Classical Dynamics

A Modern Perspective

Classical Dynamics

A Modern Perspective

E C G Sudarshan
University of Texas at Austin, USA

N Mukunda
formerly of Indian Institute of Science, India

HINDUSTAN BOOK AGENCY

World Scientific

NEW JERSEY • LONDON • SINGAPORE • BEIJING • SHANGHAI • HONG KONG • TAIPEI • CHENNAI • TOKYO

Published by

World Scientific Publishing Co. Pte. Ltd.
5 Toh Tuck Link, Singapore 596224
USA office: 27 Warren Street, Suite 401-402, Hackensack, NJ 07601
UK office: 57 Shelton Street, Covent Garden, London WC2H 9HE

British Library Cataloguing-in-Publication Data
A catalogue record for this book is available from the British Library.

CLASSICAL DYNAMICS
A Modern Perspective

ISBN 978-981-4713-87-0
ISBN 978-981-4730-01-3 (pbk)

Printed in Singapore

निरास्थ सर्वसन्देहं एकीकृत्य सुदर्शनं।
रहस्यं यो दर्शयति भजामि गुरुमीश्वरं॥

He, Who eliminates all doubts
He, Who welds all insights into knowledge
Who thus enables me to behold the Reality that was hidden
To Him, the Guru, my adoration.

अखंडमंडलाकारं व्याप्तं येन चराचरं।
तत्पदं दर्शितं येन तस्मै श्री गुरवे नमः॥

He, Who permeates all entities static and dynamic
He, Who gives form to the entire universe
He, Who reveals the knowledge of the Absolute
The Preceptor, to Him, homage!

Classical Dynamics Redux

The first version of this book appeared about forty years ago. The present reprinting is based on a complete retyping and reformatting to make for easier reading. In our preface to the 1974 edition of *Classical Dynamics: A Modern Perspective*, we had described our motivations in writing our book in some detail. All those motivations remain valid and could well be repeated today.

Classical mechanics is the codification of direct observations of the physical phenomena around us. As in elementary geometry a few observations such as the principle of inertia are treated as self-evident. The variational method, together with the action principle, leads to the equations of motion. The comparison of the equations of motion and variations thereof lead to algebraic relations between the dynamical variables. All the dynamics can be formulated in terms of these.

In simple systems there are no further restrictions except the equations of motion. But in more interesting cases, there are constraints implied by the equations of motion, and the truly independent dynamical variables have to be disentangled. A method of dealing with such difficulties was proposed by Dirac: he used the formalism of Dirac brackets, namely modified Lie bracket relations between the true dynamical variables. In this book we have followed this method. An early example of this complication is gauge invariance of classical electrodynamics.

Schwinger introduced the action principle as a fundamental framework for formulating dynamics. In this formalism the Lie algebraic structure is implicit and applies to the full relativistic transformations and internal symmetries. We have followed the formulations of Schwinger and Dirac as the guiding principle.

Our work has also been influenced by Wigner's approach to relativistic dynamics and the usual group representations. We have followed as closely as possible a re-formulation of classical mechanics in the spirit of quantum mechanics.

Around the time this book was first written, two important developments in classical mechanics were under way: the move towards the more intrinsic and global language of differential geometry, and the study of nonlinear systems with the accompanying arsenal of new concepts and methods. While the inclusion of such developments in a new version of our book seemed out of the question, over the years many physicists have told us how useful they found our treatment; in particular the coverage of basic Lie group theory from the physics perspective, the discussion of the space-time symmetry groups, and the probably first ever comprehensive treatment of Dirac's constrained Hamiltonian dynamics already mentioned above. Much of what we wrote was new in spirit and content, inspired, as mentioned, by our understanding of quantum mechanics. It is this that explains the paucity of references in the text. All things considered, the reappearance of this book, after being unavailable for some time, seems welcome.

Our sincere and heartfelt thanks are due to Mr. K. Srinivas who has prepared the manuscript in LaTeX, and to Professors Ram Ramaswamy and Subhash Chaturvedi, without whose initiative, encouragement and hard work, this reprinting would not have seen the light of day.

Austin, Texas, USA
Bengaluru, India
May 2015

E. C. G. Sudarshan
N. Mukunda

Preface

This book is the public declaration of an "affair of the heart" that we have had with classical dynamics for most of our adult lives. Indian tradition recognizes ten stages for love beginning with the beauty of form, through stages of closeness and of agony, to ultimate bliss. With classical dynamics we have progressed through many of the stages. It is our belief that the beauty of form of classical dynamics is only her calling card and recognition of this beauty but the first step.

We see classical dynamics not as a part of physics, but as physics itself. We see her form pervading all of physics, and we have endeavored to bring out this aspect. Classical dynamics is traditionally treated as if it were an earlier stage of development of physics now superceded by more ambitious theories: as if the interest in classical dynamics is only justified for historical reasons. We have found her to be otherwise. If true beauty implies that she is ever new, then classical dynamics is truly beautiful. And the dozen or so years of work represented by this book is an offering to this ever new, ever beautiful object of our love.

This book treats classical dynamics as a subject in its own right. In the initial eleven chapters, we deal with the general formalism of classical dynamics culminating in the Hamiltonian formulation. The next four chapters deal with group theory and its relevance to classical dynamics. The remaining chapters are devoted again to questions of dynamics with particular emphasis on questions of invariance groups and structure analysis of specific dynamical systems.

We are keenly aware of many omissions. Much of the beautiful work on topological questions in dynamics has not been dealt with because a satisfying treatment would have made the size of this book unresonable. We very much wished to add a chapter on "Applications of Classical Mechanics to Group Theory" but abandoned it as too ambitious. No discussion has been given of the interrelationship of classical and quantum mechanics. But we believe that the most serious omission of all is the question of measurement and measurability in classical dynamics — perhaps in another edition?

Although much of the material is "classical," much of it is new in the manner in which it is viewed and presented. It is our hope that even the most learned of our friends who read this book will find something to intrigue and to amuse them.

We have tried to make the presentation reasonably self-contained and entirely accessible to anyone with adequate undergraduate preparation. Mathematical skill and sophistication would be helpful, but a sense of adventure in ideas is even more essential. We think that the book could form the basis of a one-year course on classical dynamics; with skill and careful selection of topics it could also be used for a one-semester course.

This book is a research monograph that reviews the most recent developments in classical mechanics. It includes several new results and many new applications being developed in classical mechanics. Its primary interest is in problems of formalism and method, and it is not meant to be a class-room text. It reviews and summarizes the development of classical mechanics in a formalization that is especially intended to bring out the aspects of symmetry and invariance. It does this in the modern context of brackets, including the recently developed Dirac brackets. It is also discusses recent developments in relativistic action-at-a-distance theories.

The preparation of the book has taken us many years. One or the other author has been associated with the University of Rochester, Princeton University, Syracuse University, Brandeis University, Tata Institute of Fundamental Research and The University of Texas during this period. The constructive criticism of numerous friends is gratefully acknowledged.

We owe much inspiration to three masters: Edmund Whittaker, whose book Analytical Dynamics of Particles and Rigid Bodies has been the well-spring of our understanding of classical dynamics: Paul Dirac, whose writings and words are an inspiration in all other endeavors in physics as well as in this: and Julian Schwinger, mentor to the senior author and the most extraordinary teacher of the spirit of dynamics. To them, we must add Robert Marshak, from whom both the authors received so much, and Austin Glesson for his friendship and unflagging encouragement in this work.

Austin, Texas, USA.
Bangalore, India
December 1972

E.C.G. Sudarshan
N. Mukunda

Acknowledgements

The major part of the writing of this book was done while the senior author (ECGS) was at the University of Texas at Austin, and the junior author (NM) was at the Tata Institute of Fundamental Research at Bombay. The latter acknowledges with gratitude the appreciation bestowed on this work by Prof. M.G.K. Menon, Director of the Tata Institute of Fundamental Research.

The material of Chapters 12 and 13, and part of Chapter 14, is based on notes taken by the junior author (NM) during a course on Group Theory taught by Prof. V. Bargmann at Princeton University in the academic year 1964-65. We have tried to bring our discussion in these chapters to the level of clarity and elegance characteristic of Prof. Bargmann's presentation.

For the typing and correcting of a sometimes difficult manuscript, our thanks go to several people at John Wiley and at the University Texas. For much helpful advice during the production of the book, we thank Beatrice Shube of the house of Wiley.

Finally, the junior author (NM) wishes to record his gratitude to his wife, Usha, for her constant encouragement which has been responsible in no small way for completion of this book in a reasonable time. To the senior author (ECGS) his wife Lalita is a source of unfailing inspiration and unbounded sustenance.

Austin, Texas, USA	*E.C.G. Sudarshan*
Bangalore, India	*N. Mukunda*
Date: 1973	

Acknowledgements

Contents

Chapter 1

Introduction: Newtonian Mechanics

Physics is the science of change and of measurement. The description, contemplation, analysis, and esthetics of change manifested as motion within a classical framework are classical dynamics.

Change is seen against the changeless. How then do we make a theory of change without a discussion of the changeless background? It comes about by virtue of two circumstances. First we presuppose the observing intelligence, unaffected by the changes and yet cognizant of them. This intelligence transcends the physical body of the observer that is part of the "external universe." Second, the external universe considered as a single system does not provide an objective changeless background: we must further divide it into two parts, one of which is the experimental system and the other the background universe. It is this background universe that provides the standard of comparison and it must necessarily include the measuring apparatus. This separation of the external universe into the background universe and the experimental system is an essential step in the creation of any science, even more so than the earlier division of the universe of our experience into an external universe and an internal observing intelligence. It is the background universe that provides the passive background that gives meaning to coordinates and labels for the experimental system.

The transfer of dynamical data from the experimental system to the measuring apparatus is not describable within classical mechanics, but such measuring apparatus must be presupposed before we can give meaning to classical mechanics. True, the apparatus considered as a dynamical system has its own laws; but as soon as it is included in the investigation the experimental system gets enlarged, in turn surrounded by yet other apparati. This chasm will always remain *in any well-defined framework.* Within any such decomposition the transference of data from the experimental system to the measuring instruments, and hence to the comprehending intelligence, is beyond description, analysis or verification!

We must therefore realize that the systematization of our experience in terms of classical dynamics presupposes certain conventions, and as these conventions change the physical system also changes. With this understanding we may embark on a discussion of classical dynamics.

The foundations of classical dynamics were laid by Galileo and Newton, the latter giving them a definitive mathematical form. In the mechanics of Galileo and Newton, it was postulated that there are special frames in nature, equivalently special observers, with respect to whom the laws of motion are particularly simple. For all practical purposes it was assumed that the "fixed stars" provided such a frame. Such special frames are called "inertial frames." The equations of particle motion set up by Newton were gradually generalized step by step until ultimately a very beautiful and flexible form was reached, a form capable of describing the most diverse types of systems. The most important names to be mentioned in this development are Lagrange, Hamilton, and Poisson. The formalism, in a sense, acquired a life of its own; and it could describe in a unified way both the dynamics of particles and of fields, two very different kinds of entities in the classical domain. Thus even Maxwell's description of the electromagnetic field becomes a part of classical dynamics.

The existence of inertial frames and the ways in which these frames are related to one another are properly described by means of the notions of group theory. A most remarkable fact is that the form that classical dynamics achieved at the hands of Hamilton, Poisson, Lagrange, Jacobi and others should turn out to be so well suited to the expression of the action of groups on dynamical systems. It is this feature of classical dynamics that is retained in quantum dynamics. Even at the stage when one dispenses with the idea that there exist special observers in nature, as one does in gravitation theory today, the ideas of group action and their expression in the dynamics of systems remain and can still be handled by the previously built formalism. Unfortunately we cannot carry our discussion in this book that far.

In the initial chapters of this book, namely upto and including Chapter 11, we study the construction of the general formalism of classical dynamics, what has come to be called the *Hamiltonian* or *canonical* formalism. Following this, we devote Chapters 12 to 15 inclusive to learning some aspects of group theory and the structures of those groups that are relevant to classical dynamics. We then return to classical dynamics, and combine the formalism of the initial chapters with the groups one by one, to study the types of dynamical systems that emerge.

Newtonian Mechanics

For the time being, we shall take an orthodox, limited view of what the building blocks in dynamics are. We shall assume that the statements to follow are relative to one of the special "inertial" observers of Newtonian theory. Let us start with a discussion of the basic unit of elementary mechanics— the material point. A material point has attached to it an intrinsic parameter called its mass, m; and an instantaneous description of it amounts to specifying its location in space. A moving material point is described by variable parameters of position, and the motion is completely specified by giving the trajectory,

$$\boldsymbol{r} = \boldsymbol{r}(t), \tag{1.1}$$

t being the time. Such descriptions of motion are, however, not dynamical, they are simply observations. Thus we have, for example, the statements

$$\begin{aligned} s &= ut + \frac{1}{2}at^2 , \\ x &= A\cos(\omega t + \phi), \end{aligned} \tag{1.2}$$

familiar from uniformly accelerated motion and simple harmonic motion, respectively. They are not yet the statements of dynamics. An interesting class of such trajectory statements is that obtained by considering $\boldsymbol{r}(t)$ to give the parametric equations for an "orbit," t being the parameter. As an example we write the equation for motion under gravity as a vector equation,

$$\boldsymbol{r}(t) = \boldsymbol{u}t + \frac{1}{2}\boldsymbol{a}t^2 , \tag{1.3}$$

$\boldsymbol{u}$ being an arbitrary vector whereas $\boldsymbol{a}$ is in the negative z direction. Componentwise, we have

$$x(t) = u_x t, \qquad y(t) = u_y t, \qquad z(t) = u_z t - \frac{1}{2}gt^2 . \tag{1.4}$$

Eliminating t we get the two equations:

$$x = \frac{u_x}{u_y}y , \qquad z = \frac{u_z}{u_x}x - \frac{1}{2}\left(\frac{g}{u_x^2}\right)x^2 , \tag{1.5}$$

which are the equations to a parabola in a vertical plane. The most celebrated of such orbit theorems are those of Kepler on planetary motion.

To convert this elementary situation into a problem in mechanics, we must try to predict the trajectory $\boldsymbol{r}(t)$ from a knowledge of the mass of the particle, the initial conditions, and a specification of the background universe. It is assumed that the relevant initial conditions are simply the initial position and the initial velocity, and that the background universe is equivalent to a "force" acting on the particle and depending on its instantaneous position and time only. Newton's laws of motion are then:

$$m\frac{d^2}{dt^2}\boldsymbol{r}(t) = \boldsymbol{F}(\boldsymbol{r}(t), t) . \tag{1.6}$$

This is a second-order differential equation that can, in principle, be integrated to provide $\boldsymbol{r}(t)$ as a function of t, involving the initial position $\boldsymbol{r}(0)$ and velocity $\dot{\boldsymbol{r}}(0)$ as integration constants. (The dot over a function of t signifies the derivative with respect to t.) Conversely, given a family of trajectories $\{\boldsymbol{r}(t)\}$ that are parametrized by the initial conditions $\boldsymbol{r}(0)$, $\dot{\boldsymbol{r}}(0)$, we can compute the mass times the acceleration, $m(d^2/dt^2)\boldsymbol{r}(t)$; if this can be expressed as a function of $\boldsymbol{r}(t)$ alone (and perhaps t) through elimination of $\boldsymbol{r}(0)$ and $\dot{\boldsymbol{r}}(0)$, we can say that the force

$$\boldsymbol{F}(\boldsymbol{r}(t),t) \equiv m\frac{d^2}{dt^2}\,\boldsymbol{r}(t)$$

is acting on the material point in the region of space covered by the trajectories. Needless to say, this may not necessarily be possible. Note also that the expression for $\boldsymbol{F}(\boldsymbol{r}(t),t)$ is supposed to depend only on the instantaneous value of $\boldsymbol{r}$ and not on the entire trajectory: Newtonian mechanics contemplates only forces that are local in time.

This apparently circular definition of force in terms of motion, and vice versa, acquires content in two ways: 1. there do exist in nature families of trajectories satisfying the requirements to give rise to a force: 2. there are cases when in two *different* mechanical problems we recognize the background universe to be the same. In this instance, the differential equations describing the motion acquire an empirical content because they say that for two different particles with masses m_1 and m_2 and trajectories $\boldsymbol{r}_1(t)$ and $\boldsymbol{r}_2(t)$ we find an equality of the forces $\boldsymbol{F}_1(\boldsymbol{r},t) = \boldsymbol{F}_2(\boldsymbol{r},t)$, where

$$m_i\frac{d^2}{dt^2}\,\boldsymbol{r}_i(t) = \boldsymbol{F}_i(\boldsymbol{r}_i(t),t) \qquad i = 1,2. \tag{1.7}$$

(The equality $\boldsymbol{F}_1 = \boldsymbol{F}_2$ may have to be replaced by the equality of a characteristic multiple of $\boldsymbol{F}_1$ and a similar multiple of $\boldsymbol{F}_2$.)

Two kinds of generalizations are possible at this stage. On the one hand, apart from the position dependence, we can allow for a velocity dependence of the forces. For example, we know empirically that the force on a charged particle in an electromagnetic field is given by the Lorentz force law:

$$\boldsymbol{F} = e\left(\boldsymbol{E} + \frac{1}{c}\,\boldsymbol{v}\times\boldsymbol{B}\right)\,, \qquad \boldsymbol{v} = \dot{\boldsymbol{r}}\,. \tag{1.8}$$

Thus the equations of motion could be generalized to read

$$\frac{d^2}{dt^2}(m\boldsymbol{r}(t)) = \boldsymbol{F}(\boldsymbol{r},\dot{\boldsymbol{r}},t), \tag{1.9}$$

which in general is a nonlinear second-order differential equation. The second generalization enlarges the definition of the momentum, $m\dot{\boldsymbol{r}}$. We can have an equation of motion in the form

$$\frac{d}{dt}\,\boldsymbol{p} = \boldsymbol{F}(\boldsymbol{r},\dot{\boldsymbol{r}},t), \tag{1.10}$$

where the momentum $\boldsymbol{p}$ need not necessarily be $m\dot{\boldsymbol{r}}$ but can be a more general function of $\dot{\boldsymbol{r}}$ and other things. For example, in relativistic mechanics we have

$$\boldsymbol{p} = \frac{m_0\dot{\boldsymbol{r}}}{(1-\dot{\boldsymbol{r}}\cdot\dot{\boldsymbol{r}}/c^2)^{\frac{1}{2}}}\,, \tag{1.11}$$

which is not simply proportional by a constant factor to the velocity $\dot{\boldsymbol{r}}$. We can equally well regard the mass m as a function of velocity and write

$$\boldsymbol{p} = m(\dot{\boldsymbol{r}})\dot{\boldsymbol{r}} \,, \qquad m(\dot{\boldsymbol{r}}) = m_0(1 - \dot{\boldsymbol{r}} \cdot \dot{\boldsymbol{r}}/c^2)^{-\frac{1}{2}} \,. \tag{1.12}$$

These statements are equivalent.

Chapter 2

Generalized Coordinates and Lagrange's Equations

We now generalize the considerations of the preceding chapter to cover many-particle systems, then write the equations of motion in a form valid for any choice of coordinates.

In a many-particle system the force on each particle is made up of two parts: that originating from outside the system, and that caused by other particles of the system. We need three Cartesian coordinates to determine the position of a single particle at one instant of time, so that for an N-particle system we need $3N$ coordinates. Often, however, these $3N$ coordinates are not all independent but have to obey equations of constraint. In such a case, a "force of constraint" acts on each particle; this force is in general made up of the two types described earlier. The role of the "forces of constraint" is to ensure that the equations of constraint are satisfied during the motion. However, these forces do no work on the system (they are "passive"), and for many purposes one is not interested in them. Therefore, one eliminates these forces from the equations of motion for the $3N$ Cartesian coordinates and obtains a set of equations for a smaller number of independent "generalized coordinates."

Constraint equations are of three types:

1. vanishing of certain functions of the $3N$ Cartesian coordinates;
2. inequalities (for example, positivity) for certain functions of these coordinates;
3. linear nonintegrable relations among the differentials of these coordinates.

Constraints of type 1 are called "holonomic"; those of types 2 and 3 "nonholonomic." In each type, the time variable could appear explicitly. The constraints of type 2 can be handled by introducing forces of constraint that prevent the inequality from being violated. This force of constraint can be derived from a potential that is infinite in the domain of coordinates violating the inequalities of constraint and zero elsewhere. The constraints of type 3 are more complicated to deal with because they are dynamical constraints. We deal with constrained dynamical systems in chapter 8. At this stage we restrict ourselves to holonomic constraints, allowing for a time dependence. Rigid body constraints, for example, are holonomic and time independent.

Let us write x_r, $r = 1, 2, \cdots 3N$, for the Cartesian coordinates of the N particles. Suppose we have $(3N - k)$ holonomic constraints to be obeyed:

$$\phi_\lambda(x_r; t) = 0, \qquad \lambda = 1, 2 \cdots 3N - k, \tag{2.1}$$

so that only k of the coordinates x_r are independent. Choose k independent functions of the x_r,

$$q_s = q_s(x_r; t). \qquad s = 1, 2, \cdots, k, \tag{2.2}$$

so that, on using (2.1), we can write each x_r as a function of the variables q_s and the time t:

$$x_r = x_r(q_1, \cdots, q_k; t), \qquad r = 1, 2, \cdots, 3N. \tag{2.3}$$

In other words, using the representation of the x_r given by Eq.(2.3), the constraints of Eq. (2.1) are obeyed identically, and the variables q_s can all be treated as independent. The variables q_s are called "generalized coordinates." Differentiating Eq. (2.3) with respect to t, we get:

$$\dot{x}_r \equiv \frac{d}{dt} x_r = \sum_{s=1}^{k} \frac{\partial x_r}{\partial q_s} \dot{q}_s + \frac{\partial x_r}{\partial t}. \tag{2.4}$$

The quantities $\dot{q}_s$ are called "generalized velocities," and the above connection between them and the Cartesian velocities $\dot{x}_r$ is valid when the constraints are obeyed. Regarding the generalized coordinates and their velocities as $2k$ independent variables for purposes of partial differentiation, we obtain from Eq.(2.4) the following:

$$\frac{\partial \dot{x}_r}{\partial \dot{q}_s} = \frac{\partial x_r}{\partial q_s}, \qquad \frac{d}{dt} \frac{\partial x_r}{\partial q_s} = \frac{\partial \dot{x}_r}{\partial q_s}. \tag{2.5}$$

At each instant of time, any set of values for the coordinates q_s describes a configuration of the system consistent with the constraints. Consider two such configurations of the system, corresponding to coordinates q_s and $q_s + \delta q_s$, where δq_s are small quantities of the first order. The passage from the configuration q_s to the configuration $q_s + \delta q_s$, at one instant of time, is called a virtual, infinitesimal, instantaneous displacement of the system. The changes in the Cartesian coordinates in such a displacement are given by

$$\delta x_r = \sum_{s=1}^{k} \frac{\partial x_r}{\partial q_s} \delta q_s. \tag{2.6}$$

The equations of motion for the q_s must be obtained from those of x_r and the statement that in a displacement of the type described above, the forces of constraint do no work. The Cartesian component of the force corresponding to the coordinate x_r is split up into a force of constraint, C_r, and the *rest* denoted by F_r. So the Newtonian equations of motion for the coordinates x_r read:

$$m_r\ddot{x}_r = F_r + C_r \ , \qquad r = 1, 2, \cdots, 3N. \tag{2.7}$$

For $r = 1, 2, 3, m_r$ denotes the mass of the first particle, and so on. Since the virtual displacements δq_s are all independent, and since the force C_r performs the work $C_r\delta x_r$ in a small displacement satisfying the constraints, the equations of motion for q_s can be obtained from the following equations:

$$\sum_{r=1}^{3N} m_r\ddot{x}_r\frac{\partial x_r}{\partial q_s} = \sum_{r=1}^{3N} F_r\frac{\partial x_r}{\partial q_s} \ , \qquad s = 1, 2, \cdots, k. \tag{2.8}$$

The combination of forces $\sum_{r=1}^{3N} F_r(\partial x_r/\partial q_s)$ appearing here on the right hand side is called the "generalized force" Q_s corresponding to the coordinate q_s. To the extent that F_r is given as some function of the Cartesian coordinates and velocities, each Q_s can be written as some function of the generalized coordinates and their velocities. The expression on the left-hand side of Eq.(2.8), involving the generalized accelerations $\ddot{q}_s$, can be rewritten as follows:

$$\begin{aligned}\sum_r m_r\ddot{x}_r\frac{\partial x_r}{\partial q_s} &= \sum_r m_r\left[\frac{d}{dt}\left(\dot{x}_r\frac{\partial \dot{x}_r}{\partial \dot{q}_s}\right) - \dot{x}_r\frac{d}{dt}\frac{\partial x_r}{\partial q_s}\right] \\ &= \sum_r\left[\frac{d}{dt}\frac{\partial}{\partial \dot{q}_s}\left(\frac{1}{2}m_r\dot{x}_r\dot{x}_r\right) - \frac{\partial}{\partial q_s}\left(\frac{1}{2}m_r\dot{x}_r\dot{x}_r\right)\right] = \frac{d}{dt}\frac{\partial T}{\partial \dot{q}_s} - \frac{\partial T}{\partial q_s} \ ,\end{aligned} \tag{2.9}$$

where the total kinetic energy T is defined by

$$T = \sum_{r=1}^{3N}\frac{1}{2}m_r\dot{x}_r\dot{x}_r = T(q_s, \dot{q}_s) \ , \tag{2.10}$$

and, with the help of Eq.(2.4) has been written explicitly in terms of q_s and $\dot{q}_s$. In obtaining Eq.(2.9), use has been made of Eq.(2.5). We thus obtain the equations of motion for the generalized coordinates q_s:

$$\frac{d}{dt}\left(\frac{\partial T}{\partial \dot{q}_s}\right) - \frac{\partial T}{\partial q_s} = Q_s \ . \tag{2.11}$$

These are the Lagrange equations of motion and are generalizations of the Newtonian equations. The generalized forces Q_s are determined by those external forces acting on the system, and those acting between particles of the system, that are not forces of constraint. Note that in the Lagrangian equations of motion the coordinates q_s and the velocities $\dot{q}_s$ are treated as independent variables when taking partial derivatives and that the generalized force Q_s has a dimension determined by q_s in such a way that the product q_sQ_s is an energy.

In general, the work done by the forces F_r in an infinitesimal, virtual displacement, namely,

$$\delta W = \sum_r F_r \delta x_r = \sum_s Q_s \delta q_s, \tag{2.12}$$

will not be a "perfect differential." that is, a quantity determined solely by the initial and final configurations. If, however, the work done by the F_r as the configuration of the system changes by a finite amount from an initial to a final configuration depends only on these "end-point" configurations and not on the particular path followed between these configurations, then there exists a function V of the coordinates q_s and possibly the time t such that for a small instantaneous displacement we have:

$$\delta W = V(q_s;t) - V(q_s + \delta q_s;t) = -\sum_s \frac{\partial V}{\partial q_s}\delta q_s. \tag{2.13}$$

In such a case, the generalized forces Q_s are derivable from the "potential" V:

$$Q_s = -\frac{\partial V}{\partial q_s}, \qquad s = 1, 2, \cdots, k. \tag{2.14}$$

Because by assumption V depends only on the configuration and is therefore independent of the velocities $\dot{q}_s$, the equations of motion Eq.(2.11) take the form

$$\frac{d}{dt}\frac{\partial L}{\partial \dot{q}_s} - \frac{\partial L}{\partial q_s} = 0, \qquad s = 1, 2, \cdots, k, \tag{2.15}$$

where L, the "Lagrange function" (or Lagrangian) is given by

$$L(q_s, \dot{q}_s; t) = T(q_s, \dot{q}_s; t) - V(q_s; t). \tag{2.16}$$

Such dynamical systems are called "conservative systems," and their equations of motion are particularly simple in appearance. But from the structure of these equations it is clear that this formulation is possible for any dynamical system whose generalized forces Q_s are derivable from a generalized potential $V(q_s, \dot{q}_s; t)$ in the form:

$$Q_s = \frac{d}{dt}\frac{\partial V(q, \dot{q}; t)}{\partial \dot{q}_s} - \frac{\partial V(q, \dot{q}; t)}{\partial q_s}. \tag{2.17}$$

This is the most general type of generalized force allowing the introduction of a Lagrangian L and a compact set of equations of motion as in Eq.(2.15).

Let us illustrate this for the most familiar velocity-dependent force, the electromagnetic Lorentz force on a charged particle:

$$\boldsymbol{F}(\boldsymbol{r}, \dot{\boldsymbol{r}}, t) = e\left(\boldsymbol{E}(\boldsymbol{r}, t) + \frac{1}{c}\dot{\boldsymbol{r}} \times \boldsymbol{B}(\boldsymbol{r}, t)\right). \tag{2.18}$$

The coordinates q_s are the Cartesian components x, y, z of the position vector $\boldsymbol{r}$ of the particle; e is the charge of the particle, c the velocity of light, and

$\boldsymbol{E}(\boldsymbol{r},t), \boldsymbol{B}(\boldsymbol{r},t)$ are the electric and magnetic fields respectively at the position of the particle at the time t. These fields are expressed in terms of the scalar and vector potentials $\phi(\boldsymbol{r},t)$ and $\boldsymbol{A}(\boldsymbol{r},t)$ by

$$\begin{aligned} \boldsymbol{E}(\boldsymbol{r},t) &= -\boldsymbol{\nabla}\phi(\boldsymbol{r},t) - \frac{1}{c}\frac{\partial}{\partial t}\boldsymbol{A}(\boldsymbol{r},t), \\ \boldsymbol{B}(\boldsymbol{r},t) &= \boldsymbol{\nabla} \times \boldsymbol{A}(\boldsymbol{r},t)\ . \end{aligned} \tag{2.19}$$

With the use of these expressions, the force on the particle takes the form:

$$\begin{aligned} \frac{1}{e}\boldsymbol{F}(\boldsymbol{r},\dot{\boldsymbol{r}},t) &= -\boldsymbol{\nabla}\phi(\boldsymbol{r},t) - \frac{1}{c}\frac{\partial}{\partial t}\boldsymbol{A}(\boldsymbol{r},t) + \frac{1}{c}\dot{\boldsymbol{r}} \times (\boldsymbol{\nabla} \times \boldsymbol{A}(\boldsymbol{r},t)) \\ &= -\boldsymbol{\nabla}\phi(\boldsymbol{r},t) - \frac{1}{c}\left(\frac{\partial}{\partial t}\boldsymbol{A}(\boldsymbol{r},t) + \dot{\boldsymbol{r}} \cdot \boldsymbol{\nabla}\boldsymbol{A}(\boldsymbol{r},t)\right) + \frac{1}{c}\boldsymbol{\nabla}\dot{\boldsymbol{r}} \cdot \boldsymbol{A}(\boldsymbol{r},t) \\ &= -\boldsymbol{\nabla}\left(\phi(\boldsymbol{r},t) - \frac{1}{c}\dot{\boldsymbol{r}} \cdot \boldsymbol{A}(\boldsymbol{r},t)\right) - \frac{1}{c}\frac{d}{dt}\boldsymbol{A}(\boldsymbol{r},t). \end{aligned} \tag{2.20}$$

Here, we have written $(d/dt)\boldsymbol{A}(\boldsymbol{r},t)$ for the *total* time derivative of the vector potential, the dependence of $\boldsymbol{r}$ on t being given by the motion of the particle. For a nonrelativistic particle with mass m, the equation of motion is then

$$\frac{d}{dt}m\dot{\boldsymbol{x}} = \boldsymbol{F}(\boldsymbol{r},\dot{\boldsymbol{r}},t) = -e\boldsymbol{\nabla}\left(\phi - \frac{1}{c}\dot{\boldsymbol{r}} \cdot \boldsymbol{A}\right) - \frac{e}{c}\frac{d}{dt}\boldsymbol{A}\ ,$$

or

$$\frac{d}{dt}\left(m\dot{\boldsymbol{r}} + \frac{e}{c}\boldsymbol{A}\right) + e\boldsymbol{\nabla}\left(\phi - \frac{1}{c}\dot{\boldsymbol{r}} \cdot \boldsymbol{A}\right) = 0. \tag{2.21}$$

To put this in the Lagrangian form, we need a function $L(\boldsymbol{r},\dot{\boldsymbol{r}},t)$ such that

$$\frac{\partial L}{\partial \dot{\boldsymbol{r}}} = m\dot{\boldsymbol{r}} + \frac{e}{c}\boldsymbol{A}(\boldsymbol{r},t), \qquad \frac{\partial L}{\partial \boldsymbol{r}} = \boldsymbol{\nabla}\left(\frac{e}{c}\dot{\boldsymbol{r}} \cdot \boldsymbol{A}(\boldsymbol{r},t) - e\phi(\boldsymbol{r},t)\right);$$

the solution is easily seen to be

$$L(\boldsymbol{r},\dot{\boldsymbol{r}},t) = \frac{1}{2}m\dot{\boldsymbol{r}} \cdot \dot{\boldsymbol{r}} + \frac{e}{c}\dot{\boldsymbol{r}} \cdot \boldsymbol{A}(\boldsymbol{r},t) - e\phi(\boldsymbol{r},t). \tag{2.22}$$

The first term in L is the kinetic energy T for a nonrelativistic particle; one recognizes the generalized velocity-dependent potential to be

$$V(\boldsymbol{r},\dot{\boldsymbol{r}},t) = e\phi(\boldsymbol{r},t) - \frac{e}{c}\dot{\boldsymbol{r}} \cdot \boldsymbol{A}(\boldsymbol{r},t). \tag{2.23}$$

It thus appears that for a large class of systems the equations of motion are completely determined in terms of a single Lagrange function $L(\boldsymbol{q},\dot{\boldsymbol{q}},t)$. The resulting equations of motion. Eq.(2.15), are second-order differential equations for the coordinates q_s regarded as functions of t. Whereas the Newtonian equations of motion directly expressed the accelerations in terms of the forces that are functions of the positions and velocities, this is not the case with the Lagrangian equations of motion. Here, after having obtained the equations in the form Eq.(2.15), one must try to express the generalized accelerations $\ddot{q}_s$ as functions of the generalized positions and velocities q_s and $\dot{q}_s$, and whether this can be done depends entirely on the functional form of the Lagrangian. The case wherein one can solve for $\ddot{q}_s$, that is, rewrite Eq.(2.15) to express $\ddot{q}_s$ in terms of all the q's and $\dot{q}$'s , is called the "standard case." In such a case, in principle, the resulting equations can be integrated. By giving the "initial data." namely a set of physically admissible but otherwise independent values for the variables q_s and $\dot{q}_s$ at a time $t = t_0$, the functions $q_s(t)$, and thus the motion of the system, are determined for all later times.

If the Lagrangian for a dynamical system is of such a functional form that we are not able to obtain all the accelerations $\ddot{q}_s$ from the Lagrangian equations as functions of q's and $\dot{q}$'s, then the system is said to be "subject to constraints." The constraints referred to here are not the same as those that led to the introduction of generalized coordinates earlier in this chapter, but have to do with the peculiar functional form of the Lagrangian. Such systems are discussed in detail in Chapter 8 and 9.

Chapter 3

The Hamilton and Weiss Variational Principles and the Hamilton Equations of Motion

One can visualize the development in time of a system described by the Lagrangian equations of motion in the following way (unless otherwise stated, we are considering the standard case). The number of generalized coordinates, equivalently the number of "degrees of freedom," being k, the k-dimensional configuration space of the system is defined as a real space whose points are labeled by the k generalized coordinates $q_1, q_2, \cdots, q_k$. Each q_s goes over a range determined by the physical meaning of that coordinate. Each point in configuration space corresponds to a possible configuration of the system at one instant of time. If the configuration of the system at the time t_0 corresponds to the point Q_0 in configuration space, and if we start the system going in a particular direction at Q_0 by specifying the velocities $\dot{q}_s$ at t_0, then the Lagrangian equations of motion will completely determine the generalized coordinates $q_s(t)$ for all later times; that is, the configuration of the system at each instant t is determined. The representative point Q_t in configuration space describes a certain class of trajectories in that space; the trajectory is fixed once one knows the starting point and direction and magnitude of velocity, that is, the initial configuration and velocities. Alternatively, we could specify the configurations Q_1 and Q_2 at an initial time t_1 and a later time t_2, and then the equations of motion (in general) pick out a unique trajectory leading from Q_1 at t_1 to Q_2 at t_2; this trajectory tells us with what velocities we should start the system at Q_1 at time t_1 to ensure its arrival at Q_2 at time t_2. In either case the velocities at each interior point on the trajectory are determined.

Now, the Lagrangian equations of motion are differential equations related only to local properties of the trajectories. The two ways of determining a trajectory described above correspond to alternative ways of specifying integration constants in solving these differential equations. The Hamilton and Weiss variational principles are global characterizations of the equations of motion: they consider the motion of the dynamical system over finite intervals of time and determine directly the trajectory of the representative point in configuration space.

Consider two points Q_1, Q_2 in configuration space, with coordinates $q_s^{(1)}$ and $q_s^{(2)}$ respectively. Let C be any path connecting Q_1 to Q_2. We can imagine the system starting from Q_1 at a time t_1, traveling along C at a certain variable speed, and reaching Q_2 at the time t_2. This would correspond to giving the coordinates of points on C as functions of time t, $q_s(t)$, such that

$$q_s(t_1) = q_s^{(1)}, \qquad q_s(t_2) = q_s^{(2)} . \tag{3.1}$$

Because the time of arrival at each point on C is given, the velocities $\dot{q}_s$ at each point on C are also determined:

$$\dot{q}_s(t) = \frac{d}{dt} q_s(t); \tag{3.2}$$

and the Lagrangian L of the system, being a function of the variables q_s, $\dot{q}_s$, t, has a definite numerical value at each point on C. We then set up a functional Φ of the path C, called the "action functional," by

$$\Phi[C] = \int_{t_1}^{t_2} dt L(q_s(t), \dot{q}_s(t), t) . \tag{3.3}$$

This expression Φ is called a functional of the functions $q_s(t)$ and L because it depends on the functional forms of these quantities expressed in terms of t, in the same way in which an ordinary function depends on the numerical values of its independent arguments. Now let C' be a path differing infinitesimally from C with end points $Q_1'(Q_2')$ infinitesimally close to $Q_1(Q_2)$. In parametrizing the points on C', let us say that t runs from t_1' at Q_1' to t_2' at Q_2', where

$$t_1' - t_1 = \Delta t_1, \qquad t_2' - t_2 = \Delta t_2 , \tag{3.4}$$

and where the time differences Δt_j are small quantities of the first order. The coordinates of points on C' are given by functions $q_s'(t)$ differing infinitesimally from $q_s(t)$:

$$q_s'(t) = q_s(t) + \delta q_s(t) . \tag{3.5}$$

(The arguments of the functions q_s' and q_s have ranges differing by infinitesimals of the first order; thus any error in Eq.(3.5) is of the second order of smallness). At corresponding times t, the velocities on C' and C are related by the derivative of Eq.(3.5):

$$\dot{q}_s'(t) \equiv \frac{d}{dt} q_s'(t) = \dot{q}_s(t) + \frac{d}{dt}(\delta q_s(t)). \tag{3.6}$$

We now compute the first-order change in the action functional Φ in going from C to C':

$$
\begin{aligned}
\Delta\Phi \equiv \Phi[C'] - \Phi[C] &= \int_{t_1'}^{t_2'} L(q_s'(t), \dot{q}_s'(t), t)dt - \int_{t_1}^{t_2} L(q_s(t), \dot{q}_s(t), t)dt \\
&= \int_{t_1}^{t_2} [L(q_s', \dot{q}_s', t) - L(q_s, \dot{q}_s, t)]dt \\
&\quad + \int_{t_2}^{t_2'} L(q_s', \dot{q}_s', t)dt - \int_{t_1}^{t_1'} L(q_s', \dot{q}_s', t)dt \\
&= \int_{t_1}^{t_2} dt \sum_s \left[\frac{\partial L}{\partial q_s}\delta q_s(t) + \frac{\partial L}{\partial \dot{q}_s}\frac{d}{dt}\delta q_s(t)\right] + [L\Delta t]\Big|_{t_1}^{t_2} \\
&= \int_{t_1}^{t_2} dt \sum_s \left[\frac{\partial L}{\partial q_s} - \frac{d}{dt}\left(\frac{\partial L}{\partial \dot{q}_s}\right)\right] \delta q_s(t) \\
&\quad + \left[L\Delta t + \sum_s \frac{\partial L}{\partial \dot{q}_s}\delta q_s(t)\right]\Bigg|_{t_1}^{t_2} .
\end{aligned} \tag{3.7}
$$

Note that even though L might be explicitly time dependent, no term of the form $\partial L/\partial t$ appears. Now we define the "total variations" Δq_s in the coordinates q_s at the end points of the path C as the differences in the coordinates of $Q_1'(Q_2')$ and $Q_1(Q_2)$:

$$\Delta q_s(t_j) = q_s'(t_j') - q_s(t_j) = \delta q_s(t_j) + \dot{q}_s(t_j)\Delta t_j\ , \quad j = 1, 2. \tag{3.8}$$

In terms of these total variations, we can write $\Delta\Phi$ in the form:

$$\Delta\Phi = \int_{t_1}^{t_2} dt \sum_s \left[\frac{\partial L}{\partial q_s} - \frac{d}{dt}\frac{\partial L}{\partial \dot{q}_s}\right] \delta q_s(t) + \left[\sum_s p_s \Delta q_s - H\Delta t\right]\Bigg|_{t_1}^{t_2} . \tag{3.9}$$

We have written p_s and $-H$ for the coefficients of the total variations:

$$p_s = \frac{\partial L}{\partial \dot{q}_s}, \quad H = \sum_s p_s \dot{q}_s - L(q, \dot{q}, t). \tag{3.10}$$

Notice that the integral in $\Delta\Phi$ is to be evaluated along the path C. This expression for $\Delta\Phi$ leads to the two variational principles of Hamilton and Weiss.

Hamilton's Principle

Consider first the special class of variations of path that leave the end points and terminal times unchanged:

$$\Delta t_j = \Delta q_s(t_j) = 0. \tag{3.11}$$

For such variations we have

$$\Delta_\sigma \Phi = \int_{t_1}^{t_2} dt \sum_s \left[\frac{\partial L}{\partial q_s} - \frac{d}{dt}\frac{\partial L}{\partial \dot{q}_s}\right] \delta q_s(t). \tag{3.12}$$

(The subscript σ on Δ indicates a special variation). Comparing this with the Lagrangian equations of motion, we see that the first-order variation in Φ vanishes for an arbitrary, infinitesimal change in the path C that leaves the end points and terminal times fixed if, along C, Lagrange's equations of motion are obeyed, that is, if the path C together with the rate at which the system moves along C is the actual trajectory that the system would follow in accordance with the equations of motion in going from Q_1 at time t_1 to Q_2 at time t_2. The action functional is stationary, and possibly has an extremum, for such a path.

Thus we have an important and elegant characterisation of the dynamical path followed by a system in configuration space, obeying the equations of motion. To use this characterisation to determine the dynamical path, we can alter our point of view and express the laws of motion in the following manner: given the configurations Q_1,Q_2 at times $t_1, t_2 (t_2 > t_1)$, the actual dynamical path C followed by the system is that for which the action functional Φ is stationary and possibly has an extremum. Equivalently, C is that path, about which infinitesimal special variations produce no change in Φ:

$$\Delta_\sigma \int_{t_1}^{t_2} L(q_s(t), \dot{q}_s(t), t)dt = 0 . \tag{3.13}$$

This is Hamilton's principle, and Lagrange's equations follow from it. We note that among the variations of path contemplated in Hamilton's principle we include those in which the system passes through the same points of configuration space but at a different speed, namely, variations in which the coordinates $q_s(t)$ take on the same set of values but at altered times.

Let us pause at this point and analyze the "deduction"; we have derived a set of differential equations from statements about the values of a functional of finite paths. But the important thing is that we have allowed *arbitrary* infinitesimal variations in the path to be made; that is, the change in $q_s(t)$ in any small interval of time is quite independent of what happens in the complementary interval of time. Thus in the limit one could consider a $\delta q_s(t)$ which is nonzero only for an infinitesimally small interval of time, which would lead to differential equations of motion.

It is important to note that throughout the variation, the relation

$$\dot{q}_s(t) = \frac{d}{dt} q_s(t)$$

is maintained, so that the change in the velocity is the time derivative of the change in the coordinate; in other words, q_s and $\dot{q}_s$ are not varied independently. This constraint is characteristic of Lagrangian mechanics, where the velocities are always the time derivatives of the coordinates. Another point is that in any variation, the Lagrangian considered as a function of $q_s, \dot{q}_s$, and t preserves its functional form.

The Weiss Action Principle

Let us now go back and consider the case of a general variation in the path C:

the effect on the action functional Φ is given in Eq.(3.9). Once again we see that if the trajectory C is that on which the Lagrange equations of motion are satisfied, then the integral on the right-hand side of Eq.(3.9) vanishes and we get

$$\Delta\Phi[C] = \left[\sum_s p_s \Delta q_s - H\Delta t\right]\Bigg|_{t_1}^{t_2} \tag{3.14}$$

Thus the variation $\Delta\Phi$ in the action receives contributions only from the end points of the trajectory C, and these contributions can be expressed in terms of the total variations Δq_s, Δt. One calls the coefficient p_s of the total variation Δq_s the "generalized momentum" corresponding to the coordinate q_s, whereas the coefficient H of $-\Delta t$ is called the "Hamiltonian."

We now extend Hamilton's principle to the Weiss action principle by formulating the latter as follows: given a system with the action functional $\Phi[C]$, the dynamical path followed by the system in configuration space is that path, about which general variations produce only "end point contributions" to $\Delta\Phi$:

$$\Delta\Phi[C] \equiv \Delta \int_{t_1}^{t_2} L(q_s(t), \dot{q}_s(t), t)dt = \left[\sum_s p_s \Delta q_s - H\Delta t\right]\Bigg|_{t_1}^{t_2} \tag{3.15}$$

The coefficients p_s and $-H$ in $\Delta\Phi$ are said to be "conjugate" to the variables q_s and t, respectively. We see that the Weiss action principle leads both to the Lagrangian equations of motion, as well as to the identification of the important dynamical variables p_s and H. With the variables p_s, the Lagrangian equations of motion can be written as

$$p_s = \frac{\partial L}{\partial \dot{q}_s}\,, \tag{3.16a}$$

$$\dot{p}_s = \frac{\partial L}{\partial q_s}\,; \tag{3.16b}$$

it is clear, of course, that Eq.(3.16a) is the defining equation for p_s, whereas Eq.(3.16b) is the true equation of motion in the Newtonian and Lagrangian sense.

We have derived the Weiss action principle in configuration space from a Lagrangian starting point, with the generalized coordinates and velocities as the basic variables. A complete specification of the dynamics of a system consists, in this framework, in the choice of the Lagrange function and the postulation of the Weiss action principle. This is a generalization of Newton's second law of motion and specifies a general pattern for the equations of dynamics.

There is an alternative form of the Weiss action principle, which requires us to pass from the Lagrangian form with its use of coordinates and velocities to a form in which coordinates q_s and the conjugate momenta p_s appear on an equal footing. This framework for the discussion of dynamics is due to Hamilton, and the resulting form of dynamics is called "Hamiltonian dynamics." It is this formulation of classical mechanics that has, historically, been found appropriate as a starting point for quantum mechanics. From the manner in which the dynamical variables p_s and H appeared in the Lagrangian form of the Weiss action principle, it is plausible that in eliminating the velocities $\dot{q}_s$ in favor of the momenta p_s, the dominant role played by the Lagrangian will be taken over by the Hamiltonian. We now proceed to develop this new form of the action principle.

According to the definition of the momenta p_s in Eq.(3.10), these variables are given explicitly as functions of the variables q_s and $\dot{q}_s$, and possibly t. It is shown in Chapter 8 that in the standard case it is possible to invert the equations defining p_s and express the velocities $\dot{q}_s$ as functions of q_s and p_s:

$$\dot{q}_s = \dot{q}_s(q, p, t) \ . \tag{3.17}$$

In fact, being able to solve for the accelerations from the Lagrangian equations of motion and being able to solve for $\dot{q}_s$ in terms of q_s, p_s are precisely equivalent. In order to treat the $2k$ variables q_s, p_s as independent ones, we define the $2k$-dimensional "phase space" of the system to be a real space whose points are labeled by the q's and p's. The range of each q and each p is again determined by the physical meaning of the relevant variable. Any given configuration of the system at an instant of time (set of values for the q's) together with any given momenta (and hence velocities) at that instant of time collectively determine the "phase" of the system at that instant and correspond to a single point P in phase space. As in the case of configuration space, here again we can visualize trajectories: given two points P_1 and P_2 in phase space and initial and final times t_1 and t_2, we imagine the system traveling from P_1 at time t_1 to P_2 at time t_2 along some path $\mathcal{C}$ in phase space at some (variable) speed. This trajectory corresponds to specifying the points on $\mathcal{C}$ as functions of time; that is, giving the functions $q_s(t), p_s(t)$ subject to the boundary conditions

$$q_s(t_j) = q_s^{(j)} \ , \qquad p_s(t_j) = p_s^{(j)} \ , \qquad j = 1, 2, \tag{3.18}$$

$q_s^{(j)}, p_s^{(j)}$ being the coordinates of P_j. Now, before we define the corresponding action functional, we must clarify the following point: Because on an arbitrary trajectory $\mathcal{C}$ both q_s and p_s are specified as functions of t independently of one another, first, the velocities $\dot{q}_s$ will be determined at each point on $\mathcal{C}$; second, the relations among the variables q_s, $\dot{q}_s$, p_s, namely Eqs.(3.10) or (3.17), will, in general, not be obeyed on $\mathcal{C}$. On the actual trajectory determined by the equations of motion, these relations among $q_s, \dot{q}_s$, and p_s must be valid. The situation, then, is the following: in setting up the action principle in phase space, we ignore these relations among the three sets of variables and consider general trajectories $\mathcal{C}$ of the type described above. When we finally derive the equations of motion from the new form of the action principle, we find that one-half of the equations restore the proper connection among q_s, $\dot{q}_s$, and p_s, whereas the rest embody the Newtonian (or Lagrangian) equations of motion.

The relation between the Lagrangian and the Hamiltonian is given in Eq.(3.10). Using Eq.(3.17), we can express the Hamiltonian as an explicit function of q_s,

p_s, and t. We then have

$$L(q_s, \dot{q}_s, t) = \sum_{s=1}^{k} p_s \dot{q}_s - H(q_s, p_s, t) \ . \tag{3.19}$$

With the help of this expression for the Lagrangian, we define the action functional $\Psi(\mathcal{C})$ corresponding to a trajectory $\mathcal{C}$ in phase space:

$$\Psi(\mathcal{C}) = \int_{t_1}^{t_2} dt \left[\sum_s p_s(t)\dot{q}_s(t) - H(q(t), p(t), t) \right] \ . \tag{3.20}$$

[Remember that the trajectory $\mathcal{C}$ is given once one has the independent functions $q_s(t)$ and $p_s(t)$]. We now consider a small variation of the trajectory $\mathcal{C}$ in which we alter the coordinates of the end points by small quantities of the first order, alter the functions $q_s(t), p_s(t)$ independently by small amounts $\delta q_s(t)$ and $\delta p_s(t)$, and alter the terminal times by amounts Δt_j. The variation in the velocities is as before, given by

$$\delta \dot{q}_s(t) = \frac{d}{dt} \delta q_s(t) \ . \tag{3.6}$$

The Weiss action principle in phase space is postulated in the following form: for general variations of path about the actual path $\mathcal{C}$ of a system in phase space, the variation in the action functional Ψ shall receive only "end-point contributions" and shall be expressible in terms of the total variations Δq_s, Δt_j as

$$\Delta\Psi(\mathcal{C}) \equiv \Delta \int_{t_1}^{t_2} \left[\sum_s p_s \dot{q}_s - H(q, p, t) \right] dt = \left[\sum_s p_s \Delta q_s - H \Delta t \right] \Bigg|_{t_1}^{t_2} \tag{3.21}$$

Let us deduce the equations of motion from this postulate. Using primes for the quantities referring to the path $\mathcal{C}'$ obtained from $\mathcal{C}$ by a general variation, we have:

$$\begin{aligned}
\Delta\Psi \equiv & \Psi(\mathcal{C}') - \Psi(\mathcal{C}) = \int_{t_1'}^{t_2'} dt \left[\sum_s p_s'(t)\dot{q}_s'(t) - H(q'(t), p'(t), t) \right] \\
& - \int_{t_1}^{t_2} dt \left[\sum_s p_s(t)\dot{q}_s(t) - H(q(t), p(t), t) \right] \\
= & \int_{t_1}^{t_2} dt \left[\sum_s \dot{q}_s(t)\delta p_s(t) + p_s(t)\delta\dot{q}_s(t) - \frac{\partial H}{\partial p_s}\delta p_s(t) \right. \\
& \left. - \frac{\partial H}{\partial q_s}\delta q_s(t) \right] + \left[\left(\sum_s p_s \dot{q}_s - H(q, p, t) \right) \Delta t \right] \Bigg|_{t_1}^{t_2} \\
= & \int_{t_1}^{t_2} dt \left[\sum_s \left(\dot{q}_s - \frac{\partial H}{\partial p_s} \right) \delta p_s(t) - \sum_s \left(\dot{p}_s + \frac{\partial H}{\partial q_s} \right) \delta q_s(t) \right]
\end{aligned}$$

$$+\left[\sum_s p_s\delta q_s + \left(\sum_s p_s\dot{q}_s - H\right)\Delta t\right]\Bigg|_{t_1}^{t_2}$$

$$= \int_{t_1}^{t_2} dt\left[\sum_s\left(\dot{q}_s - \frac{\partial H}{\partial p_s}\right)\delta p_s(t)\right.$$

$$\left. - \sum_s\left(\dot{p}_s + \frac{\partial H}{\partial q_s}\right)\delta q_s(t)\right] + \left[\sum_s p_s\Delta q_s - H\Delta t\right]\Bigg|_{t_1}^{t_2} \tag{3.22}$$

The fact that the total variations in the momenta, Δp_s, do not appear here is caused by the particular form of the integrand in the action functional $\Psi[\mathcal{C}]$, Eq.(3.20). This form is, essentially, the mathematical relationship between the Lagrangian and the Hamiltonian and is explained in the next chapter. Comparing the above form for $\Delta\Psi$ with the postulated form given in Eq.(3.21) we get the differential equations of motion of the actual trajectory in phase space:

$$\dot{q}_s(t) = \frac{\partial H}{\partial p_s}\,, \qquad \dot{p}_s = -\frac{\partial H}{\partial q_s}\,. \qquad s = 1, 2, \cdots, k\,. \tag{3.23}$$

These are the Hamiltonian equations of motion and could have been obtained equally well by considering special variations that leave the end points and terminal times of the path $\mathcal{C}$ fixed. We also verify from the general expression for $\Delta\Psi$ that p_s and $-H$ are conjugate to q_s and t, respectively.

It is in this Hamiltonian form that the Weiss action principle finds its most succinct formulation. However, we have yet to check explicitly that the Lagrangian and Hamiltonian forms of the action principle are equivalent, in other words, that the Lagrangian equations in the form of Eq.(3.16) and the Hamiltonian equations in the form of Eq.(3.23) are identical in content. This demonstration, and the clarification of the relation between the two formalisms, is given in the next chapter. Assuming this equivalence for the moment, we can appreciate the differences between the trajectories and the action functional in configuration space on the one hand and in phase space on the other. The Hamiltonian equations of motion are, by construction, the Euler-Lagrange variational equations for the phase-space form of the action principle. They are first-order time-derivative equations, in contrast to the second-order time-derivative Lagrangian equations, but now we have twice as many equations. If we are given the phase (configuration plus momenta) of the system at time t_0, corresponding to the point P_0 in phase space, the ensuing motion, and hence the phase space trajectory of the representative point, is fully determined. In contrast, the dynamical trajectory in configuration space requires for its determination both the starting configuration and the starting direction (the velocities $\dot{q}_s$). This translates into properties of the two action functionals Φ and Ψ. Given initial and final configurations Q_1, Q_2 and times t_1, t_2, one can, in general, always find a connecting configuration space trajectory C for which $\Phi[C]$ is stationary; that is, in general one will be able to arrange the velocities at the initial time in such a way that the system is in the configuration Q_2 at the later time t_2.

However, in phase space, given the initial and final phases P_1, P_2 and times t_1, t_2, for *almost all* choices of P_2 and t_2 there is no trajectory $\mathcal{C}$ that makes $\Psi[\mathcal{C}]$ stationary. Even if P_2 is "properly" chosen, for *almost all* choices of t_2 there will still be no path $\mathcal{C}$ making $\Psi[\mathcal{C}]$ stationary!

Chapter 4

The Relation Between the Lagrangian and the Hamiltonian Descriptions

We now consider the problem of proving the equivalence of the configuration space and phase space versions of the Weiss action principle. We do this indirectly by deriving the Hamiltonian equations of motion from the Lagrangian ones. Recall that the latter have been put into the following form:

$$p_s = \frac{\partial L}{\partial \dot{q}_s}\,, \qquad \dot{p}_s = \frac{\partial L}{\partial q_s} \qquad s = 1, 2, \cdots, k\,. \tag{4.1}$$

Here the first set of equations define the p's as functions of $q, \dot{q}$, and t; whereas the second set of equations alone specify the accelerations of the system. In the Lagrangian form, when we take the partial derivative of a function with respect to a certain coordinate q_s, we are supposed to keep the *other* coordinates q_r and *all* the velocities $\dot{q}_r$ fixed: conversely, to differentiate with respect to a particular velocity $\dot{q}_s$ we keep *all* the coordinates q_r and the *remaining* velocities $\dot{q}_r$ fixed. We want to change from this scheme to one in which the coordinates q_s and momenta p_s are the fundamental independent variables; this is to be done by eliminating the velocities $\dot{q}_s$ by making use of the first half of Eq.(4.1). To be able to express the right-hand side of the "true" equations of motion in this new scheme, we look for a function of q_s and p_s (and t) whose partial derivative with respect to a particular coordinate q_s, keeping the other coordinates and all the momenta fixed, is equal to the partial derivative of the Lagrangian L with respect to that coordinate evaluated according to the rules of the Lagrangian scheme. Let us denote this function by $-\hat{H}(q, p, t)$; thus we require that

$$\left.\frac{\partial L}{\partial q_s}(q, \dot{q}, t)\right|_{\dot{q}\text{ fixed}} = -\left.\frac{\partial \hat{H}}{\partial q_s}(q, p, t)\right|_{p\text{ fixed}} \tag{4.2}$$

Let us write out the differentials of $\hat{H}$ and L:

$$dH = \sum_s \frac{\partial \hat{H}}{\partial q_s} dq_s + \sum_s \frac{\partial \hat{H}}{\partial p_s} dp_s + \frac{\partial \hat{H}}{\partial t} dt \; ,$$
$$dL = \sum_s \frac{\partial L}{\partial q_s} dq_s + \sum_s \frac{\partial L}{\partial \dot{q}_s} d\dot{q}_s + \frac{\partial L}{\partial t} dt \; . \tag{4.3}$$

Of course, not all the differentials dq_s, dp_s, and $d\dot{q}_s$ are independent. Using the definitions of the momenta p_s and the requirement of Eq.(4.2), we obtain

$$d(\hat{H} + L) = \sum_s \left(\frac{\partial \hat{H}}{\partial p_s} dp_s + p_s d\dot{q}_s \right) + \frac{\partial}{\partial t}(\hat{H} + L)dt \; ,$$

that is

$$d(\hat{H} + L - \sum_s p_s \dot{q}_s) = \sum_s \left(\frac{\partial \hat{H}}{\partial p_s} - \dot{q}_s \right) dp_s + \frac{\partial}{\partial t}(\hat{H} + L)dt \; . \tag{4.4}$$

Now the left-hand side is a perfect differential, whereas on the right the differentials dq_s are absent. Regarding the dq_s and dp_s as independent, it follows that a function $\hat{H}$ satisfying Eq.(4.2) can be found if and only if a function f of p_s alone and t can be found such that:

$$\hat{H} + L - \sum_s p_s \dot{q}_s = f(p_s, t) \; , \tag{4.5a}$$
$$\frac{\partial \hat{H}}{\partial p_s} - \dot{q}_s = \frac{\partial f}{\partial p_s} \; , \tag{4.5b}$$
$$\frac{\partial}{\partial t}(\hat{H} + L) = \frac{\partial f}{\partial t} \; . \tag{4.5c}$$

These conditions can all be rewritten in terms of the function $H = \hat{H} - f$; they read

$$H(q_s, p_s, t) = \sum_s p_s \dot{q}_s - L(q_s, \dot{q}_s, t) \; , \tag{4.6a}$$
$$\frac{\partial H}{\partial p_s} = \dot{q}_s \; , \tag{4.6b}$$
$$\frac{\partial H}{\partial t} = -\frac{\partial L}{\partial t} \; . \tag{4.6c}$$

Because f does not depend on q_s, in Eq.(4.2) $\hat{H}$ can be replaced by H. Thus a function $H(q, p, t)$ obeying Eq.(4.2) exists, provided that H can be made to satisfy Eq.(4.6). However, if we just use Eq.(4.6a) to *define* $H(q, p, t)$, then Eq.(4.6b) *is* verified by the definition of p_s; Eq.(4.6c) is obviously also satisfied. (When verifying Eq.(4.6b), we must remember that in evaluating the derivative of the right-hand side of Eq.(4.6a) with respect to a particular momentum p_s, we keep the other momenta and all the coordinates fixed, so that the velocities $\dot{q}_s$ vary). Therefore with the function H defined by Eq.(4.6a), we have secured both Eq.(4.2) (with $\hat{H}$ replaced by H), and Eqs.(4.6b) and (4.6c). Because up to now we have only made use of the definitions of p_s in terms of q_s and $\dot{q}_s$, it follows that Eq.(4.6b), which we recognize as one set of the Hamiltonian equations of motion, is in fact identical to the coordinate-momentum-velocity connection; the velocities have been "solved for" and expressed explicitly in terms of the coordinates and momenta. The "genuine" equation of motion is obtained by combining Eqs.(4.1) and (4.2). All in all, then, we have the definition of the Hamiltonian:

$$H(q,p,t) = \sum_s p_s \dot{q}_s - L(q,\dot{q},t) \ , \tag{4.6a}$$

and the set of equations

$$\frac{\partial H}{\partial p_s} = \dot{q}_s \ , \tag{4.7a}$$

$$-\frac{\partial H}{\partial q_s} = \dot{p}_s \ . \tag{4.7b}$$

The Hamiltonian form of the Weiss action principle is thus justified, and the passage from the Lagrangian to the Hamiltonian is now clear. Equation (4.6a) is an example of what is called a Legendre transformation. It should be pointed out that though from the Newtonian and Lagrangian view-point, Eq.(4.7a) corresponds just to the coordinate-momentum-velocity relationship and only Eq.(4.7b) is interpreted as a "true" equation of motion, from the Hamiltonian viewpoint all of these equations are to be treated on the same footing, and all are to be regarded as "equations of motion".

The inverse transformation can also be accomplished: given a Hamiltonian $H(q,p,t)$, and the equations of motion Eq.(4.7), in the standard case we can solve for the momenta p_s in terms of q_s, $\dot{q}_s$, and t. This is done by "inverting" Eq.(4.7a). In the remaining Eq.(4.7b), we look for a function $-L$ of q_s, $\dot{q}_s$, and t whose partial derivatives with respect to q_s, keeping $\dot{q}_s$ fixed, equal those of H with respect to q_s, keeping p_s fixed. It is obvious that we should define

$$L(q,\dot{q},t) = \sum_s p_s \dot{q}_s - H(q,p,t) \ . \tag{4.8}$$

Then, with the help of Eq.(4.7a) alone, we get

$$\begin{aligned} dL &= \sum_s \left(dp_s \dot{q}_s + p_s d\dot{q}_s - \frac{\partial H}{\partial p_s} dp_s - \frac{\partial H}{\partial q_s} dq_s \right) - \frac{\partial H}{\partial t} dt \\ &= \sum_s \left(p_s d\dot{q}_s - \frac{\partial H}{\partial q_s} dq_s \right) - \frac{\partial H}{\partial t} dt \ , \end{aligned}$$

that is,

$$\frac{\partial L}{\partial q_s} = -\frac{\partial H}{\partial q_s} , \tag{4.9a}$$

$$\frac{\partial L}{\partial \dot{q}_s} = p_s , \tag{4.9b}$$

$$\frac{\partial L}{\partial t} = -\frac{\partial H}{\partial t} . \tag{4.9c}$$

Equation (4.9b) expresses p_s in terms of q_s and $\dot{q}_s$, and the remaining Hamiltonian equations of motion, Eq.(4.7b), turn into the Lagrangian equations of motion:

$$\frac{d}{dt}\left(\frac{\partial L}{\partial \dot{q}_s}\right) = \frac{\partial L}{\partial q_s} . \tag{4.10}$$

This completes the demonstration, in the standard case, of the complete equivalence of the Lagrangian and the Hamiltonian forms of dynamics.

We can now throw some light on the physical interpretation of the Hamiltonian H. For the usual nonrelativistic Lagrangian, the dependence on the velocities $\dot{q}_s$ is at most quadratic. If we write

$$L(q, \dot{q}, t) = \sum_{rs} \dot{q}_r \dot{q}_s L^{(2)}_{rs}(q, t) + \sum_{s} \dot{q}_s L^{(1)}_s(q, t) + L^{(0)}(q, t) , \tag{4.11}$$

we find

$$p_s = 2\sum_{r} \dot{q}_r L^{(2)}_{rs}(q, t) + L^{(1)}_s(q, t) ,$$

$$H(q, p, t) = \sum_{rs} \dot{q}_r \dot{q}_s L^{(2)}_{rs}(q, t) - L^{(0)}(q, t) . \tag{4.12}$$

In particular, if the kinetic energy T is quadratic in the velocities $\dot{q}_s$ and the potential energy V is velocity independent, it follows that

$$L = T - V , \quad H = T + V , \tag{4.13}$$

so that the Hamiltonian represents the total energy of the system.

Conservation Laws and the Action Principle

Aside from serving as a compact and elegant statement of the fundamental laws of motion of a dynamical system, the action principle can be used to deduce conservation laws characteristic of the system. These are consequences of properties of the Lagrangian that follow from its functional form, and they are general aids in the problem of solving the equations of motion.

The basic idea is to combine the action principle statement of the laws of motion with knowledge about the behavior of the Lagrangian under various transformations. Let us begin with the simplest situation, in which the Lagrangian happens to be invariant under some geometrical transformation in configuration space. It will suffice to consider infinitesimal transformations. By an infinitesimal geometrical transformation we mean a rule that, applied to any curve C in configuration space, generates another curve C' differing infinitesimally from C. Each point Q on C, with coordinates q_s, is taken into a point Q' on C' with the coordinates

$$q'_s = q_s + \epsilon\phi_s(q) \equiv q_s + \bar{\delta}q_s \; ; \tag{4.14}$$

ϵ is a small parameter, and $\phi_s(q)$ are specific functions of q_r that characterize the transformation. (We write $\bar{\delta}q$ and not δq to stress that this is a specific, not an arbitrary, variation of C.) In effect, we have here a mapping of the points of configuration space onto themselves, which when applied to the points comprising C yields C'. If we imagine the system point traveling along C, the points on C are parametrized by time and have coordinates $q_s(t)$: at each point the velocities $\dot{q}_s(t)$ are determined. Applying the transformation, we obtain a conceivable motion along C', points on which have coordinates $q'_s(t)$; at corresponding points Q and Q' on C and C', the velocities differ by the time derivative of Eq.(4.14):

$$\bar{\delta}\dot{q}_s(t) = \frac{d}{dt}\bar{\delta}q_s(t) = \sum_r \epsilon\frac{\partial\phi_s}{\partial q_r}\dot{q}_r(t) \; . \tag{4.15}$$

Let us suppose the Lagrangian is invariant under this transformation; that is, with the change in the q's given by Eq.(4.14) and in the $\dot{q}$'s by Eq.(4.15), there is no change in L:

$$\bar{\delta}L \equiv L(q_s + \epsilon\phi_s, \dot{q}_s + \epsilon\dot{\phi}_s, t) - L(q_s, \dot{q}_s, t) = 0 \tag{4.16}$$

Then we have a conservation law. We know that, under the most general infinitesimal variation of the trajectory in configuration space, the variation in the action functional is given by Eq.(3.9):

$$\begin{aligned} \Delta\Phi[C] &\equiv \sum_s \int_{t_1}^{t_2} dt L_s(q, \dot{q}, \ddot{q}, t)\delta q_s(t) + \left[\sum_s p_s \delta q_s(t)\right]\Bigg|_{t_1}^{t_2} , \\ L_s &\equiv \frac{\partial L}{\partial q_s} - \frac{d}{dt}\frac{\partial L}{\partial \dot{q}_s} \; . \end{aligned} \tag{4.17}$$

[It is enough to consider variations that leave the terminal times t_1, t_2 fixed]. Here $\delta q_s(t)$ describes an arbitrary variation of C. This expression for $\Delta\Phi[C]$ in particular is true for the specific variation described by $\bar{\delta}q_s$ when we write $\bar{\Delta}\Phi[C]$. On the other hand, because t_1 and t_2 are kept fixed, $\bar{\Delta}\Phi[C]$ is also equal to the integral along C of the variation $\bar{\delta}L$ of L, but this happens to vanish:

$$\bar{\Delta}\Phi[C] \equiv \int_{t_1}^{t_2} dt\bar{\delta}L(q, \dot{q}, t) = 0 \; . \tag{4.18}$$

Equating the results obtained from these two ways of computing $\overline{\Delta}\Phi[C]$ for the particular variation $\overline{\delta}q_s$, we obtain an identity valid for all paths C:

$$-\sum_s \int_{t_1}^{t_2} dt L_s(q,\dot{q},\ddot{q},t)\overline{\delta}q_s(t) = \left[\sum_s p_s\overline{\delta}q_s(t)\right]\Bigg|_{t_1}^{t_2} . \tag{4.19}$$

The content of this identity is no more and no less than the statement that L is invariant under the transformation considered: it has only been expressed in a special way. If now C is a dynamical path on which the equations of motion are obeyed, the expressions L_s vanish; therefore the right-hand side above must also vanish. In other words the bracketed factor has the same values at t_1 and t_2 and thus at all times for a given state of motion. With the use of Eq.(4.14), we thus get

$$\sum_s \phi_s(q)p_s \equiv \sum_s \phi_s(q)\frac{\partial L}{\partial \dot{q}_s} = \text{constant of motion} . \tag{4.20}$$

Thus the characteristic of an invariance property of the Lagrangian under a geometric, or point, transformation in configuration space is that it leads to a constant of motion that is *linear* in the conjugate momenta p_s; the coefficients are functions of the q's, determined by the transformation. We can also consider, as a simple generalization, invariance under a geometric transformation that is time dependent: this would involve a set of functions $\phi_s(q,t)$ in Eq.(4.14), and an extra term would appear in Eq.(4.15) for $\overline{\delta}\dot{q}_s(t)$. But as long as $\overline{\delta}L$ vanishes, we still have the constant of motion (Eq(4.20)), with $\phi_s(q,t)$ in place of $\phi_s(q)$.

If we examine the foregoing derivation, which is essentially to establish an identity of the form of Eq.(4.19) in which one side involves an integral with L_s in the integrand and the other side some expression evaluated at the terminal times, we can see two generalizations that still lead to a conservation law. The first is to consider in place of a geometric transformation a more general transformation that applies to a trajectory C to generate another trajectory C'. (A trajectory is a curve with a specified rate of travel along it, that is, with specified velocities.). If at time t the system point passes through Q on C, or through Q' on C', we will permit the coordinate differences $\overline{\delta}q_s$ between Q and Q' to depend on the velocities at Q:

$$\overline{\delta}q_s = \epsilon\phi_s(q,\dot{q}) . \tag{4.21}$$

The expression for $\overline{\delta}\dot{q}_s$ is just the time derivative of $\overline{\delta}q_s$; thus it can depend (linearly) on the accelerations at Q. Under such a transformation of trajectory, suppose that L is not quite invariant, but that the change in L can be expressed as the total time derivative of a suitable function; this is the second generalization, and allows for $\overline{\delta}L$ the form:

$$\begin{aligned}\overline{\delta}L &\equiv L(q_s+\overline{\delta}q_s, \dot{q}_s+\overline{\delta}\dot{q}_s, t) - L(q_s,\dot{q}_s,t)\\ &= \epsilon\frac{d}{dt}F(q_s,\dot{q}_s,t) .\end{aligned} \tag{4.22}$$

If $F = 0$, we would have said that L is invariant under the transformation considered; if $F \neq 0$, we shall say that L is *quasi-invariant* under this transformation. Even with such behavior of L we can get a conservation law. We again have two ways of computing $\bar{\Delta}\Phi[C]$ for the particular variation $\bar{\delta}q_s$; one is to specialize Eq.(4.17) by substituting $\bar{\delta}q_s$ is place of δq_s, and the other is to invoke the quasi-invariance of L to get:

$$\bar{\Delta}\Phi[C] \equiv \int_{t_1}^{t_2} dt\bar{\delta}L(q,\dot{q},t) \equiv \epsilon[F(q,\dot{q},t)]\Big|_{t_1}^{t_2} . \tag{4.23}$$

Thus we end up with the identity:

$$-\sum_s \int_{t_1}^{t_2} dt L_s(q,\dot{q},\ddot{q},t)\bar{\delta}q_s(t) = \left[\sum_s p_s\bar{\delta}q_s(t) - \epsilon F(q,\dot{q},t)\right]\Bigg|_{t_1}^{t_2} \tag{4.24}$$

which is valid for all trajectories C and generalizes Eq.(4.19). Again, this identity is just an expression of the quasi-invariance of L, no more and no less. And again, if C is a dynamical path on which $L_s = 0$, the right-hand side gives a conserved quantity:

$$\sum_s \phi_s(q,\dot{q})p_s - F(q,\dot{q},t) = \text{constant of motion} . \tag{4.25}$$

Such a constant of motion can have a dependence on the p's that is other than linear, when the $\dot{q}$'s are expressed in terms of the q's and p's. Incorporating an explicit time dependence in the ϕ_s is straightforward. Thus the generalization consists in making use of quasi-invariance properties of L under transformations that act directly on a conceivable "history" (sequence of configurations labeled by time) to yield other such.

Invariance and quasi-invariance properties of the Lagrangian can also be interpreted in a slightly different fashion. Starting with some trajectory C and applying the transformation Eq.(4.21) under which L is quasi-invariant, we arrive at a slightly altered but definite trajectory $C' = C_\epsilon$, say, ϵ being small and kept fixed. We can write Eq.(4.23) in the form

$$\Phi[C_\epsilon] = \Phi[C] + [\epsilon F]\Big|_{t_1}^{t_2} . \tag{4.26}$$

Let us assume that the transformation that takes C into C_ϵ is invertible. Now the Weiss form of the action principle says that C is a dynamical path if and only if an arbitrary infinitesimal variation in C produces only end-point contributions in $\Delta\Phi[C]$. Starting with an arbitrary variation about C and applying to it the transformation that takes C to C_ϵ, we obtain some variation about C_ϵ; conversely because the transformation is invertible, an arbitrary variation about C_ϵ is the image of a suitably chosen one about C. Because according to Eq.(4.26) the difference between $\Phi[C_\epsilon]$ and $\Phi[C]$ is already made up of end-point terms only, we conclude that an arbitrary variation about C_ϵ will produce only end-point contributions in $\Delta\Phi[C_\epsilon]$ if a corresponding statement about C is true: that is, C_ϵ will be a dynamical path if C is. Stated in another way, the functions $q_s(t) + \epsilon\phi_s(q, \dot{q})$ will obey the equations of motion (to first order in ϵ) if the functions $q_s(t)$ do. Thus every transformation under which L is either invariant or quasi invariant is a transformation under which the equations of motion are invariant. The point to note is that it is enough to have *quasi-invariance* of L in order to get invariance of the equations of motion.

Returning to the problem of conservation laws, by now the reader will recognize that there are weaker conditions than quasi-invariance that can lead to a constant of motion. In fact, if under the transformation Eq.(4.21) we have for $\overline{\delta}L$ the form

$$\overline{\delta}L = \epsilon \sum_s L_s(q, \dot{q}, \ddot{q}, t) g_s(q, \dot{q}, t) + \epsilon \frac{d}{dt} F(q, \dot{q}, t) , \qquad (4.27)$$

where the functions g_s remain finite on a dynamical path (or at any rate where $L_s g_s$ vanishes on a dynamical path), then the identity of Eq.(4.24) will hold in a slightly altered form, and the expression appearing in Eq.(4.25) will be conserved. In practice, however, quasi-invariances and invariances of the Lagrangian are quite common, but behavior such as in Eq.(4.27) is not. We have pointed out this possibility for the sake of completeness.

The discussion we have been giving is completely general. A constant of motion that arises by any of the methods given above may have an explicit time dependence, or it may not. We might be inclined to think that it is a "useful" constant of motion only if it has no explicit time dependence; but even if it does, it is useful to the extent that the expression involved depends only on the q's and $\dot{q}$'s, whereas the equations of motion involve the accelerations. The expression in Eq.(4.25) will have no explicit time dependence if both ϕ_s and F have none or if both do and there is a cancellation.

The character of these conservation laws changes greatly if the functions ϕ_s that determine the transformations turn out to involve *arbitrary functions of time*, for then, instead of arriving at quantities that are conserved in each state of motion, we obtain quantities that are conserved by being forced to vanish in all states of motion! These are called *constraints*: we study in Chapter 8 in more detail how constraints can arise in this way.

Chapter 5

Invariance Properties of the Lagrangian & Hamiltonian Descriptions, Poisson and Lagrange Brackets, and Canonical Transformations

The study of the invariance properties of the Lagrangian and the Hamiltonian equations of motion is a very important aspect of the formal structure of classical dynamics. Because both systems of equations are equivalent to variational principles that can be stated to a large extent in a coordinate-independent way, we must expect these invariance properties to be present. In specific cases the invariance properties can be used to find particularly suitable coordinates in terms of which the equations of motion are more tractable. From a more general point of view it is desirable to display and understand the full transformation structure of the classical theory because it has many analogies with the structure of quantum mechanics. However, we are not able to explain this latter relationship in any detail here. We deal, first, with the Lagrangian description, and later with the Hamiltonian one.

The Lagrangian equations of motion are derived in Chapter 2 from the Newtonian equations by first introducing generalized coordinates in terms of which the holonomic constraints of the system were automatically taken care of, and then determining the differential equations of motion for these coordinates. Beyond requiring that the constraint equations be identically obeyed, the choice of the generalized coordinates was left unspecified. Therefore, if one set of coordinates $q_1, q_2, \cdots, q_k$ were to be replaced by another one, $Q_1, Q_2, \cdots, Q_k$, the latter being given as independent functions of the former, then the equations of motion in terms of Q_s must be of exactly the same form as before, with the functional form of the Lagrangian possibly altered because it is reexpressed in terms of the new coordinates and velocities. That this must be so is also evident from the fact that the equations of motion in the coordinates q_s are just the Euler-Lagrange equations coming from the Hamilton variational principle: it follows that all that needs to be done is to write the Lagrangian in terms of the new variables, then the new equations of motion will be just the Euler-Lagrange equations corresponding to the new Lagrangian. Although, as we say, this is self-evident, nevertheless it is instructive to go through the explicit demonstration of this invariance.

Let us suppose, then, that we have the Lagrangian equations of motion given for the coordinates q_s, with a certain Lagrangian $L(q, \dot{q}, t)$. The new coordinates Q_s would be expressed as k independent functions of the q's and, being independent functions, would permit us to express each q_s in terms of the Q's:

$$Q_s = Q_s(q), \qquad q_s = q_s(Q). \tag{5.1}$$

(For simplicity, we ignore a possible t dependence in these equations.) The two sets of velocities are linearly related,

$$\dot{Q}_s = \sum_r \frac{\partial Q_s}{\partial q_r}\dot{q}_r \ , \qquad \dot{q}_s = \sum_r \frac{\partial q_s}{\partial Q_r}\dot{Q}_r \ , \tag{5.2}$$

and following the rules of partial differentiation appropriate to the Lagrangian form, we also get:

$$\frac{\partial \dot{Q}_s}{\partial \dot{q}_r} = \frac{\partial Q_s}{\partial q_r} \ , \qquad \frac{\partial \dot{q}_s}{\partial \dot{Q}_r} = \frac{\partial q_s}{\partial Q_r} \ . \tag{5.3}$$

Notice that the two sets of coordinates appear quite symmetrically in all these equations. One more pair of equations quite easily derived is:

$$\frac{d}{dt}\frac{\partial Q_s}{\partial q_r} = \frac{\partial \dot{Q}_s}{\partial q_r} \ , \qquad \frac{d}{dt}\frac{\partial q_s}{\partial Q_r} = \frac{\partial \dot{q}_s}{\partial Q_r} \ . \tag{5.4}$$

Let the new Lagrangian, $L'(Q, \dot{Q}, t)$ be obtained from L by substituting for q and $\dot{q}$ their expressions in terms of Q and $\dot{Q}$:

$$L(q, \dot{q}, t) \equiv L'(Q, \dot{Q}, t) \ . \tag{5.5}$$

Then, using the foregoing equations connecting the two sets of variables, as well as the equations of motion for the q's, we find:

$$\begin{aligned} \frac{d}{dt}\frac{\partial L'}{\partial \dot{Q}_r}(Q, \dot{Q}, t) &= \frac{d}{dt}\sum_s \frac{\partial L}{\partial \dot{q}_s}\frac{\partial \dot{q}_s}{\partial \dot{Q}_r} = \frac{d}{dt}\sum_s \frac{\partial L}{\partial \dot{q}_s}\frac{\partial q_s}{\partial Q_r} \\ &= \sum_s \left(\frac{\partial L}{\partial q_s}\frac{\partial q_s}{\partial Q_r} + \frac{\partial L}{\partial \dot{q}_s}\frac{\partial \dot{q}_s}{\partial Q_r} \right) = \frac{\partial L'}{\partial Q_r}(Q, \dot{Q}, t) \ , \\ \text{i.e.} \qquad \frac{d}{dt}\frac{\partial L'}{\partial \dot{Q}_r} &= \frac{\partial L'}{\partial Q_r} \ . \end{aligned} \tag{5.6}$$

Thus we see that the old Lagrangian equations of motion imply the new ones, and, because the two sets of coordinates enter symmetrically, the new Lagrangian equations imply the old ones. In other words, the two sets of equations are equivalent, and we have explicitly verified that the Lagrangian formulation *preserves* its *form* under arbitrary invertible changes of coordinates in configuration space. We should note that in such transformations, the transformation law for the coordinates determines that for the velocities and also that the numerical value of the Lagrangian is unaltered, although as we said earlier, its *functional form would change in general.*

Let us write the transformation equations given above in terms of the q's and p's instead of the q's and $\dot{q}$'s. From the definitions of the two sets of momenta:

$$p_s = \frac{\partial L}{\partial \dot{q}_s} , \qquad P_s = \frac{\partial L'}{\partial \dot{Q}_s} , \tag{5.7}$$

and from Eq.(5.3) we derive:

$$Q_s = Q_s(q) , \qquad P_s = \sum_r p_r \frac{\partial q_r}{\partial Q_s} . \tag{5.8}$$

Thus even though the p's and P's are defined in terms of specific Lagrangians, the transformation equations connecting them do not refer to anything other than the relation between the q's and Q's. It is clear that because the two sets of Lagrangian equations of motion are equivalent, and because each set could be rewritten as a system of Hamilton's equations in terms of its own Hamiltonian, the two sets of Hamiltonian equations must be equivalent; that is, under the phase space transformation defined by Eq.(5.8), the form of the Hamiltonian equations of motion is preserved with the new Hamiltonian suitably defined. Because, as we said before, the transformation equations make no reference to the particular Lagrangian, they have the property of leaving *any* Hamiltonian set of equations form invariant. The particular type of phase-space transformation we have obtained here is a special case of what is called a "contact transformation." It is characterized by the invariance of a certain differential expression:

$$\sum_s P_s dQ_s = \sum_s p_s dq_s . \tag{5.9}$$

A discussion of some properties of contact transformations, which form a subset (and subgroup) of all transformations leaving the Hamiltonian equations form invariant, is given in the next chapter.

Let us now proceed to an examination of the Hamiltonian form of the equations of motion, namely, from Eq.(3.23) the system:

$$\dot{q}_s = \frac{\partial H}{\partial p_s} , \qquad \dot{p}_s = -\frac{\partial H}{\partial q_s} , \qquad s = 1, 2 \cdots , k. \tag{5.10}$$

The first thing to notice is that the time derivatives of the basic variables are directly given as functions of the same variables, unlike the Lagrangian form where the accelerations $\ddot{q}_s$ are "buried" in the equations of motion. Therefore, we are able to write down an equation of motion for any function f of the q's, the p's, and t, for we have:

$$\begin{aligned}\dot{f} \equiv \frac{df}{dt} &= \sum_s \left\{ \frac{\partial f}{\partial q_s}\dot{q}_s + \frac{\partial f}{\partial p_s}\dot{p}_s \right\} + \frac{\partial f}{\partial t} \\ &= \sum_s \left\{ \frac{\partial f}{\partial q_s}\frac{\partial H}{\partial p_s} - \frac{\partial f}{\partial p_s}\frac{\partial H}{\partial q_s} \right\} + \frac{\partial f}{\partial t} \,. \end{aligned} \tag{5.11}$$

The algebraic expression involving f and H that appears here is of fundamental importance and is called the "Poisson bracket" (PB) of f with H. More generally, for any two (differentiable) functions $A(q,p,t)$ and $B(q,p,t)$ on phase space, their PB is a third function C given by

$$C = \{A, B\} = \sum_s \left\{ \frac{\partial A}{\partial q_s}\frac{\partial B}{\partial p_s} - \frac{\partial A}{\partial p_s}\frac{\partial B}{\partial q_s} \right\} . \tag{5.12}$$

Apart from noticing the obvious facts that $\{A, B\}$ is linear in A as well as in B and changes sign under their interchange, we postpone till later an examination of the properties of the PB. Now we use it to write the general Hamiltonian equation of motion in a concise and elegant way:

$$\dot{f} = \{f, H\} + \frac{\partial f}{\partial t} \,. \tag{5.13}$$

If the function f has no explicit time dependence, the term $\partial f/\partial t$ drops out. The basic equations of motion for the q's and p's are simply special cases of Eq.(5.13), resulting from special choices of f:

$$\dot{q}_s = \{q_s, H\} \,, \qquad \dot{p}_s = \{p_s, H\} \,. \tag{5.14}$$

Because of the antisymmetry property, the PB of H with itself vanishes, and the equation of motion for the Hamiltonian is:

$$\frac{dH}{dt} = \frac{\partial H}{\partial t} \,. \tag{5.15}$$

Therefore in the course of the motion, the time dependence of the Hamiltonian stems only from its explicit time dependence; if it has none, the Hamiltonian is a "constant of motion." More generally, if a function $A(q,p)$ has a vanishing PB with H, then it is a constant of motion for the system.

The PB form of the general equation of motion emphasizes the equivalent roles played by the coordinates q_s and the momenta p_s in Hamiltonian dynamics. This suggests that the elementary identification of a generalized coordinate with position and of a generalized momentum with mass times velocity is not essential in formulating Hamiltonian dynamics. In view of this, one is led to the consideration of general transformations of phase-space coordinates, (q,p) to (Q,P), under which the Hamiltonian equations of motion retain their form.

A general coordinate transformation in phase space has the following form:

$$Q_s = Q_s(q,p,t), \qquad P_s = P_s(q,p,t)\ , \qquad s = 1,2,\cdots,k \tag{5.16}$$

It involves $2k$ *independent* functions, allowing us to express the q's and p's in terms of the Q's, P's, and t. We interpret Eq.(5.16) as providing a new set of labels for the points of phase space, which were originally labeled by the q's and p's. We now want to derive conditions on the transformation to ensure that it leaves the Hamiltonian equations of motion invariant for all possible Hamiltonians. In other words, given Eq.(5.10) for some H, there should exist a new Hamiltonian $H'(Q,P,t)$, determined jointly by H and the transformation equations, such that in the new coordinates the equations of motion should appear as:

$$\dot{Q}_s = \frac{\partial H'}{\partial P_s}\ , \qquad \dot{P}_s = -\frac{\partial H'}{\partial Q_s}\ , \qquad s = 1,2,\cdots,k\ . \tag{5.17}$$

We first analyze this requirement via the Weiss action principle and later on in this chapter, by a more direct calculation. The phase space functional Ψ leading to the original Hamiltonian equations is:

$$\Psi[\mathcal{C}] = \int_{t_1}^{t_2} \left[\sum_s p_s\dot{q}_s - H(q,p,t)\right] dt\ , \tag{5.18}$$

the integration being along the trajectory $\mathcal{C}$. Similarly, the new equations of motion. Eq.(5.17), would result from an application of the Weiss action principle to the functional Ψ' defined as:

$$\Psi'[\mathcal{C}] = \int_{t_1}^{t_2} \left[\sum_s P_s\dot{Q}_s - H'(Q,P,t)\right] dt\ . \tag{5.19}$$

The same general trajectory $\mathcal{C}$ appears as the argument both of Ψ and of Ψ': in the one case its points are given by functions $q(t)$, $p(t)$, in the other case by functions $Q(t)$, $P(t)$. It is clear that the old and the new Hamiltonian equations will be equivalent if the relation between the functionals Ψ and Ψ' is such that whenever a general variation about a particular trajectory $\mathcal{C}$ produces only end-point contributions in $\Delta\Psi$, the same variation about this same trajectory likewise produces only end-point contributions in $\Delta\Psi'$. Then, on those trajectories in phase space along which the old equations of motion are valid, the new ones will be valid too, and vice versa. A particularly simple way of ensuring this is to require that for an *arbitrary* trajectory $\mathcal{C}$, the difference between Ψ and Ψ' depends on the end points of $\mathcal{C}$ only:

$$\Psi'[\mathcal{C}] - \Psi[\mathcal{C}] = F(\Pi_2,t_2) - F(\Pi_1,t_1)\ . \tag{5.20}$$

Here, $F(\Pi,t)$ is a (possibly time-dependent) function on phase space. If Eq.(5.20) is to hold for all $\mathcal{C}$, we can identify the integrands and obtain, in general,

$$\sum_s P_s\dot{Q}_s - H'(Q,P,t) = \sum_s p_s\dot{q}_s - H(q,p,t) + \frac{dF}{dt}\ . \tag{5.21}$$

This is a sufficient condition for form invariance of the Hamiltonian equations of motion; as we shall see, it will also tell us how to define the new Hamiltonian H'. This equation can be used in two ways, either to derive conditions on the transformation of coordinates or actually to construct allowed transformations. We do the former here and discuss the latter aspect in the next chapter.

Because the new coordinates Q, P are all independent, we can imagine the function F, as well as the q's and p's, expressed in terms of them; in that case, Eq.(5.21) can be rewritten as follows:

$$\sum_s P_s\dot{Q}_s - H' = \sum_{s,r} p_s\left(\frac{\partial q_s}{\partial Q_r}\dot{Q}_r + \frac{\partial q_s}{\partial P_r}\dot{P}_r\right) + \sum_s p_s\frac{\partial q_s}{\partial t} - H$$
$$+\sum_r\left(\frac{\partial F}{\partial Q_r}\dot{Q}_r + \frac{\partial F}{\partial P_r}\dot{P}_r\right) + \frac{\partial F}{\partial t}\ . \tag{5.22}$$

Identifying the coefficients of $\dot{Q}_r$, $\dot{P}_r$ gives us:

$$P_r - \sum_s p_s\frac{\partial q_s}{\partial Q_r} = \frac{\partial F}{\partial Q_r}(Q,P,t)\ , \tag{5.23a}$$

$$-\sum_s p_s\frac{\partial q_s}{\partial P_r} = \frac{\partial F}{\partial P_r}(Q,P,t)\ , \tag{5.23b}$$

$$H'(Q,P,t) = H(q,p,t) - \sum_s p_s\frac{\partial q_s}{\partial t} - \frac{\partial F}{\partial t}(Q,P,t)\ . \tag{5.23c}$$

If the function F can be found, satisfying Eqs.(5.23a) and (5.23b) then we can use it in Eq.(5.23c) to obtain H'. The necessary and sufficient condition for the existence of F is a set of three integrability conditions:

$$\frac{\partial}{\partial Q_{r'}}\left(P_r - \sum_s p_s\frac{\partial q_s}{\partial Q_r}\right) = \frac{\partial}{\partial Q_r}\left(P_{r'} - \sum_s p_s\frac{\partial q_s}{\partial Q_{r'}}\right)\ , \tag{5.24a}$$

$$\frac{\partial}{\partial P_{r'}}\left(\sum_s p_s\frac{\partial q_s}{\partial P_r}\right) = \frac{\partial}{\partial P_r}\left(\sum_s p_s\frac{\partial q_s}{\partial P_{r'}}\right)\ , \tag{5.24b}$$

$$\frac{\partial}{\partial P_{r'}}\left(P_r - \sum_s p_s\frac{\partial q_s}{\partial Q_r}\right) = \frac{-\partial}{\partial Q_r}\left(\sum_s p_s\frac{\partial q_s}{\partial P_{r'}}\right)\ . \tag{5.24c}$$

These simplify immediately to the following form:

$$\sum_s\left(\frac{\partial q_s}{\partial Q_r}\frac{\partial p_s}{\partial Q_{r'}} - \frac{\partial q_s}{\partial Q_{r'}}\frac{\partial p_s}{\partial Q_r}\right) = 0\ , \tag{5.25a}$$

$$\sum_s\left(\frac{\partial q_s}{\partial P_r}\frac{\partial p_s}{\partial P_{r'}} - \frac{\partial q_s}{\partial P_{r'}}\frac{\partial p_s}{\partial P_r}\right) = 0\ , \tag{5.25b}$$

$$\sum_s\left(\frac{\partial q_s}{\partial Q_r}\frac{\partial p_s}{\partial P_{r'}} - \frac{\partial q_s}{\partial P_{r'}}\frac{\partial p_s}{\partial Q_r}\right) = \delta_{rr'}\ . \tag{5.25c}$$

We show later that, except for a rather uninteresting class of transformations, Eq.(5.25) contains not only sufficient but necessary conditions for a transformation to leave the Hamiltonian form of dynamics invariant.

The characteristic structure of the expressions appearing in Eq.(5.25) leads us to define a new bracket called the "Lagrange bracket" (LB). If $A_1(q,p)$, $\cdots, A_{2k}(q,p)$ is a set of $2k$ independent functions on phase space so that the q's and p's can be written as functions of the A's, then the LB of two A's computed with respect to the phase space coordinates q, p is defined by

$$[A_j, A_l] = \sum_{s=1}^{k} \left(\frac{\partial q_s}{\partial A_j} \frac{\partial p_s}{\partial A_l} - \frac{\partial q_s}{\partial A_l} \frac{\partial p_s}{\partial A_j} \right) , \quad j, l = 1, 2, \cdots, 2k . \tag{5.26}$$

Of course, after evaluation of the partial derivatives, everything could be reexpressed in terms of the q's and p's; thus $[A_j, A_l]$ is also a phase-space function. Just like the PB, the LB is antisymmetric in its two arguments; however, we need to be given $2k$ independent functions before we can define the LB of any two of them. The LB's and PB's of a set of $2k$ independent functions A_j form matrices that are essentially inverses of one another:

$$\sum_{l=1}^{2k} [A_j, A_l]\{A_l, A_m\} = -\delta_{jm} , \quad j, m = 1, 2, \cdots, 2k . \tag{5.27}$$

The proof of this fact is easy enough, but we shall give it later, along with other important properties of LB's using a simpler notation. We can now write Eq.(5.25) in LB form:

$$[Q_r, Q_{r'}] = [P_r, P_{r'}] = 0 , \qquad [Q_r, P_{r'}] = \delta_{rr'} , \tag{5.28}$$

or equivalently, on using Eq.(5.27), in PB form:

$$\{Q_r, Q_{r'}\} = \{P_r, P_{r'}\} = 0 , \qquad \{Q_r, P_{r'}\} = \delta_{rr'} . \tag{5.29}$$

In both these equations, the relevant brackets are computed with respect to the original variables q_s, p_s.

Transformations $q, p \to Q, P$ satisfying Eq.(5.28) or equivalently Eq.(5.29) are called "canonical transformations". Only such transformations will leave the Hamiltonian scheme invariant. The PB's (LB's) of the basic variables Q, P, with one another are called the fundamental PB's (LB's). If one considers the identity transformation, $Q_s = q_s$ and $P_s = p_s$, Eqs.(5.28) and (5.29) are trivially obeyed. Thus one characterizes a canonical transformation by saying that it preserves the values of the fundamental PB's and LB's. Later on we find several alternative and equivalent ways of defining canonical transformations.

If one considers those transformations that are not explicitly time dependent, Eq.(5.23c) shows that the old and the new Hamiltonians are numerically equal, but generally differ in functional form because they are written in different variables. Thus under such transformations, the Hamiltonian retains its physical significance of being the total energy of the system. One can also see that, because of Eq.(5.9), Eq. (5.8) defines a particular type of canonical transformation, corresponding to taking $H' = H, F = 0$ in Eq.(5.21).

Tensor Notation

To be able to handle PB's and LB's more easily, and to understand their properties and relationships more transparently, we now introduce a compact notation, treating all the coordinates q_s, p_s as components of a single entity. (However, it may sometimes still be worthwhile to distinguish between the q's and p's).

Denote the $2k$ variables q_s, p_s by ω^μ, $\mu = 1, 2, \cdots, 2k$. For $\mu = 1, 2, \cdots, k$, ω^μ is the same as q_μ, and for $\mu \geqq k+1$, ω^μ equals $p_{\mu-k}$. A function $f(q,p)$ can be written $f(\omega)$. To express the PB of a function f with a function g concisely, let us define a $2k$ dimensional antisymmetric matrix $\|\epsilon^{\mu\nu}\|$ with the following matrix elements:

$$\epsilon^{\mu\nu} = \begin{cases} 0 & \text{if } \mu, \nu \leqq k \text{ or } \mu, \nu > k; \\ 1 & \text{if } \mu \leqq k,\ \nu = \mu + k; \\ -1 & \text{if } \nu \leqq k,\ \mu = \nu + k\ . \end{cases} \tag{5.30}$$

This matrix is clearly nonsingular; its inverse, denoted by $\|\epsilon_{\mu\nu}\|$, obeys

$$\epsilon_{\mu\nu}\epsilon^{\nu\lambda} = \delta_\mu^\lambda, \tag{5.31}$$

and has the following matrix elements:

$$\epsilon_{\mu\nu} = \begin{cases} 0 & \text{if } \mu, \nu \leqq k \text{ or } \mu, \nu > k; \\ -1 & \text{if } \mu \leqq k,\ \nu = \mu + k; \\ +1 & \text{if } \nu \leqq k,\ \mu = \nu + k\ . \end{cases} \tag{5.32}$$

(For Greek indices going over the range $1, 2, \cdots, 2k$, the summation convention is adopted.) Now the PB of $f(\omega)$ with $g(\omega)$ assumes a very simple form:

$$\{f, g\} = \epsilon^{\mu\nu} \frac{\partial f}{\partial \omega^\mu} \frac{\partial g}{\partial \omega^\nu}\ . \tag{5.33}$$

Similarly for the LB: given the set of independent functions $A_j(\omega)$, $j = 1, 2, \cdots, 2k$, the LB of any two of these A's is:

$$[A_j, A_l] = -\epsilon_{\mu\nu} \frac{\partial \omega^\mu}{\partial A_j} \frac{\partial \omega^\nu}{\partial A_l}\ . \tag{5.34}$$

It is now very easy to verify Eq.(5.27): it depends only on the fact that the matrices $\|\epsilon^{\mu\nu}\|$ and $\|\epsilon_{\mu\nu}\|$ are inverses of one another, and that the same is true for $\|\partial A_j/\partial\omega^\mu\|$ and $\|\partial\omega^\mu/\partial A_j\|$:

$$\sum_{l=1}^{2k}[A_j, A_l]\{A_l, A_m\} = -\sum_l \epsilon_{\mu\nu}\frac{\partial\omega^\mu}{\partial A_j}\frac{\partial\omega^\nu}{\partial A_l}\,\epsilon^{\rho\sigma}\,\frac{\partial A_l}{\partial\omega^\rho}\frac{\partial A_m}{\partial\omega^\sigma}$$
$$= -\epsilon_{\mu\nu}\epsilon^{\rho\sigma}\,\frac{\partial\omega^\mu}{\partial A_j}\,\delta^\nu_\rho\,\frac{\partial A_m}{\partial\omega^\sigma}$$
$$= -\frac{\partial\omega^\mu}{\partial A_j}\frac{\partial A_m}{\partial\omega^\mu} = -\frac{\partial A_m}{\partial A_j} = -\delta_{jm}\ . \quad (5.35)$$

Actually, as far as this property is concerned, the antisymmetry of the ϵ matrices is irrelevant.

We next use Eq.(5.33) to establish an important identity for PB's. Let f, g, and h be any three functions of ω, and consider the following cyclic sum:

$$\sum\{\{f,g\},h\} = \{\{f,g\},h\} + \{\{g,h\},f\} + \{\{h,f\},g\}\ ;$$

using Eq.(5.33) twice we write:

$$\sum\{\{f,g\},h\} = \sum \epsilon^{\mu\nu}\epsilon^{\rho\sigma}\frac{\partial}{\partial\omega^\mu}\left(\frac{\partial f}{\partial\omega^\rho}\frac{\partial g}{\partial\omega^\sigma}\right)\frac{\partial h}{\partial\omega^\nu}$$
$$= \sum \epsilon^{\mu\nu}\epsilon^{\rho\sigma}\left(\frac{\partial^2 f}{\partial\omega^\mu\partial\omega^\rho}\frac{\partial g}{\partial\omega^\sigma}\frac{\partial h}{\partial\omega^\nu} + \frac{\partial f}{\partial\omega^\rho}\frac{\partial^2 g}{\partial\omega^\mu\partial\omega^\sigma}\frac{\partial h}{\partial\omega^\nu}\right)\ .$$

Because the sum is cyclic and because the Greek indices are summed over, we can interchange $f \to h \to g \to f$ as well as replace $\rho \to \nu \to \sigma \to \mu \to \rho$ in the second term to obtain:

$$\sum\{\{f,g\},h\} = \sum(\epsilon^{\mu\nu}\epsilon^{\rho\sigma} + \epsilon^{\rho\sigma}\epsilon^{\nu\mu})\frac{\partial^2 f}{\partial\omega^\mu\partial\omega^\rho}\frac{\partial g}{\partial\omega^\sigma}\frac{\partial h}{\partial\omega^\nu}\ .$$

Because of the antisymmetry of the ϵ-matrix, the cyclic sum is seen to vanish identically,

$$\sum\{\{f,g\},h\} = 0\ . \quad (5.36)$$

This basic property of the PB's is called the Jacobi identity. It is a consequence of the antisymmetry of the matrix $\|\epsilon^{\mu\nu}\|$; neither the nonsingular nature nor the specific form of this matrix are needed to prove the identity.

An analogous identity, following from the antisymmetry of the matrix $\|\epsilon_{\mu\nu}\|$, can be established without difficulty for LB's:

$$\sum_{(jlm)}\frac{\partial}{\partial A_j}[A_l, A_m] \equiv \frac{\partial}{\partial A_j}[A_l, A_m] + \frac{\partial}{\partial A_l}[A_m, A_j] + \frac{\partial}{\partial A_m}[A_j, A_l] \quad (5.37)$$
$$= 0.$$

(Here, A_j, A_l, and A_m are three out of a set of $2k$ independent functions).

Let us now collect together the main properties of PB's and LB's. For the former we have:

(i) *Antisymmetry:* $\{f, g\} = -\{g, f\}$.

(ii) *Linearity:* $\{c_1 f_1 + c_2 f_2, g\} = c_1\{f_1, g\} + c_2\{f_2, g\}$, c_1 and c_2 being numbers.

(iii) *For Any Number c and Any f:* $\{c, f\} = 0$; c is a"neutral" element.

(iv) *Jacobi Identity:* $\sum\{\{f, g\}, h\} = 0$.
To these may be added the special properties:

(v) *Fundamental PB's:* $\{\omega^\mu, \omega^\nu\} = \epsilon^{\mu\nu}$.

(vi) *Product Rule:* $\{f_1 f_2, g\} = f_1\{f_2, g\} + \{f_1, g\} f_2$.

In an abstract scheme of definition, the first four properties define a "generalized PB", and if items 5 and 6 also hold, we have "standard PB's".
For LB's we have these main properties:

(i) *Antisymmetry:* $[A_j, A_l] = -[A_l, A_j]$.

(ii) *If c Is a Number:* $[cA_j, A_l] = (1/c)[A_j, A_l]$.

(iii) : $\sum_{(jlm)}(\partial/\partial A_j)[A_l, A_m] = 0$.

The conditions for a transformation $\omega \to \omega'$ to be canonical can be given in elegant form using the new notation. Thus the fundamental PB condition, Eq.(5.29), appears as:

$$\epsilon^{\mu\nu} \frac{\partial \omega'^\rho}{\partial \omega^\mu} \frac{\partial \omega'^\sigma}{\partial \omega^\nu} = \epsilon^{\rho\sigma} \ , \tag{5.38}$$

whereas the fundamental LB condition, Eq.(5.28), is

$$\epsilon_{\mu\nu} \frac{\partial \omega^\mu}{\partial \omega'^\rho} \frac{\partial \omega^\nu}{\partial \omega'^\sigma} = \epsilon_{\rho\sigma} \ . \tag{5.39}$$

These equations allow us to interpret canonical transformations as those transformations that leave $\epsilon^{\mu\nu}$ and $\epsilon_{\mu\nu}$, defined to transform as second-rank contravariant and covariant tensors, respectively, numerically invariant.

What can be said now about the behavior of general functions on phase space, and PB's and LB's among them, with respect to a canonical transformation? If $f(\omega)$ is a given function and the transformation $\omega \to \omega'$ is canonical, we obtain a function f' of ω' by the definition

$$f'(\omega') = f(\omega) \ . \tag{5.40}$$

In general, f and f' have different functional forms; this definition for going from f to f' is characteristic of a "scalar field". If f and g are two functions of ω, and f' and g' their transforms according to Eq.(5.40), we can compute the PB of f' and g' with respect to ω'. Using Eq.(5.38) but with ω and ω'

interchanged (both Eqs. (5.38) and (5.39) hold with ω and ω' interchanged, as is easily verified), we find:

$$\begin{aligned}\{f',g'\}(\omega') &= \epsilon^{\mu\nu}\,\frac{\partial f'(\omega')}{\partial\omega'^{\mu}}\,\frac{\partial g'(\omega')}{\partial\omega'^{\nu}}\\ &= \epsilon^{\mu\nu}\,\frac{\partial f(\omega)}{\partial\omega^{\rho}}\,\frac{\partial g(\omega)}{\partial\omega^{\sigma}}\,\frac{\partial\omega^{\rho}}{\partial\omega'^{\mu}}\,\frac{\partial\omega^{\sigma}}{\partial\omega'^{\nu}} = \epsilon^{\rho\sigma}\,\frac{\partial f(\omega)}{\partial\omega^{\rho}}\,\frac{\partial g(\omega)}{\partial\omega^{\sigma}}\,,\end{aligned}\tag{5.41}$$

that is,

$$\{f',g'\}(\omega') = \{f,g\}(\omega)\,.$$

This means that the PB of any two functions is invariant under canonical transformations and may be evaluated with the use of any system of canonical coordinates, it being understood of course that when we alter the coordinates, the functions transform according to Eq.(5.40). Another way of expressing this is to say that the PB of two scalar fields is also a scalar field, and one can give Eq.(5.41) the concise form:

$$\{f,g\}' = \{f',g'\}\,.\tag{5.42}$$

Because the LB's are always inverse to the PB's, they must share this invariance property. Using Eq.(5.39), with ω and ω' interchanged, we find:

$$\begin{aligned}[A_j,A_l]_{\omega'} &= \epsilon_{\mu\nu}\,\frac{\partial\omega'^{\mu}}{\partial A_j}\,\frac{\partial\omega'^{\nu}}{\partial A_l} = \epsilon_{\mu\nu}\,\frac{\partial\omega'^{\mu}}{\partial\omega^{\rho}}\,\frac{\partial\omega'^{\nu}}{\partial\omega^{\sigma}}\,\frac{\partial\omega^{\rho}}{\partial A_j}\,\frac{\partial\omega^{\sigma}}{\partial A_l}\\ &= \epsilon_{\rho\sigma}\,\frac{\partial\omega^{\rho}}{\partial A_j}\,\frac{\partial\omega^{\sigma}}{\partial A_l}\,.\end{aligned}\tag{5.43}$$

In words: the LB's of a given set of $2k$ independent functions on phase space are independent of the particular canonical coordinate system used for evaluating them:

$$[A_j,A_l]_{\omega'} = [A_j,A_l]_{\omega}\,.$$

These invariance properties of the brackets of arbitrary functions can be used as a definition of canonical transformations, since the fundamental brackets, both Poisson and Lagrange, correspond to special choices of the functions.

Having used the tensor notation for a simple derivation of many properties of canonical transformations and the two kinds of brackets, we conclude this chapter by using it to investigate directly the nature of a transformation that preserves the Hamiltonian form of equations, thereby giving a treatment independent of the Weiss action principle. For the sake of clarity, in this part we use the symbols ϕ and ψ to indicate explicitly the functional forms of ω and ω', the two sets of coordinates, expressed in terms of one another:

$$\omega'^{\mu} = \phi^{\mu}(\omega,t)\,,\qquad \omega^{\mu} = \psi^{\mu}(\omega',t)\,.\tag{5.44}$$

By definition, one has:

$$\frac{\partial\psi^{\rho}(\omega',t)}{\partial\omega'^{\mu}}\,\frac{\partial\phi^{\mu}(\omega,t)}{\partial\omega^{\sigma}} = \delta^{\rho}_{\sigma}\,.\tag{5.45}$$

For a given Hamiltonian $H(\omega, t)$, the equations of motion have the standard PB form obtained from Eqs.(5.13) and (5.33):

$$\dot{\omega}^\mu = \epsilon^{\mu\nu} \frac{\partial H}{\partial \omega^\nu} . \tag{5.46}$$

Using Eq.(5.44), the equations for ω'^μ will then be:

$$\begin{aligned} \dot{\omega}'^\mu &= \{\phi^\mu, H\} + \frac{\partial \phi^\mu}{\partial t} \\ &= \epsilon^{\gamma\beta} \frac{\partial \phi^\mu}{\partial \omega^\gamma} \frac{\partial H}{\partial \omega^\beta} + \frac{\partial \phi^\mu}{\partial t} . \end{aligned} \tag{5.47}$$

We want the transformation to be such that, for every H, there should be a corresponding $H'(\omega', t)$ in terms of which we can write

$$\epsilon^{\gamma\beta} \frac{\partial \phi^\mu}{\partial \omega^\gamma} \frac{\partial H}{\partial \omega^\beta} + \frac{\partial \phi^\mu}{\partial t} = \epsilon^{\mu\nu} \frac{\partial H'(\omega', t)}{\partial \omega'^\nu} . \tag{5.48}$$

Then the new equations of motion would again assume the Hamiltonian form. Rewriting Eq.(5.48) as

$$\frac{\partial H'}{\partial \omega'^\nu} = \epsilon_{\nu\mu} \epsilon^{\gamma\beta} \frac{\partial \phi^\mu}{\partial \omega^\gamma} \frac{\partial H}{\partial \omega^\beta} + \epsilon_{\nu\mu} \frac{\partial \phi^\mu}{\partial t} ,$$

the necessary and sufficient conditions for the existence of H' are the following integrability conditions:

$$\begin{aligned} &\frac{\partial}{\partial \omega'^\lambda} \left(\epsilon_{\nu\mu} \epsilon^{\gamma\beta} \frac{\partial \phi^\mu}{\partial \omega^\gamma} \frac{\partial H}{\partial \omega^\beta} + \epsilon_{\nu\mu} \frac{\partial \phi^\mu}{\partial t} \right) \\ &\qquad = \frac{\partial}{\partial \omega'^\nu} \left(\epsilon_{\lambda\mu} \epsilon^{\gamma\beta} \frac{\partial \phi^\mu}{\partial \omega^\gamma} \frac{\partial H}{\partial \omega^\beta} + \epsilon_{\lambda\mu} \frac{\partial \phi^\mu}{\partial t} \right) . \end{aligned} \tag{5.49}$$

To save on writing, let us indicate the partial derivatives of ϕ^μ and H with respect to the ω's by subscripts:

$$\frac{\partial \phi^\mu}{\partial \omega^\gamma} = \phi^\mu_{,\gamma} ; \qquad \frac{\partial^2 \phi^\mu}{\partial \omega^\gamma \partial \omega^\rho} = \phi^\mu_{,\gamma\beta} , \text{ and so on} . \tag{5.50}$$

Then the basic condition, Eq.(5.49), reads:

$$\left\{ \epsilon_{\nu\mu} \epsilon^{\gamma\beta} \phi^\mu_{,\gamma} H_{,\beta} + \epsilon_{\nu\mu} \frac{\partial \phi^\mu}{\partial t} \right\}_{,\rho} \frac{\partial \psi^\rho}{\partial \omega'^\lambda} = \left\{ \epsilon_{\lambda\mu} \epsilon^{\gamma\beta} \phi^\mu_{,\gamma} H_{,\beta} + \epsilon_{\lambda\mu} \frac{\partial \phi^\mu}{\partial t} \right\}_{,\rho} \frac{\partial \psi^\rho}{\partial \omega'^\nu} .$$

Multiply both sides by $\phi^\lambda_{,\sigma} \phi^\nu_{,\tau}$ to get, on using Eq.(5.45):

$$\begin{aligned} &\left\{ \epsilon_{\nu\mu} \epsilon^{\gamma\beta} \phi^\mu_{,\gamma\rho} H_{,\beta} + \epsilon_{\nu\mu} \epsilon^{\gamma\beta} \phi^\mu_{,\gamma} H_{,\beta\rho} + \epsilon_{\nu\mu} \frac{\partial \phi^\mu_{,\rho}}{\partial t} \right\} \delta^\rho_\sigma \phi^\nu_{,\tau} \\ &= \left\{ \epsilon_{\lambda\mu} \epsilon^{\gamma\beta} \phi^\mu_{,\gamma\rho} H_{,\beta} + \epsilon_{\lambda\mu} \epsilon^{\gamma\beta} \phi^\mu_{,\gamma} H_{,\beta\rho} + \epsilon_{\lambda\mu} \frac{\partial \phi^\mu_{,\rho}}{\partial t} \right\} \delta^\rho_\tau \phi^\lambda_{,\sigma} . \end{aligned}$$

Collecting terms with different dependences on H, we get:

$$\epsilon^{\gamma\beta}\{\phi^{\nu}_{,\tau}\epsilon_{\nu\mu}\phi^{\mu}_{,\gamma\sigma} - \phi^{\lambda}_{,\sigma}\epsilon_{\lambda\mu}\phi^{\mu}_{,\gamma\tau}\}\, H_{,\beta} + \phi^{\nu}_{,\tau}\epsilon_{\nu\mu}\phi^{\mu}_{,\gamma}\epsilon^{\gamma\beta}H_{,\beta\sigma}$$
$$- \phi^{\mu}_{,\gamma}\epsilon_{\lambda\mu}\phi^{\lambda}_{,\sigma}\epsilon^{\gamma\beta}H_{,\beta\tau} + \phi^{\nu}_{,\tau}\epsilon_{\nu\mu}\frac{\partial\phi^{\mu}_{,\sigma}}{\partial t} - \phi^{\lambda}_{,\sigma}\epsilon_{\lambda\mu}\frac{\partial\phi^{\mu}_{,\tau}}{\partial t} = 0$$

By changing some of the summation indices, we can put this into a simpler form:

$$\epsilon^{\gamma\beta}\{\phi^{\nu}_{,\tau}\epsilon_{\nu\mu}\phi^{\mu}_{,\sigma\gamma} + \phi^{\nu}_{,\tau\gamma}\epsilon_{\nu\mu}\phi^{\mu}_{,\sigma}\}\, H_{,\beta} + \phi^{\nu}_{,\tau}\epsilon_{\nu\mu}\phi^{\mu}_{,\gamma}\epsilon^{\gamma\beta}H_{,\beta\sigma}$$
$$- H_{,\tau\beta}\epsilon^{\beta\gamma}\phi^{\mu}_{,\gamma}\epsilon_{\mu\lambda}\phi^{\lambda}_{,\sigma} + \frac{\partial}{\partial t}(\phi^{\nu}_{,\tau}\epsilon_{\nu\mu}\phi^{\mu}_{,\sigma}) = 0\,. \tag{5.51}$$

The basic quantity appearing here is the expression

$$L_{\tau\sigma} = \phi^{\nu}_{,\tau}\epsilon_{\nu\mu}\phi^{\mu}_{,\sigma} = -L_{\sigma\tau} = \epsilon_{\nu\mu}\frac{\partial\omega'^{\nu}}{\partial\omega^{\tau}}\frac{\partial\omega'^{\mu}}{\partial\omega^{\sigma}}\,, \tag{5.52}$$

which we immediately recognize as the LB's of the ω with respect to the ω'. Using $L_{\tau\sigma}$, Eq.(5.51) assumes a simpler form:

$$\{L_{\tau\sigma}, H\} + L_{\tau\alpha}\epsilon^{\alpha\beta}H_{,\beta\sigma} - H_{,\tau\alpha}\epsilon^{\alpha\beta}L_{\beta\sigma} + \frac{\partial}{\partial t}L_{\tau\sigma} = 0\,. \tag{5.53}$$

Here, the first term is the PB of $L_{\tau\sigma}$ with H. To make this basic integrability condition even easier to handle, define two matrices $\hat{L}$ and $\hat{H}$ with matrix elements given by

$$\hat{L}^{\rho}_{\tau} = L_{\tau\alpha}\epsilon^{\alpha\rho}\,, \quad \hat{H}^{\rho}_{\tau} = H_{,\tau\alpha}\epsilon^{\alpha\rho}\,, \tag{5.54}$$

the subscript indicating the row and the superscript the column. We then obtain the elegant form

$$\{\hat{L}, H\} + (\hat{L}\hat{H} - \hat{H}\hat{L}) + \frac{\partial\hat{L}}{\partial t} = 0 \tag{5.55}$$

for the condition that the transformation of coordinates preserve the structure of the Hamiltonian equations. The first term here is a matrix made up of the PB's of the elements of $\hat{L}$ with H; the second term is a matrix commutator. If Eq.(5.55) must be true for all possible Hamiltonians H, then the three terms must separately vanish as they involve different order derivatives of H. Thus we must have:

$$\{\hat{L}, H\} = 0\,, \tag{5.56a}$$

$$\hat{L}\hat{H} = \hat{H}\hat{L}\,, \tag{5.56b}$$

$$\frac{\partial\hat{L}}{\partial t} = 0\,. \tag{5.56c}$$

Remembering that H is arbitrary, from Eq.(5.56a) we conclude that $\hat{L}$ is independent of ω, from Eq.(5.56b) that it is a multiple of the unit matrix, and from

Eq.(5.56c) that it is time independent. All in all, the conditions for the integrability of Eq.(5.48), and so the existence of the new Hamiltonian H', become the LB conditions:

$$L_{\alpha\beta} \equiv [\omega^\alpha, \omega^\beta]_{\omega'} \equiv \epsilon_{\mu\nu} \frac{\partial \omega'^\mu}{\partial \omega^\alpha} \frac{\partial \omega'^\nu}{\partial \omega^\beta} = a\epsilon_{\alpha\beta} \tag{5.57}$$

where a is a real constant. If $a = 1$, this condition is that the transformation be canonical, as we have derived earlier. The only other possibilities are essentially given by:

1. *Dilation:* $\omega'^\mu = \sqrt{a}\omega^\mu$, $\quad a > 0;\quad H' = aH$;
2. *Interchange of* q_s *and* p_s*, with* $a = -1$*:* $Q_s = p_s$, $P_s = q_s$; $H' = -H$.

It is trivial to check that the Hamiltonian equations are indeed preserved under these two transformations, which are not canonical transformations by our definition. However, if H is physically identified as the energy of the system, then either changing its scale or its sign is not particularly interesting and thus we do not consider these transformations further. Our direct analysis has thus led us back to the canonical transformations as the most general ones consistent with the equations of motion in Hamiltonian form.

Chapter 6

Group Properties and Methods of Constructing Canonical Transformations

Consider those canonical transformations on a fixed $2k$-dimensional phase space that are not explicitly time dependent. All such transformations taken together form a group with an infinite number of elements. That is to say, the following laws, which define a group, hold:

I. *Group Composition Law:* Given two canonical transformations, $\omega \to \omega'$ and $\omega' \to \omega''$, we can define a product transformation, namely $\omega \to \omega''$, which is also canonical. From Eq.(5.38), we have

$$\epsilon^{\alpha\beta} \frac{\partial \omega'^{\mu}}{\partial \omega^{\alpha}} \frac{\partial \omega'^{\nu}}{\partial \omega^{\beta}} = \epsilon^{\alpha\beta} \frac{\partial \omega''^{\mu}}{\partial \omega'^{\alpha}} \frac{\partial \omega''^{\nu}}{\partial \omega'^{\beta}} = \epsilon^{\mu\nu} ,$$

so that

$$\begin{aligned} \epsilon^{\alpha\beta} \frac{\partial \omega''^{\mu}}{\partial \omega^{\alpha}} \frac{\partial \omega''^{\nu}}{\partial \omega^{\beta}} &= \epsilon^{\alpha\beta} \frac{\partial \omega''^{\mu}}{\partial \omega'^{\rho}} \frac{\partial \omega''^{\nu}}{\partial \omega'^{\sigma}} \frac{\partial \omega'^{\rho}}{\partial \omega^{\alpha}} \frac{\partial \omega'^{\sigma}}{\partial \omega^{\beta}} \\ &= \epsilon^{\rho\sigma} \frac{\partial \omega''^{\mu}}{\partial \omega'^{\rho}} \frac{\partial \omega''^{\nu}}{\partial \omega'^{\sigma}} = \epsilon^{\mu\nu} . \end{aligned}$$

II. *Associativity of Composition Law*: Let T_1, T_2, T_3 be three successive canonical transformations, $\omega' = \phi(\omega)$, $\omega'' = \psi(\omega')$, and $\omega''' = \chi(\omega'')$, respectively. By I above, the transformation $T_2 T_1$ is given by

$$\omega'' = \psi(\phi(\omega))$$

and $T_3(T_2 T_1)$ is then given by, say,

$$\omega''' = \chi(\psi(\phi(\omega))) = \Omega(\omega) .$$

On the other hand, if we combine T_3 and T_2 first, we get for T_3T_2:

$$\omega''' = \chi(\psi(\omega')),$$

and then $(T_3T_2)T_1$ is given by

$$\omega''' = \chi(\psi(\phi(\omega))) = \Omega(\omega) \ .$$

exactly as before. Thus, it does not matter in what sequence the products are evaluated: the result of $T_3T_2T_1$ is to express ω''' in terms of ω, and this is unique.

III. *Identity:* The identity transformation $\omega' = \omega$ obviously obeys the conditions for being canonical.

IV. *Inverses:* If $\omega \to \omega'$ is canonical, then so is the transformation $\omega' \to \omega$. Because

$$\epsilon^{\alpha\beta} \frac{\partial \omega'^{\mu}}{\partial \omega^{\alpha}} \frac{\partial \omega'^{\nu}}{\partial \omega^{\beta}} = \epsilon^{\mu\nu} \ ,$$

multiplying both sides by $(\partial\omega^{\rho}/\partial\omega'^{\mu})(\partial\omega^{\sigma}/\partial\omega'^{\nu})$ gives

$$\epsilon^{\rho\sigma} = \epsilon^{\mu\nu} \frac{\partial \omega^{\rho}}{\partial \omega'^{\mu}} \frac{\partial \omega^{\sigma}}{\partial \omega'^{\nu}} \ ,$$

and thus $\omega' \to \omega$ is canonical. The product of these two transformations is obviously the identity transformation.

Thus all the properties needed to define a group are trivially obeyed. (In Chapter 12 we discuss group structure in a little more detail). We can call this group the canonical group. It is characterised by the dimension, $2k$, of the phase space. (It is understood that all along we deal with the phase space of a given system with a fixed number, k, of degrees of freedom). In order to maintain consistently the interpretation of canonical transformations as constituting elements of a group, we must think of each transformation as being specified completely by the functional forms of the functions ϕ^{μ} that express one set of coordinates in terms of another: $\omega'^{\mu} = \phi^{\mu}(\omega)$, and not pay attention to the ranges of values of the ω's and ω''s: indeed it does not matter *which* are the old coordinates and *which* are the new ones. All the group operations deal directly with these functional forms, combining two functional forms to give a third, and so on.

The canonical group is a very large group. In the later chapters of this book we shall see how in the description of physical systems possessing geometrical and other kinds of symmetry, geometrical and other operations are made to correspond to particular canonical transformations and thus pick out various types of subgroups of the full group. However, certain subgroups of the full group can be easily identified right away: we give three examples here, and comment on the main properties of each of them.

We consider first the *identity component* of the full group: it consists of all those canonical transformations that can be connected continuously to the identity. That is, the transformation ϕ belongs to this component if a one-parameter family of canonical transformations ϕ_τ, $0 \leqq \tau \leqq 1$, can be found such that ϕ_τ is continuous in τ, reduces to the identity transformation for $\tau = 0$ and to the given ϕ for $\tau = 1$. It is easy to see that all such canonical transformations obey the group laws among themselves; that is, they form a subgroup of the full group. In practice, although it may be easy to check whether a given transformation is canonical, it may be difficult to find out whether it lies in the identity component.

Another interesting subgroup is formed by the *contact transformations* mentioned briefly in the last chapter. Reverting to the p, q language, contact transformations are those canonical transformations for which one has the following relationship between the new and the old variables:

$$\sum_{s=1}^{k} P_s dQ_s = \sum_{s=1}^{k} p_s dq_s \ . \tag{6.1}$$

From this definition, it is obvious that these transformations do form a subgroup of the full group. The example given in Eq.(5.8) is a particularly simple example of a contact transformation, in which Q depended only on q and P depended on p in a linear homogeneous fashion. More generally we can show that if Eq.(6.1) is obeyed, then Q_s is a homogeneous function of the p's of degree zero, and P_s is a homogeneous function of the p's of degree one. The proof is quite simple: expressing P_s and Q_s as functions of q and p, Eq.(6.1) leads to:

$$\sum_s P_s \frac{\partial Q_s}{\partial q_r} = p_r \ , \tag{6.2a}$$

$$\sum_s P_s \frac{\partial Q_s}{\partial p_r} = 0 \ , \qquad r = 1, 2, \cdots k \ . \tag{6.2b}$$

In any event, the PB relations characterizing a canonical transformation are obeyed by Q and P. Thus we have:

$$\sum_r \frac{\partial Q_s}{\partial q_r}\frac{\partial Q_{s'}}{\partial p_r} = \sum_r \frac{\partial Q_s}{\partial p_r}\frac{\partial Q_{s'}}{\partial q_r} \ , \tag{6.3a}$$

$$\sum_r \frac{\partial Q_{s'}}{\partial q_r}\frac{\partial P_s}{\partial p_r} = \sum_r \frac{\partial Q_{s'}}{\partial p_r}\frac{\partial P_s}{\partial q_r} + \delta_{ss'} \ . \tag{6.3b}$$

If we multiply Eq.(6.3a) by $P_{s'}$, sum on s', and use both parts of Eq.(6.2) we get:

$$\sum_r p_r \frac{\partial Q_s}{\partial p_r} = 0, \qquad s = 1, 2, \cdots, k \ ; \tag{6.4}$$

by the same process, Eq.(6.3b) leads to

$$\sum_r p_r \frac{\partial P_s}{\partial p_r} = P_s, \qquad s = 1, 2, \cdots, k. \tag{6.5}$$

Euler's theorem on homogeneous functions then leads to the properties of Q_s and P_s mentioned earlier. One more property of the Q's which was evident in the example of Eq.(5.8) generalizes to a general contact transformation. This is that the Q_s considered as k functions of the q_s, and ignoring their dependence on the p's, must, in fact, be independent functions of the q_s. This follows from Eq.(6.2a) and the fact that p_s are independent variables: the Jacobian matrix $\partial Q_s/\partial q_r$ must be nonsingular and cannot possess any purely q_s-dependent null eigenvectors. Conversely, a canonical transformation in which these three properties hold, namely the homogeneity properties of the Q's and P's in their dependence on p_s, and the independence of Q's as functions of the q's can be shown to be a contact transformation. Thus these properties can be used to define this subgroup of the canonical group.

A third kind of subgroup of the canonical group arises in the following way. Several of the physically interesting operations, such as rotations of space or translations, have the property that the initial and the final variables go over precisely the same range of values in "charting" the whole of phase space. The clearest example in the case of rotations and translations is to use Cartesian position and momentum variables for each particle in a many-particle system. Every such coordinate, q or p, goes over the entire real line, both before and after the transformation. Here we are interested not only in the functional form of the canonical transformation, but also in the range of values the new variables take as the old ones go over their physically admissible ranges. Let ω^μ be a specific set of coordinates for a $2k$-dimensional phase space, and let these coordinates have certain ranges of values determined by the physical interpretation of these coordinates. Consider all those canonical transformations $\phi : \omega^\mu \to \omega'^\mu = \phi^\mu(\omega)$ such that as ω goes over its range once, ω'^μ also goes over the *same* range once. It is obvious that such "range-preserving" canonical transformations form a subgroup of the full group. Now, it might appear that this subgroup depends on the particular set of coordinates ω^μ that was chosen in the definition above, but this is not so because such transformations can be given an alternative interpretation that frees them from the particular coordinates used. This is to interpret the mapping $\omega \to \omega' = \phi(\omega)$ as a *mapping of phase space onto itself* in a one-to-one reversible fashion. Under the canonical transformation ϕ, the phase-space point with coordinates ω is mapped into the point with coordinate values ω', such that the functions ϕ^μ expressing the coordinates of the image point in terms of the original point obey the PB condition:

$$\epsilon^{\alpha\beta}\frac{\partial\phi^\mu(\omega)}{\partial\omega^\alpha}\frac{\partial\phi^\nu(\omega)}{\partial\omega^\beta} = \epsilon^{\mu\nu} .$$

This characterization of such a canonical transformation has a meaning in every coordinate system and we shall call the corresponding subgroup of the full group the regular subgroup and its elements as *regular canonical transformations*. To repeat, a regular canonical transformation is a canonical transformation that can be interpreted as a one-to-one, invertible mapping of the points of phase space onto themselves. Expressed in any canonical coordinate system ω, the functions ϕ^μ expressing the coordinates of the image of the point ω in terms of ω^μ satisfy the conditions required of a canonical transformation, and as ω^μ vary over their physical ranges, the $\phi^\mu(\omega)$ go over precisely the same range. In general, the regular subgroup is determined by the dimensionality of the phase space as well as the particular physical system. (Let us add the following: in order to define "canonical coordinate systems" we first pick a specific coordinate system in phase space and agree once and for all that this one is canonical. Any other coordinate system is canonical if it is obtained from this one by any element of the full canonical group.)

We must stress that not every canonical transformation can be reinterpreted as corresponding to a certain mapping of phase space onto itself. The interpretation that is common to all canonical transformations is the one where we regard the points of phase space as "given", but can use different sets of coordinates to label them. Regular canonical transformations admit an alternative, and sometimes more appropriate, interpretation. We give a simple example of a transformation that is canonical but *not* regular. Take the case of a single q and a single p, each going over the entire real line so that the phase space consists of the two-dimensional plane in these variables. The transformation $q, p \to Q, P$ where

$$Q = \frac{1}{2}(q^2 + p^2), \qquad P = -\tan^{-1}\frac{q}{p} \tag{6.6}$$

is easily checked to be canonical. However it only provides a new coordinate system (radial and polar variables in the plane) and cannot be thought of as a mapping of the phase space onto itself.

After this discussion of the canonical group and some of its natural subgroups, we describe in the remainder of this chapter some of the ways of actually constructing canonical transformations.

Infinitesimal and One Parameter Subgroups of Canonical Transformations

From the way in which the identity component of the canonical group is defined, we are naturally led to the study of canonical transformations that are close to the identity, that is, are infinitesimal. (We assume for the present that we are dealing with regular canonical transformations as well.) For an infinitesimal transformation, one can write:

$$\omega'^\mu = \omega^\mu + \delta\theta\,\phi^\mu(\omega), \tag{6.7}$$

where $\delta\theta$ is a small quantity of the first order and the ϕ^μ are some functions of ω. Then the condition for a canonical transformation,

$$\epsilon^{\alpha\beta}\frac{\partial}{\partial\omega^\alpha}(\omega^\mu + \delta\theta\,\phi^\mu)\frac{\partial}{\partial\omega^\beta}(\omega^\nu + \delta\theta\,\phi^\nu) = \epsilon^{\mu\nu}\,, \tag{6.8}$$

gives to lowest order in $\delta\theta$ the condition:

$$\frac{\partial}{\partial\omega^\alpha}(\epsilon_{\beta\mu}\phi^\mu) = \frac{\partial}{\partial\omega^\beta}(\epsilon_{\alpha\mu}\phi^\mu). \tag{6.9}$$

This means that there is a function, $\phi(\omega)$ say, of which $\epsilon_{\alpha\mu}\phi^\mu$ is the gradient:

$$\epsilon_{\alpha\mu}\phi^\mu(\omega) = \frac{\partial}{\partial\omega^\alpha}\phi(\omega) \ . \tag{6.10}$$

Using this in Eq.(6.7), the general form of an infinitesimal canonical transformation turns out to be

$$\omega'^\mu = \omega^\mu + \delta\theta\epsilon^{\mu\nu}\frac{\partial\phi(\omega)}{\partial\omega^\nu} = \omega^\mu + \delta\theta\{\omega^\mu, \phi\} \ . \tag{6.11}$$

The $\phi(\omega)$ is an arbitrary function (except for our requirement that the transformation be regular). We can also write this equation in the form

$$\delta\omega^\mu = \omega'^\mu - \omega^\mu = \delta\theta\{\omega^\mu, \phi\} \ . \tag{6.12}$$

If we interpret the transformation as a mapping of phase space onto itself, then the infinitesimal quantities $\delta\omega^\mu$ represent the differences in the coordinates of the image of the point ω and ω itself.

Let us next define and compute the effect of an infinitesimal regular canonical transformation on a phase-space function $f(\omega)$. According to Eq.(5.40), a new functional form, f', is naturally defined by the rule:

$$f'(\omega') = f(\omega) \ . \tag{6.13}$$

This definition makes sense whether the transformation is regular or not. However, if it is regular, then ω' goes over the same range as ω, and then it becomes meaningful to ask for the change in the functional form of f induced by the regular canonical transformation $\omega \to \omega'$. By this is meant the *difference* of f' and f, for the same argument ω, considered as a function of ω:

$$\delta' f(\omega) \equiv f'(\omega) - f(\omega) \ . \tag{6.14}$$

(We must use different symbols δ, δ' in Eqs.(6.12) and (6.14); their meanings are quite different). Working to lowest order in $\delta\theta$, we find:

$$\begin{aligned} \delta' f(\omega) &= f'(\omega) - f(\omega) = f'(\omega) - f'(\omega') \simeq f(\omega) - f(\omega') \\ &= f(\omega^\mu) - f(\omega^\mu + \delta\theta\{\omega^\mu, \phi\}) = -\delta\theta\{\omega^\mu, \phi\}\frac{\partial f}{\partial\omega^\mu} \ , \end{aligned}$$

i.e.,

$$\delta' f(\omega) = -\delta\theta\{f, \phi\} \ . \tag{6.15}$$

The difference in sign in Eqs.(6.12) and (6.15) follows from their different interpretations.

Intuitively speaking, a finite canonical transformation that can be connected continuously to the identity should be built up as a succession of infinitesimal transformations. Up to now we used PB's to *define* canonical transformations. Equations (6.12) and (6.15) show the deep relationship between PB's and canonical transformations in a new light: the former can be used in the process of constructing the latter. This statement can be made more precise if we convert Eq.(6.12) into a differential equation with respect to θ:

$$\frac{d\omega^\mu}{d\theta} = \{\omega^\mu, \phi(\omega)\}_\omega = \epsilon^{\mu\nu}\frac{\partial\phi(\omega)}{\partial\omega^\nu} . \tag{6.16}$$

Let us first clarify the meaning of this equation. We look upon the ω^μ as functions of θ; ϕ is a function with a fixed functional form, and ω^μ are the unknowns in the differential equation. Thus on the right-hand side of this equation, the arguments of ϕ are just the quantities we are trying to solve for. From the theory of differential equations, we know that there is a unique solution for ω^μ if we are given the values of ω^μ at $\theta = 0$. Calling these boundary values of ω^μ as ω_0^μ, the solution of Eq.(6.16) can be written as

$$\omega^\mu = \psi^\mu(\omega_0;\theta), \qquad \psi^\mu(\omega_0;0) = \omega_0^\mu . \tag{6.17}$$

Because we will treat the ω_0^μ themselves as variable parameters, we write Eq.(6.16) in the form

$$\begin{aligned}\frac{\partial}{\partial\theta}\psi^\mu(\omega_0;\theta) &= \epsilon^{\mu\nu}\frac{\partial\phi(\omega)}{\partial\omega^\nu}\bigg|_{\omega^\mu=\psi^\mu(\omega_0;\theta)}\\ &= \epsilon^{\mu\nu}\frac{\partial\phi}{\partial\psi^\nu}(\psi(\omega_0;\theta)) .\end{aligned} \tag{6.18}$$

We wish to demonstrate that the *transformation* $\omega_0^\mu \to \omega^\mu$ *is a canonical transformation.* To do this, we must investigate the dependence of ψ^μ on ω_0. Consider the LB's:

$$\eta_{\mu\nu}(\omega_0;\theta) = \epsilon^{\alpha\beta}\frac{\partial\psi^\alpha(\omega_0;\theta)}{\partial\omega_0^\mu}\frac{\partial\psi^\beta(\omega_0;\theta)}{\partial\omega_0^\nu} . \tag{6.19}$$

Using Eq.(6.18), we find

$$\begin{aligned}\frac{\partial}{\partial\theta}\eta_{\mu\nu}(\omega_0;\theta) &= \epsilon_{\alpha\beta}\frac{\partial}{\partial\omega_0^\mu}\frac{\partial\psi^\alpha}{\partial\theta}\frac{\partial\psi^\beta}{\partial\omega_0^\nu} + \epsilon_{\alpha\beta}\frac{\partial\psi^\alpha}{\partial\omega_0^\mu}\frac{\partial}{\partial\omega_0^\nu}\frac{\partial\psi^\beta}{\partial\theta}\\ &= \epsilon_{\alpha\beta}\frac{\partial}{\partial\omega_0^\mu}\left(\epsilon^{\alpha\lambda}\frac{\partial\phi(\psi)}{\partial\psi^\lambda}\right)\frac{\partial\psi^\beta}{\partial\omega_0^\nu} + \epsilon_{\alpha\beta}\frac{\partial\psi^\alpha}{\partial\omega_0^\mu}\frac{\partial}{\partial\omega_0^\nu}\left(\epsilon^{\beta\lambda}\frac{\partial\phi(\psi)}{\partial\psi^\lambda}\right)\\ &= -\delta_\beta^\lambda\frac{\partial^2\phi(\psi)}{\partial\psi^\lambda\partial\psi^\rho}\frac{\partial\psi^\rho}{\partial\omega_0^\mu}\frac{\partial\psi^\beta}{\partial\omega_0^\nu} + \delta_\alpha^\lambda\frac{\partial^2\phi(\psi)}{\partial\psi^\lambda\partial\psi^\rho}\frac{\partial\psi^\rho}{\partial\omega_0^\nu}\frac{\partial\psi^\alpha}{\partial\omega_0^\mu}\\ &= 0 .\end{aligned} \tag{6.20}$$

Thus these LB's are constant with respect to θ and could be evaluated at $\theta = 0$. We thus get the important result

$$\epsilon_{\alpha\beta}\frac{\partial\psi^\alpha(\omega_0;\theta)}{\partial\omega_0^\mu}\frac{\partial\psi^\beta(\omega_0;\theta)}{\partial\omega_0^\nu} = \epsilon_{\mu\nu} , \tag{6.21}$$

which can also be translated into the equivalent PB form:

$$\{\psi^\alpha, \psi^\beta\}_{\omega_0} = \epsilon^{\mu\nu} \frac{\partial \psi^\alpha(\omega_0;\theta)}{\partial \omega_0^\mu} \frac{\partial \psi^\beta(\omega_0;\theta)}{\partial \omega_0^\nu} = \epsilon^{\alpha\beta} \,. \tag{6.22}$$

These equations display the fundamental property of the solutions of Eq.(6.16): if $\omega^\mu(\theta)$ is a set of functions obeying Eq.(6.16), then the ω^μ for any θ are obtained from their boundary values ω_0^μ at $\theta = 0$ by a canonical transformation. More generally, for any two values θ_1, θ_2 of $\theta, \omega^\mu(\theta_1)$ and $\omega^\mu(\theta_2)$ are obtainable one from the other by a canonical transformation. Continuing, as we have up to now, to deal with a function $\phi(\omega)$ not explicitly dependent on θ, we can demonstrate a composition law for these canonical transformations:

$$\psi^\mu(\psi(\omega_0;\theta_1);\theta_2) = \psi^\mu(\omega_0;\theta_1+\theta_2) \,. \tag{6.23}$$

The proof consists in noting that by virtue of Eq.(6.17), the two sides agree at $\theta_2 = 0$, whereas with respect to θ_2 they both obey the same differential equation:

$$\frac{\partial}{\partial \theta_2} \psi^\mu(\psi(\omega_0;\theta_1);\theta_2) = \epsilon^{\mu\nu} \frac{\partial \phi(\xi)}{\partial \xi^\nu}\bigg|_{\xi^\mu = \psi^\mu(\psi(\omega_0;\theta_1);\theta_2)} , \tag{6.24a}$$

$$\frac{\partial}{\partial \theta_2} \psi^\mu(\omega_0;\theta_1+\theta_2) = \epsilon^{\mu\nu} \frac{\partial \phi(\xi)}{\partial \xi^\nu}\bigg|_{\xi^\mu = \psi^\mu(\omega_0;\theta_1+\theta_2)} . \tag{6.24b}$$

From the uniqueness of the solutions to these equations with given boundary conditions, Eq.(6.23) follows.

We can summarize by saying that the solution to the differential equation (Eq.(6.16)), equivalently (Eq.(6.18)), constitutes a *one-parameter group of canonical transformations* ψ_θ. For $\theta = 0$, it is the identity transformation. When we compose the two transformations ψ_{θ_1} and ψ_{θ_2}, we obtain the transformation $\psi_{\theta_1+\theta_2}$; $\psi_{-\theta}$ is inverse to ψ_θ.

For *all* these properties to hold, it is essential that ϕ be *independent of* θ. If ϕ depends explicitly on θ, then Eqs.(6.21) and (6.22) are still valid, so that for each $\theta, \psi^\mu(\omega_0;\theta)$ arises from ω_0^μ by a canonical transformation. However, the transformations for different θ no longer obey the composition law, Eq.(6.23); what happens is that in Eq.(6.24a) we encounter on the right the quantity $\phi(\xi;\theta_2)$, whereas in (6.24b) there appears $\phi(\xi;\theta_1+\theta_2)$, so that the two differential equations are not identical.

Because the passage from ω_0^μ to $\psi^\mu(\omega_0;\theta)$ is canonical, there is an alternative way of looking upon Eqs.(6.16) and (6.18). Here we only consider the case where ϕ does not depend explicitly on θ. We then find that even the implicit dependence of ϕ on θ in Eqs.(6.16) and (6.18) vanishes:

$$\begin{aligned} \frac{d}{d\theta}\{\phi(\psi(\omega_0;\theta))\} &= \frac{\partial \phi}{\partial \psi^\alpha}(\psi(\omega_0;\theta)) \frac{\partial \psi^\alpha}{\partial \theta}(\omega_0;\theta) \\ &= \epsilon^{\alpha\beta} \frac{\partial \phi(\psi)}{\partial \psi^\alpha} \frac{\partial \phi(\psi)}{\partial \psi^\beta} = 0 \,. \end{aligned} \tag{6.25}$$

Therefore we have the following functional relation:

$$\phi(\psi(\omega_0;\theta)) = \phi(\omega_0) \ . \tag{6.26}$$

Now the right-hand side of Eq.(6.18) is just the PB of ψ^μ and $\phi(\psi)$ evaluated with respect to the variables ψ^μ,

$$\{\psi^\mu, \phi(\psi)\}_\psi ;$$

but since a PB can be evaluated in any canonical coordinate system, we can equally well evaluate it using the ω_0 system. If we do so and at the same time use Eq.(6.26), we see that the basic differential equation for $\psi(\omega_0;\theta)$ assumes the form

$$\frac{\partial \psi^\mu}{\partial \theta}(\omega_0;\theta) = \{\psi^\mu(\omega_o;\theta), \phi(\omega_0)\}_{\omega_0}, \qquad \psi^\mu(\omega_0;0) = \omega_0^\mu \ . \tag{6.27}$$

Although this is completely equivalent to Eq.(6.16) (when ϕ does not depend on θ explicitly), the interpretation of this equation is quite different from that of Eq.(6.16). The unknowns are the functions $\psi^\mu(\omega_0;\theta)$, the variables ω_0 are independent variables not dependent on θ, and it is these that enter as arguments of ϕ. For Eq.(6.27) we can write down a formal solution as a power series in θ, using the given boundary conditions:

$$\omega^\mu = \psi^\mu(\omega_0;\theta) = \omega_0^\mu + \theta\{\omega_0^\mu, \phi(\omega_0)\} + \frac{\theta^2}{2!}\{\{\omega_0^\mu, \phi\}, \phi\}, \cdots ; \tag{6.28}$$

all the P.B.'s are evaluated with respect to the ω_0's. This expression is, of course, the solution to Eq.(6.18) as well. We have thus obtained an explicit, though formal, expression for the coordinates ω^μ obtained from the ω_0^μ by means of the canonical transformation generated by the differential Eq.(6.16).

Assuming that the above transformation, $\omega_0 \to \omega$, is regular, we can develop a similar power series expression to represent functions of the ω. Let $A(\omega_0)$ be a given function of ω_0 with a specific functional form. The transformation being regular, we can consider the function $A(\omega)$ and examine its functional form expressed as a function of ω_0:

$$A(\omega) = A(\psi(\omega_0;\theta)) = A_\theta(\omega_0), \qquad A_0(\omega_0) = A(\omega_0) \ . \tag{6.29}$$

This function obeys a differential equation similar to Eq.(6.27):

$$\begin{aligned}\frac{\partial}{\partial\theta}A_\theta(\omega_0) &= \frac{\partial}{\partial\theta}A(\psi(\omega_0;\theta)) = \frac{\partial A(\psi)}{\partial\psi^\lambda}\frac{\partial\psi^\lambda}{\partial\theta} = \epsilon^{\lambda\mu}\frac{\partial A(\psi)}{\partial\psi^\lambda}\frac{\partial\phi(\psi)}{\partial\psi^\mu} \\ &= \{A(\psi), \phi(\psi)\}_\psi = \{A_\theta(\omega_0), \phi(\omega_0)\}_{\omega_0} \ .\end{aligned} \tag{6.30}$$

Here, we first used Eq.(6.18) to evaluate $(\partial\psi^\lambda/\partial\theta)$, then switched from ψ to ω_0 in evaluating the PB's and finally made use of the invariance of ϕ, Eq.(6.26). We can now express $A_\theta(\omega_0)$ as a power series in θ just as in Eq.(6.28):

$$\begin{aligned}A(\omega) = A_\theta(\omega_0) &= A(\omega_0) + \theta\{A(\omega_0), \phi(\omega_0)\} \\ &+ \frac{\theta^2}{2!}\{\{A(\omega_0), \phi(\omega_0)\}, \phi(\omega_0)\} + \cdots \ ,\end{aligned} \tag{6.31}$$

all brackets being computed with respect to ω_0. This is a generalization of Eq.(6.28); in fact the earlier equation is a special case of this one, corresponding to taking $A(\omega_0) = \omega_0^\mu$.

[Because there is a danger of confusion here, it may be useful to distinguish between the two ways of "producing new functions from old" that we have used. The original method of defining a function of the new variables, $f'(\omega')$, given a function $f(\omega)$ of the old variables by equating them numerically, $f'(\omega') = f(\omega)$, is meaningful for all transformations of coordinates $\omega \to \omega'$, even if the ω' have a different physical meaning and different ranges from the ω. Here this law is called the law for a "scalar field." If in particular the transformation is *regular*, then it is meaningful to consider the difference in functional form of f' and f by evaluating them at the same argument. This is what is done in Eq.(6.14). However, precisely when the transformation is regular, there is a second way to obtain a new function from $f(\omega)$: we can consider meaningfully the function f for the new argument ω', $f(\omega')$, and then ask how this would appear if expressed in terms of ω. One would have $f(\omega') = \bar{f}(\omega)$, say, and f and $\bar{f}$ would have different functional forms. It is this latter change in functional form, $f \to \bar{f}$, that is involved in Eqs.(6.29) and (6.30), and explicitly displayed in Eq.(6.31). The two processes $f \to f'$ and $f \to \bar{f}$, which are both available for regular transformations, are in opposite directions and complementary to one another.]

There is a compact and elegant notation for displaying these results concerning one-parameter subgroups of canonical transformations, which is appropriately introduced at this point. We will continue to assume that the function ϕ is not explicitly dependent on the parameter θ and also that the finite canonical transformations resulting from the basic differential equation Eq.(6.16) are regular. The function ϕ is called the *generator* of this one-parameter subgroup of canonical transformations. In terms of $\phi(\omega_0)$, let us define the linear partial differential operator $\tilde{\phi}(\omega_0)$ as follows:

$$\tilde{\phi}(\omega_0) = \epsilon^{\mu\nu} \frac{\partial \phi(\omega_0)}{\partial \omega_0^\mu} \frac{\partial}{\partial \omega_0^\nu} \,. \tag{6.32}$$

This operator is designed to act on functions of ω_0, and the PB of $\phi(\omega_0)$ with any function of ω_0 is obtained by making $\tilde{\phi}$ act on that function:

$$\{\phi(\omega_0), f(\omega_0)\} = \tilde{\phi}(\omega_0) f(\omega_0) \,. \tag{6.33}$$

The canonical transformation generated by $\phi(\omega_0)$ takes ω_0^μ into $\omega^\mu = \psi^\mu(\omega_0;\theta)$, and the dependence of ψ on its arguments is displayed in Eq.(6.28); we see that using $\tilde{\phi}(\omega_0)$, we can write it as

$$\begin{aligned} \omega^\mu = \psi^\mu(\omega_0;\theta) &= \omega_0^\mu - \theta\tilde{\phi}(\omega_0)\omega_0^\mu \\ &+ \frac{\theta^2}{2!}\tilde{\phi}(\omega_0)\tilde{\phi}(\omega_0)\omega_0^\mu \cdots \\ &= e^{-\theta\tilde{\phi}(\omega_0)}\omega_0^\mu \,; \end{aligned} \tag{6.34}$$

more briefly,

$$\omega = e^{-\theta\tilde{\phi}}\omega_0 \,. \tag{6.35}$$

For the development in θ of any function A, namely Eq.(6.31), we get:

$$A(\omega) = A_\theta(\omega_0) = e^{-\theta\tilde{\phi}(\omega_0)}A(\omega_0) \,. \tag{6.36}$$

Combining Eqs.(6.35) and (6.36), we can write the simple equation

$$A(e^{-\theta\tilde{\phi}}\omega_0) = e^{-\theta\tilde{\phi}(\omega_0)}A(\omega_0) \,. \tag{6.37}$$

In all these equations, the exponential is an "honest" one; we really mean a power series in $\tilde{\phi}$, and a power of $\tilde{\phi}$ acts on a function $A(\omega_0)$ by making the $\tilde{\phi}$'s act one after another. The following other useful property comes from the Jacobi identity for the PB: for two functions $A(\omega_0)$, $B(\omega_0)$, we have

$$\begin{aligned}\{A(\omega), B(\omega)\}_\omega &= e^{-\theta\tilde{\phi}(\omega_0)}\{A(\omega_0), B(\omega_0)\}_{\omega_0} \\ &= \{e^{-\theta\tilde{\phi}(\omega_0)}A(\omega_0), e^{-\theta\tilde{\phi}(\omega_0)}B(\omega_0)\}_{\omega_0} \,.\end{aligned} \tag{6.38}$$

This is proved easily by first noting that the two sides agree at $\theta = 0$, then showing that they obey the same differential equation in θ. In doing the latter, one has to use Eq.(6.30), the analogous equations for $B(\omega)$ and $\{A(\omega), B(\omega)\}$, and the Jacobi identity. The meaning of this equation is that we may apply the operator $e^{-\theta\tilde{\phi}}$ either before or after evaluating a PB and get the same answer both ways.

The most important application of all this is to the Hamiltonian equations of motion themselves. Eq.(5.14) is exactly the same as Eq.(6.16), with θ replaced by t and $\phi(\omega)$ by the Hamiltonian $H(\omega)$. Similarly, Eq.(5.13), that specifies the time dependence of a function $f(\omega)$, which has a given functional form and has the time-dependent coordinates ω as arguments, is exactly the same as Eq.(6.30). (Assume that f has no explicit time dependence.) We can describe the equations of motion by saying that the Hamiltonian is the generator of a one-parameter family of canonical transformations in phase space; if H has no explicit t-dependence, this is actually a one-parameter *group* of canonical transformations.

Assume now that H is not explicitly time dependent and also that the canonical transformations it generates are regular. Then the formal solution of the equations of motion may be pictured as follows: if the phase of the system at $t = 0$ is given by the point ω_0^μ, then the phase at time t is obtained by performing a finite canonical transformation:

$$\omega^\mu(t) = e^{-t\tilde{H}(\omega_0)}\omega_0^\mu \,. \tag{6.39}$$

As we allow ω_0 to vary over all of phase space, so will $\omega(t)$, so that the operation $e^{-t\tilde{H}(\omega_0)}$ acts as a t-dependent canonical mapping of phase space onto itself. If $f(\omega)$ is a dynamical variable with a given form, its time dependence is given by

$$f(\omega(t)) = e^{-t\tilde{H}(\omega_0)}f(\omega_0) \,, \tag{6.40}$$

and this is the solution of Eq.(5.13) (with $\partial f/\partial t = 0$). Equation (6.26) for the generator ϕ coincides exactly with the statement made in Chapter 5 that a Hamiltonian with no explicit t-dependence is a constant of motion:

$$H(\omega(t)) = H(\omega_0) . \tag{6.41}$$

More generally, a function $A(\omega)$ is a constant of motion if its has vanishing PB with $H(\omega)$, then we have:

$$A(\omega(t)) = e^{-t\tilde{H}(\omega_0)} A(\omega_0) = A(\omega_0) . \tag{6.42}$$

Equation (6.38) leads to Jacobi's theorem: if A and B are constants of motion, then so is their PB. Because the vanishing of the PB of H and a constant of motion A is a symmetrical statement concerning these two variables, not only does a constant of motion preserve its functional form under the one-parameter group of canonical transformations generated by H, but conversely, the Hamiltonian preserves *its* form under the one-parameter group generated by a constant of motion. The importance of one-parameter groups of canonical transformations should be clear by now!

Finite Canonical Transformations: Other Ways of Getting Them

We return to the characterization of canonical transformations that was obtained from the Action functional, Eq.(5.21):

$$\sum_s P_s \dot{Q}_s - H'(Q, P, t) = \sum_s p_s \dot{q}_s - H(q, p, t) + \frac{dF}{dt} . \tag{6.43}$$

Earlier we analyzed this equation by imagining F to be written out in terms of the $2k$ independent variables Q_s, P_s, (and t), and then derived equations that defined canonical transformations. We now examine other ways of handling this equation. Let us suppose that the old variables q_s and the new variables Q_s together form $2k$ independent variables. Then F can be written in terms of them and t; let the corresponding functional form be called $-F_{(1)}$. Developing the right-hand side of Eq.(6.43) in terms of $\dot{q}_s$ and $\dot{Q}_s$, we obtain:

$$p_s = \frac{\partial F_{(1)}}{\partial q_s}(q, Q, t) , \tag{6.44a}$$

$$P_s = -\frac{\partial F_{(1)}}{\partial Q_s}(q, Q, t) , \tag{6.44b}$$

$$H' = H + \frac{\partial F_{(1)}}{\partial t} . \tag{6.44c}$$

The transformation is implicitly, yet completely, defined. We must solve Eq.(6.44a) to express Q_s in terms of q, p and t; then we substitute this in Eq.(6.44b) to get P_s in terms of q, p, t. To be able to carry out the first step, the necessary and sufficient condition on $F_{(1)}$ is:

$$\det \left| \frac{\partial^2 F_{(1)}}{\partial q_r \partial Q_s} \right| \neq 0 . \tag{6.45}$$

Thus given any function $F_{(1)}$ of q, Q and t obeying Eq.(6.45), a unique canonical transformation is determined by Eqs. (6.44). The function $F_{(1)}$ is called the "generating function" of the transformation. The word "generating" is used here in a somewhat different sense than in the case of one-parameter subgroups of canonical transformations.

Equations similar to the foregoing ones can be set up if, for example, the q_s and P_s form an independent set. One rewrites Eq.(6.43) as

$$\sum_s p_s \dot{q}_s - H = -\sum_s Q_s \dot{P}_s - H' + \frac{d}{dt}\left(\sum_s P_s Q_s - F\right) . \tag{6.46}$$

Writing $F_{(2)}(q, P, t)$ for the function whose total derivative appears on the right-hand side, we get the analogues to Eqs.(6.44):

$$p_s = \frac{\partial F_{(2)}}{\partial q_s}(q, P, t), \quad Q_s = \frac{\partial F_{(2)}}{\partial P_s}(q, P, t), \quad H' = H + \frac{\partial F_{(2)}}{\partial t} . \tag{6.47}$$

The first of these equations permits the P_s to be expressed in terms of q, p and t if

$$\det \left| \frac{\partial^2 F_{(2)}}{\partial q_r \partial P_s} \right| \neq 0 . \tag{6.48}$$

and any function $F_{(2)}$ satisfying this condition defines completely a canonical transformation.

The equations for generating functions of types $F_{(3)}(p, Q, t)$ and $F_{(4)}(p, P, t)$ are:

$$q_s = -\frac{\partial F_{(3)}}{\partial p_s}, \quad P_s = -\frac{\partial F_{(3)}}{\partial Q_s}, \; H' = H + \frac{\partial F_{(3)}}{\partial t}, \quad \det \left| \frac{\partial^2 F_{(3)}}{\partial p_r \partial Q_s} \right| \neq 0 ; \tag{6.49}$$

and

$$q_s = -\frac{\partial F_{(4)}}{\partial p_s}, \quad Q_s = \frac{\partial F_{(4)}}{\partial P_s}, \; H' = H + \frac{\partial F_{(4)}}{\partial t}, \quad \det \left| \frac{\partial^2 F_{(4)}}{\partial p_r \partial P_s} \right| \neq 0 . \tag{6.50}$$

Thus if one of these four combinations of old and new coordinates forms a set of $2k$ independent variables, we can write out the transformation in closed, though implicit, form. Actually, however, such closed expressions can be obtained in a more general situation. As long as we can make up a set of $2k$ independent variables, drawing one member from each canonically conjugate old pair (q_s, p_s), and one from each new pair (Q_s, P_s), by the addition of a total time derivative we can alter the terms $\sum_s p_s \dot{q}_s$ and $\sum_s P_s \dot{Q}_s$ in Eq.(6.43) so that only the time derivatives of these $2k$ independent variables appear, and then we would obtain equations similar to the four special situations considered above. But it is an interesting property of canonical transformations that such a choice of independent variables can always be made! It is a useful exercise to prove this fact. It is no loss of generality to assume that the k old variables in the set are $q_1, q_2, \cdots, q_k$. Let us start by adding to these as many of the Q's as we possibly can, with the property that they together with the q_s are all independent. Again without any loss of generality, we can assume that these Q's are $Q_1, Q_2, \cdots, Q_A$, $A \leqq k$; thus the remaining variables $Q_{A+1}, \cdots Q_k$ must be functions of q_s and the "earlier" Q's. We next show that the $2k$ variables $q_1, \cdots, q_k$, $Q_1, \cdots, Q_A$, $P_{A+1}, \cdots, P_k$ form an independent set. Taking the Jacobian of this set with respect to the q's and p's, this amounts to proving that

$$\det \begin{vmatrix} \frac{\partial Q_1}{\partial p_1} \cdots \frac{\partial Q_A}{\partial p_1} \frac{\partial P_{A+1}}{\partial p_1} \cdots \frac{\partial P_k}{\partial p_1} \\ \cdots \\ \frac{\partial Q_1}{\partial p_k} \cdots \frac{\partial Q_A}{\partial p_k} \frac{\partial P_{A+1}}{\partial p_k} \cdots \frac{\partial P_k}{\partial p_k} \end{vmatrix} \neq 0 \,. \tag{6.51}$$

Let the indices a, b go over the values $1, 2, \cdots, A$, and $A+1, \cdots, k$, respectively. Using the expression of the Q_a in terms of q_s and p_s, for a small variation in q_s and p_s we obtain:

$$\delta Q_a = \sum_s \frac{\partial Q_a}{\partial q_s} \delta q_s + \sum_s \frac{\partial Q_a}{\partial p_s} \delta p_s \,, \qquad a = 1, 2, \cdots, A \,. \tag{6.52}$$

The condition that the Q_a and the q_s are all independent means that it is impossible to eliminate the δp_s from this equation. This can be stated in the form:

$$\sum_a \xi_a \frac{\partial Q_a}{\partial p_s} = 0 \Rightarrow \xi_a = 0. \tag{6.53}$$

Next let us consider the expression of the Q_b in terms of the q_s and Q_a:

$$Q_b = f_b(q_s, Q_a), \qquad b = A+1, \cdots, k. \tag{6.54}$$

These can be put into the conditions:

$$\{Q_b, Q_a\} = 0, \qquad \{Q_b, P_{b'}\} = \delta_{bb'} \,,$$

that follow from the canonical nature of the transformation. Making use of the vanishing of the PB's $\{Q_a, Q_{a'}\}$ and $\{Q_a, P_b\}$, we get

$$\sum_s \frac{\partial f_b}{\partial q_s} \frac{\partial Q_a}{\partial p_s} = 0, \qquad a = 1, 2, \cdots, A; \qquad b = A+1, \cdots, k; \tag{6.55a}$$

$$\sum_s \frac{\partial f_b}{\partial q_s} \frac{\partial P_{b'}}{\partial p_s} = \delta_{bb'}, \qquad b, b' = A+1, \cdots, k \,. \tag{6.55b}$$

The matrix appearing here is the $k \times k$ matrix

$$\left(\frac{\partial Q_a}{\partial p_s} \frac{\partial P_b}{\partial p_s} \right) ,$$

and the determinant of this matrix appears in Eq.(6.51). If the determinant vanishes, then the matrix has a null eigenvector; that is, there are non zero quantities ξ_a, $\eta_{b'}$ such that

$$\sum_a \xi_a \frac{\partial Q_a}{\partial p_s} + \sum_{b'} \eta_{b'} \frac{\partial P_{b'}}{\partial p_s} = 0 \,. \qquad s = 1, 2, \cdots, k. \tag{6.56}$$

But if we multiply Eq.(6.55a) by ξ_a, Eq.(6.55b) by $\eta_{b'}$ and sum on a and b', Eq.(6.56) makes the left-hand side vanish, whereas on the right-hand side we get η_b. Thus the η's must be zero. Putting this into Eq.(6.56) and comparing with Eq.(6.53), we conclude that the ξ's must also be zero. Thus the matrix written above has no nontrivial null eigenvectors, the determinant in Eq.(6.51) is nonzero, and the variables $q_1 \cdots q_k$, $Q_1 \cdots Q_A$, $P_{A+1} \cdots P_k$ form an independent set. We can also easily generalize the argument to the case where we start with one member from each pair (q_s, p_s) in the independent set.

Thus we see that every finite canonical transformation can be given in a closed (but implicit) form that is a variant of the four simple forms described earlier. In each case there is a unique generating function, determining the transformation.

It may nevertheless be interesting to see how one would modify the formulae given above if, for example, we decided to use the q_s and Q_s preferentially but found that they were not all independent. Such cases are called "degenerate cases," not in any general sense, but from the point of view of using certain chosen sets of $2k$ coordinates. For example, in the case $k = 2$ the transformation

$$Q_1 = p_1, \qquad P_1 = -q_1, \qquad Q_2 = q_2, \qquad P_2 = p_2$$

is degenerate from the point of view of each of the four types of generating functions $F_{(1)}, \cdots, F_{(4)}$. We give now a discussion of these cases.

Degenerate Cases

Suppose that we wish to use a generating function of type $F_{(1)}(q, Q, t)$ but that there exist $(k - A)$ independent relations involving q, Q and t:

$$\Omega_b(q, Q, t) = 0, \qquad b = A + 1, \cdots, k. \tag{6.57}$$

(For later convenience, we have numbered these relations in this way). This implies that $F_{(1)}$ does not have a unique functional form and also that the velocities $\dot{q}_s$ and $\dot{Q}_s$ are not all independent. To handle this situation, we introduce a set of undetermined multipliers λ_b, which allow us to formally treat all the q_s and Q_s as independent; eventually the λ_b must be determined in such a way that Eq.(6.57) is obeyed. Choosing some definite functional form for $F_{(1)}$, we have to consider the system

$$\sum_s p_s \dot{q}_s - H = \sum_s P_s \dot{Q}_s - H' + \sum_s \left(\frac{\partial F_{(1)}}{\partial q_s} \dot{q}_s + \frac{\partial F_{(1)}}{\partial Q_s} \dot{Q}_s \right) + \frac{\partial F_{(1)}}{\partial t}, \tag{6.58a}$$

$$\sum_s \left(\frac{\partial \Omega_b}{\partial q_s} \dot{q}_s + \frac{\partial \Omega_b}{\partial Q_s} \dot{Q}_s \right) + \frac{\partial \Omega_b}{\partial t} = 0, \qquad b = A + 1, \cdots, k. \tag{6.58b}$$

Now multiply the latter equations by λ_b, sum on b, add the result to the first equation, and then identify the coefficients of $\dot{q}_s$ and $\dot{Q}_s$ as though they were all independent:

$$p_s = \frac{\partial F_{(1)}}{\partial q_s} + \sum_b \lambda_b \frac{\partial \Omega_b}{\partial q_s} \equiv \frac{\partial}{\partial q_s}\left(F_{(1)} + \sum_b \lambda_b \Omega_b\right); \tag{6.59a}$$

$$P_s = -\frac{\partial F_{(1)}}{\partial Q_s} - \sum_b \lambda_b \frac{\partial \Omega_b}{\partial Q_s} \equiv -\frac{\partial}{\partial Q_s}\left(F_{(1)} + \sum_b \lambda_b \Omega_b\right); \tag{6.59b}$$

$$H' = H + \frac{\partial F_{(1)}}{\partial t} + \sum_b \lambda_b \frac{\partial \Omega_b}{\partial t} \equiv H + \frac{\partial}{\partial t}\left(F_{(1)} + \sum_b \lambda_b \Omega_b\right). \tag{6.59c}$$

Equations (6.57) and (6.59) together form the generalization of Eq.(6.44). To make the resemblance greater, we can write the present set in the form

$$p_s = \frac{\partial F'}{\partial q_s}, \qquad P_s = -\frac{\partial F'}{\partial Q_s}, \qquad H' = H + \frac{\partial F'}{\partial t}; \quad \Omega_b = 0;$$
$$F'(q, Q, \lambda, t) = F_{(1)} + \sum_b \lambda_b \Omega_b(q, Q, t)\,. \tag{6.60}$$

Thus it now takes $(k - A + 1)$ functions $F_{(1)}$, Ω_b of q, Q, t to completely define the transformation. We can regard Eqs. (6.57), (6.59a), and (6.59b) as $3k - A$ equations for the unknowns P_s, Q_s, λ_b; when these are solved, λ_b is determined, and the transformation fully specified.

Actually, what we have done in this degenerate case is exactly equivalent to the kind of solution that is available in the case of any canonical transformation, which is described in the remarks following Eq.(6.50); the equations have only been worked out so that they bear a close resemblance to the case where q_s and Q_s are independent. The role of the multipliers λ_b, when they are obtained as functions on phase space, is to make up, together with the $k + A$ independent coordinates contained in q_s and Q_s, a *total* of $2k$ independent functions. We can convince ourselves of this by viewing the foregoing calculation slightly differently. Let us assume that the constraint equations, Eq.(6.57), allow us to express $Q_{A+1}, \cdots, Q_k$ as functions of $q_s, Q_1, \cdots, Q_A$, and t; there is no loss of generality here. We then write these equations explicitly in the form:

$$\Omega_b(q_s, Q_s, t) = Q_b - f_b(q_s, Q_a, t) = 0, \qquad b = A+1, \cdots, k;$$
$$a = 1, 2, \cdots, A. \tag{6.61}$$

$F_{(1)}$ has a definite functional form in terms of q_s, Q_a, and t. The starting point of the transformation is then:

$$\sum_s p_s \dot{q}_s - H = \sum_{a=1}^{A} P_a \dot{Q}_a + \sum_{b=A+1}^{k} P_b \left(\sum_s \frac{\partial f_b}{\partial q_s}\dot{q}_s + \sum_a \frac{\partial f_b}{\partial Q_a}\dot{Q}_a + \frac{\partial f_b}{\partial t}\right) - H'$$
$$+ \sum_s \frac{\partial F_{(1)}}{\partial q_s}\dot{q}_s + \sum_{a=1}^{A} \frac{\partial F_{(1)}}{\partial Q_a}\dot{Q}_a + \frac{\partial F_{(1)}}{\partial t}. \tag{6.62}$$

Identifying coefficients of $\dot{q}_s$ and $\dot{Q}_a$ we get:

$$p_s = \frac{\partial F_{(1)}}{\partial q_s} + \sum_b P_b \frac{\partial f_b}{\partial q_s}, \quad P_a = -\frac{\partial F_{(1)}}{\partial Q_a} - \sum_b P_b \frac{\partial f_b}{\partial Q_a};$$

$$H' = H + \frac{\partial F_{(1)}}{\partial t} + \sum_b P_b \frac{\partial f_b}{\partial t}. \tag{6.63}$$

All the p_s and the P_a for $a = 1, 2, \cdots, A$, are expressed in terms of all the q_s, the Q_a for $a = 1, 2, \cdots, A$, and the remaining P_b, $b = A+1, \cdots, k$. Equation (6.61) already expresses the Q_b in terms of q_s, Q_a, and t. We can nicely combine Eqs.(6.61) and (6.63) into one set if we use the generating function F defined as:

$$F(q_s, Q_a, P_b, t) = F_{(1)}(q_s, Q_a, t) + \sum_b P_b f_b(q_s, Q_a, t), \tag{6.64}$$

for we then have:

$$p_s = \frac{\partial F}{\partial q_s}, \quad P_a = -\frac{\partial F}{\partial Q_a}, \quad Q_b = \frac{\partial F}{\partial P_b};$$

$$H' = H + \frac{\partial F}{\partial t}. \tag{6.65}$$

On the one hand, comparison with the treatment of the degenerate case using the multipliers λ_b shows that essentially the λ_b are the new momenta P_b; on the other hand, Eq.(6.65) shows that the $2k$ variables q_s, Q_a, P_b have been chosen as independent ones, and these equations fit exactly into the general pattern discussed after Eq.(6.50). Thus the meaning for the λ's as well as the equivalence of the two ways of dealing with the "degenerate" cases, are both clear.

Role of the Action as a Generator of Type I

We have now studied two ways of building up canonical transformations. The first of these is to "generate" them through the solutions to a family of differential equations of the form of Eq.(6.16); this method clearly works only for canonical transformations continuously connected to the identity. The second method can be used for all cases and involves giving a single function F of $2k$ independent variables, half of them old ones and the other half new ones. In both methods some amount of work has to be done before one has the transformation spelled out explicitly. One has to solve the differential equations in the first method, whereas in the second, one has to turn one half of the equations "inside out" in order to express the new variables as functions of the old ones.

We have seen that the Hamiltonian form of the equations of motion for a dynamical system corresponds exactly to the first method of constructing canonical transformations; thus the development in time of the system is a continuous unfolding of a canonical transformation. The Hamiltonian H acts as the generator (infinitesimal generator, to be precise) of this canonical transformation, playing the role of the function ϕ in Eq.(6.16). The question naturally arises:

What is the function F, depending on half of the phase-space coordinates at an initial time t_1 and the "other" half of the coordinates at a final time t_2, that determines the finite canonical transformation taking the system from time t_1 to time t_2 according to the second method of building up a canonical transformation? We shall show that the action function Φ introduced in Eq.(3.3) plays this role.

We confine ourselves to a treatment of Lagrangians belonging to the standard type; that is to say, those Lagrangians for which the straightforward transition to the Hamiltonian outlined in Chapters 3 and 4 is applicable. The discussion we give in Chapter 8 then shows that in such cases, at a given instant of time t the coordinates q_s and velocities $\dot{q}_s$ in configuration space may be specified independently; *there can be no relations among the q_s and $\dot{q}_s$ variables.* In turn, this shows that the coordinates q_s at an initial time t_1 and the coordinates Q_s at a final time t_2 *can be chosen completely independently of one another for any t_2 except possibly for certain isolated values*; for given values of these initial and final times and configurations, q_s at t_1 and Q_s at t_2, the equations of motion pick out an (in general) unique trajectory in configuration space connecting these configurations. Suppose that this were not so and that there was one (or more) relations involving the Q_s, q_s and the times t_1, t_2:

$$\xi(q_s, t_1; Q_s, t_2) = 0. \tag{6.66}$$

If we were to set $t_2 = t_1$ here, this relation must reduce to an identity, as the q_s by themselves are certainly independent variables. However, we could first differentiate this equation with respect to t_2 once and then set $t_2 = t_1$, and we would then end up with a relation among the q_s and $\dot{q}_s$ at one time, t_1, which we said is impossible for a Lagrangian system of standard type. If we find that a single differentiation of Eq.(6.66) with respect to t_2 does not yield such a relation between q_s and $\dot{q}_s$, we could differentiate it twice (or more), substitute for the accelerations $\ddot{Q}_s$ (or higher derivatives) in terms of Q_s and $\dot{Q}_s$, because for a standard Lagrangian the $\ddot{Q}_s$ can always be solved for in this way, and then set $t_2 = t_1$. In this way, we would at some stage arrive at a nontrivial relation among the q_s and $\dot{q}_s$ at one time t_1 generated by every relation of the type in Eq.(6.66). Thus we see quite generally that if the Lagrangian is of standard type, the initial and final configurations may always be chosen independently for almost all times t_1 and t_2.

Equation (3.3) defines the action functional $\Phi[C]$ in configuration space for any trajectory C connecting given configurations at times t_1 and t_2. As we explained there, Φ is a functional because the trajectory C can be arbitrarily chosen. Let us now consider the *value* of $\Phi[C]$ for the particular dynamical trajectory C_{dyn} determined by the equations of motion in order to go from the configuration q_s at t_1 to Q_s at t_2: for this choice of $C, \Phi[C_{\text{dyn}}]$ is stationary (according to the action principle of Hamilton), but more importantly for our present purposes, $\Phi[C_{\text{dyn}}]$ *can be thought of as an ordinary function of the $2k$ independent variables q_s, Q_s, $s = 1, \cdots, k$ and the times t_1, t_2.* By specializing the *functional* $\Phi[C]$ to a unique trajectory C that is completely determined by its end points, we have obtained a *function* of the initial and final configurations. It is customary to call this function S

$$S(q_s, t_1; Q_s, t_2) = \int_{\substack{t_1 \\ \text{along } C_{dyn}}}^{t_2} dt L(q_s(t), \dot{q}_s(t), t);$$

$$q_s(t_1) = q_s, \qquad q_s(t_2) = Q_s. \tag{6.67}$$

In words, S is the time integral of the Lagrangian along the dynamical path of the system.

The most important properties of S are contained in Eq.(3.9), which expresses the change in $\Phi[C]$ for an *arbitrary* variation in the trajectory C. We are now interested in small variations in the initial and final configurations and times but with the trajectory C always taken to be the dynamical one. Thus we will use Eq.(3.9) for the case in which the variation in the trajectory C is determined by the (small and independent) variations in q_s, Q_s, t_1 and t_2. With this specialization, we directly obtain:

$$\sum_{s=1}^{k} \left(\frac{\partial S}{\partial q_s} \Delta q_s + \frac{\partial S}{\partial Q_s} \Delta Q_s \right) + \frac{\partial S}{\partial t_1} \Delta t_1 + \frac{\partial S}{\partial t_2} \Delta t_2$$

$$= \sum_{s=1}^{k} (P_s \Delta Q_s - p_s \Delta q_s) - (H(Q, P, t_2)\Delta t_2 - H(q, p, t_1)\Delta t_1). \tag{6.68}$$

With respect to the variables q_s and Q_s we see that S obeys:

$$p_s = \frac{-\partial S}{\partial q_s}, \qquad P_s = \frac{\partial S}{\partial Q_s}, \qquad s = 1, 2, \cdots, k; \tag{6.69}$$

whereas with respect to the times t_1, t_2 we find:

$$\frac{\partial S}{\partial t_2} + H(Q, P, t_2) = 0, \tag{6.70a}$$

$$\frac{\partial S}{\partial t_1} - H(q, p, t_1) = 0. \tag{6.70b}$$

Keeping the initial time t_1 fixed for the moment and considering the dependence of S on q_s, Q_s, and $t = (t_2)$, we compare Eqs.(6.44a) and (6.44b) with (6.69) and arrive at the conclusion that $-S$ now plays the role of the (finite) generating function $F_{(1)}(q, Q, t)$ of type I of the earlier treatment. Thus if the function S were known and if Eq.(6.69) could be solved to express the Q_s, P_s as functions of q_s, p_s, t (and t_1), these expressions would directly give us the finite canonical transformation describing the motion of the system in phase space over the time interval t_1 to t.

One does not expect these results to be of much use in actually solving the equations of motion for a classical Lagrangian system, because in computing the function S we need to know the dynamical trajectory of the system in configuration space, and this presupposes a knowledge of the solution of the equations of motion! Rather, the interest lies in the insight gained into the relationship between the Hamiltonian and the Lagrangian forms of dynamics, in particular between the Hamiltonian and the Lagrangian. We see that whereas the Hamiltonian describes the *infinitesimal* form of the canonical transformation generated by the motion in phase space, the time integral of the Lagrangian describes the *finite* form of this canonical transformation. This reciprocal relationship has an analogy in quantum mechanics as well, and is most clearly seen in the formulation due to Feynman.

It is natural to expect that both parts of Eq.(6.70) are analogous to the previous Eq.(6.44c). We explain how this comes about by considering the case of (6.70a). By now it should be clear that if somehow the function S were known, and if the Eq.(6.69) could be solved for the Q_s, P_s as functions of q_s, p_s, t (and also t_1), say, in the form

$$Q_s = \phi_s(q_s, p_s, t; t_1), \qquad P_s = \psi_s(q_s, p_s, t; t_1), \tag{6.71}$$

then these expressions are in fact the solutions to the equations of motion

$$\frac{dQ_s}{dt} = \frac{\partial H(Q,P,t)}{\partial P_s}, \qquad \frac{dP_s}{dt} = -\frac{\partial H(Q,P,t)}{\partial Q_s} \tag{6.72}$$

subject to the initial conditions

$$Q_s\Big|_{t=t_1} = q_s, \; P_s\Big|_{t=t_1} = p_s. \tag{6.73}$$

Equation (6.71) in fact exhibits the solution in terms of time t and the given initial values q_s, p_s. Now it is clear that the q_s, p_s do not depend on t at all, because they are the fixed initial values of Q_s, P_s. In fact if any t dependence were ascribed to them, Eq.(6.72) would not be obeyed any more! Now let us consider the canonical transformation taking us *from* the Q_s, P_s set *to* the q_s, p_s set; we now think of the former as the old, the latter as the new variables. The finite generating function is seen from Eq.(6.69) to be S, not $-S$, because the roles of old and new variables have been interchanged; the passage from Q_s, P_s to q_s, p_s is a t-dependent canonical transformation. But this time-dependent transformation must be such as to make the "new" variables q_s, p_s independent of t, given that the "old" ones obey Eq.(6.72). The time dependence, if any, of the q_s, p_s would be governed by the "new" Hamiltonian determined according to Eq.(6.44c):

$$H'(q,p,t) = H(Q,P,t) + \frac{\partial S}{\partial t}\,; \tag{6.74}$$

by Eq.(6.70a) this vanishes,

$$H'(q,p,t) = 0, \tag{6.75}$$

thus ensuring that the q_s, p_s are indeed t independent. A parallel interpretation clearly can be given for Eq.(6.70b) as well.

Chapter 7

Invariant Measures in Phase Space and Various Forms of Development in Time

The two equivalent ways of defining canonical transformations using either PB's or LB's. Eqs.(5.28) and (5.29), are both differential in character. These conditions involving the fundamental brackets have been shown to be completely equivalent to the invariance properties, under canonical transformations, of the PB's and LB's among arbitrary functions on phase space. There is another type of entity, namely, the collection of volume and surface elements of various dimensions in phase space, which are also invariant under canonical transformations. We turn now to a discussion of these expressions.

Let us restrict ourselves to the subgroup of regular canonical transformations of a $2k$-dimensional phase space, $\omega \to \omega'$. The basic $2k$-dimensional volume element is defined as

$$d\omega \equiv d\omega^1 \cdots d\omega^{2k} \equiv dq_1 \cdots dq_k dp_1 \cdots dp_k, \tag{7.1}$$

and if $A(\omega)$ is a certain function of ω, its volume integral over the space is the multiple integral:

$$\int \cdots \int A(\omega) d\omega^1 \cdots d\omega^{2k} \equiv \int A(\omega) d\omega. \tag{7.2}$$

The ranges of the integration variables are defined by their physical meanings. For a canonical transformation, we have the condition:

$$\epsilon^{\mu\nu} \frac{\partial \omega'^\rho}{\partial \omega^\mu} \frac{\partial \omega'^\sigma}{\partial \omega^\nu} = \epsilon^{\rho\sigma} \tag{7.3}$$

from Eq.(5.38). Taking the determinant of both sides of this equation and interpreting the left-hand side as the product of three matrices, we get the following condition for the Jacobian of the transformation:

$$J = \det\left|\frac{\partial\omega'^{\rho}}{\partial\omega^{\mu}}\right|, \qquad J^2 = 1. \tag{7.4}$$

Now in general we cannot conclude that $J = 1$, but if the canonical transformation is connected continuously to the identity, we can draw this conclusion. Thus at least for such transformations we have immediately the result that the volume element in phase space is invariant:

$$d\omega' = J\, d\omega = d\omega. \tag{7.5}$$

More generally, because the volume element is an intrinsically positive quantity, we should really use $|J|$ in Eq.(7.5), so that the canonical invariance of $d\omega$ is always true. (In fact, it is true for any canonical transformation, regular or otherwise.) For a regular canonical transformation, we can consider the new function $\bar{A}(\omega)$ determined by a given function $A(\omega)$ by the equation

$$A(\omega') = \bar{A}(\omega). \tag{7.6}$$

(See the remarks following Eq.(6.31). From Eq.(7.5) we derive the invariance of the phase space integral of A:

$$\int A(\omega)d\omega \equiv \int A(\omega')d\omega' \equiv \int \bar{A}(\omega)d\omega' \equiv \int \bar{A}(\omega)d\omega. \tag{7.7}$$

If we are dealing with a pair of functions $A(\omega)$, $B(\omega)$ and each defines a new function according to Eq.(7.6), and if we think of the integral of the product $A(\omega)B(\omega)$ over all phase space as the "scalar product" of A and B, then we conclude that this scalar product, too, is an invariant under regular canonical transformations:

$$(\bar{A}, \bar{B}) \equiv \int \bar{A}(\omega)\bar{B}(\omega)d\omega = \int A(\omega)B(\omega)d\omega = (A, B). \tag{7.8}$$

This invariant integration over phase space is of great importance in connection with the picture of the Hamiltonian equations of motion as depicting a "continuous unfolding" of a group of canonical transformations. This aspect of the equations of motion is discussed in the previous chapter. Let us consider only Hamiltonians not explicitly dependent on time. Then we see that the phase at time t is related to the initial phase, ω_0 at $t = 0$ by

$$\omega^{\mu}(t) \equiv \omega^{\mu} = e^{-t\tilde{H}(\omega_0)}\omega_0^{\mu}, \tag{7.9}$$

whereas for a general dynamical variable $f(\omega)$:

$$f(\omega(t)) = e^{-t\tilde{H}(\omega_0)}f(\omega_0). \tag{7.10}$$

Now let us imagine that the initial phase ω_0 is not precisely known, but that there is a certain probability density $\rho(\omega_0)$ for having various initial phases. This function obeys

$$\rho(\omega_0) \geqq 0, \qquad \int \rho(\omega_0) d\omega_0 = 1; \tag{7.11}$$

and for a small volume $\delta\omega_0$, $\rho(\omega_0)\delta\omega_0$ is the probability that the point representing the initial phase lies in the volume $\delta\omega_0$. If we want to know the expected value of a coordinate, or a general dynamical variable, at time $t = 0$, we must introduce the "expectation value" or "average value" at $t = 0$:

$$\begin{aligned} \langle \omega^\mu \rangle_0 &\equiv \int \omega_0^\mu \rho(\omega_0) d\omega_0, \\ \langle f \rangle_0 &\equiv \int f(\omega_0) \rho(\omega_0) d\omega_0. \end{aligned} \tag{7.12}$$

Naturally, an uncertainty in the values of the dynamical variables at time $t = 0$ will introduce an uncertainty at any later time t; to get the expected value of a variable at time $t > 0$, we must suitably average the solutions of the equations of motion over the initial conditions. Clearly we must have

$$\begin{aligned} \langle \omega^\mu \rangle_t &= \int \omega^\mu(t, \omega_0) \rho(\omega_0) d\omega_0 = \int (e^{-t\tilde{H}(\omega_0)} \omega_0^\mu) \rho(\omega_0) d\omega_0, \\ \langle f \rangle_t &= \int (e^{-t\tilde{H}(\omega_0)} f(\omega_0)) \rho(\omega_0) d\omega_0. \end{aligned} \tag{7.13}$$

The distribution function $\rho(\omega_0)$ describes the "state of the system" at time $t = 0$ and is given once and for all for a specific state of motion. In the formulation developed above, it has no intrinsic time-dependence. The time dependence of the expected values of various dynamical variables $f(\omega)$ arises from the time dependence of the basic variables ω^μ, the functional form of $f(\omega)$ remaining unaltered in time. We recall that this time dependence, which is displayed in explicit form in Eqs.(7.9) and (7.10), arose from the basic Hamiltonian equations of motion:

$$\begin{aligned} \dot{\omega}^\mu &= \{\omega^\mu, H(\omega)\}_\omega, \\ \dot{f}(\omega) &= \{f(\omega), H(\omega)\}_\omega, \end{aligned} \tag{7.14}$$

which, as we saw in the last chapter, can be rewritten in a form in which the "initial value" variables ω_0 are used in place of ω both in $H(\omega)$ and in the process of computing the PB's.

Now the quantities of real interest are the expected values of dynamical variables at any time t, $\langle f \rangle_t$. From its expression in Eq.(7.13) and the canonical invariance of the phase-space integral, we see immediately that we can write it in the alternative form:

$$\begin{aligned}\langle f\rangle_t &= \int (e^{-t\tilde{H}(\omega_0)} f(\omega_0))\rho(\omega_0)d\omega_0 \\ &= \int f(\omega_0)(e^{t\tilde{H}(\omega_0)}\rho(\omega_0))d\omega_0,\end{aligned} \tag{7.15}$$

and in particular, for the coordinates themselves,

$$\langle \omega^\mu\rangle_t = \int \omega_0^\mu (e^{t\tilde{H}(\omega_0)}\rho(\omega_0))d\omega_0. \tag{7.16}$$

In this form we see that the time dependence has been transferred from the dynamical variable to the distribution function! The time-dependent distribution function obeys an equation of motion:

$$\begin{aligned}\rho_t(\omega_0) &= e^{t\tilde{H}(\omega_0)}\rho(\omega_0), \\ \frac{\partial}{\partial t}\rho_t(\omega_0) &= -\{\rho_t(\omega_0), H(\omega_0)\}_{\omega_0},\end{aligned} \tag{7.17}$$

and this equation naturally differs in sign on the right-hand side from Eq.(7.14).

In analogy with the situation in quantum mechanics, these two forms of dynamics are called the Heisenberg and the Schrödinger forms, respectively. In the Heisenberg form, the "state of the system" is specified by a time-independent density function $\rho(\omega_0)$ in phase space, determined by the probabilities of various phases at time $t = 0$. Each dynamical variable, such as the momentum or the kinetic energy of a single particle in a many-particle system, obeys the equation of motion in Eq.(7.14); it is *represented* by a function with a time-dependent functional form when expressed in terms of the variables ω_0, but by a function with a time-independent functional form when expressed in terms of the variables $\omega(t)$. (In most cases, the latter expression in terms of $\omega(t)$ actually serves as the *definition* of the dynamical variable.) In the Schrödinger form, on the other hand, the state of the system is specified by the time-dependent density function $\rho_t(\omega_0)$ obeying the equation of motion in Eq.(7.17); thus, expressed in terms of ω_0, it has a varying functional form, whereas the functional forms $f(\omega_0)$ of dynamical variables do not change in time. In both forms, we get the same values for $\langle f\rangle_t$. The basic time-dependent entity in the Heisenberg picture is the dynamical variable, and in the Schrödinger picture it is the density function ρ. We said above that in the Heisenberg picture the varying functional form $f_t(\omega_0)$ is related to a fixed functional form in terms of the variables $\omega(t)$:

$$f_t(\omega_0) = f(\omega(t)), \qquad \omega(t) = e^{-i\tilde{H}(\omega_0)}\omega_0. \tag{7.18}$$

There is an analogue to this in the Schrödinger picture; from Eq.(7.17) we see the varying functional form $\rho_t(\omega_0)$ obeys

$$\rho_t(\omega_0) = \rho(\omega(-t)), \qquad \omega(-t) = e^{t\tilde{H}(\omega_0)}\omega_0. \tag{7.19}$$

One can see from this discussion that what really is of significance is the "relative motion" between distribution functions and dynamical variables; either $\rho(\omega_0)$ is held fixed and the $f_t(\omega_0)$ move "one" way, or the $f(\omega_0)$ stay fixed while $\rho_t(\omega_0)$ moves the "other" way. In both forms, we get identical results for expected values of measurements. The essential ingredients of the theory are dynamical variables and density functions; average values of the former are given by canonical invariant phase-space integrals with the latter. The dependence on time of these average values must come from the time variation of the quantities entering the phase space integral. Different pictures of dynamics arise solely because of the canonical invariance of the phase space integral and are related by time-dependent canonical transformations. The Heisenberg and Schrödinger pictures are two extreme pictures. The Dirac or "Interaction" picture is an intermediate one, assigning time dependence to both dynamical variables and density functions, and we describe it shortly.

The Schrödinger form of dynamics is actually the one normally used in statistical mechanics. Here, one sets up an ensemble, represented by a cloud of points in phase space; that is, a collection of identical systems with identical Hamiltonians, but distributed at time $t = 0$ in phase space with a density given by $\rho(\omega_0)$. One then lets each system develop in time on its own; the whole cloud of points moves in phase space in a manner completely determined by the Hamiltonian. This motion is just the mapping of phase space onto itself corresponding to the one-parameter group of canonical transformations generated by H and gives rise at each time t to a time-dependent distribution function $\rho_t(\omega_0)$. The equation of motion for $\rho_t(\omega_0)$, Eq.(7.17) above, is called Liouville's equation in statistical mechanics. One also recalls that in defining average values of dynamical variables in statistical mechanics, one uses functions of ω_0 that are fixed in their form and need no equations of motion themselves. Thus we see that the usual phase-space formulation of classical statistical mechanics is in reality the Schrödinger form of mechanics.

Let us now explain briefly the Dirac or Interaction picture form of classical dynamics. This arises when we split the total Hamiltonian into two parts and assign one part to control the time dependence of dynamical variables, whereas the other part acts on the distribution function. We can obtain this picture most simply by starting from the Schrödinger form and making a time-dependent canonical transformation. Let us write the total Hamiltonian $H(\omega_0)$ as a sum of a "free" part H_0 and an interaction term V:

$$H(\omega_0) = H_0(\omega_0) + V(\omega_0). \tag{7.20}$$

Then, starting from the time-dependent distribution function $\rho_t(\omega_0)$ of the Schrödinger picture, we define an "Interaction picture distribution function" $\rho_t^{(I)}(\omega_0)$:

$$\rho_t^{(I)}(\omega_0) = e^{-t\tilde{H}_0(\omega_0)}\rho_t(\omega_0). \tag{7.21}$$

At the same time each dynamical variable f is represented by a time-dependent function of ω_0:

$$f_t^{(I)}(\omega_0) = e^{-t\tilde{H}_0(\omega_0)}f(\omega_0). \tag{7.22}$$

The effect is to leave unchanged the expression for $\langle f \rangle_t$:

$$\langle f \rangle_t = \int f_t^{(I)}(\omega_0)\rho_t^{(I)}(\omega_0)d\omega_0. \tag{7.23}$$

Now the explicit time dependence of $\rho_t^{(I)}(\omega_0)$ is different from that of $\rho_t(\omega_0)$ of the Schrödinger picture; it obeys the equation of motion

$$\begin{aligned}
\frac{\partial}{\partial t}\rho_t^{(I)}(\omega_0) &= -e^{-t\tilde{H}_0(\omega_0)}(\tilde{H}_0(\omega_0)\rho_t(\omega_0)) + e^{-t\tilde{H}_0(\omega_0)}\frac{\partial}{\partial t}\rho_t(\omega_0) \\
&= e^{-t\tilde{H}_0(\omega_0)}\{\rho_t(\omega_0), H_0(\omega_0)\}_{\omega_0} + e^{-t\tilde{H}_0(\omega_0)}\{\rho_t(\omega_0), -H(\omega_0)\}_{\omega_0} \\
&= -e^{-t\tilde{H}_0(\omega_0)}\{\rho_t(\omega_0), V(\omega_0)\}_{\omega_0} \\
&= -\{\rho_t^{(I)}(\omega_0), e^{-t\tilde{H}_0(\omega_0)}V(\omega_0)\}_{\omega_0}.
\end{aligned} \tag{7.24}$$

In working out this law for $\rho_t^{(I)}(\omega_0)$, we used the Schrödinger equation of motion for $\rho_t(\omega_0)$, Eq.(7.17), as well as the canonical invariance of PB's in the form given in Eq.(6.38). We also assume that both $H(\omega_0)$ and $H_0(\omega_0)$ generate regular canonical transformations. Now if we compare this equation of motion with that of $\rho_t(\omega_0)$, we see that whereas the latter involves a Hamiltonian $H(\omega_0)$ not explicitly dependent on time, the former involves one that does have an explicit time dependence. (We assume that $H(\omega_0)$, $H_0(\omega_0)$, and hence $V(\omega_0)$ all have no explicit t-dependence.) This interaction Hamiltonian is given by

$$\begin{aligned}
V_t(\omega_0) &= e^{-t\tilde{H}_0(\omega_0)}V(\omega_0) = V(\omega(t)), \\
\omega(t) &= e^{-t\tilde{H}_0(\omega_0)}\omega_0,
\end{aligned} \tag{7.25}$$

and in terms of it the equation of motion is

$$\frac{\partial}{\partial t}\rho_t^{(I)}(\omega_0) = -\{\rho_t^{(I)}(\omega_0), V_t(\omega_0)\}_{\omega_0}. \tag{7.26}$$

On the other hand, the dynamical variables $f_t^{(I)}(\omega_0)$ obey

$$\frac{\partial}{\partial t}f_t^{(I)}(\omega_0) = \{f_t^{(I)}(\omega_0), H_0(\omega_0)\}_{\omega_0}; \tag{7.27}$$

"their Hamiltonian" has no explicit t-dependence. Equations (7.26) and (7.27) are the basic equations of the Dirac picture, along with Eq.(7.23) defining $\langle f \rangle_t$. The dynamical variables behave as though we were in a Heisenberg picture with just a free Hamiltonian H_0; the distribution function behaves as though we were in a Schrödinger picture with just an interaction Hamiltonian, but this has an explicit time dependence. This formulation is useful if we are able to solve exactly Eq.(7.27), say, and wish to use a perturbative approach in taking account of the interaction V.

It may be helpful to remark that in explaining the three forms of dynamics, and appreciating their relations to one another, we have always used a fixed system of phase-space coordinates ω_0 in the entire discussion. This coordinate system does not change with time; indeed such a change is undefined. When we refer to the time dependence of any function on phase space, whether it is a state or a dynamical variable, we always refer to the *explicit* time dependence obtained when that entity is written out in terms of the coordinates ω_0; the several equations of motion always refer to the changing functional form of the corresponding quantities written in terms of ω_0. It is possible to discuss the effects of using a *time-dependent coordinate system* in phase space, although we shall not do it in any detail here. The equations of motion that one would then derive would refer to the *explicit* time dependences that are present in various quantities *after* they have been expressed in terms of the *time-dependent* coordinates. The only reason for making this point is that one can view the passage from the Heisenberg to the Schrödinger form of mechanics as corresponding to the replacement of the fixed phase-space coordinate system ω_0 by the time-dependent coordinate system $\omega^\mu(t) = e^{-t\tilde{H}(\omega_0)}\omega_0^\mu$, rewriting dynamical variables and distribution functions $f_t(\omega_0)$ and $\rho(\omega_0)$ of the Heisenberg picture in terms of $\omega(t)$ and then examining the explicit time dependences of the resulting expressions.

The Poincaré Invariants

For a dynamical system with one degree of freedom in configuration space, $k = 1$, a transformation $q, p \to Q, P$ in phase space that preserves the phase-space measure is already a canonical transformation. This happens because in this case the Jacobian J of the transformation is the same as the PB of Q with P:

$$J = \begin{vmatrix} \frac{\partial Q}{\partial q} & \frac{\partial Q}{\partial p} \\ \frac{\partial P}{\partial q} & \frac{\partial P}{\partial p} \end{vmatrix} = \{Q, P\}_{q,p}; \tag{7.28}$$

hence, if $J = 1$, the transformation is canonical, because the other two PB's $\{Q, Q\}$ and $\{P, P\}$ vanish in any case. However, if k is $\geqq 2$, then a measure-preserving transformation is not necessarily canonical, and the question arises whether there is some measure in phase space whose invariance is equivalent to the transformation being canonical. In fact, there is a whole sequence of measures of increasing (even) dimensions, all of which are invariant under canonical transformations, and conversely the invariance of the lowest of them, the two-dimensional one, under a transformation forces the transformation to be canonical. These invariants are called the Poincaré Invariants. In the case $k = 1$, the basic two dimensional measure coincides with the volume element in the two dimensional phase space. For $k \geqq 2$, the invariance of the four, six and higher dimensional measures by themselves is not sufficient to guarantee the canonical nature of the transformation. The canonical invariance of the two-dimensional measure can be developed from the phase-space form of the Weiss action principle. However, we first give a direct description of this invariant and then explain its relation to the action principle.

Consider a set S of points in phase space that form a two-dimensional surface in that space. That means that the coordinates ω^μ of the points of S can be given as continuous functions of two real variables u_1, u_2 with these variables going over prescribed ranges. We then define the elemental "area" dS on the surface S by means of

$$dS = -\epsilon_{\mu\nu}\frac{\partial\omega^\mu}{\partial u_1}\frac{\partial\omega^\nu}{\partial u_2}\, du_1\, du_2 = [u_1, u_2]du_1\, du_2 \tag{7.29}$$

and the "area" of the surface S as

$$\int dS = -\int du_1\, du_2\, \epsilon_{\mu\nu}\frac{\partial\omega^\mu}{\partial u_1}\frac{\partial\omega^\nu}{\partial u_2} = \int du_1\, du_2 [u_1, u_2]. \tag{7.30}$$

Thus dS is defined relative to a canonical coordinate system, ω^μ, and a parametrisation of S by u_1, u_2. But we are going to see that it does not depend on the choice of u_1, u_2, and is also independent of the canonical coordinate system used. We have expressed things here in terms of the LB of u_1 with u_2, because on S all the coordinates ω^μ are functions of the u's. We must make it clear that this definition of "area" is not quite the same as a naive geometrical idea of area. In the first place, the definition of dS cannot specify its sign because, for example, the interchange of u_1 and u_2 reverses the sign of dS. (We, of course, regard the expression $du_1\, du_2$ as intrinsically positive.) It is even possible that the LB $[u_1, u_2]$ changes sign from one region to another on S, so that dS also changes sign. We can even consider surfaces S for which the elementary area dS identically vanishes, so that S is assigned zero area. For this we can, for example, take the two-dimensional region obtained by varying ω^1 and ω^2 (q_1 and q_2), keeping all the remaining ω's fixed; the parameters u_1 and u_2 can be taken to be ω^1, ω^2 themselves, and the vanishing of dS follows.

First we prove that up to a sign, the definition of dS does not change if we replace the parameters u_1, u_2 by any other pair v_1, v_2:

$$\begin{aligned}
[u_1, u_2] &= -\epsilon_{\mu\nu}\frac{\partial\omega^\mu}{\partial u_1}\frac{\partial\omega^\nu}{\partial u_2} = -\sum_{a,b}\epsilon_{\mu\nu}\frac{\partial\omega^\mu}{\partial v_a}\frac{\partial\omega^\nu}{\partial v_b}\frac{\partial v_a}{\partial u_1}\frac{\partial v_b}{\partial u_2} \\
&= -\epsilon_{\mu\nu}\frac{\partial\omega^\mu}{\partial v_1}\frac{\partial\omega^\nu}{\partial v_2}\left(\frac{\partial v_1}{\partial u_1}\frac{\partial v_2}{\partial u_2} - \frac{\partial v_2}{\partial u_1}\frac{\partial v_1}{\partial u_2}\right) \\
&= [v_1, v_2]\begin{vmatrix} \frac{\partial v_1}{\partial u_1} & \frac{\partial v_1}{\partial u_2} \\ \frac{\partial v_2}{\partial u_1} & \frac{\partial v_2}{\partial u_2} \end{vmatrix}
\end{aligned} \tag{7.31}$$

Here we used the antisymmetry of $\epsilon_{\mu\nu}$ as well as the expressions for the v's in terms of the u's. Now the last factor in Eq.(7.31), the determinant, is just the Jacobian for the transformation $u \to v$, so that we do get

$$[u_1, u_2]du_1\, du_2 = \pm[v_1, v_2]dv_1\, dv_2 \; ; \tag{7.32}$$

this shows that, the overall sign of dS apart, the definition of area is intrinsic and does not depend on the variables we use to parametrize the surface. The proof of the invariance of dS under a canonical transformation $\omega \to \omega'$ is equally simple; indicating the two LB's explicitly, we just fall upon the invariance of these brackets under canonical transformations:

$$\begin{aligned}[u_1, u_2]_{\omega'} &= -\epsilon_{\mu\nu} \frac{\partial \omega'^{\mu}}{\partial u_1} \frac{\partial \omega'^{\nu}}{\partial u_2} = -\epsilon_{\mu\nu} \frac{\partial \omega'^{\mu}}{\partial \omega^{\alpha}} \frac{\partial \omega'^{\nu}}{\partial \omega^{\beta}} \frac{\partial \omega^{\alpha}}{\partial u_1} \frac{\partial \omega^{\beta}}{\partial u_2} \\ &= -\epsilon_{\alpha\beta} \frac{\partial \omega^{\alpha}}{\partial u_1} \frac{\partial \omega^{\beta}}{\partial u_2} = [u_1, u_2]_{\omega} \, . \end{aligned} \tag{7.33}$$

Here we first used the expression for the new coordinates ω' of points on S in terms of u_1 and u_2, then their expressions in terms of the ω^{μ} given by the canonical transformation equations. Thus we have proved that the definition of area is independent of the parameters u_1, u_2 as well as invariant under canonical transformations; that is, it is a canonical invariant.

The converse is also easy to prove: if we have a coordinate transformation $\omega \to \omega'$ in phase space and if for *every* two-dimensional surface S in phase space the area is the same whether computed, according to Eq.(7.30), in terms of the ω^{μ} or the ω'^{μ}, then the transformation must be canonical. Given the parameters u_1, u_2 for the points on a two-dimensional surface S, locally we can extend the pair u_1, u_2 by adding $(2k-2)$ functions $u_3, u_4, \cdots, u_{2k}$, such that the whole set of u's forms $2k$ independent functions of the ω^{μ}, S being characterized by the fact that on it the function $u_3, u_4, \cdots, u_{2k}$ are constant. The invariance of dS under the transformation $\omega \to \omega'$, because it must be true for all surfaces S, obviously implies the invariance of LB's of functions chosen from any independent set of $2k$ functions; this in turn can happen only if the transformation is canonical. Thus we have found a characterisation of canonical transformations in terms of invariant measures. Of course, it is just a reformulation of the requirement of the invariance of LB's.

It is more usual to write the element of area dS in the following form:

$$\begin{aligned} dS &= \sum_s dq_s \, dp_s, \\ \text{Area of } S &= \int \sum_s dq_s \, dp_s. \end{aligned} \tag{7.34}$$

Here each individual term in the sum is actually *defined* by working out its Jacobian with respect to the variables u_1, u_2 on S, because each q and each p is a function of u_1, u_2 on the surface:

$$dq_s \, dp_s = \begin{vmatrix} \frac{\partial q_s}{\partial u_1} & \frac{\partial q_s}{\partial u_2} \\ \frac{\partial p_s}{\partial u_1} & \frac{\partial p_s}{\partial u_2} \end{vmatrix} du_1 \, du_2 = \left(\frac{\partial q_s}{\partial u_1} \frac{\partial p_s}{\partial u_2} - \frac{\partial q_s}{\partial u_2} \frac{\partial p_s}{\partial u_1} \right) du_1 \, du_2 \, . \tag{7.35}$$

The summation on s clearly reproduces the LB of u_1 and u_2, thus Eqs. (7.30) and (7.34) are really the same. Here again each term of the sum in Eq.(7.34) is not intrinsically positive; the s-th term $dq_s \, dp_s$ denotes the projection on the $q_s - p_s$ plane of the naive geometrical element of area $du_1 du_2$, but the relative signs of the projections for various values of s are important. As in the example given earlier, namely the surface on which q_1, q_2 vary and the remaining q's and p's stay fixed, each term in Eq.(7.34) may vanish!

Let us now connect the canonical invariance of the two-dimensional measure in Eq.(7.34) to the action principle. We make use of Eq.(3.22), which gives the change $\Delta\Psi$ in the phase-space action functional $\Psi[\mathcal{C}]$ caused by a variation in the trajectory $\mathcal{C}$. Let $\mathcal{C}$ be chosen to be the dynamical trajectory $\mathcal{C}_{\text{dyn}}$ on which Hamilton's equations hold. This $\mathcal{C}_{\text{dyn}}$ is determined by any one point on it (phase at one time); thus $\Psi[\mathcal{C}_{\text{dyn}}]$ becomes an ordinary function $S(q, p, t_1; t_2)$ of the phase $q_s = q_s(t_1), p_s = p_s(t_1)$ at t_1, and of the times t_1 and t_2. We keep t_1, t_2 fixed in the sequel. Making a small variation $\delta q_s, \delta p_s$ in the initial phase, the change in $\mathcal{C}_{\text{dyn}}$ is determined, and in principle Eq.(3.22) tells us the change in S. (All this is quite similar to the discussion in Chapter 6, pages (63-64)). We have, in fact,

$$\delta S(q, p, t_1; t_2) = \sum_s P_s \delta Q_s - \sum_s p_s \delta q_s \,. \tag{7.36}$$

The notation here is the following: Q_s and P_s are the values of the dynamical variables at time $t_2, Q_s = q_s(t_2), P_s = p_s(t_2)$, and we know that Q_s, P_s arise from q_s, p_s by means of a canonical transformation, determined by the Hamiltonian and t_1, t_2. The variation δQ_s induced in Q_s, similarly δP_s, is fully determined by δq_s, δp_s, but δp_s and δP_s do not appear explicitly in Eq.(7.36). Imagine now that we have a continuous two-parameter family of initial conditions described by functions $q_s(u_1, u_2)$, $p_s(u_1, u_2)$; this turns Q_s, P_s, and S into functions of u_1, u_2 as well. The variations δq, δp in initial phase will now be given by a variation δu_1, δu_2 in u_1 and u_2, and by the passage of time this will induce the variation δQ, δP. Let us now take the initial phase q, p round a *closed curve* C in the $u_1 - u_2$ plane; at the end of the circuit there should be no change in S; thus we must have:

$$\oint_C dS(q(u), p(u), t_1; t_2) = 0,$$

that is,

$$\sum_s \oint_C P_s dQ_s = \sum_s \oint_C p_s \, dq_s \,. \tag{7.37}$$

Here, $dq_s = (\partial q_s/\partial u_1)du_1 + (\partial q_s/\partial u_2)du_2$, where du_1 and du_2 are the components of an infinitesimal arc of C. If C encloses the area A in the $u_1 - u_2$ plane, we can use Gauss, theorem to give Eq.(7.37) the form

$$\sum_s \int_A du_1 \, du_2 \left(\frac{\partial}{\partial u_1} \left(P_s \frac{\partial Q_s}{\partial u_2} \right) - \frac{\partial}{\partial u_2} \left(P_s \frac{\partial Q_s}{\partial u_1} \right) \right)$$
$$= \sum_s \int_A du_1 \, du_2 \left(\frac{\partial}{\partial u_1} \left(p_s \frac{\partial q_s}{\partial u_2} \right) - \frac{\partial}{\partial u_2} \left(p_s \frac{\partial q_s}{\partial u_1} \right) \right),$$

that is,

$$\int_A \sum_s dQ_s \, dP_s = \int_A \sum_s dq_s \, dp_s. \tag{7.38}$$

Remembering that the dependence of Q, P on u_1, u_2 arises by performing a canonical transformation on q, p to arrive at Q, P and then substituting $q(u), p(u)$ as functions of u_1, u_2, what we have arrived at via the action principle is the canonical invariance of the measure (Eq.(7.34)).

The two-dimensional Poincaré integral invariant generalizes to invariant measures associated with four, six, $\cdots$ dimensional surfaces in phase space. Again it is the basic properties of the LB that lead to the invariance properties of these measures. However, we will forego a discussion of these higher-dimensional objects here.

Chapter 8

Theory of Systems with Constraints

We have explained in Chapter 4 how one can pass from the Lagrangian to the Hamiltonian form of mechanics, and vice versa, in the case of systems that belong to the standard case. The preceding three chapters have indicated the power of the machinery of canonical transformations to which one is led from the Hamiltonian formulation of the theory. The first systematic discussion of systems whose Lagrangians are not of the standard type was given by Dirac. It is interesting to examine these systems in a general way from two different points of view. First, the nature of the solutions of the Lagrangian equations of motion can be strikingly different from the standard case. Secondly, one would like to see how much of the structure of the Hamiltonian formulation can be retained, and even how the Lagrangian form can be recast in the Hamiltonian form. Indeed the primary aim of Dirac was to develop a standard technique for "Hamiltonising" a non-standard Lagrangian and to use the new Hamiltonian form for developing the quantum mechanics of such systems. Even within the framework of classical theory, however, it is satisfying to see how far one can carry the analysis for an arbitrary Lagrangian. We devote this chapter and part of the next one to an exposition of the Dirac theory.

General Considerations

For the sake of simplicity, we consider systems with a finite number, k, of degrees of freedom. Given the Lagrangian $L(q_s, \dot{q}_s, t)$, define the matrix W of second partial derivatives with respect to the velocities:

$$W_{rs}(q, \dot{q}, t) = \frac{\partial^2 L}{\partial \dot{q}_r \partial \dot{q}_s}. \tag{8.1}$$

With its help, we can write the Lagrangian equations of motion in such a form as to expose the accelerations $\ddot{q}_s$:

$$\sum_s W_{rs}(q,\dot{q},t)\ddot{q}_s = \frac{\partial L}{\partial q_r} - \sum_s \frac{\partial^2 L}{\partial \dot{q}_r \partial q_s}\dot{q}_s \equiv \alpha_r(q,\dot{q},t). \tag{8.2}$$

The accelerations can be solved for, that is, the Lagrangian is standard, if and only if the matrix $\|W\|$ is nonsingular, that is,

$$\Delta(q,\dot{q}) = \det\ |W_{rs}(q,\dot{q},t)| \neq 0. \tag{8.3}$$

If this determinant vanishes, then the Lagrangian is nonstandard. Consider now the standard technique for passing to the Hamiltonian form: we define the momenta p_s by

$$p_s = \frac{\partial L}{\partial \dot{q}_s}, \qquad s = 1, 2, \cdots, k, \tag{8.4}$$

and then try to solve these equations to express the $\dot{q}_s$ in terms of q, p, and t. But this can be done precisely when the condition in Eq.(8.3) holds! If $\det|W| = 0$, we cannot eliminate the velocities in terms of q, p, t either. Thus, as we stated in Chapter 3 (see the remark immediately following Eq.(3.17), these two problems go together; if the determinant Δ vanishes identically, the Lagrangian equations of motion do not give us all the accelerations as functions of positions and velocities, and *at the same time* we are unable to eliminate the velocity variables completely.

As we saw in Chapter 5, the natural group of invariance of the Lagrangian formulation is the group of coordinate transformations in configuration space; a general transformation leads from the coordinates q_s to new ones, Q_s, the latter being k independent functions of the former. It is an easy matter to verify the vanishing of Δ to be an invariant property under these transformations; hence the standard or nonstandard nature of L is also independent of the coordinates used in configuration space. This happens because the transformation law for the velocities is fixed by the one for the coordinates.

On the other hand, the situation is different when we consider the passage from a Hamiltonian to a Lagrangian. In the first place, in contrast to the Lagrangian equations, the Hamiltonian equations directly express $\dot{q}_s, \dot{p}_s$ as functions of q_s, p_s, and t, whatever the Hamiltonian may be. (We have already noted this in Chapter 5, in the remark following Eq.(5.10).) If we want to pass to a Lagrangian, we should solve the first set of the Hamiltonian equations,

$$\dot{q}_s = \frac{\partial H}{\partial p_s}, \tag{8.5}$$

to get the p_s in terms of the $q, \dot{q}$ and t; this is possible if and only if

$$D(q,p) = \det\left|\frac{\partial^2 H}{\partial p_r \partial p_s}\right| \neq 0. \tag{8.6}$$

However, if we were to use the vanishing or nonvanishing of this determinant to call the Hamiltonian nonstandard or standard, respectively, this property would not be invariant under canonical transformations, which form the natural invariance group of the Hamilton equations! One can see this as follows: given any function $f(q,p)$ on phase space, it can be shown that one can choose a canonical set of coordinates Q_s, P_s such that, for example,

$$P_1 = f(q,p) .$$

[We will not prove this statement here, but refer the reader to the book by Eisenhart. (L.P. Eisenhart, *Continuous Groups of Transformations*, Dover Publications, Inc., New York, 1961.). We note though that the theorem is true at least locally, and also that the canonical transformation involved here is in general not a regular one.] Setting $f = H$, assuming H is not explicitly time dependent, for simplicity, the new Hamiltonian H' is

$$H'(Q,P) = H(q,p) = P_1,$$

and clearly the standard procedure for passing to the Lagrangian fails; in the Q, P system the determinant D in Eq.(8.6) is zero. But this theorem also shows that even if in a particular coordinate system $D(q,p)$ vanishes, by a combination of two canonical transformations we can pass to a coordinate system $\bar{Q}, \bar{P}$ in which the Hamiltonian is some preassigned function $\bar{H}(\bar{Q}, \bar{P})$ such that

$$\det \left| \frac{\partial^2 \bar{H}}{\partial \bar{P}_r \partial \bar{P}_s} \right| \neq 0.$$

In this coordinate system the passage to the Lagrangian is possible. Though these arguments are local in nature, we can conclude that there is no canonical invariant characterization of a Hamiltonian being standard or not. One can always find coordinate systems in which a given Hamiltonian is standard.

The Lagrangian Equations for Constrained Systems

Let us now examine in some detail the Lagrangian equations for the non-standard case to get a general idea of the nature of the solutions of the equations of motion. This also helps in understanding the Dirac theory, which is described later. We assume that the Lagrangian is such that the equations of motion do not lead to any inconsistencies at any stage in the argument.

Because the matrix $\|W_{rs}\|$ in Eq.(8.2) is singular, its rank R computed by treating all the variables $q_1, \cdots, q_k$, $\dot{q}_1, \cdots, \dot{q}_k$ as independent is less than k. Consequently there exist $k - R$ linearly independent null eigenvectors $\lambda_s^{(a)}(q,\dot{q})$ for this matrix:

$$\sum_r \lambda_r^{(a)}(q,\dot{q}) W_{rs}(q,\dot{q}) = 0, \qquad a = 1, 2, \cdots, k - R;$$
$$s = 1, 2, \cdots, k. \tag{8.7}$$

(For ease in writing, we omit all explicit time dependences.) The null eigenvectors $\lambda_r^{(a)}$ are linearly independent with the understanding that all the q_s and $\dot{q}_s$

are treated as independent variables, at this stage. Using Eq.(8.7) in Eq. (8.2), we get $k - R$ relations involving the q's and $\dot{q}$'s:

$$\sum_r \lambda_r^{(a)}(q,\dot{q})\alpha_r(q,\dot{q}) = 0, \qquad \alpha = 1, 2, \cdots, k - R. \tag{8.8}$$

These relations, and all other relations among the q's and $\dot{q}$'s that may turn up in the analysis, are called constraints in the Lagrangian sense. They are consequences of the equations of motion, and in general place restrictions on the choice of the initial values of the q's and $\dot{q}$'s.

We must now analyze Eq.(8.8). As the first (and simplest) possibility, let us suppose the structure of the Lagrangian is such that they are identically satisfied. Then there are no constraints, and there are no equations beyond Eq.(8.2) to determine the motion. Because the rank of $\|W\|$ is R, we can use Eq.(8.2) to express R of the accelerations in terms of the rest of the accelerations, all the coordinates, and all the velocities. It is no loss of generality to suppose that we can solve for $\ddot{q}_1, \ddot{q}_2, \cdots, \ddot{q}_R$. We then have equations of the form

$$\ddot{q}_j = f_j(q_1 \cdots q_R;\ \dot{q}_1 \cdots \dot{q}_R | q_{R+1} \cdots q_k;\ \dot{q}_{R+1} \cdots \dot{q}_k;\ \ddot{q}_{R+1} \cdots \ddot{q}_k). \qquad j = 1, \cdots, R. \tag{8.9}$$

The solution of the equations of motion can then be described as follows: choose a set of functions of time in any way you like for the coordinates $q_{R+1}, \cdots, q_k$; assign an arbitrary (but, of course, physically meaningful) set of initial values at $t = 0$ for $q_1, \cdots, q_R$ and $\dot{q}_1, \cdots \dot{q}_R$; then the $q_1 \cdots q_R$ are uniquely determined for all later times.

The most important qualitative feature is the appearance of arbitrary functions of time in the general solution of the equations of motion. This is a general feature of the dynamics of constrained systems, and we will encounter it again.

Returning to Eqs.(8.8), these are in general not identically obeyed. Let us imagine that out of these $k - R$ equations, a certain number K, $0 \leqq K \leqq k - R$, are functionally independent (at this point we treat all the q's and $\dot{q}$'s as independent variables), and let us write these constraints as

$$\gamma_a(q,\dot{q}) = 0. \qquad a = 1, 2, \cdots, K \leqq k - R. \tag{8.10}$$

If we imagine a $2k$ dimensional space S with independent coordinates $q_s, \dot{q}_s$, then the Eq.(8.10) defines a $(2k - K)$ dimensional surface V in S, to which the motion is constrained. In dealing with a set of independent equations such as Eq.(8.10) to define a surface V in S, we follow a convention in the choice of the form of the functions γ_a; we assume that the $K \times 2k$ matrix of partial derivatives

$$\left\| \frac{\partial \gamma_a}{\partial q_s} \; \frac{\partial \gamma_a}{\partial \dot{q}_s} \right\|, \tag{8.11}$$

where one first treats all the q_s and $\dot{q}_s$ as independent, and after differentiation restricts them to V, has finite matrix elements and is of rank K. This has the consequence that if the function γ_1 obeys this condition, then neither $(\gamma_1)^2$ nor $(\gamma_1)^{\frac{1}{2}}$, for example, will obey it.

The rank R of $\|W_{rs}\|$ was first computed in the space S, that is, with all the q's and $\dot{q}$'s being independent. But the equations of motion have restricted the motion to the surface V of lower dimensionality, so that we must go back and recompute the rank of $\|W_{rs}\|$ after restricting the variables to the surface V. When this is done, although the rank cannot increase, it could, in principle, decrease. That means that with the variables constrained to V we may find some more null eigenvectors for the matrix $\|W_{rs}\|$, and these in turn may introduce more independent constraints among the q_s and $\dot{q}_s$; the motion then becomes restricted to a surface of lower dimensionality than V. We have to keep doing this series of operations until the following situation is reached. We have the equations of motion (Eq.(8.2)); the motion is restricted to a surface V' of dimensionality $(2k - K')$ defined by K' independent constraint equations

$$\gamma_a(q, \dot{q}) = 0, \qquad a = 1, 2, \cdots, K'. \tag{8.12}$$

On V', the rank of $\|W_{rs}\|$ is R', and $K' \leqq k - R'$. For every null eigenvector $\lambda_r(q, \dot{q})$ of the matrix $\|W_{rs}\|$ on V', the equation

$$\sum_r \lambda_r(q, \dot{q})\alpha_r(q, \dot{q}) = 0$$

is obeyed automatically on V' as a result of Eq.(8.12).

Now we separate (by algebraic manipulations) the constraint equations defining V' into a maximum number involving the q_s alone, and a remainder, which also involves the $\dot{q}_s$ in an essential manner. Let us suppose we have K'_1 of the first type and K'_2 of the second, with $K' = K'_1 + K'_2$. We call these constraints of type A and type B, respectively; Eq.(8.12) then appears in the following equivalent form:

$$\gamma_a^{(1)}(q) = 0, \qquad a = 1, 2, \cdots, K'_1; \qquad \text{(A type)}; \tag{8.13a}$$

$$\gamma_b^{(2)}(q, \dot{q}) = 0, \qquad b = 1, 2, \cdots, K'_2; \qquad \text{(B type)}. \tag{8.13b}$$

The number K'_2 is just the rank of the $K' \times k$ matrix

$$\left\| \frac{\partial \gamma_a}{\partial \dot{q}_s} \right\|$$

evaluated on V', whereas the essential dependence of the B-type constraints on the velocities simply means that the rank of the $K'_2 \times k$ matrix

$$\left\| \frac{\partial \gamma_b^{(2)}}{\partial \dot{q}_s} \right\|$$

evaluated on V' is K'_2. This operation of separating a set of independent constraints into A-type and B-type ones will be done repeatedly.

Up to this point, the working out of the constraints, Eq.(8.13), and the restriction of the motion to V' has all been done by starting with the equations of motion and carrying out algebraic operations, only; it did not involve differentiating any equation with respect to time. Now we have to check whether the constraints are preserved in time. First of all, we add to Eq.(8.13) the time derivatives of the A-type constraint equations:

$$\sum_s \frac{\partial \gamma_a^{(1)}}{\partial q_s} \dot{q}_s = 0, \qquad a = 1, 2, \cdots, K_1'. \tag{8.14}$$

The combined set, Eqs. (8.13) and (8.14), taken together may increase the number of independent A-type constraints and/or the number of independent B-type constraints. In the former case, we must next consider the time derivatives of the new A-type constraints, add them to all the existing constraints, and reconsider the situation. We keep doing this until we end up with K_1'' and K_2'' constraints of types A and B, respectively:

$$\gamma_a^{(1)}(q) = 0, \qquad a = 1, 2, \cdots, K_1''; \tag{8.15a}$$

$$\gamma_b^{(2)}(q, \dot{q}) = 0, \qquad b = 1, 2, \cdots, K_2''; \tag{8.15b}$$

such that the equations

$$\sum_s \dot{q}_s \frac{\partial \gamma_a^{(1)}}{\partial q_s} = 0, \qquad a = 1, 2, \cdots, K_1'',$$

are implied by Eq.(8.15). These equations, namely Eq.(8.15), define a surface V'' to which the motion is restricted; it also follows from the fact that Eq.(8.15a) give rise to no new B-type constraints that $K_2'' \geqq K_1''$.

The next step is to examine those constraints whose time derivatives involve the accelerations. We must add to the analysis the second derivatives of Eq.(8.15a) and the first derivatives of Eq.(8.15b):

$$\sum_{r,s} \frac{\partial^2 \gamma_a^{(1)}}{\partial q_r \partial q_s} \dot{q}_r \dot{q}_s + \sum_s \frac{\partial \gamma_a^{(1)}}{\partial q_s} \ddot{q}_s = 0, \qquad a = 1, 2, \cdots, K_1''; \tag{8.16a}$$

$$\sum_s \left(\frac{\partial \gamma_b^{(2)}}{\partial q_s} \dot{q}_s + \frac{\partial \gamma_b^{(2)}}{\partial \dot{q}_s} \ddot{q}_s \right) = 0, \qquad b = 1, 2, \cdots, K_2''. \tag{8.16b}$$

We may thus generate more equations to determine the accelerations, independent of the original equations of motion. We have now to consider the entire system of Eqs.(8.2),(8.15), and (8.16), and go through each of the steps described in detail above. Whenever a new set of equations involving the accelerations turns up, more constraints may be generated by algebraic operations. Every new A-type constraint may lead to more of both types A and B by means of a single differentiation with respect to time. Differentiating an A-type constrain twice, and a B-type constraint once can lead to new equations for the accelerations.

For a system with a finite number of degrees of freedom, for which genuine motion is possible, this iterative process must end after a finite number of steps. The final state of affairs can be described as follows. There will be R''' independent equations for the accelerations,

$$\sum_r W'_{jr}(q,\dot{q})\ddot{q}_r = \alpha'_j(q,\dot{q}); \qquad j = 1,2,\cdots,R''' \leqq k. \tag{8.17}$$

The motion is restricted to a surface V''' defined by K_1''' independent A-type and K_2''' independent B-type constraints:

$$\gamma_a^{(1)}(q) = 0, \qquad a = 1,2\cdots,K_1'''; \tag{8.18a}$$

$$\gamma_b^{(2)}(q,\dot{q}) = 0, \qquad b = 1,2\cdots,K_2'''. \tag{8.18b}$$

The equations

$$\frac{d}{dt}\,\gamma_a^{(1)}(q) = 0, \qquad a = 1,2,\cdots,K_1'''$$

are derivable by algebraic operations alone from Eq.(8.18); that is, they are obeyed on V'''. The equations

$$\frac{d^2}{dt^2}\,\gamma_a^{(1)}(q) = 0, \qquad a = 1,\cdots,K_1''',$$

$$\frac{d}{dt}\,\gamma_b^{(2)}(q,\dot{q}) = 0, \qquad b = 1,2,\cdots,K_2'''$$

are derivable by algebraic operations alone from Eqs.(8.17) and (8.18). This situation can be described by saying that any operation which could, in principle, yield more constraints or more equations for the accelerations does not in fact do so. From the relationships among the equations described above we easily derive the inequalities:

$$K_1''' \leqq K_2''' \leqq R''' \leqq k. \tag{8.19}$$

This is the general picture for the Lagrangian description of a constrained system. To get some idea of the nature of the solutions of these equations, we can proceed as follows. The functional independence of the A-type constraints implies that on V''' the matrix

$$\left\| \frac{\partial \gamma_a^{(1)}}{\partial q_s} \right\| \tag{8.20}$$

has maximal rank, namely K_1'''. Thus we can in principle rewrite the B-type constraints so that the first K_1''' of them read

$$\sum_s \dot{q}_s\,\frac{\partial \gamma_a^{(1)}}{\partial q_s} = 0, \qquad a = 1,2,\cdots,K_1'''. \tag{8.21}$$

Let the remaining B-type constraints be written as

$$\psi_b^{(2)}(q,\dot{q}) = 0, \qquad b = 1,2,\cdots,K_2''' - K_1'''. \tag{8.22}$$

The whole set of B-type constraints taken in the form of Eqs.(8.21) together with (8.22) must still depend in an essential way on the velocities; that is, the matrix

$$\left\| \begin{matrix} \dfrac{\partial \gamma_a^{(1)}}{\partial q_s} \\ \dfrac{\partial \psi_b^{(2)}}{\partial \dot{q}_s} \end{matrix} \right\| . \tag{8.23}$$

is also of maximal rank, K_2''', on V'''. Thus we can in principle arrange that the first K_2''' equations for the accelerations take the form

$$\sum_s \frac{\partial \gamma_a^{(1)}}{\partial q_s} \ddot{q}_s + \sum_{rs} \frac{\partial \gamma_a^{(1)}}{\partial q_r \partial q_s} \dot{q}_r \dot{q}_s = 0, \quad a = 1, 2, \cdots K_1'''; \tag{8.24a}$$

$$\sum_s \frac{\partial \psi_b^{(2)}}{\partial \dot{q}_s} \ddot{q}_s + \sum_s \frac{\partial \psi_b^{(2)}}{\partial q_s} \dot{q}_s = 0, \quad b = 1, 2, \cdots K_2''' - K_1'''. \tag{8.24b}$$

(The point is that the rank of the matrix of Eq.(8.20) being maximal ensures that the velocities cannot be eliminated from Eq.(8.21), and similarly the maximal rank of the matrix of Eq.(8.23) makes Eqs.(8.24) independent equations for the accelerations). The remaining $(R''' - K_2''')$ equations for the $\ddot{q}_s$ can be written as

$$\sum_r W_{jr}''(q, \dot{q}) \ddot{q}_r = \alpha_j''(q, \dot{q}), \qquad j = 1, 2, \cdots, (R''' - K_2'''). \tag{8.25}$$

With this rewriting, the basic equations are in the form of Eqs.(8.18a), (8.21), (8.22), (8.24), and (8.25). The A-type constraints Eq.(8.18a) can be used to express K_1''' of the q's as functions of the remaining q's. We may number the degrees of freedom in such a way that these q's are $q_1, q_2, \cdots, q_{K_1'''}$; denoting the remaining q's by q_σ, $\sigma = K_1''' + 1, \cdots, k$, the A-type constraints are:

$$q_1 = f_1(q_\sigma), q_2 = f_2(q_\sigma), \cdots, q_{K_1'''} = f_{K_1'''}(q_\sigma). \tag{8.26}$$

Now we can use these equations, together with Eqs. (8.21) and (8.24a), to eliminate these K_1''' coordinates completely from the theory; they and their velocities as well as accelerations are all given in terms of the $q_\sigma, \dot{q}_\sigma$, and $\ddot{q}_\sigma$. We will then be dealing with a system with $(k - K_1''')$ coordinates, governed by a set of $(K_2''' - K_1''')$ B-type constraints:

$$\psi_b^{(2)}(q_s, \dot{q}_s) = \phi_b^{(2)}(q_\sigma, \dot{q}_\sigma) = 0, \qquad b = 1, \cdots, (K_2''' - K_1''') \tag{8.27}$$

and a set of $(R''' - K_1''')$ independent equations for the $\ddot{q}_\sigma$:

$$\sum_\sigma \left(\frac{\partial \phi_b^{(2)}}{\partial \dot{q}_\sigma} \ddot{q}_\sigma + \frac{\partial \phi_b^{(2)}}{\partial q_\sigma} \dot{q}_\sigma \right) = 0, \quad b = 1, 2, \cdots (K_2''' - K_1'''); \tag{8.28a}$$

$$\sum_\sigma W_{j\sigma}''''(q_\rho, \dot{q}_\rho) \, \ddot{q}_\sigma = \alpha_j'''(q_\rho, \dot{q}_\rho), \quad j = 1, 2, \cdots (R''' - K_2'''). \tag{8.28b}$$

(Equation (8.28b) is Eq.(8.25) rewritten using only the q_σ's.) The motion takes place on a $(2k - K_1''' - K_2''')$ dimensional surface in the $2(k - K_1''')$ dimensional space of $q_\sigma, \dot{q}_\sigma$.

The equations of motion, Eq.(8.28), guarantee that if the B-type constraints of Eq.(8.27) are obeyed at time $t = 0$, they will be obeyed for all t. Thus we need to use the constraint equations only in restricting the allowed initial values, but we can confine ourselves to the acceleration equations for solving the equations of motion. Thus considering only Eq.(8.28) allows us to express$(R''' - K_1''')$ accelerations, the $\ddot{q}_\alpha$'s, in terms of the rest, the $\ddot{q}_A$'s:

$$\ddot{q}_\alpha = g_\alpha(q_\beta, \dot{q}_\beta | q_A, \dot{q}_A, \ddot{q}_A). \tag{8.29}$$

The number of q_A's is $(k - R''')$ (which could be zero!), and the number of q_α's is $(R''' - K_1''')$; the set q_σ is broken up into q_α's and q_A's. We can easily convince ourselves that no B-type constraint can involve only the q_A and $\dot{q}_A$, if the accelerations $\ddot{q}_\alpha$ could be solved for. Thus one has a situation similar to Eq.(8.9); choose an arbitrary set of functions of time for the q_A, assign any initial values at $t = 0$ for $q_\alpha, \dot{q}_\alpha$ obeying Eq.(8.27), then the q_α are determined uniquely for all t and will always obey Eq.(8.27). The number of arbitrary functions of time appearing in the general solution of the equations of motion is simply the number of accelerations that could not be solved for. For those coordinates whose accelerations could be solved for, we have as many restrictions on their initial values and velocities as we have B-type constraints. In principle, both these numbers could be zero. An interesting point is that after elimination of variables, no A-type constraints are left. In this sense, the most general constrained Lagrangian system is one involving only B-type constraints.

In this discussion of the treatment of nonstandard Lagrangians, we assume at various stages that certain desirable algebraic manipulations can be carried out. Thus, for example, we rewrite several equations in equivalent but special forms that expose some of the variables as explicit functions of the others. These manipulations, although always possible in principle, may be awkward in practice; in that case, our analysis tells us just what we may expect to happen, although the forms of the equations may be different. We note that this general pattern for handling constraints can also be used if one is given a Lagrangian system of equations and, independently, a set of nonholonomic constraints involving the coordinates and the velocities; that is, we can handle constraints that do not follow from the action principle but are added on "from the outside."

We conclude this part of the discussion of constrained systems by indicating by means of special examples certain situations that naturally involve singular Lagrangians. In fact, the development of a general approach for handling such systems was stimulated by the study of certain kinds of field theories that possess special invariance properties called *gauge invariances*. An example of this is the treatment of the electromagnetic field in Chapters 20 and 21. But the ideas involved can be exhibited equally well by considering Lagrangians involving finite numbers of degrees of freedom and having special invariance properties.

This draws upon the discussion of conservation laws and the action principle given in Chapter 4.

Consider then a Lagrangian $L(q_s, \dot{q}_s)$ that for simplicity has no explicit time dependence. Suppose that there is a class of infinitesimal transformations of trajectory in configuration space that involves a certain number of *arbitrary* functions of time, $f_a(t)$ under which L is *quasi-invariant* (L being invariant is a special case). That is, we have the infinitesimal transformation:

$$\bar{\delta} q_s = \epsilon \sum_a f_a(t) \phi_{as}(q, \dot{q}) \tag{8.30}$$

with ϵ a small parameter, f_a arbitrary functions of t, ϕ_{as} specific functions of $q, \dot{q}$, and the associated equation for $\bar{\delta}\dot{q}_s$; under this transformation the change in L takes the form:

$$\bar{\delta} L = \epsilon \frac{d}{dt}\left(\sum_a f_a(t) F_a(q, \dot{q})\right). \tag{8.31}$$

Under these circumstances we are automatically led to a particular case of the Lagrangian theory of constraints: the essential point is (quasi) invariance of L under transformations allowing *arbitrary* functions of time. If we look upon Eq.(8.30) as defining a particular variation of a configuration space trajectory C, the variation of the action functional $\Phi[C]$ can be evaluated in two ways: equating these results in an identity valid for all C:

$$\begin{aligned} -\sum_s \int_{t_1}^{t_2} dt L_s(q, \dot{q}, \ddot{q}) \bar{\delta} q_s(t) \\ = \epsilon \sum_a \left[f_a(t) \left(\sum_s p_s \phi_{as}(q, \dot{q}) - F_a(q, \dot{q}) \right) \right] \Bigg|_{t_1}^{t_2}. \end{aligned} \tag{8.32}$$

This is Eq.(4.24). But consider the meaning of this identity: on the left we have an integral that depends on the values of the $f_a(t)$ for t in the range $t_1 \leqq t \leqq t_2$, whereas on the right only the values of $f_a(t)$ at t_1, t_2 appear. Because these are arbitrary functions of time, the only way this identity can be true is that both sides vanish identically. Therefore, the form of the Lagrangian and of Eq.(8.30) must be such that the equations

$$\sum_s \phi_{as}(q, \dot{q}) L_s(q, \dot{q}, \ddot{q}) = 0, \tag{8.33a}$$

$$\sum_s \phi_{as}(q, \dot{q}) p_s - F_a(q, \dot{q}) = 0, \tag{8.33b}$$

$$\tag{8.33c}$$

hold *as identities*. It is not necessary to invoke the equations of motion for the validity of the equations. Recalling the definition of L_s, namely,

$$L_s(q, \dot{q}, \ddot{q}) = \frac{\partial L}{\partial q_s} - \frac{d}{dt}\frac{\partial L}{\partial \dot{q}_s} \tag{8.34}$$

and the quantities $W_{rs}(q,\dot{q})$, $\alpha_r(q,\dot{q})$ from Eqs.(8.1) and (8.2), the identity in Eq. (8.33a) takes the form

$$\sum_{rs} \phi_{as}(q,\dot{q}) W_{sr}(q,\dot{q}) \ddot{q}_r = \sum_{s} \phi_{as}(q,\dot{q}) \alpha_s(q,\dot{q}) \,. \tag{8.35}$$

Because the accelerations $\ddot{q}_s$ appear only on the left and because this equation is true without imposition of the equations of motion, both sides must vanish. Therefore the matrix $\|W_{rs}\|$ possesses null eigenvectors, and we have

$$\sum_{s} \phi_{as} W_{sr} = 0, \qquad \sum_{s} \phi_{as} \alpha_s = 0. \tag{8.36}$$

This establishes that we have here a nonstandard Lagrangian. As for the equations of motion, these may be written as $L_s = 0$. But the left-hand sides of these equations satisfy the identities of Eq.(8.33a); thus the equations of motion are not independent of one another. If the matrix $\|W_{rs}\|$ has no more null eigenvectors independent of the ϕ_{as}, we have here the simplest case of a nonstandard Lagrangian system; we do not have enough equations to determine all the accelerations, but there are no constraints, either. All the equations such as Eq.(8.8), which might have led to constraints, do not, in fact, do so, because according to Eq.(8.36) they are identically satisfied. This simplest possibility is the one discussed on page 80.

According to the discussion in Chapter 4 (see page 26 onward), any quasi invariance property of the Lagrangian leads, in principle, to a constant of motion. What is the situation in the present case? From Eqs. (4.25), (8.30), and (8.31), the conserved quantity should be:

$$\sum_{a} f_a(t) \left(\sum_{s} \phi_{as}(q,\dot{q}) p_s - F_a(q,\dot{q}) \right) . \tag{8.37}$$

But we have arbitrary time-dependent factors here! Thus this "constant of motion" must somehow vanish identically; Eq. (8.33b) assures us that it does. This is the way in which the intimate relation between the action principle, conservation laws, and quasi-invariance properties is maintained.

There is a more physical way of understanding why quasi-invariance properties of the kind discussed here can occur only for nonstandard Lagrangians. We know that any quasi-invariance property of L implies an invariance property of the equations of motion (see Chapter 4, page 29). Given any solution of the equations of motion obeying some prescribed boundary conditions at $t = t_0$, and applying the transformation Eq.(8.30), we generate a new solution. Because the $f_a(t)$ are at our disposal, we can choose them in such a way that the new solution obeys the same boundary conditions as the old one at $t = t_0$; then the two solutions differ only for $t > t_0$. This shows that

1. the boundary conditions at $t = t_0$ do not determine the solution for $t > t_0$ uniquely, and

2. the *general* solution will involve arbitrary functions of time. Therefore, not all the accelerations can be determinate.

All these features can be seen without using the action functional, just by combining the equations of motion and the quasi-invariance of L; however, the route we have followed expresses things more compactly.

A slight generalization of the foregoing treatment is worth recording; it is actually this form of the theory that is needed in the Lagrangian treatment of the electromagnetic field. The generalization consists of admitting on the right-hand side of Eq.(8.30) terms involving the time derivatives of the arbitrary functions $f_a(t)$; we consider

$$\bar{\delta} q_s = \epsilon \sum_a (f_a(t)\phi_{as}(q,\dot{q}) + \dot{f}_a(t)\psi_{as}(q,\dot{q})), \tag{8.38}$$

where both ϕ_{as} and ψ_{as} are specific functions of $q, \dot{q}$. Supposing that L is quasi-invariant again, we write:

$$\bar{\delta} L = \epsilon \frac{d}{dt} \sum_a (f_a(t)F_a(q,\dot{q}) + \dot{f}_a(t)G_a(q,\dot{q})). \tag{8.39}$$

In place of the identity Eq.(8.32) we have:

$$\begin{aligned} &-\sum_{a,s} \int_{t_1}^{t_2} dt\ L_s(q,\dot{q},\ddot{q})(f_a(t)\phi_{as}(q,\dot{q}) + \dot{f}_a(t)\psi_{as}(q,\dot{q})) \\ &\equiv \sum_a \left[f_a(t) \left(\sum_s \phi_{as}(q,\dot{q})p_s - F_a(q,\dot{q}) \right) \right. \\ &\qquad \left. + \dot{f}_a(t) \left(\sum_s \psi_{as}(q,\dot{q})p_s - G_a(q,\dot{q}) \right) \right] \Bigg|_{t_1}^{t_2} \end{aligned} \tag{8.40}$$

We transform the term involving $\dot{f}_a$ on the left by means of partial integration and achieve the form:

$$\begin{aligned} &-\sum_{a,s} \int_{t_1}^{t_2} dt\ f_a(t) \left(\phi_{as} L_s - \frac{d}{dt}(\psi_{as} L_s) \right) \\ &\equiv \sum_a \left[f_a(t) \left(\sum_s \{\phi_{as} p_s + \psi_{as} L_s\} - F_a \right) + \dot{f}_a(t) \left(\sum_s \psi_{as} p_s - G_a \right) \right] \Bigg|_{t_1}^{t_2}. \end{aligned} \tag{8.41}$$

Now we can conclude that both sides must vanish identically, so the form of L must lead to

$$\sum_s \left(\phi_{as}(q,\dot{q})L_s(q,\dot{q},\ddot{q}) - \frac{d}{dt}(\psi_{as}(q,\dot{q})L_s(q,\dot{q},\ddot{q})) \right) = 0; \tag{8.42a}$$

$$\sum_s (\phi_{as}(q,\dot{q})p_s + \psi_{as}(q,\dot{q})L_s(q,\dot{q},\ddot{q})) - F_a(q,\dot{q}) = 0; \tag{8.42b}$$

$$\sum_s \psi_{as}(q,\dot{q})p_s - G_a(q,\dot{q}) = 0. \tag{8.42c}$$

The nonstandard nature of L now follows from looking at the terms in Eq.(8.42a) involving $\dddot{q}$, or the $\ddot{q}$ terms in Eq.(8.42b): either way we get a set of null eigenvectors for $\|W_{rs}\|$:

$$\sum_s \psi_{as}(q,\dot{q})W_{sr}(q,\dot{q}) = 0. \tag{8.43}$$

In contrast to the earlier situation, however, these null eigenvectors of $\|W_{rs}\|$ not only prevent us from determining all the accelerations, they also lead to some constraints, because we cannot say that the expressions $\sum_s \psi_{as}\alpha_s$ vanish identically. In fact, when the equations of motion $L_s = 0$ are imposed, the identity (8.42b) turns into a constraint equation:

$$\sum_s \phi_{as}(q,\dot{q})p_s - F_a(q,\dot{q}) = 0. \tag{8.44}$$

We repeat that this is a consequence of the equations of motion; thus it must be handled appropriately (Eq.(8.42c) is, however, an identity). In the present situation, the constant of motion supplied by the action principle takes the form:

$$\frac{1}{\epsilon}\sum_s p_s\bar{\delta}q_s - \sum_a (f_aF_a + \dot{f}_aG_a);$$

that is,

$$\sum_a f_a(t)\left(\sum_s \phi_{as}p_s - F_a\right) + \sum_a \dot{f}_a(t)\left(\sum_s \psi_{as}p_s - G_a\right), \tag{8.45}$$

and the only way this can be conserved is by its vanishing for all time. The coefficient of $\dot{f}_a$ does vanish identically, because of Eq.(8.42c), that of f_a vanishes as a consequence of the equations of motion which led to the constraints Eq.(8.44). Alternatively, we can say that because the connections between the action principle, quasi-invariances and conservation laws must be maintained in any case, we end up with a set of constraints. To put it picturesquely, too much of a (quasi) invariance property leads to a constant of motion that stays constant by always vanishing; it either vanishes identically or by virtue of the equations of motion, and in the latter event it is a constraint.

Hamiltonian Treatment — The Dirac Theory

The point of departure for the Hamiltonian theory is, as always, the definition of the momentum variables

$$p_s = \frac{\partial L}{\partial \dot{q}_s}\,, \qquad s = 1, 2, \cdots, k. \tag{8.4}$$

The nonmaximal rank, R, of the matrix $\|W_{rs}\|$ determines that only R velocity variables can be expressed as functions of the q's, p's, and the remaining $(k-R)$ velocities, and also that there exist $(k-R)$ independent relations among the p's and q's. These relations, which are direct consequences of the definition of the p's, can be written as:

$$\Omega_\rho(p_s, q_s) = 0, \qquad \rho = R+1, \cdots, k. \tag{8.46}$$

They show that of the $2k$ variables q_s, p_s, only $(k+R)$ are truly independent. The relations in Eq.(8.46) are called "primary constraints," the word primary denoting the fact that the equations of motion were not used to obtain them. The functions Ω_ρ must actually be independent functions of the p's by themselves, for otherwise we could eliminate the p's from Eq.(8.46) to obtain constraints involving the q's alone, which is impossible, at least at this stage. In principle, then, and without loss of generality, we suppose that Eq.(8.46) allows us to express the momenta $p_{R+1}, \cdots, p_k$ in terms of $p_1, \cdots, p_R$ and the q's:

$$p_\rho = \psi_\rho(q_s, p_j), \qquad \rho = R+1, \cdots, k. \tag{8.47}$$

(The index j takes the values $1, 2 \cdots, R$) Now going back to Eq.(8.4) let the R velocities that can be solved for be written $\dot{q}_\alpha$, the rest $\dot{q}_A$: α takes on R values out of $1, 2, \cdots, k$, A takes on the rest, thus we write:

$$\dot{q}_\alpha = \zeta_\alpha(q_s, p_j, \dot{q}_A), \qquad \text{number of values of } \alpha = R. \tag{8.48}$$

Equations (8.47) and (8.48), which are k in all, are identical in content to Eq. (8.4). For the $2k$-dimensional space S with independent coordinates $q_s, \dot{q}_s$ we now have an alternative coordinate system made up of the variables q_s, p_j, $\dot{q}_A$. It follows that the functions ψ_ρ and ζ_α have unambiguous functional forms.

In the standard case, the Hamiltonian is obtained as a function of q_s and p_s by rewriting the expression:

$$\sum_s p_s \dot{q}_s - L(q, \dot{q}) \tag{8.49}$$

in terms of q_s and p_s. This expression is a well-defined function on the space S; in the present case, we can rewrite it in terms of the $2k$ independent variables q_s, p_j, $\dot{q}_A$, and it then has a specific functional form:

$$\sum_s p_s \dot{q}_s - L(q, \dot{q}) = \tilde{W}(q_s, p_j, \dot{q}_A). \tag{8.50}$$

Let us compute the partial derivatives of $\tilde{W}$ with respect to its independent arguments. We have:

$$\frac{\partial\tilde{W}}{\partial q_s} = \sum_\rho \frac{\partial\psi_\rho}{\partial q_s}\dot{q}_\rho + \sum_\alpha p_\alpha \frac{\partial\zeta_\alpha}{\partial q_s} - \frac{\partial L}{\partial q_s} - \sum_\alpha \frac{\partial L}{\partial\dot{q}_\alpha}\frac{\partial\zeta_\alpha}{\partial q_s} = \sum_\rho \frac{\partial\psi_\rho}{\partial q_s}\dot{q}_\rho - \frac{\partial L}{\partial q_s}; \tag{8.51a}$$

$$\frac{\partial\tilde{W}}{\partial p_j} = \dot{q}_j + \sum_\alpha p_\alpha \frac{\partial\zeta_\alpha}{\partial p_j} - \sum_\alpha \frac{\partial L}{\partial\dot{q}_\alpha}\frac{\partial\zeta_\alpha}{\partial p_j} + \sum_\rho \dot{q}_\rho\frac{\partial\psi_\rho}{\partial p_j} = \dot{q}_j + \sum_\rho \dot{q}_\rho \frac{\partial\psi_\rho}{\partial p_j}; \tag{8.51b}$$

$$\frac{\partial\tilde{W}}{\partial\dot{q}_A} = p_A + \sum_\alpha p_\alpha \frac{\partial\zeta_\alpha}{\partial\dot{q}_A} - \frac{\partial L}{\partial\dot{q}_A} - \sum_\alpha \frac{\partial L}{\partial\dot{q}_\alpha}\frac{\partial\zeta_\alpha}{\partial\dot{q}_A} = 0. \tag{8.51c}$$

Thus although, in principle, the function $\tilde{W}$ could be a function of all the $2k$ independent variables $q_s, p_j, \dot{q}_A$, we see that as a result of the structure of Eq.(8.49), $\tilde{W}$ does not depend on the "unsolved" velocity variables $\dot{q}_A$. (This is the property of the Legendre transformation.) Thus we have:

$$\sum_s p_s\dot{q}_s - L(q,\dot{q}) = \tilde{W}(q_s, p_j); \tag{8.52a}$$

$$\frac{-\partial L}{\partial q_s} = \frac{\partial\tilde{W}}{\partial q_s} - \sum_\rho \dot{q}_\rho\frac{\partial\psi_\rho}{\partial q_s}; \tag{8.52b}$$

$$\dot{q}_j = \frac{\partial\tilde{W}}{\partial p_j} - \sum_\rho \dot{q}_\rho\frac{\partial\psi_\rho}{\partial p_j}, \qquad j = 1, 2, \cdots, R. \tag{8.52c}$$

In writing Eq.(8.47) we make the *convention* that the variables $q_1 \cdots q_k$, $p_1 \cdots p_R$ form a *maximal* independent set out of the q_s and p_s, and thus $p_{R+1} \cdots p_k$ are functions of them. Having numbered the degrees of freedom in this way, we cannot, to start with, pin down the precise values that the subscript α in Eq.(8.48) takes; equivalently, we cannot say *which* $(k - R)$ velocity variables $\dot{q}_A$ cannot be solved for. But now we see from Eq.(8.52c) that we can assume the $\dot{q}_\alpha$'s to be just the $\dot{q}_j$, and the $\dot{q}_A$ to be the $\dot{q}_\rho$! We may, in fact, identify Eq. (8.52c) with Eq.(8.48). Thus it is no loss of generality to say that if the momenta p_ρ became dependent upon the remaining phase space variables q_s, p_j, then the *corresponding* velocities $\dot{q}_\rho$ remain independent. It is interesting that in both Eqs.(8.52b) and (8.52c), the right-hand side depends on these velocities $\dot{q}_\rho$ only linearly.

Equations (8.52b) and (8.52c) are exactly analogous to Eqs. (4.2) and (4.6b), respectively. They are consequences of the way in which the function $\tilde{W}$ is related to the Lagrangian. Making use of the Lagrangian equations of motion, the equivalent Hamiltonian set can now be written down:

$$\dot{q}_j = \frac{\partial \tilde{W}}{\partial p_j} - \sum_\rho \dot{q}_\rho \frac{\partial \psi_\rho}{\partial p_j}, \qquad j = 1, 2, \cdots, R; \tag{8.53a}$$

$$\dot{p}_s = \frac{-\partial \tilde{W}}{\partial q_s} + \sum_\rho \dot{q}_\rho \frac{\partial \psi_\rho}{\partial q_s}, \qquad s = 1, 2, \cdots, k. \tag{8.53b}$$

These equations show to what extent we have succeeded in expressing the velocities $\dot{q}_s$ and the generalized forces $\partial L/\partial q_s$ of the Lagrangian equations of motion in terms of the q's and p's. We must now analyze and develop these equations further; they do not represent the final form of the Hamiltonian theory.

In the Lagrangian treatment, in the first instance there are a certain number of undetermined accelerations and a certain number of constraints. By considering the time derivatives of the constraints, new independent equations for the accelerations as well as new constraints are generated. Finally, a set of constraints and a set of equations for the accelerations are arrived at, with the property that the time derivatives of the former do not give rise to any new equations of either variety. Every acceleration that remains undetermined at this final stage gives rise to one arbitrary function of time in the general solution to the equations of motion. All this is repeated in the Hamiltonian form of the theory. To start with, we have a certain number of undetermined velocities, $\dot{q}_\rho$, and a certain number of constraints on the Hamiltonian variables, Eq.(8.46). These initially undetermined velocities correspond to the initially undetermined accelerations of the Lagrangian theory. When we decide to work with the p's and q's, *we have to carry along these velocities $\dot{q}_\rho$ as extra variables.* Just as in the Lagrangian case, when we examine the time derivatives of the (primary) constraints, we may get more constraints on the p's and q's as well as equations connecting the $\dot{q}_\rho$ to the p's and q's. A final form will be reached in which the full set of constraint equations and the set of equations connecting q's, p's, and $\dot{q}_\rho$ are such that the time derivatives of the former do not lead to any new equations of either sort. Only a subset of the initially undetermined velocities $\dot{q}_\rho$ remain undetermined at this stage; the others are expressed in terms of these, the q's and the p's. Every velocity variable out of the set $\dot{q}_\rho$ that remains undetermined in this final form of the theory appears as one arbitrary function in the general solution of the Hamiltonian equations of motion.

Let us now see in detail how all this comes about. As we noted, at the start we have to take account of the primary constraints of Eq.(8.46). This means that even if we started with a $2k$-dimensional phase space defined by $2k$ independent coordinates q_s, p_s, the motion is going to be confined, in the first instance, to a surface of lower dimensionality, defined by the constraint equations. Nevertheless, to give the theory a more flexible form and to allow us to use constructs like PB's, it is desirable to work in the full $2k$-dimensional phase space. This gives us the freedom to think of functions defined over all of phase space, compute their partial derivatives, for example, with respect to each of the q's and p's, and delay to the end the restriction of the variables to the hypersurface defined by the constraints, the "constrained hypersurface." In fact, of course, we only need the freedom to deal with functions defined in a finite "shell" surrounding the constrained hypersurface; the shell may be as small as you like, but finite. For this purpose, Dirac has introduced the concepts of "weak" and "strong" equations, which we now describe.

Let the constrained hypersurface in phase space be called U, let $f(q,p)$, $g(q,p)$ be two functions defined in a finite neighborhood of U. The values of f and g on U are obtained by replacing the variables p_ρ by the functions $\psi_\rho(q_s,p_j)$ (see Eq.(8.47).) If after this replacement, f and g become equal, that is, if f and g are equal on U, then we say that they are "weakly equal," and write it in the form:

$$f(q_s,p_s) \approx g(q_s,p_s). \tag{8.54}$$

Now, both functions f and g possess $2k$-dimensional "gradient vectors" at each point in phase space, with components $(\partial f/\partial q_s,\, \partial f/\partial p_s)$ and $(\partial g/\partial q_s,\, \partial g/\partial p_s)$, respectively. If f equals g on U, and also if the gradient of f agrees with that of g when the arguments are restricted to U, we say that f and g are "strongly equal":

$$f(q_s,p_s) \equiv g(q_s,p_s). \tag{8.55}$$

Note that in checking the strong equality of f and g, we first compute the gradients treating all the $2k$ q's and p's as independent, then restrict them to U.

The hypersurface U can be defined by a set of weak equations. Let us define a set of functions on phase space by:

$$\phi_\rho(q_s,p_s) = p_\rho - \psi_\rho(q_s,p_j), \qquad \rho = R+1,\cdots,k. \tag{8.56}$$

The functions ψ_ρ have well-defined functional forms, and as they stand, they define functions on phase space that are independent of p_ρ. Then U can be defined by the weak equations:

$$\phi_\rho \approx 0. \tag{8.57}$$

(Clearly, ϕ_ρ do not vanish strongly because $\partial\phi_\rho/\partial p_\sigma = \delta_{\rho\sigma}$ does not vanish on U.)

If f and g are weakly equal, what can be said about their gradients evaluated on U? If we have two points on U with infinitesimal coordinate differences δq, δp, then only $\delta q_s, \delta p_j$ are independent, whereas

$$\delta p_\rho = \sum_s \frac{\partial\psi_\rho}{\partial q_s}\,\delta q_s + \sum_j \frac{\partial\psi_\rho}{\partial p_j}\,\delta p_j \tag{8.58}$$

Therefore, the weak equality of f and g leads to the following equations valid on U:

$$\frac{\partial f}{\partial q_s} + \sum_\rho \frac{\partial f}{\partial p_\rho}\frac{\partial\psi_\rho}{\partial q_s} = \frac{\partial g}{\partial q_s} + \sum_\rho \frac{\partial g}{\partial p_\rho}\frac{\partial\psi_\rho}{\partial q_s}, \qquad s = 1,\cdots,k; \tag{8.59a}$$

$$\frac{\partial f}{\partial p_j} + \sum_\rho \frac{\partial f}{\partial p_\rho}\frac{\partial\psi_\rho}{\partial p_j} = \frac{\partial g}{\partial p_j} + \sum_\rho \frac{\partial g}{\partial p_\rho}\frac{\partial\psi_\rho}{\partial p_j}, \qquad j = 1,\cdots,R. \tag{8.59b}$$

We can rewrite these equalities in terms of ϕ_ρ and convert them into a system of weak equations:

$$\frac{\partial}{\partial q_s}\left(f-\sum_\rho \phi_\rho \frac{\partial f}{\partial p_\rho}\right) \approx \frac{\partial}{\partial q_s}\left(g-\sum_\rho \phi_\rho \frac{\partial g}{\partial p_\rho}\right), \qquad s=1,\cdots,k; \quad (8.60a)$$

$$\frac{\partial}{\partial p_j}\left(f-\sum_\rho \phi_\rho \frac{\partial f}{\partial p_\rho}\right) \approx \frac{\partial}{\partial p_j}\left(g-\sum_\rho \phi_\rho \frac{\partial g}{\partial p_\rho}\right), \qquad j=1,\cdots,R. \quad (8.60b)$$

In the second set of equations, j is restricted to the values $1,\cdots,R$; however, if we replace j by σ lying in the range $R+1,\cdots,k$, the weak equality is maintained; each side becomes zero because $\partial\phi_\rho/\partial p_\sigma = \delta_{\rho\sigma}$. Thus we actually have the system of weak equations:

$$\frac{\partial}{\partial q_s}\left(f-\sum_\rho \phi_\rho \frac{\partial f}{\partial p_\rho}\right) \approx \frac{\partial}{\partial q_s}\left(g-\sum_\rho \phi_\rho \frac{\partial g}{\partial p_\rho}\right); \quad (8.61a)$$

$$\frac{\partial}{\partial p_s}\left(f-\sum_\rho \phi_\rho \frac{\partial f}{\partial p_\rho}\right) \approx \frac{\partial}{\partial p_s}\left(g-\sum_\rho \phi_\rho \frac{\partial g}{\partial p_\rho}\right), \qquad s=1,\cdots,k, \quad (8.61b)$$

it being understood that the second set is trivially true for $s > R$. We then have the theorem that the weak equality

$$f \approx g$$

implies the strong equality

$$f-\sum_\rho \phi_\rho \frac{\partial f}{\partial p_\rho} \equiv g-\sum_\rho \phi_\rho \frac{\partial g}{\partial p_\rho}.$$

In particular, if we take g to be the zero function, we have the following result: if f vanishes weakly, it is strongly equal to a linear combination of the constraints:

$$f \approx 0 \Rightarrow f \equiv \sum_\rho \phi_\rho \frac{\partial f}{\partial p_\rho}. \quad (8.62)$$

Let us now use this language in examining the Hamiltonian set, Eq.(8.53). In these equations, the role of the Hamiltonian is played by the function $\tilde{W}(q_s, p_j)$. However, our formalism gives us the freedom to use in place of $\tilde{W}$ any other function $W(q_s, p_s)$ defined in phase space that is weakly equal to $\tilde{W}$; although $\tilde{W}$ is independent of the p_ρ when we move away from the hypersurface U, W could depend on these variables. The weak equality of $\tilde{W}$ and W leads, as proven before, to the weak equations:

$$\frac{\partial \tilde{W}}{\partial q_s} \approx \frac{\partial}{\partial q_s}\left(W-\sum_\rho \phi_\rho \frac{\partial W}{\partial p_\rho}\right),$$

$$\frac{\partial \tilde{W}}{\partial p_s} \approx \frac{\partial}{\partial p_s}\left(W-\sum_\rho \phi_\rho \frac{\partial W}{\partial p_\rho}\right), \qquad s=1,2,\cdots,k. \quad (8.63)$$

If we use these in the Hamiltonian equations and at the same time use ϕ_ρ in place of ψ_ρ, we get:

$$\dot{q}_j \approx \frac{\partial}{\partial p_j}\left(W - \sum_\rho \phi_\rho \frac{\partial W}{\partial p_\rho}\right) + \sum_\rho \dot{q}_\rho \frac{\partial \phi_\rho}{\partial p_j}, \qquad j = 1, 2, \cdots, R; \tag{8.64a}$$

$$\dot{p}_s \approx \frac{-\partial}{\partial q_s}\left(W - \sum_\rho \phi_\rho \frac{\partial W}{\partial p_\rho}\right) - \sum_\rho \dot{q}_\rho \frac{\partial \phi_\rho}{\partial q_s}, \qquad s = 1, 2, \cdots, k. \tag{8.64b}$$

Just as we found in the case of Eq.(8.60b) that we could allow the index j to go over all the values $1, 2, \cdots, k$, so also here we find that in Eq.(8.64a) we can allow j to go from 1 to k. If $j = \sigma > R$, both sides simply become $\dot{q}_\sigma$, the undetermined velocities. Thus we can, in fact, write these equations as:

$$\dot{q}_s \approx \frac{\partial}{\partial p_s}\left(W - \sum_\rho \phi_\rho \frac{\partial W}{\partial p_\rho}\right) + \sum_\rho \dot{q}_\rho \frac{\partial \phi_\rho}{\partial p_s}; \tag{8.65a}$$

$$\dot{p}_s \approx \frac{-\partial}{\partial q_s}\left(W - \sum_\rho \phi_\rho \frac{\partial W}{\partial p_\rho}\right) - \sum_\rho \dot{q}_\rho \frac{\partial \phi_\rho}{\partial q_s}; \qquad s = 1, 2 \cdots, k, \tag{8.65b}$$

realizing that Eq.(8.65a) is empty for $s > R$. With the functions ϕ_ρ chosen in the form of Eq.(8.56), it is clear that the gradients $\partial\phi_\rho/\partial q_s$, $\partial\phi_\rho/\partial p_s$ are all independent of p_ρ and depend explicitly only on q_s and p_j. Therefore, actually the weak equality sign in Eq.(8.64) and (8.65) has no effect on the terms involving $\dot{q}_\rho$ and really refers only to the first term involving W. However, if in place of ϕ_ρ we use some other set of phase-space functions ϕ'_ρ whose weak vanishing defines U, the weak equality would apply to both terms in Eqs.(8.64), (8.65). In any case, the *linear* dependence on the velocity variables $\dot{q}_\rho$ remains unchanged.

If we write $H(q_s, p_s)$ for the phase-space function appearing above,

$$H(q_s, p_s) = W(q_s, p_s) - \sum_\rho \phi_\rho \frac{\partial W}{\partial p_\rho}, \tag{8.66}$$

it is characterized by the property of being strongly equal to $\tilde{W}$:

$$H \equiv \tilde{W}. \tag{8.67}$$

Beyond this, H is arbitrary. By introducing H and using the PB notation, the Hamiltonian equations of motion take the neat form:

$$\begin{aligned} \dot{q}_s &\approx \{q_s, H\} + \sum_\rho \{q_s, \phi_\rho\}\dot{q}_\rho, \\ \dot{p}_s &\approx \{p_s, H\} + \sum_\rho \{p_s, \phi_\rho\}\dot{q}_\rho, \qquad s = 1, \cdots, k. \end{aligned} \tag{8.68}$$

These must be supplemented by the primary constraints

$$\phi_\rho(q_s, p_s) \approx 0, \qquad \rho = R+1, \cdots, k. \tag{8.69}$$

The original Lagrangian equations have now been recast into a phase-space form using PB's and weak equations. The velocity variables $\dot{q}_\rho$ remain in the formalism, *but their PB's with functions on phase space are to be regarded as undefined.* In any case, they always appear multiplied by the weakly vanishing functions ϕ_ρ. With the basic equations in this form we can examine the time development of the constraints, see how they give rise to more constraints and/or restrictions on the $\dot{q}_\rho$, and so on; in short, we can carry out the analysis outlined soon after Eq.(8.53).

For any phase-space function $g(q,p)$ we have the equation:

$$\frac{d}{dt}g(q,p) \approx \{g, H\} + \sum_\rho \{g, \phi_\rho\}\dot{q}_\rho. \tag{8.70}$$

Therefore, the condition that the primary constraints must be obeyed for all time gives us a set of equations:

$$\{\phi_\sigma, H\} + \sum_\rho \{\phi_\sigma, \phi_\rho\}\dot{q}_\rho \approx 0, \qquad \sigma = R+1, \cdots, k. \tag{8.71}$$

We must analyze these conditions. The simplest possibility is that the previously undetermined velocities $\dot{q}_\rho$ get determined by these conditions! This can happen if and only if the antisymmetric matrix $\|\{\phi_\sigma, \phi_\rho\}\|$ is nonsingular, that is

$$\det|\{\phi_\sigma, \phi_\rho\}| \not\approx 0, \tag{8.72}$$

on U (thus there must be an even number of primary constraints). If $\|C_{\rho\sigma}\|$ is the weak inverse of this matrix, that is,

$$\sum_\sigma C_{\rho\sigma}(q,p)\{\phi_\sigma, \phi_{\rho'}\} \approx \delta_{\rho\rho'}, \tag{8.73}$$

then we have

$$\dot{q}_\rho \approx -\sum_\sigma C_{\rho\sigma}(q,p)\{\phi_\sigma, H\}, \tag{8.74}$$

and the general equation of motion is:

$$\frac{d}{dt}\, g(q,p) \approx \{g, H\} - \sum_{\rho,\sigma} \{g, \phi_\rho\}C_{\rho\sigma}\{\phi_\sigma, H\}. \tag{8.75}$$

This is to be supplemented, of course, by the primary constraint equations. The situation here is analogous to that case of the Lagrangian treatment where no arbitrary functions of time appear in the solution of the equations of motion. Here we can assign initial values to the q_s and p_s at $t = 0$ in any way we like as long as the primary constraints are obeyed at $t = 0$. Then the equations of motion permit us to calculate q_s and p_s for all later times, and the constraints will be obeyed for all time; naturally, the structure of the right-hand side of Eq.(8.75) is such that if g were one of the ϕ's, the whole expression vanishes weakly. Thus there are restrictions on the initial values, but no arbitrariness in the time development.

As we saw in Chapter 6, the solution of the Hamiltonian equations of motion, in the standard case, gives rise to a one-parameter group of (regular) canonical transformations in phase space. Is there any such statement that we can make in the case of Eq.(8.75)? In fact, there is, and a large part of the next chapter consists of working out the answer to this question. Here, we will content ourselves with a few remarks regarding the expression appearing in Eq.(8.75). If the functions ϕ_ρ are defined as in Eq.(8.56), the PB's $\{\phi_\rho, \phi_\sigma\}$ are actually functions only of q_s and p_j; therefore, if the matrix $\|\{\phi_\rho, \phi_\sigma\}\|$ is nonsingular on U, we can assert that it is nonsingular on the entire phase space, and the inverse matrix $\|C_{\rho\sigma}\|$ exists throughout phase space. Therefore, the expression on the right-hand side of Eq.(8.75) is defined for all q_s, p_s, and it is quite generally true that it vanishes identically if g is a ϕ. This expression is called the "Dirac bracket" of g and H:

$$\{g, H\}^* = \{g, H\} - \sum_{\rho,\sigma} \{g, \phi_\rho\} C_{\rho\sigma} \{\phi_\sigma, H\}. \tag{8.76}$$

This remarkable expression, first introduced by Dirac, is a generalization of the PB and shares many of the properties of the latter, as is shown in the next chapter. We see that it has appeared here in a very natural way, allowing us to write the equations of motion in the form:

$$\frac{d}{dt}\, g(q, p) \approx \{g, H\}^*. \tag{8.77}$$

Let us now return to the analysis of Eq.(8.71). In general, the matrix $\|\{\phi_\rho, \phi_\sigma\}\|$ is singular, and the velocities $\dot{q}_\rho$ are not all determined; if the rank of the matrix is $M(< k - R)$, there are $(k - R - M)$ linearly independent null eigenvectors $\lambda_\rho^{(a)}(q, p)$:

$$\sum_\sigma \lambda_\sigma^{(a)}(q, p)\{\phi_\sigma, \phi_\rho\} \approx 0, \qquad a = 1, 2, \cdots, (k - R - M). \tag{8.78}$$

These are weak equations, and the rank, too, has been computed on U. Combining Eqs.(8.71) and (8.78), we find the following further conditions on q_s and p_s:

$$\sum_\sigma \lambda_\sigma^{(a)}(q, p)\{\phi_\sigma, H\} \approx 0, \qquad a = 1, 2, \cdots, (k - R - M). \tag{8.79}$$

These equations may or may not be obeyed on U; if they are not, we have produced more constraints that restrict the motion in phase space to a hypersurface U' of lower dimensionality than U. Let us suppose that out of the Eq.(8.79), a certain number A are independent among themselves and of the ϕ_ρ equations. Write these constraint functions as $\chi_a(q, p)$, $a = 1, 2 \cdots, A$; the new constrained hypersurface U' is defined by the $(k - R + A)$ weak equations:

$$\phi_\rho \approx 0, \qquad \rho = R + 1, \cdots, k; \tag{8.80a}$$

$$\chi_a \approx 0, \qquad a = 1, 2, \cdots, A; \tag{8.80b}$$

and is of dimensionality $(k + R - A)$. Weak and strong equality now refer to U'; in fact, at each stage of the analysis they refer to the constraints operating at that stage. The new constraints $\chi_a \approx 0$, and all others that may yet appear, are called secondary constraints. They arise only after the equations of motion are used at least once.

As in the Lagrangian treatment, we must now go back and recompute the rank of $\|\{\phi_\sigma, \phi_\rho\}\|$ *with the variables restricted to* U'. Because fewer variables are independent on U' than on U, the rank could decrease (it certainly will not increase), giving rise to more equations of the form of Eq.(8.78) and hence more conditions of the form of Eq.(8.79). This may produce more independent secondary constraints to be added to the χ_a, and thus the constrained hypersurface is restricted further. This is all quite similar to the Lagrangian treatment. This particular process of generating secondary constraints ends when the following situation is reached. The motion is restricted to the hypersurface U'' defined by the $(k - R)$ primary and A' secondary constraints:

$$\begin{aligned} \phi_\rho &\approx 0, \qquad \rho = R+1, \cdots, k; \\ \chi_a &\approx 0, \qquad a = 1, 2, \cdots, A'; \end{aligned} \tag{8.81}$$

on U'', $\|\{\phi_\rho, \phi_\sigma\}\|$ has rank M'; $(A' \leqq k - R - M')$; and for every null eigenvector λ_ρ:

$$\sum_\rho \lambda_\rho \{\phi_\sigma, \phi_\rho\} \approx 0,$$

the condition

$$\sum_\rho \lambda_\rho \{\phi_\rho, H\} \approx 0$$

is obeyed on U''.

These A' secondary constraints arise from the requirement that the primary ones be preserved in time. We must now add the requirement that these secondary constraints, too, be preserved in time. That is, we must analyze the complete set of equations:

$$\begin{aligned} \{\phi_\rho, H\} + \sum_\sigma \{\phi_\rho, \phi_\sigma\}\dot{q}_\sigma &\approx 0, \qquad \rho = R+1, \cdots, k; \\ \{\chi_a, H\} + \sum_\sigma \{\chi_a, \phi_\sigma\}\dot{q}_\sigma &\approx 0, \qquad a = 1, 2, \cdots, A', \end{aligned} \tag{8.82}$$

as a whole, in the same way in which the original set, Eq.(8.71), was analyzed. We have here some information on the $\dot{q}_\rho$ and some constraints on the q's and p's. In place of the square matrix $\|\{\phi_\rho, \phi_\sigma\}\|$, we now deal with the rectangular one:

$$\left\| \begin{matrix} \{\phi_\rho, \phi_\sigma\} \\ \{\chi_a, \phi_\sigma\} \end{matrix} \right\| \tag{8.83}$$

with $(k - R + A')$ rows and $(k - R)$ columns. Every left null eigenvector $(\lambda_\rho, \lambda_a)$ of this matrix, obeying

$$\sum_\rho \lambda_\rho \{\phi_\rho, \phi_\sigma\} + \sum_a \lambda_a \{\chi_a, \phi_\sigma\} \approx 0$$

leads to a condition on the q's and p's:

$$\sum_{\rho} \lambda_{\rho}\{\phi_{\rho}, H\} + \sum_{a} \lambda_{a}\{\chi_{a}, H\} \approx 0.$$

Of course, this condition contains nothing new if all the λ_a are zero. If the λ_a are not all zero, it is either obeyed on U'', or it implies an independent secondary constraint. Every new secondary constraint that arises in this way gives rise to one more equation involving the velocities $\dot{q}_\rho$ when we add the requirement that it be preserved in time. This process of getting new equations involving the velocities $\dot{q}_\rho$ and new secondary constraints by considering the time derivatives of existing constraints ends after a finite number of steps; we then have $(k-R)$ primary and A'' secondary constraints defining a hypersurface U''' in phase space:

$$\phi_\rho \approx 0, \qquad \rho = R+1, \cdots, k; \qquad \chi_a \approx 0, \qquad a = 1, 2, \cdots, A''. \tag{8.84}$$

Weak equality now refers to U'''. The velocity variables obey the following system of equations:

$$\begin{aligned} &\{\phi_\rho, H\} + \sum_{\sigma} \{\phi_\rho, \phi_\sigma\}\dot{q}_\sigma \approx 0, \qquad \rho = R+1, \cdots, k; \\ &\{\chi_a, H\} + \sum_{\sigma} \{\chi_a, \phi_\sigma\}\dot{q}_\sigma \approx 0, \qquad a = 1, 2, \cdots, A'' \end{aligned} \tag{8.85}$$

For every left null eigenvector of the $(k - R + A'') \times (k - R)$ matrix

$$\mathcal{D} = \left\| \begin{matrix} \{\phi_\rho, \phi_\sigma\} \\ \{\chi_a, \phi_\sigma\} \end{matrix} \right\|, \quad \begin{matrix} \rho, \sigma = R+1, \cdots, k, \\ a = 1, 2, \cdots, A'', \end{matrix}$$

having components $(\lambda_\rho, \lambda_a)$, the condition:

$$\sum_{\rho} \lambda_\rho\{\phi_\rho, H\} + \sum_{a=1}^{A''} \lambda_a\{\chi_a, H\} \approx 0$$

is obeyed. Thus one has reached the stage where the ϕ_ρ, χ_a for $\rho = R+1, \cdots, k$ and $a = 1, \cdots, A''$ form a complete set of constraints, and where no more equations for the $\dot{q}_\rho$ can be generated.

Now we must see how much information there is in Eq.(8.85) to help determine the $\dot{q}_\rho$. If on the constrained hypersurface U''', the matrix $\mathcal{D}$ were of maximal rank, which is $(k-R)$, we would have exactly $(k-R)$ linearly independent equations for the $\dot{q}_\rho$ among the $(k - R + A'')$ equations Eq.(8.85), and this would suffice to fix all the $\dot{q}_\rho$ as functions of q's and p's. More generally, if the rank of $\mathcal{D}$ is M'', where $M'' \leqq (k-R)$, exactly M'' linearly independent combinations of the $\dot{q}_\rho$ are determined in terms of q's and p's whereas $(k - R - M'')$ linear combinations *remain completely free.* Let us see how these combinations are formed. The number M'' is both the maximum number of linearly independent rows and the maximum number of linearly independent columns of $\mathcal{D}$.

Thus there must be $(k - R - M'')$ independent relations among the columns of $\mathcal{D}$:

$$\sum_\sigma \{\phi_\rho, \phi_\sigma\}\xi_\sigma^{(\alpha)} \approx 0,$$
$$\sum_\sigma \{\chi_a, \phi_\sigma\}\xi_\sigma^{(\alpha)} \approx 0, \qquad \alpha = 1, 2, \cdots, (k - R - M''). \tag{8.86}$$

Now the primary constraints $\phi_\rho \approx 0$ can be equally well replaced by any $(k-R)$ linearly independent combinations of themselves and this will not make any change in the theory. Based on Eq.(8.86), let us replace the ϕ_ρ by the $(k - R - M'')$ independent combinations

$$\phi_\alpha = \sum_\sigma \xi_\sigma^{(\alpha)}\phi_\sigma, \qquad \alpha = 1, 2, \cdots, (k - R - M''), \tag{8.87}$$

and M'' other combinations which we call $\phi_\beta, \beta = 1, 2, \cdots, M''$. The important property of the ϕ_α's is that, according to Eq.(8.86), the PB of any ϕ_α with each of the complete set of constraints vanishes weakly:

$$\{\phi_\alpha, \phi_{\alpha'}\} \approx \{\phi_\alpha, \phi_\beta\} \approx \{\phi_\alpha, \chi_a\} \approx 0. \tag{8.88}$$

No linear combination of the ϕ_β's can have this property, because if it did, we would have more relations among the columns of $\mathcal{D}$, in addition to Eq.(8.86). The functions ϕ_α are called "first class," the ϕ_β "second class"; a first-class function is defined as one whose PB with the complete set of constraints vanishes weakly. What is the purpose of separating the ϕ_ρ into a maximum number of first class ϕ_α's, and a balance of second class ϕ_β's? *It tells us precisely which combinations of the* $\dot{q}_\rho$ *are determined and which ones are not.* To see this, let us first separate the secondary constraints χ_a also into the two classes; we take advantage of the fact that for all purposes the set χ_a could be replaced by a new set χ'_a defined as

$$\chi'_a = \sum_b S_{ab}\chi_b + \sum_\alpha S_{a\alpha}\phi_\alpha + \sum_\beta S_{a\beta}\phi_\beta,$$

as long as the matrix $\|S_{ab}\|$ is nonsingular. Using this freedom, we bring as many of the χ's, the χ_A's, into the first class as possible, and call the remaining second-class χ's the χ_B's. The basic properties of the χ_A are

$$\{\chi_A, \chi_{A'}\} \approx \{\chi_A, \chi_B\} \approx \{\chi_A, \phi_\alpha\} \approx \{\chi_A, \phi_\beta\} \approx 0; \tag{8.89}$$

no linear combination of the χ_B and ϕ_β can be first class. Now the equations for the $\dot{q}_\rho$, Eq.(8.85), can be rewritten in terms of the complete set of constraints ϕ_α, ϕ_β, χ_A, χ_B. The $\dot{q}_\rho$ always appear in the combination $\sum_\sigma \phi_\sigma \dot{q}_\sigma$; rewriting this combination, we have:

$$\sum_\sigma \phi_\sigma \dot{q}_\sigma = \sum_\alpha \phi_\alpha v_\alpha + \sum_\beta \phi_\beta v_\beta. \tag{8.90}$$

The v_α's and v_β's are linearly independent combinations of the $\dot{q}_\rho$'s. Equations (8.85) now take the equivalent form:

$$\{\phi_\alpha, H\} \approx 0, \qquad \alpha = 1, 2, \cdots, (k - R - M''); \tag{8.91a}$$

$$\{\chi_A, H\} \approx 0; \tag{8.91b}$$

$$\{\phi_\beta, H\} + \sum_{\beta'} \{\phi_\beta, \phi_{\beta'}\} v_{\beta'} \approx 0; \tag{8.91c}$$

$$\{\chi_B, H\} + \sum_{\beta'} \{\chi_B, \phi_{\beta'}\} v_{\beta'} \approx 0. \tag{8.91d}$$

We see that the v_α are absent because they were multiplied by the first-class constraints ϕ_α! We also know that no new equations for the v_α's and v_β's can be obtained, beyond what we already have. *Thus there are as many undetermined combinations of the velocities, v_α, as there are primary first-class constraints, and each v_α appears as one arbitrary function of time in the Hamiltonian equations of motion.* Let us also note that because Eqs.(8.91) and (8.85) are completely equivalent, the former cannot lead to any new constraints; thus Eqs.(8.91a) and (8.91b), which do not involve v_β, neither impose conditions on H nor lead to *new* conditions on the q's and p's; they express *properties* of the Hamiltonian as well as of the first-class constraints ϕ_α and χ_A. [One can also verify this explicitly by combining the first-class property of ϕ_α and χ_A with the property of the matrix $\mathcal{D}$ explained immediately following Eq.(8.85).]

The linear combinations v_β of the $\dot{q}_\rho$ are determined by Eqs.(8.91c) and (8.91d). This follows from the property that no linear combination of the ϕ_β's and χ_B's can be first class. Let us denote by Δ the matrix of PB's of the whole set of second-class constraints:

$$\Delta = \begin{Vmatrix} \{\phi_\beta, \phi_{\beta'}\} & \{\phi_\beta, \chi_{B'}\} \\ \{\chi_B, \phi_{\beta'}\} & \{\chi_B, \chi_{B'}\} \end{Vmatrix}. \tag{8.92}$$

This square matrix must be nonsingular on U''' and therefore of even dimension; otherwise we would have a linear relation among the rows of Δ, which would lead to a linear combination of ϕ_β and χ_B having weakly vanishing PB's with $\phi_{\beta'}$ and $\chi_{B'}$, and then this combination would be first class. Let the inverse to Δ be written as C, with

$$C = \begin{Vmatrix} C_{\beta\beta'} & C_{\beta B'} \\ C_{B\beta'} & C_{BB'} \end{Vmatrix}$$

and

$$\sum_{\beta'} C_{\beta\beta'} \{\phi_{\beta'}, \phi_{\beta''}\} + \sum_{B'} C_{\beta B'} \{\chi_{B'}, \phi_{\beta''}\} \approx \delta_{\beta\beta''},$$

$$\sum_{\beta'} C_{\beta\beta'} \{\phi_{\beta'}, \chi_{B''}\} + \sum_{B'} C_{\beta B'} \{\chi_{B'}, \chi_{B''}\} \approx 0;$$

$$\sum_{\beta'} C_{B\beta'}\{\phi_{\beta'},\phi_{\beta''}\} + \sum_{B'} C_{BB'}\{\chi_{B'},\phi_{\beta''}\} \approx 0;$$
$$\sum_{\beta'} C_{B\beta'}\{\phi_{\beta'},\chi_{B''}\} + \sum_{B'} C_{BB'}\{\chi_{B'},\chi_{B''}\} \approx \delta_{BB''}. \tag{8.93}$$

Combining the first of these equations with Eqs.(8.91c) and (8.91d) suitably, we immediately get:

$$v_\beta \approx -\sum_{\beta'} C_{\beta\beta'}\{\phi_{\beta'},H\} - \sum_{B'} C_{\beta B'}\{\chi_{B'},H\}; \tag{8.94}$$

using the third of Eq.(8.93) instead of the first we also obtain:

$$\sum_{\beta'} C_{B\beta'}\{\phi_{\beta'},H\} + \sum_{B'} C_{BB'}\{\chi_{B'},H\} \approx 0. \tag{8.95}$$

This will be used shortly. Note that like Eqs.(8.91a) and (8.91b), it is a *property* of H and the second-class combinations $\sum_{\beta'} C_{B\beta'}\phi_{\beta'} + \sum_{B'} C_{BB'}\chi_{B'}$, *not* a condition on either nor on the p's and q's.

We can now use the above expression for v_β in the general Hamilton equation of motion Eq.(8.70); it takes the form:

$$\begin{aligned}\frac{d}{dt}\,g(q,p) &\approx \{g,H\} + \sum_\alpha \{g,\phi_\alpha\}v_\alpha + \sum_\beta \{g,\phi_\beta\}v_\beta \\ &\approx \{g,H\} + \sum_\alpha \{g,\phi_\alpha\}v_\alpha - \sum_{\beta,\beta'} \{g,\phi_\beta\}C_{\beta\beta'}\{\phi_{\beta'},H\} \\ &\quad - \sum_{\beta,B'} \{g,\phi_\beta\}C_{\beta B'}\{\chi_{B'},H\}.\end{aligned}$$

However, in this form the ϕ_β and χ_B do not appear symmetrically. But Eq.(8.95) shows that we may add the quantity:

$$-\sum_{B,\beta'} \{g,\chi_B\}C_{B\beta'}\{\phi_{\beta'},H\} - \sum_{B,B'} \{g,\chi_B\}C_{BB'}\{\chi_{B'},H\}$$

which vanishes weakly, and then achieve symmetry in ϕ_β and χ_B. Let us write ζ_m for the entire collection of second-class constraints ϕ_β, χ_B, the first M'' of the ζ's being the ϕ_β, the rest χ_B. The matrix elements of Δ and C can then be denoted by $\Delta_{mm'}$, $C_{mm'}$ respectively, and the general equation of motion takes the form:

$$\frac{d}{dt}\,g(q,p) \approx \{g,H\} + \sum_\alpha \{g,\phi_\alpha\}v_\alpha - \sum_{mm'} \{g,\zeta_m\}C_{mm'}\{\zeta_{m'},H\}. \tag{8.96}$$

The final Hamiltonian form of the basic equations of the theory has now been reached. Its elements are (1) the general (weak) equation of motion, Eq.(8.96), (2) a set of first-class constraints

$$\phi_\alpha \approx 0, \qquad \chi_A \approx 0, \tag{8.97}$$

some of which are primary and some secondary; (3) a combined set of second-class constraints

$$\zeta_m \approx 0. \tag{8.98}$$

The Hamiltonian H introduces no arbitrariness into the right-hand side of Eq.(8.96), because it is strongly equal to the known function $\tilde{W}$ (see Eq.(8.67)), It has the important property that its PB's with ϕ_α and χ_A vanish weakly (see Eq.(8.91a) and (8.91b)). This property of H, together with the structure of the equation of motion, that is to say, the way in which the second-class constraints ζ_m appear there, ensure that *all* the constraints are preserved in time. Therefore, if the initial values of the q_s and p_s at time $t = 0$ are chosen in such a way that the representative point lies on the constrained hypersurface U''' in phase space, and if the functions $v_\alpha(t)$ are specified arbitrarily as functions of time, the equations of motion can be solved to yield the q_s and p_s for all later times; the constraints will be satisfied for all t, that is, the phase-space trajectory will always lie entirely in U'''. The similarity of this picture of the motion to the Lagrangian one is quite evident; in both cases, if the constraints are obeyed initially, they take care of themselves for later times; in both cases the general solution of the equations of motion involves arbitrary functions of time.

The separation of constraints into first class and second class is more meaningful than the separation into primary and secondary ones. This is because a given system of Lagrangian equations of motion can sometimes be derived from more than one Lagrangian, but which constraints are primary and which secondary depends on the functional form of the Lagrangian. In Eq.(8.96), the second-class constraints, both primary and secondary, appear on a common footing, whereas this is not true for the first-class ones. This suggests that in a purely Hamiltonian approach, one should add the term

$$\sum_A \{g, \chi_A\} v_A$$

involving the secondary first-class constraints χ_A to the right-hand side of Eq.(8.96), and allow the v_A, like the v_α, to be completely arbitrary. This would bring about complete symmetry between the ϕ's and χ's, but it must be remembered that only the v_α's correspond to the unsolved accelerations of the Lagrangian formalism.

The physical interpretation of the first-class constraints, and of the canonical transformations generated by them, has been discussed very beautifully by Dirac in his Yeshiva lectures, to which we must refer the reader (P.A.M. Dirac, *Lectures on Quantum Mechanics*, Belfer Graduate School of Science, Yeshiva University, New York, 1964). Although notions of primary and secondary are not very essential, it is clear and obvious from the equations of motion Eq. (8.96) that the first-class and second-class constraints are mathematically very different. Because the matrix $\|\{\zeta_m, \zeta_{m'}\}\|$ of PB's of the second-class constraints is nonsingular on U''', it is reasonable to assume that this matrix is nonsingular in a finite neighborhood of the surface U'''; thus the inverse matrix $\|C_{mm'}\|$ also exists in this larger region. This means that Eq.(8.93) may be taken to be strong equations. The first and third terms on the right-hand side of Eq.(8.96) combine to form the Dirac bracket of g with H, the Dirac bracket being defined with respect to the ζ_m. (As we see in the next chapter, the Dirac bracket is defined in general with respect to an even number of functions whose matrix of PB's is nonsingular). We could, therefore, write the general equation of motion in the form:

$$\frac{d}{dt}\, g(q,p) \approx \{g, H\}^* + \sum_\alpha \{g, \phi_\alpha\} v_\alpha,$$

but since the ϕ_α's are first class, they too, could be included in the Dirac bracket to enable us to write:

$$\frac{d}{dt}\, g(q,p) \approx \{g, H_T\}^*, \qquad H_T = H + \sum_\alpha \phi_\alpha v_\alpha. \tag{8.99}$$

We have introduced here the "total Hamiltonian" H_T: it includes all the arbitrariness coming from the variables v_α. Now, when we defined what we meant by a weak equation, we stated that in any such equation all partial differentiations with respect to the q's and p's should first be carried out as though all these variables were independent, and only afterwards should the variables be restricted to the constrained hypersurface. If the constraint equations were used before working out a PB, for example, in a weak equation, we would get wrong answers. All this is because if some function were only known to vanish weakly, then its PB with an arbitrary function will, in general, not vanish weakly. But the situation is different if we have a weak equation in which partial derivatives of functions appear only via Dirac brackets, as is the case in Eq.(8.99). In this case, we know that the Dirac bracket is so constructed that if one of the functions in the bracket is a second-class constraint ζ_m, the bracket vanishes identically whatever the other function may be. Thus if we work exclusively with Dirac brackets, it makes no difference whether we use the vanishing of the ζ_m before or after computing the brackets. In other words, the weak second-class constraint equations $\zeta_m \approx 0$ can actually be converted into strong equations,

$$\zeta_m \equiv 0,$$

and these equations can actually be used to eliminate the corresponding number of p's and q's completely from the theory by expressing them as functions of the remaining p's and q's. When this is done, we have the general Hamiltonian equation of motion Eq.(8.99), expressed in Dirac bracket form; we are dealing with a reduced number of independent p's and q's (this number being $2k$ minus the number of second-class constraints), and the only constraints left will be the first-class ones, given in Eq.(8.97). Weak equations are now relative to these constraints only. This final form of the Hamiltonian theory, with Dirac brackets replacing PB's and with only first-class constraints left, is to be compared with the Lagrangian theory where the A-type constraints were used to eliminate variables, and one only had B-type constraints left at the end.

Our work in the next chapter will answer the following two interesting questions:

(a) what is the nature of the finite transformations generated by Eq.(8.99), and what is its relation to canonical transformations;

(b) when we eliminate an even number of variables by using the second class constraints as algebraic equations, under what conditions will these variables be a subset of the q_s together with the canonically conjugate p's?

Chapter 9

The Generalized Poisson Bracket and Its Applications

The purpose of this chapter is to analyze the properties of the Dirac bracket and of the transformations generated by it. As we have seen in the preceding chapter, to define the Dirac bracket we need to be given an even number of functions, ζ_m, on a $2k$-dimensional phase space with canonical coordinates ω^μ, with the property that the matrix of PB's $\|\{\zeta_m, \zeta_{m'}\}\|$ is nonsingular. We assume, for simplicity, that this matrix is nonsingular in the entire phase space, and denote its inverse by $\|C_{mm'}(\omega)\|$:

$$\sum_{m'} C_{mm'}(\omega)\{\zeta_{m'}(\omega), \zeta_{m''}(\omega)\} = \delta_{mm''} . \tag{9.1}$$

Then the Dirac bracket of any two functions $f(\omega), g(\omega)$ is defined as

$$\{f, g\}^* = \{f, g\} - \sum_{m,m'} \{f, \zeta_m\} C_{mm'} \{\zeta_{m'}, g\}. \tag{9.2}$$

Thus it depends both on the PB structure and on the particular set of functions ζ_m. It is easy to see that the set of functions ζ_m must be functionally independent. If this were not so, there would be at least one relation among them that would allow us to express ζ_1, say, in terms of the rest:

$$\zeta_1 = \phi(\zeta_2, \cdots);$$

but this would make the first row of the matrix $\|\{\zeta_m, \zeta_{m'}\}\|$ linearly dependent on the remaining rows,

$$\{\zeta_1, \zeta_m\} = \sum_{m' \neq 1} \frac{\partial \phi}{\partial \zeta_{m'}} \{\zeta_{m'}, \zeta_m\},$$

which is impossible. We have also noted that the Dirac bracket of any function $f(\omega)$ with any of the ζ's vanishes identically:

$$\{f, \zeta_m\}^* = 0. \tag{9.3}$$

We can check quite easily that the Dirac bracket is defined by the ζ_m upto functional transformations of the ζ's among themselves. By this we mean that if we replace the ζ_m by a set ζ'_m where the latter are independent functions of the former, then the Dirac bracket does not change. For under this replacement, the matrix $\|C_{mm'}\|$ is replaced by $\|C'_{mm'}\|$ where

$$C'_{mm'} = \sum_{m_1 \, m_2} \frac{\partial \zeta_{m_1}}{\partial \zeta'_m} C_{m_1 m_2} \frac{\partial \zeta_{m_2}}{\partial \zeta'_{m'}}, \tag{9.4}$$

and using this, one easily finds that the second term on the right-hand side of Eq.(9.2) is indeed unaltered by the replacement $\zeta \to \zeta'$, $C \to C'$.

The Dirac bracket shares with the PB the properties of being antisymmetric in its arguments and linear in each of them, as well as the product rule:

$$\{f_1 f_2, g\}^* = f_1\{f_2, g\}^* + \{f_1, g\}^* f_2. \tag{9.5}$$

In addition, it obeys the Jacobi identity, as does the PB:

$$\{\{f, g\}^*, h\}^* + \{\{g, h\}^*, f\}^* + \{\{h, f\}^*, g\}^* = 0. \tag{9.6}$$

It is hardly obvious from the way the Dirac bracket has been defined in Eq.(9.2) that it would be found to obey the Jacobi identity. Indeed, the original proof of the identity consisted of a straight-forward but rather lengthy *verification* of the identity, without shedding much light on the structure of the bracket or suggesting any simple reason for suspecting that the identity might hold. But, as we emphasize later, the fact that the Jacobi identity holds is very important from the group theoretical point of view.

We prove the Jacobi identity (Eq.(9.6)) by showing that the Dirac bracket is a particular example of a generalized Poisson bracket (hereafter called GPB). The notion of a GPB has already been introduced in Chapter 5 (see page 40); it turns out to be worthwhile studying it in some detail not only to understand the Dirac bracket and its properties in a simple way, but also because it has some other interesting applications that we explain in this chapter as well as in a later one (see the discussion of pure-spin systems in Chapter 17).

Generalized Poisson Brackets

Let us consider a certain number, N, of real variables z^μ, varying over certain ranges, and functions of these variables. Let there be given a set of functions $\eta^{\mu\nu}(z)$, antisymmetric in μ and ν, obeying the identity Eq.(9.15) which is derived later on. We then define the GPB of any two functions $f(z)$, $g(z)$ to be a third function $h(z)$ given by

$$\{f, g\}^*(z) \equiv h(z) = \eta^{\mu\nu}(z) \frac{\partial f(z)}{\partial z^\mu} \frac{\partial g(z)}{\partial z^\nu} \, . \tag{9.7}$$

We first consider the behavior of η under arbitrary changes of the coordinates, $z^\mu \to z'^\mu$, the latter being independent functions of the former. We determine this behavior by defining f, g as well as $\{f, g\}^*$ to transform as "scalar fields." That is, f, g, and h go over into functions f', g', h' of z' according to

$$f'(z') = f(z), g'(z') = g(z), h'(z') = h(z). \tag{9.8}$$

The expression of h' in terms of f' and g' yields η in the new coordinates:

$$\begin{aligned} h'(z') &= \eta^{\mu\nu}(z)\frac{\partial f(z)}{\partial z^\mu}\,\frac{\partial g(z)}{\partial z^\nu} = \eta^{\mu\nu}(z)\frac{\partial f'(z')}{\partial z'^\rho}\,\frac{\partial z'^\rho}{\partial z^\mu}\,\frac{\partial g'(z')}{\partial z'^\sigma}\,\frac{\partial z'^\sigma}{\partial z^\nu} \\ &= \eta'^{\rho\sigma}(z')\,\frac{\partial f'(z')}{\partial z'^\rho}\,\frac{\partial g'(z')}{\partial z'^\sigma} \end{aligned} \tag{9.9}$$

with

$$\eta'^{\rho\sigma}(z') = \frac{\partial z'^\rho}{\partial z^\mu}\,\frac{\partial z'^\sigma}{\partial z^\nu}\,\eta^{\mu\nu}(z). \tag{9.10}$$

Thus η transforms as a second-rank antisymmetric tensor of contravariant type.

What we have done here is not to place any restrictions on the coordinate transformation $z \to z'$, but, given the transformation, we have defined a new η or rather, we have found η in the new coordinates. The GPB is an operation by which two scalar fields determine a third one. Given a scalar field, in each coordinate system it is represented by a particular function of those coordinates. The explicit expression of the function representing the scalar field $\{f, g\}^*$ in terms of those representing f and g depends on the coordinates system. Equation (9.10) shows how this expression changes when one goes from one set of coordinates to another.

We next define a generalized canonical transformation. Here, unlike the case studied earlier, we confine ourselves from the start to "regular" transformations; a transformation $z \to z'$ is regular if the new variables go over precisely the same ranges as the old ones, thus such a transformation can also be interpreted as a mapping of the points of the "phase space" onto themselves. (Phase space is now simply the space of points with coordinates z^μ). Under such a transformation, we have seen how to pass from a given function $f(z)$ to another one, $\bar{f}(z)$, according to the rule:

$$f(z') = \bar{f}(z). \tag{9.11}$$

We define a regular transformation to be a generalized canonical transformation if it preserves GPB relations; that is, if the relation $\{f, g\}^* = h$ goes over into the relation $\{\bar{f}, \bar{g}\}^* = \bar{h}$. This implies

$$\begin{aligned} \bar{h}(z) &= \eta^{\mu\nu}(z)\frac{\partial \bar{f}(z)}{\partial z^\mu}\,\frac{\partial \bar{g}(z)}{\partial z^\nu} = \eta^{\mu\nu}(z)\frac{\partial f(z')}{\partial z'^\rho}\,\frac{\partial g(z')}{\partial z'^\sigma}\,\frac{\partial z'^\rho}{\partial z^\mu}\,\frac{\partial z'^\sigma}{\partial z^\nu} \\ &= \eta^{\rho\sigma}(z')\frac{\partial f(z')}{\partial z'^\rho}\,\frac{\partial g(z')}{\partial z'^\sigma}\,, \end{aligned}$$

that is,

$$\eta^{\mu\nu}(z)\frac{\partial z'^\rho}{\partial z^\mu}\,\frac{\partial z'^\sigma}{\partial z^\nu} = \eta^{\rho\sigma}(z')\,. \tag{9.12}$$

A transformation obeying Eq.(9.12) is called a generalized regular canonical transformation. Notice that the same functions η appear on both sides of this equation, but evaluated at different points of the phase space; in order to be as general as possible in the choice of the η's we restrict ourselves to regular transformations. Of course, if we take the particular case when η is the constant matrix $\epsilon^{\mu\nu}$ corresponding to the PB, we recover the definition of an ordinary canonical transformation. It is also obvious that the set of all generalized regular canonical transformations forms a group. The basic Eq.(9.12) can be rewritten in the form:

$$\{z'^{\rho}, z'^{\sigma}\}^*_z = \eta^{\rho\sigma}(z'), \tag{9.13}$$

which means that whether we compute the GPB's of the new basic variables z' (the fundamental GPB's) with respect to z or to z', we get the same answers.

We discuss next the Jacobi identity. We demand that $\eta^{\mu\nu}(z)$ be such that for any three functions f, g, h the cyclic sum

$$\sum_{\text{cyclic}} \{\{f, g\}^*, h\}^*$$

should vanish,

$$\{\{f, g\}^*, h\}^* + \{\{g, h\}^*, f\}^* + \{\{h, f\}^*, g\}^* = 0. \tag{9.14}$$

If Eq.(9.7) is substituted here, two kinds of terms appear, those without derivatives of $\eta^{\mu\nu}$ and those with derivatives. As in the case of the PB's, the antisymmetry of $\eta^{\mu\nu}$ makes the former vanish. The vanishing of the latter leads to:

$$\eta^{\lambda\mu}(z)\frac{\partial \eta^{\nu\rho}(z)}{\partial z^{\mu}} + \eta^{\nu\mu}(z)\frac{\partial \eta^{\rho\lambda}(z)}{\partial z^{\mu}} + \eta^{\rho\mu}(z)\frac{\partial \eta^{\lambda\nu}(z)}{\partial z^{\mu}} = 0. \tag{9.15}$$

Thus the Jacobi identity for the GPB is equivalent to Eq.(9.15).

The GPB is called nonsingular or singular depending on whether the matrix $\|\eta^{\mu\nu}\|$ is nonsingular or singular. (Thus if the number N is odd, the GPB is necessarily singular; a discussion of some properties of singular GPB's is given later in this chapter, on pages 114 to 118). Both types of GPB's are of interest. If the GPB is nonsingular, we denote the inverse matrix by $\|\eta_{\mu\nu}(z)\|$:

$$\eta_{\mu\nu}(z)\eta^{\nu\lambda}(z) = \delta^{\lambda}_{\mu}. \tag{9.16}$$

Then Eq.(9.15) takes on a very simple form in terms of $\eta_{\mu\lambda}$:

$$\frac{\partial \eta_{\mu\nu}(z)}{\partial z^{\lambda}} + \frac{\partial \eta_{\nu\lambda}(z)}{\partial z^{\mu}} + \frac{\partial \eta_{\lambda\mu}(z)}{\partial z^{\nu}} = 0. \tag{9.17}$$

This implies that for a nonsingular GPB, the inverse matrix $\eta_{\mu\nu}(z)$ can be represented in the form:

$$\eta_{\mu\nu}(z) = \frac{\partial A_{\nu}(z)}{\partial z^{\mu}} - \frac{\partial A_{\mu}(z)}{\partial z^{\nu}} \tag{9.18}$$

in terms of some set of functions $A_\mu(z)$. Equation (9.17) bears a striking resemblance to Eq.(5.37), involving LB's in the usual sense:

$$\frac{\partial}{\partial A_j}[A_l, A_m] + \frac{\partial}{\partial A_l}[A_m, A_j] + \frac{\partial}{\partial A_m}[A_j, A_l] = 0. \tag{9.19}$$

(Recall that the A_j's are $2k$ independent functions of the phase space variables $\omega^\mu, \mu = 1, \cdots, 2k$.) The z-variables correspond to the A's, and the $\eta_{\mu\nu}$ to the LB's $[A_j, A_l]$. This shows the *close connection between the Jacobi identity for a nonsingular GPB, on the one hand, and a standard property of LB's on the other.* We use this connection later in deriving the Dirac bracket.

In the nonsingular case, we could use $\eta_{\mu\nu}$ to define generalized LB's as well as rewrite Eq.(9.12) in terms of $\eta_{\mu\nu}$. However, we do not do so here.

Let us now see how a generalized canonical transformation arises from the solution to a differential equation involving the GPB. Let $\phi(z, \epsilon)$ be a function of z and a real parameter ϵ, and consider the differential equation:

$$\frac{dz^\mu}{d\epsilon} = \{z^\mu, \phi(z, \epsilon)\}^*. \tag{9.20}$$

As in the case discussed in Chapter 6, the bracket on the right is to be evaluated with respect to the variables z, which are themselves functions of ϵ. As before, we introduce the values of z at $\epsilon = 0$, the z_0^μ, as variables, and write:

$$z^\mu(\epsilon) = \psi^\mu(z_0; \epsilon), \qquad \psi^\mu(z_0; 0) = z_0^\mu. \tag{9.21}$$

We will assume that ϕ is such that the transformation $z_0 \to z$ is regular; we now wish to prove that it is also a generalized canonical transformation for each value of ϵ. In view of later applications, it is desirable not to restrict oneself either to a ϕ that does not depend explicitly on ϵ, or to a GPB that is nonsingular. We wish to establish, in the most general case, the following property of the functions $\psi^\mu(z_0; \epsilon)$:

$$\eta^{\mu\nu}(z_0)\frac{\partial\psi^\alpha(z_0;\epsilon)}{\partial z_0^\mu}\frac{\partial\psi^\beta(z_0;\epsilon)}{\partial z_0^\nu} = \eta^{\alpha\beta}(\psi). \tag{9.22}$$

Let us denote by $P^{\alpha\beta}$ the left-hand side of this equation; it depends both on z_0 and ϵ. Equation (9.20) rewritten for ψ^μ appears as:

$$\frac{\partial\psi^\mu(z_0;\epsilon)}{\partial\epsilon} = \eta^{\mu\nu}(\psi)\frac{\partial\phi(\psi,\epsilon)}{\partial\psi^\nu}. \tag{9.23}$$

We can use this to compute the partial derivative of $P^{\alpha\beta}(z_0; \epsilon)$ with respect to

ϵ:

$$
\begin{aligned}
\frac{\partial P^{\alpha\beta}(z_0;\epsilon)}{\partial\epsilon} &= \frac{\partial}{\partial\epsilon}\{\psi^\alpha(z_0;\epsilon),\psi^\beta(z_0;\epsilon)\}^*_{z_0} \\
&= \left\{\frac{\partial\psi^\alpha}{\partial\epsilon},\psi^\beta\right\}^*_{z_0} + \left\{\psi^\alpha\cdot\frac{\partial\psi^\beta}{\partial\epsilon}\right\}^*_{z_0} \\
&= \left\{\eta^{\alpha\rho}(\psi)\frac{\partial\phi(\psi,\epsilon)}{\partial\psi^\rho},\psi^\beta\right\}^*_{z_0} + \left\{\psi^\alpha,\eta^{\beta\rho}(\psi)\frac{\partial\phi(\psi,\epsilon)}{\partial\psi^\rho}\right\}^*_{z_0} \\
&= \frac{\partial}{\partial\psi^\lambda}\left(\eta^{\alpha\rho}(\psi)\frac{\partial\phi(\psi,\epsilon)}{\partial\psi^\rho}\right)\{\psi^\lambda,\psi^\beta\}^*_{z_0} \\
&\quad + \frac{\partial}{\partial\psi^\lambda}\left(\eta^{\beta\rho}(\psi)\frac{\partial\phi(\psi,\epsilon)}{\partial\psi^\rho}\right)\{\psi^\alpha,\psi^\lambda\}^*_{z_0}, \\
\frac{\partial P^{\alpha\beta}}{\partial\epsilon} &= P^{\alpha\lambda}\frac{\partial}{\partial\psi^\lambda}\left(\eta^{\beta\rho}(\psi)\frac{\partial\phi(\psi,\epsilon)}{\partial\psi^\rho}\right) \\
&\quad - P^{\beta\lambda}\frac{\partial}{\partial\psi^\lambda}\left(\eta^{\alpha\rho}(\psi)\frac{\partial\phi(\psi,\epsilon)}{\partial\psi^\rho}\right).
\end{aligned}
\tag{9.24}
$$

Next we compute the partial derivative of $\eta^{\alpha\beta}(\psi)$ with respect to ϵ:

$$
\frac{\partial\eta^{\alpha\beta}}{\partial\epsilon}(\psi) = \frac{\partial\eta^{\alpha\beta}(\psi)}{\partial\psi^\lambda}\eta^{\lambda\rho}(\psi)\frac{\partial\phi(\psi,\epsilon)}{\partial\psi^\rho}.
$$

Here let us use the fact that the GPB obeys the Jacobi identity, Eq.(9.15); we get:

$$
\frac{\partial\eta^{\alpha\beta}(\psi)}{\partial\epsilon} = \left(\eta^{\alpha\lambda}(\psi)\frac{\partial\eta^{\beta\rho}(\psi)}{\partial\psi^\lambda} + \eta^{\beta\lambda}(\psi)\frac{\partial\eta^{\rho\alpha}(\psi)}{\partial\psi^\lambda}\right)\frac{\partial\phi(\psi,\epsilon)}{\partial\psi^\rho}.
$$

But this can be rearranged to read:

$$
\begin{aligned}
\frac{\partial\eta^{\alpha\beta}(\psi)}{\partial\epsilon} &= \eta^{\alpha\lambda}(\psi)\frac{\partial}{\partial\psi^\lambda}\left(\eta^{\beta\rho}(\psi)\frac{\partial\phi(\psi,\epsilon)}{\partial\psi^\rho}\right) \\
&\quad - \eta^{\beta\lambda}(\psi)\frac{\partial}{\partial\psi^\lambda}\left(\eta^{\alpha\rho}(\psi)\frac{\partial\phi(\psi,\epsilon)}{\partial\psi^\rho}\right).
\end{aligned}
\tag{9.25}
$$

Now take the difference of Eqs.(9.24) and (9.25); if we write $\Omega^{\alpha\beta}(z_0;\epsilon)$ for the expression $P^{\alpha\beta}(z_0;\epsilon)-\eta^{\alpha\beta}(\psi)$, we find that it obeys:

$$
\begin{aligned}
\frac{\partial\Omega^{\alpha\beta}(z_0;\epsilon)}{\partial\epsilon} &= \Omega^{\alpha\lambda}(z_0;\epsilon)\frac{\partial}{\partial\psi^\lambda}\left(\eta^{\beta\rho}(\psi)\frac{\partial\phi(\psi,\epsilon)}{\partial\psi^\rho}\right) \\
&\quad - \Omega^{\beta\lambda}(z_0;\epsilon)\frac{\partial}{\partial\psi^\lambda}\left(\eta^{\alpha\rho}(\psi)\frac{\partial\phi(\psi,\epsilon)}{\partial\psi^\rho}\right).
\end{aligned}
\tag{9.26}
$$

We have thus obtained a recursion relation for the derivatives of $\Omega^{\alpha\beta}$ with respect to ϵ. The nth derivative, $(\partial^n\Omega^{\alpha\beta}/\partial\epsilon^n)$, is a linear combination of the Ω's, and their first $(n-1)$ derivatives. But at $\epsilon=0$, $\Omega^{\alpha\beta}$ vanishes identically because the ψ become equal to the z_0's; thus recursively we see that *all* the derivatives $(\partial^n \Omega^{\alpha\beta}/\partial\epsilon^n)$ vanish at $\epsilon = 0$. This proves that $\Omega^{\alpha\beta}$ vanishes identically; that is, Eq.(9.22) is valid for all ϵ, and the solution to the basic differential equation, Eq. (9.20), defines a regular generalized canonical transformation.

We can now appreciate the role played by the Jacobi identity. We could always *define a group* of transformations by means of Eq.(9.12), whether or not the Jacobi identity is obeyed. This identifies one role of the GPB, namely, its use in defining a group of transformations that leave it invariant. But we have seen also that the solutions to the differential equations set up in terms of the GPB, Eq.(9.20), themselves give rise to such transformations; that is, via the GPB we generate transformations that leave all GPB relations among functions invariant. This latter role of the GPB depends crucially on the Jacobi identity being satisfied. We have already seen this fact in the case of ordinary canonical transformations in our work in Chapter 6. The antisymmetry of the constant matrices $\epsilon^{\mu\nu}$, $\epsilon_{\mu\nu}$ is vital in deriving Eq.(6.20); as we saw in Chapter 5 (see the remarks following Eq.(5.36)) the Jacobi identity for the ordinary PB is equivalent to the antisymmetry of $\epsilon^{\mu\nu}$. These are all special cases of the general properties of Lie algebras, as we see in later chapters of this book.

As in the case of canonical transformations, the situation where the "generator" ϕ does not depend explicitly on ϵ is particularly simple, and we can write out the transformation more explicitly. We notice first that ϕ does not change its form under the transformation it generates:

$$\frac{\partial \phi(\psi(z_0;\epsilon))}{\partial \epsilon} = \frac{\partial \phi(\psi)}{\partial \psi^\rho}\eta^{\rho\sigma}(\psi)\frac{\partial \phi(\psi)}{\partial \psi^\sigma} = 0,$$
$$\phi(\psi) = \phi(z_0). \tag{9.27}$$

With the use of this property and Eq.(9.22), the equation obeyed by ψ^μ, Eq.(9.23), can be rewritten as follows:

$$\begin{aligned}\frac{\partial \psi^\mu(z_0;\epsilon)}{\partial \epsilon} &= \eta^{\alpha\beta}(z_0)\frac{\partial \psi^\mu}{\partial z_0^\alpha}\frac{\partial \psi^\nu}{\partial z_0^\beta}\frac{\partial \phi(z_0)}{\partial z_0^\gamma}\frac{\partial z_0^\gamma}{\partial \psi^\nu} \\ &= \eta^{\alpha\beta}(z_0)\frac{\partial \psi^\mu}{\partial z_0^\alpha}\frac{\partial \phi(z_0)}{\partial z_0^\beta},\end{aligned}$$

that is,

$$\frac{\partial \psi^\mu(z_0;\epsilon)}{\partial \epsilon} = \{\psi^\mu(z_0;\epsilon), \phi(z_0)\}^*. \tag{9.28}$$

The GPB here is, of course, computed with respect to the variables z_0. We can then write the solution as a power series in ϵ, as in the previous case:

$$\psi^\mu(z_0;\epsilon) = \sum_{n=0}^{\infty} \frac{\epsilon^n}{n!} A_n^\mu(z_0),$$
$$A_0^\mu(z_0) = z_0^\mu, \quad A_{n+1}^\mu(z_0) = \{A_n^\mu(z_0), \phi(z_0)\}^*. \tag{9.29}$$

In terms of the operator $\tilde{\phi}$ defined as

$$\tilde{\phi}(z_0) = \eta^{\alpha\beta}(z_0)\frac{\partial\phi}{\partial z_0^\alpha}\,\frac{\partial}{\partial z_0^\beta}\,, \tag{9.30}$$

a more compact expression results:

$$\psi^\mu(z_0;\epsilon) = e^{-\epsilon\tilde{\phi}(z_0)}z_\circ^\mu. \tag{9.31}$$

We can prove easily that these transformations form a *one-parameter group* obeying the composition law:

$$\psi^\mu(\psi(z_0;\epsilon_1);\epsilon_2) = \psi^\mu(z_0;\epsilon_1+\epsilon_2); \tag{9.32}$$

(this is not valid if ϕ depends explicitly on ϵ!). Finally, one can give the equations that determine the change in functional form of a function, $f \to \bar{f}$, as defined in Eq.(9.11),

$$f(e^{-\epsilon\tilde{\phi}(z_0)}z_0) = e^{-\epsilon\tilde{\phi}(z_0)}f(z_0) = \bar{f}(z_0), \tag{9.33}$$

and the preservation of GPB relations.

$$\{e^{-\epsilon\tilde{\phi}(z_0)}f(z_0), e^{-\epsilon\tilde{\phi}(z_0)}g(z_0)\}^* = e^{-\epsilon\tilde{\phi}(z_0)}\{f(z_0), g(z_0)\}^*. \tag{9.34}$$

These equations are proved by exactly the same methods as in the case of ordinary canonical transformations. All these properties, which are generalizations of properties derived for ordinary canonical transformations, are independent of whether the GPB is singular or not. They follow from the laws (i), (ii), (iii), (iv), and (vi) in chapter 5, following Eq.(5.37), characteristic of GPB's in classical theory.

After this discussion of GPB's and generalized canonical transformations, let us look briefly at some properties of singular GPB's. According to the definition that we gave on page 110, the GPB given by the set of functions $\eta^{\mu\nu}(z)$ is singular (nonsingular) if the determinant $\delta(z)$ of the antisymmetric matrix $\|\eta^{\mu\nu}(z)\|$ is zero (nonzero). As we might expect, this definition coincides with one that can be stated in terms of the GPB $\{.,.\}^*$ itself. Let us say that a function $\theta(z)$ is *neutral* if the GPB $\{f,\theta\}^*$ vanishes identically for all $f(z)$; it is a nontrivial neutral if it is not a mere constant. Then the GPB is singular if nontrivial neutral functions exist; otherwise it is nonsingular. We now indicate the proof, which makes essential use of the Jacobi identity as expressed in Eq.(9.15). First, suppose $\theta(z)$ is a nontrivial neutral function. Then,

$$\{f,\theta\}^*(z) = \eta^{\mu\nu}(z)\frac{\partial f(z)}{\partial z^\mu}\,\frac{\partial\theta(z)}{\partial z^\nu} = 0 \quad \text{for all } f \Rightarrow \eta^{\mu\nu}(z)\frac{\partial\theta(z)}{\partial z^\nu} = 0$$
$$\Rightarrow \delta(z) = 0. \tag{9.35}$$

because we have established a linear relation among the columns of $\|\eta^{\mu\nu}\|$. Conversely, then, if $\delta(z) \neq 0$, there can be no nontrivial neutrals. Now, what is left is to prove that in case $\delta(z) = 0$, there do exist neutrals (nontrivial ones, of course);

we must also fix the number of such independent θ's. Let the rank of the matrix $\|\eta^{\mu\nu}\|$ be $(N-m)$, with $m>0$ because $\delta(z)=0$. It then happens that there are exactly m independent nontrivial neutral functions θ^a, $a=1,2,\cdots,m$. Any neutral θ must obey the system of partial differential equations:

$$D^\lambda\theta(z)=0\ ,\qquad D^\lambda\equiv\eta^{\lambda\mu}(z)\frac{\partial}{\partial z^\mu}. \tag{9.36}$$

Because of the validity of Eq.(9.15), these equations form a *complete* system; the *commutator* $D^\lambda D^\mu - D^\mu D^\lambda$ of D^λ and D^μ is a linear combination of the D's again:

$$\begin{aligned}D^\lambda D^\mu - D^\mu D^\lambda &= \eta^{\lambda\rho}\left(\frac{\partial\eta^{\mu\sigma}}{\partial z^\rho}\right)\frac{\partial}{\partial z^\sigma}-\eta^{\mu\sigma}\left(\frac{\partial\eta^{\lambda\rho}}{\partial z^\sigma}\right)\frac{\partial}{\partial z^\rho}\\ &=\left(\eta^{\lambda\rho}\frac{\partial\eta^{\mu\sigma}}{\partial z^\rho}+\eta^{\mu\rho}\frac{\partial\eta^{\sigma\lambda}}{\partial z^\rho}\right)\frac{\partial}{\partial z^\sigma}=-\eta^{\sigma\rho}\frac{\partial\eta^{\lambda\mu}}{\partial z^\rho}\frac{\partial}{\partial z^\sigma}=\frac{\partial\eta^{\lambda\mu}}{\partial z^\rho}D^\rho.\end{aligned} \tag{9.37}$$

Now because the matrix $\|\eta^{\lambda\nu}\|$ is antisymmetric, there is no distinction between linear relations among its rows and linear relations among its columns. Because the rank is $(N-m)$, only $(N-m)$ of the operators D^λ are independent, the remaining m of them can be written as linear combinations of the previous ones, with functions of z on the left for coefficients. Thus in Eq.(9.36) we have exactly $(N-m)$ independent partial differential equations, and these equations by themselves are complete; that is, the commutators of these $(N-m)D$'s with one another are linear combinations of themselves again. Now, there is a theorem concerning such complete systems of partial differential equations that states that the number of independent solutions to the system is the number of independent variables minus the number of independent equations in the system. (This is proved, for example, in Chapter I of L.P. Eisenhart, *Continuous Groups of Transformations*, Dover Publications, Inc. New York, 1961.) Therefore, we conclude that in the case of a singular GPB with rank $\|\eta^{\mu\nu}\|=(N-m)$, we find exactly m independent nontrivial neutral functions θ^a. Notice again the role of the Jacobi identity, which is what makes the system of Eq.(9.36) complete, allowing us to apply the theorem.

The functions $\theta^1,\theta^2,\cdots,\theta^m$ of the foregoing discussion form a *basis* for neutral elements: any neutral Φ is some function of the θ's. Each θ^a generates one independent null eigenvector for the matrix $\|\eta^{\mu\nu}\|$, because the θ's are functionally independent. These two facts are expressed by

$$\{f,\Phi\}^*=0\ ,\ \text{all } f\quad\Rightarrow\eta^{\mu\nu}(z)\frac{\partial\Phi}{\partial z^\nu}=0\Rightarrow\Phi=\Phi(\theta^a(z)); \tag{9.38a}$$

$$\eta^{\mu\nu}(z)\frac{\partial\theta^a}{\partial z^\nu}=0\ ,\qquad a=1,2,\cdots,m. \tag{9.38b}$$

More generally, any linear combination

$$\sum_a\alpha'_a(z)\frac{\partial\theta^a}{\partial z^\nu}$$

is also a null eigenvector of $\|\eta\|$. But the most general null eigenvector *is* such a linear combination! Let us prove this. In addition to the m independent θ's, let us choose $(N-m)$ independent functions $\psi^r(z)$, such that the θ's and ψ's altogether are N independent functions on phase space. The ψ's can be chosen in infinitely many ways, and any choice will do. The set θ^a, ψ^r gives a coordinate system in phase space, and any function of the z^μ can be rewritten as a function of the θ^a, ψ^r. Then Eq.(9.38a) can be expressed in this way:

$$\{f, \Phi\}^* = 0, \qquad \text{all } f \Rightarrow \frac{\partial \Phi}{\partial \psi^r} = 0. \tag{9.39}$$

We can use Eq.(9.10) to evaluate the components of η' in the coordinate system ψ^r, θ^a. Of the four types of components $\eta'^{rs}(\psi,\theta)$, $\eta'^{ra}(\psi,\theta)$, $\eta'^{ar}(\psi,\theta)$, $\eta'^{ab}(\psi,\theta)$, only the first survives, the last three vanish on account of Eq.(9.38b):

$$\begin{aligned}
\eta'^{rs}(\psi,\theta) &\equiv \{\psi^r, \psi^s\}^* = \eta^{\mu\nu}(z)\frac{\partial \psi^r}{\partial z^\mu}\frac{\partial \psi^s}{\partial z^\nu}\,; \\
\eta'^{ra}(\psi,\theta) &\equiv -\eta'^{ar}(\psi,\theta) \equiv \{\psi^r, \theta^a\}^* = 0; \\
\eta'^{ab}(\psi,\theta) &\equiv \{\theta^a, \theta^b\}^* = 0;
\end{aligned}$$

$$\|\eta'\| = \left\| \begin{array}{c:c} \eta'^{rs}(\psi,\theta) & 0 \\ \hdashline 0 & 0 \end{array} \right\| . \tag{9.40}$$

In these coordinates, the GPB takes the appearance:

$$\{f, g\}^* = \sum_{rs} \eta'^{rs}(\psi,\theta)\,\frac{\partial f}{\partial \psi^r}\,\frac{\partial g}{\partial \psi^s}\,. \tag{9.41}$$

This is just a GPB for functions of the ψ^r alone, the θ^a being suppressed. But then Eq.(9.39) has the meaning that there are no nontrivial neutral elements with respect to this GPB; in other words, this is a nonsingular GPB in the variables ψ^r. Therefore, the determinant of the $(N-m)$ dimensional submatrix $\|\eta'^{rs}\|$ does not vanish. The most general null vector (α'_r, α'_a) of the matrix $\|\eta'\|$ in Eq.(9.40) must then necessarily be of the form $(0, \alpha'_a)$; the first $(N-m)$ components must vanish, and the remaining m components are arbitrary. We can now transform back to the original coordinates and have the result that the most general null vector of the $N \times N$ matrix $\|\eta^{\mu\nu}\}$ has the form:

$$\alpha_\nu \equiv \alpha'_r\,\frac{\partial \psi^r}{\partial z^\nu} + \alpha'_a\frac{\partial \theta^a}{\partial z^\nu} = \alpha'_a\frac{\partial \theta^a}{\partial z^\nu}\,. \tag{9.42}$$

There is one other property of singular GPB's that is useful in some applications. Take an arbitrary set of N functions $A_\mu(z)$, and define another set $\beta^\nu(z)$ in terms of them via:

$$\beta^\nu(z) = A_\mu(z)\eta^{\mu\nu}(z). \tag{9.43}$$

Then Eq.(9.38) shows that

$$\beta^\nu(z)\frac{\partial \theta^a}{\partial z^\nu} = 0, \qquad a = 1, \cdots, m \tag{9.44}$$

Conversely, we can show that any set of functions $\beta^\nu(z)$ obeying Eq.(9.44) is necessarily expressible in the form of Eq.(9.43) with some $A_\mu(z)$! The proof simply involves examining Eq.(9.44) in the coordinate system ψ^r, θ^a. Treating β^ν as a contravariant vector, its transformed components (β'^r, β'^a) are:

$$\beta'^r = \frac{\partial \psi^r}{\partial z^\nu}\,\beta^\nu, \qquad \beta'^a = \frac{\partial \theta^a}{\partial z^\nu}\beta^\nu = 0. \tag{9.45}$$

Because the submatrix $\|\eta'^{rs}(\psi, \theta)\|$ is nonsingular, we can find a set of functions A'_s such that

$$\beta'^r = A'_s \eta'^{sr}(\psi, \theta). \tag{9.46}$$

Now we transform back to the original coordinates:

$$\beta^\nu = \frac{\partial z^\nu}{\partial \psi^r}\,\beta'^r = \frac{\partial z^\nu}{\partial \psi^r}\,A'_s \eta'^{sr} = A'_s\,\frac{\partial \psi^s}{\partial z^\lambda}\,\frac{\partial \psi^r}{\partial z^\mu}\,\frac{\partial z^\nu}{\partial \psi^r}\,\eta^{\lambda\mu}. \tag{9.47}$$

But using Eq.(9.38), we can simplify this expression:

$$\frac{\partial \psi^r}{\partial z^\mu}\,\frac{\partial z^\nu}{\partial \psi^r}\,\eta^{\lambda\mu} = \left(\delta^\nu_\mu - \frac{\partial \theta^a}{\partial z^\mu}\,\frac{\partial z^\nu}{\partial \theta^a}\right)\eta^{\lambda\mu} = \eta^{\lambda\nu}; \tag{9.48}$$

this yields the desired result:

$$\beta^\nu = A'_s\,\frac{\partial \psi^s}{\partial z^\lambda}\,\eta^{\lambda\nu} \equiv A_\mu \eta^{\mu\nu}. \tag{9.49}$$

We see that a singular GPB defined for functions of N variables z^μ can always be rewritten in such a way that it appears as a nonsingular GPB in a fewer number of variables (the latter number must be even). But we must point out that such a rewriting may not always be a good thing to do, because some symmetries of the original GPB may be obscured. A similar comment is in order in a slightly different connection. There is a theorem in the mathematical literature that shows that any GPB obeying the Jacobi identity can locally be brought into a standard form in which it appears exactly like the ordinary PB in a certain number of variables. That is to say, in each region in the phase space we can introduce local coordinates $\pi_r(z), \rho_s(z), \theta^a(z)$, all of which taken together form N independent functions of the z^μ, in terms of which the GPB assumes the "standard" form:

$$\{f, g\}^* = \sum_r \left(\frac{\partial f}{\partial \rho_r}\,\frac{\partial g}{\partial \pi_r} - \frac{\partial f}{\partial \pi_r}\,\frac{\partial g}{\partial \rho_r}\right). \tag{9.50}$$

Here we have assumed that each function is expressed in terms of the ρ's, π's, and θ's. The total number of ρ and π variables is just the rank of the matrix $\|\eta^{\mu\nu}\|$, and generally the transformation $z^\mu \to \rho_r,\ \pi_r,\ \theta^a$ is *not* a generalized canonical transformation. The θ's are an independent set of neutral elements; the ρ's and π's together form the ψ variables in Eqs.(9.40),(9.41). What is asserted here is that, given the transformation law Eq.(9.10) for η under an *arbitrary* change of coordinates, we can always find a new coordinate system in which η' assumes a particularly simple form; at least in each small region of phase space such a coordinate system can be found. Although one should always keep in mind this property of GPB's, one should not immediately conclude that the simplest way to deal with any GPB is to first put it into the standard form above; for example, one may lose sight of various symmetries in the problem if one does this. It is for this reason that we have chosen to derive the various properties of GPB's in their general form, and only at the end mention the possibility of transcribing it into a standard form.

Let us now proceed to a derivation of the Dirac bracket in this general framework, then discuss the transformations generated by it.

Structure of the Dirac Bracket.

We return now to the $2k$-dimensional phase space with canonical coordinates ω^μ, $\mu = 1, 2, \cdots, 2k$. Let there be given an even number of functions $\zeta_m(\omega)$ with a nonsingular matrix of PB's, and as in Eq.(9.1) let the inverse matrix be denoted by $\|C_{mm'}(\omega)\|$. Let us choose additional functions $\tau^a(\omega)$ that are independent among themselves and of the ζ's and sufficient in number so that the entire set ζ_m, τ^a consists of $2k$ independent functions. Every function $f(\omega)$ can also be written as a function of the ζ's and τ's. We can form two matrices, one made up of the PB's of the ζ's and τ's with one another, the other made up of their LB's (all these brackets are the standard ones evaluated with respect to the coordinates ω^μ). According to Eq.(5.27), these matrices are inverses to one another, apart from a sign. This may be expressed as follows;

$$\sum_{m'} \{\zeta_m, \zeta_{m'}\} L_{m'm''}(\zeta,\tau) + \sum_a \{\zeta_m, \tau^a\} L_{am''}(\zeta,\tau) = \delta_{mm''}, \tag{9.51a}$$

$$\sum_{m'} \{\zeta_m, \zeta_{m'}\} L_{m'b}(\zeta,\tau) + \sum_a \{\zeta_m, \tau^a\} L_{ab}(\zeta,\tau) = 0, \tag{9.51b}$$

$$\sum_{m'} \{\tau^a, \zeta_{m'}\} L_{m'm''}(\zeta,\tau) + \sum_b \{\tau^a, \tau^b\} L_{bm''}(\zeta,\tau) = 0, \tag{9.51c}$$

$$\sum_{m'} \{\tau^a, \zeta_{m'}\} L_{m'c}(\zeta,\tau) + \sum_b \{\tau^a, \tau^b\} L_{bc}(\zeta,\tau) = \delta_{ac}. \tag{9.51d}$$

Here the L's are the negatives of the various LB's:

$$L_{m'm''} = -[\zeta_{m'}, \zeta_{m''}], \quad L_{am} = -[\tau^a, \zeta_m] = -L_{ma},$$
$$L_{ab} = -[\tau^a, \tau^b]. \tag{9.52}$$

Now the LB's obey the identities given in Eq.(9.19). If we take the $2k$ functions A_j in Eq.(9.19) to be made up of the ζ's and τ's, then a subset of these identities involves differentiation with respect to the τ^a alone. These are:

$$\frac{\partial}{\partial \tau^a} L_{bc}(\zeta,\tau) + \frac{\partial}{\partial \tau^b} L_{ca}(\zeta,\tau) + \frac{\partial}{\partial \tau^c} L_{ab}(\zeta,\tau) = 0. \tag{9.53}$$

Compare this property of the L_{ab}'s with Eq.(9.17), the condition for the Jacobi identity to hold for a nonsingular GPB. It is immediately evident that if we set

$$\eta_{ab}(\zeta,\tau) = L_{ab}(\zeta,\tau), \tag{9.54}$$

and if $\eta_{ab}(\zeta,\tau)$ possesses an inverse $\eta^{ab}(\zeta,\tau)$

$$\sum_b \eta_{ab}(\zeta,\tau)\eta^{bc}(\zeta,\tau) = \delta_a^c, \tag{9.55}$$

then we can define a nonsingular GPB for functions of τ (with the ζ's suppressed) by

$$\{f,g\}^*(\zeta,\tau) = \sum_{ab} \eta^{ab}(\zeta,\tau)\, \frac{\partial f}{\partial \tau^a}\, \frac{\partial g}{\partial \tau^b}\,, \tag{9.56}$$

and this will automatically obey the Jacobi identity. The functions f, g are arbitrary functions of the ω^μ rewritten in terms of the ζ_m and τ^a; partial differentiation with respect to a τ^a is carried out keeping the other τ's and all the ζ's constant. Clearly,

$$\{f,\zeta_m\}^* = 0. \tag{9.57}$$

We now show that the GPB Eq.(9.56) is just the Dirac bracket, Eq.(9.2), of f with g defined relative to the given set ζ_m. For this, we must first find the matrix $\|\eta^{ab}\|$. We can show that the existence of $\|\eta^{ab}\|$ implies the existence of the matrix $\|C_{mm'}\|$ of Eq.(9.1) and vice versa. Given the existence of $\|C_{mm'}\|$, we use it in Eq.(9.51b) to solve for $L_{mb}(\zeta,\tau)$:

$$L_{mb}(\zeta,\tau) = -\sum_{m',a} C_{mm'}(\omega)\{\zeta_{m'},\tau^a\}\eta_{ab}(\zeta,\tau). \tag{9.58}$$

Substituting this in Eq.(9.51d) gives:

$$\sum_b \left[\{\tau^a,\tau^b\} - \sum_{mm'}\{\tau^a,\zeta_m\}C_{mm'}\{\zeta_{m'},\tau^b\}\right]\eta_{bc}(\zeta,\tau) = \delta_c^a. \tag{9.59}$$

This shows that η^{ab} exists and is given by:

$$\eta^{ab}(\zeta,\tau) = \{\tau^a,\tau^b\} - \sum_{mm'}\{\tau^a,\zeta_m\}C_{mm'}\{\zeta_{m'},\tau^b\}. \tag{9.60}$$

Conversely, it may be easily shown that if η^{ab} exists, then so does $C_{mm'}$. Using Eq.(9.60) in Eq.(9.56) we find:

$$\{f,g\}^* = \sum_b \left[\sum_a \left(\frac{\partial f}{\partial \tau^a}\{\tau^a,\tau^b\} - \sum_{mm'} \frac{\partial f}{\partial \tau_a}\{\tau^a,\zeta_m\}C_{mm'}(\zeta_{m'},\tau^b\}\right)\right]\frac{\partial g}{\partial \tau^b}.$$

It is easy to see that by the addition of terms that, in fact, add up to zero, we can rewrite this in the form:

$$\{f,g\}^* = \sum_b \left[\{f,\tau^b\} - \sum_{mm'}\{f,\zeta_m\}C_{mm'}\{\zeta_{m'},\tau^b\}\right]\frac{\partial g}{\partial \tau^b}.$$

Once again, by the addition of vanishing terms we get the final form:

$$\{f,g\}^* = \{f,g\} - \sum_{mm'} \{f,\zeta_m\} C_{mm'} \{\zeta_{m'},g\}. \qquad (9.2),$$

which is the Dirac bracket!

It is now obvious why the Jacobi identity holds for the Dirac bracket: it is because the Dirac bracket is a nonsingular GPB in the variables τ^a, constructed with the help of a matrix η^{ab} whose inverse η_{ab} obeys the condition in Eq.(9.17) as a consequence of standard properties of LB's. We also see that the Dirac bracket is an expression determined solely by the functions ζ_m, and does not depend on the choice of the additional functions τ^a that were originally used in Eq.(9.56) to define it. Although the Dirac bracket is a nonsingular GPB in the variables τ^a, it is a singular GPB in the original variables ω^μ, characterized by the matrix $\eta^{\mu\nu}$:

$$\eta^{\mu\nu}(\omega) = \epsilon^{\mu\nu} - \sum_{mm'} \epsilon^{\mu\rho} \frac{\partial \zeta_m}{\partial \omega^\rho} C_{mm'} \epsilon^{\sigma\nu} \frac{\partial \zeta_{m'}}{\partial \omega^\sigma} \qquad (9.61)$$

because, obviously,

$$\eta^{\mu\nu}(\omega) \frac{\partial \zeta_m}{\partial \omega^\nu} = 0. \qquad (9.62)$$

We take up next the nature of the generalized canonical transformations in the variables ω^μ corresponding to this singular GPB.

Dirac-Bracket Transformations

In this section, when we talk of a regular generalized canonical transformation in the phase space with coordinates ω^μ, the GPB we have in mind is the singular one given by the Dirac bracket. All the general results derived earlier in this chapter apply now as a special case. But the particularly interesting features are the roles played by the functions ζ_m and the relation of generalized canonical transformations to ordinary canonical transformations. This latter connection is relevant because the Dirac bracket is determined jointly by the ζ's and the PB structure.

Consider the generalized (regular) canonical transformation resulting from the differential equations

$$\begin{aligned}\frac{\partial \psi^\mu(\omega_0;\epsilon)}{\partial \epsilon} &= \{\psi^\mu, \phi(\psi,\epsilon)\}^* \\ &= \{\psi^\mu, \phi(\psi,\epsilon)\} - \sum_{mm'} \{\psi^\mu, \zeta_m(\psi)\} C_{mm'}(\psi) \{\zeta_{m'}(\psi), \phi(\psi,\epsilon)\},\end{aligned} \qquad (9.63)$$

with the usual boundary conditions at $\epsilon = 0$. (In the last line all the PB's are evaluated with respect to the ψ's). The transformation preserves all Dirac bracket relationships. We verify that the functions ζ_m do not change their functional forms in this sense:

$$\begin{aligned}\frac{\partial \zeta_m(\psi(\omega_0;\epsilon))}{\partial \epsilon} &= \frac{\partial \zeta_m}{\partial \psi^\mu}\{\psi^\mu, \phi(\psi, \epsilon)\}^* \\ &= \{\zeta_m(\psi), \phi(\psi,\epsilon)\}^* = 0;\end{aligned}$$

that is,

$$\zeta_m(\psi(\omega_0;\epsilon)) = \zeta_m(\omega_0). \tag{9.64}$$

This means that if we view the generalized canonical transformation $\omega_0 \to \psi(\omega_0;\epsilon)$, for a fixed ϵ, as a mapping of phase space onto itself, the values of the ζ's at the image point are always equal to their values at the source points; that is, surfaces of constant ζ_m are preserved by the transformation. This is, of course, to be expected, because the Dirac bracket was designed to give zero when one of its arguments was a ζ_m.

That the groups of generalized and ordinary (regular) canonical transformations do in general possess common elements can be seen by choosing in the differential equation (Eq.(9.63)), a function ϕ that has zero PB with the ζ_m's:

$$\{\zeta_m(\psi), \phi(\psi,\epsilon)\} = 0, \tag{9.65}$$

for then the Dirac bracket of ϕ with any other function f is the same as the PB of ϕ with f. In fact, this displays the following connection between the two kinds of brackets:

$$\{g, \zeta_m\} = 0, \qquad \text{all } m \Rightarrow \{f, g\}^* = \{f, g\}, \quad \text{all } f. \tag{9.66}$$

For such a function $\phi(\psi,\epsilon)$, the solution to Eq.(9.63) is the same as the solution to

$$\frac{\partial \psi^\mu(\omega_0;\epsilon)}{\partial \epsilon} = \{\psi^\mu, \phi(\psi,\epsilon)\}, \tag{9.67}$$

and the resulting transformation $\omega_0 \to \pi(\omega_0;\epsilon)$, for each ϵ, is both a generalized and an ordinary canonical transformation. Of course for this result to hold, Eq.(9.65) must hold in *the entire phase space.*

What happens if the generator does not obey Eq.(9.65) everywhere but only on the hypersurface U in phase space defined by the equations $\zeta_m = 0$? We can in this case prove the following result: given the generalized canonical transformation generated by the function $\phi(\psi,\epsilon)$ via the differential equation Eq.(9.63), there exists a function $\bar{\phi}(\psi,\epsilon)$ that generates an ordinary canonical transformation such that the two transformations coincide in their effects on points of the hypersurface U, although in general their effects on points not on U are different. This agreement for points on U is valid for each value of ϵ. The function $\phi(\psi,\epsilon)$ is arbitrary; that is, its PB's with the $\zeta_m(\psi)$ are not required to vanish anywhere, not even on U, but $\bar{\phi}$ is constructed from ϕ in such a way that *its* PB's with the ζ_m's vanish on U:

$$\bar{\phi}(\psi,\epsilon) = \phi(\psi,\epsilon) - \sum_{m,m'} \zeta_m(\psi) C_{mm'}(\psi)\{\zeta_{m'}(\psi), \phi(\psi,\epsilon)\},$$

$$\{\zeta_{m''}(\psi), \bar{\phi}(\psi,\epsilon)\} = -\sum_{m,m'} \zeta_m(\psi)\{\zeta_{m''}(\psi), C_{mm'}(\psi)\{\zeta_{m'}(\psi), \phi(\psi,\epsilon)\}\}. \tag{9.68}$$

Notice also that $\bar{\phi}$ and ϕ agree numerically on the hypersurface U. We prove this result first for the case when ϕ, and thus $\bar{\phi}$, does not depend explicitly on ϵ. Because the same general point ω_0 is mapped into two distinct points by the two transformations, we denote the two image points by $\psi(\omega_0;\epsilon)$ and $\xi(\omega_0;\epsilon)$; that is, $\psi(\omega_0;\epsilon)$ is the solution to

$$\frac{\partial\psi^\mu(\omega_0;\epsilon)}{\partial\epsilon} = \{\psi^\mu, \phi(\psi)\}^*, \qquad \psi^\mu(\omega_0;0) = \omega_0^\mu; \tag{9.69}$$

whereas $\xi(\omega_0;\epsilon)$ is the solution to

$$\frac{\partial\xi^\mu(\omega_0;\epsilon)}{\partial\epsilon} = \{\xi^\mu, \bar{\phi}(\xi)\}, \qquad \xi^\mu(\omega_0;0) = \omega_0^\mu\,. \tag{9.70}$$

We also use the following identity: for any function $f(\omega_0)$,

$$\begin{aligned}\{f(\omega_0), \bar{\phi}(\omega_0)\} = \{f(\omega_0), \phi(\omega_0)\}^* - \sum_{mm'} \zeta_m(\omega_0)\\ \cdot \{f(\omega_0), C_{mm'}(\omega_0)\{\zeta_{m'}(\omega_0), \phi(\omega_0)\}\}.\end{aligned} \tag{9.71}$$

The generalized canonical transformation Eq.(9.69) can be given as the power series (see Eq.(9.29))

$$\begin{aligned}\psi^\mu(\omega_0;\epsilon) = \sum_{n=0}^{\infty} \frac{\epsilon^n}{n!} A_n^\mu(\omega_0),\\ A_0^\mu(\omega_0) = \omega_0^\mu,\ A_{n+1}^\mu(\omega_0) = \{A_n^\mu(\omega_0), \phi(\omega_0)\}^*;\end{aligned} \tag{9.72}$$

correspondingly, for the ordinary canonical transformation Eq.(9.70) we have

$$\begin{aligned}\xi^\mu(\omega_0;\epsilon) = \sum_{n=0}^{\infty} \frac{\epsilon^n}{n!} B_n^\mu(\omega_0),\\ B_0^\mu(\omega_0) = \omega_0^\mu,\ B_{n+1}^\mu(\omega_0) = \{B_n^\mu(\omega_0), \bar{\phi}(\omega_0)\}.\end{aligned} \tag{9.73}$$

We now show that for each n, the difference $B_n^\mu(\omega_0) - A_n^\mu(\omega_0)$ is a linear combination of the $\zeta_m(\omega_0)$. This is obviously true for $n = 0$, because $B_0^\mu(\omega_0) = A_0^\mu(\omega) = \omega_0^\mu$. Assuming that it is true for $B_n^\mu - A_n^\mu$,

$$B_n^\mu(\omega_0) = A_n^\mu(\omega_0) + \sum_n \zeta_m(\omega_0) C_{m,n}^\mu(\omega_0), \tag{9.74}$$

we calculate $B^{\mu}_{n+1}(\omega_0)$ using Eqs.(9.73) and (9.71):

$$\begin{aligned}B^{\mu}_{n+1}(\omega_0) &= \left\{ A^{\mu}_n(\omega_0) + \sum_m \zeta_m(\omega_0) C^{\mu}_{m,n}(\omega_0), \bar{\phi}(\omega_0) \right\} \\ &= \{A^{\mu}_n(\omega_0), \phi(\omega_0)\}^* + \sum_m \zeta_m(\omega_0)\{C^{\mu}_{m,n}(\omega_0), \phi(\omega_0)\}^* \\ &\quad - \sum_{mm'} \zeta_m(\omega_0)\{B^{\mu}_n(\omega_0), C_{mm'}(\omega_0)\{\zeta_{m'}(\omega_0), \phi(\omega_0)\}\} \\ &= A^{\mu}_{n+1}(\omega_0) + \sum_m \zeta_m(\omega_0) C^{\mu}_{m,n+1}(\omega_0), \\ C^{\mu}_{m,n+1}(\omega_0) &= \{C^{\mu}_{m,n}(\omega_0), \phi(\omega_0)\}^* \\ &\quad - \sum_{m'} \{B^{\mu}_n(\omega_0), C_{mm'}(\omega_o)\{\zeta_{m'}(\omega_0), \phi(\omega_0)\}\}. \end{aligned} \tag{9.75}$$

Therefore, the proof follows by induction, and one then has:

$$\xi^{\mu}(\omega_0;\epsilon) = \psi^{\mu}(\omega_0;\epsilon) + \sum_m \zeta_m(\omega_0) \sum_{n=1}^{\infty} \frac{\epsilon^n}{n!} C^{\mu}_{m,n}(\omega_0). \tag{9.76}$$

Restricting ω_0 to U means setting $\zeta_m = 0$; thus we get;

$$\xi^{\mu}(\omega_0;\epsilon)|_{\omega_0 \text{ on } U} = \psi^{\mu}(\omega_0;\epsilon)|_{\omega_0 \text{ on } U} \tag{9.77}$$

which is the stated result.

Thus to every generalized canonical transformation there corresponds an associated ordinary canonical transformation that agrees with the former on the constrained hypersurface U. In general, of course, the associated ordinary canonical transformation does not preserve the form of the functions ζ_m, that is, in general,

$$\zeta_m(\xi(\omega_0;\epsilon)) \neq \zeta_m(\omega_0);$$

thus except for the $\zeta_m = 0$ hypersurface, the transformation $\omega_0 \to \xi(\omega_0;\epsilon)$ does not leave the hypersurfaces with constant ζ_m invariant.

For the case that ϕ depends explicitly on ϵ, it is again true that for each ϵ, the solutions to

$$\frac{\partial \psi^{\mu}(\omega_0;\epsilon)}{\partial \epsilon} = \{\psi^{\mu}, \phi(\psi,\epsilon)\}^* \tag{9.78}$$

and

$$\frac{\partial \xi^{\mu}(\omega_0;\epsilon)}{\partial \epsilon} = \{\xi^{\mu}, \bar{\phi}(\xi,\epsilon)\} \tag{9.79}$$

coincide when ω_0 lies on U. The general idea of the proof is similar to the way in which we proved that Eq.(9.23) gives rise to a generalized canonical transformation: we consider the partial derivatives $(\partial^n \psi^{\mu}/\partial \epsilon^n)$ and $(\partial^n \xi^{\mu}/\partial \epsilon^n)$ at $\epsilon = 0$, and show that the difference vanishes if ω_0 lies on U. Because the algebra is somewhat more complicated than before, we just give the outline of the proof. First we consider the effect of the transformation $\omega_0 \to \xi$ on the functions ζ_m. The functions $\zeta_m(\xi)$ depend on ϵ through ξ; for the partial derivatives with respect to ϵ we have:

$$\frac{\partial \zeta_m(\xi(\omega_0;\epsilon))}{\partial \epsilon} = \frac{\partial \zeta_m(\xi)}{\partial \xi^\mu}\{\xi^\mu, \bar{\phi}(\xi,\epsilon)\} = \{\zeta_m(\xi), \bar{\phi}(\xi,\epsilon)\}$$
$$= -\sum_{m',m''} \zeta_{m'}(\xi)\{\zeta_m(\xi), C_{m'm''}(\xi)\{\zeta_{m''}(\xi), \phi(\xi,\epsilon)\}\};$$

that is, we have a general result of the form:

$$\frac{\partial \zeta_m(\xi)}{\partial \epsilon} = \sum_{m'} \zeta_{m'}(\xi) d_{m'm}(\xi,\epsilon).$$

In deriving this result, we used Eq.(9.71). But this shows that each derivative $(\partial^n \zeta_m(\xi)/\partial \epsilon^n)$ is a linear combination of all the derivatives of lower order. Because $\zeta_m(\xi)$ vanishes at $\epsilon = 0$ if ω_0 is restricted to U, we conclude by a recursive process that

$$\frac{\partial^n \zeta_m(\xi)}{\partial \epsilon^n}\Big|_{\epsilon=0}\Big|_{\omega_0 \text{ on } U} = 0, \tag{9.80a}$$

$$\zeta_m(\xi(\omega_0;\epsilon))|_{\omega_0 \text{ on } U} = \zeta_m(\omega_0) = 0. \tag{9.80b}$$

Thus the ordinary canonical transformation $\omega_0 \to \xi(\omega_0;\epsilon)$ does leave the hypersurface U invariant. We use Eq.(9.80a) in the sequel. Returning to Eqs.((9.78). (9.79)) the computation of the partial derivatives of ψ and ξ with respect to ϵ involves the derivatives of ϕ with respect to ϵ. Now, ϕ has both an explicit dependence on ϵ and an implicit one via ψ or ξ as the case may be. We denote the explicit and the total derivatives of ϕ with respect to ϵ by $\partial/\partial\epsilon$ and $d/d\epsilon$, respectively. (Of course, both are taken at constant values of ω_0.) In the case of the equation for ψ, we need the following general expression:

$$\frac{d}{d\epsilon}\frac{\partial^n \phi}{\partial \epsilon^n} = \frac{\partial^{n+1}\phi}{\partial \epsilon^{n+1}} + \left\{\frac{\partial^n \phi}{\partial \epsilon^n}, \phi\right\}^*, \qquad n = 0, 1, 2, \cdots; \tag{9.81}$$

here ϕ denotes $\phi(\psi,\epsilon)$. Rewriting the determining equation for ξ^μ as

$$\frac{\partial \xi^\mu}{\partial \epsilon} = \{\xi^\mu, \phi(\xi,\epsilon)\}^* - \sum_{mm'} \zeta_m(\xi)\{\xi^\mu, C_{mm'}(\xi)\{\zeta_{m'}(\xi), \phi(\xi,\epsilon)\}\} \tag{9.82}$$

with the help of (9.71), the general expression needed in this case is

$$\frac{d}{d\epsilon}\frac{\partial^n \phi}{\partial \epsilon^n} = \frac{\partial^{n+1}\phi}{\partial \epsilon^{n+1}} + \left\{\frac{\partial^n \phi}{\partial \epsilon^n}, \phi\right\}^*$$
$$- \sum_{mm'} \zeta_m(\xi)\left\{\frac{\partial^n \phi}{\partial \epsilon^n}, C_{mm'}(\xi)\{\zeta_{m'}(\xi), \phi(\xi,\epsilon)\}\right\}, \tag{9.83}$$

with ϕ now denoting $\phi(\xi, \epsilon)$. Thus we see that Eq.(9.83) arises from Eq.(9.81) by the replacement $\psi \to \xi$ everywhere, together with the addition of terms linear in $\zeta_m(\xi)$. The important point is that in Eq.(9.83) no terms involving any of the partial derivatives $(\partial^r \zeta_m(\xi)/\partial \epsilon^r)$, have appeared. Now, the general expression for $(\partial^n \psi^\mu / \partial \epsilon^n)$ consists, after the use of Eq.(9.81), of a sum of terms each of which is a multiple Dirac bracket; one of the arguments in these multiple brackets is a ψ^μ, whereas the rest are all various partial derivatives ϕ, $(\partial\phi/\partial\epsilon)$, $(\partial^2\phi/\partial\epsilon^2)$, and so on, and all the brackets are computed with respect to ψ. For the first few derivatives we get:

$$\frac{\partial \psi^\mu}{\partial \epsilon} = \{\psi^\mu, \phi\}^*; \frac{\partial^2 \psi^\mu}{\partial \epsilon^2} = \{\{\psi^\mu, \phi\}^*, \phi\}^* + \left\{\psi^\mu, \frac{\partial \phi}{\partial \epsilon}\right\}^*;$$

$$\begin{aligned} \frac{\partial^3 \psi^\mu}{\partial \epsilon^3} &= \{\{\{\psi^\mu, \phi\}^*, \phi\}^*, \phi\}^* + \left\{\left\{\psi^\mu, \frac{\partial \phi}{\partial \epsilon}\right\}^*, \phi\right\}^* \\ &+ \left\{\{\psi^\mu, \phi\}^*, \frac{\partial \phi}{\partial \epsilon}\right\}^* + \left\{\{\psi^\mu, \phi\}^*, \frac{\partial \phi}{\partial \epsilon}\right\}^* \\ &+ \left\{\psi^\mu, \frac{\partial^2 \phi}{\partial \epsilon^2}\right\}^* + \left\{\psi^\mu, \left\{\frac{\partial \phi}{\partial \epsilon}, \phi\right\}^*\right\}^*. \end{aligned} \tag{9.84}$$

(ϕ is here $\phi(\psi, \epsilon)$). Using the facts that Eq.(9.83) does not contain partial derivatives of ζ_m and that the Dirac brackets of ζ_m with all other functions vanish, we find that the general expression for $(\partial^n \xi^\mu / \partial \epsilon^n)$ is obtained from that for $(\partial^n \psi^\mu / \partial \epsilon^n)$ by the replacement everywhere of ψ by ξ, and then the addition of terms linear in

$$\zeta_m(\xi), \quad \frac{\partial \zeta_m(\xi)}{\partial \epsilon}, \cdots, \frac{\partial^{n-1} \zeta_m(\xi)}{\partial \epsilon^{n-1}}.$$

(The point is that one never has to compute the Dirac bracket of

$$\frac{\partial^r \zeta_m(\xi)}{\partial \epsilon^r}$$

with any other function; these need not vanish.) This statement is obviously true for $n = 0, 1$ and an explicit calculation for $n = 2, 3$ will convince the reader of the truth of the statement for all n. Because at $\epsilon = 0$ it does not matter whether ψ or ξ appears in the terms of the type of Eq.(9.84), we need only use the previously derived Eq.(9.80) to conclude that:

$$\begin{gathered} \frac{\partial^n \xi^\mu(\omega_0; \epsilon)}{\partial \epsilon^n} |_{\epsilon=0}|_{\omega_0 \text{ on } U} = \frac{\partial^n \psi^\mu(\omega_0; \epsilon)}{\partial \epsilon^n} |_{\epsilon=0}|_{\omega_0 \text{ on } U} \cdot \\ \xi^\mu(\omega_0; \epsilon)|_{\omega_0 \text{ on } U} = \psi^\mu(\omega_0; \epsilon)|_{\omega_0 \text{ on } U} \cdot \end{gathered} \tag{9.85}$$

Thus we see in general, that the two transformations $\omega_0 \to \psi(\omega_o; \epsilon)$, $\omega_0 \to \xi(\omega_0; \epsilon)$, the first a generalized and the second an ordinary canonical transformation, coincide in their action on the constrained hypersurface U, even if the generator ϕ and the associated $\bar{\phi}$ depend explicitly on ϵ.

We can use these properties of the Dirac bracket transformations now to give a group theoretical description of the solution of the Hamiltonian equation of motion for constrained systems. For the moment let us suppose that there are no first-class constraints (either primary or secondary); then the final form of the general equation of motion is:

$$\frac{d}{dt}\, g(g,p) = \{g, H\}^*, \tag{9.86}$$

which is just Eq.(8.99). The motion is determined by the Hamiltonian H (which has no arbitrariness in its time dependence) and is confined to the constrained hypersurface U defined in phase space by the collection of second-class constraints $\zeta_m = 0$. Thus the dynamical mapping $\omega_0^\mu \to \omega^\mu(t)$ occurring in the time interval 0 to t can be viewed either as a generalized canonical transformation acting on the surface U and generated in the familiar way by the Hamiltonian H via the equation of motion Eq.(9.86), or, equivalently, as the restriction to U of an *ordinary* canonical transformation generated by the associated Hamiltonian:

$$\bar{H} = H - \sum_{mm'} \zeta_m\, C_{mm'} \{\zeta_{m'}, H\}. \tag{9.87}$$

If H is not explicitly time dependent, both form one-parameter groups of canonical transformations, one generalized, the other ordinary. (This assumes, of course, that the functions ζ_m have no explicit time dependence!) As far as the interpretation of the Hamiltonian as the energy of the system is concerned, we get the same numerical values whether we use H or $\bar{H}$ because physically the q's and p's are at all times restricted to U. It is only when we use these two Hamiltonians in the analytical process of generating the two kinds of transformations that we must keep track of their differences. The possibility of using $\bar{H}$ and not using the Dirac bracket is encountered already in the equation of motion in the form given in Eq.(8.96), because it could have been rewritten in the form:

$$\frac{dg}{dt} \approx \{g, \bar{H}\}. \tag{9.88}$$

However, if the equations $\zeta_m = 0$ are to be used to actually eliminate variables, then we must use the Dirac brackets and picture the motion as the "gradual unfolding" of a generalized canonical transformation.

If we now go back and consider the general case in which first-class constraints are present, the above description of the motion remains valid, with two additional features: first, the constrained hypersurface U is defined by the first-class constraint functions ϕ_α, χ_A also vanishing; second, the Hamiltonian has a part, $\sum_\alpha v_\alpha(t)\phi_\alpha$, which has an arbitrary time dependence. This uncertainty in the Hamiltonian does not mean that there is any uncertainty in the numerical value of the energy at each instant of time, because during the motion the ϕ_α's vanish; however, it does mean that the generalized canonical transformation describing the motion for a finite time interval t is arbitrary to a certain extent. This is because the ϕ_α's being first class only ensures the vanishing of their PB's *with the* ζ_m (and themselves) on U, but not the vanishing of their PB's with *all* the q's and p's. This point is made clear by writing out the changes in the q's and p's over a small interval of time δt, starting from Eq.(8.99):

$$\delta q_s \approx \{q_s, H\}^* \delta t + \sum_\alpha \{q_s, \phi_\alpha\} v_\alpha(t) \delta t,$$

$$\delta p_s \approx \{p_s, H\}^* \delta t + \sum_\alpha \{p_s, \phi_\alpha\} v_\alpha(t) \delta t.$$

We see that these small changes in the coordinates consists of a part that has no arbitrariness in it and is simply the effect of an infinitesimal generalized canonical transformation generated by H, and of another part that can be interpreted either as a generalized or as an ordinary infinitesimal canonical transformation produced by the ϕ_s's but of an arbitrary amount. The presence of first-class constraints has the following general effect: the numerical values of an observable quantity do not determine any unique way of writing it as a function of the basic dynamical variables q, p because one can always add a linear combination of the ϕ_α's and χ_A's without changing the value; however, the generalized canonical transformation generated on the surface U by a given function of the q's and p's changes if we add a linear combination of the ϕ_α's and χ_A's to the function. Thus one weakens the correspondence between physical observables and the transformations that they generate (and we might state that, as we see later on in this book, at least for some physical observables this is a very useful connection). This is the characteristic feature of first-class constraints. In practical cases where first-class constraints appear, for example, in the theories of electromagnetism and general relativity, it has happened that the first-class functions ϕ_α, χ_A generate (generalized or ordinary) canonical transformations that are *physically* unobservable. However, going back to the foregoing expressions for δq_s and δp_s, this clearly means that two distinct points in the surface U do not correspond to two distinct physical situations if they are mapped into one another by a transformation generated by the ϕ_α's and χ_A's and therefore the terms in δq_s and δp_s proportional to the v_α's represent *unobservable changes* in the q_s and p_s. One must then reformulate the theory in such a way that equivalence classes of points on U correspond to physical states in a one-to-one way, and one must also suitably modify the way in which an observable generates a transformation on these classes. For further developments along these directions we must refer the reader to the original papers by Dirac and others in the literature.

Structure of the Group of Dirac Bracket Transformations

There are two further mathematically interesting points regarding the Dirac bracket that permit a discussion in a general way.

The first concerns the structure of the group of Dirac bracket transformations as a whole. We have mentioned earlier the fact that with a suitable choice of coordinates, locally, any GPB can be brought into the standard form of ordinary PB's. In deriving the Dirac bracket, we first choose additional independent functions τ^a to supplement the given second-class constraint functions ζ_m, and then set up a GPB that is nonsingular in the variables τ^a. The general mathematical theorem then implies that the set of functions $\tau^a(\omega)$ can always be chosen in such a way that it is made up of two sets of functions, ρ_r and Π_r, say, equal in number, in terms of which the Dirac bracket has the form:

$$\{f(\rho,\Pi,\zeta)g(\rho,\Pi,\zeta)\}^* = \sum_{r=1}^{M}\left(\frac{\partial f}{\partial\rho_r}\frac{\partial g}{\partial\Pi_r} - \frac{\partial f}{\partial\Pi_r}\frac{\partial g}{\partial\rho_r}\right). \tag{9.89}$$

(We have supposed that the number of ζ's is $2(k-M)$). At any rate, such a form can be achieved in small regions of the original phase space. Stated in another way, it means that given $2(k-M)$ independent functions $\zeta_m(\omega)$ obeying the second-class condition, namely $\det|\{\zeta_m,\zeta_{m'}\}| \neq 0$, we can find $2M$ additional independent functions $\rho_r(\omega)$, $\Pi_r(\omega)$ so that we have specially simple values for some of the LB's between the functions of the independent set ζ_m,ρ_r,Π_r:

$$[\rho_r,\rho_{r'}]_\omega = [\Pi_r,\Pi_{r'}]_\omega = 0, \qquad [\rho_r,\Pi_{r'}]_\omega = \delta_{rr'};\ r,r' = 1,2,\cdots,M. \tag{9.90}$$

If we can choose the ρ's and Π's for the entire phase space, we see that the full group of Dirac bracket transformations has exactly the same structure as the full group of ordinary canonical transformations in a $2M$-dimensional phase space. Of course, remember that, in general, the group of Dirac bracket transformations is *not a subgroup* of the full canonical group in the original $2k$-dimensional phase space because the variables ζ_m,ρ_r,Π_r do not generally form a canonical system of coordinates. If the ρ's and Π's cannot be chosen properly for the entire phase space, we can only conclude that for the elements close to the identity, the structures of the group of Dirac bracket transformations and of the $2M$-dimensional canonical group coincide. These general arguments should not be taken to imply that rewriting things in terms of ρ's and Π's is either easy or desirable in practice!

The properties just mentioned lead us to the second point: under what conditions does the group of Dirac bracket transformations form a subgroup of the canonical group in the full $2k$-dimensional phase space? There is one particularly simple case when it does, when the ζ_m are a subset of the original p's and q's:

$$\zeta_m \equiv q_{M+1},q_{M+2},\cdots,q_k,p_{M+1},p_{M+2},\cdots,p_k. \tag{9.91}$$

Then we easily find:

$$\{f,g\}^* = \sum_{r=1}^{M}\left(\frac{\partial f}{\partial q_r}\frac{\partial g}{\partial p_r} - \frac{\partial f}{\partial p_r}\frac{\partial g}{\partial q_r}\right); \tag{9.92}$$

we may describe the Dirac bracket in such circumstances as a restricted form of the PB, in which the canonical variables $q_{M+1}\cdots q_k$, $p_{M+1}\cdots p_k$ have been "frozen." In this case a transformation generated by a function $\phi(q_1\cdots q_M, p_1\cdots p_M)$ via the Dirac bracket,

$$f(q,p) \to \bar{f}(q,p) = f(q,p) + \{f,\phi\}^* + \frac{1}{2!}\{\{f,\phi\}^*,\phi\}^* + \cdots$$

is also an ordinary canonical one in which $q_1 \cdots q_M$, $p_1 \cdots p_M$ transform among themselves, whereas the rest of the q's and p's remain unchanged. (It should, however, be noted that if ϕ also depends on the "frozen" variables, then the Dirac bracket transformation will not be canonical in the full set of q's and p's). Even if the ζ's are not as simple as in Eq.(9.91) we know that the Dirac bracket is unchanged if we replace the ζ's by independent functions of themselves; thus in some cases it may be possible to find functions ζ'_m of the ζ_m that are in fact a subset of $(k - M)$ canonical pairs of q's and p's Precisely when this can be done is made clear by the following: the necessary and sufficient condition that the Dirac bracket determined by the $2(k - M)$ ζ's corresponds to 'freezing' $(k - M)$ pairs of canonical variables in the ordinary PB is that the functions ζ_m form a *function group* of rank $2(k - M)$. (A set of functions ζ_m forms a function group if the PB's of the ζ's with each other can be expressed as functions of the ζ's alone; the rank of the function group is the number of independent functions in the group.) The necessity of the condition is obvious, because a set of $(k - M)$ canonical pairs of q's and p's does form a function group of rank $2(k - M)$. On the other hand, it can be shown that in a function group one can choose a set of independent functions consisting of Q's and P's with PB's equal to zero or ± 1, and the second-class condition on the ζ_m ensures that we will have equal numbers of Q's and *corresponding* P's so that the condition is also sufficient.

We conclude our study of the Dirac bracket by considering a question suggested by the fact that this bracket can be put into the standard form of Eq.(9.89). Given the independent ζ's, suppose we adjoin *any* $2M$ additional independent functions ρ'_r, Π'_r and define the bracket:

$$\{f, g\}^{\#} = \sum_{r=1}^{M} \left(\frac{\partial f}{\partial \rho'_r} \frac{\partial g}{\partial \Pi'_r} - \frac{\partial f}{\partial \Pi'_r} \frac{\partial g}{\partial \rho'_r} \right).$$

The Jacobi identity is certainly obeyed, and the ζ's are treated as neutrals:

$$\{f, \zeta_m\}^{\#} = 0.$$

What is then special about the Dirac bracket? The answer is, of course, clear from all the preceding work. First, even though the Dirac bracket is a special case of this new bracket, with the ρ'_r, Π'_r taken to be the ρ_r, Π_r of Eq.(9.89), *the Dirac bracket is actually determined by the ζ's alone*, whereas the bracket $\{,\}^{\#}$ is not. Second, the Dirac bracket has many special properties relating it to the PB in many ways, for example, Eq.(9.66), and this is not so for the bracket $\{,\}^{\#}$. This comment only shows us again the close connections between Dirac and Poisson brackets.

Theory of Lagrangians Linear in the Velocities

The theory we have developed for the GPB finds an immediate and interesting application in the treatment of systems whose Lagrangians have the specially simple form:

$$L(q_s, \dot{q}_s) = \sum_{s=1}^{k} A_s(q)\dot{q}_s - V(q). \tag{9.93}$$

(For simplicity we assume that L has no explicit time dependence.) Such Lagrangians are nonstandard in the usual terminology; in fact the Lagrangian equations of motion, which are

$$\sum_{r=1}^{k} \eta_{sr}(q)\dot{q}_r = \frac{\partial V}{\partial q_s}, \qquad s = 1, \cdots, k,$$

$$\eta_{sr}(q) = \frac{\partial A_r}{\partial q_s} - \frac{\partial A_s}{\partial q_r}, \tag{9.94}$$

do not involve the accelerations at all! Although the general procedure for dealing with nonstandard Lagrangians, explained in detail in the previous chapter, is certainly available for the present case, it would naturally be better to treat this rather simple situation on its own, without any equations for the accelerations being introduced at any stage. Such a treatment is immediately available. Let us suppose that the matrix $\|\eta_{rs}\|$ is nonsingular and has the inverse $\|\eta^{rs}\|$. Then comparison of Eq.(9.94) with Eqs. (9.17) and (9.18) shows that the equations of motion written as:

$$\dot{q}_s = \sum_{r=1}^{k} \eta^{sr}(q)\frac{\partial V}{\partial q_r}$$

are already expressed in terms of a nonsingular GPB in the variables q_s:

$$\dot{q}_s = \{q_s, V\}^*, \qquad \{f(q), g(q)\}^* = \sum_{r,s} \eta^{rs}(q)\frac{\partial f}{\partial q_r}\frac{\partial g}{\partial q_s}. \tag{9.95}$$

It is as though the GPB was meant for the treatment of such Lagrangians! The "potential" $V(q)$ acts as the Hamiltonian, and the Jacobi identity is valid for the GPB. There are no constraints in the problem, and the motion in the k-dimensional "state-space" with coordinates $q_1 \cdots q_k$ is a one-parameter group of generalized canonical transformations.

If the matrix $\|\eta_{rs}\|$ is singular, we encounter a problem with constraints. However, we do not go into the detailed treatment of this situation here, because we have discussed the problem of handling constraints sufficiently to leave it to the reader to supply the details in the present case. We must only mention that when the constraint analysis is fully worked out, one again has a GPB form for the equations of motion, and a generalized canonical description of the motion itself.

The Poisson Bracket in Lagrangian form

The way in which the general theory is developed in the previous chapters for standard Lagrangians has been such that a group theoretical interpretation of the motion emerged only in the Hamiltonian framework in phase space. We now have at hand the formalism appropriate to the demonstration of the fact that a group theoretical analysis via GPB's exists in the Lagrangian framework also. This is achieved by just *rewriting* the ordinary PB in terms of the generalized positions and velocities $q_s, \dot{q}_s$ of the Lagrangian theory. The processes of partial differentiation with respect to the q's and p's of phase space are replaced by differentiation with respect to the q's and $\dot{q}$'s. The Lagrangian being of standard type, the matrix $\|W_{rs}\|$ of Eq.(8.1), possesses an inverse, which we write as $\|\Lambda_{rs}\|$:

$$W_{rs}(q,\dot{q}) = \frac{\partial^2 L}{\partial \dot{q}_r \partial \dot{q}_s}, \qquad \sum_r \Lambda_{sr}(q,\dot{q}) W_{rs'}(q,\dot{q}) = \delta_{ss'}. \tag{9.96}$$

Starting with the definition of the p_s,

$$p_s = \frac{\partial L}{\partial \dot{q}_s},$$

and differentiating both sides with respect to a p_r, keeping the other p's and all the q's fixed, yields

$$\delta_{rs} = \sum_u \frac{\partial^2 L}{\partial \dot{q}_s \partial \dot{q}_u} \frac{\partial \dot{q}_u}{\partial p_r} = \sum_u W_{su} \frac{\partial \dot{q}_u}{\partial p_r};$$

that is,

$$\frac{\partial \dot{q}_u}{\partial p_r} - \Lambda_{ur}(q,\dot{q}). \tag{9.97}$$

This tells us how to transcribe the operation $\partial/\partial p_s$ from phase space to configuration-plus-velocity space:

$$\frac{\partial}{\partial p_s} \longrightarrow \sum_r \frac{\partial \dot{q}_r}{\partial p_s} \frac{\partial}{\partial \dot{q}_r} = \sum_r \Lambda_{sr}(q,\dot{q}) \frac{\partial}{\partial \dot{q}_r}. \tag{9.98}$$

If we differentiate the relation defining p_s, with respect to a q_r, keeping the other q's and all the p's fixed, we get

$$\sum_u \frac{\partial^2 L}{\partial \dot{q}_s \partial \dot{q}_u} \frac{\partial \dot{q}_u}{\partial q_r} + \frac{\partial^2 L}{\partial \dot{q}_s \partial q_r} = 0.$$

that is,

$$\frac{\partial \dot{q}_u}{\partial q_r} = -\sum_s \Lambda_{us}(q,\dot{q}) \frac{\partial^2 L}{\partial \dot{q}_s \partial q_r}. \tag{9.99}$$

(What is being done in the left-hand sides of Eqs.(9.97) and (9.99) is to differentiate the phase space expression of the $\dot{q}$'s in terms of q's and p's with respect to the q's and p's, and the right-hand sides give the results back in terms of q's and $\dot{q}$'s). Thus for $\partial/\partial q_s$ we have the rule:

$$\frac{\partial}{\partial q_s} \rightarrow \frac{\partial}{\partial q_s} - \sum_{u,r} \Lambda_{ur}(q,\dot{q}) \frac{\partial^2 L}{\partial \dot{q}_u \partial q_s} \frac{\partial}{\partial \dot{q}_r}. \tag{9.100}$$

Now, given any two phase-space functions f, g, we write them in terms of $q_s, \dot{q}_s$, and their PB in phase space becomes the following bracket in Lagrangian variables:

$$\{f,g\}^* = \sum_s \left\{ \left(\frac{\partial f}{\partial q_s} - \sum_{u,r} \Lambda_{ur} \frac{\partial^2 L}{\partial \dot{q}_u \partial q_s} \frac{\partial f}{\partial \dot{q}_r} \right) \sum_v \Lambda_{sv} \frac{\partial g}{\partial \dot{q}_v} \right\} - (f \leftrightarrow g). \tag{9.101}$$

We have put an asterisk here because although it is numerically equal to the PB of phase space, it is a GPB in the q's and $\dot{q}$'s. Let us write ξ_A collectively for the q's and $\dot{q}$'s: the values of A will be denoted by $r = 1, \cdots, k$, and $\dot{r} = 1, \cdots, k$:

$$\xi_r = q_r, \qquad \xi_{\dot{r}} = \dot{q}_r.$$

Then the GPB takes the form:

$$\begin{aligned} \{f, g\}^* &= \sum_{A,B} \eta^{AB}(q, \dot{q}) \frac{\partial f}{\partial \xi_A} \frac{\partial g}{\partial \xi_B}, \\ \eta^{rs} &= 0, \\ \eta^{r\dot{s}} &= -\eta^{\dot{r}s} = \Lambda_{rs}, \\ \eta^{\dot{r}\dot{s}} = -\eta^{\dot{s}\dot{r}} &= \sum_{u,v} \frac{\partial^2 L}{\partial \dot{q}_v \partial q_u} (\Lambda_{ru}\Lambda_{sv} - \Lambda_{rv}\Lambda_{su}). \end{aligned} \tag{9.102}$$

The matrix $\|\eta^{AB}\|$ has a complicated appearance; the simplicity lies in the appearance of the inverse matrix $\|\eta_{AB}\|$, which can be easily worked out. We find

$$\begin{aligned} \eta_{AB}(q, \dot{q}) &= \frac{\partial}{\partial \xi_A} \mathcal{A}_B(q, \dot{q}) - \frac{\partial}{\partial \xi_B} \mathcal{A}_A(q, \dot{q}), \\ \mathcal{A}_s(q, \dot{q}) &= \frac{\partial L}{\partial \dot{q}_s}, \qquad \mathcal{A}_{\dot{s}} = 0. \end{aligned} \tag{9.103}$$

This discloses the real structure of the GPB we have here! Instead of using the expressions $\partial L / \partial \dot{q}_s$ to define the momenta p_s, and then passing to the Hamiltonian form, we use these expressions to define the first set of components of the vector field $\mathcal{A}_A$; the second set is zero. We then form the "curl" of $\mathcal{A}_A$ according to Eq.(9.18) to get the tensor η_{AB}, and finally define the GPB with the inverse tensor η^{AB}. It is straightforward algebra to verify that the Lagrangian equations of motion are reexpressed in the form:

$$\begin{aligned} \frac{d}{dt} f(q, \dot{q}) &= \{f, H\}^*, \\ H(q, \dot{q}) &= \sum_s \dot{q}_s \frac{\partial L}{\partial \dot{q}_s} - L(q, \dot{q}). \end{aligned} \tag{9.104}$$

The Hamiltonian H is retained as a function of $q, \dot{q}$. If we take $f = q_s$ in this equation of motion, we get an identity; by taking $f = \dot{q}_s$, we get the Lagrangian equations for the accelerations.

We have, then, the description of the solutions of the Lagrangian system of equations in the $2k$-dimensional space of the q_s and $\dot{q}_s$ as constituting a generalized canonical transformation, and if L has no explicit time dependence, a one-parameter group of such transformations. This is an exact transcription of the picture of the motion in phase space. But the main drawback in this Lagrangian reformulation of the ordinary canonical framework is precisely the fact that the GPB Eq.(9.102) has a form dependent on the Lagrangian, that is, on the particular system considered. In contrast, the canonical framework is fixed by the number of degrees of freedom and the physical interpretation and allowed values of the canonical variables, and can therefore be common to a large family of systems.

Chapter 10

Dynamical Systems with Infinitely Many Degrees of Freedom and Theory of Fields

In the discussion so far we have considered only systems that have arbitrarily many, but a finite number of, degrees of freedom. Now we extend these considerations to cover some aspects of the treatment of dynamical systems with an infinite number of degrees of freedom. Many of the important and interesting dynamical systems of this type involve *classical fields* as the basic dynamical variables. Examples are the classical electromagnetic and gravitational fields. In Hamiltonian field theory the infinity of the degrees of freedom is a *continuous infinity.* All the formal developments of the preceding chapters, such as the Lagrangian and Hamiltonian descriptions, the PB and its generalizations, canonical transformation theory, and even the Dirac theory of constraints, can *formally* be carried over to the description of field systems. However, it is natural to expect that the very fact that we have an infinite number of degrees of freedom, countable or not, introduces new and novel mathematical (and hence physical!) features that were completely absent previously. Most of these new features have been encountered for the first time and studied from the quantum field theory point of view, but we would like to indicate here that they are equally relevant for classical mechanics. It turns out that the easiest way to exhibit these features, and the most transparent way, is to consider dynamical systems with a *countable* infinity of degrees of freedom. For this reason we adopt the following order of presentation of the material in this chapter. We first exhibit by means of simple examples some characteristic differences between systems with a finite and countably infinite number of degrees of freedom. Then we show formally that there is a correspondence between the cases of countably infinite and continuously infinite degrees of freedom. We then consider the treatment of fields by Lagrangian and Hamiltonian methods, and conclude by discussing the example of sound vibrations in an ideal gas.

Let us begin by restating in a new way the main structural properties of the Hamiltonian description of systems with a finite number, k, of degrees of freedom. Such a description uses a $2k$-dimensional phase space with k canonical pairs of coordinates (q_s, p_s), $s = 1, \cdots k$, obeying the fundamental PB relations

$$\{q_r, q_s\} = \{p_r, p_s\} = 0, \qquad \{q_r, p_s\} = \delta_{rs}. \tag{10.1}$$

Dynamical variables $f(q,p)$, $g(q,p)$, $\cdots$ are functions of the basic variables, and their PB's are defined according to

$$\{f, g\}(q,p) = \sum_{s=1}^{k} \left(\frac{\partial f}{\partial q_s} \frac{\partial g}{\partial p_s} - \frac{\partial f}{\partial p_s} \frac{\partial g}{\partial q_s} \right). \tag{10.2}$$

A regular canonical transformation, interpretable both as a change of coordinates in phase space as well as a mapping of the points of phase space onto themselves, consists in the passage from the variables $(q_1, \cdots q_k,\ p_1, \cdots p_k)$ to a new set $(Q_1 \cdots Q_k,\ P_1 \cdots P_k)$, the latter being $2k$ independent functions of the former preserving the fundamental PB relations:

$$\{Q_r, Q_s\} = \{P_r, P_s\} = 0, \qquad \{Q_r, P_s\} = \delta_{rs}. \tag{10.3}$$

We have seen that such a transformation also determines a mapping of the collection of all dynamical variables onto itself: the variable $f(q,p)$ is mapped into the variable $\bar{f}(q,p)$ by the definition

$$f(Q, P) = \bar{f}(q,p). \tag{10.4}$$

Given the collection of all dynamical variables, we know that there are certain operations that have a *canonical invariant* significance. These are:

1. The operation of taking linear combinations of a given set of functions of (q,p) to produce a new function of (q,p), the coefficients in the combination being pure numbers.

2. The operation of *pointwise multiplication* of two, or any number of functions of (q,p) to yield a new function of (q,p).

3. The operation of taking the PB of two functions of (q,p).

Operations (1) and (2) can be generalized to that of constructing functions of a given set of functions in an algebraic manner. (1) and (2) together give the set of all dynamical variables the structure of an *associative algebra*, whereas we shall see in later chapters that (3) converts it into a *Lie algebra*. We shall here content ourselves with saying that the dynamical variables form an *algebraic system* with the *defining operations* (1),(2), and (3), which operations give the system a certain *structure*. The canonical invariance of these operations, and hence of the structure of the system, means the following: if we have a relation among a set of functions $f_a(q,p)$, and the only operations involved in stating this relation are the operations enumerated above, each one occurring any number of times and in any sequence but always at one and the same point (q,p), then *precisely the same relation* will hold among the functions $\bar{f}_a(q,p)$ into which the $f_a(q,p)$ are mapped by a regular canonical transformation. Thus every regular canonical transformation acts as a structure-preserving mapping of the algebraic system of dynamical variables onto itself; every relation among the elements of the algebraic system, involving the defining operations only, is preserved. Such a mapping is called an *automorphism* of the algebraic system.

For the case of finitely many degrees of freedom, the converse of the statements just made is also true. That is, we have the fundamental result: every automorphism of the algebraic system of dynamical variables on a finite-dimensional phase space can be implemented as a regular canonical transformation. Let us prove this result. Under the given automorphism, let the coordinates q_s, p_s, each of which is a dynamical variable, be mapped into the functions $Q_s(q,p)$, $P_s(q,p)$ and let this mapping be regular. The effect of the automorphism on the q's and p's already determines its effect on any algebraic function $f(q,p)$, because the operations (1) and (2) are the only ones involved in defining such a function. Thus Eq.(10.4) is obeyed by the automorphism. Because pure numbers are unchanged by an automorphism, we can next use the fact that all PB relations are preserved by the automorphism to conclude that the fundamental relations

$$\{Q_r, Q_s\} = \{P_r, P_s\} = 0, \qquad \{Q_r, P_s\} = \delta_{rs}, \qquad r,s = 1,\ldots,k \tag{10.3}$$

are valid. This already means that the transformation $q,p \to Q,P$ is a canonical one, but we can go on to actually exhibit the generating function for the transformation. From our work in Chapter 6 we know that if Eq.(10.3) is obeyed by the Q's and P's, then there is an integer $A, 0 \leqq A \leqq k$, such that the variables $q_1, \cdots, q_k$, $Q_1, \cdots, Q_A$, $P_{A+1}, \cdots, P_k$ form an independent set. (We make this simplifying assumption on the numbering of the degrees of freedom. This is an assumption because the transformation $(qp) \to (QP)$ is taken to be regular; basically, A is the maximum number of the Q's that can be adjoined to the q_s in building up an independent set of $2k$ variables.) Let now the indices a, a' go over the range $1, 2, \cdots, A$; b, b' over the range $A+1, \cdots, k$; and r, s as usual from 1 to k. Thus q_r, Q_a, and P_b are $2k$ independent variables, and we may therefore imagine p_s, Q_b and P_a exhibited as functions of them:

$$\begin{aligned} p_s &= \psi_s(q_r, Q_a, P_b), \\ Q_{b'} &= \theta_{b'}(q_r, Q_a, P_b), \\ P_{a'} &= \chi_{a'}(q_r, Q_a, P_b), \end{aligned} \tag{10.5}$$

Then our work in Chapter 6 tells us that there exists a single function $F(q_r, Q_a, P_b)$

of which the ψ_s, θ_b and χ_a are partial derivatives:

$$\psi_s = \frac{\partial F}{\partial q_s}\ ,\ \theta_b = \frac{\partial F}{\partial P_b}\ ,\ \chi_a = \frac{-\partial F}{\partial Q_a}. \tag{10.6}$$

This function F is the generating function of the (regular) canonical transformation $q, p \to Q, P$, and can be determined (up to an additive constant) once the automorphism is known.

It is easily seen that the set of all invertible automorphisms of an algebraic system forms a group; the product of two structure-preserving mappings, defined as the effect of one followed by the other, is also structure preserving, and this way of "multiplying" mappings clearly obeys the associative law. Thus for systems with a finite number of degrees of freedom, the group of automorphisms of the algebraic system of dynamical variables is identical with the group of regular canonical transformations. For systems with infinitely many degrees of freedom, we find that not every automorphism can always be implemented as a canonical transformation!

We have considered in Chapter 6 the special set of one-parameter families of canonical transformations:

$$\begin{aligned} Q_s &= e^{-\epsilon\tilde{\phi}} q_s = q_s + \epsilon\{q_s, \phi\} + \frac{\epsilon^2}{2!}\{\{q_s, \phi\}, \phi\} + \cdots , \\ P_s &= e^{-\epsilon\tilde{\phi}} p_s = p_s + \epsilon\{p_s, \phi\} + \frac{\epsilon^2}{2!}\{\{p_s, \phi\}, \phi\} + \cdots . \end{aligned} \tag{10.7}$$

The quantity $\phi(q, p)$ is the (infinitesimal) generator of this canonical transformation and is defined to within an additive constant by the transformation. All such one-parameter groups of automorphisms are canonically implemented by a generator that is (to within a finite additive constant) a well-defined dynamical variable.

Let us now have a look at some of the problems that arise when we generalize to systems with infinitely many degrees of freedom. If we label the dynamical variables countably, such a system is associated with the variables:

$$q_s, p_s, \qquad 1 \leqq s < \infty.$$

The fundamental PB's are formally unchanged:

$$\{q_r, q_s\} = \{p_r, p_s\} = 0, \quad \{q_r, p_s\} = \delta_{rs}, \qquad 1 \leqq r, s < \infty. \tag{10.8}$$

The expression for the PB is now an infinite series,

$$\{f(q,p), g(q,p)\} = \sum_{s=1}^{\infty} \left(\frac{\partial f}{\partial q_s} \frac{\partial g}{\partial p_s} - \frac{\partial f}{\partial p_s} \frac{\partial g}{\partial q_s} \right), \tag{10.9}$$

which poses the problem that even if f and g may be "well defined" in a suitable sense, the right-hand side in Eq.(10.9) may not exist! To be able to carry out the generalization of Hamiltonian mechanics to such systems, we are thus forced to restrict the set of dynamical variables to those for which the right-hand side of Eq.(10.9) is well-defined. A simple (though by no means the only) choice is to consider only finitely generated polynomials in the fundamental q's and p's. The general dynamical variable is thus restricted to be a finite linear combination of terms of the type

$$\prod_{s=1}^{\infty} q_s^{a_s} p_s^{b_s}, \qquad \sum_{s=1}^{\infty}(a_s + b_s) < \infty, \tag{10.10}$$

with the exponents a_s, b_s being nonnegative integers. The convergence of the infinite series here means that in any monomial of the above type, all but a finite number of the a's and b's vanish. This means that a finitely generated polynomial is nothing but a finite polynomial in a finite subset of the q_s and p_s. Given any two such "finite polynomials," their PB is also a "finite polynomial." Actually one can consider a larger class of dynamical variables, namely those functions $f(q,p)$ that depend only on a finite number of the q_s and p_s (but on these their dependence is unrestricted); the PB $\{f, g\}$ in which at least one element is of this type involves only a finite sum on the right-hand side of Eq.(10.9). If both f and g are of this type, then so is $\{f, g\}$.

An *automorphism* of the algebra generated by the infinite number of q_s and p_s is completely specified if we are given the functions $Q_s(q,p)$, $P_s(q,p)$ that the basic variables q_s, p_s are mapped into; we will require that Q_s and P_s, $1 \leqq s < \infty$, be "well-defined" functions of q, p, that they permit one to evaluate their PB's with one another, and that these have the standard values:

$$\{Q_r, Q_s\} = \{P_r, P_s\} = 0, \qquad \{Q_r, P_s\} = \delta_{rs}, \qquad 1 \leqq r, s < \infty. \tag{10.11}$$

We say that the automorphism is *canonically implementable* if it can be given in terms of a well-defined generating function F in one of the four forms developed in Chapter 6 (or some variant thereof, for example, using all the q_s, some of the Q_s and the rest of the P_s as independent variables), or in case we have a one-parameter group of automorphisms, if we can exhibit a well-defined infinitesimal generator for the automorphisms.

Suppose we define a one-parameter group of automorphisms by using Eq. (10.7) with ϕ chosen to be a "finite polynomial." In such a case, there exists an integer N such that ϕ does not depend on q_s, p_s for $s > N$; it is then clear that under this automorphism, the variables $q_1, \cdots, q_N, p_1, \cdots, p_N$ undergo a canonical transformation among themselves while the remaining q's and p's are unaffected. For such functions ϕ, the group of automorphisms is canonically implementable. However, we may in some cases have to go beyond generators of the "finite polynomial" type, or it may even happen that some one-parameter groups of automorphisms cannot be canonically implemented by *any* dynamical variable $\phi(q,p)$. Examples of this follow.

A very important case of a dynamical system with infinitely many degrees of freedom is given by a system of harmonic oscillators. The Hamiltonian may be taken to be

$$H(q,p) = \frac{1}{2}\sum_{s=1}^{\infty}(p_s^2 + \omega_s^2 q_s^2). \tag{10.12}$$

We note that it is *not* a "finite polynomial"! Nevertheless, we can formally work out the Hamiltonian equations of motion; they are

$$\dot{q}_s = \{q_s, H\} = p_s, \quad \dot{p}_s = \{p_s, H\} = -\omega_s^2 q_s, \qquad s = 1, 2, \cdots, \infty. \tag{10.13}$$

The solution of these equations is immediate:

$$\begin{aligned} q_s(t) &= q_s(0)\cos\omega_s t + p_s(0)\frac{\sin\omega_s t}{\omega_s}, \\ p_s(t) &= p_s(0)\cos\omega_s t - q_s(0)\omega_s \sin\omega_s t, \qquad s = 1, 2, \cdots, \infty \end{aligned} \tag{10.14}$$

We easily verify that the time-dependent mapping

$$q_s(0) \to Q_s = q_s(t), \qquad p_s(0) \to P_s = p_s(t) \tag{10.15}$$

forms a one-parameter group of automorphisms. Formally, the Hamiltonian $H(q,p)$ acts as the generator of this group of automorphisms, but it is not a finite polynomial. This has the consequence that although there are some configurations for which the energy is a finite constant of the motion, there are also other configurations (other states of motion) that can be properly described by the Hamiltonian equations of motion (Eq.(10.13)) but for which the energy as defined by Eq.(10.12) does not exist! We only have to consider states of motion in which each oscillator mode involving the pair (q_s, p_s) makes a minimum contribution to the energy! For such states of motion, the time development is properly described as the gradual unfolding of a time-dependent family of automorphisms, but they are *not* canonical transformations generated by the Hamiltonian Eq.(10.12), because the latter does not exist.

As another related example, consider the Hamiltonian resulting from "loading" each of the oscillators in the previous example:

$$H'(q,p) = \frac{1}{2}\sum_{s=1}^{\infty}(p_s^2 + \omega_s^2 q_s^2 + 2g_s q_s), \tag{10.16}$$

where the g_s are a set of coupling constants. In analogy with the case of a finite number of degrees of freedom, we can reduce this Hamiltonian to the earlier "unloaded" one by the substitution

$$Q_s = q_s + \frac{g_s}{\omega_s^2}, \qquad P_s = p_s, \qquad s = 1, 2, \cdots, \infty \tag{10.17}$$

The mapping $q, p \to Q, P$ is obviously an automorphism preserving the fundamental PB relations. In terms of the new variables, $H'(q,p)$ can be rewritten as

$$H'(q,p) = \frac{1}{2}\sum_{s=1}^{\infty}(P_s^2 + \omega_s^2 Q_s^2) - \frac{1}{2}\sum_{s=1}^{\infty}\frac{g_s^2}{\omega_s^2} = H(Q,P) - \Delta, \tag{10.18}$$

but the procedure is or is not well-defined according as the energy-shift:

$$\Delta = \frac{1}{2}\sum_{s=1}^{\infty}\frac{g_s^2}{\omega_s^2} \tag{10.19}$$

is or is not finite. At the same time, we see that the automorphism Eq.(10.17) is formally generated by:

$$\phi(q,p) = \sum_{s=1}^{\infty} \frac{1}{\omega_s^2} q_s p_s \tag{10.20}$$

via Eq.(10.7) with $\epsilon = 1$, but this is not a finite polynomial and thus will not exist in certain configurations.

These simple examples indicate the essential new features arising from the existence of an infinite number of degrees of freedom. Although all our work in the preceding chapters can be *formally* taken over for systems with infinitely many degrees of freedom, we must keep in mind these essential differences, especially the distinction between canonically implementable and nonimplementable automorphisms, and the possibility that certain dynamical variables may be defined in some states of motion but not in others. We recognize the possibility that whether or not a given automorphism is canonically implementable can depend on the state of motion of the system!

Theory of Fields

For a system with the countably infinite set of basic dynamical variables $q_s, p_s, s = 1, 2 \cdots$, the phase or state at an instant of time involves specifying the values of the countable infinity of variables q_s, p_s at that instant. There are many dynamical systems with infinitely many degrees of freedom, systems of fields, which are not most naturally visualized in this way. For example, if we have a dynamical system involving a certain set of fields on three-dimensional Euclidean space, the phase of the system at an instant of time is determined by giving the values of a certain set of functions at each point of space at that instant of time. Some of these functions, the $\phi_a(\boldsymbol{x})$, replace the q's of the earlier treatment, and the functions $\pi_a(\boldsymbol{x})$ replace the p's. Here the points of three-dimensional space have been labeled by their Cartesian coordinates $\boldsymbol{x}$ relative to some fixed origin and a set of orthogonal axes at that origin; the continuous three-vector label $\boldsymbol{x}$ now enumerates the infinity of degrees of freedom of the system, instead of the discrete index s used previously (of course in addition to $\boldsymbol{x}$ we may have the index a to number the fields). The general time-dependent variables $q_s(t), p_s(t)$ are now replaced by $\phi_a(\boldsymbol{x}, t)$ and $\pi_a(\boldsymbol{x}, t)$, respectively. From the Hamiltonian viewpoint, the ϕ_a are the basic coordinates, and the π_a the canonically conjugate momenta.

Let us first satisfy ourselves that these two ways of counting the degrees of freedom can (at least formally) be put in correspondence. Given a description of a system in terms of the countable set q_s, p_s $1 \leqq s < \infty$, let us define new variables $\phi(\boldsymbol{x})$, $\pi(\boldsymbol{x})$ as the real linear combinations

$$\phi(\boldsymbol{x}) = \sum_{s=1}^{\infty} u_s(\boldsymbol{x}) q_s, \qquad \pi(\boldsymbol{x}) = \sum_{s=1}^{\infty} v_s(\boldsymbol{x}) p_s, \tag{10.21}$$

in terms of a set of functions $u_s(\boldsymbol{x})$, $v_s(\boldsymbol{x})$. Then we have the PB relations

$$\{\phi(\boldsymbol{x}), \phi(\boldsymbol{y})\} = \{\pi(\boldsymbol{x}), \pi(\boldsymbol{y})\} = 0, \qquad \{\phi(\boldsymbol{x}), \pi(\boldsymbol{y})\} = \sum_{s=1}^{\infty} u_s(\boldsymbol{x}) v_s(\boldsymbol{y}). \quad (10.22)$$

Let us demand that the linear combinations (Eq.(10.21)) are nonsingular in the sense that we can solve them for q_s and p_s in the form:

$$q_s = \int d^3x U_s(\boldsymbol{x})\phi(\boldsymbol{x}), \qquad p_s = \int d^3x V_s(\boldsymbol{x})\pi(\boldsymbol{x}); \quad (10.23)$$

then the two sets of functions u_s, v_s, U_s, V_s must obey:

$$\phi(\boldsymbol{x}) = \int d^3y \sum_{s=1}^{\infty} u_s(\boldsymbol{x}) U_s(\boldsymbol{y})\phi(\boldsymbol{y}), \quad \pi(\boldsymbol{x}) = \int d^3y \sum_{s=1}^{\infty} v_s(\boldsymbol{x}) V_s(\boldsymbol{y})\pi(\boldsymbol{y});$$

$$q_s = \sum_{r=1}^{\infty} \int d^3x U_s(\boldsymbol{x}) u_r(\boldsymbol{x}) q_r, \qquad p_s = \sum_{r=1}^{\infty} \int d^3x V_s(\boldsymbol{x}) v_r(\boldsymbol{x}) p_r. \quad (10.24)$$

In view of these relations, it is natural to choose the functions u_s, v_s to be complete reciprocal bases for functions of $\boldsymbol{x}$ in the sense

$$U_s(\boldsymbol{x}) = v_s(\boldsymbol{x}), \qquad V_s(\boldsymbol{x}) = u_s(\boldsymbol{x});$$

$$\sum_{s=1}^{\infty} u_s(\boldsymbol{x}) v_s(\boldsymbol{y}) = \delta^{(3)}(\boldsymbol{x}-\boldsymbol{y}), \qquad \int d^3x\, u_r(\boldsymbol{x}) v_s(\boldsymbol{x}) = \delta_{rs}. \quad (10.25)$$

Then the original fundamental PB relations among the q's and p's are replaced by

$$\{\phi(\boldsymbol{x}), \phi(\boldsymbol{y})\} = \{\pi(\boldsymbol{x}), \pi(\boldsymbol{y})\} = 0, \qquad \{\phi(\boldsymbol{x}), \pi(\boldsymbol{y})\} = \delta^{(3)}(\boldsymbol{x}-\boldsymbol{y}). \quad (10.26)$$

This shows that a Hamiltonian *field* system is really a system with an infinite number of degrees of freedom. The linear connection between the countable and continuous infinities in the degrees of freedom, contained in Eqs.(10.21) and (10.23), can be thought of as a generalized form of a linear matrix transformation. Thus to go from the q_s to the $\phi(\boldsymbol{x})$ we use a "matrix" with "matrix elements" $u_s(\boldsymbol{x})$; the rows of this "matrix" are labeled continuously by $\boldsymbol{x}$, whereas its columns are countable and labeled by s. The inverse matrix has matrix elements $v_s(\boldsymbol{x})$; now the rows are labeled countably, and the columns continuously. Equation (10.25) tells us that these two matrices are inverses of one another; the matrix $\|v_s(\boldsymbol{x})\|$ is both the right and left inverse of the matrix $\|u_s(\boldsymbol{x})\|$.

From a mathematical point of view the fundamental PB's (Eq.(10.26)) may be somewhat unsatisfactory because of the appearance of the singular function $\delta^{(3)}(\boldsymbol{x}-\boldsymbol{y})$. But this can be resolved by taking Eq.(10.26) to be the symbolic expression for the PB relations

$$\{\phi(a), \pi(b)\} = \int d^3x\, a(\boldsymbol{x}) b(\boldsymbol{x}),$$

$$\phi(a) = \int d^3x\, a(\boldsymbol{x})\phi(\boldsymbol{x}), \quad \pi(b) = \int d^3x\, b(\boldsymbol{x})\pi(\boldsymbol{x}), \quad (10.27)$$

and requiring $a(\boldsymbol{x})$ and $b(\boldsymbol{x})$ to be such that the right-hand side of the PB in Eq.(10.27) exists. But this just means that $\phi(a)$ and $\pi(b)$ are like the countably labeled coordinates q_s, p_s!

In dealing with dynamical systems involving fields, the techniques for defining generalized momenta, and for passing from a Lagrangian to a Hamiltonian, that were appropriate for cases where the degrees of freedom were countable, must be suitably generalized. For simplicity let us consider a system involving just one field variable $\phi(\boldsymbol{x}, t)$, and let us suppose it is described by a (standard) Lagrangian depending on $\phi(\boldsymbol{x}, t)$ and $(\partial\phi/\partial t)(\boldsymbol{x}, t)$ for all $\boldsymbol{x}$ and the instant t:

$$L\left[\phi(\boldsymbol{x},t),\ \frac{\partial\phi}{\partial t}(\boldsymbol{x},t)\right]. \tag{10.28}$$

Because we have a continuous infinity of generalized coordinates, the $\phi(\boldsymbol{x}, t)$ for all $\boldsymbol{x}$, and their generalized velocities $(\partial\phi/\partial t)(\boldsymbol{x}, t)$, the Lagrangian is a *functional* of these variables. Here it must be made clear that although the time derivative of $\phi(\boldsymbol{x}, t)$ is treated as independent of $\phi(\boldsymbol{x}, t)$, being the velocity variable, *the spatial derivatives $\partial\phi/\partial\boldsymbol{x}$ are not new independent quantities but are, in a generalized sense, functions of the basic coordinates $\phi(\boldsymbol{x}, t)$*. Clearly, if the numerical values of $\phi(\boldsymbol{x}, t)$ were specified for all $\boldsymbol{x}$ at a certain time t, one would also know the values of $\partial\phi/\partial\boldsymbol{x}$ at that time. It is important to realize this in order that we may be able to define the "partial derivatives" of the Lagrangian L with respect to its arguments properly. For this purpose we imagine altering the $\phi(\boldsymbol{x}, t)$ and $(\partial\phi/\partial t)(\boldsymbol{x}, t)$ in L by small amounts, $\delta\phi(\boldsymbol{x}, t)$ and $\delta\dot{\phi}(\boldsymbol{x}, t)$, these increments being independent of one another at any one time, and compute the change in L without changing the functional form of L. We can always express the change in L in the form

$$\delta L[\phi,\dot{\phi}] = \int d^3x \left(\frac{\delta L}{\delta\phi(\boldsymbol{x},t)}\,\delta\phi(\boldsymbol{x},t) + \frac{\delta L}{\delta\dot{\phi}(\boldsymbol{x},t)}\,\delta\dot{\phi}(\boldsymbol{x},t)\right). \tag{10.29}$$

If L depends on the spatial derivatives of ϕ, we first encounter in δL terms involving the gradient $(\partial/\partial\boldsymbol{x})\delta\phi$ of $\delta\phi$, but by means of integration by parts all such terms can be finally put into the above form. *The coefficients $(\delta L/\delta\phi)(\boldsymbol{x}, t)$, $(\delta L/\delta\dot{\phi})(\boldsymbol{x}, t)$ of the independent increments $\delta\phi$, $\delta\dot{\phi}$ are defined to be the partial functional derivatives of L with respect to $\phi(\boldsymbol{x}, t)$ and $\dot{\phi}(\boldsymbol{x}, t)$ respectively.* The Euler Lagrange equations of motion then take the form:

$$\frac{d}{dt}\,\frac{\delta L}{\delta\dot{\phi}(\boldsymbol{x},t)} - \frac{\delta L}{\delta\phi(\boldsymbol{x},t)} = 0. \tag{10.30}$$

We have here a continuous infinity of equations of motion, one for each value of $\boldsymbol{x}$; in other words, we have a field equation.

To pass to the Hamiltonian, we define the canonical momentum field conjugate to ϕ as

$$\pi(\boldsymbol{x},t) = \frac{\delta L}{\delta\dot{\phi}(\boldsymbol{x},t)}, \tag{10.31}$$

and the Hamiltonian H is a functional to be written out in terms of the canonically conjugate fields ϕ, π:

$$H[\phi, \pi] = \int d^3x \, \pi(\boldsymbol{x}, t)\dot{\phi}(\boldsymbol{x}, t) - L[\phi, \dot{\phi}] \, . \tag{10.32}$$

The basic PB's have been written down in Eq.(10.26). The general dynamical variable in the Hamiltonian scheme is a functional F of ϕ, π; for such variables, the PB takes the form:

$$\{F[\phi, \pi], G[\phi, \pi]\} = \int d^3x \left(\frac{\delta F}{\delta\phi(\boldsymbol{x}, t)} \frac{\delta G}{\delta\pi(\boldsymbol{x}, t)} - \frac{\delta F}{\delta\pi(\boldsymbol{x}, t)} \frac{\delta G}{\delta\phi(\boldsymbol{x}, t)} \right) . \tag{10.33}$$

The general Hamiltonian equation of motion is, of course,

$$\frac{d}{dt} F[\phi, \pi] = \{F[\phi, \pi], H[\phi, \pi]\}. \tag{10.34}$$

In all these expressions the *functional* derivatives with respect to ϕ and π appear, and these are defined in the same way in which $\delta L/\delta\phi$, $\delta L/\delta\dot{\phi}$ were defined.

If the Lagrangian L is not of standard type, then of course the Dirac theory of constraints applies to it; of course, this theory must be suitably rewritten, just as we have rewritten PB's and so on before.

For many systems of interest, it happens that the Lagrangian L is "local"; that is, it is expressed as an integral over three-dimensional space of a function $\mathcal{L}$ of the field ϕ, and a finite number of its partial derivatives with respect both to $\boldsymbol{x}$ and to t (although, of course, no partial derivative higher than the first with respect to t appears):

$$L[\phi, \dot{\phi}] = \int d^3x \; \mathcal{L}(\phi, \phi_j, \phi_{jk}, \cdots ; \dot{\phi}, \dot{\phi}_j, \dot{\phi}_{jk}, \cdots). \tag{10.35}$$

Here we have indicated partial derivatives of ϕ with respect to the space variables by subscripts, and it is understood that only derivatives up to some finite order appear in $\mathcal{L}$:

$$\begin{aligned} \phi_j &= \frac{\partial\phi}{\partial x_j} \, , \; \phi_{jk} = \frac{\partial^2\phi}{\partial x_j \partial x_k} \, , \cdots ; \\ \dot{\phi} &= \frac{\partial\phi}{\partial t} \, , \; \dot{\phi}_j = \frac{\partial^2\phi}{\partial x_j \partial t} \, , \dot{\phi}_{jk} = \frac{\partial^3\phi}{\partial x_j \partial x_k \partial t} , \cdots \, . \end{aligned} \tag{10.36}$$

In such a case, $\mathcal{L}$ is called the *Lagrangian density* of the system; one sees from Eq.(10.31) that the canonically conjugate momentum field $\pi(\boldsymbol{x}, t)$ will also be a *local* function of ϕ and its partial derivatives. For example, if $\mathcal{L}$ does not depend on $\dot{\phi}_j, \dot{\phi}_{jk}, \cdots$, we simply get

$$\pi(\boldsymbol{x}, t) = \frac{\partial\mathcal{L}}{\partial\dot{\phi}} \, (\phi, \phi_j, \cdots ; \dot{\phi}); \tag{10.37}$$

otherwise we have the more general, but still local, expression

$$\pi(\boldsymbol{x},t) = \frac{\partial \mathcal{L}}{\partial \dot{\phi}} - \sum_j \frac{\partial}{\partial x_j}\frac{\partial \mathcal{L}}{\partial \dot{\phi}_j} + \sum_{jk} \frac{\partial^2}{\partial x_j \partial x_k}\frac{\partial \mathcal{L}}{\partial \dot{\phi}_{jk}} - \cdots \tag{10.38}$$

[The relations $\delta\dot{\phi}_j = (\partial/\partial x_j)\delta\dot{\phi}$ and so on have been used here.] But, in general, $\dot{\phi}$ need not be a local function of π and its spatial derivatives up to some finite order. For the cases when Eq.(10.37) is adequate, though, we get $\dot{\phi}$ by algebraic operations from Eq.(10.37) and can express it as a local function of π and ϕ. (This is not necessarily true in case Eq.(10.38) has to be used.) For such local Lagrangians it follows that the Hamiltonian H also appears as the volume integral of a local *Hamiltonian density*; using Eq.(10.37) we have:

$$H[\phi,\pi] = \int d^3x\, \mathcal{H}(\phi,\pi), \quad \mathcal{H}(\phi,\pi) = \pi(\boldsymbol{x},t)\dot{\phi}(\boldsymbol{x},t) - \mathcal{L} = \dot{\phi}\,\frac{\partial \mathcal{L}}{\partial \dot{\phi}} - \mathcal{L}. \tag{10.39}$$

The total Hamiltonian H is formally well-defined and represents the total energy of the system. We could interpret $\mathcal{H}$ as the *energy density* of the system, but we must remember that the addition of the divergence of a local vectorial density (with suitable boundary conditions at infinity) to $\mathcal{H}$ would affect the energy density without changing the total energy at all!

The extension of the above formalism to the case where we have a set of fields ϕ_a at each point $\boldsymbol{x}$, labeled by an index, a, is quite easy. It is appropriate now to give here an example of a classical field theory so that one can see the above formalism "in action."

Sound Waves in an Ideal Gas-an Example

Consider sound vibrations in an ideal classical gas. To simplify matters we assume that the vibrations are sufficiently small in amplitude that the fractional changes in density and pressure are also treated as small quantities. Denoting the pressure and the density at each point $\boldsymbol{x}$ at time t by $S(\boldsymbol{x},t)$, $\rho(\boldsymbol{x},t)$, respectively, we have the adiabatic equation:

$$\frac{\rho^\gamma}{S} = \text{constant}; \tag{10.40}$$

γ is the ratio of the specific heat at constant pressure to that at constant volume, and the constant on the right-hand side of Eq.(10.40) depends on the local temperature. The S and ρ differ from their uniform equilibrium values S_0 and ρ_0 by small functions of $\boldsymbol{x}$. The potential energy density of the gas at a density ρ (relative to the potential energy density at density ρ_0 being chosen zero) is given by

$$\begin{aligned} U(\boldsymbol{x},t) &= \frac{1}{\rho_0}\int_{\rho_0}^{\rho} S'\,d\rho' = S_0 \int_{\rho_0}^{\rho}\left(\frac{\rho'}{\rho_0}\right)^\gamma \frac{d\rho'}{\rho_0} \\ &= \frac{S_0}{\gamma+1}\left\{\left(\frac{\rho}{\rho_0}\right)^{\gamma+1} - 1\right\}. \end{aligned} \tag{10.41}$$

If $\xi(\boldsymbol{x}, t)$ is the amplitude of vibration, then

$$\frac{\rho}{\rho_0} = 1 - \boldsymbol{\nabla} \cdot \boldsymbol{\xi}, \tag{10.42}$$

and the potential energy density can be approximated by a quadratic expression:

$$U(\boldsymbol{x}, t) = S_0 \left\{ -\boldsymbol{\nabla} \cdot \xi(\boldsymbol{x}, t) + \frac{\gamma}{2}(\boldsymbol{\nabla} \cdot \xi)^2 + \cdots \right\}. \tag{10.43}$$

The kinetic energy density is given by:

$$T(\boldsymbol{x}, t) = \frac{1}{2} \rho \dot{\boldsymbol{\xi}} \cdot \dot{\boldsymbol{\xi}} = \left\{ \frac{1}{2} \rho_0 \dot{\boldsymbol{\xi}} \cdot \dot{\boldsymbol{\xi}} + \cdots \right\}. \tag{10.44}$$

Retaining terms no higher than quadratic, we are led to consider a Lagrangian density $\mathcal{L}$:

$$\begin{aligned}\mathcal{L}(\boldsymbol{x}, t) &= T(\boldsymbol{x}, t) - U(\boldsymbol{x}, t) \\ &= \frac{1}{2} \rho_0 \dot{\boldsymbol{\xi}}^2(\boldsymbol{x}, t) - S_0 \left\{ -\boldsymbol{\nabla} \cdot \boldsymbol{\xi} + \frac{\gamma}{2}(\boldsymbol{\nabla} \cdot \boldsymbol{\xi})^2 \right\},\end{aligned} \tag{10.45}$$

and the total Lagrangian L, which is a functional of the independent coordinates $\boldsymbol{\xi}$ and velocities $\dot{\boldsymbol{\xi}}$:

$$L[\boldsymbol{\xi}, \dot{\boldsymbol{\xi}}] = \int \mathcal{L}(\boldsymbol{x}, t) d^3x. \tag{10.46}$$

Here we have a field system involving three fields at each point $\boldsymbol{x}$, correspondingly we will have a canonical momentum field $\boldsymbol{\pi}$ which is also a vector. Computing the functional derivatives of L according to the definitions given earlier, we find

$$\frac{\delta L}{\delta \boldsymbol{\xi}(\boldsymbol{x}, t)} = \gamma S_0 \boldsymbol{\nabla}(\boldsymbol{\nabla} \cdot \boldsymbol{\xi}), \qquad \frac{\delta L}{\delta \dot{\boldsymbol{\xi}}(\boldsymbol{x}, t)} = \rho_0 \dot{\boldsymbol{\xi}}(\boldsymbol{x}, t); \tag{10.47}$$

and the Lagrangian field equation is

$$\rho_0 \ddot{\boldsymbol{\xi}}(\boldsymbol{x}, t) - \gamma S_0 \boldsymbol{\nabla}(\boldsymbol{\nabla} \cdot \boldsymbol{\xi}(\boldsymbol{x}, t)) = 0. \tag{10.48}$$

The canonical momentum $\boldsymbol{\pi}(\boldsymbol{x}, t)$ and the Hamiltonian H are easily obtained:

$$\begin{aligned}\boldsymbol{\pi}(\boldsymbol{x}, t) &= \frac{\delta L}{\delta \dot{\boldsymbol{\xi}}} = \rho_0 \dot{\boldsymbol{\xi}}(\boldsymbol{x}, t); \\ H &= \int d^3x \, \mathcal{H}'(\boldsymbol{x}, t), \qquad \mathcal{H}'(\boldsymbol{x}, t) = \frac{1}{2} \rho_0 \dot{\boldsymbol{\xi}}^2 + \frac{1}{2} S_0 \gamma (\boldsymbol{\nabla} \cdot \boldsymbol{\xi})^2 - S_0 \boldsymbol{\nabla} \cdot \boldsymbol{\xi}.\end{aligned} \tag{10.49}$$

We can rewrite the Hamiltonian density in terms of the new variables $\boldsymbol{\xi}, \boldsymbol{\pi}$; at the same time we drop the last term $\boldsymbol{\nabla} \cdot \boldsymbol{\xi}$ in $\mathcal{H}'$ (this term contributes nothing to

H unless the acoustic field $\boldsymbol{\xi}(\boldsymbol{x},t)$ fails to vanish at infinity in spatial directions) and thus we get:

$$H = H[\boldsymbol{\xi},\boldsymbol{\pi}] = \int d^2x \mathcal{H}(\boldsymbol{x},t)\,,$$
$$\mathcal{H}(\boldsymbol{x},t) = \frac{1}{2\rho_0}\,\boldsymbol{\pi}^2(\boldsymbol{x},t) + \frac{\gamma}{2}\,S_0(\boldsymbol{\nabla}\cdot\boldsymbol{\xi}(\boldsymbol{x},t))^2. \tag{10.50}$$

It is an interesting exercise to apply the Weiss action principle in Lagrangian form, developed in Chapter 3 for systems with finite numbers of degrees of freedom, directly to the Lagrangian field theory described by $L[\boldsymbol{\xi},\dot{\boldsymbol{\xi}}]$. By considering the variation of the action (defined as before as the time integral of $L[\boldsymbol{\xi},\dot{\boldsymbol{\xi}}]$ along a sequence of configurations described by a certain set of values for all the independent coordinates $\boldsymbol{\xi}(\boldsymbol{x},t)$ for each instant of time) resulting from variations $\delta\boldsymbol{\xi}(\boldsymbol{x},t)$ in the $\boldsymbol{\xi}$'s and variations $\boldsymbol{\Delta}t_1,\boldsymbol{\Delta}t_2$ in the terminal points of the time integration, and expressing the result in terms of the total variations $\boldsymbol{\Delta}t_j$ and

$$\boldsymbol{\Delta\xi}(\boldsymbol{x},t_j) = \delta\boldsymbol{\xi}(\boldsymbol{x},t_j) + \dot{\boldsymbol{\xi}}(\boldsymbol{x},t_j)\boldsymbol{\Delta}t_j, \qquad j=1,2,$$

we obtain the Lagrangian equations of motion (Eq.(10.48)), the definition of the momenta $\boldsymbol{\pi}(\boldsymbol{x},t)$, as well as the Hamiltonian H. It is important to realize that the quantities varied are the *time dependences* of the $\boldsymbol{\xi}(\boldsymbol{x},t)$ at each point $\boldsymbol{x}$, and that $\boldsymbol{x}$ undergoes no variation of any kind. The label $\boldsymbol{x}$ is here the exact analogue of the indices $r,s,\cdots$ appearing on the coordinates $q_r,q_s,\cdots$, previously. We leave the details to the interested reader.

The fundamental PB relations take the form:

$$\{\xi_j(\boldsymbol{x},t),\xi_k(\boldsymbol{x}',t)\} = \{\pi_j(\boldsymbol{x},t),\pi_k(\boldsymbol{x}',t)\} = 0,$$
$$\{\xi_j(\boldsymbol{x},t),\pi_k(\boldsymbol{x}',t)\} = \delta_{jk}\delta^{(3)}(\boldsymbol{x}-\boldsymbol{x}'), \qquad j,k=1,2,3. \tag{10.51}$$

The Hamiltonian equations of motion are easily derived:

$$\dot{\boldsymbol{\xi}}(\boldsymbol{x},t) = \{\boldsymbol{\xi}(\boldsymbol{x},t),H[\boldsymbol{\xi},\boldsymbol{\pi}]\} = \frac{1}{\rho_0}\boldsymbol{\pi}(\boldsymbol{x},t),$$
$$\dot{\boldsymbol{\pi}}(\boldsymbol{x},t) = \{\boldsymbol{\pi}(\boldsymbol{x},t),H[\boldsymbol{\xi},\boldsymbol{\pi}]\} = -\gamma S_0\boldsymbol{\nabla}(\boldsymbol{\nabla}\cdot\boldsymbol{\xi}(\boldsymbol{x},t)), \tag{10.52}$$

and they are together equivalent to Eq.(10.48).

The Hamiltonian Eq.(10.50) shows some similarity to the Hamiltonian given in Eq.(10.12); however they are not quite the same because in the present case we do not have a collection of *uncoupled* oscillators in the variables $\boldsymbol{\xi}(\boldsymbol{x},t)$, $\boldsymbol{\pi}(\boldsymbol{x},t)$. The presence of the gradient operator $\boldsymbol{\nabla}$ in the Hamiltonian and in the equations of motion shows that the time evolution of the coordinate $\boldsymbol{\xi}(\boldsymbol{x},t)$ at the point $\boldsymbol{x}$ depends on the coordinates $\boldsymbol{\xi}$ at "neighboring" points, thus the $\boldsymbol{\xi}(\boldsymbol{x},t)$ do not represent independent uncoupled coordinates. This is as it should be because it leads to propagation, the excitation at each point being passed on to the "neighboring points" and they, in turn, to their neighbors, and so on. (Remember that $\boldsymbol{x}$ plays here the role of the indices $r,s,\cdots$ on q_r,p_s used previously.)

Let us first show what kind of wave motion results from the Lagrangian equations of motion, and then show how to obtain the uncoupled modes of motion. For the former purpose, we separate $\boldsymbol{\xi}(\boldsymbol{x},t)$ and $\boldsymbol{\pi}(\boldsymbol{x},t)$ into their divergence-free (solenoidal) and curl-free (lamellar) components:

$$\begin{aligned}\boldsymbol{\xi} &= \boldsymbol{\xi}^T + \boldsymbol{\xi}^L, \qquad & \boldsymbol{\pi} &= \boldsymbol{\pi}^T + \boldsymbol{\pi}^L;\\ \nabla\cdot\boldsymbol{\xi}^T &= \nabla\cdot\boldsymbol{\pi}^T = 0, \qquad & \nabla\times\boldsymbol{\xi}^L &= \nabla\times\boldsymbol{\pi}^L = 0.\end{aligned} \tag{10.53}$$

(This separation is *not* a local operation.) Then the Hamiltonian equations of motion break up into the form:

$$\dot{\boldsymbol{\xi}}^T = \frac{1}{\rho_0}\boldsymbol{\pi}^T, \quad \dot{\boldsymbol{\pi}}^T = 0; \quad \dot{\boldsymbol{\xi}}^L = \frac{1}{\rho_0}\boldsymbol{\pi}^L, \quad \dot{\boldsymbol{\pi}}^L = -\gamma S_0 \nabla^2 \boldsymbol{\xi}^L. \tag{10.54}$$

The equations of motion for the solenoidal parts $\boldsymbol{\xi}^T, \boldsymbol{\pi}^T$ are reminiscent of the free-particle equations of motion and lead to a secular nonperiodic time development. The momentum density $\boldsymbol{\pi}^T$, and thus its contribution to the density of kinetic energy, remains constant. If the displacement $\boldsymbol{\xi}$ is to remain small and bounded at all times, we must consider $\boldsymbol{\pi}^T = 0$; this is necessary only because we have neglected nonlinear effects. But within this approximation, the solenoidal components do not contribute to sound propagation. The lamellar part $\boldsymbol{\xi}^L$ obeys a wave equation:

$$\frac{\rho_0}{\gamma S_0}\ddot{\boldsymbol{\xi}}^L - \nabla^2\boldsymbol{\xi}^L = 0; \tag{10.55}$$

we have purely longitudinal waves with velocity $c = \sqrt{\gamma S_0/\rho_o}$.

In order to exhibit the Hamiltonian as a sum of *uncoupled* oscillators, we pass to the Fourier transforms of $\boldsymbol{\xi}(\boldsymbol{x},t)$ and $\boldsymbol{\pi}(\boldsymbol{x},t)$:

$$\begin{aligned}\tilde{\boldsymbol{\xi}}(\boldsymbol{k},t) &= (2\pi)^{-3/2}\int d^3x\; e^{-i\boldsymbol{k}\cdot\boldsymbol{x}}\boldsymbol{\xi}(\boldsymbol{x},t),\\ \tilde{\boldsymbol{\pi}}(\boldsymbol{k},t) &= (2\pi)^{-3/2}\int d^3x\; e^{i\boldsymbol{k}\cdot\boldsymbol{x}}\boldsymbol{\pi}(\boldsymbol{x},t).\end{aligned} \tag{10.56}$$

These variables are complex; complex conjugation of $\tilde{\boldsymbol{\xi}}$ or $\tilde{\boldsymbol{\pi}}$ amounts to replacing $\boldsymbol{k}$ by $-\boldsymbol{k}$:

$$\tilde{\boldsymbol{\xi}}(\boldsymbol{k},t)^* = \tilde{\boldsymbol{\xi}}(-\boldsymbol{k},t); \qquad \tilde{\boldsymbol{\pi}}(\boldsymbol{k},t)^* = \tilde{\boldsymbol{\pi}}(-\boldsymbol{k},t). \tag{10.57}$$

In terms of them, the nonzero PB's, and the Hamiltonian are

$$\begin{aligned}&\{\tilde{\xi}_j(\boldsymbol{k},t), \tilde{\pi}_k(\boldsymbol{k}',t)\} = \delta_{jk}\delta^{(3)}(\boldsymbol{k}-\boldsymbol{k}');\\ &H = \int d^3k\left(\frac{1}{2\rho_0}\sum_j \tilde{\pi}_j(\boldsymbol{k},t)\tilde{\pi}_j(\boldsymbol{k},t)^*\right.\\ &\qquad\left. + \frac{\gamma}{2}\,S_0\left(\sum_j k_j\tilde{\xi}_j(\boldsymbol{k},t)\right)\left(\sum_l k_l\tilde{\xi}_l(\boldsymbol{k},t)\right)^*\right).\end{aligned} \tag{10.58}$$

If we split the $\tilde{\xi}$ and $\tilde{\pi}$ into their real and imaginary parts,

$$\begin{aligned}
\tilde{\boldsymbol{\xi}}^{(1)}(\boldsymbol{k},t) &= \frac{1}{\sqrt{2}}(\tilde{\boldsymbol{\xi}}(\boldsymbol{k},t)+\tilde{\boldsymbol{\xi}}(-\boldsymbol{k},t)),\\
\tilde{\boldsymbol{\xi}}^{(2)}(\boldsymbol{k},t) &= \frac{i}{\sqrt{2}}(\tilde{\boldsymbol{\xi}}(\boldsymbol{k},t)-\tilde{\boldsymbol{\xi}}(-\boldsymbol{k},t)),\\
\tilde{\boldsymbol{\pi}}^{(1)}(\boldsymbol{k},t) &= \frac{1}{\sqrt{2}}(\tilde{\boldsymbol{\pi}}(\boldsymbol{k},t)+\tilde{\boldsymbol{\pi}}(-\boldsymbol{k},t)),\\
\tilde{\boldsymbol{\pi}}^{(2)}(\boldsymbol{k},t) &= \frac{-i}{\sqrt{2}}(\tilde{\boldsymbol{\pi}}(\boldsymbol{k},t)-\tilde{\boldsymbol{\pi}}(-\boldsymbol{k},t)),
\end{aligned} \tag{10.59}$$

then because of their symmetry properties in $\boldsymbol{k}$ space,

$$\tilde{\boldsymbol{\xi}}^{(1)}(\boldsymbol{k},t)=\tilde{\boldsymbol{\xi}}^{(1)}(-\boldsymbol{k},t),\quad \tilde{\boldsymbol{\xi}}^{(2)}(\boldsymbol{k},t)=-\tilde{\boldsymbol{\xi}}^{(2)}(-\boldsymbol{k},t),\qquad \text{and so on,}$$

we can restrict ourselves to the region $k_3 \geqq 0$, say, in $\boldsymbol{k}$ space. With this understanding, the basic nonvanishing PB's are

$$\{\tilde{\xi}_j^{(\alpha)}(\boldsymbol{k},t),\tilde{\pi}_l^{(\beta)}(\boldsymbol{k}',t)\}=\delta_{\alpha\beta}\delta_{jl}\delta^{(3)}(\boldsymbol{k}-\boldsymbol{k}'),\qquad k_3>0, k_3'>0, \tag{10.60}$$

and the Hamiltonian is

$$H=\int_{k_3>0} d^3k\left(\frac{1}{2\rho_0}\sum_{j,\alpha}\left(\tilde{\pi}_j^{(\alpha)}(\boldsymbol{k},t)\right)^2+\frac{\gamma S_0}{2}\sum_{\alpha}\left(\sum_j k_j\tilde{\xi}_j^{(\alpha)}(\boldsymbol{k},t)\right)^2\right). \tag{10.61}$$

These equations exhibit the system as a collection of a continuous infinity of uncoupled harmonic oscillators (after the further separation of $\tilde{\boldsymbol{\xi}}^{(\alpha)}$ and $\tilde{\boldsymbol{\pi}}^{(\alpha)}$ into their longitudinal and transverse parts and the disregard of the latter). But we must remember that the Fourier transformation Eq.(10.56) is not a transformation of the type discussed in connection with Eq.(10.27). This means that even though Eqs.(10.60) and (10.61) describe a collection of harmonic oscillators there are no genuine normal modes for the system; Eq.(10.61) represents the degree to which one can simulate normal modes.

Chapter 11

Linear and Angular Momentum Dynamical Variables and Their Significance

We have mentioned in previous chapters the important connection between dynamical variables in phase space and the one-parameter groups of canonical transformations that they generate. This connection is a feature of the Hamiltonian form of dynamics. For example (see the discussion in Chapter 6 on page 56) a constant of motion is a function $A(q,p)$ on phase space that generates a group of canonical transformations under which the Hamiltonian of the system does not change. We have noted in Chapter 9 that for a constrained system, the presence of first-class constraints weakens the reciprocal relationship between dynamical variables and the corresponding groups of canonical transformations. It is worthwhile to discuss in some detail the transformation-theoretical significance of two very important dynamical variables that are defined for most systems of physical interest, these being the variables of linear and angular momentum. Such a discussion serves as a model for the later treatment of other important dynamical variables (introduced by the theories of Galilean and Poincaré relativity). In Newtonian particle mechanics, the primary significance of the linear and angular momenta is that they are constants of motion for certain kinds of forces between the particles, and their forms are derived from the form of the basic equations of motion. On the other hand, it turns out that in Hamiltonian dynamics it is more natural to *define* these variables by means of the transformations that they generate on the basic dynamical variables, the q's and p's, of the theory. One can also appreciate this relationship from the following point of view. In discussing Lagrangian and Hamiltonian field theory in the previous chapter, we noted that the continuous three-vector

$\boldsymbol{x}$ that on the one hand enumerates the continuous infinity of coordinates $\phi(\boldsymbol{x}, t)$ and canonically conjugate momenta $\pi(\boldsymbol{x}, t)$, is on the other hand to be identified (in most physically interesting examples) with points in three-dimensional Euclidean space. Thus we spoke of Lagrangian and Hamiltonian *densities* and of the energy density. This geometrical interpretation of the label $\boldsymbol{x}$ naturally leads to important consequences for a dynamical system of fields. Euclidean space and the relations of Euclidean geometry are unchanged under a certain set of operations on points in that space; these are the translations and rotations of three-dimensional space and all these operations taken together comprise the *Euclidean group.* If a physical system embedded in Euclidean space is described in the Hamiltonian form of dynamics, then the effects of translations and rotations on the system must be adequately described in the mathematical framework. We see in this chapter how the linear and angular momenta are intimately connected with the translational and rotational properties for systems permitting a Lagrangian and Hamiltonian description. We find that both for multiparticle and field systems we can in fact *define* these dynamical variables in terms of the effects of translations and rotations. But the meaning of the algebraic relationship between the different components of linear and angular momenta, as well as the complete group-theoretical relationship between them and the Euclidean group, becomes clear in later chapters.

Let us begin by briefly recapitulating the situation in Newtonian mechanics. Consider a system of N interacting nonrelativistic particles with masses $m_r, r = 1, 2, \cdots, N$. With respect to a given origin of coordinates and an orthogonal set of axes at that origin, let us denote the *Cartesian* position vectors of the particles by $\boldsymbol{q}_r$. If we assume that the forces on the particles are all derivable in the usual way from a potential function V that depends only on the magnitudes of the interparticle distances, then the Newtonian equations of motion are:

$$\begin{aligned} \frac{d}{dt}\,\boldsymbol{p}_r &= -\frac{\partial}{\partial \boldsymbol{q}_r}\,V(|\boldsymbol{q}_1 - \boldsymbol{q}_2|,\ |\boldsymbol{q}_1 - \boldsymbol{q}_3|, \cdots), \\ \boldsymbol{p}_r &= m_r \frac{d}{dt}\,\boldsymbol{q}_r, \qquad r = 1, 2, \cdots, N. \end{aligned} \tag{11.1}$$

It follows immediately that the *total linear momentum*, defined as the vector sum of the individual Cartesian momentum vectors $\boldsymbol{p}_r$, is a constant of motion:

$$\boldsymbol{P} = \sum_{r=1}^{N} \boldsymbol{p}_r, \qquad \frac{d}{dt}\,\boldsymbol{P} = 0. \tag{11.2}$$

This is a consequence of the fact that the potential V depends only on the *coordinate differences* $(\boldsymbol{q}_r - \boldsymbol{q}_s)$, in other words of the fact that if we apply a uniform translation $\boldsymbol{q}_r \to \boldsymbol{q}_r + \boldsymbol{a}$ to all the coordinate vectors, V does not change. Under such a translation, all the velocities and accelerations are unaltered. But such a uniform translation of all the Cartesian position vectors by $\boldsymbol{a}$ is precisely what would result if we shifted the origin of the coordinate system by the amount $-\boldsymbol{a}$, leaving the directions of the orthogonal axes unchanged, and described the configuration of the system from the new coordinate system. Let us now recover the result of Eq.(11.2) from the Lagrangian version of the equations of motion. We can use the connection between conservation laws and the action principle given in Chapter 4. Omitting for simplicity an explicit time dependence, we assume that the Lagrangian $L(\boldsymbol{q}, \dot{\boldsymbol{q}})$ is unchanged under the above-described translation of coordinates:

$$L(\boldsymbol{q}_r + \boldsymbol{a}, \dot{\boldsymbol{q}}_r) = L(\boldsymbol{q}_r, \dot{\boldsymbol{q}}_r). \tag{11.3}$$

The transformation $\boldsymbol{q}_r \to \boldsymbol{q}_r + \boldsymbol{a}$ is an example of a geometric or point transformation in configuration space, and thus leads to a constant of motion linear in the canonical momenta. Following Eq.(4.20), we thus get:

$$\boldsymbol{P} = \sum_{r=1}^{N} \boldsymbol{p}_r = \sum_{r=1}^{N} \frac{\partial L}{\partial \dot{\boldsymbol{q}}_r} = \text{constant}. \tag{11.4}$$

Conversely, if this conservation law is to be an automatic consequence of the equations of motion for all states of motion, Eq.(11.3) must hold and the translational invariance of L follows.

Thus we see that for a many-particle system describable in terms of a Lagrangian, conservation of three-dimensional linear momentum follows from invariance of the Lagrangian under translations of the Cartesian coordinate system, and vice-versa. Actually the proof using the Lagrangian is more general than the one using the Newtonian equations, because it tells us how to define the total momentum even if the generalized potential in the Lagrangian is velocity-dependent. The point to stress is that in this more general situation the canonical momentum $\boldsymbol{p}_r$ need not be just the kinetic momentum $m_r\dot{\boldsymbol{q}}_r$; it has to be defined as in Eq.(11.4), and it is the sum of these Cartesian canonical momenta that is conserved if L obeys Eq.(11.3).

Again for the class of Lagrangian systems, let us next show that the conservation of three-dimensional angular momentum follows from invariance of L under rotations of the Cartesian axes, and vice versa. For simplicity we consider the effects of infinitesimal rotations. Under the small rotation of the coordinate axes by the amount $\delta\theta$ around the axis $\boldsymbol{n}$ ($\boldsymbol{n}$ a unit vector), the geometrical identification of $\boldsymbol{q}_r$ and $\dot{\boldsymbol{q}}_r$ determines the changes they undergo:

$$\boldsymbol{q}_r \to \boldsymbol{q}_r + \delta\theta\, \boldsymbol{n} \times \boldsymbol{q}_r, \qquad \dot{\boldsymbol{q}}_r \to \dot{\boldsymbol{q}}_r + \delta\theta\, \boldsymbol{n} \times \dot{\boldsymbol{q}}_r, \qquad |\delta\theta| \ll 1. \tag{11.5}$$

This again is a geometric transformation in configuration space, and if L is invariant under it, the action principle gives a constant of motion linear in the canonical momenta; because $\delta\theta$ and $\boldsymbol{n}$ are arbitrary, we get from Eq.(4.20),

$$\boldsymbol{J} = \sum_{r=1}^{N} \boldsymbol{q}_r \times \boldsymbol{p}_r = \sum_{r=1}^{N} \boldsymbol{q}_r \times \frac{\partial L}{\partial \dot{\boldsymbol{q}}_r} = \text{constant}. \tag{11.6}$$

(Once again we must keep in mind that the contribution of each particle to the total angular momentum $\boldsymbol{J}$ is not necessarily its "kinetic" angular momentum

$m_r\boldsymbol{q}_r \times \dot{\boldsymbol{q}}_r$, if velocity-dependent potentials are present.) Conversely, if the conservation of $\boldsymbol{J}$ is to follow automatically from the equations of motion, L must be invariant under Eq.(11.5).

We can now transcribe these properties to the Hamiltonian form. The phase-space variables are the Cartesian coordinates $\boldsymbol{q}_r$ and Cartesian canonical momenta $\boldsymbol{p}_r$, obeying the basic PB rules:

$$\{q_{rj}, q_{sk}\} = \{p_{rj}, p_{sk}\} = 0, \qquad \{q_{rj}, p_{sk}\} = \delta_{rs}\delta_{jk}; \\ 1 \leqq r,\ s \leqq N;\ 1 \leqq j,\ k \leqq 3. \tag{11.7}$$

The translational invariance of L and of $\dot{\boldsymbol{q}}_r$ imply that of $\boldsymbol{p}_r$, whereas the rotational invariance of L and the three-vector nature of $\dot{\boldsymbol{q}}_r$ imply that $\boldsymbol{p}_r$ is a vector. Consequently, the expression

$$\sum_{r=1}^{N} \boldsymbol{p}_r \cdot \dot{\boldsymbol{q}}_r$$

is both translationally and rotationally invariant, and the invariance properties of the Lagrangian become invariance properties for the Hamiltonian:

$$H(\boldsymbol{q}_r, \boldsymbol{p}_r) = H(\boldsymbol{q}_r + \boldsymbol{a}, \boldsymbol{p}_r); \tag{11.8a}$$

$$H(\boldsymbol{q}_r + \delta\theta\, \boldsymbol{n} \times \boldsymbol{q}_r, \boldsymbol{p}_r + \delta\theta\, \boldsymbol{n} \times \boldsymbol{p}_r) = H(\boldsymbol{q}_r, \boldsymbol{p}_r)\,. \tag{11.8b}$$

The crucial point now is that these invariance properties of the Hamiltonian under the geometrical transformations of the basic variables can be reinterpreted as the invariance of the Hamiltonian with respect to the groups of canonical transformations generated by $\boldsymbol{P}$ and $\boldsymbol{J}$, respectively! Let us examine the case of $\boldsymbol{P}$ first. The effect of a translation in the coordinate system is to transform the basic variables as follows:

$$\boldsymbol{q}_r \to \boldsymbol{q}_r + \boldsymbol{a}, \qquad \boldsymbol{p}_r \to \boldsymbol{p}_r, \qquad r = 1, \cdots, N. \tag{11.9}$$

Obviously, the basic PB relations are preserved by this transformation, so that in the terminology of Chapter 10, a translation acts as an automorphism on the dynamical variables of the system. Making use of the especially simple form of $\boldsymbol{P}$ in terms of $\boldsymbol{p}_r$, we can easily work out the effect of a finite canonical transformation generated by the linear combination $\boldsymbol{a} \cdot \boldsymbol{P}$:

$$e^{-\boldsymbol{a}\cdot\tilde{\boldsymbol{P}}}\boldsymbol{p}_r = \boldsymbol{p}_r, \qquad e^{-\boldsymbol{a}\cdot\tilde{\boldsymbol{P}}}\boldsymbol{q}_r = \boldsymbol{q}_r + \boldsymbol{a}, \qquad r = 1, \cdots, N, \tag{11.10}$$

which coincides with Eq.(11.9). We know from the work of the previous chapter that the automorphism Eq.(11.9) must be canonically implemented and that the infinitesimal generators should be uniquely determined up to additive constants, and now we see that the infinitesimal generators are just the linear momenta $\boldsymbol{P}$. Equation (11.8a) can be given in the new form:

$$e^{-\boldsymbol{a}\cdot\tilde{\boldsymbol{P}}}H(\boldsymbol{q}_r, \boldsymbol{p}_r) = H(\boldsymbol{q}_r, \boldsymbol{p}_r). \tag{11.11}$$

Thus the transformation-theoretical significance of the total linear momentum $\boldsymbol{P}$ can be expressed as follows: it is that (three-vector) dynamical variable which generates a group of canonical transformations whose effect on the basic variables $\boldsymbol{q}_r$, $\boldsymbol{p}_r$ is exactly the same as the effect of a translation of the spatial coordinate system. Conversely, this property can be used to define $\boldsymbol{P}$. For any Hamiltonian system for which there is a *natural action* of three-dimensional translations as a group of automorphisms on the algebraic system of dynamical variables, we define the total linear momentum $\boldsymbol{P}$ as the infinitesimal generator of these automorphisms. This determines $\boldsymbol{P}$ for such systems uniquely up to an additive constant. But the real value of this definition of $\boldsymbol{P}$ lies in the fact that one need not necessarily use Cartesian $\boldsymbol{q}_r$'s and $\boldsymbol{p}_r$'s in phase space; any set of generalized q's and p's is admissible, as long as one is sure that translations act as automorphisms on these variables, and as long as one is able to write down the way these general q's and p's change under a translation. It is also clear that there may be cases where $\boldsymbol{P}$ can be defined by the foregoing considerations, but the Hamiltonian does not obey Eq.(11.11); thus $\boldsymbol{P}$ may not be conserved. For example, this is so if in addition to interparticle forces in a many-particle system, the particles were to experience a uniform gravitational field.

Let us next examine the properties of the total angular momentum $\boldsymbol{J}$. We have seen that it is associated with the effects of rotations on the system. Because the group of rotations has a more complicated structure than the group of translations, we here restrict ourselves to infinitesimal rotations. Using Cartesian variables to start with, namely $\boldsymbol{q}_r$ and $\boldsymbol{p}_r$, one easily verifies that the transformed variables obtained as a result of a small rotation of amount $\delta\theta$ about the direction $\boldsymbol{n}$,

$$\boldsymbol{q}_r \to \boldsymbol{q}_r + \delta\theta\ \boldsymbol{n} \times \boldsymbol{q}_r, \qquad \boldsymbol{p}_r \to \boldsymbol{p}_r + \delta\theta\ \boldsymbol{n} \times \boldsymbol{p}_r \tag{11.12}$$

obey the same basic PB relations as the $\boldsymbol{q}_r$ and $\boldsymbol{p}_r$ (up to first order in $\delta\theta$, but later chapters show this for finite rotations) namely Eq.(11.7). Thus (small) rotations of the Cartesian coordinate system also act as automorphisms of the algebra of the $\boldsymbol{q}_r$'s and $\boldsymbol{p}_r$'s. Using now the expression for $\boldsymbol{J}$ given in Eq.(11.6), we can work out the effect of an infinitesimal canonical transformation generated by $\boldsymbol{J}$. We find:

$$e^{-\delta\theta \boldsymbol{n}\cdot\tilde{\boldsymbol{J}}}\boldsymbol{q}_r \simeq \boldsymbol{q}_r - \delta\theta\{\boldsymbol{n}\cdot\boldsymbol{J}, \boldsymbol{q}_r\} = \boldsymbol{q}_r + \delta\theta\ \boldsymbol{n} \times \boldsymbol{q}_r,$$
$$e^{-\delta\theta \boldsymbol{n}\cdot\tilde{\boldsymbol{J}}}\boldsymbol{p}_r \simeq \boldsymbol{p}_r - \delta\theta\{\boldsymbol{n}\cdot\boldsymbol{J}, \boldsymbol{p}_r\} = \boldsymbol{p}_r + \delta\theta\ \boldsymbol{n} \times \boldsymbol{p}_r, \tag{11.13}$$

which coincides exactly with Eq.(11.12). Here, we have used the basic PB's Eq.(11.7) to evaluate those between $\boldsymbol{J}$ and $\boldsymbol{q}_r$ and $\boldsymbol{J}$ and $\boldsymbol{p}_r$. We can now use these properties of $\boldsymbol{J}$ as definitions. For any Hamiltonian system for which there is a natural action of three-dimensional orthogonal rotations as a group of automorphisms on the dynamical variables, we define the total angular momentum as the infinitesimal generator of these automorphisms. In practice we would evaluate $\boldsymbol{J}$ by examining small rotations. Once more, this definition of $\boldsymbol{J}$ determines it up to an additive constant; however, it can be used, whatever be the generalized q's and p's one is dealing with, as long as one knows the effects of (small) rotations on these q's and p's. Once $\boldsymbol{J}$ has been defined, the Hamiltonian will obey Eq.(11.8b), which we now rewrite as

$$e^{-\delta\theta\boldsymbol{n}\cdot\tilde{\boldsymbol{J}}}H(q,p) = H(q,p) \tag{11.14}$$

if it is rotationally invariant, otherwise not. Only in the former case will $\boldsymbol{J}$ be conserved in time. A simple example in which $\boldsymbol{J}$ can be defined but is not a constant of motion is the case of charged particles in a uniform magnetic field.

The characterisation of total linear and angular momenta that we have now achieved, by examining the mechanics of a system of many particles, is a basic and general one, and is the one that we carry over to the general case. Recall that the Lagrangian description of mechanics was originally derived in Chapter 2 by starting from the Newtonian mechanics of point particles. But we have seen thereafter that the Lagrangian and Hamiltonian theories are very general forms of dynamics, and we have developed the appropriate techniques for dealing with Lagrangians and Hamiltonians of almost any functional forms. For example, a Lagrangian linear in the velocities is not most naturally interpreted as a collection of point particles! But it is the value of associating the linear and angular momenta with the Euclidean group that these important dynamical quantities can be set up for any system on which the transformations of the Euclidean group act as automorphisms.

As another important example, let us next discuss these variables in the case of a system involving a single field $\phi(\boldsymbol{x},t)$ on Euclidean space. Let us assume for simplicity that the Lagrangian is local and can be given in the form:

$$L[\phi,\dot{\phi}] = \int \mathcal{L}(\phi,\phi_j,\cdots;\dot{\phi})d^3x. \tag{11.15}$$

We have also assumed that the Lagrangian density $\mathcal{L}$ does not depend on the spatial derivatives $\dot{\phi}_j, \dot{\phi}_{jk}\cdots$ of the velocity variable $\dot{\phi}$. As we explained in the last chapter, this makes the canonically conjugate momentum $\pi(\boldsymbol{x},t)$ and $\dot{\phi}(\boldsymbol{x},t)$ local functions of one another. We have $\pi(\boldsymbol{x},t)$ defined by

$$\pi(\boldsymbol{x},t) = \frac{\delta L[\phi,\dot{\phi}]}{\delta\dot{\phi}(\boldsymbol{x},t)} = \frac{\partial\mathcal{L}}{\partial\dot{\phi}}, \tag{11.16}$$

and the Lagrangian equation of motion is

$$\dot{\pi}(\boldsymbol{x},t) = \frac{\delta L[\phi,\dot{\phi}]}{\delta\phi(\boldsymbol{x},t)}. \tag{11.17}$$

Now the action of a space translation on the field variables $\phi,\dot{\phi}$ is as follows:

$$\phi(\boldsymbol{x},t)\to\phi(\boldsymbol{x}-\boldsymbol{a},t), \quad \dot{\phi}(\boldsymbol{x},t)\to\dot{\phi}(\boldsymbol{x}-\boldsymbol{a},t). \tag{11.18}$$

This transformation law is different in structure from the one given in Eq.(11.9). It follows from the geometrical identification of $\boldsymbol{x}$ in a given coordinate system.

Thus the corresponding velocity variable $\dot{\phi}$ also changes. Nevertheless, this is again a geometrical transformation in configuration space, so if L is invariant under it we will find a constant of motion linear in $\pi(\boldsymbol{x}, t)$. Because for infinitesimal $\boldsymbol{a}$, $|\boldsymbol{a}| \ll 1$, we have

$$\delta\phi(\boldsymbol{x}, t) = -\boldsymbol{a} \cdot \boldsymbol{\nabla}\phi(\boldsymbol{x}, t),$$

the action principle yields:

$$\boldsymbol{P} = -\int d^3x \ \pi(\boldsymbol{x}, t)\boldsymbol{\nabla}\phi(\boldsymbol{x}, t) = \text{constant}. \tag{11.19}$$

In the Lagrangian treatment this shows us how to define the total linear momentum $\boldsymbol{P}$. Similarly if L is rotationally invariant, that is, unchanged when ϕ and $\dot{\phi}$ change as follows:

$$\phi(\boldsymbol{x}, t) \to \phi(\boldsymbol{x} - \delta\theta \ \boldsymbol{n} \times \boldsymbol{x}, t), \quad \dot{\phi}(\boldsymbol{x}, t) \to \dot{\phi}(\boldsymbol{x} - \delta\theta \ \boldsymbol{n} \times \boldsymbol{x}, t). \tag{11.20}$$

then we have the law:

$$\boldsymbol{J} = -\int d^3x \ \pi(\boldsymbol{x}, t) \ \boldsymbol{x} \times \boldsymbol{\nabla} \ \phi(\boldsymbol{x}, t) = \text{constant}. \tag{11.21}$$

This, too, is linear in π. This shows us how to define the total angular momentum for a simple field system; in this case we have simply a single field variable $\phi(\boldsymbol{x}, t)$ and the effect of a small rotation is to merely change the argument $\boldsymbol{x}$ appropriately.

Going over to the Hamiltonian treatment, we have the basic variables $\phi(\boldsymbol{x}, t)$, $\pi(\boldsymbol{x}, t)$ obeying the fundamental PB relations:

$$\begin{aligned} \{\phi(\boldsymbol{x}, t), \phi(\boldsymbol{y}, t)\} &= \{\pi(\boldsymbol{x}, t), \pi(\boldsymbol{y}, t)\} = 0, \\ \{\phi(\boldsymbol{x}, t), \pi(\boldsymbol{y}, t)\} &= \delta^{(3)}(\boldsymbol{x} - \boldsymbol{y}), \end{aligned} \tag{11.22}$$

and the Hamiltonian H is the integral of a density $\mathcal{H}$:

$$H[\phi, \pi] = \int d^3x \mathcal{H}(\phi, \pi), \qquad \mathcal{H}(\phi, \pi) = \pi(\boldsymbol{x}, t)\dot{\phi}(\boldsymbol{x}, t) - \mathcal{L}(\phi, \cdots; \dot{\phi}). \tag{11.23}$$

The field equations of motion are

$$\dot{\phi}(\boldsymbol{x}, t) = \{\phi(\boldsymbol{x}, t), H\} = \frac{\delta H}{\delta\pi(\boldsymbol{x}, t)}, \ \dot{\pi}(\boldsymbol{x}, t) = \{\pi(\boldsymbol{x}, t), H\} = -\frac{\delta H}{\delta\phi(\boldsymbol{x}, t)}. \tag{11.24}$$

In terms of these variables, translations and rotations act in the following fashion:

$$\phi(\boldsymbol{x}, t) \to \phi(\boldsymbol{x} - \boldsymbol{a}, t), \quad \pi(\boldsymbol{x}, t) \to \pi(\boldsymbol{x} - \boldsymbol{a}, t); \tag{11.25a}$$
$$\phi(\boldsymbol{x}, t) \to \phi(\boldsymbol{x} - \delta\theta \ \boldsymbol{n} \times \boldsymbol{x}, t), \quad \pi(\boldsymbol{x}, t) \to \pi(\boldsymbol{x} - \delta\theta \ \boldsymbol{n} \times \boldsymbol{x}, t). \tag{11.25b}$$

These are trivially seen to be automorphisms of the variables, because Eq.(11.22) is preserved. Formally, we can attempt to set up the linear and angular momenta $\boldsymbol{P}, \boldsymbol{J}$ as the infinitesimal generators of Eq.(11.25a), (11.25b), respectively. That is, we seek functionals P_j of ϕ and π with the properties:

$$\begin{aligned}
\{\phi(\boldsymbol{x},t), P_j\} &\equiv \frac{\delta P_j}{\delta\pi(\boldsymbol{x},t)} = -\frac{\partial\phi(\boldsymbol{x},t)}{\partial x_j}, \\
\{\pi(\boldsymbol{x},t), P_j\} &\equiv -\frac{\delta P_j}{\delta\phi(\boldsymbol{x},t)} = -\frac{\partial}{\partial x_j}\pi(\boldsymbol{x},t);
\end{aligned} \tag{11.26}$$

and P_j can then be identified as the components of the total linear momentum. Similarly, we can identify the total angular momentum with functionals J_j having the properties:

$$\begin{aligned}
\{\phi(\boldsymbol{x},t), J_j\} &\equiv \frac{\delta J_j}{\delta\pi(\boldsymbol{x},t)} = -(\boldsymbol{x}\times\boldsymbol{\nabla})_j\,\phi(\boldsymbol{x},t), \\
\{\pi(\boldsymbol{x},t), J_j\} &\equiv -\frac{\delta J_j}{\delta\phi(\boldsymbol{x},t)} = -(\boldsymbol{x}\times\boldsymbol{\nabla})_j\,\pi(\boldsymbol{x},t)\,.
\end{aligned} \tag{11.27}$$

We may check that the expressions obtained in Eqs. (11.19) and (11.21) have the desired properties; it is also quite easy to directly set up these functionals using the basic relations Eq.(11.22) and the conditions imposed on them. We can write them as follows:

$$\begin{aligned}
P_j &= \int T_j(\boldsymbol{x},t)d^3x, \qquad T_j = -\pi(\boldsymbol{x},t)\frac{\partial}{\partial x_j}\phi(\boldsymbol{x},t); \\
J_1 &= \int M_j(\boldsymbol{x},t)d^3x, \qquad M_j = -\pi(\boldsymbol{x},t)(\boldsymbol{x}\times\boldsymbol{\nabla})_j\phi(\boldsymbol{x},t).
\end{aligned} \tag{11.28}$$

Thus in the case of a field system also we see how we can formally set up these dynamical variables from considering the action of the Euclidean group on the fields, remembering of course that here, too, $\boldsymbol{P}$ as well as $\boldsymbol{J}$ are determined up to additive constants.

Just as the variables $\boldsymbol{P}$ and $\boldsymbol{J}$ are interpreted as generators of space translations and space rotations on the fundamental field variables, a comparison of Eqs.(11.26) and (11.27) with Eq.(11.24) tells us that the Hamiltonian acts as the generator of the group of automorphisms corresponding to *time translations.* Indeed, because the basic PB relations Eq.(11.22) are valid at each instant of time, time translations do act as automorphisms, but these are *dynamical* automorphisms, in contrast to the *kinematical* automorphisms generated by $\boldsymbol{P}$ and $\boldsymbol{J}$. It would be natural to think of the quantities $T_j(\boldsymbol{x},t)$, $M_j(\boldsymbol{x},t)$ and $\mathcal{H}(\boldsymbol{x},t)$ as the *densities* of linear momentum, angular momentum, and energy, respectively, but one must keep in mind that one can add exact differentials to these without altering the total quantities $\boldsymbol{P}, \boldsymbol{J}, H$ so that the densities are arbitrary to this extent.

There are two comments regarding these dynamical variables for a "local" field system that are of interest. First, these densities $T_j(\boldsymbol{x}), M_j(\boldsymbol{x}), \mathcal{H}(\boldsymbol{x})$ are all "localized," which means that the PB of any two of them at distinct space points vanishes, namely,

$$\{T_j(\boldsymbol{x},t),T_k(\boldsymbol{y},t)\} = \{T_j(\boldsymbol{x},t),\mathcal{H}(\boldsymbol{y},t)\} = \{\mathcal{H}(\boldsymbol{x},t),\mathcal{H}(\boldsymbol{y},t)\} = 0\,, \quad \boldsymbol{x} \neq \boldsymbol{y}\,. \tag{11.29}$$

We can show this by computing the PB's of the basic fields ϕ, π with the densities themselves. We obtain:

$$\begin{aligned}
\{\phi(\boldsymbol{x},t),T_j(\boldsymbol{y},t)\} &= -\delta^{(3)}(\boldsymbol{x}-\boldsymbol{y})\frac{\partial}{\partial x_j}\phi(\boldsymbol{x},t),\\
\{\pi(\boldsymbol{x},t),T_j(\boldsymbol{y},t)\} &= -\delta^{(3)}(\boldsymbol{x}-\boldsymbol{y})\frac{\partial}{\partial x_j}\pi(\boldsymbol{x},t)\,;
\end{aligned} \tag{11.30a}$$

$$\begin{aligned}
\{\phi(\boldsymbol{x},t),M_j(\boldsymbol{y},t)\} &= -\delta^{(3)}(\boldsymbol{x}-\boldsymbol{y})(\boldsymbol{x}\times\boldsymbol{\nabla})_j\phi(\boldsymbol{x},t),\\
\{\pi(\boldsymbol{x},t),M_j(\boldsymbol{y},t)\} &= -\delta^{(3)}(\boldsymbol{x}-\boldsymbol{y})(\boldsymbol{x}\times\boldsymbol{\nabla})_j\pi(\boldsymbol{x},t)\,;
\end{aligned} \tag{11.30b}$$

$$\begin{aligned}
&\{\phi(\boldsymbol{x},t),\mathcal{H}(\boldsymbol{y},t)\} = \{\phi(\boldsymbol{x},t),\pi(\boldsymbol{y},t)\dot{\phi}(\boldsymbol{y},t)-\mathcal{L}(\boldsymbol{y},t)\}\\
&= \delta^{(3)}(\boldsymbol{x}-\boldsymbol{y})\dot{\phi}(\boldsymbol{y},t)+\pi(\boldsymbol{y},t)\,\frac{\delta\dot{\phi}(\boldsymbol{y},t)}{\delta\pi(\boldsymbol{x},t)}-\frac{\partial\mathcal{L}(\boldsymbol{y},t)}{\partial\dot{\phi}(\boldsymbol{y},t)}\,\frac{\delta\dot{\phi}(\boldsymbol{y},t)}{\delta\pi(\boldsymbol{x},t)}\\
&= \delta^{(3)}(\boldsymbol{x}-\boldsymbol{y})\dot{\phi}(\boldsymbol{x},t),\\
&\{\pi(\boldsymbol{x},t),\mathcal{H}(\boldsymbol{y},t)\} = -\pi(\boldsymbol{y},t)\frac{\delta\dot{\phi}(\boldsymbol{y},t)}{\delta\phi(\boldsymbol{x},t)}+\frac{\delta\mathcal{L}(\boldsymbol{y},t)}{\delta\phi(\boldsymbol{x},t)}+\frac{\partial\mathcal{L}(\boldsymbol{y},t)}{\partial\dot{\phi}(\boldsymbol{y},t)}\frac{\delta\dot{\phi}(\boldsymbol{y},t)}{\delta\phi(\boldsymbol{x},t)}\\
&= \frac{\delta\mathcal{L}(\boldsymbol{y},t)}{\delta\phi(\boldsymbol{x},t)}\\
&= \frac{\delta L(t)}{\delta\phi(\boldsymbol{x},t)}\delta^{(3)}(\boldsymbol{x}-\boldsymbol{y})+\frac{\partial}{\partial y_j}\left(\frac{\partial\mathcal{L}(\boldsymbol{y},t)}{\partial\phi_j(\boldsymbol{y},t)}\delta^{(3)}(\boldsymbol{x}-\boldsymbol{y})\right)\\
&-\frac{\partial}{\partial y_j}\left(\delta^{(3)}(\boldsymbol{x}-\boldsymbol{y})\frac{\partial}{\partial y_k}\frac{\partial\mathcal{L}(\boldsymbol{y},t)}{\partial\phi_{jk}(\boldsymbol{y},t)}-\frac{\partial\mathcal{L}(\boldsymbol{y},t)}{\partial\phi_{jk}(\boldsymbol{y},t)}\frac{\partial}{\partial y_k}\delta^{(3)}(\boldsymbol{x}-\boldsymbol{y})\right)\cdots\\
&= \dot{\pi}(\boldsymbol{x},t)\delta^{(3)}(\boldsymbol{x}-\boldsymbol{y})+\frac{\partial}{\partial y_j}\left(\frac{\partial\mathcal{L}(\boldsymbol{y},t)}{\partial\phi_j(\boldsymbol{y},t)}\delta^{(3)}(\boldsymbol{x}-\boldsymbol{y})\right.\\
&\left.+\frac{\partial\mathcal{L}(\boldsymbol{y},t)}{\partial\phi_{jk}(\boldsymbol{y},t)}\frac{\partial}{\partial y_k}\delta^{(3)}(\boldsymbol{x}-\boldsymbol{y})-\delta^{(3)}(\boldsymbol{x}-\boldsymbol{y})\frac{\partial}{\partial y_k}\,\frac{\partial\mathcal{L}(\boldsymbol{y},t)}{\partial\phi_{jk}(\boldsymbol{y},t)}\cdots\right).
\end{aligned} \tag{11.30c}$$

Here we have had to introduce the variational derivatives of $\dot{\phi}(\boldsymbol{y},t)$ with respect to $\phi(\boldsymbol{x},t)$, $\pi(\boldsymbol{x},t)$, and of $\mathcal{L}(\boldsymbol{y},t)$ with respect to $\phi(\boldsymbol{x},t)$. The Lagrangian equation of motion is also used. Equation (11.30) shows that all the PB's of the form of Eq.(11.29) vanish as long as the space points $\boldsymbol{x}$ and $\boldsymbol{y}$ are distinct, so that each of the seven important dynamical variables P_j, J_j and H is an integral of a "localized density."

The second interesting comment has to do with the fact that the field system has an infinite number of degrees of freedom. Using the terminology of the previous chapter, we see that $\boldsymbol{P}, \boldsymbol{J}$ and H are not "finite polynomials" in the variables $\phi(\boldsymbol{x},t)$, $\pi(\boldsymbol{x},t)$ because they involve integrals over all space with uniform measure. There is, therefore, the possibility that configurations may exist in which the localized densities $T_j(\boldsymbol{x},t)$, $M_j(\boldsymbol{x},t)$, $\mathcal{H}(\boldsymbol{x},t)$, are well defined, but the integrated quantities do not exist! It is, therefore, natural to separate the dynamical configurations of a field system into those in which the seven quantities P_j, J_j, H are defined (the "standard" configurations) and those in which they are not defined (the "nonstandard" configurations). By any canonical transformation generated by any finite polynomial we cannot pass from a standard configuration to a nonstandard one, or vice versa. Hence, as long as we restrict ourselves to transformations generated by finite polynomials, the standard configurations form a closed set by themselves. The nonstandard configurations form a residual set that can be further separated into distinct closed (infinite) subsets of configurations with no transformation generated by finite polynomials being able to connect one subset to another. But it would take us beyond the scope of this book to attempt to give a proof of these statements, and we must content ourselves with these qualitative statements.

Summarizing the work of this chapter, we have seen the intimate connection between the dynamical variables of linear and angular momentum and the action of translations and rotations on a system. These quantities play a dual role; on the one hand they act as the infinitesimal generators of the automorphisms that reflect the (essentially kinematical) behavior of the system under translations and rotations; on the other hand for an isolated system they are constants of motion reflecting the independence of the Hamiltonian of the absolute origin or the absolute orientation of the coordinate system. We have seen this connection in detail for a system of interacting particles, as well as for a simple single-field system. We now abstract this property of linear and angular momenta and always define them to be the generators of the automorphisms of the Euclidean group,whatever be the detailed nature of the system. Thus this definition (which, of course, only determines them up to additive constants) makes sense for a system of particles alone, or a system of fields alone, or a system involving particles as well as fields, in fact, any system on which the operations of the Euclidean group have a meaning.

This work must now be generalized and carried further, in many ways. First, in regard to the treatment of fields, we have considered only the case of a single field, a *scalar* field ϕ, in which case there is the following relationship between the densities of angular and linear momentum, following from Eq.(11.28):

$$M_{jk}(\boldsymbol{x},t) = x_j T_k(\boldsymbol{x},t) - x_k T_j(\boldsymbol{x},t). \tag{11.31}$$

(We have expressed the density of angular momentum here as an anti-symmetric second-rank tensor.) This must be generalized to the case where we have several field functions $\phi_a(\boldsymbol{x},t)$ that undergo a "mixing" under a rotation, thereby introducing extra terms into the density $M_{jk}(\boldsymbol{x},t)$. This is a more or less technical point, but one that has an interesting interpretation. More important, we have to analyze the relationships that exist *between* the variables P_j, J_j (and H), that is, the values of their PB's with one another; we must also understand the details of the relationship between these PB relations and the structure of the Euclidean group. We have also to use the work of this chapter as a model for describing the Hamiltonian dynamics of systems subject to larger groups of transformations, like the transformations of Galilean and Poincaré relativity; these will bring in transformations in which both space and time coordinates are transformed in certain ways. As a preparation for all this work, it is now to our advantage to devote the next three chapters to an exposition of the relevant properties of Lie groups and Lie algebras, and their realizations. We will then have at hand the necessary mathematical tools for carrying out the foregoing tasks.

Chapter 12

Sets, Topological Spaces, Groups

The primary purpose of this and the two succeeding chapters is to provide the reader with an adequate understanding of the properties of Lie groups, Lie algebras, and their realizations so that he can then follow the relevance and use of these things in classical Hamiltonian mechanics, some aspects of which are discussed in the last few chapters of this book. The treatment that we give here of these mathematical topics certainly cannot aspire to be a very complete one, because that would be out of place in a book devoted more to classical mechanics than to group theory. On the other hand, too brief a treatment of the mathematics would have made these chapters useless except for those who know the subjects already. In the choice of material covered, therefore, a compromise has to be made; we have included all that is needed for our later applications, in addition to those topics that make for cohesion and continuity of presentation. In this chapter, we discuss sets, groups, and topological groups and give the definition of a Lie group. Chapter 13 describes the structure of Lie groups in more detail and explains how one goes from a Lie group to its Lie algebra and vice versa. Chapter 14 describes the kinds of realizations of these abstract objects that are relevant in Hamiltonian mechanics, namely realizations of Lie algebras and Lie groups via PB's and canonical transformations, respectively.

As he goes along, the reader will recognize that some of the ideas introduced in these three chapters have appeared already in some of the previous ones. For example, the idea of a group was used when we spoke of the group properties of canonical transformations, and the main properties of a Lie algebra have been met with in discussing the properties of PB's. However, there should be no harm in some repetition of these topics; it should only lead to a better understanding of them!

Sets

The notion of a set is a basic one. We can consider a set to be a collection of objects selected according to some criterion or generated by some process, or a set may sometimes be defined by directly identifying the objects it is made up of. The objects contained in a set S are called "elements" of S; if an element x belongs to a set S, we express this fact by the notation $x \in S$; if not, we write $x \notin S$. Given a set S, another set S_1 is called a "subset" of S if every $x \in S_1$ also obeys $x \in S$; we write $S_1 \subseteq S$ to express this relationship. If it happens that in addition to containing S_1 as a subset, S contains at least one element not belonging to S_1, then S_1 is a "proper" subset of S, and we write $S_1 \subset S$. Given two sets S_1, S_2, their "union" is a set S written $S = S_1 \cup S_2$, consisting of all elements of S_1 together with all those of S_2; that is, $x \in S$ if and only if $x \in S_1$ and/or $x \in S_2$. The "intersection" of the sets S_1, S_2 is a set S' written $S' = S_1 \cap S_2$, consisting of all elements belonging to S_1 as well as to S_2. It is clear that one can take unions and intersections of an arbitrary number of sets by a straightforward generalization of the foregoing definitions.

Starting with the idea of sets, the next important concept is that of a "mapping." Let there be given two sets S and T. A mapping $f : S \to T$ is a rule that associates with each element $x \in S$ a unique element $f(x) \in T$: we say that x "goes into" $f(x)$ under the mapping f or that $f(x)$ is "the image of x" under f. As we have defined a mapping, two distinct elements in S need not necessarily possess distinct images, and not every element in T need arise as the image of some element in S. The collection of all those elements in T that do arise as images of elements of S forms a subset of T called the "range" of the mapping, written $f(S)$. If $f(S) = T$, we say that the mapping is "onto." In the particular case in which the sets S and T are the same, the general mapping f is a mapping of S into S, and in case $f(S) = S$, we have a mapping of S onto S.

When can one talk of an inverse to a mapping? If f is one to one, and onto, from S to T, then it is easy to see that, in fact, f establishes a unique correspondence between the elements of S and those of T. (A mapping f is one to one if each image element $f(x)$ in T is the image of one and only one initial element x in S.) In this case, we can define the inverse mapping f^{-1}, *from T to S*, as follows: given $y \in T$, $f^{-1}(y)$ is that unique element in S such that

$$f(f^{-1}(y)) = y. \tag{12.1}$$

It is trivial to modify the definition of f^{-1} in case f is one to one, but fails to be onto. One merely restricts the elements y in Eq.(12.1) to the subset $f(S)$ in T; that is, the "domain of definition" of f^{-1} is not T but $f(S)$.

By their very nature, mappings can be "multiplied" into one another to yield new mappings. Consider three sets S, T, U and two mappings f, g : f mapping S into T and g mapping T into U. We can then define a mapping h from S into U by the rule:

$$x \in S: \qquad h(x) \in U; \qquad h(x) = g(f(x)). \tag{12.2}$$

We write $h = gf$ and call it the product mapping of g with f; it is merely the result of applying f and g in succession, of course, f first and g second. It is important to write g and f in a particular order; if S, T, and U are all distinct sets, gf has meaning whereas fg does not. If the sets T, U are identical with S, then we have defined the product of two mappings of S into S, the result being yet another mapping $S \to S$. In this case gf and fg both have meaning, but they need not coincide.

The crucial property of the rule for multiplying mappings is that these products are *associative*. Let S, T, U, V be four sets, and let us have three mappings $f : S \to T, g : T \to U, h : U \to V$. Then the product gf maps S into U, and this multiplied by h yields a mapping $h(gf)$ from $S \to V$. Similarly, the product hg is a mapping from T to V, and thus the product $(hg)f$ is also a mapping from S to V. Thus we have two mappings, $h(gf)$ and $(hg)f$, both from S to V. But they are identical! Let us write $F = h(gf)$, $G = (hg)f$. For any $x \in S$, write $f(x) = y \in T$, $g(y) = z \in U$, $h(z) = w \in V$. Clearly, $(gf)(x) = g(f(x)) = g(y) = z \in U$, and $(hg)(y) = h(g(y)) = h(z) = w \in V$. Thus we find, for any $x \in S$, $F(x) = h(z) = w$, $G(x) = (hg)(y) = w$, and the mappings $F = h(gf)$ and $G = (hg)f$ are identical:

$$h(gf) = (hg)f. \tag{12.3}$$

This is the law of associativity for product of mappings. As long as the maps are written in a specific sequence, we can multiply them out in any order we please; the answer is the same and we can unambiguously denote it by hgf. This associative law clearly extends to products of any number of mappings.

Groups

The notion of a group arises by abstraction from some of the properties of mappings between sets that we have just described. A group G is a set with a binary composition defined on its elements. If a, b are any two elements in G, then to the ordered pair a, b corresponds a unique third element $c \in G$, which may be written $c = ab$ and called the product of a with b. The law that determines c from a and b is called the group composition law. It must have the following properties:

(i) *Associativity:* for any three elements $a, b, c \in G$, we must have

$$(ab)c = a(bc). \tag{12.4}$$

(ii) *Identity:* there must exist a unique element $e \in G$, called the *identity*, such that for all $a \in G$, we have

$$ae = ea = a. \tag{12.5}$$

(iii) *Inverses:* to each $a \in G$, there must correspond a unique element called "the inverse of a" and written a^{-1}, such that

$$aa^{-1} = a^{-1}a = e. \tag{12.6}$$

Any set G with a binary composition law obeying (i), (ii), and (iii) is called a group.

We have stated the properties defining a group so that they appear as simple as possible. One could weaken properties (ii) and (iii) relating to the existence of the identity and inverses and demand, for example, only the existence of a left-identity e_L and a left-inverse a_L^{-1} to each $a \in G$. That is, in place of Eq.(12.5), we postulate that e_L has the property:

$$e_L a = a, \qquad \text{all} \quad a \in G; \tag{12.5$'$}$$

and in place of Eq.(12.6), we require that to each $a \in G$, there correspond a unique element a_L^{-1} obeying

$$a_L^{-1} a = e_L. \tag{12.6$'$}$$

However, it is easy to demonstrate that from Eqs.(12.5)′ and (12-6)′ we are led back to Eqs.(12.5) and (12.6)! First let us prove from Eqs. (12.5)′ and (12.6)′ alone that a_L^{-1} serves as a right inverse to a: consider the element:

$$x = (a_L^{-1})_L^{-1} a_L^{-1} a a_L^{-1},$$

the first factor being the left inverse of a_L^{-1}. By the associative law (i), we have two ways of evaluating x. Indicating the sequence of products by brackets, we have the possibility:

$$x = (\{(a_L^{-1})_L^{-1} a_L^{-1}\} a) a_L^{-1} = (e_L a) a_L^{-1} = a a_L^{-1};$$

an alternative possibility is:

$$x = (a_L^{-1})_L^{-1} \{(a_L^{-1} a) a_L^{-1}\} = (a_L^{-1})_L^{-1} \{e_L a_L^{-1}\} = (a_L^{-1})_L^{-1} a_L^{-1} = e_L.$$

Note that in both methods of computing x, only Eq.(12.5)′ and (12.6)′ have been used. Thus we find:

$$a_L^{-1} a = a a_L^{-1} = e_L\,,$$

and we may now simply write a^{-1} instead of a_L^{-1}. Next we show that e_L is a right identity as well, by using what we have just proven: for any a we have:

$$a e_L = a(a^{-1} a) = \{a a^{-1}) a = e_L a = a.$$

and we are justified in writing e in place of e_L. The uniqueness of the element e, and of the inverse to a given element, can also be proved quite easily so that we are fully justified in adopting the simple form (i), (ii), (iii) for the laws that define a group.

Even though the group composition law is required to be associative, it is not required to be commutative; that is to say, in general the elements ab and ba need not be the same. If a group G is such that for all pairs a, b the two products ab and ba are identical, then G is called an "Abelian group"; otherwise it is a non-Abelian group. Again, a group G may be made up of a finite number of distinct elements, in which case it is called a finite group; on the other hand, if G

has an infinite number of distinct elements, it is an infinite group. In the latter case, two possibilities arise: the number of elements in G may be denumerably infinite, or it may be nondenumerably infinite, in which case we may loosely call G a continuous group.

Given a set G that is also a group, there may be a subset $G_1 \subseteq G$ whose elements form a group by themselves, of course with respect to the composition law given in G. That is, the identity e of G may be contained in G_1, and products and inverses of elements in G_1 may again be contained in G_1. In that case G_1 is called a subgroup of G. Here is a succinct test to determine whether a subset G_1 is a subgroup: for any two elements $a, b \in G_1$, the product $a^{-1}b$ must also belong to G_1. If G_1 is a proper subset of G, $G_1 \subset G$, and is also a subgroup, we call it a proper subgroup of G. There is a particularly important kind of subgroup, called an invariant, or a normal subgroup, that arises in the following way. Given a subgroup G_1 of G and any element $b \in G$, it is an easy matter to verify that the subset of elements of the form bxb^{-1}, for all $x \in G_1$, which may be denoted bG_1b^{-1}, is also a subgroup of G. If for all $b \in G$, we have the equality $bG_1b^{-1} = G_1$, then G_1 is called a normal, or an invariant, subgroup of G. (To be precise, this means that each element in bG_1b^{-1} belongs to G_1 and vice versa, *not* that $bxb^{-1} = x$ for all $x \in G_1$.)

Several of the groups that are of interest in classical (as well as in quantum!) mechanics are infinite groups with a nondenumerably infinite set of elements. The ones we have in mind are the groups of rotations of three-dimensional space, translations in three-dimensional space, the Euclidean and Galilean groups, and, turning to special relativity, the homogeneous as well as the inhomogeneous Lorentz groups. As we said before, groups of this type are, loosely speaking, continuous groups in the sense that in order to "enumerate" their elements, use must be made of (a finite number of) real parameters or coordinates varying continuously in certain ranges. It is clear that we have here a combination of the idea of a group with the mathematically distinct idea of continuity; in order to understand this combination properly, it is worthwhile devoting some space to a precise definition of the notion of continuity.

Topological Spaces

Let S be a set with elements $a, b, \cdots, x, \cdots$. We say that a *topology* θ has been defined on S if we are given a collection of subsets $N, N', N'', \cdots$ of S with the following properties: to each $x \in S$ must be associated some of these subsets, which will be called "neighborhoods of x" and denoted by $N_x, N'_x, N''_x, \cdots$, and we must have:

(i) $x \in N_x$ for N_x any neighborhood of x.

(ii) If N_x, N'_x be neighborhoods of x, there must exist a neighborhood N''_x of x obeying $N''_x \subset N_x \cap N'_x$.

(iii) If N_x be a neighborhood of x, and $y \in N_x$, there must exist a neighborhood N_y of y such that $N_y \subset N_x$.

Thus in a topology θ on S, each element x comes with the set of its neighborhoods, and these neighborhoods must satisfy these three conditions. Two different topologies θ, θ' will differ in what subsets of S are called neighborhoods of an element x. It should also be stressed that even if an element x is contained in one of the subsets N defined when the topology θ is given, this N need not be a neighborhood of x; for example, if x and y are two distinct elements of S, and N_y a neighborhood of y such that $x \in N_y$, then N_y need not be a neighborhood of x.

A simple example is to take the set S as the points of real, n-dimensional Euclidean space and the neighborhoods of a point x as the open spheres of nonzero radius centred on x. That is, the neighborhoods of x can be indexed by the nonzero radius ρ, $N_{x,\rho}$, and are given by: $N_{x,\rho}$ =set of all points y such that $|x-y| < \rho$, $0 < \rho < \infty$. (The $|x-y|$ is just the Euclidean distance between x and y). With this definition of neighborhoods, one can easily verify that the three laws defining a topology θ are all obeyed. One can vary this example a bit by defining the neighborhoods $N_{x,\rho}$ of x to be the open spheres described before, but with ρ restricted by, say, $0 < \rho \leqq 1$; the three laws for a topology are again obeyed, and we obtain another possible topology θ' on S. The difference between θ and θ' lies in the fact that if x and y are any two points in S, y is always contained in some neighborhood of x in the topology θ, but this need not be so in θ'. However, in a sense that is described later, these two topologies θ, θ' are, in fact, equivalent.

A set S with a topology θ defined on it is called a "topological space". Consider now a subset O of a topological space S. We say that O is "open" if the following is satisfied: if $x \in O$, then there exists a neighborhood N_x of x such that $N_x \subset O$. The decision whether a certain subset O of S is open or not will depend on the given topology, that is, on the given definition of neighborhoods. If O is open, the complement of O, written CO, is said to be "closed". (The complement CS_1 of a subset S_1 of a set S is the subset of S consisting of all those elements which are not in S_1.) We can see by property (iii) in the definition of a topology that every neighborhood N is an open set. Of course, a given subset O of S need be neither open nor closed! There are some properties of open and closed sets that are easily proved, which we shall simply state. The union of an arbitrary collection of open sets and the intersection of a finite number of open sets are both open sets. On the other hand, the union of a finite number of closed sets and the intersection of any collection of closed sets, are both closed sets. We can also see easily that an open set O is the union of a collection of neighborhoods. For, to each element $x \in O$, there is a neighborhood $N_x \subset O$, and we can check that the union of these neighborhoods N_x, $\bigcup_x N_x$, is precisely O.

The definition of a topological space has been given to provide a framework in which to discuss the continuity properties of mappings. It turns out that the definition of continuity depends essentially on that of open sets. Therefore, two topologies θ, θ' on S that lead to the same open sets lead to the same notion of continuity, and can therefore be considered equivalent. More precisely, two topologies θ, θ' are called equivalent if a subset O that is open in the topology θ is also open in θ', and vice versa. It is very easy to given necessary and sufficient conditions for θ and θ' to be equivalent. Let the neighborhoods of a point x in the topology θ be written $N_x, N'_x, \cdots$ and in the topology θ' be written $M_x, M'_x, \cdots$. First assume that θ and θ' are equivalent. Now, in any given topology each neighborhood is an open set. The equivalence of θ and θ' implies that any N_x, open in θ, must also be open in θ'; because x is an element of N_x, this implies the existence of an M_x obeying $M_x \subset N_x$. Interchanging the roles of θ and θ' we conclude that any neighborhood M_x in θ' must contain some neighborhood N_x in θ, $N_x \subset M_x$. Therefore, we have the necessary conditions for equivalence of θ and θ': given any neighborhood N_x of x in θ, there exists a neighborhood M_x of x in θ' such that $M_x \subset N_x$; given any neighborhood M'_x of x in θ', there exists a neighborhood N'_x of x in θ such that $N'_x \subset M'_x$. These conditions are also sufficient to ensure that θ and θ' are equivalent. If a subset O is open in θ, and $x \in O$, then there is an $N_x \subset O$; then there is an $M_x \subset N_x$, and thus $M_x \subset O$; it follows that O is also open in θ'. The converse is also obviously true: an open set in θ' is open in θ as well.

We are now prepared to define what is meant by a continuous mapping. Let S and T be two topological spaces, with neighborhoods $N_x, N'_x, \cdots$ for any $x \in S$, $M_y, M'_y, \cdots$ for any $y \in T$. (If S and T coincide, we of course identify the neighborhoods $M, M', \cdots$ with $N, N', \cdots$) Let $f : S \to T$ be a mapping from S to T. We say that f is a continuous mapping if, $y = f(x)$ being the image of x and given a neighborhood M_y of y, there exists a neighborhood N_x of x such that $f(N_x) \subset M_y$. That is, given $y = f(x)$ and M_y a neighborhood of y, there is a neighborhood N_x of x such that for any $x' \in N_x$, $f(x') \in M_y$. The analogy of this definition of continuity to that for real functions of a real variable is evident. It is interesting that this definition of continuity can be reformulated in terms of open sets alone. Let us first define the "inverse image" of a subset of T. If $T_0 \subset T$ is a subset of T, the inverse image $f^{-1}(T_0)$ of T_0 is the subset of S consisting of all those elements x for which $f(x) \in T_0$. Then we have the alternative definition of continuity: the mapping $f : S \to T$ is continuous if the inverse image $f^{-1}(O)$ of every open subset $O \subset T$ is an open subset of S.

Let us prove that these two definitions of continuity are identical. Let f be continuous in the first sense and let O be an open subset of T. If $x \in f^{-1}(O)$, then $y = f(x) \in O$. Because O is open in T, there exists an M_y, a neighborhood of y, such that $M_y \subset O$. By the continuity of f in the first sense, we can find a neighborhood N_x of x fulfilling $f(N_x) \subset M_y$. From $f(N_x) \subset O$, it then follows that $N_x \subset f^{-1}(O)$. Thus for any element $x \in f^{-1}(O)$, we have succeeded in finding an $N_x \subset f^{-1}(O)$, which means that $f^{-1}(O)$ is an open subset of S. Therefore, f is continuous according to the second definition. Conversely, let f be continuous in the second sense, let x be an element in S, and $y = f(x)$ it image in T. Any neighborhood M_y of y is an open set in T, thus its inverse image $f^{-1}(M_y)$ is an open set in S containing the element x. Consequently, we can find a neighborhood N_x of x fulfilling $N_x \subset f^{-1}(M_y)$, in other words, $f(N_x) \subset M_y$. Thus f is continuous in the first sense, and the identity of the two definitions of continuity is established. The reader can now also satisfy himself that if f is a continuous mapping from S to T with respect to given topologies in S and T and if these are replaced by equivalent topologies, then f remains continuous.

Because open and closed sets are complements of one another, we can state the property of continuity in terms of closed sets as well. For any subset $T_0 \subset T$, we have the equality $f^{-1}(CT_0) = Cf^{-1}(T_0)$ relating the inverse image of the complement of T_0 and the complement of the inverse image of T_0. If then $A \subset T$ is a closed subset of T, we have that $f^{-1}(A) = Cf^{-1}(CA)$, and because $f^{-1}(CA)$ is open in S, being the inverse image of the open set CA in T, $f^{-1}(A)$ is closed in S. Thus for a continuous mapping $f : S \to T$, *the inverse image of a closed set in* T *is a closed set in* S. This definition of continuity is easily verified to be equivalent to the one in terms of open sets. All in all, under a continuous mapping $f : S \to T$, every open set in T arises from, or is the image of, an open set in S, and every closed set in T arises from a closed set in S; to be more precise, the *inverse images* of open (closed) sets are open (closed) sets. However, we cannot interchange the roles of S and T here! Namely, if we are given an open set $A \subset S$, we *cannot* assert from the continuity of the mapping f that $f(A)$ is an open set in T. In general, because $f^{-1}(f(A)) \supseteq A$ and because we cannot be sure that the equality sign in always valid here, it is not possible to compare $f^{-1}(f(A))$ with A directly. If f were a one-to-one mapping, then for any set $A \subset S$, we certainly have $f^{-1}(f(A)) = A$, but even then we cannot conclude from the continuity of f, A being open, that $f(A)$ is also open; all we can assert is that $f(A)$ cannot be closed, because if it were, its inverse image A would also have to be closed. This means that in case f is a continuous mapping from S onto T and is also one to one so that f^{-1} exists as a mapping from T onto S, f^{-1} *need not be a continuous mapping.* Thus the continuity of a mapping does not imply that of its inverse, assuming the latter exists!

A specially important type of mapping $f : S \to T$ from one topological space S to another T arises if f is one to one and continuous and if f^{-1} *happens to be continuous from* T *to* S. Such a mapping is called a *homeomorphism* between S and T. Thus a homeomorphism between S and T is a one-to-one correspondence between the sets S and T *that is continuous in both directions.* The importance of such a mapping is evident from the following: let us be given a mapping g of S into S, and let it be continuous (with respect to the topology given in S). Using the one-to-one correspondence f from S onto T, we can define a mapping g' of T into T: for any $y \in T$, we just set $g'(y) = f(g\{f^{-1}(y)\})$. Then if f is a homeomorphism, it follows that g' is a continuous mapping in T. Thus homeomorphic mappings preserve notions of continuity. (The proof of the statement made here is left to the reader).

We are now in a position to combine the structure of a group and of a topological space to define a topological group.

Topological Groups

Let G be a set that is a group and a topological space at the same time. We call G a *topological group* if the group operations of taking inverses of given elements and taking the products of pairs of elements are both continuous with respect to the given topology. Let us spell out these conditions in detail. If $a \in G$ is an arbitrary element in G, we write $N_a, N'_a, \cdots$ for the neighborhoods of a in a given topology. Then we must have:

(i) *Continuity of Inverses:* Given any neighborhood $N_{a^{-1}}$ of the element a^{-1} there must exist a neighborhood N_a of a such that $b \in N_a$ implies $b^{-1} \in N_{a^{-1}}$. We can write this as follows: to each $N_{a^{-1}}$ there exists some N_a obeying $(N_a)^{-1} \subset N_{a^{-1}}$. [$(N_a)^{-1}$ is the set of inverses of all elements in N_a].

(ii) *Continuity of Products:* Let a, b be any two elements in G, and $c = ab$ their product. Then given any neighborhood N_c of c, there must exist neighborhoods N_a, N_b of a, b, respectively such that $a' \in N_a$, $b' \subset N_b$ implies $a'b' \in N_c$, or that $N_a N_b \subset N_c$ in an obvious notation.

Let us make a few comments regarding this definition. The topological space G comes with a set of neighborhoods $N_a, N'_a, \cdots$ for each element a; there is no reason to expect, however, that each neighborhood $N_{a^{-1}}$ of a^{-1} is obtained by taking the inverses of all the elements in some neighborhood N_a of a! That is, there need be no N_a obeying $(N_a)^{-1} = N_{a^{-1}}$. Similarly, given a, b and $c = ab$, each neighborhood N_c of c need not arise from multiplying all the elements in some N_a with all those in some N_b!

Another point is that the definition of continuity of a mapping that we gave earlier can be applied as it stands to the operation of taking inverses of elements of G, because the correspondence $a \to a^{-1}$ is a mapping of G onto G. Condition (i) simply requires this mapping to be continuous. As for the product law in G, we must first rewrite it as a mapping from the set $G \times G$ consisting of all *pairs* of elements (a, b), a and $b \in G$ onto G, define in an obvious way a topology for $G \times G$, and then require continuity for this mapping. But because we have been quite explicit in stating the requirements under (ii), there is no need here for this additional development.

Because for a topological group G the group operations are continuous, it turns out to be possible to define two new topologies on G that are equivalent to each other and to the original one. Let the given topology be called θ, and in it let the neighborhoods of the identity e be $N_e, N'_e, \cdots$. Let us define a new system of neighborhoods $N_a^L, N_a^{L'}, \cdots$ for any element a by taking them to be of the form $aN_e, aN'_e, \cdots$. First we prove that for each element a, any new neighborhood N_a^L contains an old one, N_a, and vice versa. From the equation $a^{-1}a = e$ and the continuity of group multiplication (in θ), given any N_e, there exist (old) neighborhoods $N_{a^{-1}}, N_a$ such that $N_{a^{-1}} N_a \subset N_e$. That is, taking the element $a^{-1} \in N_{a^{-1}}$, for any N_e we can find a N_a obeying $a^{-1}N_a \subset N_e$ or $N_a \subset aN_e$. But every new neighborhood of a is just a subset of the form aN_e for some N_e; thus one-half of our statement is proved. Conversely, applying continuity of group multiplication to the product $ae = a$, given any N_a we can find an N_e and an N'_a obeying $N'_a N_e \subset N_a$. Choosing the element $a \in N'_a$ here, we conclude that $aN_e \subset N_a$, or that any old neighborhood of a contains a new one. Using these properties, we can next prove that the new neighborhoods $N_a^L, N_a^{L'}, \cdots$ have the properties necessary to define a new topology θ^L. First, given two of these new neighborhoods $N_a^L = aN_e$, $N_a^{L'} = aN'_e$, we wish to assert the existence of an $N_a^{L''} = aN''_e$, say, such that $aN''_e \subset aN_e \cap aN'_e$. But this is the same as $N''_e \subset N_e \cap N'_e$, and such an N''_e can certainly be found because the N_e and N'_e are neighborhoods of e in θ. Second, given the new neighborhood aN_e of a and an element $b \in aN_e$, we wish to demonstrate the existence of a new neighborhood bN'_e of b such that $bN'_e \subset aN_e$. This is the same as the requirement $a^{-1}bN'_e \subset N_e$. Now, because $a^{-1}b \in N_e$ and N_e is a neighborhood in θ of e, we can find an old neighborhood $N_{a^{-1}b}$ of $a^{-1}b$ within $N_e : N_{a^{-1}b} \subset N_e$. Next, by what we proved a little while ago, we can find a new neighborhood $a^{-1}bN'_e$ of $a^{-1}b$ contained in this old one: $a^{-1}bN'_e \subset N_{a^{-1}b}$. This N'_e fulfills our conditions, as we have $a^{-1}bN'_e \subset N_e$ or $bN'_e \subset aN_e$. Thus we see that with the help of the neighborhoods of the identity given by the original topology, $N_e, N'_e, \cdots$, we are able to define a new system of neighborhoods for any element a, namely aN_e, $aN'_e, \cdots$, and from the continuity of the group operations we can see that these new neighborhoods do form a topology θ^L, and that θ^L and θ are equivalent. It is obvious that we can also define a second system of new neighborhoods by $N_e a, N'_e a \cdots$ for any a, that these define a new topology θ^R, and that θ^R and θ are also equivalent. Finally, all three topologies $\theta, \theta^L, \theta^R$ are equivalent, and one can satisfy oneself that the group operations, being continuous in θ, are continuous in θ^L and θ^R as well.

With these facts, it is natural to redefine the concept of a topological group in terms of neighborhoods of the identity alone and to generate neighborhoods for an arbitrary element a from these by definition. It is worthwhile going through the details of this reformulation, and checking that the earlier conditions defining a topological group are obeyed. We now define a topological group G to be a group in which certain subsets $N, N', N'', \cdots$ are defined as "neighborhoods of the identity", and in which two sets of conditions are imposed:

Set I:

(i) $e \in N$, N any neighborhood of e;

(ii) Given N, N', there exists an $N'' \subset N \cap N'$

(iii) For any $a \in N$, there exists an N' such that $aN' \subset N$.

Set II:

(iv) Given N, there is an N' such that $(N')^{-1} \subset N$;

(v) Given N, there is an N' such that $N'N' \subset N$;

(vi) Given N, and any $a \in G$, there is an N' such that $aN'a^{-1} \subset N$.

The role of conditions (i), (ii), and (iii) is to ensure that if we define the neighborhoods of any element a to be the subsets $aN, aN', \cdots$ obtained via group multiplication, then these neighborhoods do form a topology; conditions (iv),(v), and (vi) then ensure that the group operations are continuous in this topology. We now prove these two statements, even though some of the arguments are very similar to ones given earlier.

First, let us prove that we do have a topology. Clearly, $a \in aN$ for any N. If we are given two neighborhoods aN and aN', because by (ii) there is an $N'' \subset N \cap N'$, with this N'' we have a neighborhood aN'' obeying $aN'' \subset aN \cap aN'$. Finally, let $b \in aN$, that is, $a^{-1}b \in N$. By (iii), we can find an N' obeying $a^{-1}bN' \subset N$ or $bN' \subset aN$ so that the neighborhood bN' of b is contained within aN. Thus we do have a topological space.

Second, we consider the continuity of the group operations. As for inverses: let $a^{-1}N$ be a neighborhood of a^{-1}. By (vi) we can find an N' obeying $aN'a^{-1} \subset N$. By (iv) we can then find an N'' obeying $(N'')^{-1} \subset N'$. Thus with N'' we have $a(N'')^{-1}a^{-1} \subset N$, $(N'')^{-1}a^{-1} \subset a^{-1}N$, or $(aN'')^{-1} \subset a^{-1}N$. Thus we have succeeded in finding a neighborhood of a, namely, aN'', such that the inverses of the elements in this neighborhood are all contained in the given neighborhood of a^{-1}, and the continuity of the operation of taking inverses is proved. Finally for products: let a, b be any two elements of G, and abN some neighborhood of their product. By (v) we can find an N' such that $N'N' \subset N$, and by (vi) we can then find an N'' such that $b^{-1}N''b \subset N'$. Thus we have $b^{-1}N''bN' \subset N$, or $aN''bN' \subset abN$. Because aN'' and bN' are neighborhoods of a, b, respectively, the continuity of taking products is also proved.

We have thus seen that because the group multiplication law can carry us from the identity element to any other one, the definition of a topological group can be given entirely in terms of conditions on neighborhoods of the identity. We leave it as an exercise for the reader to check that the alternative choice of the subsets $Na, N'a, \cdots$ as neighborhoods of a leads to a topology equivalent to the one used above; use must be made here of the conditions (i) to (vi) imposed previously. As another exercise, he may satisfy himself that a topological group as defined originally is such that its neighborhoods of the identity do obey conditions (i) to (vi). (We have only demonstrated that if (i) to (vi) are satisfied, then the earlier conditions for a topological group are obeyed). There is one other condition on the structure of a topological group that is usually added to conditions (i) to (vi) given previously, but which was not needed in our analysis up to now. This is the condition of "separability" and reads:

Condition (vii): Given any element $a \in G$ other than e, there exists a neighborhood N of e such that $a \notin N$.

There are a couple of interesting and easy-to-understand properties of a general topological group G that we would like to describe now, before specializing to the case of Lie groups. Because we have a topology on G, we can speak of "continuous curves" in G. Let τ be a real variable in the range $0 \leq \tau \leq 1$, and for each τ let $Z(\tau)$ be an element of G such that $Z(\tau)$ is a continuous function of τ. Stated precisely, this amounts to the following: if for $\tau = \tau_0$, $Z(\tau_0) = a \in G$ and N_a is any neighborhood of a, then we can find an interval of length 2δ,

$\delta > 0$, around τ_0 such that for $|\tau - \tau_0| < \delta$, $Z(\tau) \in N_a$. Because any such N_a arises from some neighborhood N of e as $N_a = aN$, we can also state the definition of a continuous curve as follows: the curve $\tau \to Z(\tau) \in G$, $0 \leqq \tau \leqq 1$, is continuous if at every point τ_0 and for any neighborhood N of e we can find a $\delta(\tau_0) > 0$ such that if $|\tau - \tau_0| < \delta(\tau_0)$ then $Z^{-1}(\tau)Z(\tau_0) \in N$. Using this notion of continuous curves in G, we can split up G into disjoint subsets. Given two elements $a, b \in G$, we will call them *equivalent* if we can find a continuous curve $\tau \to Z(\tau)$, $0 \leqq \tau \leqq 1$ such that $Z(0) = a$ and $Z(1) = b$. One can easily convince onself that this idea of equivalence has all the properties required of an *equivalence relation*:

1. every element $a \in G$ is equivalent to itself;
2. if a is equivalent to b, then b is equivalent to a; and
3. if a is equivalent to b, and b to c, then a is equivalent to c.

Consequently, one sees that the group G splits up into disjoint subsets: any two elements in one subset are equivalent, and no two elements taken from different subsets are equivalent. When the group G has been split up this way into disjoint subsets or "equivalence classes" (with respect to the relation of equivalence given above), there is one particular equivalence class that can be singled out: this is the class of elements equivalent to the identity element e. This class is called the "identity component" of G and is written G_0. We can easily convince ourselves that G_0 is, in fact, a *subgroup* of G! The basic property of the elements of G_0 is that any one of them can be continuously connected to the identity e. But if a, b are any two elements in G_0 and $Z_1(\tau)$, $Z_2(\tau)$ are two continuous curves leading from e to a, b, respectively, the continuity of the two operations of taking inverses and products leads to the conclusion that the curve $Z(\tau) = Z_1(\tau)^{-1}Z_2(\tau)$ is also continuous. Because for $\tau = 0$, $Z(0) = e$ and for $\tau = 1$, $Z(\tau) = a^{-1}b$, along with a and b, $a^{-1}b$ also belongs to G_0, thus establishing that G_0 is a subgroup. One can go even further and show that G_0 is an *invariant subgroup*. If $Z(\tau)$ is continuous and leads from e to $a \in G_0$, and $c \in G$ is any other element, the continuity of the curve $cZ(\tau)c^{-1}$ is immediate, hence $cac^{-1} \in G_0$ along with a.

In the definition of a continuous curve, we demanded that to each neighborhood N of e we could find a positive number $\delta(\tau_0) > 0$ such that $|\tau - \tau_0| < \delta(\tau_0)$ implied $Z(\tau)^{-1}Z(\tau_0) \in N$. In this form we allowed δ to depend on τ_0 (and, of course, also on N!). It is possible to show quite easily, although we do not do so here, that given N, a choice of δ can be made *that is independent of* τ_0. Thus one can *prove* the uniform continuity of a curve that obeys the definition of continuity we gave earlier; this happens because the interval $0 \leqq \tau \leqq 1$ on the real line is closed. An interesting consequence of this property of uniform continuity is the following:given any fixed neighborhood N of the identity e, any element $a \in G_0$ can be expressed as a *finite product* of elements in N; that is, given a, we can find elements $a_1, a_2, \cdots, a_n$, all $a_j \in N$ and $n < \infty$ such that

$a = a_1 a_2 \cdots a_n$. In this sense, the entire identity component G_0 of a topological group G may be finitely generated from any neighborhood of the identity element.

We conclude this chapter with the definition of a Lie group. A topological group G is called a *Lie group* if there exists some neighborhood N_0 of the identity that can be mapped homeomorphically on to an open, bounded subset of the real Euclidean space E_n for some n. That is to say, we can establish a one-to-one correspondence between the group elements $a \in N_0 \subset G$ and points lying in an open bounded region in E_n in such a way that this correspondence is a continuous mapping in both directions.The topology to be used for E_n is the customary one given by open spheres. Thus we are able in a Lie group to assign n real variables as coordinates to the group elements in N_0, such that the coordinates of a^{-1} are continuous functions of the coordinates of a, and similarly those of the product ab are continuous functions of those of a and b. The number n is called the dimension of the Lie group; it is assumed that it is finite. The groups connected with geometrical properties of space and time, such as the three-dimensional rotation group $R(3)$, the Euclidean group $E(3)$, and the Galilean and Lorentz groups, are all Lie groups. In the following chapter we analyze some of the properties of Lie groups that are relevant for our purposes; one makes an *assumption* that the functions involved in computing inverses and products are *differentiable sufficiently often*, in addition to being continuous. This then leads to the possibility of making a rather detailed study of Lie groups through the study of certain kinds of partial differential equations.

Chapter 13

Lie Groups and Lie Algebras

Consider a Lie group G in which coordinates have been introduced in some neighborhood N_0 of the identity e. Thus, if a is an element of G in N_0, it corresponds to a point in n-dimensional real Euclidean space with coordinates

$$a \to \alpha^1,\ \alpha^2, \cdots, \alpha^n\ ; \tag{13.1}$$

and as a varies over N_0, the points α^j vary over some open bounded region in E_n. It is customary to assume, and there is no loss of generality here, that to the identity e corresponds the origin in E_n:

$$e \to \alpha^j = 0,\ j = 1, \cdots, n\ . \tag{13.2}$$

The correspondence between $a \in N_0$ and points in the relevant region in E_n is one to one. Call this latter region R. We know from the properties of G that there exist neighborhoods N, N' of e such that $(N)^{-1} \subset N_0$ and $N'N' \subset N_0$. Consequently, if $a \in N$, then $a^{-1} \subset N_0$, and a^{-1}, too, can be assigned coordinates in R. If $a, b \subset N'$, then $ab \subset N_0$ and ab also can be assigned coordinates in R. Proceeding in this way, we can easily convince ourselves that there is a suitable neighborhood N'' such that if three elements a, b, c are picked, all from N'', their product abc lies in N_0 and, hence, has coordinates. In the sequel we will deal with products of, at most, three elements at a time (exceptions, if any, are dealt with suitably); thus if all the group elements considered are taken from N'' we are free to describe them and their products in terms of coordinates. We shall from now on not mention explicitly the restriction of elements to N''.

Let $ab = c$, and let the coordinates of a, b, and c be written α^j, β^j, and γ^j, respectively. The group composition law is expressed by giving the γ^j as functions of α, β:

$$c = ab \Rightarrow \gamma^j = f^j(\alpha^1, \cdots, \alpha^n; \beta^1, \cdots, \beta^n) = f^j(\alpha, \beta), \quad j = 1, 2, \cdots, n. \tag{13.3}$$

We know that the functions f^j are continuous in their $2n$ arguments. We *assume that they possess continuous partial derivatives with respect to the α's and β's up to derivatives of order three.* We can call the f^j the group-composition functions; they are functions of two group elements, equivalently of two points in the space of the α's and β's.

Some properties of the composition functions f^j follow immediately from the convention of Eq.(13.2). These are:

$$\begin{aligned} f^j(\alpha,0) &= \alpha^j\ , \qquad \text{because } ae = a, \\ f^j(0,\alpha) &= \alpha^j\ , \qquad \text{because } ea = a. \end{aligned} \tag{13.4}$$

Similarly, the coordinates of a^{-1} can be obtained from those of a by solving the equations

$$f^j(\alpha,\alpha') = 0, \qquad a \to \alpha^j\ , \qquad a^{-1} \to \alpha'^j. \tag{13.5}$$

Certainly for sufficiently small α, that is, for α^j close enough to the origin, the solution for α'^j will be unique and will be the coordinates of a^{-1}.

Clearly there is the possibility of changing the coordinates used to describe the elements $a, b, \cdots$. If we start with the coordinate system $a \to \alpha^j$ and introduce a new system $a \to \bar{\alpha}^j$, where the $\bar{\alpha}^j$ are continuous and (sufficiently often) differentiable functions of α^j and vice versa, then the group composition will be described by a different set of composition functions, and these will have the same differentiability properties as the original $f^j(\alpha,\beta)$. Such a change of coordinates, consistent with the convention of Eq.(13.2), is described by the system of equations:

$$\begin{aligned} a \to \bar{\alpha}^j, \quad \bar{\alpha}^j &= \phi^j(\alpha), \quad \alpha^j = \psi^j(\bar{\alpha}); \\ ab = c \to \bar{\gamma}^j &= \bar{f}^j(\bar{\alpha},\bar{\beta}); \\ \phi^j(0) &= \psi^j(0) = 0; \\ \bar{f}^j(\bar{\alpha},0) &= \bar{f}^j(0,\bar{\alpha}) = \bar{\alpha}^j, \end{aligned} \tag{13.6}$$

and all we insist upon is the continuity and differentiability of the functions ϕ^j and ψ^j. The functional forms of $\bar{f}^j$, of course, follow from those of f^j and ϕ^j, whereas those of ψ^j are determined by ϕ^j. On occasion we must examine the manner in which various objects of interest change when such a change of coordinates is made.

Let us now give a very brief outline of the contents of the rest of this chapter. As a result of the assumption of differentiability of the composition functions f^j, it turns out to be possible to translate the associative property of group multiplication into differential form, namely into a system of partial differential equations for the f^j. Although the f^j are functions of two group elements, these equations will introduce a set of functions of one group element only, such that the f^j can be determined in terms of them. The condition that the partial differential equations for the f^j be soluble leads to a set of integrability conditions. It turns out that these conditions show that the group multiplication law is entirely determined in terms of a set of constants (pure numbers) called the structure constants of the group. That is, the Lie group determines the set of structure constants and vice versa. Actually, this last statement must be modified: there are in general several Lie groups, all of which give rise to the same set of structure constants. What is common to them is the group structure in a sufficiently small neighborhood of the identity, and it is this structure that can be reconstructed from the structure constants. In the course of the analysis the following important concepts are introduced: one-parameter subgroups of Lie groups, canonical coordinate systems, and Lie algebras.

Associativity in Differential Form

Let a, b, c be three elements in a Lie group G, and consider the associativity property of group multiplication:

$$c(ab) = (ca)b. \tag{13.7}$$

In a given coordinate system, $a \to \alpha^j$, this can be expressed as a property of the composition functions $f^j(\alpha, \beta)$. As long as no confusion is likely to arise, we shall sometimes write $f^j(a, b)$ with the group elements a, b as arguments instead of their coordinates α, β as arguments. then, Eq.(13.7) can be written as:

$$f^j(c, ab) = f^j(ca, b). \tag{13.8}$$

Let us write γ^j for the coordinates of c and δ^j for those of ca. Keeping a and b fixed, if we differentiate both sides of Eq.(13.8) with respect to γ^k, we obtain

$$\frac{\partial f^j(c, ab)}{\partial \gamma^k} = \frac{\partial f^j(ca, b)}{\partial \delta^l} \frac{\partial f^l(c, a)}{\partial \gamma^k}, \tag{13.9}$$

a summation on a repeated index being assumed. If we now set c equal to the identity element e in Eq.(13.9) we arrive at:

$$\left(\frac{\partial f^j(c, ab)}{\partial \gamma^k}\right)_{\gamma=0} = \frac{\partial f^j(a, b)}{\partial \alpha^l} \left(\frac{\partial f^l(c, a)}{\partial \gamma^k}\right)_{\gamma=0}. \tag{13.10}$$

This is a differential expression of the associativity property and is a property of the functions $f^j(\alpha, \beta)$. We can put it into a more transparent form if we define two matrices $F(a, b)$ and $H(a)$ as follows:

$$\begin{aligned} F(a, b) &\equiv \|f^j_k(a, b)\|, \qquad f^j_k(a, b) \equiv \frac{\partial f^j(a, b)}{\partial \alpha^k}, \\ H(a) &= \|\eta^j_k(a)\|, \qquad \eta^j_k(a) \equiv f^j_k(e, a). \end{aligned} \tag{13.11}$$

By interpreting superscripts (subscripts) as row (column) indices, Eq.(13.10) can be put into the matrix form

$$F(a, b)H(a) = H(ab). \tag{13.12}$$

Now, when a is the identity e, Eq.(13.4) shows that $H(a)$ reduces to the unit matrix so that $H(e)$ is certainly nonsingular. Continuity implies, then, that there is some neighborhood of e such that for a within this neighborhood $H(a)$ remains nonsingular and possesses an inverse, which we write as $\Xi(a)$:

$$H(a)^{-1} = \Xi(a) = \|\xi^j_k(a)\|, \qquad \xi^j_k(e) = \delta^j_k. \tag{13.13}$$

Then we can write Eq.(13.12) as:

$$F(a,b) = H(ab)\Xi(a). \tag{13.14}$$

The interesting point about this equation is that with the help of associativity, we have been able to express $F(a,b)$, which is a function of *two* group elements, in terms of the matrix H, which is a function of just *one* element. Actually, if we regard the functions $\eta^j_k(a)$ as "given", we can look upon Eq.(13.14) as a system of partial differential equations for $f^j(\alpha,\beta)$; we see this by writing out Eq.(13.14) in detail:

$$\frac{\partial f^j(\alpha,\beta)}{\partial \alpha^k} = \eta^j_l(f(\alpha,\beta))\xi^l_k(\alpha). \tag{13.15}$$

Here the coordinates β^j of b are held fixed, and for $\alpha^j = 0$ we have the boundary condition:

$$f^j(0,\beta) = \beta^j. \tag{13.16}$$

Clearly an alternative differential expression of associativity can be obtained by considering the partial derivatives of $f^j(\alpha,\beta)$ with respect to the β's. We define matrices $\hat{F}, \hat{H}, \hat{\Xi}$ analogous to F, H, Ξ, by

$$\begin{aligned} \hat{F}(\alpha,\beta) &= \|\hat{f}^j_k(\alpha,\beta)\|, & \hat{f}^j_k(\alpha,\beta) &= \frac{\partial f^j(\alpha,\beta)}{\partial \beta^k} \ ; \\ \hat{H}(\alpha) &= \|\hat{\eta}^j_k(\alpha)\|, & \hat{\eta}^j_k(\alpha) &= \hat{f}^j_k(\alpha,0) \ ; \\ \hat{\Xi}(\alpha) &= \hat{H}(\alpha)^{-1} = \|\hat{\xi}^j_k(\alpha)\|; & \hat{H}(0) &= \hat{\Xi}(0) = 1 \ . \end{aligned} \tag{13.17}$$

Then we find from Eq.(13.8) the analogues to Eqs.(13.14), (13.15), and (13.16):

$$\begin{aligned} &\hat{F}(a,b) = \hat{H}(ab)\hat{\Xi}(b) \ ; \\ &\frac{\partial f^j(\alpha,\beta)}{\partial \beta^k} = \hat{\eta}^j_l(f(\alpha,\beta))\hat{\xi}^l_k(\beta), \ f^j(\alpha,0) = \alpha^j \ . \end{aligned} \tag{13.18}$$

We have, then, two systems of partial differential equations, each of which is a consequence of the associativity property. Conversely, we can now ask if these equations, together with the appropriate boundary conditions, are enough to *determine* the composition functions $f^j(\alpha,\beta)$! Let us pose this question more precisely: suppose we are given the functions $\eta^j_k(\alpha)$, $\xi^j_k(\alpha)$, each of which is a function of n real variables, such that the matrix formed by the ξ's is inverse to that formed by the η's, and for $\alpha = 0$ both matrices reduce to the unit matrix. Can we then solve the partial differential equations Eq.(13.15) for the functions $f^j(\alpha,\beta)$ depending on $2n$ real variables, the β's being introduced by the boundary conditions Eq.(13.16), and if we then use these $f^j(\alpha,\beta)$ to define a group composition law, will the associative property be present? The answer to these questions is as follows: in the first place, given a general set of functions $\eta^j_k(\alpha)$, *there will be no solutions at all to Eq.(13.15)!* That is, the requirement that Eq.(13.15) be solvable for $f^j(\alpha,\beta)$ imposes a set of conditions on the functions $\eta^j_k(\alpha)$; these are the familiar *integrability conditions* connected with any set of partial differential equations. But, interestingly enough, once one knows that the η's obey the integrability conditions, the structure of the system of Eq.(13.15) guarantees that the associative law will be valid! It is not necessary to know the precise forms of the functions $f^j(\alpha,\beta)$; it is enough to know that they exist and satisfy Eqs.(13.15) and (13.16). Let us prove this. We take the solution $f^j(\alpha,\beta)$ to Eqs. (13.15) and (13.16), having been given an admissible set of η's, and wish to prove the relation

$$f^j(\gamma, f(\alpha,\beta)) = f^j(f(\gamma,\alpha),\beta)\ , \tag{13.19}$$

corresponding to the relation $c(ab) = (ca)b$. Consider the two sides of Eq.(13.19) as functions of γ^j, keeping α and β fixed. If we denote the expression on the left by $L^j(\gamma;\alpha,\beta)$, use of Eqs.(13.15) and (13.16) immediately shows that L is a solution of the system:

$$\frac{\partial L^j}{\partial \gamma^k} = \eta^j_l(L)\xi^l_k(\gamma)\ ,\ L^j(0;\alpha,\beta) = f^j(\alpha,\beta)\ . \tag{13.20}$$

On the other hand, writing $R^j(\gamma;\alpha,\beta)$ for the right-hand side of Eq.(13.19), we only need to use Eq.(13.15) twice and Eq.(13.16) once to see that R obeys the same system of equations as L:

$$\begin{aligned}\frac{\partial R^j}{\partial \gamma^k} =& \eta^j_m(R)\xi^m_l(f(\gamma,\alpha))\eta^l_p(f(\gamma,\alpha))\xi^p_k(\gamma) = \eta^j_l(R)\xi^l_k(\gamma),\\ R^j(0;\alpha,\beta) =& f^j(\alpha,\beta)\ .\end{aligned} \tag{13.21}$$

The important property used here is that the matrices $\|\eta\|$ and $\|\xi\|$ are inverse to one another. Hence from the uniqueness of the solutions to Eqs.(13.20) and (13.21), we conclude that Eq.(13.19) is indeed valid.

For the sake of completeness, let us satisfy ourselves that the functions $f^j(\alpha,\beta)$ resulting from Eqs.(13.15) and (13.16), given an admissible set of η's, lead to inverses with the expected properties. The boundary condition Eq.(13.16) already guarantees the property $ea = a$ of the identity for any a, because the coordinates of the identity are $\alpha^j = 0$. On the other hand, if we set $\beta^j = 0$ in Eq.(13.16) we easily check that the solution to Eq.(13.15) is simply $f^j(\alpha,0) = \alpha^j$, so that we are assured that $ae = a$ for any a. Now, given an element a with coordinates α^j, we can solve for the coordinates α'^j of its (left) inverse a^{-1} from the equations:

$$f^j(\alpha',\alpha) = 0\ , \qquad a \to \alpha, \qquad a^{-1} \to \alpha'\ . \tag{13.22}$$

Because the matrices $\|\eta\|$ and $\|\xi\|$ are nonsingular, we see from Eq.(13.15) that the Jacobian matrix

$$\left\| \frac{\partial f^j(\alpha', \alpha)}{\partial \alpha'^k} \right\|$$

is also nonsingular, thus Eq.(13.22) has a unique solution for a^{-1}, given a, and this obeys $a^{-1}a = e$. All we now have to check is whether with this a^{-1} we have $aa^{-1} = e$, that is whether

$$f^j(\alpha, \alpha') = 0 . \tag{13.23}$$

But this follows from the associative law (just proved!) applied to the product $(a^{-1})^{-1}a^{-1}a$, where the (left) inverse $(a^{-1})^{-1}$ of a^{-1} is obtained via an equation just like Eq.(13.22):

$$(a^{-1})^{-1}a^{-1}a = ea = a = (a^{-1})^{-1}e = (a^{-1})^{-1} . \tag{13.24}$$

Thus the inverses computed using $f^j(\alpha, \beta)$ have correct properties.

Summarizing, we can say that the laws of group structure are completely equivalent to the integrability conditions for the partial differential Eq.(13.15) being satisfied. If the group structure is given, the composition functions $f^j(\alpha, \beta)$ obey Eqs.(13.15) and (13.16); thus the integrability conditions must be satisfied obviously. Conversely, if these conditions are obeyed, the functions $f^j(\alpha, \beta)$ can be solved for, and the form of Eqs.(13.15) and (13.16) guarantees the proper group structure. The form of the integrability conditions is taken up later in this chapter.

Let us conclude this part with some remarks concerning the matrices $H(\alpha)$, $\hat{H}(\alpha)$ and their inverse $\Xi(\alpha)$, $\hat{\Xi}(\alpha)$. These matrices obviously depend on the coordinate system $a \to \alpha^j$ used in G, although all that we have done up to now holds good in every coordinate system. If from the system $a \to \alpha^j$ we switch over to another one, $a \to \bar{\alpha}^j$, according to the general scheme given in Eq.(13.6), we will encounter new matrices $\bar{\eta}^j_k(\bar{\alpha})$, $\bar{\xi}^j_k(\bar{\alpha})$, and so on. It is possible to give a meaning to the matrix $H(\alpha)$ in terms of infinitesimal transformations. If a, b are two elements of G and $c = ab$ is their product, and if we keep b fixed and consider a to be close to e, the coordinates of c can be expanded in a Taylor series in those of a to yield:

$$\begin{aligned}\gamma^j = f^j(\alpha, \beta) &= f^j(0, \beta) + \left(\frac{\partial f^j(\alpha', \beta)}{\partial \alpha'^k} \right)_{\alpha'=0} \alpha^k + \cdots \\ &= \beta^j + \eta^j_k(\beta)\alpha^k + \cdots = \beta^j + (H(\beta)\alpha)^j + \cdots .\end{aligned} \tag{13.25}$$

Thus the matrix $H(\beta)$ is related to the small change that the coordinates β^j of b undergo when b is multiplied on the left by an element close to the identity. A parallel meaning for $\hat{H}(\beta)$ can easily be developed.

One-Parameter Subgroups

Let a Lie group G with composition functions $f^j(\alpha, \beta)$ (in a specific coordinate system) obeying Eqs. (13.15) and (13.16) be given. A *continuous, one-*

parameter subgroup of G is a continuous, differentiable curve $\sigma \to a(\sigma) \in G$, $-\infty < \sigma < \infty$, in G, whose elements are closed under multiplication:

$$a(\sigma)a(\tau) = a(\tau)a(\sigma) = a(\sigma + \tau), \; -\infty < \sigma, \; \tau < \infty . \tag{13.26}$$

Such a subgroup is automatically Abelian, because group composition corresponds to adding the parameters σ and τ. Differentiability means that the coordinates $\alpha^j(\sigma)$ of $a(\sigma)$ are differentiable functions of σ. [Note that although the elements $a(\sigma)$ must be defined for all real values of σ, we may be able to assign coordinates $a(\sigma) \to \alpha^j(\sigma)$ only for σ in some neighborhood of the origin, say $-\sigma_0 < \sigma < \sigma_0$. Note also that we have by convention made $\sigma = 0$ correspond to the identity: $a(0) = e$.] The property of being elements of a one-parameter subgroup can thus be given a differential expression. From Eq.(13.26), we get:

$$f^j(\alpha(\tau), \alpha(\sigma)) = \alpha^j(\tau + \sigma) \; , \tag{13.27}$$

and if we differentiate both sides with respect to τ and use Eq.(13.15) we obtain:

$$f^j_k(\alpha(\tau), \alpha(\sigma))\frac{d\alpha^k(\tau)}{d\tau} = \frac{d\alpha^j(\tau + \sigma)}{d\tau} = \frac{d\alpha^j(\tau + \sigma)}{d\sigma} \; . \tag{13.28}$$

Now setting $\tau = 0$ on both sides, we find:

$$\frac{d\alpha^j(\sigma)}{d\sigma} = \eta^j_k(\alpha(\sigma))t^k \; , t^k = \left(\frac{d\alpha^k(\sigma)}{d\sigma}\right)_{\sigma=0} \; . \tag{13.29}$$

Here we have written t^k for the derivatives of the $\alpha^k(\sigma)$ at $\sigma = 0$. Thus the coordinates $\alpha^j(\sigma)$ of the elements on a continuous, differentiable, one-parameter subgroup $\sigma \to a(\sigma)$ obey this simple differential equation in the variable σ. The numbers t^k are said to form the "tangent vector" to the one-parameter subgroup. To the differential equation we can add the boundary condition:

$$\alpha^j(0) = 0 \; . \tag{13.30}$$

Instead of differentiating the composition function f in Eq.(13.27) with respect to the first argument, we could equally well have done so with respect to the second one. In that case, we would have established the simple differential equation:

$$\frac{d\alpha^j(\sigma)}{d\sigma} = \hat{\eta}^j_k(\alpha(\sigma))t^k \; , \tag{13.31}$$

with no change in the tangent vector t^k or in the boundary condition Eq.(13.30).

Let us now ask whether the differential Eq.(13.29) and boundary condition Eq.(13.30) can serve as a complete characterisation of a one-parameter subgroup. More precisely, given the group G and the functions $f^j(\alpha, \beta)$, $\eta^j_k(\alpha)$, suppose we take an arbitrary set of numbers t^k and solve Eqs.(13.29) and (13.30) and thus obtain a set of functions $\alpha^j(\sigma)$; will these be the coordinates of elements $a(\sigma)$ on some one-parameter subgroup of G with tangent vector t^k? The answer is in the affirmative, and the proof is simple. Having solved for $\alpha^j(\sigma)$, let us use the given composition functions $f^j(\alpha, \beta)$ to define:

$$L^j(\sigma; \tau) = f^j(\alpha(\sigma); \alpha(\tau)) \, . \tag{13.32}$$

By virtue of Eqs.(13.15), (13.16), (13.29), and (13.30), we find that $L^j(\sigma;\tau)$ obeys the system:

$$\frac{dL^j(\sigma;\tau)}{d\sigma} = \eta^j_k(L)\xi^k_l(\alpha(\sigma))\eta^l_m(\alpha(\sigma))t^m = \eta^j_k(L)t^k, \; L^j(0;\tau) = \alpha^j(\tau) \, . \tag{13.33}$$

But $\alpha^j(\sigma + \tau)$ obviously obeys the same differential equation and boundary condition:

$$\frac{d\alpha^j(\sigma+\tau)}{d\sigma} = \eta^j_k(\alpha(\sigma+\tau))t^k \; , \; \alpha^j(\sigma+\tau)\Big|_{\sigma=0} = \alpha^j(\tau) \, . \tag{13.34}$$

Therefore, $L^j(\sigma;\tau)$ and $\alpha^j(\sigma+\tau)$ must be the same; that is, for any choice of the tangent vector t^k, the solution $\alpha^j(\sigma)$ to Eqs.(13.29) and (13.30) obeys

$$f^j(\alpha(\sigma), \alpha(\tau)) = \alpha^j(\sigma + \tau) \tag{13.35}$$

and, therefore, corresponds to a one-parameter subgroup. That the t^k we started with is the tangent vector to this subgroup follows from Eq.(13.29) because the matrix $\|\eta\|$ is the unit matrix at the origin. Thus every differentiable one-parameter subgroup determines a unique tangent vector t^k and conversely.

There is one point that needs clarification. In determining the one-parameter subgroup corresponding to a given tangent vector t^k, the functions $\alpha^k(\sigma)$ can be solved for and Eq.(13.35) established only for σ in some neighborhood of $\sigma = 0$. Suppose that in the first instance the elements $a(\sigma)$ for $-2\sigma_0 < \sigma < 2\sigma_0$ alone have been obtained by knowing their coordinates $\alpha^j(\sigma)$, and that the multiplication law has only been established for $|\sigma|$, $|\tau| < \sigma_0$. Thus, in coordinate-free form the property $a(\sigma)a(\tau) = a(\sigma+\tau)$ has been verified only in this range. Then we can build up all the other elements in this subgroup by taking products of elements of the form $a(\sigma)$ with $|\sigma| < \sigma_0$. If σ is any positive real number whatsoever, we can express it in the form:

$$\sigma = m\frac{\sigma_0}{3} + \rho \; , \; m = \text{integer} \; , \; 0 \leqq \rho < \frac{\sigma_0}{3}, \tag{13.36}$$

and we then *define* $a(\sigma)$ directly in coordinate-free form as

$$a(\sigma) = \underbrace{a\left(\frac{\sigma_0}{3}\right) a\left(\frac{\sigma_0}{3}\right) \cdots a\left(\frac{\sigma_0}{3}\right)}_{\longleftarrow\; m \text{ factors} \;\longrightarrow} a(\rho) = \left[a\left(\frac{\sigma_0}{3}\right)\right]^m a(\rho) \, . \tag{13.37}$$

All the factors here commute with one another. If $\sigma < 0$, we set $a(\sigma) = [a(-\sigma)]^{-1}$. One can then easily check that the composition law Eq.(13.26) is valid for all σ and τ so that the *entire* one-parameter subgroup has been constructed.

Some simple properties of the solution to Eqs.(13.29) and (13.30) are easy to discover. In order to indicate explicitly the dependence of the solution on the vector t^k, let us write it in the form:

$$\alpha^j = \alpha^j(\sigma; t^1, t^2, \cdots, t^n) . \tag{13.38}$$

Now, it is an obvious property of Eqs.(13.29) and (13.30) that if we change the scale of σ by replacing it by σ/λ, $\lambda > 0$, and at the same time replace t^k by λt^k, the solution is left unchanged. Thus we have

$$\alpha^j(\sigma; t) = \alpha^j(\sigma/\lambda; \lambda t) = \alpha^j(1; \sigma t) \equiv \chi^j(\sigma t^1, \sigma t^2, \cdots, \sigma t^n) , \tag{13.39}$$

the χ^j being functions of n arguments only. We then obtain from Eqs.(13.29) and (13.30) two useful properties of the χ^j:

$$t^k \frac{\partial \chi^j(t)}{\partial t^k} = \eta^j_k(\chi) t^k , \ \chi^j(0) = 0 , \ \left(\frac{\partial \chi^j(t)}{\partial t^k}\right)_{t=0} = \delta^j_k . \tag{13.40}$$

All the foregoing analysis is valid in every coordinate system for G because although we used a specific one, $a \to \alpha^j$, we did not specify which one it was. Thus in every coordinate system we have a one-to-one correspondence between one-parameter subgroups and tangent vectors, which is the same thing as saying that this correspondence is *intrinsically coordinate independent*. The set of all tangent vectors t is defined to form a real n-dimensional linear vector space, the *tangent vector space*; and to each one-parameter subgroup $\sigma \to a(\sigma)$ in G there corresponds an element t of this vector space and conversely. If a coordinate system is introduced in G, at the same time a *corresponding basis* is thereby introduced in the tangent vector space; the description of the elements $a(\sigma)$ by the *coordinates* $\alpha^j(\sigma)$ leads at once to the description of the associated tangent vector t in terms of its *components* t^k in the corresponding basis. If we change from the coordinates $a \to \alpha^j$ in G to a new system $a \to \bar{\alpha}^j$, according to Eq.(13.6), then at the same time a new basis is introduced in the tangent vector space, related to the old one by a linear transformation. Because Eqs.(13.29) and (13.30) are valid in any coordinate system, it is immediate that the new coordinates $\bar{\alpha}^j(\sigma)$ of a one-parameter subgroup $\sigma \to a(\sigma)$ will obey:

$$\frac{d\bar{\alpha}^j(\sigma)}{d\sigma} = \bar{\eta}^j_k(\bar{\alpha}(\sigma))\bar{t}^k , \ \bar{t}^k = \left(\frac{d\bar{\alpha}^k(\sigma)}{d\sigma}\right)_{\sigma=0} , \ \bar{\alpha}_k(0) = 0 . \tag{13.41}$$

The linear relation between t^k and $\bar{t}^k$ follows from their definitions and the transformation equations Eq.(13.6):

$$\bar{t}^k = \left(\frac{\partial \phi^k(\alpha)}{\partial \alpha^l}\right)_{\alpha=0} t^l \equiv \phi^k_l(0) t^l , \tag{13.42}$$

where the $\phi^k_l(0)$ denote the derivatives of $\phi^k(\alpha)$ at the origin. Thus we can say that the one-parameter subgroup $\sigma \to a(\sigma)$ and the corresponding tangent vector t remain the same; only the coordinates $\alpha^j(\sigma)$ of $a(\sigma)$ and the components t^k of t are replaced by the new coordinates and components $a(\sigma) \to \bar{\alpha}^j(\sigma)$, $t \to \bar{t}^k$.

Equation (13.42) determines the change in basis in the tangent vector space consequent upon a change in coordinates in G. However, the fact that only the first derivatives of the transformation functions, $\phi^j_k(0)$, at the origin appear in Eq.(13.42) means that there will be many systems of coordinates in G, all of which lead to the same basis in the tangent vector space. The general condition that this should happen is that the first term in a Taylor expansion of $\bar{\alpha}^k$ in terms of the α's should be α^k, and conversely, because in that case $\phi^j_k(0) = \delta^j_k$.

The general connection between one-parameter subgroups of a Lie group and tangent vectors leads to a new and expressive notation for group elements. If $\sigma \to a(\sigma)$ is such a subgroup and t is its tangent vector, *we denote the element* $a(\sigma)$ by $\exp(\sigma t)$. We know that only the product σt is relevant; thus this is an adequate notation. The composition law for the group elements $\exp(\sigma t)$ is expressed, as is to be expected, by

$$\exp(\sigma t)\exp(\tau t) = \exp((\sigma + \tau)t) \ . \tag{13.43}$$

This is the same as Eq.(13.26), the Abelian composition law for two elements on the same one-parameter subgroup. In a coordinate system $a \to \alpha^j$ for G, t has correspondingly components t^k, and the coordinates of $\exp(t)$ are given by the functions χ^j of Eq.(13.39):

$$\exp(t) \to \chi^j(t^1, \cdots, t^n) \ . \tag{13.44}$$

Canonical Coordinates of the First kind

As we have emphasized, all the analysis carried out up to now is valid in every coordinate system in G. However, Eq.(13.44) naturally suggests a way of determining a specially simple class of coordinate systems for G. Let us ask the following question: starting from the coordinate system $a \to \alpha^j$ in G, in which the element $\exp(t)$ is assigned the coordinates $\chi^j(t^1 t^2 \cdots t^n)$, is it possible to pass to a new system of coordinates in which $\exp(t)$ is just assigned the components t^j of t as coordinates? This would certainly be the case if, in some neighborhood of the identity, *every* element $a \in G$ did lie on *some* one-parameter subgroup, for then all we would have to do is to associate with every $a \in G$ (in this neighborhood) the tangent vector t in terms of which we can write $a = \exp(t)$, and then assign to a the coordinates t^j, which are the components of t. But this can be done! If we take the n functions $\chi^j(t^1 \cdots t^n)$ that are determined once we are given the coordinate system $a \to \alpha^j$ in G, Eq.(13.40) shows that the Jacobian matrix $\|\partial\chi^j/\partial t^k\|$ is nonsingular at the origin $t^j = 0$, and so remains nonsingular in some neighborhood of the origin. Consequently, in this neighborhood we can turn the functions $\chi^j(t)$ "inside out", and express all the t^j as functions of the χ's:

$$\chi^j = \chi^j(t^1, \cdots, t^n) \to t^j = \zeta^j(\chi^1, \cdots \chi^n) \ . \tag{13.45}$$

This proves that given any element a in this neighborhood (any set of values for the χ^j), we obtain a unique tangent vector t^j whose associated one-parameter subgroup passes through a at $\sigma = 1$.

The coordinate system that has been obtained in this way, starting from a given coordinate system $a \to \alpha^j$ and working out the corresponding functions $\chi^j(t^1 \cdots t^n)$ and the functions $\zeta^j(\chi^1 \cdots \chi^n)$ of Eq.(13.45), is called a *canonical coordinate system of the first kind.* The element $a \in G$ with coordinates α^j in the original system, is assigned the coordinates

$$a \to \zeta^j(\alpha^1, \cdots \alpha^n) \tag{13.46}$$

in the new system. Elements $a(\sigma)$ on a one-parameter subgroup $\exp(\sigma t)$ with tangent vector t now have the coordinates

$$a(\sigma) = \exp(\sigma t) \to \sigma t^1, \sigma t^2, \cdots, \sigma t^n \ . \tag{13.47}$$

Because by considering all possible one-parameter subgroups we got all possible tangent vectors, the converse is also true: in a canonical coordinate system (of the first kind), a family of elements $a(\sigma)$ with coordinates $\sigma t^1, \sigma t^2, \cdots, \sigma t^n$ automatically forms a one-parameter subgroup!

It is clear that there is a whole family of canonical coordinate systems of the first kind, all related to one another by linear transformations. The crucial, coordinate-independent fact is that in some neighborhood of the identity, every element lies on some one-parameter subgroup. As we saw earlier, each way of assigning coordinates $a \to \alpha^j$ in G at the same time is a way of choosing a basis for tangent vector space, and we see now that it also determines a unique canonical coordinate system of the first kind. We can say that the canonical coordinate system that assigns the coordinates $\zeta^j(\alpha^1 \cdots \alpha^n)$ to the element a previously assigned the coordinates α^j is *associated* with the latter coordinate system. Thus we have a well-defined way of passing from any coordinate system to its unique, associated canonical coordinate system. If we have two coordinate systems $a \to \alpha^j$, $a \to \bar{\alpha}^j$ for G, related by Eq.(13.6), the two associated canonical coordinate systems are linearly related: the general element $\exp(t)$ has coordinates t^j in the former, $\bar{t}^j$ in the latter, and these obey Eq.(13.42). Of course, it can sometimes happen that two different coordinate systems α^j, $\bar{\alpha}^j$ for G determine the same basis in tangent vector space, in which case their associated canonical coordinate systems coincide; this happens if $\phi^j_k(0)$, which appears in Eq.(13.42), is δ^j_k,. As a particular example of this situation, it is obvious that the system $a \to \alpha^j$ in G and its associated canonical system determine the same basis in tangent vector space; this follows from the property

$$\left(\frac{\partial \chi^j(t)}{\partial t^k}\right)_{t=0} = \delta^j_k$$

of the functions χ^j that connect these two coordinate systems (see Eq.(13.40)).

The basic properties of a canonical coordinate system of the first kind can also be expressed in terms of its associated system of functions $\eta^j_k(a)$. Suppose the coordinates $a \to \alpha^j$ are canonical. Then the solution $\alpha^j(\sigma;t)$ to Eqs.(13.29) and (13.30) must be

$$\alpha^j(\sigma;t) \stackrel{*}{=} \sigma t^j \ . \tag{13.48}$$

[We indicate by an asterisk any equation valid only in canonical coordinates.] Putting this known solution into Eq.(13.29), we get a characteristic property of the functions η:

$$\eta^j_k(\alpha)\alpha^k \stackrel{*}{=} \alpha^j \ . \tag{13.49}$$

Conversely, if in some coordinate system Eq.(13.49) is obeyed, we see that Eq.(13.48) is a solution to Eqs.(13.29) and (13.30), so the system is a canonical one. Thus Eq.(13.49) is a complete characterization of a canonical coordinate system of the first kind. An alternative and completely equivalent characterization is

$$\hat{\eta}^j_k(\alpha)\alpha^k \stackrel{*}{=} \alpha^j \ . \tag{13.50}$$

The Integrability Conditions

Let us now go back to the system of partial differential equations (Eq.(13.15)) and ask under what conditions they permit us to solve for the $f^j(\alpha,\beta)$. As is well known from the theory of partial differential equations, the necessary and sufficient conditions on the η's for Eq.(13.15) to be solvable are the *integrability conditions*

$$\frac{\partial}{\partial\alpha^m}\left[\eta^j_l(f(\alpha,\beta))\xi^l_k(\alpha)\right] = \frac{\partial}{\partial\alpha^k}\left[\eta^j_l(f(\alpha,\beta))\xi^l_m(\alpha)\right] \ . \tag{13.51}$$

Here the parameters β are kept constant. Let us indicate the partial derivatives of the functions η, ξ by a subscript preceded by a comma, the partial differentiation always being carried out with respect to the argument of η or ξ. In computing the derivatives of the η's on both sides of Eq.(13.51) with respect to the α's, use must be made of Eq.(13.15), because the arguments of the η's are the functions $f(\alpha,\beta)$, and it is these latter that carry the α dependence. In this way, Eq.(13.51) leads to the following equation:

$$\begin{aligned}\eta^j_{l,p}(f)\eta^p_q(f)\xi^q_m(\alpha)\xi^l_k(\alpha) &+ \eta^j_l(f)\xi^l_{k,m}(\alpha)\\ =\eta^j_{l,p}(f)\eta^p_q(f)\xi^q_k(\alpha)\xi^l_m(\alpha) &+ \eta^j_l(f)\xi^l_{m,k}(\alpha) \, .\end{aligned} \tag{13.52}$$

Now interchange the dummy indices q, l in the first term on the right, bring the terms involving derivatives of the η's to one side, and take the other terms to the other side: Eq. (13.52) then appears as

$$\left[\eta^j_{l,p}(f)\eta^p_q(f) - \eta^j_{q,p}(f)\eta^p_l(f)\right]\xi^q_m(\alpha)\xi^l_k(\alpha) = \left[\xi^l_{m,k}(\alpha) - \xi^l_{k,m}(\alpha)\right]\eta^j_l(f) \, . \tag{13.53}$$

Now because the matrices ξ, η are inverse to one another, we can rewrite Eq. (13.53) in such a way that only quantities with f as argument appear on one side and only those with α as argument appear on the other. All we have to do is to multiply both sides of Eq.(13.53) by the factor $\eta^m_r(\alpha)\eta^k_s(\alpha)\xi^u_j(f)$, and we get:

$$\xi^u_j(f)\left[\eta^p_r(f)\eta^j_{s,p}(f) - \eta^p_s(f)\eta^j_{r,p}(f)\right] = \eta^m_r(\alpha)\eta^k_s(\alpha)\left[\xi^u_{m,k}(\alpha) - \xi^u_{k,m}(\alpha)\right] . \tag{13.54}$$

This is the most convenient form of the integrability conditions. Because Eq. (13.54) has been obtained from Eq. (13.51) by nonsingular operations only, it is completely equivalent to Eq.(13.51). Now, the two sides of Eq.(13.54) are functions evaluated at two different and independent points α, f in the group; thus the two sides can be equal only if each of them is a number independent of α and f. Therefore the necessary and sufficient conditions that Eq.(13.15) be solvable are that there exist a set of constants c^u_{rs} such that

$$\eta^m_r(\alpha)\eta^k_s(\alpha)\left[\xi^u_{m,k}(\alpha) - \xi^u_{k,m}(\alpha)\right] = -c^u_{rs} , \tag{13.55a}$$

$$\xi^u_j(\alpha)\left[\eta^p_r(\alpha)\eta^j_{s,p}(\alpha) - \eta^p_s(\alpha)\eta^j_{r,p}(\alpha)\right] = -c^u_{rs} . \tag{13.55b}$$

Actually, there is only one set of conditions here: because the matrices η, ξ are inverse to one another, we establish easily that

$$\eta^j_{s,p}(\alpha) = -\eta^j_r(\alpha)\eta^k_s(\alpha)\xi^r_{k,p}(\alpha) ; \tag{13.56}$$

if this is used on the left-hand side of Eq.(13.55b), it becomes identical to the expression on the left-hand side of Eq.(13.55a). The complete integrability conditions are thus contained in Eq.(13.55a). We can also write them in the following convenient forms:

$$\Delta^u_{mk}(\alpha) \equiv \xi^u_{m,k}(\alpha) - \xi^u_{k,m}(\alpha) + c^u_{rs}\xi^r_m(\alpha)\xi^s_k(\alpha) , \ \Delta^u_{mk}(\alpha) = 0 ; \tag{13.57a}$$

$$\eta^p_s(\alpha)\eta^j_{r,p}(\alpha) - \eta^p_r(\alpha)\eta^j_{s,p}(\alpha) = c^u_{rs}\eta^j_u(\alpha) . \tag{13.57b}$$

What we mean by writing the integrability conditions in these various forms is that there should exist a certain set of constants c^u_{rs} such that Δ^u_{mk}, defined in terms of the ξ's and these constants in Eq.(13.57a), vanish, and similarly such that the η's obey Eq.(13.57b). As in the case of Eq.(13.55), it is evident that Eqs.(13.57a) and (13.57b) are equivalent.

The constants c^u_{rs} are called the "structure constants" of the Lie group that we are studying. One such set of structure constants is determined by each Lie group; although the actual numerical values depend on the coordinate system used, we see later that the values in two different coordinates are very simply related. Now, the remarkable thing is that a given set of structure constants uniquely determines a corresponding Lie group, at least in some neighborhood of the identity. Of course, not any set of constants c^u_{rs} can serve as the structure constants of some Lie group; for example, there is the obvious property evident from Eq.(13.55) that these constants must be antisymmetric with respect to the two subscripts. We will find that the particular form that the left-hand side of Eq.(13.55a) takes leads to one more bilinear relation for the c^u_{rs}, called the *Jacobi identity*. Apart from these two conditions, the antisymmetry in the subscripts and the Jacobi identity, the c^u_{rs} are completely arbitrary; any set satisfying these two conditions is acceptable and gives rise to a Lie group in some neighborhood of the identity.

Properties of the Structure Constants

We assume that the Lie group G has been given, and we now examine the properties of the corresponding set of structure constants.

1. One property of the c^u_{rs} has been mentioned already; it is the antisymmetry property
$$c^u_{rs} = -c^u_{sr} \; . \tag{13.58}$$
All that needs to be added is that this property is present whatever be the coordinate system used to arrive at the c^u_{rs}.

2. From Eq.(13.57b) we can given a simple expression for the c^u_{rs} in terms of the second derivatives of the composition functions $f^j(\alpha,\beta)$; we set $\alpha = 0$ in Eq.(13.57b) and recall the definition of the functions $\eta^j_k(\alpha)$ to obtain:
$$c^j_{rs} = \left(\frac{\partial^2 f^j(\alpha,\beta)}{\partial\alpha^r \partial\beta^s} - \frac{\partial^2 f^j(\alpha,\beta)}{\partial\alpha^s \partial\beta^r} \right)_{\alpha=\beta=0} \tag{13.59}$$
Now, this expression for the c^j_{rs} has been arrived at by examining the integrability conditions for the Eqs.(13.15) and (13.16). Clearly, we could equally well have started out by examining the integrability requirements for Eq.(13.18), because these equations also guarantee the proper group structure. In that case, we would have been led to the existence of a set of constants $\hat{c}^j_{rs}$ that obeyed the analogue to Eq.(13.57b):
$$\hat{\eta}^p_s(\alpha)\hat{\eta}^j_{r,p}(\alpha) - \hat{\eta}^p_r(\alpha)\hat{\eta}^j_{s,p}(\alpha) = \hat{c}^u_{rs}\hat{\eta}^j_u(\alpha) \; . \tag{13.60}$$
However, if we set $\alpha = 0$ here, we find that the left-hand side of Eq.(13.60) is the same as the expression in Eq.(13.59) except for a sign; that is the two sets of structure constants obey
$$\hat{c}^j_{rs} = -c^j_{rs} \; . \tag{13.61}$$
This is a useful relation in our work later on.

Equation (13.59) can be rewritten in a slightly different form. If we make a Taylor series expansion of $f^j(\alpha,\beta)$ in α and β, the boundary conditions at $\alpha = 0$ and $\beta = 0$ imply a structure of the following form:
$$f^j(\alpha,\beta) = \alpha^j + \beta^j + a^j_{rs}\alpha^r\beta^s + \cdots \; . \tag{13.62}$$
In particular, all terms higher than linear must involve both α and β. Then, the structure constants are just the antisymmetric parts of the coefficients a^j_{rs}:
$$c^j_{rs} = -\hat{c}^j_{rs} = a^j_{rs} - a^j_{sr} \; . \tag{13.63}$$

3. The expression (Eq.(13.59)) for c^j_{rs} in terms of the composition functions $f^j(\alpha,\beta)$ makes it very easy to determine the way these constants change

under a change of coordinate system in G. Let us pass from the coordinates $a \to \alpha^j$ to the new ones $a \to \bar{\alpha}^j$ according to Eq.(13.6). This transformation involves the set of functions ϕ^j and the inverse set ψ^j. As before, we indicate partial derivatives of these functions by appropriate subscripts. Then we easily find that the structure constants $\bar{c}^j_{rs}$ evaluated in the new coordinates are related to c^j_{rs} by

$$\bar{c}^j_{rs} = \phi^j_{,l}(0)\psi^m_{,r}(0)\psi^n_{,s}(0)c^l_{mn} \ . \tag{13.64}$$

This is a linear reversible transformation.

Earlier we had seen that the two coordinate systems $a \to \alpha^j$, $a \to \bar{\alpha}^j$ determine the same coordinate system in tangent vector space if and only if the transformation functions ϕ^j and ψ^j obey:

$$\phi^j_{,k}(0) = \psi^j_{,k}(0) = \delta^j_k \ . \tag{13.65}$$

Although it is certainly true that if Eq.(13.65) is obeyed the two sets of structure constants are also identical, it must be clear that the classes of coordinate systems that lead to the same structure constants are, in general, larger than and contain several classes of coordinate systems leading to the same basis in tangent vector space. This is simply because the requirement that the right-hand side of Eq.(13.64) equal c^j_{rs} is a much weaker condition on the ϕ's and ψ's than Eq.(13.65). However, one thing is clear: if we pass from a given coordinate system $a \to \alpha^j$ to its associated canonical coordinate system of the first kind, Eq.(13.65) is valid; thus neither the basis for tangent vectors nor the set of structure constants will change (of course, in this case the ϕ's and ψ's are just the functions ζ^j, χ^j of Eqs.(13.45) and (13.39), respectively).

4. The structure constants are closely related to what is called the *commutator* $q(a, b)$ of any two elements $a, b \in G$, which is defined as

$$q(a, b) = aba^{-1}b^{-1} \ . \tag{13.66}$$

If either a or b is the identity, then so is q; consequently, if we expand the coordinates w^j of $q(a, b)$ as a power series in the coordinates α^j, β^j of a, b, the leading terms will be linear both in α and in β. Let us rewrite Eq.(13.66) in the form $ab = q(a, b)ba$ and then use Eq.(13.62) to compute the coordinates on both sides. We get

$$\alpha^j + \beta^j + a^j_{rs}\alpha^r\beta^s + \cdots = w^j + \beta^j + \alpha^j + a^j_{rs}\beta^r\alpha^s + \cdots \ ,$$

so that the leading terms in w^j are

$$w^j = (a^j_{rs} - a^j_{sr})\alpha^r\beta^s + \cdots = c^j_{rs}\alpha^r\beta^s + \cdots \ , \tag{13.67}$$

because of Eq.(13.63). Therefore, the structure constants are the coefficients of the leading terms in the expansion of the coordinates of the commutator $q(a, b)$ in terms of the coordinates of a and b.

If the group G is Abelian, $q(a,b)$ is the identity always, so that its coordinates vanish. Therefore, *for an Abelian Lie group all the structure constants vanish.* This is a coordinate-independent statement, by virtue of the linearity of the transformation laws (Eq.(13.64)) for these constants. Our later work shows that the converse is also true: if the structure constants vanish, the corresponding Lie group must be Abelian.

5. The last property of the c^j_{rs} to be derived is the Jacobi identity, which arises from a reinterpretation of Eq.(13.57b). Imagine some function $\Phi(a)$ defined on the group G, or more precisely in some neighborhood of the identity; in the coordinate system $a \to \alpha^j$, let us continue to use the symbol Φ for the function and write it as $\Phi(\alpha)$. Assume that $\Phi(\alpha)$ is differentiable. Earlier we saw that the functions $\eta^j_k(\alpha)$ come in very naturally when we consider the small changes in the coordinates of an element a that are caused by multiplying a on the left by an element close to the identity (cf.Eq.(13.25)). We can now ask for the change in the value of Φ when we change its argument a little bit in this way. Now any element close to e lies on some one-parameter subgroup, so it can be written as $\exp(\tau t)$ where τ is small and t is some tangent vector. The product $\exp(\tau t)a$ arises from a by "shifting" it a little bit, and we can shift a in various "directions" by considering various tangent vectors t. For a given direction, that is, for a given t, we can compute the rate of change of Φ in that direction, in other words, the derivative of $\Phi(\exp(\tau t)a)$ with respect to τ. With the help of Eqs.(13.15) and (13.29) we get

$$\begin{aligned}\frac{d}{d\tau}\Phi(\exp(\tau t)a) &= \frac{\partial \Phi(f)}{\partial f^j}\eta^j_k(f)\xi^k_l(\exp(\tau t))\eta^l_m(\exp(\tau t))t^m \\ &= t^k\eta^j_k(f)\frac{\partial\Phi(f)}{\partial f^j}\ , \quad f^j = f^j(\exp(\tau t), a)\ . \qquad (13.68)\end{aligned}$$

If we set $\tau = 0$ here, we obtain

$$\left.\frac{d}{d\tau}\Phi(\exp(\tau t)a)\right)_{\tau=0} = t^k\eta^j_k(a)\frac{\partial}{\partial\alpha^j}\Phi(a)\ . \qquad (13.69)$$

Thus the gradient of $\Phi(a)$ at the point a in the direction of the tangent vector t is obtained by applying to Φ a certain first-order linear differential operator; this operator is completely fixed once one knows the tangent vector t^k, and we write it as $\eta_{[t]}$:

$$\eta_{[t]}(a) \equiv t^k\eta^j_k(a)\frac{\partial}{\partial\alpha^j}\ . \qquad (13.70)$$

This operator $\eta_{[t]}$ depends linearly on t in the sense that

$$\eta_{[t_1]} + \eta_{[t_2]} = \eta_{[t_1+t_2]}\ , \qquad (13.71)$$

and $\eta_{[t]}$ vanishes identically if and only if t vanishes identically (recall that $\|\eta\|$ is a nonsingular matrix.) Even though this differential operator $\eta_{[t]}$

has been constructed in a specific coordinate system for G, it is obvious that it has a coordinate-independent meaning because the left-hand side of Eq.(13.69) has a coordinate-independent meaning and value. Consequently, if in a new coordinate system $a \to \bar{\alpha}^j$ we have the functions $\bar{\eta}^j_k(\bar{\alpha})$ and if t has components $\bar{t}^k$, we have the equality

$$\bar{t}^k \bar{\eta}^j_k(a) \frac{\partial}{\partial \bar{\alpha}^j} \equiv t^k \eta^j_k(a) \frac{\partial}{\partial \alpha^j} = \eta_{[t]} \ . \tag{13.72}$$

Now if D_1 and D_2 are any two first-order linear differential operators in any set of variables, the operator $D_1 D_2 - D_2 D_1$, also called the *commutator* of D_1 and D_2, is, as is well known, another first-order linear differential operator. Thus if we take two different tangent vectors t and u, say, the commutator $\eta_{[t]}\eta_{[u]} - \eta_{[u]}\eta_{[t]}$ will turn out to be a first-order linear differential operator. However, the operators of the form in Eq.(13.70) are rather special and are not the most general ones that could be written down; they are special because they are built up from the functions $\eta^j_k(a)$ and pure numbers t^k. Interestingly enough, Eq.(13.57b) tells us that the commutator of two of these $\eta_{[t]}$'s is another operator of the same structure! For, if we calculate the commutator of $\eta_{[t]}$ and $\eta_{[u]}$, we obtain:

$$\begin{aligned}
\eta_{[t]}\eta_{[u]} - \eta_{[u]}\eta_{[t]} &= t^k u^l \left(\eta^j_k(a) \frac{\partial}{\partial \alpha^j} \eta^m_l(a) \frac{\partial}{\partial \alpha^m} - \eta^m_l(a) \frac{\partial}{\partial \alpha^m} \eta^j_k(a) \frac{\partial}{\partial \alpha^j} \right) \\
&= t^k u^l \left(\eta^j_k(a) \eta^m_{l,j}(a) \frac{\partial}{\partial \alpha^m} - \eta^m_l(a) \eta^j_{k,m}(a) \frac{\partial}{\partial \alpha^j} \right) \\
&= t^k u^l \left(\eta^m_k(a) \eta^j_{l,m}(a) - \eta^m_l(a) \eta^j_{k,m}(a) \right) \frac{\partial}{\partial \alpha^j} \\
&= c^p_{lk} t^k u^l \eta^j_p(a) \frac{\partial}{\partial \alpha^j} \equiv -v^p \eta^j_p(a) \frac{\partial}{\partial \alpha^j} = -\eta_{[v]} \ , \\
v^p &= c^p_{kl} t^k u^l \ .
\end{aligned} \tag{13.73}$$

Thus we find that if t and u are any two tangent vectors, then they determine a third tangent vector v such that the commutator of $\eta_{[t]}$ and $\eta_{[u]}$ is just $\eta_{[v]}$! This is the content of the integrability condition in Eq.(13.57b); in fact this property of these first-order linear differential operators is completely equivalent to Eq.(13.57b) and thus to the integrability conditions.

It is useful to introduce the following notation for summarizing the way in which the tangent vector v has been obtained from t and u:

$$v = [t, u] \Leftrightarrow v^p = c^p_{kl} t^k u^l \ . \tag{13.74}$$

Because the operators $\eta_{[t]}$ and $\eta_{[u]}$ have coordinate-independent significance, so must their commutator. In other words, the process by which the vector v has been obtained from t and u is an intrinsic, coordinate-independent one. It is trivial to verify this directly: if t^k, u^l and v^p are transformed to $\bar{t}^k$, $\bar{u}^l$ and $\bar{v}^p$, respectively, and if the c's are transformed to the $\bar{c}$'s, using Eqs.(13.42) and (13.64) in the two cases, we find that the relation

$$\bar{v}^p = \bar{c}^p_{kl}\bar{t}^k\bar{u}^l \tag{13.75}$$

is satisfied. Now, the commutator of the two $\eta_{[\cdot]}$'s appearing on the left-hand side of Eq.(13.73) is usually denoted by $[\eta, \eta]_-$, thus we can write Eq.(13.73) in the concise form

$$[\eta_{[t]}, \eta_{[u]}]_- = -\eta_{[t,u]} \ . \tag{13.76}$$

This expression, together with a standard property of the commutator on the left-hand side, leads directly to the Jacobi identity. Take three independent tangent vectors t, u, w and construct a cyclic sum of double commutators of the associated η's:

$$\begin{aligned}\chi_{t,u,w} \equiv & [[\eta_{[t]}, \eta_{[u]}]_-, \eta_{[w]}]_- \\ & + [[\eta_{[u]}, \eta_{[w]}]_-, \eta_{[t]}]_- + [[\eta_{[w]}, \eta_{[t]}]_-, \eta_{[u]}]_- \ .\end{aligned}$$

One property that can be seen immediately is that $\chi_{t,u,w}$ is totally anti-symmetric in t, u, and w. Further, if all the terms in $\chi_{t,u,w}$ are written out (and there are 12 in all) they all cancel against one another; thus $\chi_{t,u,w}$ vanishes identically. On the other hand, use of Eqs.(13.71) and (13.76) permits us to express $\chi_{t,u,w}$ as a single $\eta_{[\cdot]}$:

$$\begin{aligned}\chi_{t,u,w} & \equiv \sum_{\text{cyclic}} \left[[\eta_{[t]}, \eta_{[u]}]_-, \eta_{[w]}\right]_- = \sum_{\text{cyclic}} \left[\eta_{[t,u]}, \eta_{[w]}\right]_- \\ & = \sum_{\text{cyclic}} \eta_{[[t,u],w]} = \eta_{\sum_{\text{cyclic}} [[t,u],w]} \ .\end{aligned} \tag{13.77}$$

But we know that in general, $\eta_{[v]}$ vanishes if and only if the corresponding vector v itself vanishes! Thus we derive from the vanishing of $\chi_{t,u,w}$ the result that if t, u, w are any three tangent vectors, the combination rule (Eq.(13.74)) obeys

$$\sum_{\text{cyclic}} [[t, u], w] = 0 \ . \tag{13.78}$$

This is the coordinate-independent expression of the Jacobi identity. If we use Eq.(13.74) to rewrite Eq.(13.78) in a coordinate-system, we get

$$c^j_{pq}c^k_{jr} + c^j_{qr}c^k_{jp} + c^j_{rp}c^k_{jq} = 0 \ . \tag{13.79}$$

This is the Jacobi identity for the structure constants. Just like the anti-symmetry property (Eq.(13.58)), we see that the identity (Eq.(13.79)) is a consequence of the particular analytic forms that the left-hand sides of Eqs.(13.55a) and (13.55b) take.

Let us trace the main steps of the development up to now. First of all we saw that the associativity property of group multiplication is completely equivalent to the possibility of solving the partial differential equations, Eqs.(13.15) and (13.16), for the composition functions $f^j(\alpha,\beta)$, regarding the functions $\eta^j_k(\alpha)$ as something given. Second, we derived a system of *integrability conditions* on the functions $\eta^j_k(\alpha)$, $\xi^j_k(\alpha)$; these can be given in several forms, namely, Eqs.(13.55), (13.57), (13.73), or (13.76), and they are the necessary and sufficient conditions for the solvability of Eqs.(13.15) and (13.16). These integrability conditions require a set of constants c^j_{kl} to exist, obeying, for example, Eq.(13.55). Thirdly, we saw that if any such set of constants c^j_{kl} do exist, then they cannot be arbitrary but must obey two conditions: these are the antisymmetry requirement (Eq.(13.58)) and the Jacobi identity (Eq.(13.79)). Therefore the existence and these two properties of the structure constants c^j_{kl} are direct consequences of the associativity property of group multiplication. The converse is also true, and all the foregoing steps can be retraced. Namely, we shall prove that given any set of constants c^j_{kl} obeying Eqs.(13.58) and (13.79), we can *solve* the equations, for example, $\Delta^j_{kl} = 0$ (see Eq.(13.57a)), and obtain the functions $\xi^j_k(\alpha)$, $\eta^j_k(\alpha)$, and from these solve Eqs.(13.15) and (13.16) to get the group composition functions $f^j(\alpha,\beta)$. The important point here is that the properties of Eqs.(13.58) and (13.79) of the structure constants are sufficient to allow us to carry out this reconstruction; *no other restrictions need be placed on them.* This will mean that the Jacobi identity for the structure constants is not just a consequence of associativity of group multiplication but is even equivalent to it: the Jacobi identity is just the expression of associativity for elements of G infinitesimally close to e. As we have stated previously, the reconstruction of the group G from the knowledge of its structure constants only leads to the determination of the properties of G in some neighbourhood of the identity, and one cannot obtain the properties of G "in the large". This is because, in general, there are several Lie groups, all of which have the same structure in some neighborhood of the identity and thus lead to the same set of structure constants, but which differ from one another globally. Before we carry out our program of reconstructing the group from its structure constants, we formalize the important properties of tangent vectors and structure constants in the concept of a *Lie algebra.*

Lie Algebras

It is sufficient for our purposes to define a real Lie algebra over the field of real numbers. What we shall do is to formalize the properties of tangent vectors that we have found and make them the basis of the definition of a new mathematical object. A Lie algebra L is a finite (n) dimensional real linear vector space with elements $x, y, u, v, \cdots$ in which a "Lie bracket" is defined, which associates with every ordered pair of vectors $x, y \in L$, a third vector $z \in L$, written $z = [x, y]$, with the following properties:

1. $[x, y] = -[y, x]$.

2. $[\lambda x + \lambda' x', y] = \lambda[x, y] + \lambda'[x', y]$, λ, λ' real numbers.

3. For any three vectors x, y, w, the Jacobi identity holds, namely:

$$[[x,y],w]+[[y,w],x]+[[w,x],y]=0\ .$$

These three properties of a Lie bracket are contained in the words "antisymmetry, linearity, Jacobi identity". We see that the space of tangent vectors associated with a Lie group G of dimension n, with the combination law, Eq.(13.74), as the Lie bracket, forms a Lie algebra of dimension n. If e_i, $i = 1, 2, \cdots, n$, is a basis for L as a real vector space, the linearity property (2) above shows that the Lie bracket of any two vectors $x, y \in L$ can be determined once we know the brackets of the basis vectors e_i with one another. Because these can be expanded again in terms of the e_i, we can write

$$[e_k, e_l] = c^j_{kl} e_j\ , \tag{13.80}$$

and with the expansions $x = x^k e_k$, $y = y^k e_k$ for x and y, we have

$$z = [x,y]\ , \qquad z^k = c^k_{lm} x^l y^m\ . \tag{13.81}$$

Properties (1) and (3) of a Lie algebra lead us back to properties of Eqs.(13.58) and (13.79), respectively, for the structure constants c^j_{kl}.

We have seen that a given Lie group G leads to its associated Lie algebra L, and now we show that a given Lie algebra L leads to a unique associated Lie group, at least in some neighborhood of the identity.

From a Lie Algebra to its (Local) Lie Group

Let us be given a set of structure constants c^j_{kl} obeying Eqs.(13.58) and (13.79), and let us attempt to construct the functions $\xi^j_k(\alpha), \eta^j_k(\alpha)$ by solving the integrability conditions in the form Eq.(13.57a):

$$\Delta^u_{mk}(\alpha) \equiv \xi^u_{m,k}(\alpha) - \xi^u_{k,m}(\alpha) + c^u_{rs}\xi^r_m(\alpha)\xi^s_k(\alpha)\ , \ \Delta^u_{mk}(\alpha) = 0\ ,$$
$$\xi^j_k(0) = \delta^j_k\ . \tag{13.82}$$

Actually, we know in advance that there will be ambiguities in attempting to solve Eq.(13.82) for $\xi^j_k(\alpha)$. Thus, if we were already given some Lie group G, we have seen that there will be several coordinate systems in G, all of which lead to the same set of structure constants, but each of which has its own set of functions $\xi^j_k(\alpha)$; and in each coordinate system Eq.(13.82) will be obeyed, with different functions $\xi^j_k(\alpha)$ in different coordinates, but all with the same c^u_{rs}! Conversely, the invariance properties of the system of Eq.(13.82) prevent us from finding a unique solution $\xi^j_k(\alpha)$ given the c^u_{rs}, *unless we impose from outside some extra relations that amount to selecting one (or a few) special coordinate systems.* We can see how to go about this problem by trying to solve for the $\xi^j_k(\alpha)$ in a power series in the α's and looking at the leading terms. Suppose, to start with, we do not assume any subsidiary relations among the ξ's, and we make an expansion of the form

$$\xi^j_k(\alpha) = \delta^l_k + b^j_{kl}\alpha^l + \cdots\ . \tag{13.83}$$

If we now substitute this into the equation $\Delta^j_{mk} = 0$, and examine the leading terms, we get

$$b^j_{mk} - b^j_{km} + c^j_{mk} = 0 \,. \tag{13.84}$$

Thus only the part of b^j_{mk} antisymmetric in m and k is determined by Eq.(13.82); the symmetric part is quite arbitrary! On the other hand, let us impose from outside the condition that we are working in a canonical coordinate system of the first kind; this gives us the additional relations

$$\xi^j_k(\alpha)\alpha^k = \alpha^j \tag{13.85}$$

(see Eq.(13.49)), and if we put Eq.(13.83) into Eq.(13.85) we get the restriction that the b^j_{mk} must be antisymmetric in m and k:

$$b^j_{mk}\alpha^m\alpha^k = 0 \Rightarrow b^j_{mk} = -b^j_{km} \,. \tag{13.86}$$

We now have enough information to determine these coefficients uniquely, the result being

$$b^j_{mk} = \frac{1}{2}c^j_{km} \,. \tag{13.87}$$

Here we have an indication that the ambiguities in the solution to Eq.(13.82) can be resolved by choosing to work in special coordinate systems, namely, canonical coordinates of the first kind. We will show that this indeed is the case; the entire set of Eq.(13.82) and (13.85) has a consistent and unique solution for the functions $\xi^j_k(\alpha)$, $\eta^j_k(\alpha)$.

The assumption just made concerning the use of canonical coordinates has the following implications. The arguments $\alpha^j, \beta^j \cdots$ that will appear in the functions $\xi, \eta, \Delta, \cdots$ in the analysis are coordinates of group elements $a, b, \cdots$ belonging to the group that we will construct. But since the coordinates are canonical, the tangent vector t in terms of which we would write $a = \exp(t)$ also has α^j for its components! Therefore, the parameters α^j will serve as the canonical coordinates of some element a in the group and at the same time as the components of the associated tangent vector t in the Lie algebra, in terms of which we have $a = \exp(t)$. But this can be stated in another way. The structure constants c^j_{kl} that we are starting from arise by choosing a basis in the given Lie algebra L, and the α^j are the components of t *in this basis*. From a coordinate-independent standpoint, what we are given is the Lie bracket $[u, v]$ of any two vectors u, v in L. From this same standpoint, then, we will find that the elements in some neighborhood of the identity of the Lie group G that we are going to construct from L can be written $a = \exp(u)$, $b = \exp(v), \cdots$, with $u, v, \cdots$ vectors lying in some neighborhood of the origin in L. Specifying the composition law for G will amount to a rule for writing the product $\exp(u)\exp(v)$ as $\exp(w)$ and computing w from u and v. The expression of w in terms of u and v can involve Lie brackets and nothing else. If we choose some basis for L, we get at the same time a canonical coordinate system of the first kind for G, with the components α^j of t in L as the coordinates of $a = \exp(t)$ in G. With this explanation regarding the significance of the use of canonical coordinates, we return to the problem of solving Eqs.(13.82) and (13.85).

We begin by deriving from Eqs.(13.82) and (13.85) a simpler system of equations that are already sufficient to find the functions $\xi^j_k(\alpha)$. These are obtained by multiplying the equations $\Delta^j_{mk} = 0$ by α^m and then making use of Eq.(13.85):

$$W^j_k(\alpha) \equiv \alpha^m \Delta^j_{mk}(\alpha) \equiv \alpha^m \xi^j_{m,k}(\alpha) - \alpha^m \xi^j_{k,m}(\alpha) + c^j_{rs}\alpha^r \xi^s_k(\alpha) = 0 \,. \quad (13.88)$$

[Since we use canonical coordinates exclusively from now on, we dispense with the convention of adding a $*$ to every equation valid only in these coordinates.] The first term in $W^j_k(\alpha)$ can be simplified by using a relation obtained by differentiating Eq.(13.85) with respect to α^l:

$$\alpha^k \xi^j_{k,l}(\alpha) = \delta^j_l - \xi^j_l(\alpha) \,. \quad (13.89)$$

Consequently, Eq.(13.88) takes the form

$$\left(\alpha^m \frac{\partial}{\partial \alpha^m} + 1\right) \xi^j_k(\alpha) = \delta^j_k + c^j_{rs}\alpha^r \xi^s_k(\alpha) \,. \quad (13.90)$$

This equation determines the "radial" dependence of the ξ's with respect to the α's; that is, it determines the way the ξ's vary along a "line" on which the ratios of the components α^j to one another are held fixed. In other words, if we consider a line of the form $\alpha^j = \tau\alpha^j_0$, where τ varies and the α^j_0 are kept fixed, Eq.(13.90) assumes the form

$$\left(\tau \frac{\partial}{\partial \tau} + 1\right) \xi^j_k(\tau\alpha_0) = \delta^j_k + c^j_{rs}\alpha^r_0 \tau \xi^s_k(\tau\alpha_0) \,. \quad (13.91)$$

This can be rewritten in matrix notation:

$$\frac{\partial}{\partial \tau}(\tau\Xi(\tau\alpha_0)) = \mathbb{1} + C(\alpha_0)(\tau\Xi(\tau\alpha_0)) \,,$$
$$C(\alpha_0) = \|C^j_s(\alpha_0)\| \,,\ C^j_s(\alpha_0) = c^j_{rs}\alpha^r_0 \,. \quad (13.92)$$

The fact that $C(\alpha_0)$ is independent of τ allows us to solve this equation and obtain an infinite power series for $\Xi(\tau\alpha_0)$, the result being

$$\Xi(\tau\alpha_0) = \mathbb{1} + \sum_{n=1}^{\infty} \frac{[\tau C(\alpha_0)]^n}{(n+1)!} \,. \quad (13.93)$$

Thus using only part of the information contained in Eqs.(13.82) and (13.85), we have been able to solve for the matrix $\Xi(\alpha)$. In order to write the solution more compactly, we introduce an entire analytic function $\phi_0(\zeta)$ of a complex variable ζ via

$$\phi_0(\zeta) = \frac{e^\zeta - 1}{\zeta} = 1 + \sum_{n=1}^{\infty} \frac{\zeta^n}{(n+1)!} \,. \quad (13.94)$$

Then we have

$$\Xi(\alpha) = \phi_0(C(\alpha)) = \mathbb{1} + \sum_{n=1}^{\infty} \frac{[C(\alpha)]^n}{(n+1)!} \,. \tag{13.95}$$

Clearly, the power series for $\Xi(\alpha)$ converges for all values of the α^j; in fact, in canonical coordinates we find that the $\xi^j_k(\alpha)$ are entire analytic functions of α!

Having obtained $\Xi(\alpha)$ by solving the equations $W^j_k = 0$ (cf. Eq.(13.88)), we must make sure that Eqs.(13.82) and (13.85) are really satisfied. This is done not by using the result that we got for $\Xi(\alpha)$ but by appealing to certain properties of the Δ^j_{kl} that are consequences of the way they are defined in terms of the ξ's. Of course, Eq.(13.85) is trivially satisfied, because the matrix $C(\alpha)$ annihilates α:

$$[C(\alpha)\alpha]^j = C(\alpha)^j_k \alpha^k = c^j_{lk}\alpha^l\alpha^k = 0 \,. \tag{13.96}$$

To check Eq.(13.82) we need more work. Denoting, as usual, a partial derivative by a comma followed by a subscript, the definition of Δ^j_{kl} leads to

$$\begin{aligned}\Delta^j_{kl,m} + \Delta^j_{lm,k} + \Delta^j_{mk,l} =& c^j_{rs}\left(\xi^r_{k,m}\xi^s_l + \xi^r_k\xi^s_{l,m} + \xi^r_{l,k}\xi^s_m + \xi^r_l\xi^s_{m,k}\right.\\ &\left.+\xi^r_{m,l}\xi^s_k + \xi^r_m\xi^s_{k,l}\right)\end{aligned} \tag{13.97}$$

The particular combination of derivatives of Δ that appears on the left in Eq.(13.97) is such that all the contributions from the first two terms in the definition of Δ in Eq.(13.82) cancel out, leaving only the contributions from the third term. The terms on the right in Eq.(13.97) can be rearranged as follows:

$$\begin{aligned}\Delta^j_{kl,m} + \Delta^j_{lm,k} + \Delta^j_{mk,l} =& c^j_{rs}\left(\xi^r_k\left(\xi^s_{l,m} - \xi^s_{m,l}\right) + \xi^r_l\left(\xi^s_{m,k} - \xi^s_{k,m}\right)\right.\\ &\left.+\xi^r_m\left(\xi^s_{k,l} - \xi^s_{l,k}\right)\right)\\ =& c^j_{rs}\left(\xi^r_k\Delta^s_{lm} + \xi^r_l\Delta^s_{mk} + \xi^r_m\Delta^s_{kl}\right) - J^j_{rsu}\xi^r_k\xi^s_l\xi^u_m \,,\\ J^j_{rsu} \equiv& c^j_{rv}c^v_{su} + c^j_{sv}c^v_{ur} + c^j_{uv}c^v_{rs} \,.\end{aligned} \tag{13.98}$$

In the first step, we used the antisymmetry of c^j_{rs} in r and s to regroup the terms, and then we used the definition of Δ. This second step brought in the terms bilinear in the structure constants and denoted by J^j_{rsu}; the Jacobi identity would make these terms vanish, but for the moment let us carry them along. Now we multiply both sides of Eq.(13.98) by α^k, and, using Eq.(13.85) wherever possible, we get:

$$\begin{aligned}\alpha^k\Delta^j_{lm,k} + \alpha^k\frac{\partial}{\partial\alpha^m}\Delta^j_{kl} - \alpha^k\frac{\partial}{\partial\alpha^l}\Delta^j_{km} =& c^j_{rs}\alpha^r\Delta^s_{lm}\\ &+ c^j_{rs}(\xi^r_m W^s_l - \xi^r_l W^s_m) - J^j_{rsu}\alpha^r\xi^s_l\xi^u_m \,,\end{aligned}$$

that is,

$$\begin{aligned}\left(\alpha^k\frac{\partial}{\partial\alpha^k} + 2\right)\Delta^j_{lm} =& c^j_{rs}\alpha^r\Delta^s_{lm} + W^j_{m,l} - W^j_{l,m}\\ &+ c^j_{rs}(\xi^r_m W^s_l - \xi^r_l W^s_m) - J^j_{rsu}\alpha^r\xi^s_l\xi^u_m \,.\end{aligned} \tag{13.99}$$

This, then, is an identity involving the Δ's following directly from the way they were defined, with occasional use of Eq.(13.85). Now, the important point is this: what we have already solved in order to obtain $\Xi(\alpha)$ are the equations $W^j_m = 0$; thus the identity of Eq.(13.99) assures us that if W^j_m vanishes and if the Jacobi identity is valid so that the J^j_{rsu} also vanish, then every Δ^j_{lm} obeys the equation

$$\left(\alpha^k \frac{\partial}{\partial \alpha^k} + 2\right) \Delta^j_{lm} = [C(\alpha)]^j_s \Delta^s_{lm} \; . \tag{13.100}$$

As in the case of Eq.(13.90), if we choose the line $\alpha = \tau\alpha_0$ with variable τ, we get

$$\frac{\partial}{\partial \tau}\left(\tau^2 \Delta^j_{lm}(\tau\alpha_0)\right) = [C(\alpha_0)]^j_s \left(\tau^2 \Delta^s_{lm}(\tau\alpha_0)\right) . \tag{13.101}$$

Since $C(\alpha_0)$ is independent of τ, and since at $\tau = 0$ the expressions $\tau^2\Delta^j_{lm}(\tau\alpha_0)$ vanish, Eq.(13.101) guarantees that $\tau^2\Delta^j_{lm}(\tau\alpha_0)$, and thus $\Delta^j_{lm}(\tau\alpha_0)$ vanishes identically. Thus the equations $\Delta^j_{kl}(\alpha) = 0$ are all satisfied if the simpler equations $W^j_l(\alpha) = 0$ are satisfied and if the c^j_{rs} obey the Jacobi identity; *no further restrictions need be placed on the structure constants.* In canonical coordinates, the solution for $\Xi(\alpha)$ is explicit and is given by Eq.(13.95).

To obtain the matrix $H(\alpha)$, we define the analytic function $\phi_1(\zeta)$ by

$$\phi_1(\zeta) = \frac{1}{\phi_0(\zeta)} = \frac{\zeta}{e^\zeta - 1} \; . \tag{13.102}$$

This function has a power series expansion in ζ around the origin, with radius of convergence 2π. It also has the property that it can be represented in the form

$$\phi_1(\zeta) = -\frac{\zeta}{2} + \psi(\zeta) \; , \qquad \psi(\zeta) = \frac{\zeta}{2}\coth\frac{\zeta}{2} = \text{even function of } \zeta \; . \tag{13.103}$$

This will be used later on. The matrix $H(\alpha)$ can now be written as

$$H(\alpha) = \phi_1(C(\alpha)) \; . \tag{13.104}$$

Just like $\Xi(\alpha)$, $H(\alpha)$ depends analytically on the α^j in canonical coordinates, and the power series expansion for $H(\alpha)$ in terms of $C(\alpha)$ converges for small enough α^j. These formulae for $\Xi(\alpha)$ and $H(\alpha)$ lead at once to expressions for the matrices $\hat{\Xi}(\alpha)$, $\hat{H}(\alpha)$ as well, because the only difference in the two cases is that the structure constants $\hat{c}^j_{kl}$ associated with the latter matrices just differ in sign from the c^j_{kl} (see (Eq.(13.61)). Therefore, the solutions for $\hat{\Xi}(\alpha)$ and $\hat{H}(\alpha)$ following from the alternative set of integrability conditions in Eq.(13.60), in canonical coordinates, are

$$\hat{\Xi}(\alpha) = \phi_0(-C(\alpha)) \; , \qquad \hat{H}(\alpha) = \phi_1(-C(\alpha)) \; . \tag{13.105}$$

In Eqs.(13.95),(13.104) and (13.105), we have expressions for the full set of matrices $\Xi, H, \hat{\Xi}, \hat{H}$. We now introduce an item of notation, before using them for obtaining the composition functions $f^j(\alpha,\beta)$. This involves the use of the Lie brackets in expressing the effect of any of these matrices, $\hat{\Xi}(\alpha)$ for example, on a vector. For any two vectors α^j, t^j, the components of $C(\alpha)t$ are

$$[C(\alpha)t]^j = C(\alpha)^j_k t^k = c^j_{lk}\alpha^l t^k = ([\alpha,t])^j \ , \tag{13.106}$$

where $[\alpha, t]$ is just the Lie bracket of α and t (cf.(13.81)). Since we have to deal with powers of $C(\alpha)$, we do the following: to each vector α^j we associate an operator Ω_α that acts linearly on vectors t^j to produce new vectors:

$$\Omega_\alpha t = [\alpha, t] \ ; \qquad \Omega_\alpha(\lambda t + \lambda' t') = \lambda\Omega_\alpha t + \lambda'\Omega_\alpha t' \ ; \tag{13.107a}$$

$$(\Omega_\alpha t)^j = [C(\alpha)t]^j = C(\alpha)^j_k t^k \ . \tag{13.107b}$$

Given a Lie algebra L, Eq.(13.107a) defines in a coordinate-independent way a linear operator Ω_α associated with any $\alpha \in L$, acting on L; the dependence of Ω_α on α is also linear, and Eq.(13.107b) relates Ω_α to the matrix $C(\alpha)$ in a basis for L. Now we can write:

$$\begin{aligned}
\Xi(\alpha)t &= \phi_0(\Omega_\alpha)t = \left[1 + \sum_{n=1}^{\infty}\frac{(\Omega_\alpha)^n}{(n+1)!}\right] t \\
&= t + \sum_{n=1}^{\infty}\frac{1}{(n+1)!}[\alpha,[\alpha,\cdots[\alpha,t]]\cdots] \\
&= t + \frac{[\alpha,t]}{2!} + \frac{[\alpha,[\alpha,t]]}{3!} + \cdots \ ; \\
H(\alpha)t &= \phi_1(\Omega_\alpha)t = -\frac{1}{2}\Omega_\alpha t + \psi(\Omega_\alpha)t = -\frac{1}{2}[\alpha,t] + \psi(\Omega_\alpha)t; \\
\hat{\Xi}(\alpha)t &= \phi_0(-\Omega_\alpha)t \ ; \qquad \hat{H}(\alpha)t = \phi_1(-\Omega_\alpha)t \ .
\end{aligned} \tag{13.108}$$

These formulae become completely coordinate-independent if in place of the arguments α^j in $\Xi(\alpha)$, $H(\alpha)$ and so on we write the abstract group element $a = \exp(u)$, u being some vector in L, for we then obtain:

$$\Xi(\exp(u))t = t + \sum_{n=1}^{\infty}\frac{1}{(n+1)!}[u,[u,\cdots[u,t]]\cdots] \tag{13.109}$$

and so on. All the brackets appearing in Eqs. (13.108) and (13.109) are Lie brackets; the expression for $\Xi(\exp(u))t$, for example, is an infinite series, the general term written in Eq.(13.109) involving an n-fold Lie bracket.

Returning to the problem of determining the composition functions $f^j(\alpha,\beta)$, it turns out to be more convenient to use both Eqs.(13.15) and (13.18) simultaneously. Keeping the elements α^j, β^j of L fixed, consider $f^j(\tau\alpha,\tau\beta)$. This obeys the differential equation:

$$\begin{aligned}
\frac{d}{d\tau}f^j(\tau\alpha,\tau\beta) &= f^j_k(\tau\alpha,\tau\beta)\alpha^k + \hat{f}^j_k(\tau\alpha,\tau\beta)\beta^k \\
&= [H(f)\Xi(\tau\alpha)\alpha]^j + [\hat{H}(f)\hat{\Xi}(\tau\beta)\beta]^j \\
&= [H(f)\alpha]^j + [\hat{H}(f)\beta]^j = [\phi_1(\Omega_f)\alpha + \phi_1(-\Omega_f)\beta]^j \ .
\end{aligned} \tag{13.110}$$

By use of Eq.(13.103) here, this becomes

$$\begin{aligned}\frac{d}{d\tau}f(\tau\alpha,\tau\beta) &= \frac{1}{2}\Omega_f(\beta-\alpha)+\psi(\Omega_f)(\alpha+\beta)\\ &= \frac{1}{2}[\alpha-\beta,f]+\psi(\Omega_f)(\alpha+\beta)\ .\end{aligned} \tag{13.111}$$

This equation, along with the boundary condition that f vanishes at $\tau=0$, serves as a generating equation for f. In coordinate-free form, we would rewrite Eq.(13.111) as follows: we think of $\tau\alpha^j$ as being the canonical coordinates of $a(\tau)=\exp(\tau u)$, $\tau\beta^j$ being those of $b(\tau)=\exp(\tau v)$, and $f^j(\tau\alpha,\tau\beta)$ as those of $a(\tau)b(\tau)=\exp(w(\tau))$, so that α^j, β^j, f^j are the components of u, v, and w, respectively; then we have

$$\begin{aligned}&\exp(\tau u)\exp(\tau v)=\exp(w(\tau))\ ,\\ \frac{dw(\tau)}{d\tau} &= \frac{1}{2}[u-v,w]+\psi(\Omega_w)(u+v)\,;\qquad w(0)=0\ .\end{aligned} \tag{13.112}$$

Because they are complicated and nonlinear Eqs. (13.111) and (13.112) do not permit us to write down the solution in a simple form; however, we can substitute a power series in τ for f or w and solve these equations term by term. The leading terms in the solution can be obtained fairly simply. We have

$$\psi(\Omega_f)=1+\sum_{s=1}^{\infty}\beta_s(\Omega_f)^{2s}=1+\frac{1}{12}(\Omega_f)^2+\cdots\ . \tag{13.113}$$

Substituting

$$f(\tau\alpha,\tau\beta)=\sum_{n=1}^{\infty}\tau^n f_n(\alpha,\beta)\ , \tag{13.114}$$

we get:

$$\begin{aligned}&f_1(\alpha,\beta)+2\tau f_2(\alpha,\beta)+3\tau^2 f_3(\alpha,\beta)+\cdots\\ &\quad=\frac{1}{2}[\alpha-\beta,\ \tau f_1(\alpha,\beta)+\tau^2 f_2(\alpha,\beta)+\cdots]+\alpha+\beta\\ &\qquad+\frac{1}{12}[\tau f_1(\alpha,\beta)+\cdots,[\tau f_1(\alpha,\beta)+\cdots,\alpha+\beta]]+\cdots\ .\end{aligned} \tag{13.115}$$

Thus we obtain:

$$\begin{aligned}f_1(\alpha,\beta)&=\alpha+\beta\,;\qquad f_2(\alpha,\beta)=\frac{1}{4}[\alpha-\beta,\alpha+\beta]=\frac{1}{2}[\alpha,\beta];\\ f_3(\alpha,\beta)&=\frac{1}{3}\left\{\frac{1}{2}\left[\alpha-\beta,\frac{1}{2}[\alpha,\beta]\right]+\frac{1}{12}[\alpha+\beta,[\alpha+\beta,\alpha+\beta]]\right\}\\ &=\frac{1}{12}[\alpha-\beta,[\alpha,\beta]]\ .\end{aligned} \tag{13.116}$$

The coordinate-free form for the composition law for G is then:

$$\exp(u)\exp(v) = \exp(w) \; ,$$
$$w = u + v + \frac{1}{2}[u,v] + \frac{1}{12}[u - v, [u,v]] + \cdots \; . \qquad (13.117)$$

The higher terms involve repeated Lie brackets of increasing order, and the series converges for u and v in some finite neighborhood of the origin in L. This expression for w in terms of u and v is called the *Baker-Campbell-Hausdorff formula.*

Let us now summarize what has been done. In the first part of this chapter we saw that every Lie group G leads to an associated Lie algebra L, the Lie bracket in L being determined by the structure of G "in the small". We saw that this Lie bracket is antisymmetric and obeys the Jacobi identity as a consequence of the associativity of group multiplication in G. In the second part, we have seen that these properties of the Lie bracket are sufficient to allow a reconstruction of the group G, given the Lie algebra L, at least in some neighborhood of the identity in G. If $u, v, \cdots$ are vectors in L, then the elements of G are $\exp(u)\exp(v), \cdots$ The composition law tells us that the product $\exp(u)\exp(v)$ is $\exp(w)$, w being given in terms of u and v by the Baker-Campbell-Hausdorff formula. This formula converges and gives a finite value for w if u and v are sufficiently close to the origin in L. That the Lie group G so obtained from L does have L for its Lie algebra is trivially checked from the Baker-Campbell-Hausdorff formula and Eq.(13.63).

We have pointed out repeatedly that many Lie groups G possess the same Lie algebra L. Conversely, what L determines uniquely is a certain *simply connected* Lie group $\bar{G}$, and $\bar{G}$ is called the *universal covering group* of all those Lie groups G that have L for their Lie algebra. However, it would take us too far afield to go into these matters in any detail here, and when the occasion arises in the sequel, we shall be content with pointing out examples of this situation.

We conclude this chapter with a few more comments concerning the reciprocal relationship between Lie groups and Lie algebras. In the previous chapter we defined the concepts of subgroups and invariant subgroups of any group. These can be applied to Lie groups as well, and they then lead to corresponding definitions in the case of a Lie algebra. If L is a given Lie algebra, we say L_1 is a *subalgebra* of L if L_1 is a linear subspace of L in the vector space sense and if, in addition, the Lie bracket of any two vectors in L_1 is also contained in L_1. It turns out that if G is a Lie group and it possesses a sub-Lie group G_1, then the tangent vectors to all one-parameter subgroups of G lying within G_1 determine a subalgebra L_1 in the Lie algebra L of G, and conversely. The Lie algebra notion corresponding to an invariant subgroup is that of an invariant subalgebra: L_1 is an invariant subalgebra of L if the conditions for L_1 to be a subalgebra are obeyed, and, in addition, if the Lie bracket of any element in L_1 with any element in L is contained in L_1. Here again, the tangent vectors to all one-parameter subgroups in an invariant Lie subgroup of a Lie group G determine an invariant subalgebra in the Lie algebra L of G, and conversely.

Thus $L_1 \subset L$ is a subalgebra if it is a subspace of L and also:

$$u \in L_1 \ , \ v \in L_1 \Rightarrow [u, v] \in L_1 \ ; \tag{13.118}$$

it is an invariant subalgebra if

$$u \in L_1 \ , \ v \in L \Rightarrow [u, v] \in L_1 \ . \tag{13.119}$$

We make repeated use of these ideas in our considerations.

Chapter 14

Realizations of Lie Groups and Lie Algebras

Our major interest in Lie algebras and Lie groups is in their relation to classical mechanics. We are interested in utilizing the structural similarities by defining a correspondence of dynamics and Lie algebras, as well as in the expression of various kinds of symmetries in a given dynamical framework. We are thus led to introduce the concepts of realizations and representations.

In the previous two chapters we studied some properties of Lie groups and Lie algebras as abstract mathematical structures. Any specific mathematical system, with suitable properties and operations defined in it, is said to furnish a "realization" of an abstract system (be it a group or a Lie group or a Lie algebra) if we can define a mapping or correspondence of the abstract system into (or onto) the specific system. Under this mapping, the relations (or operations) defined in the abstract system must be mirrored by the operations defined in the mathematical system. In any particular realization, it may well happen that the specific system has more properties or has more meaningful operations defined than are actually necessary for the realization; such "extra" features become a way of characterizing the nature of the realization. Correspondingly, it is clear that a given abstract system may be realized in several essentially different ways.

We saw in Chapter 12 that the concept of a group arises by abstraction from the properties of mappings among sets. The concept of the realizations of a group takes us back to these mappings. Let a group G be given, and let us denote its elements by $a, b, \cdots, g, \cdots$. Let S be some set with elements $x, y, \cdots$. If to each $a \in G$ we can associate a mapping T_a of S onto S that is one to one and invertible, such that the product of two of these mappings $T_a T_b$ is the mapping T_{ab}:

$$T_a T_b = T_{ab} \ , \tag{14.1}$$

then we say that we have a realization of G by means of transformations on S. Here, the multiplication of mappings is the natural one explained in Chapter 12, whereas the product of a and b is given by the group structure. Thus we may express the content of Eq.(14.1) by saying that the group multiplication law is preserved by the product of the corresponding mappings. We can also see that because the multiplication law for mappings automatically obeys the associative law, we are able to speak consistently of realizations of a group in this way only because the associativity property of the group multiplication was made one of the defining properties of a group. One important class of realizations of a group goes by the name of *representation*, or more precisely *linear representation.* This arises when the set S is a (real or complex) linear vector space, and the mappings $T_a, T_b, \cdots$ are non-singular linear transformations on this space. Such representations are the quantities that are of direct application in quantum mechanics, though a particular representation, called the adjoint representation of a Lie group, will be of importance in our considerations as well. Of course, the realizations of groups that are immediately relevant to classical Hamiltonian mechanics are realizations by canonical transformations: here the set S would be the phase space of some dynamical system, and the mappings $T_a, T_b \cdots$ canonical transformations (assumed to be regular). Alternatively, if S were taken to be the set of all functions on the phase space, then the canonical transformations acting as point transformations in phase space would induce corresponding linear transformations on S.

As for Lie algebras, we have already come across two examples of concrete realizations of this kind of structure. Our discussion in Chapter 5 shows that if we define the Lie bracket of two functions on phase space to be their PB, then all the properties of a Lie algebra are satisfied: we have the properties of linearity, antisymmetry as well as the Jacobi identity. Thus, the PB is a realization of Lie brackets and is the realization relevant to classical mechanics. As one might expect, we shall see that this realization of Lie brackets corresponds to realizations of a Lie group by canonical transformations. Other Lie brackets relevant in classical mechanics are the realizations via GPBs and Dirac brackets discussed in Chapters 8 and 9, and these go with realizations of Lie groups via generalized or Dirac canonical transformations. Yet another realization of Lie brackets has appeared in the previous chapter; this is by means of linear first-order differential operators, with the commutator $[X, Y] \equiv XY - YX$ serving as the definition of the Lie bracket of two such operators X and Y. This use of the commutator of linear operators as the Lie bracket is the realization of Lie brackets relevant to quantum mechanics. Another form of this realization, pertaining again to quantum mechanics, is one wherein one has (finite or infinite-dimensional) matrices corresponding to linear operators on a linear vector space, and one again adopts the commutator $AB - BA$ of two matrices A, B as their Lie bracket, AB representing, of course, the matrix product of A and B. As is to be expected, this realization of the Lie bracket goes with realizations of a Lie group by means of linear transformations in a vector space, in other words with linear representations of a Lie group.

After these general comments on realizations, we now discuss in more detail some specific realizations, namely, the adjoint representation, realization via a transformation group, and via canonical transformations.

Adjoint Representation of a Lie Group

We saw in the preceding chapter that given a Lie algebra L it is possible to reconstruct a corresponding Lie group G in some neighborhood of the identity, which has L for its Lie algebra. If we write $u, v, w, \cdots$ for the vectors in L, we can denote the elements of G by $\exp(u), \exp(v), \cdots$. The identity in G is $\exp(0)$, the 0 being the null vector in L; the inverse of $\exp(u)$ is $\exp(-u)$; and for small enough u and v, the product $\exp(u)\exp(v)$ can be written as $\exp(w)$, with w given in terms of u and v by the Baker-Campbell-Hausdorff formula, Eq.(13.117). A basis for L, in which the vector u has components u^j, gives rise to a corresponding canonical coordinate system of the first kind for G, with the element $\exp(u)$ being assigned these same u^j as coordinates. In such a coordinate system for G, a family of elements $a(\tau)$, τ a real variable, with coordinates τu^j, automatically forms a one-parameter subgroup of G with u as its tangent vector.

Let a one-parameter subgroup $\sigma \to a(\sigma)$ in G be given, and let t be its tangent vector. For some $u \in L$, let $b = \exp(u)$ be a fixed element in G. It is obvious that the family of elements $\sigma \to a'(\sigma) = ba(\sigma)b^{-1}$ also forms a one-parameter subgroup of G, with some tangent vector t', say. How is t' related to t and b? This is easy enough to calculate; in a general coordinate system for G, with $\alpha^j(\sigma)$, $\alpha'^j(\sigma)$ being the coordinates of $a(\sigma)$, $a'(\sigma)$, respectively. Eqs.(13.15), (13.18), and (13.29) lead to:

$$
\begin{aligned}
t'^j &= \left(\frac{d\alpha'^j(\sigma)}{d\sigma}\right)_{\sigma=0} = \left(\frac{d}{d\sigma} f^j(ba(\sigma), b^{-1})\right)_{\sigma=0} \\
&= f^j_k(b, b^{-1}) \left(\frac{d}{d\sigma} f^k(b, a(\sigma))\right)_{\sigma=0} \\
&= \eta^j_l(e)\xi^l_k(b)\hat{f}^k_m(b,e)\eta^m_n(e)t^n = [\Xi(b)\hat{H}(b)t]^j \ .
\end{aligned} \tag{14.2}
$$

If we use canonical coordinates, we set $b = \exp(u)$ and then use Eq.(13.108), to compute the effect of the matrix $\Xi(b)\hat{H}(b)$ on t:

$$
\begin{aligned}
\exp(u) : t \to t' &= \phi_0(\Omega_u)\phi_1(-\Omega_u)t = \left(\frac{e^\zeta - 1}{\zeta} \cdot \frac{\zeta}{1 - e^{-\zeta}}\right)_{\zeta=\Omega_u} \cdot t \\
&= e^{\Omega_u} t = t + \frac{1}{1!}[u, t] + \frac{1}{2!}[u, [u, t]] + \cdots \ .
\end{aligned} \tag{14.3}
$$

This passage from t to t' is a linear transformation acting on L viewed as a linear vector space. Thus each element $b = \exp(u)$ in G determines a corresponding linear transformation in L, which we shall write as R_b:

$$
b = \exp(u) \to R_b = R_{\exp(u)} : R_{\exp(u)} t = e^{\Omega_u} t \ . \tag{14.4}
$$

These linear transformations $R_a, R_b, \cdots$ automatically give us a representation of the group G. If t is any vector in L, tangent to the one-parameter subgroup $\exp(\sigma t)$, $R_b t$ is the vector tangent to the subgroup $b\exp(\sigma t)b^{-1}$, and $R_a R_b t$ is the tangent vector to the subgroup $a(b\exp(\sigma t)b^{-1})a^{-1}$, but the elements of this last subgroup can also be written $(ab)\exp(\sigma t)(ab)^{-1}$ so that the corresponding tangent vector is $R_{ab}t$; thus we have proved the representation property:

$$R_a R_b = R_{ab} \,. \tag{14.5}$$

This representation is called the adjoint representation of G. We can summarize the situation as follows: given an abstract Lie algebra L, it determines locally its associated Lie group G, and then it also serves as a vector space on which we obtain a realization of G by means of linear transformations. Both the multiplication rule for G and the action of the linear transformations corresponding to elements of G are given explicitly in terms of the Lie bracket in L, namely by Eq.(13.117), and Eqs. (14.3), (14.4), respectively. All that is needed for this entire development is the set of laws that define a Lie bracket, namely the linearity, antisymmetry, and the Jacobi identity. We can write Eqs(14.4) and (14.5) in the following way: for any $u, v \in L$ (sufficiently close to the origin), we have:

$$e^{\Omega_u} e^{\Omega_v} = e^{\Omega_w} \;, \; w = u + v + \frac{1}{2}[u,v] + \frac{1}{12}[u - v, [u,v]] + \cdots \,. \tag{14.6}$$

In addition to the fact that the Lie algebra L of G serves in this way as the linear space in which the adjoint representation of G is obtained, we should note that at the same time we have a linear representation of L by means of the linear transformations $\Omega_u, \Omega_v, \cdots$ acting on L! This "self-realization" of L goes naturally with the adjoint representation of G. That the correspondence $u \to \Omega_u$ between elements of L and linear transformations on L is a representation of L is easy to see. Clearly, Ω_u is a linear function of u. If we adopt the commutator $[\Omega_u, \Omega_v]_-$ of two of these Ω's as the representative of the Lie bracket in L, the Jacobi identity for L guarantees that L is properly represented:

$$\begin{aligned}[\Omega_u, \Omega_v]_- t &= \Omega_u(\Omega_v t) - \Omega_v(\Omega_u t) = [u, [v,t]] - [v, [u,t]] \\ &= [[u,v],t] = \Omega_{[u,v]} t, \;\; [\Omega_u, \Omega_v]_- = \Omega_{[u,v]} \,. \end{aligned} \tag{14.7}$$

This feature that we have just seen, namely that a realization of G is accompanied by a corresponding realization of L with a suitable definition of Lie bracket, is a general one, as we shall see.

The exponential notation $\exp(u)$ for elements of G is a formal one; it is possible to introduce this notation because there is some neighborhood of the identity e in G in which every element lies on some one-parameter subgroup. Writing $\exp(u)$ for an element $a \in G$ only means that a is that element that is reached for the value $\sigma = 1$ of the parameter labeling the elements of the one-parameter subgroup having u for its tangent vector. At this level $\exp(u)$ does not stand for the exponential function of something or other. But the appropriateness of this notation becomes clear from the structure of the adjoint representation: Eq.(14.4) shows that the abstract element $\exp(u)$ is realized, in this representation, by the exponential function of the linear operator Ω_u determined by u and acting on L! In other words, the quantities $e^{\Omega_u}, e^{\Omega_v}, \cdots$

appearing above are "honest exponentials," and are defined by the exponential series in the arguments $\Omega_u, \Omega_v, \cdots$. This correspondence is a general feature; we see that in realizations of Lie groups by transformation groups or by canonical transformations, the abstract element $\exp(u)$ gets represented by the "honest exponential" of a first-order linear differential operator determined by u. Similarly, in a linear representation of G by matrices, the ordinary exponential of a matrix determined by u corresponds to $\exp(u)$.

There are several important properties of the adjoint representation that are worth explaining. The first concerns one-parameter subgroups. Just as we saw that the elements lying on such a subgroup in G are subject to the differential equation Eq.(13.29) it turns out that the *representatives* of such a subgroup also obey a characteristic differential equation and can in fact be defined to be the solutions of this differential equation. Let $\exp(\sigma u)$ be such a subgroup, $R_{\exp(\sigma u)}$ the representative of a general element, t a general vector in L, and let us write:

$$t(\sigma) = R_{\exp(\sigma u)}t\,, \qquad t(0) = t\,. \tag{14.8}$$

Then $t(\sigma)$ obeys the differential equation

$$\frac{dt(\sigma)}{d\sigma} = \Omega_u t(\sigma) = [u, t(\sigma)]\,, \tag{14.9}$$

and it can be recovered by solving this equation subject to the initial conditions specified in Eq.(14.8). Equivalently, one could write down a differential equation plus boundary condition for $R_{\exp(\sigma u)}$ itself; it reads:

$$\frac{d}{d\sigma}R_{\exp(\sigma u)} = \Omega_u R_{\exp(\sigma u)}\; R_{\exp(0)} = \text{unit operator on } L\,. \tag{14.10}$$

Another useful and important property of the adjoint representation is a consequence of the Jacobi identity and the differential Eq.(14.9) we have just set up. Let t and v be any two vectors in L, let $t(\sigma)$ be defined as in Eq.(14.8) and $v(\sigma)$ analogously. For each σ, we can compute the Lie bracket of $t(\sigma)$ and $v(\sigma)$; call it $w(\sigma)$:

$$\begin{gathered} t(\sigma) = R_{\exp(\sigma u)}t = e^{\sigma\Omega_u}t,\; v(\sigma) = R_{\exp(\sigma u)}v = e^{\sigma\Omega_u}v\,, \\ [t(\sigma), v(\sigma)] = w(\sigma)\,. \end{gathered} \tag{14.11}$$

We can compute the derivative of $w(\sigma)$ with respect to σ; we find:

$$\begin{aligned} \frac{dw(\sigma)}{d\sigma} &= \left[\frac{dt(\sigma)}{d\sigma}\,, v(\sigma)\right] + \left[t(\sigma)\,, \frac{dv(\sigma)}{d\sigma}\right] \\ &= [[u, t(\sigma)], v(\sigma)] + [[v(\sigma), u], t(\sigma)] = -[[t(\sigma), v(\sigma)], u] \\ &= [u, w(\sigma)] = \Omega_u w(\sigma)\,, \end{aligned} \tag{14.12}$$

by using the Jacobi identity for the Lie bracket. Thus $w(\sigma)$ obeys the same kind of differential equation with respect to σ as do $t(\sigma)$ and $v(\sigma)$, and it can be determined in terms of its value at $\sigma = 0$ exactly as $t(\sigma)$ and $v(\sigma)$ in Eq.(14.11). We obtain:

$$w(\sigma) = e^{\sigma\Omega_u} w(0) = e^{\sigma\Omega_u}[t, v] \; . \tag{14.13}$$

Combining Eqs.(14.13) and (14.11), and dropping σ, we can write our result in the following forms:

$$[e^{\Omega_u} t,\, e^{\Omega_u} v] = e^{\Omega_u}[t, v] \; , \; [R_{\exp(u)} t,\, R_{\exp(u)} v] = R_{\exp(u)}[t, v] \; . \tag{14.14}$$

We can describe this result in words by saying that under the transformations of L induced by the adjoint representation of G, the values of Lie brackets are preserved; more generally, any relation among the elements of L involving Lie brackets alone (apart from the processes of taking linear combinations and numerical multiples of elements of L) will be preserved.

The similarity of much of what we have been doing here to our earlier work on the connection between PB's and canonical transformations, in Chapters 5 and 9, must be obvious to the reader. We return to an explanation of this relationship later; now we proceed to study another class of realizations of Lie groups, which again leads to the particular case of realizations by canonical transformations.

Realizations via Transformation Groups

Let us go back to the definition of a general realization of a Lie group G by means of mappings of a set S onto itself. We are now interested in the case in which the set S is a manifold of points that can be described by means of a finite number of real coordinates such that we can think of functions defined on S and be able to apply the processes of differentiation and analysis on these functions. Ultimately, we are interested in the case where S is the phase space of some classical system, but we consider this restriction later. A set of mappings of such a set S onto itself, forming a realization of a Lie group G — such that the functions that express the coordinates of an image point in terms of those of the initial point as well as the coordinates of the group element are continuous and differentiable with respect to these arguments — is said to be a *transformation group*. We can say that we have a realization of G by means of a transformation group. The work of Lie started originally from a study of the structure of transformation groups and introduced the ideas of structure constants, tangent vectors, and so on in this context. As is usual now, we have described the abstract structure of Lie groups first, and we look upon transformation groups as an example of realizations of abstract groups.

The analysis of such a realization of a Lie group is very similar to the analysis of the preceding chapter. Let us denote by $\omega, \omega', \cdots$ the points of the manifold S, and in a specific coordinate system for S use $\omega^\rho, \omega'^\rho, \cdots$ to denote the coordinates of these points. To an element a in G, there corresponds a mapping T_a of S onto S that takes the point ω into the point $\omega' = T_a\omega$. We express the coordinates of ω' in terms of those of a and ω (some coordinate system having been chosen for G) in the form

$$a \to T_a : \omega \to \omega' = T_a\omega, \qquad \omega'^{\rho} = h^{\rho}(\alpha, \omega) \; . \tag{14.15}$$

(Greek superscripts will indicate coordinates for S, latin ones for G). The facts that T_e is the identity transformation and that the T's obey Eq.(14.1), lead to two conditions on the functions h:

$$h^{\rho}(0, \omega) = \omega^{\rho} \; ; \tag{14.16a}$$
$$h^{\rho}(\alpha, h(\beta, \omega)) = h^{\rho}(f(\alpha, \beta), \omega) \; . \tag{14.16b}$$

Here, the f's are the group composition functions. We can refer to Eq.(14.16b) as the *representation property* for the functions h. Because the h's describe mappings, there is no need to write down separately an associative law; it is automatically valid, as was already explained.

Actually, the law of group multiplication studied in Chapter 13 is already a special example of a transformation group-type realization of the group G; Eq.(14.16b) reduces to Eq.(13.8) if we identify the set S with the group G itself and also replace the h's by f's. When we consider the product ab of two elements a and b in G, by keeping a fixed and letting b vary over G, we obtain all elements of G, and we can think of this as a transformation acting on G. This transformation, F_a say, is determined by a and sends a general element b into ab. The associative law for G, which can be written either as $(ab)c = a(bc)$ or $f(f(\alpha, \beta), \gamma) = f(\alpha, f(\beta, \gamma))$, can now be written as $F_a(F_b c) = F_{ab}c$, or, dropping the variable element c, as $F_a F_b = F_{ab}$. Thus the associative law for group multiplication in G could equally well be thought of as the representation property for these mappings F_a acting on G itself. Notice that yet another way of writing this relation is $F_a F_b = F_{F_a b}$. But the point is that in this self-realization of G we *prefer* to think of the element $F_a b$ as *the product of a and b*, and thus it is more natural to think of the law $(ab)c = a(bc)$ as the associative law for the product rule rather than as the representation property for the F's, although these two interpretations are equivalent.

As in the case of the group composition law, here too we can set up a system of partial differential equations to determine the transformation functions h. The only difference is that now the group G and hence the functions $f(\alpha, \beta)$ are supposed to be given, and we are interested in sets of functions h^{ρ} obeying Eq.(14.16). Let us differentiate both sides of Eq.(14.16b) with respect to α^j, the coordinates of a; we get:

$$\frac{\partial h^{\rho}(\alpha, h(\beta, \omega))}{\partial \alpha^j} = \left(\frac{\partial h^{\rho}(f, \omega)}{\partial f^k} \right)_{f = f(\alpha, \beta)} \eta^k_l(f(\alpha, \beta))\xi^l_j(\alpha) \; , \tag{14.17}$$

use having been made of Eq.(13.15). With $\alpha = 0$, Eq.(14.17) gives us:

$$\lambda^{\rho}_j(h(\beta, \omega)) = \frac{\partial h^{\rho}(\beta, \omega)}{\partial \beta^k} \eta^k_j(\beta) \; ,$$

or equivalently,

$$\frac{\partial h^\rho(\alpha,\omega)}{\partial\alpha^k} = \lambda^\rho_j(h)\xi^j_k(\alpha) \ , \tag{14.18}$$

where we have introduced a set of functions $\lambda^\rho_j(\omega)$ defined on S by

$$\lambda^\rho_j(\omega) = \left(\frac{\partial h^\rho(\alpha,\omega)}{\partial\alpha^j}\right)_{\alpha=0} \ . \tag{14.19}$$

These functions λ^ρ_j constitute generalizations of the η's of the previous chapter, and we can look upon Eq.(14.18) as a system of partial differential equations for the transformation functions h^ρ. Thus the representation property Eq.(14.16b) when translated into differential form has allowed us to express a system of functions depending on two arguments α, ω in terms of simpler systems of functions, one depending on ω alone and the other on α alone. The h's and λ's belong to the transformation group, the ξ's to the abstract group.

Just as we did for the η's, it is possible to give a meaning to the λ's in terms of infinitesimal transformations. Namely, if a is an element of G close to e and ω any point in S, the coordinates of the image point $a\omega$ may be expanded in a Taylor series in the coordinates of a to yield:

$$\begin{aligned} \omega' = a\omega : \ \omega'^\rho = h^\rho(\alpha,\omega) &= h^\rho(0,\omega) + \left(\frac{\partial h^\rho(\alpha',\omega)}{\partial\alpha'^j}\right)_{\alpha'=0} \alpha^j + \cdots \\ &= \omega^\rho + \lambda^\rho_j(\omega)\alpha^j + \cdots \end{aligned} \tag{14.20}$$

Thus the functions $\lambda^\rho_j(\omega)$ are related to the small changes that the coordinates ω^ρ of ω undergo when the transformation T_a corresponding to an element a close to e is applied to ω.

Let us now reverse the argument and see whether a solution to Eq.(14.18) and the boundary condition Eq.(14.16a) will lead to functions $h^\rho(\alpha,\omega)$ satisfying Eq.(14.16b) and hence giving rise to a transformation group. As in the case of the abstract group, the condition that Eq.(14.18) be solvable leads to a set of integrability conditions on the λ's. We examine them in a moment, but let us now suppose the λ's do obey them, and let us prove that the h's that arise by solving Eqs.(14.18) and (14.16a) do possess the representation property. We set:

$$L^\rho(\alpha) \equiv h^\rho(\alpha, h(\beta,\omega)) \ , \qquad R^\rho(\alpha) \equiv h^\rho(f(\alpha,\beta),\omega) \ , \tag{14.21}$$

where the element β of G and the point ω of S are held fixed. For $L^\rho(\alpha)$, we easily establish the system of equations:

$$\frac{\partial L^\rho(\alpha)}{\partial\alpha^j} = \lambda^\rho_k(L)\xi^k_j(\alpha) \ , \qquad L^\rho(0) = h^\rho(\beta,\omega) \ , \tag{14.22}$$

use having been made of Eq.(14.18) and (14.16b) only. But with the help of the same equations, together with known properties of the composition functions

$f(\alpha, \beta)$, we find that $R^\rho(\alpha)$ obeys the same equations as $L^\rho(\alpha)$:

$$\frac{\partial R^\rho(\alpha)}{\partial \alpha^j} = \lambda^\rho_k(R)\xi^k_l(f)f^l_j(\alpha, \beta) = \lambda^\rho_k(R)\xi^k_l(f)\eta^l_m(f)\xi^m_j(\alpha)$$
$$= \lambda^\rho_k(R)\xi^k_j(\alpha)\,, \qquad R^\rho(0) = h^\rho(\beta, \omega)\,. \tag{14.23}$$

Because they obey the same differential equations and boundary conditions with respect to α, the equality of $L^\rho(\alpha)$ and $R^\rho(\alpha)$, and thus the validity of the representation property of Eq. (14.16b), follows.

Now let us examine the integrability conditions for Eq.(14.18). These arise by demanding that the partial derivative of the right-hand side of Eq.(14.18) with respect to α^l, say, should give an expression symmetrical in k and l. Let us denote derivatives with respect to coordinates of points of S by commas followed by greek subscripts, and with respect to coordinates of elements of G by commas followed by latin subscripts. Then the conditions we are after read:

$$\lambda^\rho_{j,\sigma}(h)\lambda^\sigma_m(h)\xi^m_l(\alpha)\xi^j_k(\alpha) + \lambda^\rho_j(h)\xi^j_{k,l}(\alpha)$$
$$= \lambda^\rho_{j,\sigma}(h)\lambda^\sigma_m(h)\xi^m_k(\alpha)\xi^j_l(\alpha) + \lambda^\rho_j(h)\xi^j_{l,k}(\alpha)\,. \tag{14.24}$$

Transposing terms, we get:

$$(\lambda^\sigma_m(h)\lambda^\rho_{j,\sigma}(h) - \lambda^\sigma_j(h)\lambda^\rho_{m,\sigma}(h))\xi^m_l(\alpha)\xi^j_k(\alpha) = \lambda^\rho_j(h)(\xi^j_{l,k}(\alpha) - \xi^j_{k,l}(\alpha))\,. \tag{14.25}$$

Finally, we can multiply both sides by η's to remove the ξ's on the left, and we have:

$$\lambda^\sigma_m(h)\lambda^\rho_{j,\sigma}(h) - \lambda^\sigma_j(h)\lambda^\rho_{m,\sigma}(h) = \lambda^\rho_{j'}(h)\eta^l_m(\alpha)\eta^k_j(\alpha)(\xi^{j'}_{l,k}(\alpha) - \xi^{j'}_{k,l}(\alpha))\,. \tag{14.26}$$

Let us now use a known property of the ξ's and η's, given by Eq.(13.55a), and at the same time replace the argument h by a general point ω in S, to put these conditions into the following form:

$$\lambda^\sigma_m(\omega)\lambda^\rho_{j,\sigma}(\omega) - \lambda^\sigma_j(\omega)\lambda^\rho_{m,\sigma}(\omega) = -c^k_{mj}\lambda^\rho_k(\omega)\,. \tag{14.27}$$

These are the integrability conditions on the functions $\lambda^\rho_k(\omega)$; they are a generalization to the case of a realization via a transformation group, of Eq.(13.57b). Once we have a system of functions $\lambda(\omega)$ obeying Eq.(14.27), we can solve Eqs.(14.18) and (14.16a) for the h's and thus obtain the transformation group. What we need from the group G is a knowledge of its structure constants c^k_{mj}.

We analyze Eq.(14.27) by a method similar to that used in the analysis of Eq.(13.57b). For this purpose, imagine a function $\Phi(\omega)$ defined on S, and consider the change in $\Phi(\omega)$ when we apply to ω a transformation T_a corresponding to an element a close to e. As Eq.(14.20) shows, the functions $\lambda(\omega)$ naturally appear in this calculation. An element a close to e can be written $\exp(\tau t)$ where t is some tangent vector and $|\tau|$ is very small. Let us enquire after the dependence on τ of $\Phi(T_{\exp(\tau t)}\omega)$. We find:

$$\frac{d}{dt}\,\Phi(h(\exp(\tau t),\omega)) = \Phi_{,\rho}(h)\lambda^{\rho}_{j}(h)\xi^{j}_{k}(\exp(\tau t))\eta^{k}_{l}(\exp(\tau t))t^{l}$$
$$= t^{j}\lambda^{\rho}_{j}(h)\frac{\partial}{\partial h^{\rho}}\,\Phi(h)\;, \qquad h = h(\exp(\tau t),\omega), \tag{14.28}$$

use having been made of Eq.(14.18) and (13.29). We interpret this result as follows. Keeping the point ω and the tangent vector t fixed, the image point $h(\exp(\tau t),\omega)$ traces out a curve or orbit in S as τ varies, and each point on this orbit can be labeled by the corresponding value of τ. Restricting the argument of Φ to this orbit, we get a function of τ. Equation (14.28) says that at any point τ on this orbit the rate of change of Φ with respect to τ is obtained merely by applying to Φ at this point a first-order partial differential operator determined by the tangent vector t. Calling this operator Λ_t, and replacing h by the general point ω in S, we have the association:

$$t \in L \to \Lambda_t(\omega) = t^{j}\lambda^{\rho}_{j}(\omega)\frac{\partial}{\partial\omega^{\rho}}\;, \tag{14.29}$$

and Eq.(14.28) reads:

$$\frac{d}{dt}\Phi(T_{\exp(\tau t)}\omega) = \Lambda_t(T_{\exp(\tau t)}\omega)\Phi(T_{\exp(\tau t)}\omega). \tag{14.30}$$

We return to the use of this equation in a moment. Let us now note two things. First, the association $t \to \Lambda_t$ given in Eq. (14.29) is a coordinate-independent one and has an intrinsic significance that is not lost if we make a change of coordinates either in G or in S, or in both. This is assured by the way the functions $\lambda(\omega)$ are altered by changes in coordinates in G and S and the transformation law for the t^{j}. We leave it to the reader to verify this fact. Second, by using these differential operators $\Lambda_t(\omega)$, we can put the integrability conditions Eq.(14.27) into the simpler form:

$$[\Lambda_t(\omega), \Lambda_u(\omega)]_{-} = -\Lambda_{[t,u]}(\omega)\;; \tag{14.31}$$

on the left we have the commutator of the operators Λ_t, Λ_u corresponding to the tangent vectors t, u, respectively, whereas on the right appears the operator corresponding to the Lie bracket of t and u. This form of the integrability conditions is a generalization of Eq.(13.76). The problem of finding all possible realizations of a Lie group G by means of transformation groups can thus be stated in infinitesimal terms as follows: find all possible realizations by means of linear first-order partial differential operators, in some set of variables ω, of the Lie algebra L of G; every such realization of L leads to a system of functions $\lambda^{\rho}_{j}(\omega)$ obeying the integrability conditions Eq. (14.27) and thus permits the construction of the transformation functions $h^{\rho}(\alpha,\omega)$ enjoying the representation property Eq.(14.16b). Two remarks must, however, be made. One is that a realization obtained in this way, in general, turns out to be a realization only of the universal covering group $\bar{G}$ of G, and extra conditions may have to be imposed if a realization of G itself is required. The second is that one must impose the condition on the functions $\lambda_j^\rho(\omega)$ that they lead to a solution $h^\rho(\alpha,\omega)$, which can be interpreted as a one-to-one invertible mapping of S onto S for each α. It is very difficult to translate this global requirement into a simple condition on the realization of L by the operators Λ_t, but it must be kept in mind.

The fact that Eq.(14.31) has a universal appearance for all realizations has interesting consequences. Suppose the vectors e_j form a basis for the Lie algebra L. Then Eq.(14.31) is equivalent to the requirement that there be a set of linear first-order differential operators $\Lambda_j(\omega)$ obeying the commutation rules:

$$[\Lambda_j(\omega), \Lambda_k(\omega)]_- = c^l_{kj}\Lambda_l(\omega) \ . \tag{14.32}$$

Thus the number of operators Λ_j and the commutation rules they obey are the same in all realizations! If one examines the structure of linear matrix representations of a Lie group G, one finds again that there must be a linear matrix representation of L, with one matrix X_j corresponding to each basis vector e_j in L, obeying

$$[X_j, X_k]_- \equiv X_j X_k - X_k X_j = c^l_{kj} X_l \ . \tag{14.33}$$

From a practical point of view, this means that one may compute the structure constants of a Lie group G with the help of any one of its linear representations (assuming this to be a *faithful* representation so that $X_t \neq 0$ for t any nonzero element in L) or, equally well by starting from any faithful realization of G by means of a transformation group. Once one has obtained the c^l_{jk} from a particular faithful realization or representation, one knows the equations that must be obeyed in any realization or representation. Using this property, one often defines a Lie group G, not by directly writing down its multiplication law in abstract terms, but by means of a *defining representation* or *defining realization* of it, which one asserts to be faithful. We use this method in defining the Galilei, Lorentz, and other groups.

We return now to the promised use of Eq.(14.30). Imagine that some realization of L by means of operators Λ_t has been constructed, leading therefore to a realization of G by means of transformations T_a in S, with associated transformation functions $h^\rho(\alpha,\omega)$. Let $\Phi(\omega)$ be one of a suitably chosen class of functions on S, for example, the class of infinitely differentiable functions. Then we can set up a *linear representation* $a \to \bar{R}_a$ acting on these functions $\Phi(\omega)$ (which we assume constitute a real linear vector space) by the definition:

$$a \to \bar{R}_a \ : \qquad [\bar{R}_a\Phi](\omega) = \Phi(T_{a^{-1}}\omega) \ . \tag{14.34}$$

That is, the operator $\bar{R}_a$ acts on the function Φ to yield a new function $\bar{R}_a\Phi$ whose value at ω is specified in Eq.(14.34). It is trivial to verify that the representation property $\bar{R}_a\bar{R}_b = \bar{R}_{ab}$ is valid. (We are using the symbol $\bar{R}_a$ for the representative of a in order not to confuse this representation with the adjoint representation in which the representative of a was written R_a). Now an explicit closed form can be given for $\bar{R}_a$ in terms of the operators Λ_t forming the realization of L! Let us consider a one-parameter subgroup $\sigma \to a(\sigma) = \exp(\sigma t)$ in G, and set:

$$\bar{R}_{a(\sigma)}\Phi = \Phi_\sigma \ . \tag{14.35}$$

Then combining Eq.(14.34) with Eq.(14.30) we see that:

$$\begin{aligned}\frac{\partial}{\partial\sigma}\Phi_\sigma(\omega) &= \frac{\partial}{\partial\sigma}\Phi(T_{\exp(-\sigma t)}\omega) = -\Lambda_t(T_{\exp(-\sigma t)}\omega)\Phi(T_{\exp(-\sigma t)}\omega)\\ &= -[\Lambda_t(\omega')\Phi(\omega')]_{\omega'=T_{\exp(-\sigma t)}\omega}\\ &= -[\Lambda_t\Phi]_\sigma(\omega).\end{aligned} \tag{14.36}$$

Here, what we mean is that for any given Φ, $\Lambda_t\Phi$ is some definite function on S determined by t and Φ, and $[\Lambda_t\Phi]_\sigma$ is a function obtained from $\Lambda_t\Phi$ by the general rule of Eq.(14.35). We can also express Eq.(14.36) in the form:

$$\frac{\partial}{\partial\sigma}\bar{R}_{a(\sigma)}\Phi = -\bar{R}_{a(\sigma)}\Lambda_t\Phi \ . \tag{14.37}$$

But now the only dependence on σ on the right-hand side again comes from the presence of the operator $\bar{R}_{a(\sigma)}$, just as on the left, so that we can successively determine all the higher derivatives of the left-hand side with respect to σ; we get:

$$\frac{\partial^n}{\partial\sigma^n}\bar{R}_{a(\sigma)}\Phi = \bar{R}_{a(\sigma)}(-\Lambda_t)^n\Phi \ . \tag{14.38}$$

Consequently, for small enough σ we can make a power series expansion for Φ_σ in terms of σ, and we find:

$$\begin{aligned}\Phi_\sigma(\omega) \equiv [\bar{R}_{\exp(\sigma t)}\Phi](\omega) \equiv \Phi(T_{\exp(-\sigma t)}\omega) &= \sum_{n=0}^{\infty}\frac{\sigma^n}{n!}\left(\frac{\partial^n}{\partial\sigma'^n}\Phi_{\sigma'}(\omega)\right)_{\sigma'=0}\\ &= e^{-\sigma\Lambda_t(\omega)}\Phi(\omega) \ .\end{aligned} \tag{14.39}$$

That is, the realization of G via a transformation group on the manifold S has led to a linear representation of G defined on functions on S, in which the element $a = \exp(u)$ is represented by:

$$a = \exp(u) \to \bar{R}_a = \bar{R}_{\exp(u)} = e^{-\Lambda_u(\omega)} \ . \tag{14.40}$$

As in the example of the adjoint representation, here again we see that the element of G denoted formally as $\exp(u)$ is represented by the ordinary exponential function of (the negative of) the first-order linear differential operator $\Lambda_u(\omega)$ associated with u in the realization of L. The composition law for these "honest exponentials," of course, is:

$$\begin{aligned}&e^{-\Lambda_u(\omega)}\,e^{-\Lambda_v(\omega)} = e^{-\Lambda_w(\omega)},\\ &\qquad w = u + v + \frac{1}{2}[u,v] + \frac{1}{12}[u-v,[u,v]] + \cdots\end{aligned} \tag{14.41}$$

(u, v being close enough to zero), and once again it is evident that Eq.(14.41) is valid because we have a realization of L in which the requirements of linearity, antisymmetry, and the Jacobi identity for the Lie bracket are all obeyed.

We conclude this study of the structure of realizations via transformation groups by giving an alternative derivation of the Baker-Campbell-Hausdorff formula. For this purpose, let us identify the space S with the Lie group G itself so that the general point ω in S becomes a general element in G; then let us consider the realization of G by left multiplication in which the transformation T_a corresponding to an element a is that which maps a general element b of G into ab. Then the functions $\lambda_j^\rho(\omega)$ of this realization are just the functions $\eta_j^k(\alpha)$ of the preceding chapter, and the differential operator $\Lambda_u(\omega)$ corresponding to a vector $u \in L$ coincides with the operator $\eta_{[u]}(\alpha)$ of the preceding chapter (see Eq.(13.70)). Let us also use canonical coordinates of the first kind for G, in which case we have seen that the matrix $H(\alpha) = \|\eta_j^k(\alpha)\|$ has the form:

$$a = \exp(u) \ : \qquad H(a) = \phi_1(C(u)) \ , \qquad C(u)^j_k = c^j_{lk} u^l \ . \tag{14.42}$$

(This is Eq.(13.104)). Now in this realization of G let us specialize Eq.(14.39) to the case in which the function Φ is one of the (canonical) coordinates themselves; that is, where $\Phi(\exp(u)) = u^j$ for some fixed j. Then Eq.(14.39) leads directly to the group composition law in canonical coordinates, for we find, setting $\exp(u)\exp(v) = \exp(w)$,

$$\begin{aligned} w^j &= \Phi(\exp(w)) = \Phi(T_{\exp(u)}\exp(v)) = \exp[\eta_{[u]}(v)]\Phi(\exp(v)) \\ &= \exp(\eta_{[u]}(v))v^j \ , \\ w^j &= \exp\left[u^k \eta^l_k(v) \frac{\partial}{\partial v^l}\right] v^j \ . \end{aligned} \tag{14.43}$$

(Remember that the point $\exp(v)$ at which the operator $\eta_{[u]}$ in the exponent was required, according to Eq.(14.39), has coordinates v^l because we are using canonical coordinates.) It is instructive to work out the first few terms of Eq.(14.43) and see that our new derivation gives results in agreement with the previous calculation. We retain terms in w^j that are up to trilinear in u^j and v^j combined; thus for $\eta(v)$ we must use the expansion:

$$\begin{aligned} \eta^l_k(v) &= \left(-\frac{1}{2}C(v) + 1 + \frac{1}{12}C(v)C(v) + \cdots\right)^l_k \\ &= \delta^l_k - \frac{1}{2}c^l_{mk}v^m + \frac{1}{12}c^l_{pm}c^m_{qk}v^p v^q + \cdots \end{aligned} \tag{14.44}$$

Retaining terms up to second order in u as well, Eq.(14.43) leads to:

$$
\begin{aligned}
w^j =& v^j + u^k \eta_k^l(v) \frac{\partial}{\partial v^l} v^j + \frac{1}{2} u^k u^{k'} \eta_k^l(v) \frac{\partial}{\partial v^l} \eta_{k'}^{l'}(v) \frac{\partial}{\partial v^{l'}} v^j + \cdots \\
=& v^j + u^k \eta_k^j(v) + \frac{1}{2} u^k u^{k'} \eta_k^l(v) \frac{\partial}{\partial v^l} \eta_{k'}^j(v) + \cdots \\
=& v^j + u^k \left(\delta_k^j - \frac{1}{2} c_{mk}^j v^m + \frac{1}{12} c_{pm}^j c_{qk}^m v^p v^q + \cdots \right) \\
&+ \frac{1}{2} u^k u^{k'} \left(\delta_k^l - \frac{1}{2} c_{mk}^l v^m + \cdots \right) \\
&\times \left(-\frac{1}{2} c_{lk'}^j + \frac{1}{12} c_{lm'}^j c_{qk'}^{m'} v^q + \frac{1}{12} c_{pm'}^j c_{lk'}^{m'} v^p \cdots \right) \\
=& v^j + u^j + \frac{1}{2} [u, v]^j + \frac{1}{12} [v, [v, u]]^j \\
&+ \frac{1}{2} u^k u^{k'} \left(-\frac{1}{2} c_{kk'}^j + \frac{1}{12} c_{km'}^j c_{qk'}^{m'} v^q + \frac{1}{12} c_{pm'}^j c_{kk'}^{m'} v^p + \frac{1}{4} c_{lk'}^j c_{mk}^l v^m \right) + \cdots \\
=& v^j + u^j + \frac{1}{2} [u, v]^j + \frac{1}{12} [v, [v, u]]^j + \frac{1}{24} [u, [v, u]]^j \\
&+ \frac{1}{8} [[v, u], u]^j + \cdots \\
=& v^j + u^j + \frac{1}{2} [u, v]^j + \frac{1}{12} [u - v, [u, v]]^j + \cdots , \\
w =& u + v + \frac{1}{2} [u, v] + \frac{1}{12} [u - v, [u, v]] + \cdots . \qquad (14.45)
\end{aligned}
$$

Thus we have arrived at the Baker-Campbell-Hausdorff formula by an alternative route.

Realizations of a Lie Group by Canonical Transformations

We have seen that a realization of a Lie group G by means of a transformation group on a manifold S can be set up if we can find a realization of the corresponding Lie algebra L by means of linear first-order partial differential operators acting on suitable functions on S, the commutator of these operators being chosen to represent the abstract Lie bracket operation in L (cf. Eq.(14.31) and the remarks following it). The resulting transformations acting on S are called *point-transformations*, because each element a in G is realized by a mapping T_a that takes points of S into points of S in a one-to-one invertible manner. In applications to classical Hamiltonian mechanics, we are interested in a special class of such realizations of Lie groups by means of point transformation groups, namely realizations in which each of the representative transformations T_a acts as a *regular canonical transformation* on the manifold S. This is because, in all the applications we have in mind, any relevant Lie group G arises as the expression of some invariance or symmetry property of the dynamical system under consideration with respect to some group of transformations defined on the dynamical variables of the system. The natural requirement that the Hamiltonian structure for the equations of motion should be maintained before as well as after such transformation demands that such transformations should be realized by canonical transformations. (Recall from Chapter 5 that canonical transformations are the most general point transformations that leave the *form* of Hamilton's equations invariant.) Of course, we might wonder why we should have a realization at all, namely, why the canonical transformations corresponding to various elements of G must have the representation property! As is shown in Chapter 16, the answer is that the representation property is a *consistency condition* on the entire formalism; in other words, it follows from the requirement that the use of the Hamiltonian formalism be internally consistent and have a sensible physical interpretation. Now we analyze the consequences of restricting the general point transformation group realization of a group G to a realization by regular canonical transformations. For brevity, we refer to these as *canonical realizations* of G and also omit the word "regular" when speaking of canonical transformations.

To begin with, let us suppose that the dynamical system under study is of the standard type and not subject to any constraints. The manifold S for the realization is then the even-dimensional phase space of the system, and the superscript ρ on the coordinates ω^ρ of a point ω in S takes the values $\rho = 1, 2, \cdots, 2k$. (For $\rho \leqq k$, the ω^ρ are the q's of the system, for $\rho \geqq k+1$ they are the p's). Given the realization of L by the operators $\Lambda_t(\omega)$, the functions $\lambda_k^\rho(\omega)$ characterizing the realization are thereby known, and Eq.(14.20) tells us the form of an infinitesimal transformation acting on S corresponding to an element close to e in G. Now the PB of any two functions Φ, Ψ defined on S is given in terms of the standard matrix $\|\varepsilon^{\rho\sigma}\|$ of Chapter 5:

$$\{\Phi(\omega), \Psi(\omega)\}_{PB} = \varepsilon^{\rho\sigma} \frac{\partial \Phi(\omega)}{\partial \omega^\rho} \frac{\partial \Psi(\omega)}{\partial \omega^\sigma} . \tag{14.46}$$

(See Eq.(5.33)). *The condition that the infinitesimal point transformation Eq. (14.20):*

$$a = \exp(t): \qquad \omega \to T_{\exp(t)}\omega = \omega' : \ \omega'^\rho = \omega^\rho + t^j \lambda_j^\rho(\omega) + \cdots \tag{14.47}$$

be canonical is that the fundamental PB's be preserved. But as we have shown in Chapter 6 (see Eq.(6.7) and the following development in that chapter), this means that there must exist a set of functions $\lambda_j(\omega)$, as many in number as the dimension of L, such that the small changes $\delta\omega^\rho$ in the coordinates of ω can be written in the form:

$$\delta\omega^\rho \equiv \omega'^\rho - \omega^\rho \equiv t^j \lambda_j^\rho(\omega) = t^j \varepsilon^{\rho\sigma} \frac{\partial \lambda_j(\omega)}{\partial \omega^\sigma} . \tag{14.48}$$

Thus for a canonical realization of G, the associated realization of L must be somewhat special: the functions $\lambda_j^\rho(\omega)$ and operators $\Lambda_j(\omega)$ corresponding to a basis e_j in L must take the form:

$$\lambda_j^\rho(\omega) = \varepsilon^{\rho\sigma} \frac{\partial \lambda_j(\omega)}{\partial \omega^\sigma} , \quad \Lambda_j(\omega) = \varepsilon^{\rho\sigma} \frac{\partial \lambda_j(\omega)}{\partial \omega^\sigma} \frac{\partial}{\partial \omega^\rho} \tag{14.49}$$

in terms of some set of functions $\lambda_j(\omega)$ on S. If we use the notation introduced in Eq.(6.32), which associates with each function $\phi(\omega)$ on S a corresponding linear differential operator $\tilde{\phi}(\omega)$, which we call the tilded operator corresponding to ϕ, then Eq.(14.49) says that for a canonical realization of G, the operator $\Lambda_j(\omega)$ for each j must be the tilded operator corresponding to some function $-\lambda_j(\omega)$; more generally, for any tangent vector $t \in L$, $\Lambda_t(\omega)$ must be the tilded operator corresponding to the function $-t^j\lambda_j(\omega)$:

$$\Lambda_j(\omega) = -\widetilde{\lambda_j(\omega)}, \;\; \Lambda_t(\omega) = -t^j\widetilde{\lambda_j(\omega)}\,. \tag{14.50}$$

(The minus sign introduced here, caused by the way the λ_j are defined in Eq.(14.48), is for later convenience.) Now we must convert the basic commutation relations Eq.(14.32) for the operators Λ_j into conditions on the functions $\lambda_j(\omega)$. To do this, we note an important property of tilded operators; let ϕ and ψ be two functions on phase space, and consider the commutator of $\tilde{\phi}$ and $\tilde{\psi}$. With $\chi(\omega)$ being any other function, we obtain with the use of the Jacobi identity:

$$\begin{aligned}
[\tilde{\phi}(\omega), \tilde{\psi}(\omega)]_-\,\chi(\omega) =&(\tilde{\phi}(\omega)\tilde{\psi}(\omega) - \tilde{\psi}(\omega)\tilde{\phi}(\omega))\chi(\omega)\\
=&\{\phi(\omega), \{\psi(\omega), \chi(\omega)\}\}\\
&-\{\psi(\omega), \{\phi(\omega), \chi(\omega)\}\}\\
=&-\{\chi(\omega), \{\phi(\omega), \psi(\omega)\}\} = \widetilde{\{\phi(\omega), \psi(\omega)\}}\,\chi(\omega),\\
[\tilde{\phi}(\omega), \tilde{\psi}(\omega)]_- =& \widetilde{\{\phi, \psi\}}\,.
\end{aligned} \tag{14.51}$$

(We have presented the derivation in such a way that it remains valid for a GPB as well.) That is, *the commutator of the tilded operators associated with two functions ϕ, ψ is the tilded operator associated with the PB of ϕ with ψ.* This is a basic result, and it allows us to convert the condition Eq.(14.32) on the Λ's into a statement on the λ_j's. We have:

$$\begin{aligned}
&[\Lambda_j(\omega), \Lambda_k(\omega)]_- = [\tilde{\lambda}_j(\omega), \tilde{\lambda}_k(\omega)]_- = \widetilde{\{\lambda_j(\omega), \lambda_k(\omega)\}}\,,\\
&\widetilde{\{\lambda_j(\omega), \lambda_k(\omega)\}} = c^l_{jk}\tilde{\lambda}_l(\omega)\,.
\end{aligned} \tag{14.52}$$

Can we remove the tildes on both sides of this last equation? The answer is no; the association $\phi \to \tilde{\phi}$ is not a faithful one, because it does not take every nonzero function ϕ into a nonzero operator $\tilde{\phi}$. In fact, if ϕ were a pure number, namely a constant over S, $\tilde{\phi}$ would be zero. Thus we cannot conclude that the functions under the tilde signs on the two sides of the last line in Eq.(14.52) are equal, but that they are equal up to some constant additive factors. Thus we get the fundamental conditions on the λ_j:

$$\{\lambda_j(\omega), \lambda_k(\omega)\} = c^l_{jk}\lambda_l(\omega) + d_{jk}\;, \;\; d_{jk} = \text{pure numbers;} \tag{14.53}$$

and we have the following fundamental result: in order to obtain a *canonical realization* of a Lie group G we must look for a *PB realization* of its Lie algebra

L, in which however the basic Lie bracket relations are obeyed up to some additive numerical constants on the right-hand side. By a PB realization of L we mean a realization in which the elements of L are represented by functions of ω, and the Lie bracket is represented by the PB of these functions. We have proved the statement made in the introduction to this chapter, namely that the PB is the Lie bracket relevant to classical mechanics.

Thus far we only verified that for the infinitesimal transformations to be canonical, it is necessary to have a PB realization of L. But we have seen earlier in this chapter that a finite point transformation T_a can be built up from a differential equation involving the infinitesimal form (see Eqs. (14.36), (14.37), (14.39) and (14.40)), and our work in Chapter 6 assures us that if the infinitesimal form involves the PB, the solution of the differential equation will certainly be a finite canonical transformation. The connection between a canonical realization of G and a PB realization of L (upto additive constants) is thereby complete.

At the risk of being repetitive, let us now transcribe some of our results on general transformation groups to the present case. We have a PB realization of the Lie algebra L corresponding to the given Lie group G. For a general element $t \in L$, let us write $\lambda_t(\omega)$ for the corresponding function:

$$\lambda_t(\omega) = t^j \lambda_j(\omega). \tag{14.54}$$

Then, the element $a = \exp(u)$ in G is realized as the canonical transformation T_a acting on points in phase space and as the operator $\bar{R}_a$ acting on functions on phase space, with

$$[\bar{R}_{\exp(u)}\Phi](\omega) = \Phi(T_{\exp(-u)}\omega) = \widetilde{e^{\lambda_u(\omega)}}\Phi(\omega)\,. \tag{14.55}$$

This is just Eq.(14.39) for the present case. Specializing Φ to be one of the coordinates ω^ρ themselves, the transformation T_a itself appears as:

$$\begin{aligned} T_{\exp(u)}\omega &= \omega' \, ; \\ \omega'^\rho = e^{-\widetilde{\lambda_u(\omega)}}\,\omega^\rho &= \omega^\rho - \{\lambda_u(\omega), \omega^\rho\} + \frac{1}{2}\{\lambda_u(\omega), \{\lambda_u(\omega), \omega^\rho\}\}\cdots \end{aligned} \tag{14.56}$$

Finally, we have the composition property:

$$\begin{aligned} \widetilde{e^{\lambda_u(\omega)}}\widetilde{e^{\lambda_v(\omega)}} &= \widetilde{e^{\lambda_w(\omega)}}\,, \quad w = u + v + \frac{1}{2}[u,v] + \frac{1}{12}[u - v, [u,v]] + \cdots \\ \lambda_w(\omega) &= \lambda_u(\omega) + \lambda_v(\omega) + \frac{1}{2}\{\lambda_u(\omega), \lambda_v(\omega)\} \\ &\quad + \frac{1}{12}\{\lambda_u(\omega) - \lambda_v(\omega), \{\lambda_u(\omega), \lambda_v(\omega)\}\}\cdots + \text{pure numbers}\,. \end{aligned} \tag{14.57}$$

This last form of the composition law with PB's appearing in the Baker-Campbell-Hausdorff formula is a simple consequence of combining the relation:

$$\{\lambda_t(\omega), \lambda_u(\omega)\} = \lambda_{[t,u]}(\omega) + \text{constant}, \; t, u \in L \tag{14.58}$$

with the usual Baker-Campbell-Hausdorff formula giving w in terms of u and v.

Let us derive some simple properties of the constants d_{jk} appearing in the PB realization Eq.(14.53) of L, leaving for the last section of this chapter a more formal discussion of such realizations up to constants of Lie algebras. It is clear in the first place that the d_{jk} cannot be arbitrary numbers; for example, one property obvious from Eq.(14.53) is that they must be antisymmetric in their subscripts:

$$d_{jk} = -d_{kj} \ . \tag{14.59}$$

Another property of these constants emerges if we use the fact that the PB is subject to the Jacobi identity. Starting from Eq.(14.53), take the PB of both sides with $\lambda_m(\omega)$, and sum cyclically on j, k and m:

$$\begin{aligned}\sum_{\substack{\text{cyclic}\\ jkm}} \{\{\lambda_j(\omega), \lambda_k(\omega)\}, \lambda_m(\omega)\} &= \sum_{\substack{\text{cyclic}\\ jkm}} c^l_{jk}\{\lambda_l(\omega), \lambda_m(\omega)\}\\ &= \sum_{\substack{\text{cyclic}\\ jkm}} c^l_{jk}(c^p_{lm}\lambda_p(\omega) + d_{lm}) \ .\end{aligned} \tag{14.60}$$

Now the last expression here should vanish, but the terms bilinear in the structure constants vanish by themselves, because of the Jacobi identity for these constants! It follows that we must have:

$$\sum_{\substack{\text{cyclic}\\ jkm}} c^l_{jk} d_{lm} \equiv c^l_{jk} d_{lm} + c^l_{km} d_{lj} + c^l_{mj} d_{lk} = 0 \ . \tag{14.61}$$

Equations (14.59) and (14.61) are the only restrictions that can be placed on the d's; apart from these, they are arbitrary. Now suppose a definite set of d_{jk} obeying these conditions has been found and that a set of functions $\lambda_j(\omega)$ obeying Eq.(14.53) has also been found. We now make use of the fact that the operators $\tilde{\lambda}_j(\omega)$ corresponding to the $\lambda_j(\omega)$ are unaltered if to each $\lambda_j(\omega)$ a constant were to be added, so that the canonical realization of G is also unaffected by such a change in $\lambda_j(\omega)$. That is, as far as the canonical realization of G is concerned, we could use the functions $\lambda'_j(\omega) = \lambda_j(\omega) + d_j$ in place of $\lambda_j(\omega)$, the d_j being a set of numbers. But the d_{jk} corresponding to these new functions are different from the old ones. We have:

$$\{\lambda'_j(\omega), \lambda'_k(\omega)\} = c^l_{jk}\lambda'_l(\omega) + d'_{jk} \ , \ d'_{jk} = d_{jk} - c^l_{jk} d_l \ . \tag{14.62}$$

Then we have two realizations up to constants of the Lie algebra L, one given by the system $(\lambda_j(\omega), d_{jk})$ and the other by the system $(\lambda'_j(\omega), d'_{jk})$, both leading to the *same* canonical realization of G. It follows that we must regard these two realizations of L as *equivalent*. Two realizations of L will be *inequivalent* if for no choice of constants d_j can the constants d'_{jk}, d_{jk} be related as in Eq.(14.62).

Now in many cases of practical interest, it may happen that starting with a realization up to constants of the Lie algebra L, given by $(\lambda_j(\omega), d_{jk})$, it is possible to choose the constants d_j in Eq.(14.62) in such a way that the d'_{jk} vanish identically! If this happens, then we say that the realization up to constants $(\lambda_j(\omega), d_{jk})$ is equivalent to a *true* PB realization of L, and we say that the $\lambda'_j(\omega)$ give us a true realization of L. If this cannot be done — that is, if the d_{jk} in Eq.(14.53) cannot be "transformed away" by a redefinition of the functions $\lambda_j(\omega)$— then the realization is essentially different in character from a true realization of L, although we may, of course, make use of the transformation Eq.(14.62) to put the d_{jk} into the most simple form. The structures of the Lie algebras of many familiar Lie groups such as the three-dimensional (or any n-dimensional) rotation group, the three-dimensional Euclidean group, the Lorentz and Poincaŕe groups, are such that they allow us to prove that *any* PB realizations up to constants of these Lie algebras are equivalent to true realizations; in other words, in these cases, it is possible to combine Eqs.(14.59) and (14.61) to show that d_{jk} can be written as $c^l_{jk} d_l$ in terms of some set of constants d_l, so that one need only look for true realizations. But the Galilei group is an important example of a Lie group whose Lie algebra possesses both kinds of realizations, true as well as up to constants that cannot be transformed away. We study this in more detail in Chapter 19. In the last section of this chapter, we present another simple but important instance of this phenomenon.

The extension of the theory of canonical realizations of Lie groups G to the cases where PB's and canonical transformations are replaced by GPB's and generalized canonical transformations or by Dirac brackets and Dirac bracket transformations is straightforward. All that must be remembered is the essential reason for the presence of the numbers d_{jk} in Eq.(14.53); in the case of the usual PB, pure numbers are the only candidates for the d_{jk} because they are the only quantities that have vanishing PB's with all functions on S. In general, any quantity having vanishing bracket with all other quantities is called a *neutral element.* For a nonsingular GPB, pure numbers are the only neutral elements, whereas for a singular GPB this is not so. For example, in the case of the Dirac bracket, every function of the second-class constraints is a neutral element. The general characteristic of a neutral element in all cases is that its tilded operator vanishes, so that in going from the analogue of Eq.(14.52) to the analogue of Eq.(14.53) in a general case, we must always allow for the appearance of neutral elements on the right-hand side in Eq.(14.53). We can now say that a generalized canonical realization of a Lie group G arises from a GPB realization up to neutral elements of its Lie algebra L, and vice versa, and similarly for the Dirac bracket. Of course, the notion of equivalent realizations of L and so on is applicable once again; we have only to note that the quantities d_l in Eq.(14.62) must in all cases be neutral elements so that changing the functions $\lambda_k(\omega)$ by addition of d_k does not alter the realization of G.

The Lie Algebra of all Dynamical Variables

Let us compare the important elements in the three types of realizations of a given finite-parameter Lie group G that we have studied. In the case of the adjoint representation of G, the element $u \in L$ and the group element $a = \exp(u) \in G$ are represented by the linear operators Ω_u and $R_a = e^{\Omega_u}$, respectively; both of these operate on L itself viewed as a linear vector space and form realizations of L and G, respectively, by virtue of Eqs.(14.6) and (14.7). In the case of a point-transformation group realization of G, on a manifold S, $u \in L$ and $a = \exp(u) \in G$ are, respectively, realized by the differential operator $\Lambda_u(\omega)$, and the operator $\bar{R}_a = \exp(-\Lambda_u(\omega))$, both acting on suitable functions on S. We now have Eqs.(14.31) and (14.41). We notice a formal difference in the nature of these two realizations of L: in the first case the quantities $u, v \cdots \in L$ being represented are also the quantities on which the *representors* $\Omega_u, \Omega_v, \cdots$ are defined to act; whereas in the second case the representors $\Lambda_u(\omega), \Lambda_v(\omega), \cdots$ do not act on $u, v, \cdots$ but on functions on S. But when we look at canonical realizations we find that to some extent it is like the adjoint representation: with the PB playing the role of the Lie bracket, $u \in L$ is represented by an *ordinary function* $\lambda_u(\omega)$ on S (up to a neutral element); when one takes the Lie bracket of the representor of u with something, that something is also an ordinary function on S; and the representative $\bar{R}_a = e^{\widetilde{\lambda}_u(\omega)}$ of $a = \exp(u)$ is also defined to act on ordinary functions on S. We now have Eqs.(14.53) and (14.57). [Even for the canonical realizations one could say that every $u \in L$ is represented by a differential operator $\Lambda_u(\omega)$ of a special type and that the Lie bracket is the commutator of these operators, but because the realization is canonical, we are able to write $\Lambda_u(\omega)$ as $-\ \widetilde{\lambda}_u(\omega)$, and thus associate an ordinary function $\lambda_u(\omega)$ with u]. Because of this feature, we can *formally* apply the theory of the adjoint representation to the group of regular canonical transformations that can be continuously connected to the identity, and in this way we find a close correspondence between many results derived in Chapter 6 on the one hand, and results pertaining to the adjoint representation of a Lie group on the other. This happens in the following way. We consider the set of all real functions $\phi(\omega)$ on phase space that are infinitely often differentiable and that act as generators of regular canonical transformations. All such functions are observables of the system. There is an infinite number of them, and formally we can suppose that they form an infinite-dimensional real linear vector space. We make this into an infinite-dimensional Lie algebra, L_∞, by defining the Lie bracket to be the PB! Now any Lie algebra L determines, locally, a Lie group G, and then we have the adjoint representation with both L and G represented by means of linear operators acting on L. Thus we can ask: what is the structure of the Lie group G_∞ corresponding to the Lie algebra L_∞, and what does the adjoint representation look like in this case? Formally we can say that to every $\phi(\omega) \in L_\infty$, we have an "element" $\exp(\phi(\omega))$ in G_∞! The operator Ω_u of the adjoint representation, because its action on any element in L is defined to be the same as taking the Lie bracket of u with that element, now takes the form $\Omega_{\phi(\omega)} = \widetilde{\phi(\omega)}$: it is by applying $\widetilde{\phi(\omega)}$ to any other function $\psi(\omega) \in L_\infty$ that we get $\{\phi(\omega), \psi(\omega)\}$. Again, because the adjoint representation assigns to the group element $\exp(u)$ the linear operator $R_{\exp(u)} = e^{\Omega_u}$ acting on L, we see that the element $\exp(\phi(\omega))$ in G_∞ is represented by $R_{\exp(\phi(\omega))} = e^{\widetilde{\phi(\omega)}}$, and this is meant to act on L_∞, namely, on suitable functions on S. However, we saw in the case of the adjoint representation that Lie brackets are preserved by the transformations $R_{\exp(u)}$ (that is the content of Eq.(14.14)). Thus in the present case the transformations $e^{\widetilde{\phi(\omega)}}$ must preserve PB's and thus must be canonical transformations! Thus the most important properties of the tilded operators $\widetilde{\phi(\omega)}$ that we studied in Chapter 6 are now seen to be natural consequences of the fact that we are formally dealing with the adjoint representations of L_∞ and G_∞. The analogy can be displayed by listing in adjoining columns the quantities pertaining to a finite-dimensional Lie algebra L and those pertaining to L_∞:

General element in Lie algebra:

$$u \in L \qquad\qquad \phi(\omega) \in L_\infty$$

General element in Lie group:

$$\exp(u) \in G \qquad\qquad \exp(\phi(\omega)) \in G_\infty$$

Adjoint representation:

$$\left.\begin{matrix} u \to \Omega_u = [u, \cdot] \\ \exp(u) \to e^{\Omega_u} \end{matrix}\right\} \text{acting on } L. \qquad \left.\begin{matrix} \phi(\omega) \to \Omega_\phi = \{\phi(\omega), \cdot\} = \widetilde{\phi(\omega)} \\ \exp(\phi(\omega)) \to e^{\widetilde{\phi(\omega)}} \end{matrix}\right\} \text{acting on } L_\infty$$

$$e^{\Omega_u}[v,t] = [e^{\Omega_u}v, e^{\Omega_u}t] : \qquad e^{\widetilde{\phi(\omega)}}\{\psi(\omega), \chi(\omega)\} = \{e^{\widetilde{\phi(\omega)}}\psi(\omega), e^{\widetilde{\phi(\omega)}}\chi(\omega)\} :$$

Invariance of Lie brackets under e^{Ω_u} — $e^{\widetilde{\phi(\omega)}}$is a canonical transformation.

In fact, all the equations in the column on the right are familiar from our earlier work, and our relating them to the properties of the adjoint representation only is to help us perceive them from a new point of view. There is only one new property we have now that was not derived in Chapter 6, and that is the composition law for G_∞ or the way two canonical transformations $e^{\tilde\phi}$ and $e^{\tilde\psi}$ combine to yield a third one. This is contained in Eqs.(14.6) and (14.57), but we repeat it here:

$$\begin{aligned} e^{\widetilde{\phi(\omega)}} e^{\widetilde{\psi(\omega)}} =& e^{\widetilde{\chi(\omega)}}, \\ \chi(\omega) =& \phi(\omega) + \psi(\omega) + \frac{1}{2}\{\phi(\omega), \psi(\omega)\} \\ &+ \frac{1}{12}\{\phi(\omega) - \psi(\omega), \{\phi(\omega), \psi(\omega)\}\} + \cdots \end{aligned} \tag{14.63}$$

We had introduced different symbols, R_a and $\bar{R}_a$ to denote the representative of the group element a in the adjoint representation and in a general realization by a transformation group, respectively. However, because the special case of canonical realizations has many formal similarities to the case of the adjoint representation, it is both reasonable and convenient simply to write R_a for the representative of a in a canonical realization of G.

There is one remaining property of canonical realizations (also shared by realizations by transformation groups) that we should describe before we look at the problem of neutral elements. As we said at the beginning of this chapter, a specific realization of an abstract mathematical system may have operations defined in it that are not necessary for the realization as such but nevertheless characterize it. In the case of canonical realizations of Lie groups, the objects on which the representations act are real-valued functions on phase space, and apart from the PB, which is defined for these functions, there is the more elementary operation of associative pointwise multiplication that is also defined for them. That is, given two functions $\phi(\omega)$ and $\chi(\omega)$, their product $\psi(\omega) = \phi(\omega)\chi(\omega)$ is also a function on phase space. Now if T is a regular canonical mapping $\omega \to \omega'$ and if R is the linear mapping induced on functions of ω, so that $(R\phi)(\omega)$ is equal to $\phi(T^{-1}\omega)$, the behavior of a product with respect to R is obvious:

$$R(\phi\chi) = (R\phi)(R\chi) \ . \tag{14.64}$$

This is because the effect of R on any function is to evaluate the same function at a transformed argument. This property of canonical transformations is reflected in the following behavior of products in a PB:

$$\{\phi\chi, \psi\} = \phi\{\chi, \psi\} + \{\phi, \psi\}\chi \ . \tag{14.65}$$

This is what we called the product rule in Chapter 5 (see the discussion following Eq.(5.37)). It is also called the "derivation property" for the PB. Whereas Eqs.(14.64) and (14.65) are obvious and need no proof, it is interesting to note that they are particular examples of the following more general property of the *adjoint* representation of a Lie algebra: suppose that in a Lie algebra L a rule of combination Q is given, which assigns to every pair of vectors $u, v \in L$ a third vector $Q(u, v)$ such that Q is linear in both arguments and with respect to the Lie bracket Q possesses the following derivation property:

$$[u, Q(v, w)] = Q([u, v], w) + Q(v, [u, w]) \ . \tag{14.66}$$

Then under the adjoint representation of the Lie group G corresponding to L we have:

$$e^{\Omega_u} Q(v, w) = Q(e^{\Omega_u} v, e^{\Omega_u} w) \ . \tag{14.67}$$

We leave it to the reader to supply the details of the proof. It follows the same pattern as many of the proofs we have given and consists in showing that both sides of Eq.(14.67) arise from the same differential equation and boundary condition by virtues of Eq.(14.66). It is interesting to note, however, that Q need not necessarily represent an associative multiplication law in L, although in the application to derive Eq.(14.64) it does; for example, $Q(u, v)$ could be the Lie bracket of u with v, in which case Eq. (14.66) is the Jacobi identity and

Eq.(14.67) is the same as Eq.(14.14)!

Neutral Elements in a Lie Algebra

We have seen that the canonical realizations of a Lie group G arise from PB realizations up to neutral elements of its Lie algebra L and vice versa. Such a realization of L involves the phase-space functions $\lambda_j(\omega)$, one for each basis vector e_j in L, and the set of neutral elements d_{jk}. But the important thing is that the presence of these neutral elements in Eq.(14.53) does not affect the canonical transformations representing the elements of the Lie group G corresponding to L or the composition law for these transformations. Our method of derivation of Eq. (14.53), of course, guarantees this, but it is worthwhile to see this property directly from the structure of the adjoint representation.

Let L' be some Lie algebra. An element u_0 in L' having vanishing Lie bracket with every element of L' is called a "neutral element" of L'. It is clear that the set of all neutral elements in L' forms a linear subspace L_0 in L'. Now L' can be decomposed into the direct sum of L_0 and a complementary subspace $\bar{L}$, in the vector space sense:

$$L' = L_0 \oplus \bar{L} \,. \tag{14.68}$$

$\bar{L}$ is not uniquely determined, but any choice will do. Any vector $u' \in L'$ can be split up in a unique manner into a part $u_0 \in L_0$ and a part $\bar{u} \in \bar{L}$. Let $\bar{e}_j$ be a basis for $\bar{L}$; then the Lie brackets of these vectors with each other can be written as:

$$[\bar{e}_j, \bar{e}_k] = c^l_{jk}\bar{e}_l + \text{something in } L_0 \,. \tag{14.69}$$

These are the only nonvanishing basic Lie brackets in L'. The coefficients c^l_{jk} are a subset of the full set of structure constants for L' and are completely determined once the $\bar{e}_j$ are chosen. But one can easily convince oneself that the c^l_{jk} by themselves obey both the antisymmetry requirements and the Jacobi identity so that one could define a new Lie algebra L with basis elements e_j obeying the bracket relations:

$$[e_j, e_k] = c^l_{jk} e_l \,. \tag{14.70}$$

Obviously, $\bar{L}$ and L are of the same dimension, and we have a one-to-one correspondence between vectors in $\bar{L}$ and vectors in L: for the same u^k, $\bar{u} = u^k \bar{e}_k \in \bar{L}$ and $u = u^k e_k \in L$ correspond to one another. Let now G' and G be the Lie groups determined by L' and L, respectively, and consider the adjoint representation of G' realized on L'. For any vector $u' \in L'$, we have:

$$u' = \bar{u} + u_0, \qquad \Omega_{u'} = \Omega_{\bar{u}}, \qquad \Omega_{u_0} = 0 \,. \tag{14.71}$$

That is, in the adjoint realization of L' on itself, all elements of L_0 go into zero. Correspondingly, for elements of G' of the type $\exp(u_0)$ the representing operator R is just the unit operator:

$$R_{\exp(u_0)} = \exp(\Omega_{u_0}) = 1\!\!1 \text{ on } L' \,. \tag{14.72}$$

For a more general element $u' \in L'$, $\exp(u')$ is represented by:

$$R_{\exp(u')} = R_{exp(\bar{u})} = e^{\Omega_{\bar{u}}} \ . \tag{14.73}$$

Thus as far as the operators Ω and R are concerned, we could restrict ourselves to vectors in $\bar{L}$; further, because there is a one-one correspondence between L and $\bar{L}$, we have one operator $R_{\exp(\bar{u})}$ acting on L' corresponding to each element $\exp(u)$ in G, u being the vector in L corresponding to $\bar{u}$ in $\bar{L}$. However, these operators $R_{\exp(\bar{u})}$ obey the composition law characteristic of G! If $\bar{u}, \bar{v} \in \bar{L}$, we have:

$$\begin{aligned} R_{\exp(\bar{u})} R_{\exp(\bar{v})} &= e^{\Omega_{\bar{u}}} e^{\Omega_{\bar{v}}} = R_{\exp(w')} = R_{\exp(\bar{w})} = e^{\Omega_{\bar{w}}} \ , \\ w' &= \bar{u} + \bar{v} + \frac{1}{2}[\bar{u}, \bar{v}] + \frac{1}{12}[\bar{u} - \bar{v}, [\bar{u}, \bar{v}]] + \cdots \end{aligned} \tag{14.74}$$

Now all the brackets appearing here are to be evaluated in L'. Actually, we are only interested in $\bar{w}$; in evaluating each of the multiple brackets, in each term we can compute all the brackets except the last one as though we were dealing with vectors in L because any neutral elements that appear are going to be eliminated by the last bracket. As far as the component $\bar{w}$ of w' is concerned, we can evaluate the last bracket, too, as though we were dealing with L and drop any neutral elements that appear! This means that the vector w in L corresponding to $\bar{w}$ in $\bar{L}$ arises by an application of the Baker-Campbell-Hausdorff formula to u and v in L, u and v being the vectors that correspond to $\bar{u}$ and $\bar{v}$, respectively. Stated in terms of the representatives, we find that if $R_{\exp(\bar{u})} R_{\exp(\bar{v})} = R_{\exp(\bar{w})}$, then w is such that $\exp(u)\exp(v) = \exp(w)$ for the elements of G determined by u, v, and w. Therefore, the adjoint representation of L' and G' "boils down" to a representation of L and G, although acting on L'; L and G are just what we would be dealing with if, loosely speaking, we ignored the presence of the neutral elements in L'.

Reversing the argument, we can say that the Lie algebra L' arises from the Lie algebra L by the formal addition of neutral elements to L, thus increasing the dimension of L, and at the same time adding neutral elements to the values of Lie brackets in L. A realization up to neutral elements, in terms of PB's, for instance, of L, can be reinterpreted as a true realization of L'. The locally determined Lie group G corresponding to L goes into G' corresponding to L', and G' is called an *extension* of G. The one-parameter subgroups of G' corresponding to the added neutral elements in L' are called "phase groups." Basically, these are trivially represented by the adjoint representation of G', which makes the latter act as a representation of G.

We have noted that the Galilei group is the most important practical example of such a situation — there are nontrivial realizations up to neutral elements. A much simpler example is provided, however, by the case of a pair of canonically conjugate Hamiltonian variables, namely p and q obeying the basic PB relation:

$$\{q, p\} = 1 \ . \tag{14.75}$$

The unit appearing on the right is a neutral element. Thus Eq.(14.75) provides a PB realization up to neutral elements of a two-dimensional Lie algebra L in which all Lie brackets vanish. For a two-dimensional Lie algebra with basic elements e_1, e_2, the only possibly nonvanishing basic Lie bracket is $[e_1, e_2]$, and if this too vanishes, then *all* the structure constants are zero. The corresponding Lie group G is the group of translations in two dimensions and is Abelian. Thus we have:

$$L: \quad \text{basis } e_1, e_2 \; ; \qquad [e_1, e_2] = 0$$
$$G: \quad \text{elements } \exp(u^1 e_1 + u^2 e_2) \; ;$$
$$\exp(u^1 e_1 + u^2 e_2) \exp(v^1 e_1 + v^2 e_2) = \exp((u^1 + v^1) e_1 + (u^2 + v^2) e_2) \; . \quad (14.76)$$

Because all the c^j_{kl} vanish for L, Eq.(14.62) shows that if we had a realization up to neutral elements of L, these elements can never be transformed away, and the realization will not be equivalent to a true one. The only basic bracket where a neutral element can turn up is in the representative of $[e_1, e_2]$; thus we can enlarge L to a three-dimensional Lie algebra L' by formally adding a third basic element e_3, with the only nonvanishing Lie bracket in L' being:

$$[e_1, e_2] = e_3 \; , \qquad e_3 = \text{neutral element.} \quad (14.77)$$

The composition law for G' is easily obtained because all terms in the Baker-Campbell-Hausdorff formula beyond the third one are zero. It is:

$$u = \sum_1^3 u^j e_j \; , \quad v = \sum_1^3 v^j e_j \; :$$
$$G': \quad \exp(u)\exp(v) = \exp(w), \quad w = u + v + \frac{1}{2}(u^1 v^2 - u^2 v^1) e_3 \; . \quad (14.78)$$

Thus we can describe Eq.(14.75) in these terms: the realization upto neutral elements $e_1 \to q$, $e_2 \to p$ of L is a true realization of the three-dimensional Lie algebra $L' : e_1 \to q$, $e_2 \to p$, $e_3 \to 1$, by means of PB's. The algebra L' is called the "Heisenberg algebra" because of its importance in quantum mechanics, but it also describes the basic structure of canonically conjugate variables in classical mechanics.

The mathematical preparation of this and the two preceding chapters is adequate for us to go on now to a study of specific groups and their classical realizations. In the next chapter we describe some specific Lie groups, then we proceed to examine their realizations.

Chapter 15

Some Important Lie Groups and Their Lie Algebras

As examples of physically important Lie groups, we describe in this chapter the following groups:

1. the group T_3 of translations in three-dimensional Euclidean space;
2. the group $R(3)$ or $SO(3)$ of proper orthogonal real rotations in three-dimensional space;
3. the Euclidean group $E(3)$ consisting of a combination of T_3 and $R(3)$;
4. the Galilei group, which is the group of space-time transformations describing Newtonian relativity; and
5. the proper homogeneous and inhomogeneous Lorentz groups describing Poincaŕe relativity.

These groups arise in the following way. A description of a dynamical system, be it a system of particles or fields or rigid bodies or a combination of these, is always given with reference to a definite coordinate system in space and time. It is usual to associate an "observer" with each possible coordinate system. It is possible to choose the coordinate system in infinitely many ways, and, in general, the form of the basic dynamical equations depends on the system chosen. But for a given body of physical laws, it usually happens that a special class of coordinate systems is singled out such that the basic laws take the same mathematical form in all these systems. Then one says that these laws are "invariant" with respect to changes in coordinate systems from one system to another within this class. In practice, the particular laws decide the way in which the transformation from one coordinate system, or as it is also called, one "reference frame," to another is to be carried out. As examples, we may mention that the Newtonian equations of motion under gravity do not change if we go from one inertial frame to another one, if the relation between the space-time coordinates used in one frame and another is of a particular form, namely the Galilei form. Similarly, the Maxwell equations of electrodynamics are invariant if the space-time coordinates of two inertial frames are connected by the Lorentz transformation equations of special relativity. Now in describing the way in which these various invariance properties are implemented in Hamiltonian mechanics, two things must be done. First, one must relate the members of the set of all "allowed" frames of reference to one another in appropriate mathematical terms. The set of all transformations one can make to go from one allowed frame to all others has the structure of a group, and depending on which family of frames of reference one is talking about, one has one or other of the groups we listed before. This group is an abstract object, and it has relevance in both classical and quantum mechanics. Second, coming to the problem of Hamiltonian mechanics, we must understand in what way the description of a given physical situation or state by one allowed observer is related to the description of the same physical state by another allowed observer. Given the two observers, the element of the invariance group that connects their coordinate systems is determined, and this element should then determine the way in which the description of a specific state given by one observer is to be translated into the description given by the other observer. The fact that the basic laws take the same mathematical form for both observers implies the following: if a Lagrangian, and thus a Hamiltonian, formulation of the laws exists for one observer, then it must also exist for the other. This immediately implies that the rules for transformation from one observer to another must not destroy the Hamiltonian form of the equations of motion; in other words, *this transformation must be a (regular) canonical transformation.* We see in the next chapter that what we have here is in fact a *canonical realization* of the relevant invariance group; this emerges from a careful examination of the second aspect of the general problem just outlined. In this Chapter, we examine the first part, namely the structures of the groups themselves.

The first three groups listed, namely the translation, rotation, and Euclidean groups in three-dimensional space, are concerned with what is generally called the "kinematics" of the system; they relate the various possible choices of a Cartesian reference frame in three-dimensional space and do not involve possible changes in the choice of the time coordinate. The Galilei and Lorentz groups describe two different types of space and time transformations, one relevant in the nonrelativistic domain and the other in the relativistic domain. Both contain the three "kinematic" groups T_3, $R(3)$, and $E(3)$ as subgroups. A wider family of coordinate transformations, which constitutes not a finite parameter but, loosely speaking, an infinite parameter group, arises in general relativity. The basic equations of this theory preserve their form under arbitrary changes of coordinate system in space time, the allowed coordinate systems being the most general curvilinear ones. However, we do not study this group in this book.

The Three-Dimensional Translation Group T_3

The group T_3 describes the transformations between all those Cartesian coordinate systems in three-dimensional space that differ only in the location of the origin but not in the directions of the axes. Given two such systems S and S' say, the location of S' relative to S is completely determined by the displacement vector $\boldsymbol{a}$ leading from the origin O' of S' to O of S. The components of $\boldsymbol{a}$ along the common directions of the axes in S and S' may be written a_j, $j = 1, 2, 3$. Because the transformation leading from S to S' is to be identified with an element of T_3, *elements of* T_3 *are labeled completely by three real numbers* (a_1, a_2, a_3). Considering transformations from a fixed S to all possible S', we encounter all possible displacement vectors $\boldsymbol{a}$, hence all elements of T_3. If some fixed point P in space has coordinates x_j in S, it has coordinates $x'_j = x_j + a_j$ in S'. The group composition law for T_3 is very simple. If $\boldsymbol{a}$ describes the transition from S to S' and $\boldsymbol{b}$ from S' to S'', with the displacement vectors having the components a_j, b_j respectively, then the transition from S to S'' corresponding to the product of the elements a_j and b_j in T_3, $(b_j)(a_j)$, has a displacement vector $\boldsymbol{c}$ with components $c_j = a_j + b_j$. Thus group composition is the same as vector addition of the displacement vectors, and T_3 is Abelian. One-parameter subgroups are easily identified: they are families of elements $\boldsymbol{a}(\sigma)$ with coordinates $a_j(\sigma) = \sigma t_j$, where t_j are the components of a fixed vector. This is a family of translations of the origin, all in a given direction but of varying amounts. Therefore, these are canonical coordinates of the first kind. The Lie algebra of T_3 is three-dimensional, and because T_3 is Abelian, all the structure constants vanish. We choose as basis for the Lie algebra the tangent vectors to the one-parameter subgroups of translations in the direction of the jth Cartesian axis, for $j = 1, 2, 3$. Calling these vectors d_j, we have

$$[d_j, d_k] = 0, \qquad j, k = 1, 2, 3. \tag{15.1}$$

If the displacement vector $\overrightarrow{O'O}$ leading from S' to S has components a_j, we say that the element of T_3 corresponding to this change of frame is $\exp(\boldsymbol{a} \cdot \boldsymbol{d})$ and we write $S' = \exp(\boldsymbol{a} \cdot \boldsymbol{d})S$. In this notation we can write the composition law for T_3 as:

$$\exp(u)\exp(v) = \exp(w)\ , \qquad w = u + v. \tag{15.2}$$

The u, v, w are elements of the three-dimensional Lie algebra, which is equivalent to being vectors in three-dimensional space. Because T_3 has such a simple structure, we can easily see that all elements lie on one-parameter subgroups, and canonical coordinates of the first kind can be used for the entire group.

Three-Dimensional Proper Rotation Group $R(3)$

The group $R(3)$ consists of the set of transformations that connect all those Cartesian coordinate systems $S, S', \cdots$ which share the same origin but differ in the directions of the three coordinate axes. In addition, all the systems $S, S', \cdots$, are assumed to be of the same "handedness," namely, all right-handed (or equivalently all left-handed); the group $R(3)$ contains no elements corresponding to a transition from a right-handed to a left-handed coordinate system or vice versa. There are two ways in which the action of $R(3)$ on space can be analyzed; they are called the "passive" and the " active" interpretations. In the former, which is the way we defined $R(3)$ at the beginning of this paragraph, the coordinate system S, with respect to which one measures the components of vectors, is subjected to a rotation, leading to a new system S', while points of space are held fixed. In the latter interpretation, the coordinate system S is kept unchanged, and rotations act as linear mappings of vectors in three-dimensional space onto themselves; equivalently each point in space is mapped into a new one under the action of a rotation. Let us first give the "passive" interpretation. Relative to a given system of axes S, the orientation of another, S', is completely specified by giving the direction cosines of the three axes of S' with respect to those of S. Let us write $\boldsymbol{e}_j$ for the three unit vectors in the directions of the coordinate axes in S, and similarly $\boldsymbol{e}'_j$ in S'. Then these direction cosines are the scalar products of the $\boldsymbol{e}'_j$ with the $\boldsymbol{e}_k$, and we have:

$$A_{jk} = \boldsymbol{e}'_j \cdot \boldsymbol{e}_k \ , \qquad \boldsymbol{e}'_j = A_{jk}\boldsymbol{e}_k \ , \qquad \boldsymbol{e}_k = A_{jk}\boldsymbol{e}'_j \ . \tag{15.3}$$

(Sum on repeated indices!) If $\boldsymbol{x}$ is some vector in three-dimensional space, for example the vector $\overrightarrow{OP}$ leading from the common origin O of S and S' to some given point P, then its components x'_j in S' are given in terms of the components x_j in S by a linear transformation involving the numbers A_{jk}:

$$x'_j = A_{jk}x_k \ . \tag{15.4}$$

From elementary geometry we know that the matrix A made up of the nine elements A_{jk} is an orthogonal matrix with unit determinant:

$$A = \|A_{jk}\| \ , \qquad \det A = 1 \ ;$$

$$AA^T = A^T A = 1 \Rightarrow A_{jk}A_{j'k} = \delta_{jj'}, \qquad A_{jk}A_{jk'} = \delta_{kk'} \ . \tag{15.5}$$

(The transpose of the matrix A is written A^T). We can think of these matrices A as providing us with a *defining matrix representation* of $R(3)$. We have one element a in the abstract group $R(3)$ corresponding to each distinct matrix A obeying Eq.(15.5), and conversely. The determinant condition is to ensure that the "handedness" of the coordinate system does not change. Another way of characterizing this set of transformations is to define it as the collection of all real linear transformations in the real variables x_k, as in Eq.(15.4), which have a unit determinant and preserve the squared length of the vector $\boldsymbol{x}$:

$$\sum_{k=1}^{3}(x'_k)^2 = \sum_{k=1}^{3}(x_k)^2 \ . \tag{15.6}$$

This immediately leads to the condition that A be an orthogonal matrix. Let us now work out the group multiplication law. Let a be an element corresponding to a matrix A, and in the passive interpretation let it lead us from a frame S to a frame S', which we can write as $aS : S' = aS$. Next let b be an element with matrix B, and let it take the frame S' to the frame $S'' = bS'$. Introducing the unit vectors $\boldsymbol{e}''_j$ appropriate to S'' and using Eq.(15.3) twice, we get:

$$S'' = baS : \boldsymbol{e}''_j = B_{jl}\boldsymbol{e}'_l = B_{jl}A_{lk}\boldsymbol{e}_k = C_{jk}\boldsymbol{e}_k \ , \qquad C_{jk} = B_{jl}A_{lk} \ . \tag{15.7}$$

Thus the element $c = ba$ that leads directly from S to S'' and is the group product of b with a corresponds to the matrix C, which is the matrix product of B with $A : C = BA$. Group multiplication corresponds to matrix multiplication in this defining representation. This same rule for computing the product ba of b and a could equally well be derived by taking some fixed vector $\boldsymbol{x}$ in space, having components x_j, x'_j and x''_j in S, $S' = aS$ and $S'' = bS'$, respectively, then using Eq.(15.4) to relate x''_j to x_j. In contrast to T_3, $R(3)$ is somewhat more complicated and has a more interesting structure. Because matrix multiplication is not commutative and because all orthogonal unimodular matrices are being considered, it is clear that $R(3)$ *is not Abelian.* If A and B are unimodular orthogonal matrices, so are AB and BA, but in general these two products are unequal. The inverse of the element a, namely a^{-1}, corresponds to A^T if a corresponds to A.

In the active interpretation of the action of $R(3)$, we start with some coordinate system S and keep it unchanged, and we make each rotation $a \in R(3)$ act as a linear mapping of vectors onto themselves. If $\boldsymbol{x}$ is a general vector with components x_j in S and if a is some element of $R(3)$ corresponding to the matrix A, the action of a on $\boldsymbol{x}$ yields a new vector $\boldsymbol{y}$ with components y_j (in S again) given as follows:

$$a \in R_3 : \qquad \boldsymbol{x} \to \boldsymbol{y} = a\boldsymbol{x}, \qquad y_j = A_{jk}x_k \ . \tag{15.8}$$

Although this looks exactly like Eq.(15.4) with y written in place of x', it is clear that the interpretation is different in the two cases. Because the mapping $\boldsymbol{x}$ to $a\boldsymbol{x}$ is linear, the effect of a on any $\boldsymbol{x}$ is determined once one knows the effects of a on the three basic vectors $\boldsymbol{e}_j$ in S. This is given by:

$$a : \qquad \boldsymbol{e}_j \to \hat{\boldsymbol{e}}_j = a\boldsymbol{e}_j = A_{kj}\boldsymbol{e}_k \ . \tag{15.9}$$

This is consistent with Eq.(15.8). We have written $\hat{\boldsymbol{e}}_j$ and not $\boldsymbol{e}'_j$ for $a\boldsymbol{e}_j$ because these vectors, in the active interpretation, are not supposed to define a new reference frame but are just the images of $\boldsymbol{e}_j$ under the mapping induced by a. The forms of Eq.(15.9) and (15.8) have been chosen so that the previously derived composition law Eq.(15.7) is obeyed. To verify it using Eq.(15.9), let b and a be any two elements of $R(3)$ with matrices B, A as before; we then use the linearity of the mappings to derive:

$$\begin{aligned} &\hat{\boldsymbol{e}}_j = a\boldsymbol{e}_j, \qquad \hat{\hat{\boldsymbol{e}}}_j = b\hat{\boldsymbol{e}}_j = ba\boldsymbol{e}_j \ ; \\ &\hat{\hat{\boldsymbol{e}}}_j = b(A_{lj}\boldsymbol{e}_l) = A_{lj}b\boldsymbol{e}_l = A_{lj}B_{kl}\boldsymbol{e}_k = (BA)_{kj}\boldsymbol{e}_k = C_{kj}\boldsymbol{e}_k \ . \end{aligned} \tag{15.10}$$

The point to note is that the action of *any* rotation on *any* vector is to be deduced by linearity from Eq.(15.9), which specifies the action on the fixed basic vectors $\boldsymbol{e}_j$; it must be used for b as well as a.

Now we come to the problem of choosing a coordinate system for $R(3)$. The orthogonality requirement on the matrix A imposes a certain number of conditions on the nine matrix elements of A. Because $A^TA = 1$ ensures that $\det A = \pm 1$, the unimodular condition $\det A = 1$ will not be violated while the matrix elements of A vary continuously; thus this condition need not be counted separately. Let us now work with the active interpretation, so that $a \in R(3)$ acts as a linear mapping on vectors, $\boldsymbol{x} \to a\boldsymbol{x}$, following Eqs.(15.8) and (15.9). The matrix equation $A^T A = 1$ (which implies $AA^T = 1$) can be described as consisting of six independent conditions: three of them are that the three vectors $\hat{\boldsymbol{e}}_j$ (which are the images of $\boldsymbol{e}_j$ under the rotation a and whose coordinates in S are given by the first, second, and third columns in A for $j = 1, 2, 3$, respectively) are vectors of unit length; three more conditions are that these vectors are perpendicular to one another. Therefore, there are just three free parameters in the matrix A, and $R(3)$ *is a three-parameter group.* Now there are several ways of choosing these parameters, and each has its own merits. From elementary geometrical arguments we can set up a canonical coordinate system of the first kind in the following way. Euler's theorem states that any proper rotation a in three-dimensional space leaves one direction unaltered and consists, therefore, of a right-handed rotation by some angle about this direction as axis. Thus a rotation can be specified completely by giving the axis and angle of rotation. These two things can be combined into a single vector $\boldsymbol{\alpha}$ with components α_j and magnitude $\alpha = |\boldsymbol{\alpha}|$; the vector $\boldsymbol{\alpha}$ is the axis of the rotation, and the magnitude α, obeying $0 \leqq \alpha < 2\pi$, is the amount of the rotation. These coordinates certainly are canonical; a family of rotations $a(\sigma)$ with coordinates $\sigma\boldsymbol{\alpha}$ is a family, all of whose members are rotations about one common axis $\boldsymbol{\alpha}$ but of varying amounts $\sigma\alpha$; they obviously form a one-parameter subgroup of $R(3)$. It is also evident that any continuous one-parameter subgroup arises in this way by considering all possible rotations about some common axis. In these coordinates, the tangent vector to the subgroup $a(\sigma) \to \sigma\boldsymbol{\alpha}$ is $\boldsymbol{\alpha}$; thus in the exponential notation the rotation specified by the vector $\boldsymbol{\alpha}$ can be written $\exp(\boldsymbol{\alpha})$. However, there is one point to be noted in using these coordinates globally for uniquely labeling all elements of $R(3)$. Suppose we indicate separately the unit vector $\hat{\boldsymbol{n}}$ in the direction of $\boldsymbol{\alpha}$ and the magnitude α of $\boldsymbol{\alpha}$, so that the pair $(\hat{\boldsymbol{n}}, \alpha)$ is used to define a rotation. Clearly, the rotations $(\hat{\boldsymbol{n}}, \alpha)$ and $(-\hat{\boldsymbol{n}}, 2\pi - \alpha)$ are identical. Thus if α is allowed to vary from 0 to 2π, there will generally be two ways of assigning $\hat{\boldsymbol{n}}$ and α, hence $\boldsymbol{\alpha}$, to a given rotation. But this problem can be taken care of to some extent in the following way: we allow all possible unit vectors $\hat{\boldsymbol{n}}$, hence all possible directions for $\boldsymbol{\alpha}$, but limit the magnitude α to the range $0 \leqq \alpha \leqq \pi$. Then we have a unique way of choosing coordinates $\boldsymbol{\alpha}$ for each rotation, *except for those rotations that are of amount π about some axis.* Thus for $\alpha < \pi$ we have a way of labeling rotations uniquely by vectors $\boldsymbol{\alpha}$; whereas if $\alpha = \pi$, *the vectors $\boldsymbol{\alpha}$ and $-\boldsymbol{\alpha}$ are understood to denote the same rotation.* This problem for $R(3)$ cannot be avoided; it is connected with the topological properties of the group, and it results in $R(3)$ not being simply connected and not being its own universal covering Lie group.

Now the matrix A associated with the rotation $\exp(\boldsymbol{\alpha})$ must be computed. Because we are using the active interpretation, we subject a general vector $\boldsymbol{x}$ to a rotation of amount α about $\boldsymbol{\alpha}$, thereby obtaining a new vector $\boldsymbol{y}$, and then get the matrix elements A_{jk} by expressing the components of $\boldsymbol{y}$ in terms of those of $\boldsymbol{x}$ as in Eq.(15.8). From simple geometry we then get the result:

$$\exp(\boldsymbol{\alpha}): \qquad A_{jk}(\boldsymbol{\alpha}) = \delta_{jk}\cos\alpha + \alpha_j\alpha_k\,\frac{(1-\cos\alpha)}{\alpha^2} - \epsilon_{jkl}\alpha_l\frac{\sin\alpha}{\alpha}\,. \tag{15.11}$$

Having obtained the matrix A explicitly, we could, in principle, express the group composition law in these coordinates simply by multiplying the matrices $A(\boldsymbol{\alpha})$ and $A(\boldsymbol{\beta})$ corresponding to the elements $\exp(\boldsymbol{\alpha})$ and $\exp(\boldsymbol{\beta})$ and writing the result in the form $A(\boldsymbol{\gamma})$ for a suitable $\boldsymbol{\gamma}$. However, the result is unwieldy (and so is the computation); thus we do not go into it. (The interested reader can make use of the $R(3) - SU(2)$ connection explained in Chapter 17 and derive the $R(3)$ composition law in these coordinates).

The Lie algebra of $R(3)$ is three dimensional. A natural basis for it consists of the tangent vectors l_1, l_2, l_3 to the one parameter subgroups of rotations about the first, second, and third axis, respectively. Using Eq.(15.11), we can work out the basic Lie brackets in a straightforward way. It is clear that the components of the general tangent vector $\boldsymbol{\alpha}$ of the previous paragraph, in the basis l_j we have chosen, are just the numbers α_j. Therefore Eq.(15.11) gives us the matrix A representing the element $\exp(\alpha_j l_j)$. (We sometimes write $\exp(\boldsymbol{\alpha}\cdot\boldsymbol{l})$ for this element; it is the same as $\exp(\boldsymbol{\alpha})$ in Eq.(15.11)). Consider now, the Lie bracket $[l_1, l_2]$. To obtain it, we consider the one parameter subgroups $\exp(\sigma l_1)$, $\exp(\rho l_2)$, for small σ and ρ. From the Baker-Campbell-Hausdorff formula, we can write:

$$\exp(\sigma l_1)\exp(\rho l_2) = \exp(\boldsymbol{\gamma}), \qquad \boldsymbol{\gamma} = \sigma l_1 + \rho l_2 + \frac{\sigma\rho}{2}[l_1, l_2] + \cdots \tag{15.12}$$

It is enough for our purposes to retain just the terms that are linear in σ and ρ, and among the terms of second order, only the mixed one involving $\sigma\rho$. If we write $\boldsymbol{\beta}$ for the bracket $[l_1, l_2]$, we have:

$$\gamma_j = \sigma\delta_{j1} + \rho\delta_{j2} + \frac{\sigma\rho}{2}\beta_j + \cdots, \qquad [l_1, l_2] = \beta_j l_j\,. \tag{15.13}$$

By working to this accuracy, the matrices $A(\sigma l_1)$ and $A(\rho l_2)$ representing $\exp(\sigma l_1)$ and $\exp(\rho l_2)$ are

$$\begin{aligned} A_{jm}(\sigma l_1) &\simeq \delta_{jm} - \sigma\epsilon_{jm1}\cdots, \\ A_{mk}(\rho l_2) &\simeq \delta_{mk} - \rho\epsilon_{mk2}\cdots, \end{aligned} \tag{15.14}$$

so that $\exp(\sigma l_1)\exp(\rho l_2)$ is represented by:

$$\begin{aligned} (A(\sigma l_1)A(\rho l_2))_{jk} &\simeq \delta_{jk} - \rho\epsilon_{jk2} - \sigma\epsilon_{jk1} + \sigma\rho\epsilon_{jm1}\epsilon_{mk2}\cdots \\ &= \delta_{jk} - \rho\epsilon_{jk2} - \sigma\epsilon_{jk1} + \sigma\rho\delta_{j2}\delta_{k1}\,. \end{aligned} \tag{15.15}$$

On the other hand, we expect this to be the same as $A_{jk}(\boldsymbol{\gamma})$, which is, to the same accuracy, given by:

$$\begin{aligned}A_{jk}(\boldsymbol{\gamma}) =&\delta_{jk}\left(1-\frac{\gamma^2}{2!}+\cdots\right)+\gamma_j\gamma_k\left(\frac{1}{2!}-\frac{\gamma^2}{4!}\cdots\right)-\epsilon_{jkm}\gamma_m\left(1-\frac{\gamma^2}{3!}\cdots\right)\\ \simeq&\delta_{jk}+\frac{1}{2}(\sigma\delta_{j1}+\rho\delta_{j2}+\cdots)(\sigma\delta_{k1}+\rho\delta_{k2}\cdots)\\ &-\epsilon_{jkm}\left(\sigma\delta_{m1}+\rho\delta_{m2}+\frac{\sigma\rho}{2}\beta_m+\cdots\right)\\ \simeq&\delta_{jk}-\rho\epsilon_{jk2}-\sigma\epsilon_{jk1}+\frac{\sigma\rho}{2}(\delta_{j1}\delta_{k2}+\delta_{j2}\delta_{k1}-\epsilon_{jkm}\beta_m)+\cdots \end{aligned}\tag{15.16}$$

Equating Eqs.(15.15) and (15.16) we get:

$$\epsilon_{jkm}\beta_m=\delta_{j1}\delta_{k2}-\delta_{j2}\delta_{k1}\,,\tag{15.17}$$

or $\beta_1=\beta_2=0$, $\beta_3=1$. Therefore, the Lie bracket of l_1 and l_2 is l_3; by cyclically permuting the indices we get the full set of basic Lie brackets for $R(3)$:

$$\begin{aligned}&[l_1,l_2]=l_3,\qquad [l_2,l_3]=l_1,\qquad [l_3,l_1]=l_2\,;\\ &\qquad\qquad [l_j,l_k]=\epsilon_{jkm}l_m\,.\end{aligned}\tag{15.18}$$

If we take any two vectors $\boldsymbol{\alpha}=\alpha_j l_j$ and $\boldsymbol{\beta}=\beta_j l_j$ in the Lie algebra of $R(3)$, we find their Lie bracket to be:

$$[\boldsymbol{\alpha},\boldsymbol{\beta}]\equiv\alpha_j\beta_k[l_j,l_k]=\epsilon_{jkm}\alpha_j\beta_k l_m=(\boldsymbol{\alpha}\times\boldsymbol{\beta})_j l_j\,.\tag{15.19}$$

Thus the usual process of taking the vector product of two vectors in three-dimensional space turns out to be the same as taking the Lie bracket of two vectors in the Lie algebra of $R(3)$! Also, the familiar identity that one studies in elementary geometry:

$$\boldsymbol{\alpha}\times(\boldsymbol{\beta}\times\boldsymbol{\gamma})+\boldsymbol{\beta}\times(\boldsymbol{\gamma}\times\boldsymbol{\alpha})+\boldsymbol{\gamma}\times(\boldsymbol{\alpha}\times\boldsymbol{\beta})=0\tag{15.20}$$

is just the Jacobi identity for the Lie algebra of $R(3)$!

All one-parameter subgroups of $R(3)$ are basically similar. That is to say, given the one-parameter subgroups $a(\sigma)\to\exp(\sigma\boldsymbol{\alpha})$ and $b(\sigma)\to\exp(\sigma\boldsymbol{\beta})$, there is an element of $R(3)$ that takes the tangent vector $\boldsymbol{\alpha}$ into the tangent vector $\boldsymbol{\beta}$ in the adjoint representation of $R(3)$. (We are assuming $|\boldsymbol{\alpha}|=|\boldsymbol{\beta}|$.) In fact, the three-dimensional defining matrix representation of $R(3)$ $\exp(\boldsymbol{\alpha})\to A(\boldsymbol{\alpha})$ is the adjoint representation! From Eq.(15.19), the operation $\Omega_{\boldsymbol{\alpha}}$ of taking the Lie bracket of $\boldsymbol{\alpha}$ with something can be written as

$$\boldsymbol{\Omega}_{\boldsymbol{\alpha}}\cdot=\boldsymbol{\alpha}\times\cdot\,.\tag{15.21}$$

Then,from the work of the previous chapter (see Eq.(14.4)),we know that in the adjoint representation, $\exp(\boldsymbol{\alpha})$ will be represented by $e^{\Omega_{\boldsymbol{\alpha}}}$. If we work out the effect of this operator on a general vector $\boldsymbol{\beta}$ in the Lie algebra of $R(3)$, we get:

$$\begin{aligned}
R_{exp(\boldsymbol{\alpha})}\boldsymbol{\beta} =& e^{\Omega_\alpha}\boldsymbol{\beta} = \boldsymbol{\beta} + \boldsymbol{\alpha}\times\boldsymbol{\beta} + \frac{1}{2!}\boldsymbol{\alpha}\times(\boldsymbol{\alpha}\times\boldsymbol{\beta}) \\
&+ \frac{1}{3!}\boldsymbol{\alpha}\times(\boldsymbol{\alpha}\times(\boldsymbol{\alpha}\times\boldsymbol{\beta})) + \cdots \\
=& \boldsymbol{\beta} + \boldsymbol{\alpha}\times\boldsymbol{\beta}\left(1 - \frac{\alpha^2}{3!} + \frac{\alpha^4}{5!} - \cdots\right) \\
&+ \boldsymbol{\alpha}\times(\boldsymbol{\alpha}\times\boldsymbol{\beta})\left(\frac{1}{2!} - \frac{\alpha^2}{4!} + \cdots\right) \\
=& \boldsymbol{\beta} + \boldsymbol{\alpha}\times(\boldsymbol{\alpha}\times\boldsymbol{\beta})\frac{(1-\cos\alpha)}{\alpha^2} + \boldsymbol{\alpha}\times\boldsymbol{\beta}\frac{\sin\alpha}{\alpha} \\
=& \boldsymbol{\beta}\cos\alpha + \boldsymbol{\alpha}\cdot\boldsymbol{\beta}\,\boldsymbol{\alpha}\frac{(1-\cos\alpha)}{\alpha^2} + \boldsymbol{\alpha}\times\boldsymbol{\beta}\frac{\sin\alpha}{\alpha}\,, \\
[R_{\exp(\boldsymbol{\alpha})}\boldsymbol{\beta}]_j =& \left(\delta_{jk}\cos\alpha + \alpha_j\alpha_k\frac{1-\cos\alpha}{\alpha^2} - \epsilon_{jkm}\alpha_m\frac{\sin\alpha}{\alpha}\right)\beta_k \\
=& A_{jk}(\boldsymbol{\alpha})\beta_k\,. \qquad (15.22)
\end{aligned}$$

Comparing this with the definition of the action of a rotation on a vector given in Eq.(15.8), and with Eq.(15.11),we see that the way in which the element $a = \exp(\boldsymbol{\alpha})$ acts in the adjoint representation is to just subject the general vector $\boldsymbol{\beta}$ in the Lie algebra to the rotation a, and thus take it into the vector $a\boldsymbol{\beta}$. Now, by definition, the tangent vector to the subgroup $a\exp(\sigma\boldsymbol{\beta})a^{-1}$, when a is some fixed element of $R(3)$, is $R_a\boldsymbol{\beta}$; R_a being the representative of a in the adjoint representation. Therefore, we can combine what we have learned about the form of the adjoint representation with this fact to write:

$$\exp(\boldsymbol{\alpha})\exp(\sigma\boldsymbol{\beta})\exp(-\boldsymbol{\alpha}) = \exp(\sigma(a\boldsymbol{\beta}), \qquad a = \exp(\boldsymbol{\alpha})\,. \qquad (15.23)$$

The action of a on $\boldsymbol{\beta}$ is defined as in equation (15.8). Because any two vectors of the same magnitude but different directions can be rotated into one another,Eq.(15.23) shows that any one-parameter subgroup of $R(3)$ can be transformed into any other by multiplying all its elements on the left by an appropriate element a and on the right by a^{-1}. It is in this sense that all one-parameter subgroups of $R(3)$ are basically similar.

An alternative, commonly used coordinate system for $R(3)$ consists of the *Euler angles.* This is *not* a canonical coordinate system. We explain it by using the passive interpretation of $R(3)$. This new coordinate system for $R(3)$ arises by showing that a Cartesian frame S' in space can be obtained from another one, S, by performing three relatively simple rotations in a specified sequence. The $\boldsymbol{e}_j$ and $\boldsymbol{e}'_j$ being the basic unit vectors in S and S', respectively, we introduce two intermediate frames S_1 and S_2 with basic vectors $\boldsymbol{e}_j^{(1)}$ and $\boldsymbol{e}_j^{(2)}$, respectively. Starting with S, the frame S_1 is reached by making a rotation of amount ϕ around the third axis in S so that we have:

$$\boldsymbol{e}_1^{(1)} = \boldsymbol{e}_1\cos\phi + \boldsymbol{e}_2\sin\phi\,, \qquad \boldsymbol{e}_2^{(1)} = \boldsymbol{e}_2\cos\phi - \boldsymbol{e}_1\sin\phi\,, \qquad \boldsymbol{e}_3^{(1)} = \boldsymbol{e}_3\,. \quad (15.24)$$

Next, we perform a rotation of amount θ about the second axis in S_1 to attain S_2:

$$\boldsymbol{e}_1^{(2)} = \boldsymbol{e}_1^{(1)} \cos\theta - \boldsymbol{e}_3^{(1)} \sin\theta\,, \quad \boldsymbol{e}_2^{(2)} = \boldsymbol{e}_2^{(1)}\,, \quad \boldsymbol{e}_3^{(2)} = \boldsymbol{e}_3^{(1)} \cos\theta + \boldsymbol{e}_1^{(1)} \sin\theta\,. \tag{15.25}$$

Finally, by means of a rotation of amount ψ about the third axis in S_2, we reach the desired system of axes S':

$$\boldsymbol{e}_1' = \boldsymbol{e}_1^{(2)} \cos\psi + \boldsymbol{e}_2^{(2)} \sin\psi\,, \quad \boldsymbol{e}_2' = \boldsymbol{e}_2^{(2)} \cos\psi - \boldsymbol{e}_1^{(2)} \sin\psi\,, \quad \boldsymbol{e}_3' = \boldsymbol{e}_3^{(2)}\,. \tag{15.26}$$

These three operations can be depicted in the following diagram:

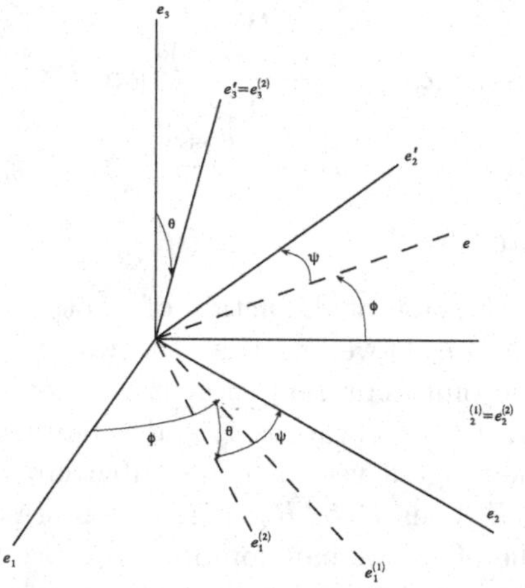

It is clear that any frame S' can be reached in this way from a given initial frame S for a suitable choice of the angles ϕ, θ, and ψ. The angles θ and ϕ are simply the spherical polar angles of the vector $\boldsymbol{e}_3'$ measured in the frame S : θ is the polar angle, ϕ the azimuth, and they obey $0 \leqq \theta \leqq \pi$, $0 \leq \phi < 2\pi$. Thus from a knowledge of $\boldsymbol{e}_3'$, θ and ϕ are determined, and one can carry out the first two rotations described above. One then arrives at the frame S_2 with the vector $\boldsymbol{e}_3^{(2)} = \boldsymbol{e}_3'$ already in the desired position and with $\boldsymbol{e}_1^{(2)}$ and $\boldsymbol{e}_2^{(2)}$ in the plane perpendicular to $\boldsymbol{e}_3'$. But the final vectors $\boldsymbol{e}_1'$ and $\boldsymbol{e}_2'$ are also in this same plane and so they must be obtained from $\boldsymbol{e}_1^{(2)}$ and $\boldsymbol{e}_2^{(2)}$ by some righthanded rotation about $\boldsymbol{e}_3'$; this fixes ψ, and it obeys $0 \leqq \psi < 2\pi$. The full set of ranges for the Euler angles is

$$0 \leqq \phi, \quad \psi < 2\pi, \qquad 0 \leqq \theta \leqq \pi\,. \tag{15.27}$$

We now have a unique way of associating the triplet (ϕ, θ, ψ) with every rotation a, except in the following special circumstances: if $\boldsymbol{e}_3' = \pm \boldsymbol{e}_3$, in which case

$\theta = 0$ or π, ϕ is indeterminate; but by convention we shall choose $\phi = 0$. The representative matrix A as a function of ϕ, θ and ψ can be worked out quite easily. We can express the $\boldsymbol{e}'_j$ in terms of the $\boldsymbol{e}_j$, using Eqs. (15.24), (15.25), and (15.26):

$$\begin{aligned}
\boldsymbol{e}'_1 =&(\cos\psi\cos\theta\cos\phi - \sin\psi\sin\phi)\boldsymbol{e}_1 \\
&+ (\cos\psi\cos\theta\sin\phi + \sin\psi\cos\phi)\boldsymbol{e}_2 - \cos\psi\sin\theta\,\boldsymbol{e}_3\,, \\
\boldsymbol{e}'_2 =&(-\sin\psi\cos\theta\cos\phi - \cos\psi\sin\phi)\boldsymbol{e}_1 \\
&+ (-\sin\psi\cos\theta\sin\phi + \cos\psi\cos\phi)\boldsymbol{e}_2 + \sin\psi\sin\theta\,\boldsymbol{e}_3\,, \\
\boldsymbol{e}'_3 =& \cos\phi\sin\theta\,\boldsymbol{e}_1 + \sin\phi\sin\theta\,\boldsymbol{e}_2 + \cos\theta\,\boldsymbol{e}_3\,.
\end{aligned} \tag{15.28}$$

Comparing this with Eq.(15.3), we get the matrix $A(\phi,\theta,\psi)$:

$$\begin{aligned}
&A(\phi,\theta,\psi) \\
&= \begin{pmatrix}
\cos\psi\cos\theta\cos\phi - \sin\psi\sin\phi; & \cos\psi\cos\theta\sin\phi + \sin\psi\cos\phi; & -\cos\psi\sin\theta \\
-\sin\psi\cos\theta\cos\phi - \cos\psi\sin\phi; & -\sin\psi\cos\theta\sin\phi + \cos\psi\cos\phi; & \sin\psi\sin\theta \\
\cos\phi\sin\theta; & \sin\phi\sin\theta; & \cos\theta
\end{pmatrix}
\end{aligned} \tag{15.29}$$

This represents the element $a(\phi,\theta,\psi)$.We can trivially relate this to the earlier parametrization. By the way $A(\phi,\theta,\psi)$ has been defined, we can factor it into the following form:

$$A(\phi,\theta,\psi) = A(0,0,\psi)A(0,\theta,0)A(\phi,0,0)\,. \tag{15.30}$$

But now the factors here can be compared with Eq.(15.11); we see that $A(0,0,\psi)$ is the same as $A(\psi,0,0)$, and this is the representative of the element $\exp(-\psi l_3)$; $A(0,\theta,0)$ is the representative of $\exp(-\theta l_2)$. Therefore Eq.(15.30) leads to the *Euler decomposition*

$$a(\phi,\theta,\psi) = \exp(-\psi l_3)\exp(-\theta l_2)\exp(-\phi l_3)\,, \tag{15.31}$$

valid for all elements of $R(3)$.

There is one last point regarding the group $R(3)$ that we will explain here, because it also helps us in understanding the form of the result Eq.(15.31). This involves reinterpreting the result of Eq.(15.11),which gives the matrix $A(\alpha)$ corresponding to the element $\exp(\alpha_j l_j)$, in the passive interpretation of $R(3)$. (Recall that Eq.(15.11) was derived using the active interpretation.) Suppose we subject the axes of S themselves to the rotation by amount α about the axis $\boldsymbol{\alpha}$; then the unit vectors $\boldsymbol{e}_j$ in S get rotated into new positions that are the basic unit vectors $\boldsymbol{e}'_j$ of the new frame S'. The $\boldsymbol{e}'_j$ will be given in terms of the $\boldsymbol{e}_j$ by $\boldsymbol{e}'_j = A_{kj}(\boldsymbol{\alpha})\boldsymbol{e}_k$, which is Eq.(15.9) but with $\boldsymbol{e}'$ written in place of $\hat{\mathbf{e}}$. We must now put this into the form of Eq.(15.3) before we can identify the element a in $R(3)$ in terms of which we can write $S' = aS$. But Eq.(15.11) shows that taking the transpose of the matrix $A(\boldsymbol{\alpha})$ amounts to reversing the sign of $\boldsymbol{\alpha}$, so that we have $\boldsymbol{e}'_j = A_{jk}(-\boldsymbol{\alpha})\boldsymbol{e}_k$, and then we can write $S' = \exp(-\alpha_j l_j)S$. Thus the reference frame S' obtained by subjecting the axes of the frame S to the rotation of amount α about $\boldsymbol{\alpha}$ is to be identified as the frame $\exp(-\alpha_j l_j)S$. This explains the negative signs in the exponents in the three factors in Eq.(15.31).

Direct and Semidirect Products of Lie Groups

The Lie algebras of the two groups T_3 and $R(3)$ are both three-dimensional. In both cases the forms of the basic Lie brackets are easy to write down,and we are able to introduce and interpret geometrically canonical coordinates of the first kind.We are also able to define both these groups quite simply, in terms of defining representations or realizations. The next three groups we wish to deal with have somewhat more complex structures. In their cases, it turns out that it is not the most natural thing, from the viewpoint of physical interpretation, to introduce canonical coordinates at all, but rather to use coordinates that have a direct and simple physical meaning; although of course one could introduce canonical coordinates if one wanted to. Also, one understands their mathematical structures better if one first understands a particular way of combining two groups to produce a third one: that is the process of forming the *semi-direct product.* It is therefore worthwhile to devote some space to an explanation of these ideas, which enables us to see how they apply to the Euclidean, Galilean and inhomogeneous Lorentz groups in turn.

Let G be a group with elements $a, b, \cdots$. An *automorphism of* G is a one-to-one, invertible mapping τ of G onto G that preserves the structure of G. The use of the word automorphism here is quite similar to its use in Chapter 10: it is a structure-preserving mapping of any given mathematical system onto itself. We will write $\tau(a)$ for the image of $a \in G$ under the automorphism τ; for each a, $\tau(a)$ is an element of G. The defining properties that make the mapping τ an automorphism are

$$\begin{aligned} &1. \quad \tau(e) = e; \\ &2. \quad \tau(a)^{-1} = \tau(a^{-1}); \\ &3. \quad \tau(a)\tau(b) = \tau(ab)\,. \end{aligned} \tag{15.32}$$

This is what we mean by saying that the structure of G is preserved by τ. Though we have listed three conditions, one can easily see that (3) leads to (1) and (2). In addition we have the requirement that $\tau(G) = G$, or that τ possess an inverse. A group G possesses many kinds of automorphisms. The simplest, most trivial one is the one that maps each element a into itself: $\tau(a) = a$. This is the *identity automorphism.* Another class of automorphisms is called the class of *inner automorphisms*; if b is some fixed element in G, the mapping τ_b which takes a into bab^{-1} for any $a \in G$ is easily verified to be an automorphism; it is called an *inner automorphism*, or the *conjugation* by b. Thus each element in G gives rise to an inner automorphism on G, although,of course, two distinct elements need not give rise to distinct inner automorphisms. An arbitrary automorphism τ is called inner if it coincides with the inner automorphism generated by some element in G; otherwise it is called *outer.* Automorphisms,being mappings, can be composed in the natural way to yield new automorphisms, and one can see that the set of all automorphisms of a group G onto itself has the structure of a group. This group that G determines is called aut G; the interesting point that emerges, and the reader can provide the proof quite easily, is that the inner automorphisms of G form a subgroup, and in fact, an invariant subgroup, of aut G.

Let now G and G' be two groups, with elements $a, b, \cdots$ and $a', b', \cdots$, respectively; each has its own law of group composition. There is one rather simple way in which one can combine G and G' to define a new group called the *direct product*, or *outer product*, $G \otimes G'$ of G and G'. One defines elements of $G \otimes G'$ to be pairs, (a, a'), the first element being drawn from G and the second from G'. If G and G' were both finite groups, with m and m' elements, respectively, then $G \otimes G'$ has mm' elements. The unit element of $G \otimes G'$ is taken to be (e, e'), e and e' being the unit elements of G and G', respectively; the product of two pairs (a, a') and (b, b') is defined to be the pair $(ab, a'b')$. This defines the composition law in $G \otimes G'$; ab is the product of a and b in G, $a'b'$ the product of a' and b' in G'. The inverse of (a, a') is (a^{-1}, a'^{-1}). One "recovers" G and G' from $G \otimes G'$ in this way: the subset of pairs (a, e') in $G \otimes G'$ obviously forms a subgroup with the same structure as G; that is, the elements of this subset can be put in one-to-one correspondence with the elements of G in such a way that products go into products and inverses into inverses. When such a relationship can be set up between any two groups one says that one has an isomorphism, or identity of structure, between the two groups, and no distinction can be made between them on the basis of the group operations alone. Thus one has recovered G from $G \otimes G'$ in the sense that the subgroup of $G \otimes G'$ consisting of elements (a, e') is isomorphic to G; in the same way the subgroup made up of the pairs (e, a') is isomorphic to G'. One can also satisfy oneself that these two natural subgroups of $G \otimes G'$ are invariant subgroups.

A slightly more complicated way of combining two groups to produce a third arises in the following way. Let H and A be two groups, whose elements will be written as $h, h', \cdots$ and $a, a', \cdots$, respectively. (The letters H and A are chosen because of the nature of these two groups in the cases that we shall be studying later). Suppose that for each element $h \in H$, we can find an automorphism τ_h acting on A such that these automorphisms provide a realization of H. That is, for each h, τ_h is a mapping of A onto itself taking a into $\tau_h(a)$, such that Eq.(15.32) is obey, and further these τ_h's obey the composition law

$$\tau_{h'}\tau_h = \tau_{h'h} \ , \qquad h', h \in H \ . \tag{15.33}$$

Then one can define a *semidirect product* of H and A, written $H \times A$, as follows: the elements of $H \times A$ are once again pairs (h, a), h being drawn from H and a from A; the composition law is given by

$$(h', a')(h, a) = (h'h, a'\tau_{h'}(a)) \, . \tag{15.34}$$

The elements from H are multiplied just by themselves, as in the case of the direct product; but as for the elements from A, one first acts on a with the automorphism $\tau_{h'}$ determined by h', then takes the product with the element a'. The facts that for each h, τ_h is an automorphism on A and that these automorphisms obey Eq. (15.33) and thus provide a realization of H make Eq.(15.34) an acceptable group composition law. For, to check the associative property, let us compute the product of these pairs in two different orders:

$$\begin{aligned}
\{(h'',a'')(h',a')\}(h,a) =&(h''h',a''\tau_{h''}(a'))(h,a)\\
=&(h''h'h,a''\tau_{h''}(a')\tau_{h''h'}(a));\\
(h'',a'')\{(h',a')(h,a)\} =&(h'',a'')(h'h,a'\tau_{h'}(a))\\
=&(h''h'h,a''\tau_{h''}(a'\tau_{h'}(a)))\\
=&(h''h'h,a''\tau_{h''}(a')\tau_{h''}(\tau_{h'}(a)))\\
=&(h''h'h,a''\tau_{h''}(a')\tau_{h''h'}(a));\\
\{(h'',a'')(h',a')\}(h,a) =&(h'',a'')\{(h'a')(h,a)\} \qquad (15.35)
\end{aligned}$$

Thus the associative law is verified. The identity element in $H \times A$ is (e, e'), e being the identity in H and e' in A. The inverse of (h, a) is found to be $(h^{-1}, \tau_{h^{-1}}(a^{-1}))$:

$$\begin{aligned}
(h,a)(h^{-1},\tau_{h^{-1}}(a^{-1})) =&(hh^{-1},a\tau_h(\tau_{h^{-1}}(a^{-1}))) = (hh^{-1},a\tau_{hh^{-1}}(a^{-1}))\\
=&(e,e')\,. \qquad (15.36)
\end{aligned}$$

The roles of the two groups H and A entering the semidirect product are quite different, so we must call $H \times A$ the *semi-direct product of H by A*. Different ways of choosing the automorphisms τ_h realizing H can lead to different ways of forming the semidirect product; thus there is, a priori, no unique semidirect product of H by A. In this sense, the notation $H \times A$ is deficient, and one must indicate somehow the choice that has been made for the τ's. But it rarely happens that in the course of a discussion one has to deal with two different semidirect products of H by A; hence in practice the notation $H \times A$ proves adequate.

Given any two groups H, A there is one trivial way to construct a semidirect product: if one chooses τ_h to be the identity automorphism on A for each h, one has a trivial realization of H by these τ's, and one then does have a semidirect product, but it collapses to the ordinary direct product $H \otimes A$ of H and A! Whether or not a more interesting choice can be made for the τ's depends on the structures of H and aut A. One must succeed in picking one element from aut A for each $h \in H$, though not necessarily distinct elements for distinct h's, such that the condition of Eq.(15.33) is obeyed, and at least for some h's the corresponding τ's should differ from the identity automorphism on A. In the practical cases that we are dealing with, H and A and the τ_h's are all easily identified from the definitions of the transformations involved.

In the case of the direct product $G \otimes G'$ of G and G', it is easy to exhibit two subgroups in $G \otimes G'$, each of which is an invariant subgroup, with one isomorphic to G, the other to G'. In the case of the semidirect product $H \times A$, one can again exhibit two natural subgroups, one isomorphic to H and the other to A, but now only the latter forms an invariant subgroup of $H \times A$. This shows up again the asymmetric way in which H and A enter the semidirect product. The subgroup isomorphic to H is made up of the pairs (h, e') in $H \times A$, and the one isomorphic to A consists of the pairs (e, a). The invariant nature of the latter is easy to check:if (h', a') is any element in $H \times A$, then

$$\begin{aligned}(h', a')(e, a)(h', a')^{-1} &= (h', a')(e, a)(h'^{-1}, \tau_{h'^{-1}}(a'^{-1})) \\ &= (h', a'\tau_{h'}(a))(h'^{-1}, \tau_{h'^{-1}}(a'^{-1})) \\ &= (e, a'\tau_{h'}(a)\tau_{h'}(\tau_{h'^{-1}}(a^{-1}))) \\ &= (e, a'\tau_{h'}(a)a'^{-1}) . \end{aligned} \tag{15.37}$$

Therefore the subgroup of elements (e, a), isomorphic to A, is carried into itself under conjugation by any element of $H \times A$ and is an invariant subgroup.For the subgroup isomorphic to H, we find:

$$\begin{aligned}(h', a')(h, e')(h', a')^{-1} &= (h'h, a')(h'^{-1}, \tau_{h'^{-1}}(a'^{-1})) \\ &= (h'hh'^{-1}, a'\tau_{h'h}(\tau_{h'^{-1}}(a'^{-1}))) \\ &= (h'hh'^{-1}, a'\tau_{h'hh'^{-1}}(a'^{-1})); \end{aligned} \tag{15.38}$$

if the semidirect product is a nontrivial one, so that for some h, τ_h is not the identity automorphism, it follows that the subgroup of elements (h, e'), isomorphic to H, is not an invariant subgroup.

There is one further process connected with formation of groups that we should explain, before we can look at the processes in Lie algebras that are analogous to the formation of direct and semidirect products of Lie groups. This is the process by which a group G possessing an invariant subgroup G_0 leads to the *factor group* G/G_0. Let us write $a, b, \cdots$ for elements in G, and $x, y, \cdots$ for those in G_0. The invariant nature of the subgroup G_0 implies that if $a \in G$ and $x \in G_0$, then axa^{-1} is some element $y \in G_0$. In set theoretic notation, this means that the subset aG_0a^{-1}, consisting of all elements axa^{-1} where a is kept fixed and x runs over all of G_0, always coincides with G_0, whatever a is chosen. However, this is the same thing as saying that for each a the subsets aG_0 and G_0a are the same: given a, for each $x \in G_0$ there is a unique $y \in G_0$ such that $ax = ya$. Now the elements of G can be broken up into disjoint subsets in the following way: we will say that two elements a, b in G are *equivalent* if we can find an element $x \in G_0$ such that $a = xb$; otherwise we will say that they are *inequivalent*. If an x can be found to obey $a = xb$, then surely an element $y \in G_0$ can also be found such that $a = by$; G_0 is an invariant subgroup. This definition of equivalence satisfies all the laws required of an equivalence relation:

1. every a is equivalent to itself;
2. if a is equivalent to b, then b is equivalent to a;
3. the equivalence of a and b, and of b and c, implies that of a and c.

[All these laws are obeyed even if G_0 is not an invariant subgroup, but then we have two distinct notions of equivalence: we could say that a and b are left equivalent (with respect to G_0) if there is some $x \in G_0$ such that $a = xb$; we could say that they are right equivalent if for some $y \in G_0$ we had $a = by$. Each would be an equivalence relation, but they would not agree. If G_0 is an invariant subgroup, then left and right equivalence become the same, and we speak of equivalence with respect to G_0]. We can now split up G into disjoint subsets, each subset consisting of elements equivalent to one another, with no element in one subset being equivalent to any element in a distinct subset. *We can then make a group out of these equivalence classes in* G, equivalence classes being the name given to these subsets. This is done as follows. It is clear that any one of these equivalence classes can be generated by picking one element in it, say a, and then taking the products ax for all $x \in G_0$, or equivalently xa for all $x \in G_0$; thus an equivalence class can be written as aG_0, a being any member of it. The group formed by these classes is called the *factor group of* G *by* G_0, G/G_0. Elements of G/G_0 are the equivalence classes with respect to G_0 in G. The unit element in G/G_0 is the equivalence class made up of G_0 itself; any other element in G/G_0 is the equivalence class aG_0, for some $a \notin G_0$; the inverse to the element aG_0 is $a^{-1}G_0$; and finally, the product of the two elements aG_0, bG_0 in G/G_0 is the equivalence class abG_0. That the definition of group multiplication given here is an acceptable one is ensured by the invariant nature of $G_0 : aG_0 = G_0a$. We leave it to the reader to check the details.

The importance of this concept of a factor group in the case of the semidirect product $H \times A$ of a group H by a group A is this: we saw that we have two natural subgroups in $H \times A$, one of which is isomorphic to H and the other to A. Let us write $\tilde{H}$ for the former and $\tilde{A}$ for the latter. We saw that while $\tilde{A}$ is an invariant subgroup of $H \times A$, $\tilde{H}$ was not. As an invariant subgroup, $\tilde{A}$ allows us now to form the factor group $H \times A/\tilde{A}$: this is isomorphic to H! The proof is quite elementary. $\tilde{A}$ is formed by the elements (e, a) in $H \times A$. If we take a fixed element (h', a') in $H \times A$ and let the element a run over all of A, we generate by the products $(h', a')(e, a)$ one equivalence class in $H \times A$. All these elements have the first factor in the pair common, for $(h', a')(e, a)$ is $(h', a'\tau_{h'}(a))$; because we allow a to vary over all of A, the element $a'\tau_{h'}(a)$ will also vary over all of A. Thus each equivalence class in $H \times A$ with respect to $\tilde{A}$ is uniquely labeled by an element $h' \in H$; these is a one-to-one correspondence between elements in H and these equivalence classes. That this correspondence is an isomorphism is obvious: the equivalence class $(h', a)\tilde{A}$ is the same as the equivalence class $(h', e)\tilde{A}$, and we know that $\tilde{H}$ and H are isomorphic, thus the definition of group multiplication in the factor group proves the assertion.

We will now give the analogues in the case of a Lie algebra, to all these concepts. (Subalgebras and invariant subalgebras of a Lie algebra, which are analogues of subgroups and invariant subgroups of a Lie group, are defined on page 204 at the end of Chapter 13.) We will be content with defining the processes involved, and will forego proofs that they are directly related to the corresponding processes with groups. An automorphism of a Lie algebra is, as the reader should by now expect, a linear mapping of the Lie algebra onto itself, which preserves the values of Lie brackets. We have already used this terminology in Chapter 10. If L is a Lie algebra with elements $u, v, \cdots$, an automorphism τ is a one-to-one, invertible mapping of L onto itself, obeying:

$$\begin{aligned} &1.\ \tau(\lambda u + \mu v) = \lambda\tau(u) + \mu\tau(v)\,, \qquad \lambda, \mu \text{ real numbers;} \\ &2.\ [\tau(u),\, \tau(v)] = \tau([u,v])\,. \end{aligned} \tag{15.39}$$

Thus, as examples, we have that the linear transformations that arise in the adjoint representation of the Lie group G determined by L act as automorphisms on L: see Eq.(14.14); canonical transformations are automorphisms when acting on functions in phase space, with the Lie bracket taken to be the PB, and so on. The infinitesimal form of an automorphism is a *derivation*: this arises when one has a continuous family of automorphisms, and one examines the effect on L of those automorphisms that are close to the identity automorphism. A *derivation* D on L is a linear mapping of L onto itself, which associates with each $u \in L$ an element $D(u) \in L$, such that:

$$\begin{aligned} &1. \qquad D(\lambda u + \mu v) = \lambda D(u) + \mu D(v)\,, \\ &2. \qquad D([u,v]) = [D(u), v] + [u, D(v)]\,. \end{aligned} \tag{15.40}$$

Starting with a derivation D, one can build up an automorphism on L by the familiar process of solving a differential equation. *Inner derivations* are derivations that correspond to taking the Lie bracket with some element in L. Thus a derivation D is inner if it equals Ω_w for some w, otherwise it is *outer*. The automorphisms of L generated by inner derivations are the transformations of the adjoint representation.

To the direct product $G \otimes G'$ of two Lie groups corresponds the *direct sum* of their Lie algebras L and L'. This is written $L \oplus L'$; as a vector space it is the direct sum of the vector spaces L and L'. Each vector in $L \oplus L'$ can be written as $u + u'$, for some $u \in L$ and some $u' \in L'$; conversely, considering all u and u' in this way, we exhaust $L \oplus L'$. We must now define the Lie bracket in $L \oplus L'$ that is the bracket of $u + u'$ and $v + v'$. For the bracket of u and v we use the one given in L, for that of u' and v' we use the one given in L', whereas elements in L are defined to have zero brackets with elements in L', and then we use linearity:

$$[u + u',\, v + v']_{L \oplus L'} = [u, v]_L + [u', v']_{L'}\,. \tag{15.41}$$

Because there is never any danger of confusion, we omit the subscripts here and write it simply as

$$[u + u',\, v + v'] = [u, v] + [u', v']\,. \tag{15.42}$$

It is evident trivially that L and L' are contained as invariant subalgebras in $L \oplus L'$. (The more precise way to state this would be to say that there are invariant subalgebras $\tilde{L}$ and $\tilde{L}'$ in $L \oplus L'$ isomorphic to L and L', respectively, but there is no real need to observe these niceties!)

Turning now to the semidirect product: let H and A be two Lie groups with corresponding Lie algebras $\mathcal{H}$ and $\mathcal{A}$, respectively, and let $H \times A$ be the semidirect product of H by A. Denote the Lie algebra of $H \times A$ by L. Then as a vector space, L is the direct sum of $\mathcal{H}$ and $\mathcal{A}$. Further, $\mathcal{H}$ forms a subalgebra and $\mathcal{A}$ an *invariant* subalgebra of L. In more detail: let us write $u, u', \cdots$ for vectors in $\mathcal{H}$ and $\alpha, \alpha', \cdots$ for those in $\mathcal{A}$ so that a general element of L is the sum $u + \alpha$. Then, the u's form a subalgebra by themselves, so that the Lie bracket of u and u' is another u'', determined by the Lie bracket in $\mathcal{H}$; similarly the α's form a subalgebra with the bracket of α and α' being another α'' determined by the Lie bracket in $\mathcal{A}$; finally, the bracket of u and α is some α' in $\mathcal{A}$. Thus the operation of taking the bracket of a fixed u with various α in $\mathcal{A}$ acts as a derivation on $\mathcal{A}$: to each $u \in \mathcal{H}$, we have a derivation D_u on $\mathcal{A}$, and $[u, \alpha] = D_u(\alpha)$. These derivations D_u corresponding to the elements in the Lie algebra $\mathcal{H}$ of H are just the infinitesimal forms of the automorphisms τ_h corresponding to elements $h \in H$ that act on A, in the definition of $H \times A$.

The notion of the factor group G/G_0 goes over into the notion of the *factor algebra*. If G is a Lie group with Lie algebra L, and G_0 is a (Lie) subgroup of G corresponding to the subalgebra L_0 in L, then L_0 is an invariant subalgebra in case G_0 is an invariant subgroup. Then the Lie algebra corresponding to G/G_0 is obtained merely by *projecting L with respect to L_0*, or in other words, by *consistently ignoring elements of L_0*. We think of two elements of L as being equivalent if they differ by an element in L_0; that is, u and v in L are equivalent if $u - v \in L_0$. In this way, L breaks up into equivalence classes of vectors; these classes then form in a natural way a vector space whose dimension is the dimension of L minus that of L_0. Any vector u in one of these classes can be chosen to represent that class; then all the other vectors in that class are gotten by adding to u all vectors u_0 in L_0. Now if we take two classes in L and a representative element from each, say, u and v, their bracket $[u, v]$ determines a vector w belonging to a certain class: $[u, v] = w$. But the class that w belongs to is determined by the classes to which u and v belong, and does not depend on which u and v were chosen to represent their classes! This is the consequence of L_0 being an invariant subalgebra. For any other representatives of the classes that u and v belong to are some vectors $u + u_0$, $v + v_0$, respectively, for some u_0, $v_0 \in L_0$; and then $[u + u_0, v + v_0] = [u, v] +$ terms in $L_0 = w + w_0$, some $w_0 \in L_0$, since if at least one element in a Lie bracket belongs to L_0, the bracket has a value in L_0. Therefore, changing the representatives u and v of two equivalence classes of vectors in L only alters the Lie bracket by a vector in L_0, and so does not alter the equivalence class to which the bracket belongs. This turns the vector space of equivalence classes in L, also called the vector space obtained by projection with respect to L_0, into a Lie algebra; this is the Lie algebra corresponding to the group G/G_0. In the particular case in which G is the semidirect product $H \times A$ with Lie algebra L, by projecting L with respect to its invariant subalgebra $\mathcal{A}$ (the Lie algebra of A) we just recover the Lie algebra $\mathcal{H}$ corresponding to H.

Now we go on to a study of the Euclidean, Galilei, and homogeneous and inhomogeneous Lorentz groups, all but the third of which are semidirect products in one way or another.

The Euclidean Group $E(3)$

The group $E(3)$ arises by combining the operations of the proper rotation group $R(3)$ and the translation group T_3; that is, it is the set of transformations connecting all possible Cartesian coordinate systems $S, S', \cdots$ in three-dimensional space, all of these being of the same handedness. If S is one Cartesian coordinate system and S' is another, it is clear from simple geometry that S can be brought into coincidence with S' in two steps. First we apply a suitable rotation a corresponding to the matrix A to go from S to S_1, S_1 having axes parallel to the axes in S'. We then write $S_1 = aS$. Next we apply a translation to S_1 to make its origin coincide with that in S'; we can then write $S' = \exp(\boldsymbol{a} \cdot \boldsymbol{d})S_1$, $\exp(\boldsymbol{a} \cdot \boldsymbol{d})$ being the appropriate element of T_3. (With our conventions, this means that the vector $\overrightarrow{O_1O'}$ from the origin in S_1 to that in S' is $-\boldsymbol{a}$). This pair of operations, first an element from $R(3)$ and then one from T_3, can be written $(A, \boldsymbol{a})$, and it is a general element of the group $E(3)$. Because it connects S with S', we write $S' = (A, \boldsymbol{a})S$. By considering all possible A and all possible $\boldsymbol{a}$, we get all the elements of $E(3)$. Let P be some fixed point in space, and let its coordinates in S be x_j, and in S' be x'_j. We relate these in the following manner. Because $S_1 = aS$ and to a corresponds the matrix A, the coordinates of P in S_1 are $x_j^{(1)}$ given by:

$$S_1 = aS : \quad x_j^{(1)} = A_{jk}x_k \,. \tag{15.43}$$

(This follows from Eq.(15.4) and the meaning of the notation $S_1 = aS$.) Then, x'_j are related to $x_j^{(1)}$ by $x'_j = x_j^{(1)} + a_j$, because $S' = \exp(\boldsymbol{a} \cdot \boldsymbol{d})S_1$; thus, all in all, we get

$$S' = (A, \boldsymbol{a})S \,, \quad x'_j = A_{jk}x_k + a_j \,, \tag{15.44}$$

relating the coordinates of a fixed point P in S' to those in S (a_j are the components of $\boldsymbol{a}$ in S_1 and S').

Because it takes three parameters to specify an element in $R(3)$ and three more to specify an element in T_3, it follows that $E(3)$ *is a six-parameter group.* We can work out the group composition law for $E(3)$ quite simply from Eq.(15.44). Let $S' = (A, \boldsymbol{a})S$ and $S'' = (B, \boldsymbol{b})S'$. Then for a fixed point P, we use Eq.(15.44) twice to relate the coordinates x''_j in S'' to x_j in S:

$$x''_j = B_{jl}x'_l + b_j = B_{jl}A_{lk}x_k + B_{jl}a_l + b_j = (BA)_{jk}x_k + B_{jl}a_l + b_j \,. \tag{15.45}$$

Thus the element of $E(3)$ taking us from S to S'' is $(BA, B_{jl}a_l + b_j)$ where the components of the translation are written explicitly. We must now identify it with the product $(B, b_j)\ (A, a_j)$, so that the composition law is

$$(B, b_j)(A, a_j) = (BA,\ b_j + B_{jl}a_l) \,. \tag{15.46}$$

This exhibits the structure of $E(3)$: it is the semidirect product of $R(3)$ by T_3. First, the elements of $R(3)$ in the two terms on the left, B and A, get combined into BA according to the composition rule in $R(3)$, completely independently of the presence of elements from T_3. Second, the mapping $a_j \to B_{jl}a_l$ of elements of T_3 onto themselves, determined by a fixed B in $R(3)$, is obviously an automorphism on T_3; the product translation on the right in Eq.(15.46) arises by composition of the translation b_j and the image of a_j under the automorphism of T_3 corresponding to B. Thus Eq.(15.46) is a special case of Eq. (15.34); it follows from our general discussion that the elements $(A, \mathbf{0})$ form a subgroup of $E(3)$ isomorphic to $R(3)$, whereas the elements $(\mathbb{1}, \boldsymbol{a})$ form an invariant subgroup isomorphic to T_3. The unit element in $E(3)$ clearly is the pair $(\mathbb{1}, \mathbf{0})$, and from Eq.(15.46) we can work out the inverse of a general element:

$$(A, a_j)^{-1} = (A^{-1}, -A_{kj}a_k)\,. \tag{15.47}$$

Let us now consider the parametrization and the Lie algebra of $E(3)$. We can label the element A in $(A, \boldsymbol{a})$ by canonical coordinates within $R(3)$ so that we have $A \to \exp(\alpha_j l_j)$. Similarly, we can use canonical coordinates within T_3 for the translation $\boldsymbol{a}$, but these are the a_j themselves; thus this element of T_3 is $\exp(a_j d_j)$. The set of six parameters (α_j, a_j) will suffice to label elements of $E(3)$ and will stand for the element $(A(\boldsymbol{\alpha}), a_j)$. These coordinates are *not* canonical ones for $E(3)$, but we will use them because they have a simple geometrical interpretation. Now, just as a general element in $E(3)$ is arrived at by first carrying out a rotation and then a translation, we can use Eq.(15.46) to show that all of $E(3)$ is generated by taking products of elements from its rotation and translation subgroups:

$$(A, a_j) = (\mathbb{1}\,, a_j)(A, 0)\ . \tag{15.48}$$

The fact that the coordinates (α_j, a_j) are not canonical ones means that *in general* a family of elements $(\sigma\alpha_j,\ \sigma a_j)$ will not form a one-parameter subgroup in $E(3)$. However, if we consider a family of the form $(\sigma\alpha_j, 0)$ or of the form $(0, \sigma a_j)$, these certainly do form one-parameter subgroups in $E(3)$, although we do not obtain all types of one-parameter subgroups in this way. There will be no danger of confusion if we continue to write l_j for the three basic tangent vectors to the subgroups of the first type and d_j for those of the second type. These six vectors l_j, d_j form a basis for the Lie algebra of $E(3)$, and, using Eq.(15.48), we can say that the most general element in $E(3)$ is $\exp(a_j d_j)\exp(\alpha_j l_j)$. The composition law takes the form:

$$\begin{aligned} &\exp(\boldsymbol{b}\cdot\boldsymbol{d})\exp(\boldsymbol{\beta}\cdot\boldsymbol{l})\exp(\boldsymbol{a}\cdot\boldsymbol{d})\exp(\boldsymbol{\alpha}\cdot\boldsymbol{l}) = \exp((\boldsymbol{b}+\boldsymbol{a}')\cdot\boldsymbol{d})\exp(\boldsymbol{\gamma}\cdot\boldsymbol{l})\,, \\ &a'_j = A_{jm}(\boldsymbol{\beta})a_m\,, \qquad \exp(\boldsymbol{\gamma}\cdot\boldsymbol{l}) = \exp(\boldsymbol{\beta}\cdot\boldsymbol{l})\exp(\boldsymbol{\alpha}\cdot\boldsymbol{l})\,. \end{aligned} \tag{15.49}$$

(To ensure that this notation is unambiguously understood, we remind the reader that if S is a Cartesian frame, the frame $S' = \exp(\boldsymbol{a}\cdot\boldsymbol{d})\exp(\boldsymbol{\alpha}\cdot\boldsymbol{l})S$ is obtained as follows: we subject the axes in S to a rotation of amount α about the direction $-\boldsymbol{\alpha}$ to get the coordinate axes of the intermediate frame S_1; we then translate the frame S_1 by the amount $-\boldsymbol{a}$, the components of $\boldsymbol{a}$ being a_j as measured in S_1 or equivalently S'. This leads to the frame S'.)

The l_j by themselves have Lie brackets corresponding to the structure of $R(3)$; similarly, the d_j by themselves have vanishing Lie brackets because these correspond to the structure of T_3. The only remaining basic Lie brackets for $E(3)$ are those between an l and a d. We can compute these from Eq.(15.49). Take the special case in which $\boldsymbol{b} = \boldsymbol{\alpha} = 0$ to get:

$$\exp(\boldsymbol{\beta} \cdot \boldsymbol{l}) \exp(\boldsymbol{a} \cdot \boldsymbol{d}) = \exp(\boldsymbol{a}' \cdot \boldsymbol{d}) \exp(\boldsymbol{\beta} \cdot \boldsymbol{l}) . \tag{15.50}$$

We can rewrite this as

$$\exp(\boldsymbol{\beta} \cdot \boldsymbol{l}) \exp(\boldsymbol{a} \cdot \boldsymbol{d}) \exp(-\boldsymbol{\beta} \cdot \boldsymbol{l}) = \exp(\boldsymbol{a}' \cdot \boldsymbol{d}) , \qquad a'_j = A_{jm}(\boldsymbol{\beta}) a_m . \tag{15.51}$$

But this equation says that in the adjoint representation of $E(3)$, the tangent vector $\boldsymbol{a} \cdot \boldsymbol{d}$ to the one-parameter subgroup $\exp(\sigma \boldsymbol{a} \cdot \boldsymbol{d})$ is transformed, by the element $\exp(\boldsymbol{\beta} \cdot \boldsymbol{l})$ of $E(3)$, into the tangent vector $\boldsymbol{a}' \cdot \boldsymbol{d}$ of the transformed one-parameter subgroup $\exp(\sigma \boldsymbol{a}' \cdot \boldsymbol{d})$. That is, using the operator $\Omega_{\boldsymbol{\beta} \cdot \boldsymbol{l}}$,

$$e^{\Omega_{\boldsymbol{\beta} \cdot \boldsymbol{l}}} \boldsymbol{a} \cdot \boldsymbol{d} = \boldsymbol{a}' \cdot \boldsymbol{d} . \tag{15.52}$$

Because a_j is arbitrary, we get:

$$e^{\Omega_{\boldsymbol{\beta} \cdot \boldsymbol{l}}} d_j = A_{kj}(\boldsymbol{\beta}) d_k . \tag{15.53}$$

Finally, using the form of Eq.(15.11) for $A_{kj}(\boldsymbol{\beta})$, and looking at the terms of order β_j on both sides we get:

$$[\boldsymbol{\beta} \cdot \boldsymbol{l},\, d_j] = -\epsilon_{kjm} \beta_m d_k , \qquad [l_k, d_j] = \epsilon_{kjm} d_m . \tag{15.54}$$

The full list of basic Lie brackets for $E(3)$ is:

$$[l_k,\, l_j] = \epsilon_{kjm} l_m , \tag{15.55a}$$
$$[d_k,\, d_j] = 0 , \tag{15.55b}$$
$$[l_k,\, d_j] = \epsilon_{kjm} d_m . \tag{15.55c}$$

All our expectations are fulfilled: $R(3)$ being a subgroup of $E(3)$, the l_j form a subalgebra in the $E(3)$ Lie algebra; T_3 being an invariant subgroup of $E(3)$, the d_j form an invariant subalgebra of the total Lie algebra; in any Lie bracket, if at least one element is a linear combination of the d_j, so is the bracket itself. According to Eq.(15.55c), each l_k acts as a derivation on the Lie algebra of T_3 when we take its Lie bracket with arbitrary combinations of the d's. From the point of view of T_3, each of these derivations is an outer one, because no combination of the d's can have nonzero Lie brackets with the d's.

The Euclidean group has arisen by a suitable combination of the operations of the first two groups we studied, namely T_3 and $R(3)$. As far as we are concerned, it is the largest group that acts on three-dimensional space alone. (There are larger groups that act on three-dimensional space, but we do not deal with them). The propositions of Euclidean geometry are those statements regarding the relationships between the objects of geometry like points, lines, and two and three-dimensional figures, that are true in all coordinate systems $S, S', \cdots$ related by elements of the Euclidean group. This class of transformations on space alone, $E(3)$, is common to both the "relativity groups" that we shall study; namely, $E(3)$ will appear as a subgroup of both the Galilei and the inhomogeneous Lorentz groups.

The Galilei Group

The Galilei group is the group of transformations in space and time that connect those Cartesian systems that are termed "inertial frames" in Newtonian mechanics. The most general relationship between two such frames is as follows: the origin of the time scale in the inertial frame S' may be shifted compared with that in S; the orientation of the Cartesian axes in S' may be different from that in S; the origin O of the Cartesian frame in S' may be moving relative to the origin O in S at a uniform velocity. One rules out the possibilities that the axes in S' may be subject to time dependent rotations, compared with the directions of the axes in S, and that the velocity of the origin of S' in its motion with respect to the origin in S may be nonuniform. This is because in either of these cases, S' would be a noninertial frame, if S were an inertial one, and Newton's laws of mechanics would not be valid in S'. Apart from the actual form of the transformation equations, the basic assumption inherent in Newtonian relativity is that there is an absolute time scale, so that the only way in which the time variables used by two different "inertial observers" could possibly differ is that zero of time for one of them may be shifted relative to the zero of time for the other.

In the Euclidean case we related the space coordinates of a point P measured in one Cartesian frame S' to their values in another frame S from which S' was obtained by some transformation of $E(3)$. Now we speak of "events" or "world-points" or "space-time points". An event is something that "happens" at some point in space at some time, and in each inertial frame it is assigned three spatial and one temporal coordinate, (x_j, t), making up four space-time coordinates. We must see how to relate the space-time coordinates of a given event in two different inertial frames S and S'. Generalizing the procedure in the case of $E(3)$, the transition from an inertial frame S to any other one, S', can be accomplished in four steps. First, we apply a spatial rotation to S to give us a frame S_1 and leave the time coordinate unchanged; we write $S_1 = aS$, $a \in R(3)$ being the appropriate rotation with corresponding matrix A. Second, we go to a frame S_2 having its spatial axes parallel to those in S_1, but with its origin O_2 moving with a uniform velocity $\boldsymbol{v}$ relative to the origin O_1 in S_1; O_2 and O_1 coincide at the common zero of time in S_2 and S_1, and $\boldsymbol{v}$ has components v_j along the axes of S_1 or S_2. Third, we leave the space axes in S_2 alone, but change the time coordinate by an amount b, to arrive at an inertial frame S_3; time in S_3 lags behind that in S_2 by the amount b. Lastly, we go from S_3 to S' by a space translation; the vector $\overrightarrow{O_3O'}$ from the origin in S_3 to that in S' being $-\boldsymbol{a}$, we write $S' = \exp(\boldsymbol{a} \cdot \boldsymbol{d})S_3$, following our convention for interpreting elements of T_3. All in all, the transition from S to S' involves ten parameters: thus *the Galilei group is a ten-parameter group.* Following the sequence of operations to go from S to S', we can write:

$$S' = (A, v_j, b, a_j)S\,, \tag{15.56}$$

and we denote a general element of the Galilei group by (A, v_j, b, a_j). For some given event in space-time, let us write (x_j, t), $(x_j^{(1)}, t^{(1)})$, $(x_j^{(2)}, t^{(2)})$, $(x_j^{(3)}, t^{(3)})$ and (x'_j, t') for its space-time coordinates in S, S_1, S_2, S_3, and S', respectively. Then we have the system of equations:

$$\begin{aligned} x_j^{(1)} &= A_{jk}x_k\,, & t^{(1)} &= t\,;\\ x_j^{(2)} &= x_j^{(1)} - v_j t^{(1)}\,, & t^{(2)} &= t^{(1)}\,;\\ x_j^{(3)} &= x_j^{(2)}\,, & t^{(3)} &= t^{(2)} - b\,;\\ x'_j &= x_j^{(3)} + a_j\,, & t' &= t^{(3)}\,. \end{aligned} \tag{15.57}$$

Combining these, we see that Eq.(15.56) is accompanied by the following space-time transformation law:

$$x'_j = A_{jk}x_k - v_j t + a_j\,, \qquad t' = t - b\,. \tag{15.58}$$

These are generalizations of Eq.(15.44) and are the basic transformation equations of Newtonian relativity.

The group composition law is obtained by combining two transformations of the type in Eq.(15.58). If $S' = (A, v_j, b, a_j)S$ and $S'' = (A', v'_j, b', a'_j)S'$, we have:

$$\begin{aligned} x''_j &= A'_{jk}x'_k - v'_j t' + a'_j = A'_{jk}(A_{kl}x_l - v_k t + a_k) - v'_j(t-b) + a'_j\\ &= (A'A)_{jl}x_l - (v'_j + A'_{jk}v_k)t + a'_j + A'_{jk}a_k + v'_j b\,,\\ t'' &= t' - b' = t - b - b'\,. \end{aligned} \tag{15.59}$$

Thus the product rule is:

$$\begin{aligned} &(A', v'_j, b', a'_j)(A, v_j, b, a_j)\\ &\quad =(A'A, v'_j + A'_{jk}v_k, b' + b, a'_j + A'_{jk}a_k + v'_j b)\,. \end{aligned} \tag{15.60}$$

The structure of the Galilei group can be inferred from this equation. The identity element is $(\mathbb{1}, 0, 0, 0)$. Inverses are given by:

$$(A, v_j, b, a_j)^{-1} = (A^{-1}, -A_{kj}v_k, -b, -A_{kj}a_k + bA_{kj}v_k)\,. \tag{15.61}$$

There are several natural subgroups:

1. the rotation subgroup $R(3)$ made up of the elements $(A, 0, 0, 0)$ with the appropriate composition law;

2. the three-dimensional Abelian subgroup $T_3^{(v)}$ of velocity transformations, consisting of the elements $(\mathbb{1}, v_j, 0, 0)$, with group composition being addition of the velocity vectors;

3. the one-dimensional Abelian subgroup T_1 of time translations, the elements being $(\mathbb{1}, 0, b, 0)$ and group composition being addition of the b's; and

4. the subgroup T_3 of displacements in space consisting of the elements $(\mathbb{1}, 0, 0, a_j)$.

The first two subgroups, $R(3)$ and $T_3^{(v)}$, combine in semidirect-product fashion, to form a subgroup $E^{(v)}(3)$ having exactly the same structure as the Euclidean group $E(3)$. The elements of $E^{(v)}(3)$ are $(A, v_j, 0, 0)$, and Eq.(15.60) gives for them the multiplication law:

$$(A', v_j', 0, 0)(A, v_j, 0, 0) = (A'A, v_j' + A_{jk}' v_k, 0, 0)\,. \tag{15.62}$$

This is just like the composition rule in Eq. (15.46) for $E(3)$. The elements of the time-translation subgroup T_1 combine with those of the space-translation subgroup T_3 in direct-product fashion to produce the four-parameter Abelian subgroup T_4 of space-time translations; elements of T_4 are $(\mathbb{1}, 0, b, a_j)$, and for them we have the composition rule:

$$(\mathbb{1}, 0, b', a_j')(\mathbb{1}, 0, b, a_j) = (\mathbb{1}, 0, b' + b, a_j' + a_j)\,. \tag{15.63}$$

Thus far, then, we have the subgroups $E^{(v)}(3) = R(3) \times T_3^{(v)}$ and $T_4 = T_1 \otimes T_3$ in the Galilei group. But now Eq.(15.60) shows that the Galilei group itself is a semidirect product of $E^{(v)}(3)$ by T_4, namely, $E^{(v)}(3) \times T_4$, or in more detail, $[R(3) \times T_3^{(v)}] \times (T_1 \otimes T_3)$. The elements of $E^{(v)}(3)$ in the two terms on the left in Eq.(15.60), (A', v_j') and (A, v_j) get combined into $(A'A, v_j' + A_{jk}' v_k)$ according to the composition law Eq.(15.62) in $E^{(v)}(3)$, independently of the presence of elements from the space-time translation subgroup T_4. On the other hand, for a given element (A', v_j') from $E^{(v)}(3)$, the mapping of T_4 onto itself defined by

$$(A', v_j') : (b, a_j) \to (b, A_{jk}' a_k + v_j' b) \tag{15.64}$$

is obviously an automorphism of T_4. The pair $(b' + b, a_j' + A_{jk}' a_k + v_j' b)$ on the right in Eq.(15.60) arises by composition of the pair (b', a_j') and the image of the pair (b, a_j) under the automorphism of T_4 determined by (A', v_j'), according to the composition rule in T_4. It follows that T_4 is an invariant subgroup of the Galilei group, whereas $E^{(v)}(3)$ is an ordinary subgroup.

As in the Euclidean case, we parametrize the Galilei group in a way that parallels the physical interpretation. If A corresponds to the rotation $\exp(\boldsymbol{\alpha} \cdot \boldsymbol{l})$ in $R(3)$, $A = A(\boldsymbol{\alpha})$, the set of ten parameters (α_j, v_j, b, a_j) suffices to parametrize all elements of the Galilei group. The range of the α_j has been determined in the section on the group $R(3)$; the v_j, b and a_j all go over the range $-\infty$ to $+\infty$ independently. These are not canonical coordinates because, in general, the family of elements $(\sigma\alpha_j, \sigma v_j, \sigma b, \sigma a_j)$ does not form a one-parameter subgroup, except in special cases. Corresponding to the sequence of operations by which the general element (A, v_j, b, a_j) is built up, we have the following decomposition:

$$(A, v_j, b, a_j) = (\mathbb{1}, 0, 0, a_j)(\mathbb{1}, 0, b, 0)(\mathbb{1}, v_j, 0, 0)(A, 0, 0, 0)\,. \tag{15.65}$$

Now we can introduce a basis for the ten-dimensional Lie algebra of the Galilei group as follows. Even though the coordinate system we are using is noncanonical, some one-parameter subgroups still have a simple form, including:

1. the family of elements $(\sigma\alpha_j, 0, 0, 0)$ forming for each α_j a one-parameter subgroup in $R(3)$;
2. the family $(0, \sigma v_j, 0, 0)$ forming for each v_j a one-parameter subgroup of velocity transformations in a given direction and belonging to $T_3^{(v)}$;
3. the one-parameter subgroup T_1 of time translations $(0, 0, \sigma, 0)$; and
4. the one-parameter subgroup $(0, 0, 0, \sigma a_j)$ of space translations in a given direction belonging to T_3.

Of course, we do not exhaust all one-parameter subgroups of the Galilei group in this way. There is no danger of confusion if we write l_j and d_j again for the basic tangent vectors to be first and fourth types of one-parameter sub-groups. For the tangent vector to the subgroup T_1, it is customary to write h, so that the element $(\mathbb{1}, 0, b, 0)$ is denoted $\exp(bh)$. For the three basic tangent vectors to the subgroups of velocity transformations in the three directions, we write g_j, so that the element $(\mathbb{1}, v_j, 0, 0)$ is $\exp(v_j g_j)$. Therefore, by Eq.(15.65) we have:

$$(A, v_j, b, a_j) = \exp(a_j d_j)\exp(bh)\exp(v_j g_j)\exp(\alpha_j l)\,, \tag{15.66}$$

and Eq.(15.60) takes the form:

$$\begin{aligned}
\exp(a'_j d_j)\exp(b'h)\exp(v'_j g_j)\exp(\alpha'_j l_j)\exp(a_j d_j)\exp(bh)\exp(v_j g_j)\exp(\alpha_j l_j)& \\
= \exp(\bar{a}_j d_j)\exp((b'+b)h)\exp(\bar{v}_j g_j)\exp(\gamma_j l_j)\ ,& \\
\bar{a}_j = a'_j + A_{jk}(\boldsymbol{\alpha}')a_k + bv'_j\,, \quad \bar{v}_j = v'_j + A_{jk}(\boldsymbol{\alpha}')v_k\,,& \\
\exp(\boldsymbol{\gamma}\cdot\boldsymbol{l}) = \exp(\boldsymbol{\alpha}'\cdot\boldsymbol{l})\exp(\boldsymbol{\alpha}\cdot\boldsymbol{l})\,.&
\end{aligned} \tag{15.67}$$

We obtain the basic Lie brackets among the l_j, g_j, d_j and h by considering special cases of Eq.(15.67). The l_j among themselves obey the bracket rules (Eq.(15.18)) characteristic of $R(3)$, whereas the g_j have vanishing brackets with one another because $T_3^{(v)}$ is Abelian. Because $R(3)$ and $T_3^{(v)}$ are combined in a semidirect product exactly as $R(3)$ and T_3 are in $E(3)$, the brackets between the l's and g's can be read off from the $E(3)$ case, namely, Eq.(15.54):

$$[l_j, g_k] = \epsilon_{jkm} g_m\ . \tag{15.68}$$

Because T_4 is Abelian, all brackets among h and the d_j vanish. Between l_j and d_k we once again have Eq.(15.54), since $E(3)$ is contained as a subgroup of the Galilei group. Taking $b' = \boldsymbol{\alpha}' = b = \boldsymbol{\alpha} = 0$ in Eq.(15.67), we see that the elements of T_3 and $T_3^{(v)}$ commute; thus the brackets of the d_j with the g_k must vanish. The remaining brackets are $[l_j, h]$ and $[g_j, h]$. But since the subgroups $R(3)$ and T_1 commute, the former brackets vanish. To obtain $[g_j, h]$, we specialize Eq.(15.67) to the case $a'_j = b' = \alpha'_j = a_j = v_j = \alpha_j = 0$:

$$\exp(\boldsymbol{v}' \cdot \boldsymbol{g}) \exp(bh) = \exp(b\boldsymbol{v}' \cdot \boldsymbol{d}) \exp(bh) \exp(\boldsymbol{v}' \cdot \boldsymbol{g}) . \tag{15.69}$$

We drop the prime on $\boldsymbol{v}$ and use the facts that h and the d_j as well as the d_j and g_k, have vanishing brackets, to write this as:

$$\exp(-bh) \exp(\boldsymbol{v} \cdot \boldsymbol{g}) \exp(bh) = \exp(\boldsymbol{v} \cdot (\boldsymbol{g} + b\boldsymbol{d})) . \tag{15.70}$$

This means that in the adjoint representation, the tangent vector $\boldsymbol{v} \cdot \boldsymbol{g}$ to the one-parameter subgroup $\exp(\sigma \boldsymbol{v} \cdot \boldsymbol{g})$ is carried by the element $\exp(-bh)$ into the tangent vector $\boldsymbol{v} \cdot (\boldsymbol{g} + b\boldsymbol{d})$. Because v_j is arbitrary, we get:

$$e^{-\Omega_{bh}} g_j = g_j + b d_j \ , \tag{15.71}$$

and comparing the term of order b on both sides:

$$[g_j, h] = d_j \ . \tag{15.72}$$

All in all, the complete set of basic Lie brackets for the Galilei group is

$$\begin{aligned} E^{(v)}(3) = R(3) \times T_3^{(v)} :& [l_j, l_k] = \epsilon_{jkm} l_m \ , \\ & [l_j, g_k] = \epsilon_{jkm} g_m \ , \ [g_j, g_k] = 0 \ ; && (15.73a) \\ T_4 = T_1 \otimes T_3 :& [h, d_k] = [d_j, d_k] = 0 \ ; && (15.73b) \\ & [l_j, h] = 0 \ , \ [l_j, d_k] = \epsilon_{jkm} d_m \ ; \\ & [g_j, h] = d_j \ , \ [g_j, d_k] = 0 \ . && (15.73c) \end{aligned}$$

In Eq.(15.73a), we have the l_j and g_j spanning the subalgebra corresponding to the subgroup $E^{(v)}(3) = R(3) \times T_3^{(v)}$; in Eq.(15.73b), we have the d_j and h spanning the Abelian invariant subalgebra corresponding to the invariant Abelian space-time translation subgroup $T_4 = T_1 \otimes T_3$; and Eq.(15.73c) shows how each of the l_j and g_j acts as a derivation on the d's and h. Because T_4 is Abelian, these are all outer derivations.

The manner in which we have exhibited the Galilei group as a semidirect product of $R(3) \times T_3^{(v)}$ by T_4 is closely related to the structure that we will find for the inhomogeneous Lorentz group. Therefore we may say that the decomposition we have given is the "physical" decomposition of the Galilei group, and Eq.(15.73) exhibits the corresponding "physical" decomposition of the Lie algebra. There is another way in which the Galilei group can be expressed as a semidirect product of two of its subgroups, one of which is invariant. This is most easily seen in terms of the Lie algebra. It is clear that the l_j and h form a subalgebra, and because the bracket $[l_j, h]$ vanishes, this corresponds to the subgroup $R(3) \otimes T_1$. The g_j and d_j also form a subalgebra: all brackets among these six elements vanish; thus they correspond to the six-parameter Abelian subgroup $T_3^{(v)} \otimes T_3$. But the latter subalgebra is an invariant one: any Lie bracket in which one of the elements is h or an l_j and the other is one of the d's or g's is equal to one of the d's or g's! Therefore, the Galilei group can also be represented as the semidirect product of $R(3) \otimes T_1$ by $T_3^{(v)} \otimes T_3$, namely, as $[R(3) \otimes T_1] \times [T_3^{(v)} \otimes T_3]$. There is no analogous decomposition in the case of the inhomogeneous Lorentz group.

We see in Chapter 19 that the most interesting aspect of the Galilei group and the Lie algebra relations (Eq.(15.73)) is that it admits a family of nontrivial realizations up to neutral elements. The presence of neutral elements can be avoided in all the basic Lie brackets except $[g_j, d_k]$. If instead of being zero, this bracket has the value $\delta_{jk}M$, M being a neutral element, this neutral element cannot be transformed away by a redefinition of the basic elements of the algebra. That the dependence on j and k must be of the form δ_{jk} is a consequence of rotational properties of the system, namely properties of the brackets involving the l_j. The existence of such realizations is very important from the point of view of physical interpretation; we shall see this in detail in Chapter 19.

The Homogeneous and Inhomogeneous Lorentz Groups

The transformation Eq.(15.58) of Newtonian relativity does not set any upper limit on the velocity of a material object or a propagating wave disturbance in an inertial frame, or on the relative velocity between two inertial frames. As is well known, this is in conflict with the experimental finding that the velocity of transmission of electromagnetic waves, a phenomenon described by Maxwell's equations of electrodynamics, is the same in two inertial frames moving relative to one another. In other words the velocity of light is an absolute constant of nature, and measurements in all inertial frames will lead to the same value, and Maxwell's equations cannot be made invariant under the Galilean transformations (Eq.(15.58)). Whereas Eq.(15.58) may appear quite "reasonable" and just what one might expect from "everyday experience," it is really valid only for small velocities v, small compared with the velocity c of light. The true transformation equations connecting the space-time data of two "inertial" observers in uniform relative motion have a different form, and they deviate significantly from Eq.(15.58) if v is a large fraction of c. These are the Lorentz transformation equations, and they are essentially derived from two requirements:

1. the equations expressing the space-time coordinates x'_j, t' of an inertial observer S' in terms of x_j, t of another observer S must be linear in these coordinates.

2. A disturbance propagating with velocity c in S must be found to possess this same velocity c in S' as well.

This leads to the negation of the idea of an absolute time, and also has the consequence that two events at spatially separated points that appear to be simultaneous in time to S will, in general, not be simultaneous in time to S'

because the time variables used by S' and S are related in a more subtle fashion than they are in Eq.(15.58), involving the space variables x_j as well. All this development is very well explained in many excellent books; here we are more interested in the group-theoretical structure of the new transformation equations. We will develop this in two parts: first, we study the set of *homogeneous* Lorentz transformations, the transformations connecting inertial frames $S, S', \cdots$ that either are rotated with respect to one another in space or are moving uniformly with respect to one another, or some combination of both, but whose origins in space and time coincide; next, we include the possibility of space-time translations as well, leading to the set of *inhomogeneous* Lorentz transformations. The former transformations give rise to the *homogeneous Lorentz group*, abbreviated HLG, whereas the latter give rise to the *inhomogeneous Lorentz group*, also called the *Poincaré group*, abbreviated IHLG or $\mathcal{P}$.

(a) The Homogeneous Lorentz Group

Let us begin with the simplest Lorentz transformation, namely a transformation in one dimension. That is, let S and S' be two inertial observers, let the directions of the space axes in the two frames be the same, and suppose S' is moving relative to S in the positive x direction with velocity $v (v < c)$. In addition, we assume that at time $t = 0$ in S, the spatial origins O' and O in S' and S coincide, and conversely as viewed from S'. If some event in space-time has coordinates (x', y', z', t') in S' and (x, y, z, t) in S, these are connected by the *Lorentz transformation equations*:

$$x' = \frac{x - vt}{\sqrt{1 - v^2/c^2}}\ y' = y\,,\ z' = z\,,\ t' = \frac{t - vx/c^2}{\sqrt{1 - v^2/c^2}}\,. \tag{15.74}$$

As we said before, these arise by requiring linearity of the transformation equations and the constancy of the velocity of light; the latter is expressed by the demand that the combination $x^2 - c^2t^2$ be invariant under the transformation, namely,

$$(x')^2 - c^2(t')^2 = x^2 - c^2t^2\,. \tag{15.75}$$

Spatial measurements in the plane perpendicular to the direction of the relative velocity are unaffected, whereas the coordinate in the direction of this velocity and the time get transformed. There are two velocity addition theorems that follow from Eq.(15.74). The first is that if some "object" is moving uniformly along the x axis in S with a velocity u, then its velocity u' along the x' axis in S' will be

$$u' = \frac{u - v}{1 - uv/c^2}\,. \tag{15.76}$$

Thus if the "object" were a light wave, with $u = c$, then $u' = c$ as well. The second theorem is that if the observer S' is moving relative to S with velocity v and if S'' is moving relative to S' with velocity v', all these relative motions being along the common x axis, then the velocity of S'' relative to S is v'', given by

$$v'' = \frac{v + v'}{1 + vv'/c^2}\,. \tag{15.77}$$

(These two addition theorems are, of course, not independent). That is, if the variables x', t' are related to x, t as in Eq.(15.74) and if x'', t'' are related to x', t' in a similar manner but with v' replacing v, then x'', t'' are related again to x, t in this same manner, but with the velocity variable being v''. Thus, in an obvious way, these Lorentz transformations in one dimension may be composed with one another to yield new transformations of the same type, and all together they constitute an Abelian one-parameter group. That this should be so is evident from the fact that these transformations can (essentially) be defined as the set of linear transformations $(x, t) \to (x', t')$ obeying the invariance requirement of Eq.(15.75). The symmetry of the right-hand side of Eq.(15.77) with respect to v and v' guarantees the Abelian nature of this group of transformations; but with the velocity v being the parameter labeling the transformations, composition of two transformations does not correspond to addition of the parameters. It is useful to introduce a new parameter in terms of which the equivalent of Eq.(15.77) becomes simple addition of the parameters, and this can be done easily. We write

$$v = c\tanh\zeta\,,\quad \frac{1}{\sqrt{1 - v^2/c^2}} = \cosh\zeta\,,\quad \frac{v/c}{\sqrt{1 - v^2/c^2}} = \sinh\zeta\,, \tag{15.78}$$

and call ζ the "rapidity" parameter. The range $-c < v < c$ corresponds to $-\infty < \zeta < \infty$, and now if the relative rapidities of S' and S and of S'' and S' are ζ and ζ', respectively, then the relative rapidity of S'' and S is

$$\zeta'' = \zeta + \zeta'\,. \tag{15.79}$$

This is the same as Eq.(15.77) in the new parameter. The transformation Eq.(15.74) has now the following appearance:

$$x' = x\cosh\zeta - ct\,\sinh\zeta\,,\ ct' = ct\,\cosh\zeta - x\,\sinh\zeta\,,\ y' = y\,,\ z' = z\,; \tag{15.80}$$

it is clear why one calls this Lorentz transformation a "hyperbolic rotation" of amount ζ in the $x - ct$ plane.

The directions of the space axes were not changed in the family of Lorentz transformations just considered. Still maintaining these directions, we can generalize the formulae to describe Lorentz transformations in an arbitrary direction. We will now go back to writing x_1, x_2, x_3 in place of x, y, z. If the inertial frames S' and S have common directions for their space axes and if S' is moving relative to S uniformly in the direction $\boldsymbol{\zeta}$ and with rapidity $\zeta = |\boldsymbol{\zeta}|$, and if as before the two origins O and O' coincide at time $t = 0$ in S as well as at $t' = 0$

in S', we have the following transformation equations:

$$x'_j = \left(\delta_{jk} + \zeta_j\zeta_k \frac{\cosh\zeta - 1}{\zeta^2}\right) x_k - ct\zeta_j \frac{\sinh\zeta}{\zeta}\,;$$
$$ct' = ct\cosh\zeta - x_j\zeta_j \frac{\sinh\zeta}{\zeta}\,, \tag{15.81}$$

where ζ_j are the components of $\boldsymbol{\zeta}$ along the axes of S' or S. The analogue of Eq.(15.75) is:

$$x'_j x'_j - c^2(t')^2 = x_j x_j - c^2 t^2\,. \tag{15.82}$$

A family of such Lorentz transformations, all along the same direction $\hat{\boldsymbol{\zeta}}$ but of varying amounts $\zeta, \zeta', \cdots$ will again form an Abelian one-parameter group, with addition of the rapidities ζ, ζ' corresponding to composition of the transformations. However, if we combine two such "pure" Lorentz transformations in two *different* directions, $\hat{\boldsymbol{\zeta}}$ and $\hat{\boldsymbol{\zeta}}'$, the product will not be a "pure" Lorentz transformation at all, but will involve rotations in space as well!

The most general homogeneous Lorentz transformation connecting two inertial frames S and S' whose space-time origins coincide is now obtained in this way: we first apply a suitable rotation to the axes in S to go to a new system of axes S_1, the time variable remaining unchanged, such that S_1 and S' have parallel axes. Thus we write $S_1 = \exp(\boldsymbol{\alpha}\cdot\boldsymbol{l})S$, $\exp(\boldsymbol{\alpha}\cdot\boldsymbol{l})$ being the appropriate element from $R(3)$. Then we apply a "special" or "pure" Lorentz transformation corresponding to the rapidity vector $\boldsymbol{\zeta}$ to S_1 to take us to S'. The components of $\boldsymbol{\zeta}$ are ζ_j, as measured in S_1 or S'. We will denote this sequence of operations by the pair $(\boldsymbol{\zeta}, \boldsymbol{\alpha})$, and write $S' = (\boldsymbol{\zeta}, \boldsymbol{\alpha})S$. The space-time coordinates x'_j, t' in S' and x_j, t in S of some event are related in this way: we first relate the coordinates $x_j^{(1)}$ in S_1 to x_j in S using Eq.(15.4), then go from $x_j^{(1)}, t$ to x'_j, t' using Eq.(15.81). We get:

$$S_1 = \exp(\boldsymbol{\alpha}\cdot\boldsymbol{l})S : x_j^{(1)} = A_{jk}(\boldsymbol{\alpha})x_k\,, \qquad t^{(1)} = t\,;$$
$$S' = (\boldsymbol{\zeta}, \mathbf{0})S_1 = (\boldsymbol{\zeta}, \boldsymbol{\alpha})S : x'_j = \left(\delta_{jk} + \zeta_j\zeta_k \frac{\cosh\zeta - 1}{\zeta^2}\right) x_k^{(1)} - ct\zeta_j \frac{\sinh\zeta}{\zeta}\,,$$
$$ct' = ct\,\cosh\zeta - x_j^{(1)}\zeta_j \frac{\sinh\zeta}{\zeta}\,. \tag{15.83}$$

It is customary to write the total transformation $S \to S'$ in "four-dimensional notation." The space and time coordinates x_j and t of a world point P, relative to S, are united into the four components of a "four vector" leading from the space-time origin of S to the event P. We write x for this four vector and x^μ for its components; the superscript μ takes the four values 0,1,2,3, and we define $x^0 = ct$, $x^j = x_j$, $j = 1, 2, 3$. The x^μ are called the "contravariant components" of x. In order to be able to rewrite Eq.(15.82) in this notation, a metric tensor $g_{\mu\nu}$ and the "covariant components" x_μ of a four-vector whose contravariant components are x^μ are also introduced. The tensor $g_{\mu\nu}$ is symmetric in μ and

ν, and the only nonvanishing components are $g_{00} = -1$, $g_{11} = g_{22} = g_{33} = +1$; the lowering of the index on x^μ to get to x_μ is defined by

$$x_\mu = g_{\mu\nu} x^\nu \quad : \qquad x_0 = -x^0 \, , \; x_j = x^j \, . \tag{15.84}$$

The inverse to $g_{\mu\nu}$ is $g^{\mu\nu}$. This obeys $g^{\mu\nu} g_{\nu\lambda} = \delta^\mu{}_\lambda$ and is also symmetric in the superscripts: the only nonvanishing components are $g^{00} = -1$, $g^{11} = g^{22} = g^{33} = +1$. The tensor $g^{\mu\nu}$ is to be used for raising indices: $x^\mu = g^{\mu\nu} x_\nu$. The combined Lorentz transformation from S to S', performed by first subjecting S to a spatial rotation and then subjecting the resulting frame S_1 to a pure Lorentz transformation, can be expressed as follows:

$$\begin{aligned}
x'^\mu &= \Lambda^\mu{}_\nu(\boldsymbol{\zeta}, \boldsymbol{\alpha}) x^\nu \, ; \\
\Lambda^j{}_k(\boldsymbol{\zeta}, \boldsymbol{\alpha}) &= A_{jk}(\boldsymbol{\alpha}) + \zeta_j \zeta_m A_{mk}(\boldsymbol{\alpha}) \frac{\cosh\zeta - 1}{\zeta^2} \, ; \\
\Lambda^j{}_0(\boldsymbol{\zeta}, \boldsymbol{\alpha}) &= -\zeta_j \frac{\sinh\zeta}{\zeta} \, ; \\
\Lambda^0{}_j(\boldsymbol{\zeta}, \boldsymbol{\alpha}) &= -\zeta_k A_{kj}(\boldsymbol{\alpha}) \frac{\sinh\zeta}{\zeta} \, ; \\
\Lambda^0{}_0(\boldsymbol{\zeta}, \boldsymbol{\alpha}) &= \cosh\zeta \, .
\end{aligned} \tag{15.85}$$

This linear transformation from x^μ to x'^μ leaves invariant the quadratic form appearing in Eq.(15.82):

$$x'^\mu x'_\mu = x^\mu x_\mu \, . \tag{15.86}$$

Consequently, for the four-dimensional matrix $\Lambda(\boldsymbol{\zeta}, \boldsymbol{\alpha})$, we have:

$$g_{\mu\mu'} \Lambda^\mu{}_\nu \Lambda^{\mu'}{}_{\nu'} \equiv \Lambda^\mu{}_\nu \Lambda_{\mu\nu'} = g_{\nu\nu'} \, . \tag{15.87}$$

From the physical interpretation it is clear that if S and S' are any two inertial frames whose space-time origins coincide and for which the senses of increasing time and the handedness of the spatial coordinate system are the same, then the space-times variables x^μ and x'^μ are related by Eq.(15.85) for some choice of $\boldsymbol{\zeta}$ and $\boldsymbol{\alpha}$. For, to determine $\boldsymbol{\alpha}$, we only have to see what rotation needs to be applied to the space axes of S to render them parallel to the axes in S'; once this has been done, the direction and magnitude of the relative velocity between S' and $S_1 = \exp(\boldsymbol{\alpha} \cdot \boldsymbol{l}) S$ will give us $\boldsymbol{\zeta}$. The set of all these transformations $S \to S' = (\boldsymbol{\zeta}, \boldsymbol{\alpha}) S$ for all possible $\boldsymbol{\zeta}$ and $\boldsymbol{\alpha}$ constitutes the homogeneous Lorentz group, or more precisely, the proper orthochronous homogeneous Lorentz group. This is therefore a six-parameter group. One can also define this group in the following way: it is the set of all transformations from one inertial frame S to another S', such that the space-time variables x'^μ used in S' are linearly related to the x^μ used in S by some matrix Λ,

$$S \to S' = \Lambda S : \qquad x'^\mu = \Lambda^\mu{}_\nu x^\nu \, , \tag{15.88}$$

where the $\Lambda^\mu{}_\nu$ obey Eq.(15.87) so that the indefinite quadratic form $x^\mu x_\mu$ is preserved. In addition: (1) $\Lambda^0{}_0$ is positive,in fact $\geqq 1$; (2) the 3×3 sub-determinant $|\Lambda^j{}_k|$, $j,k = 1,2,3$ is positive. These two conditions on the matrix Λ ensure that the senses of increasing time and the handedness of the spatial coordinate system are the same in S and S', and then it is clear that Λ must have the form of Eq.(15.85) for some $\boldsymbol{\zeta}$ and $\boldsymbol{\alpha}$. We can use the notation $S' = \Lambda S$ in place of $S' = (\boldsymbol{\zeta}, \boldsymbol{\alpha})S$ if we wish. Because the invariant quadratic form $x^\mu x_\mu$ contains three positive terms and one negative one, the group of these transformations Λ can be thought of as the group of "pseudo-orthogonal" rotations in a real four-dimensional space; the ordinary orthogonal rotations in a four-dimensional space would have preserved a positive definite quadratic form made up of four positive terms. Consequently, the proper orthochronous homogeneous Lorentz group may also be written $SO(3,1)$, the letter S summarizing the conditions $\Lambda^0{}_0 \geqq 1$, $\det|\Lambda^j{}_k| > 0$ on Λ. The group composition law for the HLG (as we shall call it from now on) is easily worked out from Eq.(15.88): If we have three inertial frames S, S', S'' with the element Λ corresponding to the matrix $\|\Lambda^\mu{}_\nu\|$ connecting S' to S, and similarly Λ' connecting S'' to S', then the coordinates x''^μ of an event, as measured in S'', will be related to the coordinates x^μ in S as follows:

$$S'' = \Lambda'\Lambda S: \; x''^\mu = \Lambda'^\mu{}_\nu x'^\nu = \Lambda'^\nu{}_\nu \Lambda^\nu{}_\sigma x^\sigma = (\Lambda'\Lambda)^\mu{}_\sigma x^\sigma\,,$$
$$(\Lambda'\Lambda)^\mu{}_\nu = \Lambda'^\mu{}_\sigma \Lambda^\sigma{}_\nu\,. \tag{15.89}$$

Consequently, the matrix representing the element $\Lambda'\Lambda$ is just the product of the individual matrices. Similarly, by use of Eq.(15.87), we see that the matrix representing the inverse of an element Λ is given as follows:

$$\Lambda^{-1} \to (\Lambda^{-1})^\mu{}_\nu = g^{\mu\mu'} g_{\nu\nu'} \Lambda^{\nu'}{}_{\mu'} \equiv \Lambda_\nu{}^\mu\,. \tag{15.90}$$

Equations (15.89) and (15.90) relating to products and inverses can be expressed in the coordinate system $\boldsymbol{\zeta}, \boldsymbol{\alpha}$ as well. Corresponding to Eq.(15.89), we have an equation like:

$$(\zeta'_j, \alpha'_j)(\zeta_j, \alpha_j) = (\zeta''_j, \alpha''_j)\,. \tag{15.91}$$

But the equations determining ζ''_j and α''_j in terms of the elements on the left are somewhat complicated, and we shall not bother to obtain them in the general case. Special cases of interest will be computed when the need arises. The computation of inverses, however, is quite simple. One can verify, using Eq.(15.85), that one has

$$(\zeta_j, \alpha_j)^{-1} = (-A_{lj}(\boldsymbol{\alpha})\zeta_l, -\alpha_j)\,. \tag{15.92}$$

The coordinates (ζ_j, α_j), which can be used to label all elements of the HLG, are useful because they have a simple and direct physical meaning. However, they are not canonical coordinates of the first kind, because a family of elements with coordinate $(\sigma\zeta_j, \sigma\alpha_j)$ will, in general, not be a one-parameter subgroup. One-parameter subgroups that do have a simple form in these coordinates are of two kinds:

1. the family of elements $(0, \sigma\alpha_j)$, the α_j being fixed, is a one-parameter subgroup of rotations around the axis α_j;

2. the family $(\sigma\zeta_j, 0)$, the ζ_j being fixed, is a one-parameter subgroup of pure Lorentz transformations, all in the direction ζ_j but of varying rapidities.

The three basic tangent vectors to the three basic sub-groups of the first type will be the l_j; whereas for the tangent vectors to the one-parameter subgroups of pure Lorentz transformations in the first, second, and third directions we will write k_j, $j = 1, 2, 3$. As a special case of Eq.(15.91), we have

$$(\zeta_j, \alpha_j) = (\zeta_j, 0)(0, \alpha_j); \tag{15.93}$$

this is the way the transformation (ζ_j, α_j) was built up. We can then also write $\exp(\zeta_j k_j)\exp(\alpha_j l_j)$ for this element of the HLG. Thus we have that if $S' = \exp(\zeta_j k_j)\exp(\alpha_j l_j)S$, then the space-time coordinates are related by Eq.(15.85). The $R(3)$ subgroup of the HLG is, of course, the set of elements $\exp(\boldsymbol{\alpha} \cdot \boldsymbol{l})$ or $(0, \alpha_j)$. In contrast to the case of the Galilei group, the subset of elements $\exp(\boldsymbol{\zeta} \cdot \boldsymbol{k})$ consisting of pure velocity transformations does not constitute a subgroup: the product $\exp(\boldsymbol{\zeta}' \cdot \boldsymbol{k})\exp(\boldsymbol{\zeta} \cdot \boldsymbol{k})$, for nonparallel $\boldsymbol{\zeta}', \boldsymbol{\zeta}$ will involve rotations as well. We now compute the basic Lie brackets among the l_j and k_j. Among the l_j we have the brackets corresponding to the $R(3)$ structure. Next, using Eqs.(15.85) and (15.89), we can derive quite easily the following result:

$$\exp(\boldsymbol{\alpha} \cdot \boldsymbol{l})\exp(\boldsymbol{\zeta} \cdot \boldsymbol{k})\exp(-\boldsymbol{\alpha} \cdot \boldsymbol{l}) = \exp(\boldsymbol{\zeta}' \cdot \boldsymbol{k}), \quad \zeta'_j = A_{jk}(\boldsymbol{\alpha})\zeta_k . \tag{15.94}$$

This has a transparent physical meaning. Now this equation has exactly the same structure as Eq.(15.57), for example,with the d's replaced by the k's. Consequently, the analogue to Eq.(15.54) holds:

$$[l_j, k_m] = \epsilon_{jmn} k_n . \tag{15.95}$$

To compute the Lie brackets among the k's, we use a method similar to the one in the case of $R(3)$. Let us consider the bracket $[k_1, k_2]$. To evaluate it, we consider the one-parameter subgroups of pure Lorentz transformations $\exp(\sigma k_1)$ and $\exp(\rho k_2)$, σ and ρ being small. If we retain only terms linear in ρ and σ, and among terms of second order only the one involving $\rho\sigma$ in all calculations, we have

$$\begin{aligned}
\exp(\sigma k_1)\exp(\rho k_2) &= \exp\left(\sigma k_1 + \rho k_2 + \frac{\sigma\rho}{2}[k_1, k_2] + \cdots\right) \\
&= \exp(\sigma k_1 + \rho k_2)\exp\left(\frac{\sigma\rho}{2}[k_1, k_2] + \cdots\right), \\
\exp(-\sigma k_1 - \rho k_2)&\exp(\sigma k_1)\exp(\rho k_2) \\
&= \exp\left(\frac{\sigma\rho}{2}[k_1, k_2] + \cdots\right).
\end{aligned} \tag{15.96}$$

Let us introduce a sequence of inertial frames S, $S_1 = \exp(\rho k_2)S$, $S_2 = \exp(\sigma k_1)S_1$, $S' = \exp(-\sigma k_1 - \rho k_2)S_2$, and relate the space-time variables in these frames to one another. At each step, we need to use only Eq.(15.81). We have:

$$
\begin{aligned}
&S_1 = \exp(\rho k_2)S : \ x_j^{(1)} = x_j - ct\rho\delta_{j2}\cdots, ct^{(1)} = ct - \rho x_2 \cdots ; \\
&S_2 = \exp(\sigma k_1)S_1 : \ x_j^{(2)} = x_j^{(1)} - ct^{(1)}\sigma\delta_{j1}\cdots, ct^{(2)} = ct^{(1)} - \sigma x_1^{(1)} \cdots ; \\
&S' = \exp(-\sigma k_1 - \rho k_2)S_2 ; \\
&x'_j = x_j^{(2)} + \frac{1}{2}(\sigma\delta_{j1} + \rho\delta_{j2})(\sigma\delta_{k1} + \rho\delta_{k2})x_k^{(2)} + ct^{(2)}(\sigma\delta_{j1} + \rho\delta_{j2})\cdots \\
&\quad = x_j^{(2)} + \frac{1}{2}\sigma\rho(\delta_{j1}x_2^{(2)} + \delta_{j2}x_1^{(2)}) + ct^{(2)}(\sigma\delta_{j1} + \rho\delta_{j2})\cdots, \\
&ct' = ct^{(2)} + (\sigma\delta_{j1} + \rho\delta_{j2})x_j^{(2)} \cdots = ct^{(2)} + \sigma x_1^{(2)} + \rho x_2^{(2)} \cdots .
\end{aligned} \tag{15.97}
$$

We now combine these to relate S' directly to S :

$$
\begin{aligned}
S' =& \exp\left(\frac{\sigma\rho}{2}[k_1, k_2]\cdots\right) S : \\
x'_j =& x_j^{(1)} - ct^{(1)}\sigma\delta_{j1} + \frac{1}{2}\sigma\rho(\delta_{j1}x_2^{(1)} + \delta_{j2}x_1^{(1)}) \\
&+ (ct^{(1)} - \sigma x_1^{(1)})(\sigma\delta_{j1} + \rho\delta_{j2})\cdots \\
=& x_j - ct\rho\delta_{j2} - ct\sigma\delta_{j1} + \sigma\rho\delta_{j1}x_2 + \frac{1}{2}\sigma\rho(\delta_{j1}x_2 + \delta_{j2}x_1) \\
&+ ct\sigma\delta_{j1} - \sigma\rho\delta_{j1}x_2 + ct\rho\delta_{j2} - \rho\sigma\delta_{j2}x_1 \cdots \\
=& x_j + \frac{1}{2}\sigma\rho(\delta_{j1}x_2 - \delta_{j2}x_1) + \cdots, \\
ct' =& ct^{(1)} - \sigma x_1^{(1)} + \sigma x_1^{(1)} + \rho x_2^{(1)} \cdots = ct - \rho x_2 + \rho x_2 \cdots = ct \cdots .
\end{aligned} \tag{15.98}
$$

This shows that to the accuracy to which we are working, S' is obtained from S by a spatial rotation alone; thus the bracket $[k_1, k_2]$ must be a combination of the l_j. If write $[k_1, k_2] = \beta_j l_j$, then we would have

$$
\begin{aligned}
S' =& \exp\left(\frac{\sigma\rho}{2}\boldsymbol{\beta}\cdot\boldsymbol{l} + \cdots\right) S : \\
x'_j =& A_{jk}\left(\frac{\sigma\rho}{2}\boldsymbol{\beta}\right) x_k = \left(\delta_{jk} - \frac{\sigma\rho}{2}\epsilon_{jkm}\beta_m \cdots\right) x_k \\
=& x_j - \frac{1}{2}\sigma\rho\epsilon_{jkm}\beta_m x_k \cdots, \qquad ct' = ct \cdots .
\end{aligned} \tag{15.99}
$$

Comparing Eqs.(15.99) and (15.98), we see that $\beta_l = -\epsilon_{1\,2l}$, or that $[k_1, k_2] = -l_3$. Cyclically permuting the indices, we get the bracket relations:

$$[k_l, k_m] = -\epsilon_{lmn} l_n \,. \tag{15.100}$$

Thus the complete list of basic Lie brackets for the Lie algebra of the HLG is:

$$[l_j, l_m] = \epsilon_{jmn} l_n \,, \quad [l_j, k_m] = \epsilon_{jmn} k_n \,, \quad [k_j, k_m] = -\epsilon_{jmn} l_n \,. \tag{15.101}$$

A more symmetrical notation for the l's and k's can be arrived at in the following manner. Let us examine the form of an element Λ in the HLG close to the identity element. Working to first order in small quantities, we can expand the corresponding matrix $\|\Lambda^\mu{}_\nu\|$ in the form

$$\Lambda \to \Lambda^\mu{}_\nu \simeq \delta^\mu_\nu + \delta\lambda^\mu{}_\nu\,, \quad |\delta\lambda| \ll 1\,. \tag{15.102}$$

The $\delta\lambda$'s will be restricted by the basic condition of Eq.(15.87):

$$\begin{aligned} \Lambda^\mu{}_\nu \Lambda_{\mu\nu'} \simeq (\delta^\mu_\nu + \delta\lambda^\mu{}_\nu)(g_{\mu\nu'} + \delta\lambda_{\mu\nu'}) \simeq g_{\nu\nu'} + \delta\lambda_{\nu\nu'} + \delta\lambda_{\nu'\nu}\,, \\ \delta\lambda_{\mu\nu} + \delta\lambda_{\nu\mu} = 0\,. \end{aligned} \tag{15.103}$$

Thus the requirement that the matrix $\|\Lambda^\mu{}_\nu\|$ be an infinitesimal homogeneous Lorentz transformation means that the $\delta\lambda$'s be antisymmetric in the subscripts. Now this element of the group could equally well have been written as

$$\Lambda = \exp(\delta\zeta_j k_j)\exp(\delta\alpha_j l_j) \simeq \exp(\delta\zeta_j k_j + \delta\alpha_j l_j)\,, \tag{15.104}$$

where again we work to first order in the small quantities $\delta\zeta$, $\delta\alpha$. Now, from Eq.(15.85), the matrix corresponding to these values of ζ and α is

$$\begin{aligned} \Lambda^j{}_k &\simeq \delta_{jk} - \epsilon_{jkl}\delta\alpha_l\,, \\ \Lambda^j{}_0 &\simeq -\,\delta\zeta_j\,, \\ \Lambda^0{}_j &\simeq -\,\delta\zeta_j\,, \\ \Lambda^0{}_0 &\simeq 1\,. \end{aligned} \tag{15.105}$$

If we compare Eq.(15.105) with Eq.(15.102), we find that the $\delta\zeta$'s and $\delta\alpha$'s can be related to the $\delta\lambda$'s as follows:

$$\delta\alpha_l = -\frac{1}{2}\epsilon_{lmn}\delta\lambda^{mn}\,, \quad \delta\zeta_j = \delta\lambda^{j0} = -\delta\lambda^{0j}\,. \tag{15.106}$$

Let us express the infinitesimal tangent vector $\delta\boldsymbol{\zeta}\cdot\boldsymbol{k} + \delta\boldsymbol{\alpha}\cdot\boldsymbol{l}$ in terms of the $\delta\lambda$'s; it is natural to write:

$$\delta\zeta_j k_j + \delta\alpha_j l_j = -\frac{1}{2}\delta\lambda^{mn}\epsilon_{jmn} l_j + \delta\lambda^{j0} k_j = -\frac{1}{2}\delta\lambda^{\mu\nu} l_{\mu\nu}\,, \tag{15.107}$$

where we have made use of the antisymmetry of $\delta\lambda^{\mu\nu}$ in μ, ν and have defined

$$l_{mn} = \epsilon_{mnj} l_j\,, \quad l_{m0} = -k_m\,, \quad l_{\mu\nu} = -l_{\nu\mu}\,. \tag{15.108}$$

In this way the six basic tangent vectors l_j and k_j are united into the six independent components of the antisymmetric set $l_{\mu\nu}$. It is easy to transcribe Eq.(15.101);it takes the form

$$[l_{\mu\nu}, l_{\sigma\rho}] = g_{\mu\sigma} l_{\nu\rho} - g_{\nu\sigma} l_{\mu\rho} + g_{\mu\rho} l_{\sigma\nu} - g_{\nu\rho} l_{\sigma\mu}\,. \tag{15.109}$$

In this notation, then, the homogeneous Lorentz transformation $\Lambda = \exp(-\frac{1}{2}\delta\lambda^{\mu\nu} l_{\mu\nu})$ corresponds to the matrix $\|\Lambda^\mu{}_\nu\|$ given in Eq.(15.102). The behavior of the elements $l_{\mu\nu}$ under the action of the adjoint representation can be easily worked out. If Λ is some (finite) element of the HLG, and $\exp(-\frac{1}{2}\delta\lambda^{\mu\nu}l_{\mu\nu})$ is an infinitesimal one, then the element $\Lambda\exp(-\frac{1}{2}\delta\lambda^{\mu\nu}l_{\mu\nu})\Lambda^{-1}$ is also an infinitesimal one. We can get the corresponding infinitesimal tangent vector $-\frac{1}{2}\delta\lambda'^{\mu\nu}l_{\mu\nu}$ by using Eq.(15.88) successively to get the effect of $\Lambda\exp(-\frac{1}{2}\delta\lambda^{\mu\nu}l_{\mu\nu})\Lambda^{-1}$ on a frame S:

$$\begin{aligned} S' =&\Lambda\exp(-\frac{1}{2}\delta\lambda^{\mu\nu}l_{\mu\nu})\Lambda^{-1}S: \qquad x^\mu \to x'^\mu\,; \\ x'^\mu =&\Lambda^\mu{}_\sigma(\delta^\sigma_\rho + \delta\lambda^\sigma{}_\rho)(\Lambda^{-1})^\rho{}_\nu\cdot x^\nu = x^\mu + \Lambda^\mu{}_\sigma\Lambda_\nu{}^\rho\delta\lambda^\sigma{}_\rho x^\nu \\ =&x^\mu + \delta\lambda'^\mu{}_\nu x^\nu\,, \\ \delta\lambda'^{\mu\nu} =&\Lambda^\mu{}_\sigma\Lambda^\nu{}_\rho\delta\lambda^{\sigma\rho}\,. \end{aligned} \tag{15.110}$$

This means that in the adjoint representation, the element Λ takes the tangent vector $\delta\lambda^{\mu\nu}l_{\mu\nu}$ into the new combination $\delta\lambda'^{\mu\nu}l_{\mu\nu}$, or because the $\delta\lambda$'s are arbitrary subject to the antisymmetry requirement, that the operator R_Λ giving the adjoint action of Λ on the Lie algebra is

$$R_\Lambda l_{\mu\nu} = \Lambda^\rho{}_\mu\Lambda^\sigma{}_\nu l_{\rho\sigma}\,. \tag{15.111}$$

Of course we could have obtained this adjoint action directly from the basic brackets (Eq.(15.101)). The action of a rotation is already familiar:

$$e^{\Omega_{\boldsymbol{\alpha}\cdot\boldsymbol{l}}}l_j = A_{kj}(\boldsymbol{\alpha})l_k \qquad e^{\Omega_{\boldsymbol{\alpha}\cdot\boldsymbol{l}}}k_j = A_{mj}(\boldsymbol{\alpha})k_m\,. \tag{15.112}$$

Repeated use of Eq.(15.101) gives for the effect of $e^{\Omega_{\boldsymbol{\zeta}\cdot\boldsymbol{k}}}$ the result:

$$\begin{aligned} e^{\Omega_{\boldsymbol{\zeta}\cdot\boldsymbol{k}}}l_j &= \left(\delta_{jm}\cosh\zeta - \zeta_j\zeta_m\frac{\cosh\zeta - 1}{\zeta^2}\right)l_m - \epsilon_{jmn}\zeta_m\frac{\sinh\zeta}{\zeta}k_n\,, \\ e^{\Omega_{\boldsymbol{\zeta}\cdot\boldsymbol{k}}}k_j &= \left(\delta_{jm}\cosh\zeta - \zeta_j\zeta_m\frac{\cosh\zeta - 1}{\zeta^2}\right)k_m + \epsilon_{jmn}\zeta_m\frac{\sinh\zeta}{\zeta}l_n\,. \end{aligned} \tag{15.113}$$

We leave it to the reader to satisfy himself, by making use of Eqs.(15.85) and (15.108), that the form of Eq.(15.111) for the adjoint representation is the same as the form of Eqs.(15.112) and (15.113).

The generalization of the expression $\Lambda = \exp(-\frac{1}{2}\delta\lambda^{\mu\nu}l_{\mu\nu})$ for an element Λ close to the identity to a general Λ gives us a system of canonical coordinates for the HLG. That is, given any element Λ in the HLG, its canonical coordinates $\lambda^{\mu\nu} = -\lambda^{\nu\mu}$ are the parameters in terms of which we could write $\Lambda = \exp(-\frac{1}{2}\lambda^{\mu\nu}l_{\mu\nu})$. In principle, these $\lambda^{\mu\nu}$ could be related to the ζ's and α's we have been using, but we have no particular need for this connection and thus shall not obtain it.It is, however, worth saying something about the different kinds of one-parameter subgroups that the HLG contains. In the case of $R(3)$, we saw that, basically, all one-parameter subgroups are similar, namely any one of them can be transformed into any other by conjugation with a suitable element of $R(3)$. The situation in the case of the HLG is quite different. We have already seen two particular examples of very different one-parameter subgroups

— a subgroup of rotations around some axis and a subgroup of pure Lorentz transformations in a given direction. These are the two extreme members of a one-parameter family of distinct types of one-parameter subgroups. We can analyze the situation in this way. We basically need to know when two tangent vectors can be transformed into one another under the action of the adjoint representation. Suppose the tangent vector $a_j l_j + b_j k_j$, a_j and b_j being real coefficients, is subjected to one of the operators R_Λ; it will then be taken into a new combination involving coefficients a'_j, b'_j. For the two special cases — of spatial rotations and pure velocity transformations — we find from Eqs.(15.112) and (15.113):

$$e^{\Omega_{\boldsymbol{\alpha}\cdot\boldsymbol{l}}} : a_j \to a'_j = A_{jk}(\boldsymbol{\alpha}) a_k \,, \qquad b_j \to b'_j = A_{jk}(\boldsymbol{\alpha}) b_k \,; \tag{15.114a}$$

$$\begin{aligned} e^{\Omega_{\boldsymbol{\zeta}\cdot\boldsymbol{k}}} : a_j \to a'_j &= \left(\delta_{jk}\cosh\zeta - \zeta_j\zeta_k \frac{\cosh\zeta - 1}{\zeta^2}\right) a_k + \epsilon_{jkm}\zeta_m \frac{\sinh\zeta}{\zeta} b_k \,, \\ b_j \to b'_j &= \left(\delta_{jk}\cosh\zeta - \zeta_j\zeta_k \frac{\cosh\zeta - 1}{\zeta^2}\right) b_k - \epsilon_{jkm}\zeta_m \frac{\sinh\zeta}{\zeta} a_k \,. \end{aligned} \tag{15.114b}$$

Now the second set of equations takes an appearance very much like the first set, if we write in terms of the complex combinations $a_j \pm ib_j$; under $e^{\Omega_{\boldsymbol{\zeta}\cdot\boldsymbol{k}}}$,

$$\begin{aligned} a'_j \pm ib'_j &= \left(\delta_{jk}\cosh\zeta - \zeta_j\zeta_k \frac{\cosh\zeta - 1}{\zeta^2} \mp i\epsilon_{jkm}\zeta_m \frac{\sinh\zeta}{\zeta}\right)(a_k \pm ib_k) \\ &= A_{jk}(\pm i\boldsymbol{\zeta})(a_k \pm ib_k) \,. \end{aligned} \tag{15.115}$$

We used here the explicit form for the rotation matrix $A_{jk}(\boldsymbol{\alpha})$ in Eq.(15.11). Combining Eqs. (15.114a) and (15.115) for a general element $\Lambda = \exp(\boldsymbol{\zeta}\cdot\boldsymbol{k})\exp(\boldsymbol{\alpha}\cdot\boldsymbol{l})$ of the homogeneous Lorentz group, the operator R_Λ takes the pair a_j, b_j into a'_j, b'_j according to:

$$R_{\exp(\boldsymbol{\zeta}\cdot\boldsymbol{k})\exp(\boldsymbol{\alpha}\cdot\boldsymbol{l})} : \begin{aligned} a'_j + ib'_j &= A_{jl}(i\boldsymbol{\zeta})A_{lk}(\boldsymbol{\alpha})(a_k + ib_k) \,, \\ a'_j - ib'_j &= A_{jl}(-i\boldsymbol{\zeta})A_{lk}(\boldsymbol{\alpha})(a_k - ib_k) \,. \end{aligned} \tag{15.116}$$

How can one characterize these matrices $A(i\boldsymbol{\zeta})A(\boldsymbol{\alpha})$? The matrix $A(\boldsymbol{\alpha})$ is a general real orthogonal matrix of determinant unity; these properties of being orthogonal and unimodular will clearly be retained if one analytically continues the components α_j to pure imaginary values $i\zeta_j$, though of course the reality property will be lost. Thus we have that $A(i\boldsymbol{\zeta})$ is a complex orthogonal matrix of unit determinant; the same is true for $A(i\boldsymbol{\zeta})A(\boldsymbol{\alpha})$. Thus the *combinations* $a_j \pm ib_j$ *undergo complex unimodular orthogonal transformations.* The transformation law for $a_j - ib_j$ is the complex conjugate of the one for $a_j + ib_j$. But we can go further: by giving all possible values to $\boldsymbol{\zeta}$ and $\boldsymbol{\alpha}$, the matrix $A(i\boldsymbol{\zeta})A(\boldsymbol{\alpha})$ yields all possible complex orthogonal unimodular matrices in three dimensions! In fact, if O is any such matrix, we can first subject it to a polar decomposition, namely write it as the product HA of a positive hermitian matrix, H, and a unitary one, A. But the orthogonal nature of $O, O^T O = O O^T = \mathbb{1}$, implies that H is an *orthogonal* hermitian matrix and A an *orthogonal* unitary one. In other words, A is a real orthogonal matrix. Of course, both H and A are unimodular. This A is just the $A(\boldsymbol{\alpha})$ in Eq.(15.116), and H is $A(i\boldsymbol{\zeta})$: the hermiticity of $A(i\boldsymbol{\zeta})$ is obvious from Eq.(15.115).

Thus we have established that under the action of the operator R_Λ of the adjoint representation, a general tangent vector $a_j l_j + b_j k_j$ goes into the vector $a'_j l_j + b'_j k_j$ such that the complex numerical three-vector $a'_j + ib'_j$ is obtained from $a_j + ib_j$ via a 3×3 complex orthogonal unimodular matrix determined by Λ; by considering all possible homogeneous Lorentz transformations Λ we obtain all possible 3×3 complex orthogonal unimodular matrices. Now, given the tangent vector having coefficients a_j, b_j, the question arises whether we can find a complex orthogonal unimodular transformation that puts $a'_j + ib'_j$ into a particularly simple form, for example, such that two of its three components vanish. In any case, the orthogonality of the transformation tells us that $(a'_j + ib'_j)\ (a'_j + ib'_j)$ must be equal to $(a_j + ib_j)(a_j + ib_j)$. A simple analysis shows that we have just two cases to consider. If $(a_j + ib_j)(a_j + ib_j)$ is *nonzero*, it is possible to arrange things so that $a'_2 + ib'_2 = a'_3 + ib'_3 = 0$. Of course, $a'_1 + ib'_1$ will then obey

$$(a'_1 + ib'_1)^2 = (a_j + ib_j)(a_j + ib_j)\,. \tag{15.117}$$

In other words, in such a case there exists an element $(\boldsymbol{\zeta}, \boldsymbol{\alpha})$ in the HLG such that we have:

$$\begin{aligned} &e^{\Omega_{\boldsymbol{\zeta}\cdot\boldsymbol{k}}} e^{\Omega_{\boldsymbol{\alpha}\cdot\boldsymbol{l}}} (a_j l_j + b_j k_j) = a'_1 l_1 + b'_1 k_1\,, \\ &\qquad (a'_1)^2 - (b'_1)^2 = a_j a_j - b_j b_j\,, \quad a'_1 b'_1 = a_j b_j\,. \end{aligned} \tag{15.118}$$

On the other hand, if $a_j + ib_j$ is such that $(a_j + ib_j)(a_j + ib_j)$ *vanishes*, we clearly cannot arrange to have *two* components of $a'_j + ib'_j$ vanish; we can however make *one* component, $a'_3 + ib'_3$, say, vanish, and further we can put the complex vector $a'_j + ib'_j$ into the simple form $(1, i, 0)$. In this case, then, there exists an element $(\boldsymbol{\zeta}, \boldsymbol{\alpha})$ in the HLG such that we have:

$$e^{\Omega_{\boldsymbol{\zeta}\cdot\boldsymbol{k}}} e^{\Omega_{\boldsymbol{\alpha}\cdot\boldsymbol{l}}} (a_j l_j + b_j k_j) = l_1 + k_2\,. \tag{15.119}$$

Thus every one-parameter subgroup $\sigma \to \Lambda(\sigma)$ in the HLG is carried by conjugation with a suitable element, either into a subgroup of the form $\sigma \to \exp(\sigma(a_1 l_1 + b_1 k_1))$ for some values of a_1 and b_1 or into the subgroup $\sigma \to \exp(\sigma(l_1 + k_2))$; these two possibilities are mutually exclusive. The first possibility gives us then a continuous family of essentially different kinds of one-parameter subgroups, different kinds corresponding to different values for the ratio a_1/b_1. At one extreme we have subgroups of purely spatial rotations around some axis, $a_1/b_1 = \pm\infty$; at the other extreme we have subgroups of pure velocity transformations in a fixed direction, $a_1/b_1 = 0$. And then we have the one-parameter subgroup $\sigma \to \exp(\sigma(l_1 + k_2))$ and all others equivalent to it under conjugation.

(b) The Inhomogeneous Lorentz Group or Poincaré Group

The Poincaré group $\mathcal{P}$ arises by combining the HLG with the group of all space-time translations in semidirect product fashion. As in the case of the Galilei group, the space-time translations constitute an Abelian invariant subgroup T_4 in $\mathcal{P}$. Elements of $\mathcal{P}$ correspond to all possible transformations connecting two inertial frames S and S' in space-time whose space-time origins as well as orientations may differ. The most general element of $\mathcal{P}$ can be built up in this way: starting with the inertial frame S, we go to a frame S_1 related to S by a homogeneous Lorentz transformation: $S_1 = \exp(\boldsymbol{\zeta}\cdot\boldsymbol{k})\exp(\boldsymbol{\alpha}\cdot\boldsymbol{l})S = \Lambda S$. Next we go from S_1 to S' by applying a spatial translation of amount $\boldsymbol{a}$ and a translation in time of amount b : $S' = \exp(bh)\exp(\boldsymbol{a}\cdot\boldsymbol{d})S_1$. The interpretation of the operations $\exp(bh)$ and $\exp(\boldsymbol{a}\cdot\boldsymbol{d})$ on S_1 is the same as in the case of the Galilei group: the vector $\overline{O_1O'}$ from the spatial origin in S_1 to that in S' has components $-a_j$ measured in S' or S_1, and the time in S' lags behind that in S_1 by the amount b. Combining both operations, the space-time coordinates of a given event as seen in S' and S are related as follows:

$$\begin{aligned} S' &= \exp(bh)\exp(\boldsymbol{a}\cdot\boldsymbol{d})\Lambda S: \qquad x'^j = \Lambda^j{}_\mu x^\mu + a_j\,, \\ x'^0 &= \Lambda^0{}_\mu x^\mu - b\,. \end{aligned} \tag{15.120}$$

We can write the two parts of this transformation more uniformly if we introduce the translation four vector a^μ by $a^j = a_j$, $a^0 = -a_0 = -b$; at the same time the four basic tangent vectors h and d_j to the subgroups of translations in time and space can be written jointly as d_μ, with $-d_0 = d^0 = h$. Then we have

$$S' = \exp(a^\mu d_\mu)\Lambda S: \qquad x'^\mu = \Lambda^\mu{}_\nu x^\nu + a^\mu\,. \tag{15.121}$$

Thus a single element of $\mathcal{P}$ can be denoted by (Λ, a^μ); Λ is some element of the HLG, and a^μ is some space-time translation. The multiplication law is derived from Eq.(15.121):

$$\begin{aligned} S'' &= (\Lambda', a'^\mu)S' = (\Lambda', a'^\mu)(\Lambda, a^\mu)S: \\ x''^\mu &= \Lambda'^\mu{}_\nu x'^\nu + a'^\mu = \Lambda'^\mu{}_\nu(\Lambda^\nu{}_\sigma x^\sigma + a^\nu) + a'^\mu\,, \\ (\Lambda', a'^\mu)(\Lambda, a^\mu) &= (\Lambda'\Lambda, a'^\mu + \Lambda'^\mu{}_\nu a^\nu)\,. \end{aligned} \tag{15.122}$$

We now analyze the structure of $\mathcal{P}$ with the help of this equation. The unit element is $(\mathbb{1}, 0)$, and inverses are given by:

$$(\Lambda, a)^{-1} = (\Lambda^{-1}, -(\Lambda^{-1})^\mu{}_\nu a^\nu)\,. \tag{15.123}$$

The elements $(\Lambda, 0)$ constitute the subgroup of homogeneous Lorentz transformations, and the elements $(\mathbb{1}, a^\mu)$ the Abelian invariant subgroup T_4 of translations in space time. By construction, we have

$$(\Lambda, a^\mu) = (\mathbb{1}, a^\mu)(\Lambda, 0)\,, \tag{15.124}$$

and Eq.(15.122) shows that $\mathcal{P}$ is the semidirect product of the HLG by T_4. Because it takes six parameters to specify Λ and four to specify a^μ, *the Poincaré group, like the Galilei group, is a ten-parameter group.*

A basis for the Lie algebra of $\mathcal{P}$ is given by the l_j, k_j, d_j, and h. The basic Lie brackets among the l_j and k_j are given in Eq.(15.101) or in covariant notation in Eq.(15.109), whereas the brackets among the d_j and h are zero because T_4 is Abelian. The remaining brackets are obtained as follows: from Eq.(15.122) we easily find

$$(\Lambda, 0)(\mathbb{1}, a^\mu)(\Lambda^{-1}, 0) = (\mathbb{1}, \Lambda^\mu{}_\nu a^\nu)\,, \tag{15.125}$$

or in the exponential form,

$$\exp(\boldsymbol{\zeta}\cdot\boldsymbol{k})\exp(\boldsymbol{\alpha}\cdot\boldsymbol{l})\exp(a^\mu d_\mu)\exp(-\boldsymbol{\alpha}\cdot\boldsymbol{l})\exp(-\boldsymbol{\zeta}\cdot\boldsymbol{k}) = \exp(\Lambda^\mu{}_\nu a^\nu d_\mu)\,. \tag{15.126}$$

Thus in the adjoint representation of $\mathcal{P}$, we have, because a^μ is arbitrary,

$$e^{\Omega_{\boldsymbol{\zeta}\cdot\boldsymbol{k}}} e^{\Omega_{\boldsymbol{\alpha}\cdot\boldsymbol{l}}} d_\mu = \Lambda^\nu{}_\mu(\boldsymbol{\zeta}, \boldsymbol{\alpha}) d_\nu\,. \tag{15.127}$$

Setting $\boldsymbol{\zeta} = 0$, we get

$$e^{\Omega_{\boldsymbol{\alpha}\cdot\boldsymbol{l}}} d_0 = d_0\,, \qquad e^{\Omega_{\boldsymbol{\alpha}\cdot\boldsymbol{l}}} d_k = A_{jk}(\boldsymbol{\alpha}) d_j\,, \tag{15.128}$$

and by arguments familiar by now, this implies the bracket relations

$$[l_j, h] = 0\,, \qquad [l_j, d_k] = \epsilon_{jkm} d_m\,. \tag{15.129}$$

These are, therefore, the same as in the case of the Galilei group. Next, we set $\boldsymbol{\alpha} = 0$ in Eq.(15.127) and get

$$\begin{aligned} e^{\Omega_{\boldsymbol{\zeta}\cdot\boldsymbol{k}}} d_0 &= \cosh\zeta\, d_0 - \zeta_j d_j \frac{\sinh\zeta}{\zeta}\,, \\ e^{\Omega_{\boldsymbol{\zeta}\cdot\boldsymbol{k}}} d_j &= d_j + \zeta_j\zeta_k \frac{\cosh\zeta - 1}{\zeta^2} d_k - \zeta_j \frac{\sinh\zeta}{\zeta} d_0\,, \end{aligned} \tag{15.130}$$

leading to:

$$[k_j, h] = d_j\,, \qquad [k_j, d_m] = \delta_{jm} h\,. \tag{15.131}$$

The full set of basic bracket relations for $\mathcal{P}$ can now be exhibited in either the physically transparent form separating the elements into l's, k's, d's, and h or in more condensed and covariant form using four-dimensional notation. The former is

$$HLG:\ [l_j, l_m] = -[k_j, k_m] = \epsilon_{jmn} l_n\,, \quad [l_j, k_m] = \epsilon_{jmn} k_n\,; \tag{15.132a}$$

$$T_4:\ [d_j, d_m] = [d_j, h] = 0\,; \tag{15.132b}$$

$$\begin{aligned} [l_j, d_m] &= \epsilon_{jmn} d_n \quad [l_j, h] = 0\,; \\ [k_j, d_m] &= \delta_{jm} h, \quad [k_j, h] = d_j\,. \end{aligned} \tag{15.132c}$$

Equation (15.132c) exhibits how each of the l_j and k_j acts as an outer derivation on the invariant Abelian subalgebra of the d_j and h. The covariant form is

$$HLG: \ [l_{\mu\nu}, l_{\rho\sigma}] = g_{\mu\rho} l_{\nu\sigma} - g_{\nu\rho} l_{\mu\sigma} + g_{\mu\sigma} l_{\rho\nu} - g_{\nu\sigma} l_{\rho\mu}\,; \tag{15.133a}$$

$$T_4: \ [d_\mu, d_\nu] = 0\,; \tag{15.133b}$$

$$[l_{\mu\nu}, d_\rho] = g_{\mu\rho} d_\nu - g_{\nu\rho} d_\mu\,. \tag{15.133c}$$

In the (mathematical) limit in which one allows the velocity of light, c, to become indefinitely large, the Lie algebra of the Poincaré group goes over into that of the Galilei group. In this process, the k_j go over into the g_j, and the pure velocity transformations also become a subgroup. Thus the subalgebra corresponding to the HLG gives place to the subalgebra corresponding to the subgroup $E^{(v)}(3)$ of the Galilei group. It now becomes clear why the decomposition of the Galilei group as $E^{(v)}(3) \times T_4$ is the more physical description of its structure.

Incorporation of Discrete Operations: Space and time Reflections

All the Lie groups that we have examined in this chapter have consisted of just their identity components, that is, the components whose elements are continuously connected to the identity. According to the work of the previous chapter, it is enough to have realizations of Lie algebras in each case in order to build up realizations of these components. There are, however, additional discrete operations on space-time that have a simple geometrical meaning: these are the operations of space reflection and time reversal. If either or both of these operations are added on to the earlier groups, we get new groups consisting of two or more disjoint components, and each such component can be characterized by one representative element in it. Each of these elements must continue to be treated as a finite element of the group when we pass to a realization, and the problem of realizing them cannot be reduced entirely to finding a realization of the Lie algebra. However, the structure of the full group leads to some conditions: each of these discrete elements will act in a specific way as an automorphism on the Lie algebra. The form of this automorphism is easily found. We consider a general one-parameter subgroup of the Lie group, and subject it to conjugation by one of these discrete elements. This leads to a new one-parameter subgroup, and the passage from the original to the new tangent vector fixes the form of the automorphism. (Note that although the discrete element does not belong to the identity component of the group, both the original and the conjugated one-parameter sub-groups do.) In a canonical realization of the full group, we will have to find directly finite canonical transformations to represent the discrete elements, such that they have an acceptable physical meaning; at the same time, acting on the functions furnishing the PB realization of the Lie algebra, these canonical transformations must reduce to the corresponding automorphisms. It turns out that only the space inversion $\mathbb{P}$ can be treated in this way: in each case $\mathbb{P}$ *does* act as an automorphism of the Lie algebra, and in a canonical realization it *is* represented by a canonical transformation. On the other hand, the requirement of a sensible physical interpretation leads to $\mathbb{T}$, time reversal, acting as an *antiautomorphism* of the

Lie algebra, and in a realization in classical mechanics being represented as an *anti-canonical* transformation! We will discuss these notions now.

Let us discuss the space inversion $\mathbb{P}$ first. As in all the cases discussed up to now, it is possible to give a passive interpretation of this operation, or an active one. We choose the former. Let S_R be some Cartesian right-handed coordinate system with origin O in three-dimensional space. The operation $\mathbb{P}$ takes us from S_R to a left-handed system S_L whose origin O' coincides with O, but whose axes are all reversed in direction relative to S_R. That is, the positive first axis in S_L is the negative first axis in S_R, and so on for the other two axes. If P is some point in space, with coordinates x_j in S_R, its coordinates x'_j in S_L are the negatives of the x_j:

$$S_L = \mathbb{P} S_R : \qquad x'_j = -x_j \, . \tag{15.134}$$

Combining $\mathbb{P}$ with the proper rotation group $R(3)$, we get the *full orthogonal group* $O(3)$. All Cartesian frames having a common origin O can be divided into two classes: one contains all right-handed frames $S_R, S'_R, \cdots$, and the other all left-handed ones, $S_L, S'_L, \cdots$. The group $O(3)$ connects all these to one another. The identity component of $O(3)$ is just $R(3)$, consisting of elements $\exp(\boldsymbol{\alpha} \cdot \boldsymbol{l})$; these connect the frames $S_R, S'_R, \cdots$, to one another and $S_L, S'_L, \cdots$, to one another. The group $O(3)$ has just one other component, whose elements are of the form $\mathbb{P}\exp(\boldsymbol{\alpha} \cdot \boldsymbol{l})$: these connect an S_R to an S_L and vice versa. Clear $\mathbb{P}\mathbb{P}$ is the identity. In the defining representation, each element in $R(3)$ corresponds to a real orthogonal 3×3 matrix A obeying $\det A = +1$. From Eq.(15.134) we see that in the defining representation of $O(3)$, each element corresponds to a real orthogonal 3×3 matrix A, whose determinant is allowed to take both possible values ± 1. Since to $\mathbb{P}$ corresponds the negative of the unit matrix, $\mathbb{P}$ *commutes with all elements in* $O(3)$.Therefore the multiplication rules for $O(3)$ are trivial:

$$\begin{aligned}
\exp(\boldsymbol{\alpha} \cdot \boldsymbol{l}) \exp(\boldsymbol{\beta} \cdot \boldsymbol{l}) &= \exp(\boldsymbol{\gamma} \cdot \boldsymbol{l}) \cdots R(3) \, ; \\
\mathbb{P}\exp(\boldsymbol{\alpha} \cdot \boldsymbol{l}) \exp(\boldsymbol{\beta} \cdot \boldsymbol{l}) &= \exp(\boldsymbol{\alpha} \cdot \boldsymbol{l}) \mathbb{P}\exp(\boldsymbol{\beta} \cdot \boldsymbol{l}) = \mathbb{P}\exp(\boldsymbol{\gamma} \cdot \boldsymbol{l}) \, ; \\
\mathbb{P}\exp(\boldsymbol{\alpha} \cdot \boldsymbol{l}) \mathbb{P}\exp(\boldsymbol{\beta} \cdot \boldsymbol{l}) &= \exp(\boldsymbol{\gamma} \cdot \boldsymbol{l})
\end{aligned} \tag{15.135}$$

Conjugation by $\mathbb{P}$ leaves any one-parameter subgroup in $O(3)$, which necessarily lies in $R(3)$ and thus has the form $\exp(\sigma \boldsymbol{\alpha} \cdot \boldsymbol{l})$, invariant: $\mathbb{P}\exp(\sigma \boldsymbol{\alpha} \cdot \boldsymbol{l}) \mathbb{P}^{-1} = \exp(\sigma \boldsymbol{\alpha} \cdot \boldsymbol{l})$. Consequently, as an automorphism on the Lie algebra of $R(3)$, $\mathbb{P}$ acts trivially as the identity automorphism:

$$\mathbb{P} : \qquad l_j \to l_j \, . \tag{15.136}$$

Combining $\mathbb{P}$ with $E(3)$ leads to more interesting consequences. Although it will still commute with the subgroup $R(3)$, it does not commute with translations. We get the full Euclidean group $\mathcal{E}(3)$ whose elements are of two types: $\exp(\boldsymbol{a} \cdot \boldsymbol{d}) \exp(\boldsymbol{\alpha} \cdot \boldsymbol{l})$ and $\mathbb{P}\exp(\boldsymbol{a} \cdot \boldsymbol{d}) \exp(\boldsymbol{\alpha} \cdot \boldsymbol{l})$. From our study of $E(3)$ and using Eq.(15.134), we see quite easily that the rules for multiplying elements in $\mathcal{E}(3)$

follow from those for $E(3)$ together with:

$$\begin{aligned}\mathbb{P}\exp(\boldsymbol{\alpha}\cdot\boldsymbol{l})\mathbb{P}^{-1} &= \exp(\boldsymbol{\alpha}\cdot\boldsymbol{l})\,,\\ \mathbb{P}\exp(\boldsymbol{a}\cdot\boldsymbol{d})\mathbb{P}^{-1} &= \exp(-\boldsymbol{a}\cdot\boldsymbol{d})\,,\\ \mathbb{P}^2 &= 1\,.\end{aligned} \tag{15.137}$$

The action of $\mathbb{P}$ as an (outer) automorphism on the Lie algebra of $E(3)$ follows:

$$\mathbb{P}: \qquad l_j \to l_j\,, \qquad d_j \to -d_j\,. \tag{15.138}$$

By now, the method of handling the $\mathbb{P}$ operation is clear. In each case it "doubles" the size of the group it is being combined with, leading to a new group with two components. The new element of structure in each case is the way $\mathbb{P}$ acts as an automorphism, via conjugation, on the identity component, and thus as an automorphism on the Lie algebra. It is sufficient to state these latter automorphisms for the Lie algebras of the Galilei and inhomogeneous Lorentz groups:

Galilei Group:

$$\mathbb{P}: \qquad l_j \to l_j, \quad g_j \to -g_j\,; \quad h \to h\,, \quad d_j \to -d_j\,. \tag{15.139}$$

Poincaré Group:

$$\mathbb{P}: \qquad l_j \to l_j, \quad k_j \to -k_j\,; \quad h \to h\,, \quad d_j \to -d_j\,. \tag{15.140}$$

This action of $\mathbb{P}$ has an acceptable physical meaning; as we saw in simple examples in Chapter 11, in a realization the l_j are identified with the angular momenta and the d_j with the linear momenta of a system, and it is reasonable that the former should not, and the latter should, change sign under space inversion. The meaning we shall find for g_j and k_j in later chapters shows that their behaviors are in accordance with their physical interpretation.

Let us finally clarify the question of the frame-dependence of the space-reflection operation. This comes in when $\mathbb{P}$ is combined with the Euclidean, Galilei or Poincaré groups. In any one of these cases, it is physically evident that there is one operation of space-reflection associated with each distinct inertial frame, and the operations in different frames are related by conjugation. Thus, if two frames are related by $S' = gS$, g being some element of the Euclidean, Galilei or Poincaré groups, and $\mathbb{P}$ denotes space-reflection in S, then in S' this operation is represented by the element $g\mathbb{P}g^{-1}$ of the extended group. In this sense, space-reflection is not Euclidean, Galilean or Poincaré invariant. This also follows from Eqs.(15.139), (15.140): $\mathbb{P}$ acts non-trivially as an automorphism on the Lie algebra. However, for a Poincaré invariant theory for example, if we have $\mathbb{P}$ invariance in one frame we have it in all frames; and if we can define $\mathbb{P}$ consistently in one frame, so can we define it in all others.

The behavior of the time reversal operation $\mathbb{T}$ is somewhat different and more subtle. As a transformation on the space-time coordinates x_j, t, we define it as

$$\mathbb{T}: \qquad x_j \to x_j, \quad t \to -t\,. \tag{15.141}$$

However, because all physical observers have a common sense of increasing time, it is more reasonable to give $\mathbb{T}$ an *active interpretation alone*, and not to regard it as corresponding to any change of frame. We are interested in combining $\mathbb{T}$ with the Galilei and Poincaré groups; for the identity components of these groups we can choose either the passive or the active interpretation, and there is no inconsistency in choosing the former, even though $\mathbb{T}$ has only an active interpretation. Combining $\mathbb{T}$ with $\mathcal{P}$, for example, we get a new group made up of two components. Elements of the identity component are just the transformations (Λ, a^μ), which are continuously connected to the identity; such an element leads from one *physical* frame S to another *physical* S', $S' = (\Lambda, a^\mu)S$, and the coordinates x'^μ, x^μ assigned to a single space-time point by S', S are related by Eq.(15.121). Elements of the second component can be written $\mathbb{T}(\Lambda, a^\mu)$. Such an element acts on a physical frame S to first yield another physical frame $S' = (\Lambda, a^\mu)S$; then $\mathbb{T}$ *acts on space-time* to take any world point with coordinates (x'^0, x'^j) *in* S' to a new world point with coordinates $(-x'^0, x'^j)$ in S'. That is, $\mathbb{T}$ *acts in* S'. Thus under $\mathbb{T}(\Lambda, a^\mu)$, a frame S goes to a frame S', and each world point P goes into a new world point P''; the coordinates of P'' in S', written x''^μ, are related to the coordinates x^μ of P in S by:

$$\begin{aligned} \mathbb{T}(\Lambda, a^\mu): &\qquad S \to S' \quad P \to P'', \\ x''^0 = -(\Lambda^0{}_\nu x^\nu + a^0), &\quad x''^j = \Lambda^j{}_\nu x^\nu + a^j . \end{aligned} \tag{15.142}$$

To repeat, the active transformation $\mathbb{T}$ defined in Eq.(15.141) is applied in the frame S'. This is enough to determine the structure of the new group. One finds quite easily that the only new pieces of information are contained in the following statements:

$$\begin{aligned} &\mathbb{T}\,\exp(\boldsymbol{\alpha}\cdot\boldsymbol{l})\mathbb{T}^{-1} = \exp(\boldsymbol{\alpha}\cdot\boldsymbol{l}) \,,\ \mathbb{T}\,\exp(\boldsymbol{\zeta}\cdot\boldsymbol{k})\mathbb{T}^{-1} = \exp(-\boldsymbol{\zeta}\cdot\boldsymbol{k}) \,, \\ &\mathbb{T}\,\exp(\boldsymbol{a}\cdot\boldsymbol{d})\mathbb{T}^{-1} = \exp(\boldsymbol{a}\cdot\boldsymbol{d}) \,,\ \mathbb{T}\,\exp(bh)\mathbb{T}^{-1} = \exp(-bh) \,; \end{aligned} \tag{15.143}$$

these are analogues to Eq.(15.137). For the case of the Galilei group, these same equations result, with k_j replaced by g_j. Thus this information, together with $\mathbb{T}^2 = 1$, suffices to determine the *mathematical* structure of the extended groups. But we run into trouble with the physical interpretation if at this point we proceed as in the case of space inversion and just demand that there be a canonical transformation representing $\mathbb{T}$ and that it act as an automorphism on the Lie algebra of $\mathcal{P}$ in a manner determined by Eq.(15.143)! For, if we did so, we would require this automorphism to be:

$\mathbb{T}$ *as an Automorphism:*

$$l_j \to l_j \,, \quad k_j \to -k_j \,; \quad d_j \to d_j \,, \quad h \to -h \,. \tag{15.144}$$

This is certainly an automorphism, and all the Lie brackets will be preserved. But as we have said earlier, the quantities that correspond to l_j, d_j, and h in a realization are the angular momenta, linear momenta, and energy, respectively, and the "natural" behavior for these under time inversion is just the opposite of Eq.(15.144): l_j and d_j should change sign, and h should not! This is because

$\mathbb{T}$ corresponds, physically, to "reversal of the motion," in which (generalized) positions do not change sign, whereas (generalized) velocities do. We shall see in examples that the k_j and g_j are, in a realization, closely related to *position*. Thus for them the natural law is not to change sign under time inversion, and this again is the opposite of Eq.(15.144). We are thus forced by the physical interpretation to allow $\mathbb{T}$ to act as an operation that does not preserve Lie brackets but reverses their signs. We call such an operation an *antiautomorphism*. For a general Lie algebra L, a linear antiautomorphism K is a mapping of L onto itself which obeys:

$$K(\lambda u + \mu v) = \lambda K u + \mu K v\,, \qquad [Ku, Kv] = -K[u, v]\,. \tag{15.145}$$

A very simple, and trivial, example of such a K is this: $Ku = -u$. More generally, in a basis e_j for L, with corresponding structure constants c^l_{jk}, an antiautomorphism K can be characterized by a matrix $\|K^l_j\|$ as follows:

$$\begin{aligned} Ke_j = K^l_j e_l\,; \qquad [K^m_j e_m, K^n_k e_n] = c^p_{mn} K^m_j K^n_k e_p = -K c^l_{jk} e_l = -K^p_l c^l_{jk} e_p\,, \\ c^p_{mn} K^m_j K^n_k = -K^p_l c^l_{jk}\,. \end{aligned} \tag{15.146}$$

The square of an antiautomorphism, or the product of two antiautomorphisms, is an ordinary automorphism. The rules for combining an antiautomorphism with the linear transformations occurring in the adjoint representation, (and these as we know are automorphisms because they preserve brackets), are obtained from Eq.(15.145):

$$\begin{aligned} K[u, v] = -[Ku, Kv] \Rightarrow K\Omega_u v = -\Omega_{Ku} K v \Rightarrow K\Omega_u = -\Omega_{Ku} K \\ \Rightarrow K e^{\Omega_u} = e^{-\Omega_{Ku}} K\,, \\ K e^{\Omega_u} K^{-1} = e^{-\Omega_{Ku}} = e^{\Omega_{(-Ku)}}\,. \end{aligned} \tag{15.147}$$

An anticanonical transformation is an antiautomorphism corresponding to the case where the Lie bracket is the PB: it is a mapping of phase space onto itself, $\omega \to \omega' = k(\omega)$, say, such that the induced linear transformation on functions on phase space, $\phi(\omega) \to \phi'(\omega) = (K\phi)(\omega) = \phi(k^{-1}(\omega))$ changes the signs of PB's:

$$\phi \to K\phi\,, \quad \psi \to K\psi : \quad \{K\phi, K\psi\} = -K\{\phi, \psi\}\,. \tag{15.148}$$

A simple example of an anticanonical transformation for a phase space described by a set of q's and p's is this: $q \to q$, $p \to -p$. In sum, the extension of our earlier considerations so as to include the time inversion $\mathbb{T}$ in a *physical* way leads to the demand that whereas the elements in the identity component of the group should be realized by canonical transformations (both in the Galilei and Poincaré cases), the element $\mathbb{T}$ and with it all elements in the component of the group containing $\mathbb{T}$ should be realized by anticanonical transformations; and restricted to the Lie algebra this anticanonical $\mathbb{T}$ should have the effect of the following antiautomorphism:

$\mathbb{T}$ *as an anti-Automorphism:*

$$\begin{aligned} &\textit{Poincaré group:} \quad l_j \to -l_j\,, \quad k_j \to k_j\,; \quad d_j \to -d_j\,, \quad h \to h\,; \\ &\textit{Galilei group:} \quad l_j \to -l_j\,, \quad g_j \to g_j\,; \quad d_j \to -d_j\,, \quad h \to h\,. \end{aligned} \tag{15.149}$$

However, it should be stressed that although we have ensured the proper physical form for the action of $\mathbb{T}$ in this way, we have *not* lost the representation property for the full group! The point is that the relative minus sign between Eq.(15.149) and the true automorphism of Eq.(15.144) is compensated by the minus sign in the exponent in the last line in Eq.(15.147) so that the group relations *are* maintained. For example, in Eq.(15.143) we see that $\mathbb{T}$ commutes with elements of $R(3)$. This will still be the case when $\mathbb{T}$ is in a realization anticanonical because, acting on l_j, it changes the sign! That is, combining Eqs.(15.147) and (15.149) we have

$$\mathbb{T}e^{\Omega_{\alpha\cdot l}}\mathbb{T}^{-1} = e^{\Omega_{(-\mathbb{T}\alpha\cdot l)}} = e^{\Omega_{-\alpha\cdot\mathbb{T}l}} = e^{\Omega_{\alpha\cdot l}}\,. \tag{15.150}$$

One can check in the same way that all the other relationships in Eq.(15.143) will be properly represented.

One might wonder whether the Hamiltonian form of the equations of motion for a system whose Hamiltonian is unchanged under the physical operation of $\mathbb{T}$ will be preserved under time reversal even with $\mathbb{T}$ anticanonical, because we proved many chapters ago that only canonical transformations leave the Hamilton equations invariant in form. But there is no cause for worry: in the earlier proof we did not consider the possibility of changing the sign of time! One side of the general equation of motion contains a PB that will change sign under the action of $\mathbb{T}$, but on the other side appears the derivative with respect to t of some variable, and this too will change sign under the geometric transformation (Eq.(15.141))!

The combined operation $\mathbb{PT}$, called "strong reflection," must be considered if we adjoin both $\mathbb{P}$ and $\mathbb{T}$ to the groups studied earlier. Geometrically, this corresponds to changing the signs of all four space-time coordinates. Each of the groups extended in this way acquires four distinct components: the identity component, the component containing $\mathbb{P}$, the one containing $\mathbb{T}$, and the one containing $\mathbb{PT}$. It is clear that elements in the first two components should be canonically realized, and those in the third and fourth anticanonically. Correspondingly, $\mathbb{PT}$ also will act as an antiautomorphism on the Lie algebra. We leave to the reader the details of working out this action for the various Lie algebras of interest as well as analysing the frame-dependences of the operations $\mathbb{T}$ and $\mathbb{PT}$.

Chapter 16

Relativistic Symmetry in the Hamiltonian Formalism

We want to examine now the formal structure of the Hamiltonian description of a dynamical system possessing symmetry under one of the groups G that we studied in the previous chapter. For simplicity, let us call any one of these groups a "relativity group." We assume, of course, that the system being discussed does permit a Hamiltonian description; in practice this may be arrived at by starting from an action principle stated in terms of a Lagrangian to which we then apply the standard rules for going over to a Hamiltonian. When we say that a certain system "possesses symmetry under a group G," we mean the following: there is a class of observers with associated space-time coordinate systems, related to one another by elements of G; there are appropriate rules for passing from the description of the system given by one observer to that given by another observer; with the use of these rules the equations of motion take the same form for all these observers. For the cases when G is the Galilei group or the Poincaré group, the subgroup T_1 corresponding to time translations or shifts in the zero of time is already present in G. As we see in detail later, this has the consequence that the general rules that connect the descriptions given by two observers linked by a general element of G contain as a special case the equations of motion, or equations for development in time. But suppose we wish to discuss a system possessing rotational symmetry alone, that is, possessing symmetry under $R(3)$, or, say, possessing symmetry under $E(3)$. In such cases, the class of observers involved is such that they all use the same time coordinate, and only their spatial coordinate systems differ, being related by elements of $R(3)$ or $E(3)$. But we are still interested in the equations of motion in time for these cases; thus we agree to extend these groups formally by including the time translations T_1 in direct-product fashion, and agree to write G for the groups $R(3) \otimes T_1$ or $E(3) \otimes T_1$ that we obtain in this way. That is, it is convenient to allow two observers not only to use different coordinates in space, but also different origins in time, and then the position of the equations of motion in the general analysis will be the same for all groups G.

Let us begin by recalling the basic elements of the Hamiltonian description of a dynamical system, as given in Chapter 7. We have a phase space, of dimension $2k$ say, with coordinates written as ω^μ or q_s, p_s; the ranges for these are fixed by their physical significance. Dynamical variables, or observables, are suitable functions $f(\omega), f'(\omega), \cdots$, on phase space. We speak of dynamical variables *at one instant of time*, each such representing some physical property of the system at that instant of time; it is assumed that a "complete" description of the system at each point in time can be given with the help of such variables. Thus there is a specific function of the q's and p's that represents "total linear momentum of the system in the first direction at time zero," another for "energy of the system at time equal to two units," and so on. (These statements are sharpened a little bit later on.) The state of the system at one instant of time, say, at $t = 0$, is given by a non-negative probability distribution function $\rho(\omega)$ on phase space, normalized to unity:

$$\int d\omega\, \rho(\omega) = 1\,. \tag{16.1}$$

It has the interpretation that at $t = 0$, the probability that the phase of the system lies in the small volume $\Delta\omega$ around ω is $\rho(\omega)\Delta\omega$. The average value of a dynamical variable $f(\omega)$ representing some physical property at $t = 0$ is given by:

$$\langle f\rangle_0 = \int d\omega\, f(\omega)\rho(\omega)\,. \tag{16.2}$$

The same physical property measured at different instants of time will lead to different average values $\langle f\rangle_t$, and this dependence on t is governed by the equations of motion for the system. In the Heisenberg picture, the time dependence of $\langle f\rangle_t$ is ascribed to that of $f(\omega)$: the same physical property is represented at different times by different functions $f_t^{(H)}(\omega)$ on phase space, whereas the "state of the system" is always represented by the probability distribution function $\rho(\omega)$ reflecting conditions at $t = 0$. This means that in this picture the phase space coordinates ω^μ represent different physical properties at different instants of time, except for those that may turn out to be constants of motion; in fact, the ω's that we are using now are the same as the "initial value variables" ω_0 of Chapter 7. The equation of motion for a general $f_t^{(H)}(\omega)$, and its formal solution, are contained in Eq.(7.13):

$$\frac{\partial}{\partial t} f_t^{(H)}(\omega) = \{f_t^{(H)}(\omega), H(\omega)\}\,, \quad f_t^{(H)}(\omega) = e^{-\widetilde{tH(\omega)}} f(\omega)\,. \tag{16.3}$$

(We are considering here only those physical properties whose definitions themselves do not depend on time; in the earlier terminology, these are the quantities without explicit time dependence.) The average value of the physical property has at time t the expression:

$$\langle f\rangle_t = \int d\omega f_t^{(H)}(\omega)\rho(\omega) = \int d\omega (e^{-\widetilde{tH(\omega)}} f(\omega))\rho(\omega)\,. \tag{16.4}$$

In the Schrödinger picture, we allow a fixed function $f(\omega)$ to represent the same physical property for all times, but we make the distribution function $\rho(\omega)$ depend on time. Now each of the basic coordinates ω^μ has a fixed physical meaning for all times, whereas the probability density $\rho_t^{(S)}(\omega)$ for the occurrence of various values of these ω's depends on time. The equation of motion for a general $\rho_t^{(S)}(\omega)$ and its formal solution are given in Eq.(7.17),and are repeated here:

$$\frac{\partial}{\partial t}\rho_t^{(S)}(\omega) = -\{\rho_t^{(S)}(\omega), H(\omega)\}, \quad \rho_t^{(S)}(\omega) = e^{\widetilde{tH(\omega)}}\rho(\omega). \tag{16.5}$$

The $\langle f\rangle_t$ takes the form:

$$\langle f\rangle_t = \int d\omega f(\omega)\rho_t^{(S)}(\omega) = \int d\omega f(\omega)(e^{\widetilde{tH(\omega)}}\rho(\omega)), \tag{16.6}$$

and Eqs.(16.4) and (16.6) agree because of the invariance of the phase-space volume element $d\omega$ under regular canonical transformations. Intermediate between these is the Dirac picture in which the total time dependence of $\langle f\rangle_t$ comes partly from $f(\omega)$ and partly from $\rho(\omega)$. If

$$H(\omega) = H_0(\omega) + V(\omega) \tag{16.7}$$

expresses a useful splitting up of the total Hamiltonian H into a "free" part (or simple part) H_0, and an "interaction" part V, then in the Dirac picture the function $f_t^{(D)}(\omega)$ representing a given physical property at time t develops according to:

$$\frac{\partial f_t^{(D)}(\omega)}{\partial t} = \{f_t^{(D)}(\omega), H_0(\omega)\}; \quad f_t^{(D)}(\omega) = e^{-\widetilde{tH_0(\omega)}}f(\omega). \tag{16.8}$$

[Once again, then, the ω^μ denote different physical properties at different times, except for those that may be constants of motion relative to $H_0(\omega)$.] In particular, the "interaction energy" is a physical property that is given, at time t, by a function $V_t^{(D)}(\omega)$:

$$\frac{\partial V_t^{(D)}(\omega)}{\partial t} = \{V_t^{(D)}(\omega), H_0(\omega)\}, \quad V_t^{(D)}(\omega) = e^{-\widetilde{tH_0(\omega)}}V(\omega). \tag{16.9}$$

The interaction energy governs the time development of $\rho_t^{(D)}(\omega)$ in the Dirac picture:

$$\frac{\partial}{\partial t}\rho_t^{(D)}(\omega) = -\{\rho_t^{(D)}(\omega), V_t^{(D)}(\omega)\}. \tag{16.10}$$

The formal solution to this equation cannot be written in terms of $V_t^{(D)}(\omega)$ alone in as simple a form, as in the earlier cases, although of course, in terms of H and H_0 the solution to Eq.(16.10) is:

$$\rho_t^{(D)}(\omega) = e^{-\widetilde{tH_0(\omega)}}e^{\widetilde{tH(\omega)}}\rho(\omega). \tag{16.11}$$

The $\langle f \rangle_t$ is given by

$$\langle f \rangle_t = \int d\omega f_t^{(D)}(\omega) \rho_t^{(D)}(\omega) . \tag{16.12}$$

To summarize, the basic ingredients in the Hamiltonian formalism are the notions of dynamical variables representing physical properties at an instant of time and probability distributions representing states of the system at an instant of time. These are combined as in Eq.(16.2) to yield average values at one instant. Depending on the picture of dynamics adopted, either the variables or the state or both develop in time according to appropriate canonical transformations.

Implicit in the foregoing discussion is the assumption that there is some fixed observer (or coordinate system) S with respect to whom all these statements are made. It is the descriptions of the system at two different times *as seen by* S that get related by a canonical transformation generated by the Hamiltonian. We must now generalize the framework to see how the description of time development given by one observer S is to be related to that given by another S', S' and S being related by a general element g of the relativity group G under discussion: $S' = gS$. The inclusion of the time translations T_1 as a subgroup of G in all cases leads to a formal simplification as follows: suppose that the description of the system by an observer S at time zero with respect to S is known, and we wish to compute from this the description given by S at a later time $t = b$ in the coordinate system S. We can now introduce a new observer S' who uses the same spatial coordinate system as S, but whose zero of time is displaced with respect to S by the amount $b : t' = t - b$ or $S' = \exp(bh)S$. It is evident that the description of the system given *at time b in S* is exactly the same as that given *in S' at time zero in S'*; thus the problem of relating two descriptions at different times, 0 and b, in one frame S is the same as relating the description at time zero in S to the description at time zero in S'. Now in each of the relativity groups G, T_1 appears as a subgroup, implying that among all those observers S' who are related to S by all possible elements g in G, there appear observers who use the same spatial coordinates as S but whose zeros of time are shifted by all possible amounts relative to S. It is therefore sufficient for us to consider the problem: given the description of a particular state of the system by one observer S at time zero in S, how do we obtain the description of that state of the system given by another observer S' at time zero in S', S' being related to S by a general element g of G? If we know the answer to this wider question, then as a special case we can obtain the form of development in time with respect to one fixed observer S, of the system. The statement that the system possesses symmetry under G means that the form of the transformation taking the description by S into that given by S' permits the use of the Hamiltonian formalism in S' given that it is valid in S. Therefore, the description of a particular state of the system at time zero in S' is obtained from the one at time zero in S by means of a canonical transformation; this canonical transformation must be completely determined by the frame S and the element $g \in G$ taking S to S', $S' = gS$ and nothing else. In particular, it must be the same, whatever be the state of the system. Depending on whether this canonical transformation acts on dynamical variables alone, or probability distribution functions alone, or on both, we will have a *relativistic Heisenberg*, a *relativistic Schrödinger*, or a *relativistic Dirac* picture, respectively.

We must now introduce an item of notation for dynamical variables. In the description of the three forms of time development with respect to one observer S, we saw that, in general, a given physical property is represented by a different function of the phase-space coordinates ω^μ at different instants of time. This feature must be generalized to the "relativistic case." Each physical property can be thought of as being characterized by a "name," and we denote these names by the letters $A, B, \cdots$. Examples of names are "total linear momentum in the first direction" and "third Cartesian component of position of particle number five." We obtain a concrete function of the ω^μ, that is, a definite function on phase space, only when we use a name *with respect to some definite observer S and at some instant of time in S*. Thus, in general, a change in the observer S or in the instant of time causes a change in the phase-space function representing the physical property A. To exhibit clearly all these dependences, we write $A(S, t; \omega)$ for the representative of A for an observer S at time t. For example, it will become clear that if A denotes "total linear momentum in the first direction," and the observer S' differs from S only in that its spatial coordinate system arises from that in S by some rotation about the third axis, then in the generalized Heisenberg picture $A(S', t; \omega)$ *must* differ from $A(S, t; \omega)$. In fact, in this picture, $A(S', t; \omega)$ should be a linear combination of $A(S, t; \omega)$ and $B(S, t; \omega)$, B denoting the quantity "total linear momentum in the second direction"; the coefficients in the linear combination should be the cosine and sine, respectively, of the angle of rotation leading from S to S'. For the reasons given in the previous paragraph, it is sufficient to consider initially only the dynamical variables $A(S, 0; \omega)$, for all S. This represents A at time zero in S. If we know how to go from $A(S, 0; \omega)$ to $A(S', 0; \omega)$, S' being related to S by a general element of the relativity group, we will surely know how to compute $A(S, t; \omega)$ from $A(S, 0; \omega)$. In this more general context, the meaning of the phase space variables ω^μ must also be clarified. We can imagine that each of the ω^μ is the representative of some specific physical quantity at time zero in some fixed frame S_0. This frame S_0 does not enter explicitly in the analysis, but we must keep in mind the fact that some such S_0 is present and stays unchanged during the discussion. We could, of course, start off by assuming that these ω^μ represent these specific properties at time zero in some other frame S_0', or we might continue to associate the ω^μ with S_0, but decide to use as basic coordinates in phase space the functions $\omega'^\mu(\omega)$ that represent these properties at time zero in some S_0'. The ω' arise from the ω by some regular canonical transformation. All these freedoms simply reflect the overall invariance of the Hamiltonian formalism under regular canonical transformations. When we discuss the relativistic Schrödinger picture, the reader will see that the ω^μ represent some fixed set of physical properties in all frames S at all times t, but only in this picture is this so.

The various relativity groups G are Lie groups of various dimensions; $R(3) \otimes$

T_1 is a four-parameter group, $E(3) \otimes T_1$ has seven parameters, whereas the Galilei and Poincaré groups both have ten parameters. When we discussed the transformation-theoretical meaning of the linear and angular momentum dynamical variables in Chapter 11, we saw by means of examples that these generate canonical transformations that have the same effects on the phase-space variables that a translation or rotation of the Cartesian coordinate system would have had. We generalize this idea in the following way. Let the Lie algebra of G be called L, and let it be of dimension n. Relative to a basis e_j, $j = 1, 2, \cdots, n$, for L, there shall be n distinguished physical quantities X_j out of all the physical quantities A relevant to the system. (These X_j are "names".) The representers of the X_j at time zero in the frame S will be $X_j(S, 0; \omega)$, and at time t they will be $X_j(S, t; \omega)$. The $X_j(S, 0; \omega)$ are to be used as generators of the canonical transformation connecting the description at time zero in S to that at time zero in S', where $S' = gS$ and g is a general element in G. This canonical transformation is an explicit function of g and the $X_j(S, 0; \omega)$, and its form is given later. Using the notation of the preceding chapter, if we pick a basis for L from the elements d_j, l_j, g_j, k_j, h, taking whatever is needed, the three X's corresponding to the d_j are called the three Cartesian components of linear momentum, and those corresponding to the l_j the components of angular momentum. The single X corresponding to h is the Hamiltonian. For want of a better name, the X's corresponding to the g_j or k_j are simply called "the generators of transformations to moving frames" or "mechanical moments."

We now analyze the three relativistic pictures of dynamics.

The Relativistic Heisenberg Picture

This picture is defined thus: a given state of motion shall correspond to one and the same probability distribution function $\rho(\omega)$ in phase space at time zero for all frames S, and therefore at all times for all S. [To identify the $\rho(\omega)$ for a particular state of motion, we only have to examine the state of the system at time zero for the distinguished observer S_0 we spoke of earlier; at this time and for S_0 we know what physical quantity each ω^μ represents, and the probability distribution for various values of these at this time fixes $\rho(\omega)$.] On the other hand, the functions $A(S', 0; \omega)$ and $A(S, 0; \omega)$ representing the physical quantity A at time zero for two observers S' and S will, in general, be different, the former arising from the latter by a canonical transformation. Let $S' = \exp(u)S$, and in the basis e_j for L, let u have components $u^j : u = u^j e_j$. Then we write this canonical transformation as follows:

$$S' = \exp(u)S: \quad A(S', 0; \omega) = R(-u^j, X_j(S, 0; \omega))A(S, 0; \omega),$$
$$R(-u^j, X_j(S, 0; \omega)) = e^{-u \cdot \widetilde{X(S,0;\omega)}}, \quad u \cdot X(S, 0; \omega) = u^j X_j(S, 0; \omega). \quad (16.13)$$

(We have put in a minus sign before the u^j so that the formulae we get later will be simple.) This equation makes explicit the way in which the $X_j(S, 0; \omega)$ are to be used in building up the canonical transformation to go from S to S'. It specifies the dependence on the group element as well. For example, suppose $\exp(u)$ connects S' and S, as well as another pair of observers S'_1 and

$S_1 : S_1' = \exp(u)S_1$. Then, to go from dynamical variables at time zero in S to time zero in S' we must use $R(-u^j, X_j(S,0;\omega))$; but to go from time zero in S_1 to time zero in S_1' we must use $R(-u^j, X_j(S_1,0;\omega))$. The dependence on u is the same, but the generators are different; thus, in general, the two canonical transformations are not the same. The basic condition on the R transformations comes from consistency of the formalism. Let S' be related to S by $\exp(u)$ as in Eq.(16.13), and S'' be related to S' by some other element $\exp(v) : S'' = \exp(v)S'$. We can relate S'' directly to S:

$$S'' = \exp(v)S' = \exp(v)\exp(u)S = \exp(w)S,\ \exp(w) = \exp(v)\exp(u)\,. \tag{16.14}$$

There are two ways in which the canonical transformation taking dynamical variables at time zero in S to those at time zero in S'' can be written down, and *these must be the same.* We must have:

$$\begin{aligned} A(S'',0;\omega) &= R(-v^j, X_j(S',0;\omega))A(S',0;\omega) \\ &= R(-v^j, X_j(S',0;\omega))R(-u^j, X_j(S,0;\omega))A(S,0;\omega)\,, \\ R(-v^j, X_j(S',0;\omega))&R(-u^j, X_j(S,0;\omega)) = R(-w^j, X_j(S,0;\omega))\,. \end{aligned} \tag{16.15}$$

The last line gives the basic condition on the R's, corresponding to the composition law in G.

To analyze this condition, we must rewrite it in terms of the functions $X_j(S,0;\omega)$ alone. Now the $X_j(S',0;\omega)$ represent the physical quantities X_j at time zero in S'; thus as a particular case of Eq.(16.13) we have:

$$X_j(S',0;\omega) = R(-u^j, X_j(S,0;\omega))X_j(S,0;\omega)\,. \tag{16.16}$$

How do we use this to simplify Eq.(16.15)? We need a simple property of these R's that follows from the structure of the adjoint representation of a Lie group. (As we saw in Chapter 14, the formal properties of regular canonical transformations generated continuously from the identity correspond exactly to similar properties for the adjoint representation of an arbitrary Lie group). We know that Lie brackets are preserved under the action of the adjoint representation; Eq.(14.13) says:

$$e^{\Omega_u}[v,w] = [e^{\Omega_u}v, e^{\Omega_u}w]\,, \tag{16.17}$$

u, v and w being any vectors in a general, unspecified, Lie algebra. We can put Eq.(16.17) into the form

$$e^{\Omega_u}\Omega_v w = \Omega_{v'}e^{\Omega_u}w\,, \qquad v' = e^{\Omega_u}v\,. \tag{16.18}$$

Because w is arbitrary, this implies the general property:

$$e^{\Omega_u}\Omega_v = \Omega_{v'}e^{\Omega_u}\,, \qquad v' = e^{\Omega_u}v\,. \tag{16.19}$$

By considering powers of Ω_v, for instance, this means that we can move e^{Ω_u} through any function of Ω_v if we replace v by v' in the process:

$$e^{\Omega_u} f(\Omega_v) = f(\Omega_{v'}) e^{\Omega_u} . \tag{16.20}$$

Taking the exponential function for f, we have the rule:

$$e^{\Omega_u} e^{\Omega_v} = e^{\Omega_{v'}} e^{\Omega_u} , \quad e^{\Omega_u} e^{\Omega_v} e^{-\Omega_u} = e^{\Omega_{v'}} , \quad v' = e^{\Omega_u} v . \tag{16.21}$$

This last can be written in the notation $R_{\exp(u)} = e^{\Omega_u}$ as well:

$$R_{\exp(u)} R_{\exp(v)} R_{\exp(-u)} = R_{\exp(v')} , \qquad v' = R_{\exp(u)} v . \tag{16.22}$$

Equation (16.21) and (16.22) are also obvious from the very definition of the adjoint representation. An equation of this same form is valid, with obvious redefinitions, for the case of canonical transformations. We can use Eq.(16.21) with Eq.(16.16) to remove the $X_j(S', 0; \omega)$ in the basic condition of Eq.(16.15). We have, in the first place,

$$\begin{aligned} R(-v^j, X_j(S', 0; \omega)) =& R(-u^j, X_j(S, 0; \omega)) R(-v^j, X_j(S, 0; \omega)) \\ & \times R(u^j, X_j(S, 0; \omega)) ; \end{aligned} \tag{16.23}$$

and using this in Eq.(16.15) gives:

$$R(-u^j, X_j(S, 0; \omega)) R(-v^j, X_j(S, 0; \omega)) = R(-w^j, X_j(S, 0; \omega)), \tag{16.24}$$

w^j being the canonical coordinates of $\exp(v) \exp(u)$. Taking the inverses of both sides, we arrive at the final equation

$$\begin{aligned} R(v^j, X_j(S, 0; \omega)) R(u^j, X_j(S, 0; \omega)) =& R(w^j, X_j(S, 0; \omega)) , \\ e^{v \cdot \widetilde{X(S,0;\omega)}} e^{u \cdot \widetilde{X(S,0;\omega)}} =& e^{w \cdot \widetilde{X(S,0;\omega)}} . \end{aligned} \tag{16.25}$$

The same set of functions $X_j(S, 0; \omega)$ appears throughout. The two terms on the left are canonical transformations corresponding to the two elements $\exp(v)$ and $\exp(u)$ of G, respectively, whereas on the right appears the canonical transformation corresponding to $\exp(w) = \exp(v) \exp(u)$. Thus we reach the conclusion that *the canonical transformations $R(u^j, X_j(S, 0; \omega))$ for a fixed set of generators constitute a canonical realization of the group G.* (Now it is clear why a minus sign was put in front of u in Eq.(16.13).) From our work in Chapter 14 it follows that *the functions $X_j(S, 0; \omega)$ representing the quantities X_j at time zero in the frame S give us a PB realization up to neutrals of the Lie algebra L of the relativity group G under discussion*:

$$\begin{aligned} L : \quad [e_j, e_k] =& c_{jk}^l e_l , \\ \{X_j(S, 0; \omega),\, X_k(S, 0; \omega)\} =& c_{jk}^l X_l(S, 0; \omega) + d_{jk} . \end{aligned} \tag{16.26}$$

These neutral elements d_{jk} do not depend on the observer S at all, but are characteristic of the physical system! The generators associated with an S' are obtained from those of S by a canonical transformation (see Eq.(16.16)), and because all PB relations are preserved by such transformations, exactly the same equations as Eq.(16.26) will be valid with S' written in place of S on both sides.

We have made good the claim made in Chapter 14: a consistent application of the basic rule of Eq. (16.13) leads to the requirement that we have a canonical realization of the group of frame transformations G. We have many canonical realizations of G, one generated by the functions $X_j(S,0;\omega)$ representing the X_j at time zero in each frame S; all these realizations are canonically equivalent because of Eq.(16.16). In the relativistic Heisenberg picture, the representatives of physical quantities change from frame to frame according to Eq.(16.13), whereas the probability function $\rho(\omega)$ is kept unchanged. The standard notation for the X's corresponding to the basic elements d_j, l_j, g_j, k_j and h is

$$d_j \to P_j\,, \quad l_j \to J_j\,, \quad g_j \to G_j\,, \quad k_j \to K_j\,, \quad j=1,2,3\,; \; h \to H\,. \tag{16.27}$$

Keeping track of the conventions made in the previous chapter for the various elementary transformations generated by these basic elements, we can draw up Table 16.1. The matrices $A(\boldsymbol{\alpha})$ and $\Lambda(\boldsymbol{\zeta},\mathbf{0})$ are given in Eq.(15.11) and (15.85), respectively.

Table 16.1:

Transformation g, $S'=gS$	Effect on space-time coordinates	Canonical Transformation R such that (Quantity at time zero in S')$=R$ (Quantity at time zero in S)
$\exp(a_j d_j)$	$x'_j = x_j + a_j\,,\; t'=t$	$\exp[-a_j \widetilde{P_j(S,0;\omega)}]$ (a)
$\exp(\alpha_j l_j)$	$x'_j = A_{jk}(\boldsymbol{\alpha})x_k\,,\; t'=t$	$\exp[-\alpha_j \widetilde{J_j(S,0;\omega)}]$ (b)
$\exp(v_j g_j)$	$x'_j = x_j - v_j t\,,\; t'=t$	$\exp[-v_j \widetilde{G_j(S,0;\omega)}]$ (c)
$\exp(\zeta_j k_j)$	$x'^{\mu} = \Lambda^{\mu}{}_{\nu}(\boldsymbol{\zeta},\mathbf{0})x^{\nu}$	$\exp[-\zeta_j \widetilde{K_j(S,0;\omega)}]$ (d)
$\exp(bh)$	$x'_j = x_j\,,\; t'=t-b$	$\exp[-b\widetilde{H(S,0;\omega)}]$ (e)

We can now specialize to the question of time development. To examine conditions at a time b in S, we define an observer $S' = \exp(bh)S$. "Time equal to b" in S is the same as "Time zero" in S'. We then define:

$$A(S,b;\omega) = A(\exp(bh)S,0;\omega)\,. \tag{16.28}$$

In words: in the Heisenberg picture, the function representing the physical property A at time b is the same as its representative at time zero in the frame $\exp(bh)S$. Now we can apply Eqs.(16.13) and (16.27) to this definition, to obtain

$$A(S,b;\omega) = e^{-b\widetilde{H(S,0;\omega)}}A(S,0;\omega)\,. \tag{16.29}$$

$H(S,0;\omega)$ is the representative of the total energy H of the system at time zero in S. Equation (16.29) is nothing but Eq.(16.3), but now such an equation is valid for every observer S. If we consider H in place of A in Eq.(16.29), we see that *in each frame S the representative of the total energy is independent of time*:

$$H(S,b;\omega) = e^{-b\widetilde{H(S,0;\omega)}} H(S,0;\omega) = H(S,0;\omega) = H(S;\omega)\,; \tag{16.30}$$

we must allow, of course, for a possible dependence on the frame S. The transformation from time zero in S to time zero in S', given by Eq.(16.13), can be combined with the transformation from time zero in S to time b in S, given by Eq.(16.29), to obtain the transformation from time b in S to time b in S'; with $S' = \exp(u)S$, we have:

$$\begin{aligned} A(S',b;\omega) &= e^{-b\widetilde{H(S';\omega)}} A(S',0;\omega) = e^{-b\widetilde{H(S';\omega)}} e^{-u\cdot\widetilde{X(S,0;\omega)}} A(S,0;\omega) \\ &= e^{-b\widetilde{H(S';\omega)}} e^{-u\cdot\widetilde{X(S,0;\omega)}} e^{b\widetilde{H(S;\omega)}} A(S,b;\omega)\,. \end{aligned} \tag{16.31}$$

But $H(S';\omega)$ is related to $H(S;\omega)$ by

$$H(S';\omega) = e^{-u\cdot\widetilde{X(S,0;\omega)}} H(S;\omega)\,, \tag{16.32}$$

and using Eq.(19.21) this gives

$$e^{-b\widetilde{H(S';\omega)}} = e^{-u\cdot\widetilde{X(S,0;\omega)}} e^{-b\widetilde{H(S;\omega)}} e^{u\cdot\widetilde{X(S,0;\omega)}}\,. \tag{16.33}$$

Using this in Eq.(16.31), we get a rather simple result:

$$A(S',b;\omega) = e^{-u\cdot\widetilde{X(S,0;\omega)}} A(S,b;\omega) \equiv R(-u^j, X_j(S,0;\omega))A(S,b;\omega)\,. \tag{16.34}$$

The canonical transformation has turned out to be independent of b! This is a feature of the Heisenberg picture: given the two frames S and S' related by the element $\exp(u)$ in G, a definite canonical transformation dependent on u and S alone is determined. This canonical transformation takes us from the representatives $A(S,b;\omega)$ of dynamical quantities A at any time b in S to their representatives $A(S',b;\omega)$ at the same time b in S', *and the same transformation is to be used whatever be the value of b*. The dependence of the transformation on S arises because we use the generators $X_j(S,0;\omega)$. In general, we do not expect the $X_j(S,b;\omega)$ to be independent of time b; for each X_j, it depends on whether that X_j has a vanishing PB with H. All in all, we can exhibit the important properties of the relativistic Heisenberg picture in this form:

(a) *Probability Distribution Function:* $\rho(\omega)$, determined by state of motion, but independent of observer S or value of time.

(b) *Representative of Quantity A:* $A(S,b;\omega)$ represents A in the frame S at time b in S.

(c) *Time development in one Frame S:* $A(S, b; \omega) = e^{-b\widetilde{H(S;\omega)}} A(S, 0; \omega)$

(d) *Transformation law for* $A(S, b; \omega)$*:* If $S' = \exp(u)S$, then

$$A(S', b; \omega) = e^{-u\cdot\widetilde{X(S,0;\omega)}}\, A(S, b; \omega) \qquad \text{for all } b\,.$$

To these must be added the realization requirements of Eqs.(16.25) and (16.26).

There is one comment to be made regarding the form chosen in Eq.(16.13) for the canonical transformation $R(u^j, X_j(S, 0; \omega))$. We have expressly chosen to use canonical coordinates of the first kind for G. Now in most practical cases these are not the most convenient ones to use. Let us consider, for example, the case $G = E(3) \otimes T_1$. Suppose two frames S and S' differ only by an element from $E(3)$, and in the notation of the last chapter let us write $S' = \exp(\boldsymbol{a}\cdot\boldsymbol{d}) \exp(\boldsymbol{\alpha}\cdot\boldsymbol{l})S$. This corresponds to using noncanonical parameters for $E(3)$. Now one might have thought that because the elements $\exp(\boldsymbol{a}\cdot\boldsymbol{d})$ and $\exp(\boldsymbol{\alpha}\cdot\boldsymbol{l})$ appear in this definite order, one could write the canonical transformation R to go from S to S' in the form:

$$R_{S\to S'} = e^{-\boldsymbol{a}\cdot\widetilde{\boldsymbol{P}(S,0;\omega)}}\, e^{-\boldsymbol{\alpha}\cdot\widetilde{\boldsymbol{J}(S,0;\omega)}}\,. \tag{16.35}$$

But this is incorrect both mathematically and physically. Mathematically, our analysis has shown that the *inverses* of the transformations appearing in Eq. (16.13) go to make up a realization of G, not these transformations themselves. This would imply that $R_{S\to S'}$ must be

$$R_{S\to S'} = \exp[-\boldsymbol{\alpha}\cdot\widetilde{\boldsymbol{J}(S, 0; \omega)}]\exp[-\boldsymbol{a}\cdot\widetilde{\boldsymbol{P}(S, 0; \omega)}]\,. \tag{16.36}$$

Physically, if we introduce an intermediate frame $S_1 = \exp(\boldsymbol{\alpha}\cdot\boldsymbol{l})S$, then we have $S' = \exp(\boldsymbol{a}\cdot\boldsymbol{d})S_1$; the natural equations to set up are:

$$\begin{aligned} R_{S\to S_1} &= \exp[-\boldsymbol{\alpha}\cdot\widetilde{\boldsymbol{J}(S, 0; \omega)}]\,, \quad R_{S_1\to S'} = \exp[-\boldsymbol{a}\cdot\widetilde{\boldsymbol{P}(S_1, 0; \omega)}]\,,\\ R_{S\to S_1} &= R_{S_1\to S'}R_{S\to S_1} = \exp[-\boldsymbol{a}\cdot\widetilde{\boldsymbol{P}(S_1, 0; \omega)}]\exp[-\boldsymbol{\alpha}\cdot\widetilde{\boldsymbol{J}(S, 0; \omega)}]\,. \end{aligned} \tag{16.37}$$

That Eqs.(16.36) and (16.37) are in agreement follows from the relation between $\boldsymbol{P}(S_1, 0; \omega)$ and $\boldsymbol{P}(S, 0; \omega)$:

$$\boldsymbol{P}(S_1, 0; \omega) = R_{S\to S_1}\boldsymbol{P}(S, 0; \omega)\,. \tag{16.38}$$

One must keep this feature of the R transformations in mind, if one decides to use noncanonical coordinates for G and wishes to compute the canonical transformation connecting two general frames S and S'.

We promised the reader at the end of Chapter 11, where we examined the forms of the canonical transformations generated by P_j and J_j, that we would explain later the nature of the PB relations *among the* P_j *and* J_j *themselves.* We now know, from Eq.(16.26), what this should be: $P_j(\omega)$ and $J_j(\omega)$ must always obey the PB relationships characteristic of $E(3)$, allowing of course, for the presence of neutral elements. The reader can check that in the examples discussed in Chapter 11, the expressions obtained for P_j and J_j do obey the Lie relations of $E(3)$, with no neutral elements being present. The PB's of the P_j among themselves vanish, because T_3 is Abelian; this is why we had no difficulty in discussing in Chapter 11 finite transformations generated by the P_j. Now we know that the J_j must obey PB relations corresponding to the non-Abelian group $R(3)$, which explains why we were content to examine infinitesimal rotations in Chapter 11! It is better to leave a more detailed account of the various PB relations in the various cases to the later chapters when we discuss each group G in turn.

The Relativistic Schrödinger Picture

This picture is defined by the following statement: a physical property A shall be represented by one and the same phase-space function $A(\omega)$ for all observers at time zero, and thus for all observers at all times. On the other hand, for a definite state of motion of the system, the probability distribution function shall undergo a canonical transformation for every change of frame. We obtain the Schrödinger picture from the Heisenberg picture by making a frame-dependent canonical transformation. Let $A(S, 0; \omega)$ be the representative of A at time zero in S, and $\rho(\omega)$ the distribution function for a specific state, all in the Heisenberg picture. The average value of A measured in S at time zero is given by

$$\langle A \rangle(S, 0) = \int d\omega \, A(S, 0; \omega)\rho(\omega) \,. \tag{16.39}$$

(In the bracket following $\langle A \rangle$ are indicated the frame S, and time, zero, for which the average value has been computed.) Using Eq.(16.13), in another frame S' and at time zero in S', for the same state of motion, we must have

$$\begin{aligned} \langle A \rangle(S', 0) &= \int d\omega \, A(S', 0; \omega)\rho(\omega) \\ &= \int d\omega \left(e^{-u \cdot \widetilde{X(S,0;\omega)}} A(S, 0; \omega) \right) \rho(\omega) \,. \end{aligned} \tag{16.40}$$

Now these average values for specified frames and times must be independent of the "picture" being used. In Eq.(16.40), we can make use of the invariance of the volume element $d\omega$ under a canonical transformation and "shift" the exponential from the dynamical variable to ρ:

$$\langle A \rangle(S', 0) = \int d\omega \, A(S, 0; \omega) \left(e^{u \cdot \widetilde{X(S,0;\omega)}} \rho(\omega) \right) \,. \tag{16.41}$$

This expression tells us how to set up the basic equations of the Schrödinger picture, because on comparing Eq.(16.41) with Eq.(16.39) we see that the frame transformation $S \to S'$ can be represented by a change in ρ alone. We have already agreed that a physical quantity A shall be represented for all frames S and times t by one fixed function, thus in place of $A(S, t; \omega)$ we simply write

$A(\omega)$. In particular, we can simply write $X_j(\omega)$ for the representatives of the X_j. Then we have the law of change for ρ: if $\rho(S,0,\omega)$ is the probability distribution function for a particular state of motion, as seen at time zero in the frame S, and $S' = \exp(u)S$ is another frame, then $\rho(S',0;\omega)$ is given thus:

$$\begin{aligned} S' = \exp(u)S : \qquad \rho(S',0;\omega) =& R(u^j, X_j(\omega))\rho(S,0;\omega) \\ =& e^{\widetilde{u\cdot X(\omega)}}\rho(S,0;\omega) \,. \end{aligned} \tag{16.42}$$

It is clear that in the Schrödinger picture each of the basic variables ω^μ represents the same physical quantities for all observers, and thus the determination of $\rho(S,0;\omega)$ for a given state of motion is immediate.

We could carry over all the properties of the $X_j(\omega)$ from what we have learned of them in the Heisenberg picture, but it is more transparent to give an independent discussion based on Eq. (16.42). Let us derive the realization property. If $S' = \exp(u)S$ and $S'' = \exp(v)S'$, we can directly relate the ρ in S'' to that in S:

$$\begin{aligned} S'' = \exp(v)S' =& \exp(v)\exp(u)S = \exp(w)S\,, \\ & \exp(w) = \exp(v)\exp(u)\,; \\ \rho(S'',0;\omega) =& R(v^j, X_j(\omega))\rho(S',0;\omega) \\ =& R(v^j, X_j(w))R(u^j, X_j(\omega))\rho(S,0;\omega) \\ =& R(w^j, X_j(\omega))\rho(S,0;\omega)\,, \\ R(v^j, X_j(\omega))R(u^j, X_j(\omega)) =& R(w^j, X_j(\omega))\,. \end{aligned} \tag{16.43}$$

Because the same set of generators $X_j(\omega)$ appears throughout, and $\exp(v)\exp(u) = \exp(w)$, the realization property is evident. Therefore, in the Schrödinger picture, *to every dynamical system possessing symmetry under the relativity group G there corresponds a canonical realization of G and a set of functions $X_j(\omega)$ forming a PB realization up to neutrals of the Lie algebra of G:*

$$\{X_j(\omega), X_k(\omega)\} = c^l_{jk}X_l(\omega) + d_{jk}\,. \tag{16.44}$$

We notice now the interesting property that *the canonical transformation appearing in Eq.(16.42) is independent of the frame S and depends only on the element* $\exp(u)$ *connecting S' to S*. This is because the generators $X_j(\omega)$ are common to all observers S. It is worth doing a short calculation to convince ourselves that there is nothing inconsistent in this procedure. Let S' be obtained by the action of $\exp(u)$ on S, and let S_1', S_1 be two other frames also connected by $\exp(u) : S_1' = \exp(u)S_1$. We want to be sure that it is quite alright to use the same canonical transformation to go from S_1 to S_1' as to go from S to S'. If we write $\exp(v)$ for the element taking S to S_1, the relationships among these four frames are given by this diagram:

$$\begin{array}{ccccc} S & \text{———} & \exp(u) & \longrightarrow & S' = \exp(u)S \\ \exp(v)\Big\downarrow & & & & \Big\downarrow\exp(v') \\ S_1 & \text{———} & \exp(u) & \longrightarrow & S_1' = \exp(u)S_1 \end{array}$$

$$\exp(v') = \exp(u)\exp(v)\exp(-u)\,.$$

Let us use Eq.(16.42) for the top and the two sides of this diagram:

$$\begin{aligned} \rho(S',0;\omega) =& R(u^j, X_j(\omega))\rho(S,0;\omega);\ \rho(S_1,0;\omega) = R(v^j, X_j(\omega))\rho(S,0;\omega)\,; \\ \rho(S_1',0;\omega) =& R(v'^j, X_j(\omega))\rho(S',0;\omega), \qquad v' = e^{\Omega_u}v\,. \end{aligned} \tag{16.45}$$

Now we combine these to relate S_1' to S_1:

$$\rho(S_1',0;\omega) = R(v'^j, X_j(\omega))R(u^j, X_j(\omega))R(-v^j, X_j(\omega))\rho(S_1,0;\omega)\,. \tag{16.46}$$

But the R's form a realization of G, and $\exp(v')$ arises from $\exp(v)$ by conjugation with $\exp(u)$, thus:

$$R(v_j', X_j(\omega)) = R(u^j, X_j(\omega))R(v^j, X_j(\omega))R(-u^j, X_j(\omega))\,. \tag{16.47}$$

If we use this in Eq.(16.46) we come out with the expected result:

$$\rho(S_1',0;\omega) = R(u^j, X_j(\omega))\rho(S_1,0;\omega)\,. \tag{16.48}$$

Thus the canonical transformation to go from the distribution function at time zero in S to the corresponding one in S' is determined fully by the relativity transformation taking S to S', and is otherwise independent of S. This is an important characteristic of the Schrödinger picture.

Now we turn to the question of time development. Analogously to the situation in the Heisenberg picture, we agree that the probability distribution function $\rho(S,b;\omega)$ at time b in the frame S is the same as the one at time zero in the frame $S' = \exp(bh)S$:

$$\rho(S,b;\omega) = \rho(\exp(bh)S,0;\omega)\,. \tag{16.49}$$

Using Eq.(16.42) here, we get the formal expression for development in time in a given S:

$$\rho(S,b;\omega) = e^{\widetilde{bH(\omega)}}\rho(S,0;\omega)\,. \tag{16.50}$$

Thus we have recovered Eq.(16.5)! The important point is this: although Eq.(16.50) is the expression of time development in any frame S, by definition the Hamiltonian $H(\omega)$ is the same in all frames because we are using the Schrödinger picture and $H(\omega)$ is the representative of the quantity "total energy of the system." Thus the same canonical transformation $e^{\widetilde{bH(\omega)}}$ generated by $H(\omega)$ is to be used in describing time evolution in all relativistically related frames S. This is just a particular case of the statement made at the conclusion of the previous paragraph. Now we can combine Eqs.(16.50) and (16.42) to determine the transformation taking ρ at time b in a frame S to ρ at time b in $S' = \exp(u)S$: this turns out dependent on b, in contrast to the corresponding result in the Heisenberg picture given in Eq.(19.34). We find:

$$\begin{aligned}\rho(S', b; \omega) =& e^{\widetilde{bH(\omega)}} R(u^j, X_j(\omega)) e^{-\widetilde{bH(\omega)}} \rho(S, b; \omega) \\ =& R(u^j, e^{\widetilde{bH(\omega)}} X_j(\omega)) \rho(S, b; \omega), \end{aligned} \tag{16.51}$$

use having been made of Eq.(16.21). We can list the important properties of the relativistic Schrödinger picture in this form:

(a) *Probability Distribution Function*: $\rho(S, b; \omega)$, determined by state of motion and dependent on the observer S as well as the time b in S.

(b) *Representative of Quantity A*: $A(\omega)$ represents A in all frames at all times.

(c) *Time Development in One Frame S*: $\rho(S, b; \omega) = e^{\widetilde{bH(\omega)}} \rho(S, 0; \omega)$.

(d) *Transformation Law for* $\rho(S, b; \omega)$: If $S' = \exp(u)S$, then

$$\rho(S', b; \omega) = R(u^j, e^{\widetilde{bH(\omega)}} X_j(\omega)) \rho(S, b; \omega).$$

To these must be added the realization properties of Eqs.(16.43) and (16.44).

There is one aspect of the relativistic Schrödinger picture that might puzzle one at first sight, but really should not; it is worthwhile to explain this aspect. It has to do with the rule that in this picture a physical quantity shall be represented by the same phase-space function for all observers. As an example, consider two observers S and S' whose only difference is that the spatial axes in S' arise from those in S by applying to the latter a rotation of amount α around the third axis. Then we can write $S' = \exp(-\alpha l_3)S$. Now consider the quantity "total linear momentum in the first direction." According to our rules, this is represented by the function $P_1(\omega)$ for S as well as for S'. Next consider the quantity "total linear momentum in a direction in the 1-2 plane at an angle α to the first direction." It is natural that in S this quantity should be represented by $\cos\alpha P_1(\omega) + \sin\alpha P_2(\omega)$; but in S' as well this quantity is represented by this same combination of $P_1(\omega)$ and $P_2(\omega)$. Now one has the notion that "total linear momentum in the first direction, in S'" should, after all, be the same thing as "total linear momentum in a direction in the 1-2 plane at an angle α to the first direction, in S," but the former is represented by $P_1(\omega)$ and the latter by $\cos\alpha P_1(\omega) + \sin\alpha P_2(\omega)$! This emphasizes the following fact: the notion of the equality of the two quantities mentioned in the previous sentence really applies only to their *values* computed for each given state of motion of the system, and in the Schrödinger picture the entire burden of change in going from S to S' is borne by the probability distribution function $\rho(\omega)$! In fact, for a given state of motion, at time zero in both frames, we have

$$\rho(S', 0; \omega) = e^{-\widetilde{\alpha J_3(\omega)}} \rho(S, 0; \omega). \tag{16.52}$$

If P_1 and $\cos\alpha P_1 + \sin\alpha P_2$ denote the two physical quantities we are talking about — independent of frames — we get:

$$\begin{aligned}\langle P_1\rangle(S',0) &= \int d\omega\, P_1(\omega)\rho(S',0;\omega) = \int d\omega P_1(\omega)(e^{-\widetilde{\alpha J_3(\omega)}}\rho(S,0;\omega))\\ &= \int d\omega(e^{\widetilde{\alpha J_3(\omega)}}P_1(\omega))\rho(S,0;\omega)\\ &= \int d\omega(\cos\alpha P_1(\omega) + \sin\alpha P_2(\omega))\rho(S,0;\omega)\\ &=\langle\cos\alpha P_1 + \sin\alpha P_2\rangle(S,0)\\ &= \cos\alpha\langle P_1\rangle(S,0) + \sin\alpha\langle P_2\rangle(S,0)\,. \end{aligned} \tag{16.53}$$

[We assume for simplicity the absence of neutral elements in the PB realization of $E(3)$ given by $J_j(\omega)$ and $P_j(\omega)$. We also use the transformation properties of $P_j(\omega)$ under the canonical transformations representing rotations; these follow from Eq.(16.15), (16.53).] We see that our expectations are fulfilled by the average values in just the right fashion, and the whole formalism "hangs together" properly.

The Relativistic Dirac Picture

We can develop this picture by making a frame-dependent canonical transformation on the relativistic Schrödinger picture. In the latter, we have the fixed function $A(\omega)$ representing the property A in all frames; in particular the physical quantities X_j are represented by the functions $X_j(\omega)$ generating a canonical realization of G. We now choose some standard frame S_0, and keep it unchanged in what follows. If S is some other frame, it is related to S_0 by some element of $G : S = \exp(u(S))S_0$. Because S_0 is fixed, there is one element $\exp(u(S))$ for each S. According to the Schrödinger picture, the value of A at time zero in S is given by

$$\langle A\rangle(S,0) = \int d\omega A(\omega)\rho(S,0;\omega) = \int d\omega A(\omega)(e^{\widetilde{u(S)\cdot X(\omega)}}\rho(S_0,0;\omega))\,. \tag{16.54}$$

Now suppose there is a useful way of separating the functions $X_j(\omega)$ into two parts, one of which we call the "free" part and the other the "interaction" part:

$$X_j(\omega) = F_j(\omega) + V_j(\omega)\,, \qquad j = 1, 2, \cdots, n\,. \tag{16.55}$$

Notice that we have introduced new physical quantities with new "names": we have "the free part of the quantity X_j", symbolized by F_j and represented by $F_j(\omega)$ for all observers because we are using the Schrödinger picture; we have "the interaction part of the quantity X_j", symbolized by V_j and represented by $V_j(\omega)$ for all observers. The V_j's are generalized "potentials." We assume that the $F_j(\omega)$ also constitute a PB realization up to neutrals of the Lie algebra of G and generate a canonical realization of G, just as the $X_j(\omega)$ do:

$$\{F_j(\omega),\, F_k(\omega)\} = c^l_{jk}F_l(\omega) + d'_{jk}\,. \tag{16.56}$$

However, there is no need to require that the neutral elements d_{jk} in the case of the X's and d'_{jk} in the case of the F's are the same. The canonical realization of G given by the $X_j(\omega)$ corresponds to the system including interaction, the one given by the $F_j(\omega)$ is a "free" realization. Now we can perform a free-canonical transformation on the two factors in Eq.(16.54) to put it into this form:

$$\begin{aligned}\langle A\rangle(S,0) &= \int d\omega\, \bar{A}(S,0;\omega)\bar{\rho}(S,0;\omega)\,,\\ \bar{A}(S,0;\omega) &= e^{-\widetilde{u(S)\cdot F(\omega)}} A(\omega)\,,\\ \bar{\rho}(S,0;\omega) &= e^{-\widetilde{u(S)\cdot F(\omega)}} e^{\widetilde{u(S)\cdot X(\omega)}} \rho(S_0,0;\omega)\,. \end{aligned} \tag{16.57}$$

(We have put bars on the A's and ρ's so that the former are not confused with the A's of the Heisenberg picture, and the latter with the ρ's of the Schrödinger picture.) We can now define the Dirac picture: the property A, represented in the Schrödinger picture by the fixed function $A(\omega)$, is now represented in the frame S at time zero by $\bar{A}(S,0;\omega)$; a particular state of motion, given at time zero in S by $\rho(S,0;\omega)$ in the Schrödinger picture, is now represented in the same frame at the same time by $\bar{\rho}(S,0;\omega)$. In order to switch over from one picture to the other we made use of the fixed frame S_0. We must now write the basic equations of the Dirac picture in a form in which no explicit reference is made to this S_0. This must be done both for dynamical variables and distribution functions. As a particular case of a general quantity A, it is clear that in the Dirac picture "the free part of X_j" is represented at time zero in S by $\bar{F}_j(S,0;\omega)$ given by

$$\bar{F}_j(S,0;\omega) = e^{-\widetilde{u(S)\cdot F(\omega)}}\, F_j(\omega)\,. \tag{16.58}$$

Similarly, "the interaction part of X_j" and X_j itself are represented in S as follows:

$$\begin{aligned}\bar{V}_j(S,0;\omega) &= e^{-\widetilde{u(S)\cdot F(\omega)}}\, V_j(\omega)\,,\\ \bar{X}_j(S,0;\omega) &= \bar{F}_j(S,0;\omega) + \bar{V}_j(S,0;\omega) = e^{-\widetilde{u(S)\cdot F(\omega)}}\, X_j(\omega)\,. \end{aligned} \tag{16.59}$$

Now let S' be another frame, related to S by the element $\exp(u)$. We can then write $S' = \exp(u)S = \exp(u)\exp(u(S))S_0 = \exp(u(S'))S_0$ with $\exp(u(S')) = \exp(u)\exp(u(S))$. We can relate dynamical variables in S' directly to those in S:

$$\begin{aligned}\bar{A}(S',0;\omega) &= e^{-\widetilde{u(S')\cdot F(\omega)}} e^{\widetilde{u(S)\cdot F(\omega)}} \bar{A}(S,0;\omega)\\ &= e^{-\widetilde{u(S)\cdot F(\omega)}} e^{-\widetilde{u\cdot F(\omega)}} e^{\widetilde{u(S)\cdot F(\omega)}} \bar{A}(S,0;\omega)\\ &= e^{-\widetilde{u\cdot \bar{F}(S,0;\omega)}} \bar{A}(S,0;\omega)\,,\\ \bar{A}(S',0;\omega) &= R(-u^j, \bar{F}_j(S,0;\omega))\bar{A}(S,0;\omega)\,. \end{aligned} \tag{16.60}$$

The final result has the same form as Eq.(16.13). In its derivation, use has been made of the fact that the $F_j(\omega)$ generate a realization of G, and also of

Eq.(16.21) and the definition in Eq.(16.58). We see that Eq.(16.60) looks just like the transformation law for $A(S, 0; \omega)$ in the Heisenberg picture, except that only the free parts of the X_j are used in constructing the R transformation. Just as in the Heisenberg picture the consistent use of Eq.(16.13) led to the realization conditions Eqs.(16.25) and (16.26), the consistent use of Eq.(16.60) in the Dirac picture will lead to analogous conditions, but these will be satisfied by virtue of Eq.(16.56). The canonical transform of Eq.(16.56) in the Dirac picture is, of course,

$$\{\bar{F}_j(S,0;\omega),\, \bar{F}_k(S,0;\omega)\} = c^l_{jk}\bar{F}_l(S,0;\omega) + d'_{jk}\,. \tag{16.61}$$

Turning next to the distribution functions in S' and S, we have from Eq.(16.57):

$$\begin{aligned}\bar{\rho}(S',0;\omega) &= e^{-\widetilde{u(S')\cdot F(\omega)}} e^{\widetilde{u(S')\cdot X(\omega)}} e^{-\widetilde{u(S)\cdot X(\omega)}} e^{\widetilde{u(S)\cdot F(\omega)}} \bar{\rho}(S,0;\omega) \\ &= e^{-\widetilde{u(S)\cdot F(\omega)}} e^{-\widetilde{u\cdot F(\omega)}} e^{\widetilde{u\cdot X(\omega)}} e^{\widetilde{u(S)\cdot X(\omega)}} e^{-\widetilde{u(S)\cdot X(\omega)}} \\ &\quad \times e^{\widetilde{u(S)\cdot F(\omega)}} \bar{\rho}(S,0;\omega) \\ &= e^{-\widetilde{u(S)\cdot F(\omega)}} \left\{ e^{-\widetilde{u\cdot F(\omega)}} e^{\widetilde{u\cdot X(\omega)}} \right\} e^{\widetilde{u(S)\cdot F(\omega)}} \bar{\rho}(S,0;\omega)\,. \end{aligned} \tag{16.62}$$

In going from the first to the second line we used the realization properties for F and X, as well as $\exp(u(S')) = \exp(u)\exp(u(S))$. Now the first and fourth canonical transformations in the last line, the ones outside the curly brackets, simply have the effect of replacing $F_j(\omega)$ and $X_j(\omega)$ in the second and third canonical transformations by $\bar{F}_j(S, 0; \omega)$ and $\bar{X}_j(S, 0; \omega)$, respectively; this follows from Eqs.(16.21), (16.58), and (16.59). Thus we get:

$$\begin{aligned}\bar{\rho}(S',0;\omega) &= e^{-\widetilde{u\cdot \bar{F}(S,0;\omega)}} e^{\widetilde{u\cdot \bar{X}(S,0;\omega)}} \bar{\rho}(S,0;\omega) \\ &= R(-u^j, \bar{F}_j(S,0;\omega)) R(u^j, \bar{X}_j(S,0;\omega)) \bar{\rho}(S,0;\omega) \end{aligned} \tag{16.63}$$

reflecting the change of frame $S' = \exp(u)S$. Unlike the cases we have had so far, the canonical transformations involved in this equation for $\bar{\rho}$ do *not* form elements of a canonical realization of G. Actually we have the product of two R transformations, the (inverse of the) first belonging to a "free" canonical realization and the second to a "fully interacting" realization of G. Equation (16.63) is the relativistic generalization of Eq.(16.11). If one wished to make a practical use of the Dirac picture, one would be more interested in the differential form of Eq.(16.63). This is easily derived. We take an "initial" frame S and a one-parameter subgroup $\exp(\tau u)$ in G, then consider the one-parameter family of frames $S_\tau = \exp(\tau u)S$. For a fixed tangent vector u, we can then develop a differential equation in τ:

$$\begin{aligned}\frac{\partial}{\partial\tau}\bar{\rho}(S_\tau,0;\omega) &= \frac{\partial}{\partial\tau}(e^{-\tau u\cdot\widetilde{\bar{F}(S,0;\omega)}}e^{\tau u\cdot\widetilde{\bar{X}(S,0;\omega)}}\,\bar{\rho}(S,0;\omega))\\ &= e^{-\tau u\cdot\widetilde{\bar{F}(S,0;\omega)}}(u\cdot\widetilde{\bar{X}(S,0;\omega)} - u\cdot\widetilde{\bar{F}(S,0;\omega)})\\ &\qquad \times e^{\tau u\cdot\widetilde{\bar{X}(S,0;\omega)}}\bar{\rho}(S,0;\omega)\\ &= e^{-\tau u\cdot\widetilde{\bar{F}(S,0;\omega)}}\{u\cdot\bar{V}(S,0;\omega)\,,\\ &\qquad e^{\tau u\cdot\widetilde{\bar{X}(S,0;\omega)}}\bar{\rho}(S,0;\omega)\}\\ &= \left\{e^{-\tau u\cdot\widetilde{\bar{F}(S,0;\omega)}}u\cdot\bar{V}(S,0;\omega)\,,\quad \bar{\rho}(S_\tau,0;\omega)\right\}\\ &= \{u\cdot\bar{V}(S_\tau,0;\omega),\bar{\rho}(S_\tau,0;\omega)\}\,,\end{aligned}$$

$$\frac{\partial}{\partial\tau}\bar{\rho}(S_\tau,0;\omega) = -\{\bar{\rho}(S_\tau,0;\omega),\, u\cdot\bar{V}(S_\tau,0;\omega)\}\,. \tag{16.64}$$

This is the relativistic extension of Eq.(16.10).

By writing the basic equations of the Dirac picture in the form of Eqs.(16.60), (16.63) and (16.64), we have freed them from explicit reference to the standard frame S_0. What remains is to generalize them from time zero in each frame to arbitrary times. We define $\bar{A}(S,b;\omega)$ and $\bar{\rho}(S,b;\omega)$ to be the same as the corresponding quantities at time zero in the frame $S' = \exp(bh)S$. Let us write H, F, and V for the total, free, and interaction, energies respectively. Then we have, with the use of Eqs.(16.60) and (16.63):

$$\begin{aligned}\bar{A}(S,b;\omega) &= \bar{A}(\exp(bh)S,0;\omega) = e^{-b\widetilde{\bar{F}(S,0;\omega)}}\bar{A}(S,0;\omega)\,.\\ \bar{\rho}(S,b;\omega) &= \bar{\rho}(\exp(bh)S,0;\omega) = e^{-b\widetilde{\bar{F}(S,0;\omega)}}e^{b\widetilde{\bar{H}(S,0;\omega)}}\bar{\rho}(S,0;\omega)\,.\end{aligned} \tag{16.65}$$

Now we must remember that in the Dirac picture, only the free energy F is represented in a fixed frame S by a time-independent function, which we can simply write as $\bar{F}(S;\omega)$; whereas the total as well as the interaction energies are time-dependent functions of ω, $\bar{H}(S,b;\omega)$ and $\bar{V}(S,b;\omega)$, respectively. Keeping this in mind, we can generalize Eqs.(16.60) and (16.63) to time b. For Eq.(16.60), the situation is quite similar to the one in the Heisenberg picture, and we get, for $S' = \exp(u)S$,

$$\bar{A}(S',b;\omega) = R(-u^j,\bar{F}_j(S,0;\omega))\bar{A}(S,b;\omega)\,. \tag{16.66}$$

For the behavior of probability distributions, we have a slightly more complicated law of change, We first get:

$$\begin{aligned}\bar{\rho}(S',b;\omega) =& e^{-b\widetilde{\bar{F}(S';\omega)}}e^{b\widetilde{\bar{H}(S',0;\omega)}}e^{-u\cdot\widetilde{\bar{F}(S,0;\omega)}}e^{u\cdot\widetilde{\bar{X}(S,0;\omega)}}\\ &\times e^{-b\widetilde{\bar{H}(S,0;\omega)}}e^{b\widetilde{\bar{F}(S;\omega)}}\bar{\rho}(S,b;\omega)\,.\end{aligned} \tag{16.67}$$

The first two factors can be moved to the right of the third one if we replace S' in them by S : our aim is to express the right-hand side completely in terms of S and $\exp(u)$. We get:

$$\begin{aligned}\bar{\rho}(S', b; \omega) =& R(-u^j, \bar{F}_j(S, 0; \omega)) e^{-b\widetilde{\bar{F}(S;\omega)}} e^{b\widetilde{\bar{H}(S,0;\omega)}} e^{u\cdot\widetilde{\bar{X}(S,0;\omega)}} e^{-b\widetilde{\bar{H}(S,0;\omega)}} \\ & \times e^{b\widetilde{\bar{F}(S;\omega)}} \bar{\rho}(S, b; \omega) . \end{aligned} \tag{16.68}$$

The second and sixth factors involving $\bar{F}(S;\omega)$ have the effect of replacing time zero by time b in the three transformations standing between them. After this, we can use Eq.(16.21) to absorb the effect of $\bar{H}(S, b; \omega)$. Finally, after these two steps, we get:

$$\bar{\rho}(S', b; \omega) = R(-u^j, \bar{F}_j(S, 0; \omega)) R(u^j, e^{b\widetilde{\bar{H}(S,b;\omega)}} \bar{X}_j(S, b; \omega)) \bar{\rho}(S, b; \omega) . \tag{16.69}$$

The differential form of this law, gotten by considering a one-parameter family of frames $S_\tau = \exp(\tau u)S$, generalizes Eq.(16.64). It reads:

$$\frac{\partial}{\partial \tau} \bar{\rho}(S_\tau, b; \omega) = -\left\{ \bar{\rho}(S_\tau, b; \omega),\ u \cdot e^{b\widetilde{\bar{H}(S_\tau,b;\omega)}} \bar{X}_j(S_\tau, b; \omega) - u \cdot \bar{F}(S_\tau, 0; \omega) \right\} . \tag{16.70}$$

Use has been made of Eq.(16.66). One must also notice that $u \cdot \bar{F}(S_\tau, 0; \omega)$ is actually independent of τ, because the transformation from $u \cdot \bar{F}(S, 0; \omega)$ to $u \cdot \bar{F}(S_\tau, 0; \omega)$ is generated by $u \cdot \bar{F}(S, 0; \omega)$ itself.

We summarize the main features of the relativistic Dirac picture in this way:

(a) *Probability Distribution Function*: $\bar{\rho}(S, b; \omega)$ determined by state of motion and dependent on the observer S as well as the time b.

(b) *Representative of Quantity A*: $\bar{A}(S, b; \omega)$ represents A in the frame S at time b.

(c) *Time Development in one frame S*: $\bar{A}(S, b; \omega) = e^{-b\widetilde{\bar{F}(S;\omega)}} \bar{A}(S, 0; \omega)$ and $\bar{\rho}(S, b; \omega) = e^{-b\widetilde{\bar{F}(S;\omega)}} e^{b\widetilde{\bar{H}(S,0;\omega)}} \bar{\rho}(S, 0; \omega)$, $\bar{F}$ and $\bar{H}$ being the representatives of the free and the total energies, respectively. (Only the first is time independent.)

(d) *Transformations between Frames*: For $S' = \exp(u)S$, the rule for dynamical variables is as in the "free" Heisenberg picture,

$$\bar{A}(S', b; \omega) = R(-u^j, \bar{F}_j(S, 0; \omega)) \bar{A}(S, b; \omega) .$$

For states, it is:

$$\bar{\rho}(S', b; \omega) = R(-u^j, \bar{F}_j(S, 0; \omega)) R(u^j, e^{b\widetilde{\bar{H}(S,b;\omega)}} \bar{X}_j(S, b; \omega)) \bar{\rho}(S, b; \omega) .$$

To these must be added the "free" and the "interacting" realization properties, the former expressed by the Eq.(16.61) and the latter by the canonical transform of Eq.(16.44):

$$\{\bar{X}_j(S,0;\omega),\, \bar{X}_k(S,0;\omega)\} = c^l_{jk}\bar{X}_l(S,0;\omega) + d_{jk}\ . \qquad (16.71)$$

The natural circumstances under which one would use the Dirac picture are when one is able to solve and obtain explicitly the "free" canonical realization of the relativity group G, but one is unable to obtain the interacting one. In that case, the behavior of dynamical variables is completely known, and for the probability distribution function one has the differential Eqs.(16.64) and (16.70) which one can try to solve in a perturbative form in the "potentials" V_j. In fact, this is how the Dirac picture was introduced in quantum mechanics. We see that actually there are several generalized "equations of motion" for $\bar{\rho}$ in this picture, corresponding to various possible tangent vectors in Eq.(16.64). Only for the choice $u = h$ does Eq.(16.64) become a differential equation for time development in one frame. For a general "direction" u in the Lie algebra of G, what appears in Eq.(16.64) is the component of the generalized potential along u, namely $u.\ \bar{V}$. A simplification that can occur is given by the following example. Suppose G is the Poincaré group, so that the X_j can be listed in the form J_j, P_j, K_j, and H. Then it can happen that the generalized potentials V_j vanish for the quantities J_j and P_j, but are nonzero only for K_j and H. This is expressed by saying that as far as the Euclidean subgroup $E(3)$ of the Poincaré group is concerned, there is no difference between the free and the interacting systems. Thus the only nontrivial "directions" u to be chosen correspond to the combinations $u = h + v_j k_j$ of h and k_j, and further from rotational properties it suffices to consider only the case $u = h + v_3 k_3$ for example. This means that the "interactions" are fully specified by the interaction terms in H and K_3 alone (although, of course, one is not saying that these can be chosen independently!). As is clear later on, if H contains an interaction term, but J_j and P_j do not, it is essential that K_j also contain interaction terms. Under these circumstances, we have a one-parameter family of nontrivial "equations of motion" for $\bar{\rho}$, corresponding to the one-parameter set of directions $u = h + v_3 k_3$ in the Lie algebra of G.

We do not make use of the relativistic Dirac picture in this book. Having understood the steps that lead from the requirement of "symmetry under a relativistic group G" to the canonical realizations of G, our approach in the following chapters is to develop some interesting canonical realizations of each possible G and to describe the corresponding physical systems. In all the work of this chapter it is evident that we could have replaced the ordinary phase space and the PB by a generalized phase space and a GPB, respectively. A Hamiltonian formalism involving GPB's combines with symmetry under a relativity group G to yield a generalized canonical realization of G. We give some examples of this in the cases of $R(3)$ and $E(3)$. Again, for each G, we must examine the extent to which neutral elements can be avoided in a realization of the Lie algebra of G. To these tasks we now turn.

Chapter 17

The Three-Dimensional Rotation Group

We study in this chapter some examples of classical mechanical realizations of the group $R(3)$. We have already explained in Chapter 15 some of the important structural properties of this group and its Lie algebra. For applications to physics this is an important group, partly for its own sake and partly because it appears as a subgroup in the other space-time transformation groups For this reason, it is useful to include a discussion of the linear matrix representations of this group, although it would take us too far afield to prove all the interesting properties we mention. We also give a description of the connection between $R(3)$ and the group $SU(2)$ consisting of all complex unitary unimodular matrices in two dimensions. This is the group that describes the rotational behavior of "spinors" in contrast to the behaviors of vectors and tensors, which are adequately described by the group $R(3)$. These matters, which are not exclusively related to classical mechanics, are taken up in the latter part of this chapter.

The first question to be examined in connection with a canonical realization of $R(3)$, or for that matter of any Lie group, is that of neutral elements. The basic Lie brackets for the Lie algebra of $R(3)$ were obtained in Chapter 15, and we repeat them here. We have three basic vectors l_j, j=1, 2, 3 in the Lie algebra, and they obey

$$[l_j, l_k] = \epsilon_{jkm} l_m \,. \tag{17.1}$$

A PB realization up to neutrals of this algebra, on a phase space with coordinates ω, consists of three functions $J_j(\omega)$ obeying

$$\{J_j(\omega), J_k(\omega)\} = \epsilon_{jkm} J_m(\omega) + d_{jk} \,. \tag{17.2}$$

The d_{jk} are pure numbers, or more generally, in case we are dealing with a GPB which may be singular, they are such functions of ω that their brackets with all functions of ω vanish identically. For the case of $R(3)$, we can show very easily that these neutral elements can always be "transformed away" by means of a suitable redefinition of the J_j, without affecting the canonical realization of $R(3)$ at all. It is evident from Eq. (17.2) that the d_{jk} are antisymmetric in their indices, so that there are just three independent neutral elements. We can write these as $d_{12} = d_3, d_{23} = d_1$ and $d_{31} = d_2$, or equivalently,

$$d_{jk} = \epsilon_{jkm} d_m\,, \quad d_m = \frac{1}{2}\epsilon_{mjk} d_{jk}\,. \tag{17.3}$$

Then, Eq.(17.2) takes the form:

$$\{J_j(\omega), J_k(\omega)\} = \epsilon_{jkm}(J_m(\omega) + d_m)\,, \tag{17.4}$$

or in terms of $J'_j(\omega) = J_j(\omega) + d_j$, we have the true realization:

$$\{J'_j(\omega), J'_k(\omega)\} = \epsilon_{jkm} J'_m(\omega)\,. \tag{17.5}$$

Thus there is no loss of generality in supposing that every canonical realization of $R(3)$ arises from a true PB realization of its Lie algebra, and therefore that the neutral elements d_{jk} in Eq. (17.2) are in fact zero. We can now connect this result with the appearance of $R(3)$ as a subgroup of the Euclidean, Galilei, Lorentz, and Poincaré groups to draw two useful conclusions. First, in any PB realization up to neutrals of the Lie algebra of any of these larger groups, we can assume right at the start that in the subalgebra corresponding to $R(3)$ the PBs do not involve any neutral elements. Second, in attempting to reduce to a minimum the number of independent neutral elements present in the remaining PB relations, we must not resort to redefinitions of the $J_j(\omega)$ by adding neutral elements to them, because that would reintroduce neutral elements in those PB relations in which there were none to begin with!

The true realization of Eq. (17.1) in which each $J_k(\omega)$ is zero is called the *trivial* realization of the $R(3)$ Lie algebra. Any other is called a *nontrivial realization.* From the form of the basic relations in Eq. (17.1), one sees that in any nontrivial realization *none of the $J_j(\omega)$ may vanish identically.* For example, if J_3 were identically zero, the vanishing of the PBs of J_3 with J_1 and J_2 would imply that J_2 and J_1, respectively, vanish as well. More generally, as the reader can easily verify, no numerical linear combination of the three J's can vanish identically in a nontrivial realization of $R(3)$.

The most familiar canonical realization of $R(3)$ is generated by the expressions for the angular momentum of a particle moving in three-dimensional space. With respect to a definite Cartesian coordinate system located in space, a particle is described by three canonically conjugate pairs of variables (q_j, p_j), $j = 1, 2, 3$. The q's are the Cartesian coordinates and the p's the Cartesian linear momenta. Relative to this coordinate system, the Cartesian components of angular momentum are defined as:

$$J_j(q,p) = \epsilon_{jkm} q_k p_m \tag{17.6}$$

By using the basic PB's among the q's and p's we verify that we have a true realization of the Lie algebra of $R(3)$:

$$\begin{aligned}
\{J_j(q,p), J_k(q,p)\} &= \epsilon_{jmn}\epsilon_{krs}\{q_m p_n, q_r p_s\} \\
&= \epsilon_{jmn}\epsilon_{krs}(q_m(-\delta_{nr})p_s + p_n\delta_{ms}q_r) \\
&= -\epsilon_{jmn}\epsilon_{kns}q_m p_s + \epsilon_{jmn}\epsilon_{krm}q_r p_n \\
&= (\delta_{jk}\delta_{ms} - \delta_{js}\delta_{mk})q_m p_s \\
&\quad -(\delta_{jk}\delta_{nr} - \delta_{jr}\delta_{nk})q_r p_n \\
&= (\delta_{jr}\delta_{kn} - \delta_{jn}\delta_{kr})q_r p_n = \epsilon_{jkm}\epsilon_{mrn}q_r p_n\,, \\
\{J_j(q,p), J_k(q,p)\} &= \epsilon_{jkm}J_m(q,p)\,. \qquad (17.7)
\end{aligned}$$

The canonical transformations generated by these J's act in a simple way on the six-dimensional phase space of the q's and p's: they reduce to three-dimensional linear transformations acting separately on the q_j and the p_j, characteristic of the fact that they form components of Cartesian vectors. A simple computation yields:

$$\{J_j(q,p), q_k\} = \epsilon_{jkm}q_m, \quad \{J_j(q,p), p_k\} = \epsilon_{jkm}p_m\,. \qquad (17.8)$$

Using these brackets, we get the effects of a finite canonical transformation to be

$$\exp[-\widetilde{\boldsymbol{\alpha}\cdot \boldsymbol{J}(q,p)}]q_j = A_{jk}(\boldsymbol{\alpha})q_k, \quad \exp[-\widetilde{\boldsymbol{\alpha}\cdot \boldsymbol{J}(q,p)}]p_j = A_{jk}(\boldsymbol{\alpha})p_k, \qquad (17.9)$$

$A(\boldsymbol{\alpha})$ being the 3×3 orthogonal matrix defined in Eq.(15.11). The J_j themselves also transform in the same linear fashion:

$$\exp[-\widetilde{\boldsymbol{\alpha}\cdot \boldsymbol{J}(q,p)}]J_j(q,p) = A_{jk}(\boldsymbol{\alpha})J_k(q,p)\,. \qquad (17.10)$$

In fact, these linear transformation laws (Eq.(17.9) or (17.10)) under rotations are the distinguishing marks of a vector, and it should be evident by now that such behavior under finite canonical transformations corresponding to finite rotations is completely equivalent to the PB relationships of the type of Eqs.(17.7) or (17.8) between the J_j and the vectorial quantity concerned. That is, in any canonical realization of $R(3)$ given by three quantities $J_j(\omega)$, if there are three functions $V_j(\omega)$ obeying

$$\{J_j(\omega), V_k(\omega)\} = \epsilon_{jkm}V_m(\omega)\,, \qquad (17.11)$$

then under a rotation the V's transform as a Cartesian vector,

$$\exp[-\widetilde{\boldsymbol{\alpha}\cdot \boldsymbol{J}(\omega)}]V_j(\omega) = A_{jk}(\boldsymbol{\alpha})V_k(\omega)\,, \qquad (17.12)$$

and vice versa.

A very interesting application of the $R(3)$ group structure occurs in the description of rigid body rotations. We first discuss this subject along the usual Lagrangian lines, then find that the equations of motion can be very simply and elegantly formulated in terms of a GPB which is nothing but the Lie algebra of $R(3)$ itself. This leads to the idea of a family of classical dynamical systems which we call "pure spin systems".

The Description of Rigid Body Rotations

Using the idea of mass points or particles as basic, a rigid body may be defined as a collection of mass points in three-dimensional space with forces of constraint acting on these masses in such a way that the magnitude of the distance between any two mass points remains constant in time. These forces are referred to as "rigid body constraints" From a kinematical point of view, it is evident that the motion of a rigid body splits up into two independent parts. One is the translational motion of the entire body, and this is adequately described by the translational motion of one fixed representative mass point in the body. The other is the rotational motion of the body about this representative point, resulting in a time dependence of the orientation of the body in space independent of its location. Of course, it is only at the level of kinematics that one can speak of these two as independent motions, and for prescribed forces acting on the body the Newtonian equations of motion generally link one kind of motion to the other. The use of the group $R(3)$ comes in only in connection with the rotational motion; thus we shall right away confine ourselves to rigid body motion in which one representative point of the body is held fixed at some location in space, and the body is permitted only to change its orientation in space. Such motion is called "rigid body rotation". In this case, in addition to the constraint forces that cause the body to be rigid, there is a force of constraint that prevents the representative point from changing its location in space. We assume that all these constraint forces do no work. It should also be clear that our discussion is not restricted to rigid bodies made up of discrete points, but includes continuous distributions of mass. This just requires replacing discrete summations by integrations in the relevant formulae.

Let us consider now the problem of choosing generalized coordinates to describe the rotational motion of a rigid body. Denote by O the point of the body that remains fixed in space, and with O as origin let S be an inertial frame, that is a fixed set of Cartesian axes, with respect to which the motion of the body is described. As usual, we write $\boldsymbol{e}_j, j = l, 2, 3$ for three unit vectors along the directions of the three axes in S. At the same time we can imagine choosing three orthogonal axes fixed in the body and with O as origin, and moving with respect to S as the body moves. Clearly the orientation of these three body-fixed axes with respect to the space-fixed axes in S at each instant of time determines completely the configuration of the rigid body itself at that instant of time. We imagine the body-fixed axes to define at each instant of time a Cartesian coordinate system $S(t)$, which does not in general represent an inertial reference frame. It is convenient to choose S and the origin of time such that $S(0) = S$. At every subsequent instant of time,there is a uniquely determined element $a(t)$ in the group $R(3)$ that leads from S to $S(t) : S(t) = a(t)S$ in the notation of Chapter 15. *This group element $a(t)$ can be chosen as the generalized coordinate for describing rigid body rotations.* In practice, one introduces a coordinate system for $R(3)$, and then the three time-dependent coordinates of *$a(t)$* become three numerical generalized coordinates for rigid body rotation.

But it should be stressed that the most important properties of the motion must necessarily have a significance that is independent of the particular coordinates chosen for $R(3)$. In most of the usual discussions, one introduces Euler angles as coordinates for $R(3)$, and then one has three angle variables $\phi(t), \theta(t), \psi(t)$ as generalized q's. This choice is well suited to obtaining a complete solution of the motion (for certain types of forces) by starting with some given set of initial conditions and integrating the equations of motion. However, here we are less interested in exhibiting the full solution of the motion than in studying the formal structure of the equations of motion themselves and the use of the $R(3)$ group structure in describing the kinematics of rigid body rotation. We therefore find it much more convenient to use canonical coordinates of the first kind for $R(3)$, that is, we shall specify elements of $R(3)$ by the axis and angle of rotation. Let us then write $a(t) = \exp(-\alpha_j(t)l_j)$ for the element of $R(3)$ taking S to $S(t)$, so that the three functions $\alpha_j(t)$ are the generalized coordinates for the problem. Let us also write $\boldsymbol{e}_j(t)$ for the three unit vectors in the directions of the three body fixed axes in the rigid body. We then have

$$S(t) = \exp(-\alpha_j(t)l_j)S : \boldsymbol{e}_j(t) = A_{kj}(\boldsymbol{\alpha}(t))\boldsymbol{e}_k, \ \boldsymbol{e}_j(0) = \boldsymbol{e}_j\,. \tag{17.13}$$

Consider some vector $\boldsymbol{V}$ in space. We can resolve it into its components $V_j^{(S)}$ along the space-fixed axes, or at each instant t into its components $V_j^{(B)}$ along the body-fixed axes. Between these two sets of components we have a time dependent linear connection:

$$\boldsymbol{V} = V_j^{(S)}\boldsymbol{e}_j = V_j^{(B)}\boldsymbol{e}_j(t), V_j^{(S)} = A_{jk}(\boldsymbol{\alpha}(t))V_k^{(B)}\,. \tag{17.14}$$

If the vector $\boldsymbol{V}$ is "really" time independent, then the components $V_j^{(S)}$ are constant in time, and $\boldsymbol{V}$ has a fixed direction in space; but then the components $V_j^{(B)}$ become naturally time dependent. On the other hand, if $\boldsymbol{V}$ is a vector "fixed rigidly in the body", for example the vector leading from one mass point in the body to another one, then it is the "body components" $V_j^{(B)}$ that are constant in time, and the "space components" $V_j^{(S)}$ that vary with time. For a general $\boldsymbol{V}$, one has time-dependent space as well as body components.

With the $\alpha_j(t)$ as generalized coordinates, the generalized velocities are, of course, the derivatives $\dot{\alpha}_j(t)$. But it is more natural to introduce another set of variables, called the angular velocities. They arise in the following way.

The body-fixed frames $S(t), S(t+\delta t)$ at two close-by instants of time must be related by an infinitesimal element of $R(3)$, allowing us to write:

$$S(t+\delta t) = \exp(-\omega_j(t)l_j\delta t)S(t)\,. \tag{17.15}$$

The three variables $\omega_j(t)$ are the angular velocities. The *angular velocity vector* $\boldsymbol{\omega}(t)$ has $\omega_j(t)$ for its Cartesian components, and since the "initial" frame that appears on the right hand side in Eq. (17.15) is $S(t)$, *to be consistent we must interpret* $\omega_j(t)$ *to be the components of* $\boldsymbol{\omega}(t)$ *along the axes in* $S(t)$. We call these variables the angular velocities because the vector $\boldsymbol{\omega}(t)$ gives the axis and instantaneous rate of rotation of the body-fixed frame $S(t)$ at time t. But it is important to understand that these variables are not true generalized velocities, in the sense that they are not the time derivatives of some generalized position variables. Rather, the angular velocities are certain linear combinations, with $\boldsymbol{\alpha}$ dependent coefficients, of the true generalized velocities $\dot{\alpha}_j$; we can think of the ω_j as generalized velocity-type variables but not as velocities. The linear transformation connecting ω_j and $\dot{\alpha}_j$ is determined by the structure of $R(3)$ and it is an interesting exercise to work it out. First, let us show that the time derivative of the matrix $A(\boldsymbol{\alpha}(t))$ is given in a simple fashion in terms of $\omega_j(t)$. From Eq. (17.15) we have:

$$\boldsymbol{e}_j(t+\delta t) = A_{kj}(\boldsymbol{\omega}(t)\delta t)\boldsymbol{e}_k(t)\,, \tag{17.16}$$

and combining this with Eq. (17.13) we get:

$$A(\boldsymbol{\alpha}(t+\delta t)) = A(\boldsymbol{\alpha}(t))A(\boldsymbol{\omega}(t)\delta t)\,. \tag{17.17}$$

Going to the limit $\delta t \to 0$ and using the explicit form of the matrix $A(\boldsymbol{\alpha})$, we find:

$$\dot{A}_{jk}(\boldsymbol{\alpha}(t)) = -A_{km}(\boldsymbol{\alpha}(t))\epsilon_{mjl}\omega_l(t)\,. \tag{17.18}$$

This equation says the following: Since the matrix $A(\boldsymbol{\alpha}(t))$ is an orthogonal matrix for all t, by differentiation of this condition with respect to t we get the result that the matrix $A^{\sim}\dot{A}$ must always be an antisymmetric matrix. Equation (17.18) identifies the three independent elements of $A^{\sim}\dot{A}$ with the three components of $\boldsymbol{\omega}(t)$. Now to relate $\omega_j(t)$ and $\dot{\alpha}_j(t)$, we can combine Eqs. (17.13) and (17.15) to write

$$\exp(-\boldsymbol{\alpha}(t+\delta t)) = \exp(-\boldsymbol{\omega}(t)\delta t)\exp(-\boldsymbol{\alpha}(t))$$

for the relevant elements of $R(3)$, or taking inverses on both sides to get rid of the minus signs, we have the basic condition:

$$\exp(\boldsymbol{\alpha}(t+\delta t)) = \exp(\boldsymbol{\alpha}(t))\exp(\boldsymbol{\omega}(t)\delta t)\,. \tag{17.19}$$

In the limit $\delta t \to 0$, we use the group composition law for $R(3)$ and Eq.(13.25) (or more precisely its analogue for the functions $\hat{\eta}^k_j(\alpha)$) to obtain:

$$\dot{\alpha}_j(t) = \hat{\eta}_{jk}(\boldsymbol{\alpha}(t))\omega_k(t)\,. \tag{17.20}$$

Here we must remember that canonical coordinates are being used for $R(3)$, and that where we previously wrote $\hat{\eta}^j_k(\boldsymbol{\alpha})$ we have now written $\hat{\eta}_{jk}(\boldsymbol{\alpha}(t))$. This is in line with the earlier convention, adopted in Chapter 13, that the superscripts (subscripts) on the functions $\eta^j_k, \xi^j_k, \hat{\eta}^j_k, \hat{\xi}^j_k$ are to be used as row (column) indices for matrix multiplication. Now the evaluation of these matrices is a straightforward matter. Let us begin with the matrix $\Xi(\alpha) = \left\|\xi^j_k(\alpha)\right\| = \|\xi_{jk}(\alpha)\|$. According to Eq. (17.1), in the case of $R(3)$ the structure constant c^l_{jk} is now equal to ϵ_{jkl}. Given the vector α_j in the Lie algebra of $R(3)$, we must set up the corresponding matrix $C(\boldsymbol{\alpha})$. It is (see Eq. (13.92))

$$C(\boldsymbol{\alpha}) = \|C_{jk}(\boldsymbol{\alpha})\|\,, C_{jk}(\boldsymbol{\alpha}) = -\epsilon_{mjk}\alpha_m\,. \tag{17.21}$$

Writing this out in full we have:

$$C(\boldsymbol{\alpha}) = \left\| \begin{matrix} 0 & -\alpha_3 & \alpha_2 \\ \alpha_3 & 0 & -\alpha_1 \\ -\alpha_2 & \alpha_1 & 0 \end{matrix} \right\| . \tag{17.22}$$

In computing $\Xi(\boldsymbol{\alpha})$, we have to deal with powers of the matrix $C(\boldsymbol{\alpha})$, but these are easy to handle. Firstly we find that the square of $C(\boldsymbol{\alpha})$ is a projection matrix $P(\boldsymbol{\alpha})$, apart from a factor:

$$\begin{aligned} \{C(\boldsymbol{\alpha})C(\boldsymbol{\alpha})\}_{jk} &= C_{jm}(\boldsymbol{\alpha})C_{mk}(\boldsymbol{\alpha}) = +\epsilon_{jmr}\epsilon_{mks}\alpha_r\alpha_s \\ &= (\delta_{js}\delta_{rk} - \delta_{jk}\delta_{rs})\alpha_r\alpha_s = -\alpha^2 P_{jk}(\boldsymbol{\alpha}), \\ P_{jk}(\boldsymbol{\alpha}) &= \delta_{jk} - \frac{\alpha_j\alpha_k}{\alpha^2},\ \alpha^2 = \alpha_j\alpha_j\,. \end{aligned} \tag{17.23}$$

This matrix $P(\boldsymbol{\alpha})$ has the following properties:

$$P(\boldsymbol{\alpha})P(\boldsymbol{\alpha}) = P(\boldsymbol{\alpha}), P(\boldsymbol{\alpha})C(\boldsymbol{\alpha}) = C(\boldsymbol{\alpha})P(\boldsymbol{\alpha}) = C(\boldsymbol{\alpha})\,. \tag{17.24}$$

Thus any nonzero even power of $C(\boldsymbol{\alpha})$ is essentially $P(\boldsymbol{\alpha})$, whereas any odd power is essentially $C(\boldsymbol{\alpha})$:

$$[C(\boldsymbol{\alpha})]^{2n} = (-1)^n\alpha^{2n}P(\boldsymbol{\alpha}), [C(\boldsymbol{\alpha})]^{2n+1} = (-1)^n\alpha^{2n}C(\boldsymbol{\alpha})\,. \tag{17.25}$$

Using these results, we get from Eq.(13.95):

$$\begin{aligned} \Xi(\boldsymbol{\alpha}) &= 1\!\!1 + \sum_{n=1}^{\infty}\frac{[C(\boldsymbol{\alpha})]^n}{(n+1)!} = 1\!\!1 + \sum_{n=0}^{\infty}\frac{[C(\boldsymbol{\alpha})]^{2n+1}}{(2n+2)!} + \sum_{n=1}^{\infty}\frac{[C(\boldsymbol{\alpha})]^{2n}}{(2n+1)!} \\ &= 1\!\!1 + \sum_{n=0}^{\infty}\frac{(-1)^n\alpha^{2n}}{(2n+2)!}C(\boldsymbol{\alpha}) + \sum_{n=1}^{\infty}\frac{(-1)^n\alpha^{2n}}{(2n+1)!}P(\boldsymbol{\alpha}) \\ &= 1\!\!1 + \frac{1-\cos\alpha}{\alpha^2}C(\boldsymbol{\alpha}) + \left(\frac{\sin\alpha}{\alpha} - 1\right)P(\boldsymbol{\alpha})\,. \end{aligned} \tag{17.26}$$

Writing out the matrix elements, we have found that the functions $\xi_{jk}(\boldsymbol{\alpha})$, evaluated in canonical coordinates and for the group $R(3)$, are:

$$\xi_{jk}(\boldsymbol{\alpha}) = \delta_{jk}\frac{\sin\alpha}{\alpha} + \frac{\alpha_j\alpha_k}{\alpha^2}\left(1 - \frac{\sin\alpha}{\alpha}\right) - \epsilon_{jkl}\alpha_l\frac{1-\cos\alpha}{\alpha^2}\,. \tag{17.27}$$

Inverting the matrix $\Xi(\boldsymbol{\alpha})$ to find $H(\boldsymbol{\alpha})$ is quite easy. We can write $H(\boldsymbol{\alpha})$ as a linear combination of $1\!\!1, C(\boldsymbol{\alpha})$, and $P(\boldsymbol{\alpha})$ and use Eq.(17.24) and the definition of $P(\boldsymbol{\alpha})$ to solve the equation $H\Xi = 1\!\!1$. In this way we get:

$$\begin{aligned} H(\boldsymbol{\alpha}) &= 1\!\!1 - \frac{1}{2}C(\boldsymbol{\alpha}) + \left(\frac{\alpha}{2}\cot\frac{\alpha}{2} - 1\right)P(\boldsymbol{\alpha})\,, \\ \eta_{jk}(\boldsymbol{\alpha}) &= \delta_{jk}\frac{\alpha}{2}\cot\frac{\alpha}{2} + \frac{\alpha_j\alpha_k}{\alpha^2}\left(1 - \frac{\alpha}{2}\cot\frac{\alpha}{2}\right) + \epsilon_{jkl}\frac{\alpha_l}{2}\,. \end{aligned} \tag{17.28}$$

Because we are using canonical coordinates, the matrices $\hat{H}(\boldsymbol{\alpha})$ and $\hat{\Xi}(\boldsymbol{\alpha})$ arise from $H(\boldsymbol{\alpha})$ and $\Xi(\boldsymbol{\alpha})$ by changing the sign of α_j (cf Eq.(13.105)):

$$\begin{aligned}
\hat{H}(\boldsymbol{\alpha}) &= H(-\boldsymbol{\alpha}), \hat{\Xi}(\boldsymbol{\alpha}) = \Xi(-\boldsymbol{\alpha}) ; \\
\hat{\eta}_{jk}(\boldsymbol{\alpha}) &= \delta_{jk}\frac{\alpha}{2}\cot\frac{\alpha}{2} + \frac{\alpha_j\alpha_k}{\alpha^2}\left(1 - \frac{\alpha}{2}\cot\frac{\alpha}{2}\right) - \epsilon_{jkl}\frac{\alpha_l}{2} ; \\
\hat{\xi}_{jk}(\boldsymbol{\alpha}) &= \delta_{jk}\frac{\sin\alpha}{\alpha} + \frac{\alpha_j\alpha_k}{\alpha^2}\left(1 - \frac{\sin\alpha}{\alpha}\right) + \epsilon_{jkl}\alpha_l\frac{1-\cos\alpha}{\alpha^2} .
\end{aligned} \tag{17.29}$$

The connection between the angular velocities ω_j and the true generalized velocities $\dot{\alpha}_j$ is now clear: we just put $\hat{\eta}_{jk}(\boldsymbol{\alpha})$ obtained in Eq.(17.29) into Eq.(17.20).

Let us proceed to express the important dynamical quantities such as angular momentum and kinetic energy in terms of the α's and $\dot{\alpha}$'s. We enumerate the mass points constituting the rigid body with an index λ. The masses can be written m_λ. The Cartesian coordinates of the λth particle, measured in the body-fixed axes, will be time independent; write $\zeta_j^{(\lambda)}$ for them. Then the Cartesian coordinates $x_j^{(\lambda)}$ of this particle measured with respect to the space fixed frame S will be time dependent:

$$x_j^{(\lambda)} = A_{jk}(\boldsymbol{\alpha}(t))\zeta_k^{(\lambda)} . \tag{17.30}$$

This is an example of Eq.(17.14); the time dependence of the x's comes only from the time dependence of the matrix $A(\boldsymbol{\alpha}(t))$. Equation (17.6) gives the elementary contribution to the angular momentum coming from a single particle. The total angular momentum comes from a sum on λ. Remembering that in the nonrelativistic case the p's in Eq.(17.6) are the mass times the $\dot{q}$'s, the total angular momentum of the rigid body can be calculated as follows:

$$\begin{aligned}
J_j &= \sum_\lambda m_\lambda \epsilon_{jkl} x_k^{(\lambda)} \dot{x}_l^{(\lambda)} = \sum_\lambda m_\lambda \epsilon_{jkl} A_{kr}(\boldsymbol{\alpha}(t))\zeta_r^{(\lambda)} \dot{A}_{ls}(\boldsymbol{\alpha}(t))\zeta_s^{(\lambda)} \\
&= -\sum_\lambda m_\lambda \zeta_r^{(\lambda)}\zeta_s^{(\lambda)} \epsilon_{jkl}\epsilon_{msn} A_{kr}(\boldsymbol{\alpha}(t)) A_{lm}(\boldsymbol{\alpha}(t))\omega_n(t) .
\end{aligned} \tag{17.31}$$

These J_j are, of course, the components of the total angular momentum with respect to the axes in S. To simplify this expression, we use the following property of the tensor ϵ:

$$\epsilon_{jkl} A_{kr} A_{lm} = A_{jp}\epsilon_{prm} . \tag{17.32}$$

This is true for any proper orthogonal rotation matrix A, and merely expresses the invariance of Lie brackets in the Lie algebra of $R(3)$ under the adjoint representation. Then J_j takes this form:

$$\begin{aligned}
J_j &= -\sum_\lambda m_\lambda \zeta_r^{(\lambda)}\zeta_s^{(\lambda)}\epsilon_{prm}\epsilon_{msn}A_{jp}(\boldsymbol{\alpha}(t))\omega_n(t) \\
&= A_{jp}(\boldsymbol{\alpha}(t))\left\{\sum_\lambda m_\lambda(\delta_{pn}\delta_{rs}-\delta_{ps}\delta_{rn})\zeta_r^{(\lambda)}\zeta_s^{(\lambda)}\right\}\omega_n(t) \\
&= A_{jp}(\boldsymbol{\alpha}(t))I_{pn}\omega_n(t)\,, \\
I_{pn} &= \sum_\lambda m_\lambda(\delta_{pn}\zeta_r^{(\lambda)}\zeta_r^{(\lambda)}-\zeta_p^{(\lambda)}\zeta_n^{(\lambda)})\,.
\end{aligned} \tag{17.33}$$

The quantities I_{pn} are called the components of the *moment of inertia tensor* of the rigid body. They depend on the point O in the rigid body that is chosen as the origin of the body-fixed axes, as well as on the orientation of the body-fixed axes. But once these are chosen in some way, the I_{pn} are constants and do not change with time. Considered as a 3×3 matrix, $I = \|I_{jk}\|$ is both *symmetric and positive definite*. Because I is symmetric, by means of an orthogonal transformation I can be made diagonal, and then the three diagonal elements I'_{jj}, for which we simply write I'_j, are positive definite quantities. This orthogonal transformation that takes us from the symmetric I to the diagonal I' is just a transformation to a new set of axes fixed in the body, with respect to which the off-diagonal components of the moment of inertia tensor are all zero. Such a system of body-fixed axes is called a *principal axis system*. We could, if we wish, assume all along that the body-fixed axes we have chosen are principal axes, but there is no need to make this assumption for the moment. All that we must remember is that such axes are always available, and that this is so whatever be the fixed point O in the rigid body.

Going back to the expression for the angular momenta, for the case that we are discussing we can regard each of the J's as a function of the α's and $\dot{\alpha}$'s, with the simplification that they are linear in the latter; combining Eq.(17.33) with Eq.(17.20) we have:

$$J_j(\boldsymbol{\alpha},\dot{\boldsymbol{\alpha}}) = A_{jk}(\boldsymbol{\alpha})I_{km}\hat{\xi}_{mn}(\boldsymbol{\alpha})\dot{\alpha}_n \tag{17.34}$$

We can adopt matrix notation in which we regard J_j and $\dot{\alpha}_j$ as column vectors, with three rows in each case, and then Eq.(17.34) takes this simple form:

$$J(\boldsymbol{\alpha},\dot{\boldsymbol{\alpha}}) = A(\boldsymbol{\alpha})I\Xi(\boldsymbol{\alpha})\dot{\boldsymbol{\alpha}}\,. \tag{17.35}$$

It is useful to introduce also the components of the angular momentum vector $\mathbf{J}$ resolved with respect to the axes in the body frame $S(t)$. We shall write $S_j(\boldsymbol{\alpha},\dot{\boldsymbol{\alpha}})$ for these components. If we compare Eqs.(17.34) and (17.35) with Eq.(17.14), we see that the S_j are obtained from Eqs.(17.34) or (17.35) by just dropping the matrix $A(\boldsymbol{\alpha})$:

$$\begin{aligned}
S_j(\boldsymbol{\alpha},\dot{\boldsymbol{\alpha}}) &= I_{jm}\omega_m(\boldsymbol{\alpha},\dot{\boldsymbol{\alpha}}) = I_{jm}\hat{\xi}_{mn}(\boldsymbol{\alpha})\dot{\alpha}_n\,, \\
S(\boldsymbol{\alpha},\dot{\boldsymbol{\alpha}}) &= I\omega(\boldsymbol{\alpha},\dot{\boldsymbol{\alpha}}) = I\hat{\Xi}(\boldsymbol{\alpha})\dot{\boldsymbol{\alpha}}\,.
\end{aligned} \tag{17.36}$$

The components of the total angular momentum resolved with respect to the body axes are just linear combinations of the angular velocities $\omega_j(t)$, with the various components of the inertia tensor for coefficients. Next we consider the expression for the total kinetic energy of the rigid body. This again is a sum of contributions, one for each mass point in the body, and these individual terms are just the familiar nonrelativistic forms. Writing T for the kinetic energy, we have:

$$\begin{aligned} T(\boldsymbol{\alpha}, \dot{\boldsymbol{\alpha}}) &= \frac{1}{2}\sum_{\lambda} m_\lambda \dot{x}_j^{(\lambda)} \dot{x}_j^{(\lambda)} = \frac{1}{2}\sum_{\lambda} m_\lambda \zeta_r^{(\lambda)} \zeta_s^{(\lambda)} \dot{A}_{jr}(\boldsymbol{\alpha}(t)) \dot{A}_{js}(\boldsymbol{\alpha}(t)) \\ &= \frac{1}{2}\sum_{\lambda} m_\lambda \zeta_r^{(\lambda)} \zeta_s^{(\lambda)} A_{jm}(\boldsymbol{\alpha}(t)) A_{jn}(\boldsymbol{\alpha}(t)) \epsilon_{mru} \epsilon_{nsv} \omega_u(t) \omega_v(t) \\ &= \frac{1}{2}\sum_{\lambda} m_\lambda \zeta_r^{(\lambda)} \zeta_s^{(\lambda)} (\delta_{rs}\delta_{uv} - \delta_{rv}\delta_{su}) \omega_u \omega_v = \frac{1}{2}\omega_r I_{rs} \omega_s \,. \end{aligned} \tag{17.37}$$

Thus the kinetic energy T is a simple bilinear expression in the angular velocities ω, the components of the inertia tensor once again appearing as the coefficients. We can write the expression for T in several equivalent ways, depending on whether we wish to express it in terms of ω, S, J or $\dot{\alpha}$. We find:

$$\begin{aligned} T(\boldsymbol{\alpha}, \dot{\boldsymbol{\alpha}}) &= \frac{1}{2}\tilde{\omega} I \omega = \frac{1}{2}\tilde{S} I^{-1} S = \frac{1}{2}\tilde{J} A(\boldsymbol{\alpha}) I^{-1} \tilde{A}(\boldsymbol{\alpha}) J \\ &= \frac{1}{2}\tilde{\dot{\boldsymbol{\alpha}}} \tilde{\hat{\Xi}}(\boldsymbol{\alpha}) I \hat{\Xi}(\boldsymbol{\alpha}) \dot{\boldsymbol{\alpha}} \,. \end{aligned} \tag{17.38}$$

The last form gives in full the dependence of T on the generalized coordinates and velocities. (In the above, we have used the tilde sign ˜ to denote the transpose of a matrix or a column vector). The point to be emphasized is that ω_j, S_j and J_j are all generalized velocity-type quantities, whereas $\dot{\alpha}_j$ are true velocities.

The expression of the kinetic energy in terms of the spatial components J_j of the total angular momentum has an interesting property. In principle, since the matrices $A(\boldsymbol{\alpha})$ and $\tilde{A}(\boldsymbol{\alpha})$ appear explicitly in this expression, one might think that there is in general a time dependence of T stemming from two sources: one from a possible dependence on time of the J_j and the other from the time dependence of the matrices $A(\boldsymbol{\alpha})$ and $\tilde{A}(\boldsymbol{\alpha})$. *But the latter time dependences do not contribute to a time variation of T at all, and any time variation of T must arise from a time variation of the J_j.* This is a purely kinematical result and is quite independent of the forces that may be acting on the rigid body, and of the equations of motion. Let us establish this fact. We first rewrite Eq.(17.18) in matrix form, making use of the definition of Eq. (17.21) for this purpose:

$$\dot{A}(\boldsymbol{\alpha}(t)) = A(\boldsymbol{\alpha}(t)) C(\boldsymbol{\omega}(t)) \,. \tag{17.39}$$

The matrix C is defined once we are given any triplet of real numbers. Now we can express $\boldsymbol{\omega}(t)$ in terms of J, and Eq. (17.39) takes the form:

$$\dot{A}(\boldsymbol{\alpha}) = A(\boldsymbol{\alpha}) C(I^{-1} \tilde{A}(\boldsymbol{\alpha}) J) \,. \tag{17.40}$$

[Remember that I^{-1} and $A^{\sim}$ are 3×3 matrices and J a 3×1 column matrix, so that the product $I^{-1}A^{\sim}J$ is a column vector with three entries]. The possible time dependence of T that we are concerned with consists of two terms in $\dot{T}$, one gotten from differentiating A and one from differentiating $A^{\sim}$; we then use Eq. (17.40) and the transposed equation to evaluate these two terms:

$$\begin{aligned}&\frac{1}{2}J^{\sim}\dot{A}I^{-1}A^{\sim}J+\frac{1}{2}J^{\sim}AI^{-1}\dot{A}^{\sim}J\\&=\frac{1}{2}J^{\sim}AC(I^{-1}A^{\sim}J)I^{-1}A^{\sim}J-\frac{1}{2}J^{\sim}AI^{-1}C(I^{-1}A^{\sim}J)A^{\sim}J\,.\end{aligned}\tag{17.41}$$

(We have used the antisymmetry of C in the second term.) But whatever be the argument $\boldsymbol{\alpha}$ of $C(\boldsymbol{\alpha})$, we always have $C(\boldsymbol{\alpha})\alpha=0$:

$$[C(\boldsymbol{\alpha})\alpha]_j=C(\boldsymbol{\alpha})_{jk}\alpha_k=-\epsilon_{mjk}\alpha_m\alpha_k=0\,;\tag{17.42}$$

thus in particular we have:

$$C(I^{-1}A^{\sim}J)I^{-1}A^{\sim}J=0,J^{\sim}AI^{-1}C(I^{-1}A^{\sim}J)=0\,.\tag{17.43}$$

Thus the two terms in Eq. (17.41) vanish separately, proving the assertion. For the time derivative of T we can then write:

$$\frac{dT}{dt}=\frac{1}{2}\dot{J}^{\sim}AI^{-1}A^{\sim}J+\frac{1}{2}J^{\sim}AI^{-1}A^{\sim}\dot{J}=J^{\sim}AI^{-1}A^{\sim}J\,.\tag{17.44}$$

We repeat that this interesting result is a purely kinematical one.

We turn now to the derivation of the Lagrangian equations of motion for rigid body rotation. We will restrict ourselves to the simple case where no external forces, and thus no external torques, act on the body. In this case, the Lagrangian coincides with the kinetic energy T. There are in fact two ways in which the equations of motion could be obtained, one via the usual Lagrangian techniques and the other by a more physical method which is actually simpler because we have assumed the absence of external forces. Let us work with the Lagrangian first. Writing L for the Lagrangian as usual, we now have:

$$L(\boldsymbol{\alpha},\dot{\boldsymbol{\alpha}})=\frac{1}{2}I_{jk}^{-1}S_jS_k,S_j=I_{jm}\hat{\xi}_{mn}(\boldsymbol{\alpha})\dot{\alpha}_n\,.\tag{17.45}$$

Therefore,

$$\begin{aligned}\frac{\partial L}{\partial\dot{\alpha}_r}&=I_{jk}^{-1}S_j\frac{\partial S_k}{\partial\dot{\alpha}_r}=I_{jk}^{-1}S_jI_{km}\hat{\xi}_{mr}(\boldsymbol{\alpha})=S_j\hat{\xi}_{jr}(\boldsymbol{\alpha})\,;\\\frac{\partial L}{\partial\alpha_r}&=I_{jk}^{-1}S_j\frac{\partial S_k}{\partial\alpha_r}=I_{jk}^{-1}S_jI_{km}\hat{\xi}_{mn,r}(\boldsymbol{\alpha})\dot{\alpha}_n=S_j\hat{\xi}_{jn,r}(\boldsymbol{\alpha})\dot{\alpha}_n\,.\end{aligned}\tag{17.46}$$

As usual, we have denoted the partial derivative of $\hat{\xi}_{mn}$ with respect to α_r by a comma followed by the subscript r. Thus the Lagrangian equations of motion are:

$$\frac{d}{dt}(S_j\hat{\xi}_{jr}(\boldsymbol{\alpha}))\equiv\dot{S}_j\hat{\xi}_{jr}(\boldsymbol{\alpha})+S_j\hat{\xi}_{jr,n}(\boldsymbol{\alpha})\dot{\alpha}_n=S_j\hat{\xi}_{jn,r}(\boldsymbol{\alpha})\dot{\alpha}_n\,.\tag{17.47}$$

Transposing terms, this reads:

$$\dot{S}_j\hat{\xi}_{jr}(\boldsymbol{\alpha}) = S_j(\hat{\xi}_{jn,r}(\boldsymbol{\alpha}) - \hat{\xi}_{jr,n}(\boldsymbol{\alpha}))\dot{\alpha}_n\,. \tag{17.48}$$

But according to the general theory of Chapter 13, the particular combination of derivatives of the functions $\hat{\xi}$ appearing here can be evaluated in term of the $\hat{\xi}$'s themselves, and the structure constants of the relevant Lie group! The analogue of Eq.(13.55) for the functions $\hat{\xi}$, specialized to the group $R(3)$, is:

$$\hat{\xi}_{jn,r}(\boldsymbol{\alpha}) - \hat{\xi}_{jr,n}(\boldsymbol{\alpha}) = \epsilon_{mkj}\hat{\xi}_{mn}(\boldsymbol{\alpha})\hat{\xi}_{kr}(\boldsymbol{\alpha})\,. \tag{17.49}$$

(When we use the analogue of Eq.(13.55) for the $\hat{\xi}$ s, the sign of the structure constants changes. This is clear from Eq.(13.61)). Using Eq.(17.49) in Eq. (17.48), the basic Lagrangian equations of motion take the form:

$$\dot{S}_k\hat{\xi}_{kr}(\boldsymbol{\alpha}) = S_j\epsilon_{mkj}\hat{\xi}_{mn}(\boldsymbol{\alpha})\dot{\alpha}_n\hat{\xi}_{kr}(\boldsymbol{\alpha})\,. \tag{17.50}$$

These can be simplified in two ways. First, because the matrix $\hat{\Xi}(\boldsymbol{\alpha})$ is non-singular, the factors $\hat{\xi}_{kr}(\boldsymbol{\alpha})$ on the two sides can be dropped. Second, we can reexpress the velocities $\dot{\alpha}_n$ in terms of the S's, thus getting rid of the other factor $\hat{\xi}_{mn}(\boldsymbol{\alpha})$. In this way the Lagrangian equations become equivalent to

$$\dot{S}_k = S_j\epsilon_{mkj}\omega_m = \epsilon_{kjm}S_jI^{-1}_{mn}S_n\,. \tag{17.51}$$

We see that the equations of motion take a simple and elegant form when expressed in terms of the components of the angular momentum resolved with respect to the body-fixed axes. Actually, like the conventional Newtonian equations, these equations in (17.51) are second-order differential equations in time for the generalized coordinates $\alpha_j(t)$; the S_j are linear in the $\dot{\alpha}$'s, so that $\dot{S}_k$ brings in the generalised accelerations $\ddot{\alpha}_j$.

The alternative method for deriving the equations of motion is as follows. Because the space-fixed coordinate system S is an inertial one, and there are no net torques acting on the body, it follows that the components of the total angular momentum along the axes in S must be independent of time. That is, the J_m are constants of motion. This can be expressed in terms of the S_j:

$$\begin{aligned}
\dot{J}_m &= \frac{d}{dt}(A_{mk}(\boldsymbol{\alpha})S_k) = \dot{A}_{mk}(\boldsymbol{\alpha})S_k + A_{mk}(\boldsymbol{\alpha})\dot{S}_k = 0\,,\\
\dot{S}_k &= -A_{mk}(\boldsymbol{\alpha})\dot{A}_{mj}(\boldsymbol{\alpha})S_j = -C_{kj}(\boldsymbol{\omega})S_j = -C_{kj}(I^{-1}S)S_j\,,\\
\dot{S}_k &= \epsilon_{kjm}S_jI^{-1}_{mn}S_n\,.
\end{aligned} \tag{17.52}$$

This is identical to Eq. (17.51). Thus we have a rather simple way of deriving the Lagrangian equations of motion for the case of torque-free rotation. We see at the same time that for this case the second-order differential equations in Eq. (17.51) can be integrated one step to read:

$$J_m(\boldsymbol{\alpha},\dot{\boldsymbol{\alpha}}) \equiv [A(\boldsymbol{\alpha})I\hat{\Xi}(\boldsymbol{\alpha})\dot{\alpha}]_m = [I\dot{\alpha}(0)]_m = \text{constant}\,. \tag{17.53}$$

The integration constants that have appeared are the initial values, at $t = 0$, of the generalized velocities, $\dot{\alpha}(0)$. We have also made use of the other initial condition that the α's themselves vanish at t=0: this follows from the convention that at t=0 the body-fixed Cartesian frame $S(t)$ coincides with the inertial frame S. In order to solve completely for the motion, that is in order to determine the α_j as functions of time, one has to solve the first-order differential equations (Eq. (17.53)), alternatively written as

$$\dot{\alpha}(t) = \hat{H}(\boldsymbol{\alpha})I^{-1}A^{\sim}(\boldsymbol{\alpha})I\dot{\alpha}(0), \alpha(0) = 0\,. \tag{17.54}$$

It is in solving these equations that it is somewhat simpler to work with Euler angles as parameters for $R(3)$ instead of canonical coordinates of the first kind. The first-order differential equations that one obtains for the Euler angles are completely equivalent to Eq. (17.54) in content, differing from them only in their forms. We forego a discussion of this aspect of the problem and do not bother to solve Eq.(17.54). However, there is one property of the nature of the motion related to the theorem of conservation of energy that is worth mentioning. Because the Lagrangian is homogeneous of degree two in the velocities $\dot{\alpha}_j$, Euler's theorem tells us that the Hamiltonian and the Lagrangian coincide. Because neither of them has an explicit time dependence, it is a consequence of the Lagrangian equations of motion that the Hamiltonian, or equivalently the kinetic energy T, is a constant of motion. But we have another way of understanding this result. We saw that the Lagrangian equations of motion coincide with the statement that the J_j are constant in time; but in that case the result embodied in Eq. (17.44), stating that T can vary in time only if the J_j do, informs us again that T must be a constant of the motion!

We turn now to an alternative interpretation and use of the equations of motion in the form of Eq. (17.51)

Classical Pure-Spin Systems

The interesting point about the equations of motion in the form of Eq. (17.51) is that they involve only the variables S_k, and no others. (Of course, they involve the time derivatives $\dot{S}_k$). The same thing is true of the kinetic energy T: if we use the S_k as the velocity-type variables, T is a function of the S's alone:

$$T = \frac{1}{2}I^{-1}_{mn}S_m S_n\,. \tag{17.55}$$

(If we decide to use the *space* components J_k of the angular momentum as the velocity-type variables, the analogous statement is not true: T depends on the α_j as well as on the J_j, although admittedly, the dependence on the α_j is rather special, in view of Eq. (17.44)). These two facts lead to the following expectation: if to the Lagrangian $L(\alpha, \dot{\alpha})$ of Eq. (17.45) we were to apply the usual rules for passing to a Hamiltonian, by introducing generalized momenta β_j canonically conjugate to the α_j and setting up the usual basic PB's among the α's and β's, and if we then computed the PB of S_j with S_k, each regarded as a function of the α's and β's, the result should be expressible in terms of the

S's alone. Let us prove that this is so. We can get the β's using Eq. (17.46) and then write the S_j in terms of the α's and β's:

$$\beta_j \equiv \frac{\partial L}{\partial \dot{\alpha}_j} = S_k \hat{\xi}_{kj}(\boldsymbol{\alpha}), S_j = \hat{\eta}_{kj}(\boldsymbol{\alpha})\beta_k \,. \tag{17.56}$$

The basic PB's would be:

$$\{\alpha_j, \alpha_k\} = \{\beta_j, \beta_k\} = 0, \{\alpha_j, \beta_k\} = \delta_{jk}\,; \tag{17.57}$$

using these we can compute the PB $\{S_j, S_k\}$:

$$\{S_j, S_k\} = \{\hat{\eta}_{mj}(\boldsymbol{\alpha})\beta_m, \hat{\eta}_{nk}(\boldsymbol{\alpha})\beta_n\} = (\hat{\eta}_{sk}(\boldsymbol{\alpha})\hat{\eta}_{mj,s}(\boldsymbol{\alpha}) - \hat{\eta}_{sj}(\boldsymbol{\alpha})\hat{\eta}_{mk,s}(\boldsymbol{\alpha}))\beta_m \,. \tag{17.58}$$

For the combination of partial derivatives of the $\hat{\eta}$'s, we use the analogue of Eq. (13.55), written for the functions $\hat{\eta}$:

$$\hat{\eta}_{sk}\hat{\eta}_{mj,s} - \hat{\eta}_{sj}\hat{\eta}_{mk,s} = \epsilon_{kjn}\hat{\eta}_{mn} \,. \tag{17.59}$$

(This is completely equivalent to Eq. (17.49)). We then get:

$$\{S_j, S_k\} = \epsilon_{kjn}\hat{\eta}_{mn}\beta_m, \{S_j, S_k\} = -\epsilon_{jkn}S_n \,. \tag{17.60}$$

Thus it is indeed true that starting from the basic PB's among the "primitive" variables α_j, β_j, the S_j possess PB's with one another whose values are expressible in terms of themselves alone; in fact, the quantities $-S_j$ give us a true PB realization of the Lie algebra of $R(3)$. If we write the equations of motion Eq. (17.51) in this form:

$$\dot{S}_k = \epsilon_{mkj}S_j I^{-1}_{mn} S_n = \epsilon_{mkj}S_j \frac{\partial T}{\partial S_m} = -\epsilon_{mnj}S_j \frac{\partial S_k}{\partial S_m}\frac{\partial T}{\partial S_n}\,, \tag{17.61}$$

we see that they can be put into the standard Hamiltonian form:

$$\dot{S}_k = \{S_k, T(\boldsymbol{S})\} \,. \tag{17.62}$$

The PB here may either be computed by going back to the basic ones in Eq.(17.57) among the α's and β's, or more simply, because only the variables S_j appear, *they may be directly computed by just using Eq. (17.60) and the derivation property*! Thus, as far as the equations of motion are concerned, all that we need are the expression of the Hamiltonian (actually this is just the kinetic energy) as a function of the S's, and the PB's among the S's; the connection of these things to the primitive quantities α_j, β_j is *not* needed. But we must be careful here! As we have emphasized several times, in the case of a rotating rigid body the variables S_k are generalized velocity-type variables, and *not every dynamical quantity associated with this system can be expressed as a function of the S_k alone.* On the one hand, we can imagine that the rigid body is placed in a uniform gravitational field, for example, and then the Lagrangian would not be the kinetic energy alone, but would include a potential energy term that is a function of the α_j alone and cannot be written in terms of the S_j. [We describe this system in the next chapter in terms of the group $E(3)$]. On the other hand, even for the case of torque free rotation when Eq. (17.62) applies, *the physical interpretation tells us that solving Eq. (17.62) does not amount to solving fully the equations of motion.* Equation (17.62), like Eq. (17.51), is an equation that specifies the accelerations $\ddot{\alpha}_j$ of the rigid body; a single integration of it leads essentially to equations like Eq. (17.53) or Eq. (17.54), involving the velocities $\dot{\alpha}_j$, and these must be integrated one more step to yield the coordinates α_j as functions of t before we can say that the equations of motion have been solved. Both these features stem from the self-evident fact that the S_k do not form a complete set of coordinates, either kinematically or dynamically, for a description of rigid body rotation. However, we can in principle conceive of a family of classical systems, each of which is completely described by three variables S_k, each with a suitable physical interpretation, and for which the equations of motion can be written in the Hamiltonian form but using a *generalized Poisson bracket* (GPB): the basic GPB's among the S_j are postulated to have the values:

$$\{S_j, S_k\} = -\epsilon_{jkm} S_m \,, \tag{17.63}$$

and the GPB of any two functions $f(\boldsymbol{S})$ and $g(\boldsymbol{S})$ is computed from here using the derivation property

$$\{fg, h\} = f\{g, h\} + \{f, h\}g \,, \tag{17.64}$$

thus giving

$$\{f(\boldsymbol{S}), g(\boldsymbol{S})\} = -\epsilon_{mnj} S_j \frac{\partial f}{\partial S_m} \frac{\partial g}{\partial S_n} \,. \tag{17.65}$$

We call such classical dynamical systems classical pure-spin systems. Such systems differ from a rotating rigid body in these respects:

1. for them, *all* dynamical variables are defined as suitable functions of the S_k;

2. the three-dimensional space with the S_k as coordinates, each with a specified range, forms the generalized phase space of the system (the question of ranges is examined on page 323);

3. solving the first-order equations of motion that express each time derivative $\dot{S}_k$ as the GPB of S_k with a Hamiltonian $H(\boldsymbol{S})$ amounts to a complete solution of the motion and suffices to determine the "phase" of the system at any time in terms of its "phase" at an earlier time.

Does the definition of Eq. (17.65) give us an acceptable GPB? In the case when we had the primitive variables α_j, β_j, the Jacobi identity for PB's of functions of S alone would be a consequence of the Jacobi identity for the normal PB, but now we want to regard the GPB's (Eq. (17.63)) as the basic ones, and we must be sure that the Jacobi identity is valid. We do not expect it to be invalid, but let us check it anyway. The linearity in f and g of the

GPB (Eq. (17.65)) is obvious and so is the antisymmetry under exchange of f and g, because ϵ_{mnj} is antisymmetric in m and n. This property of the ϵ_{mnj} being antisymmetric in their first two indices follows from their being the structure constants of a Lie algebra. (Of course, in the case at hand, the ϵ's are antisymmetric in all three indices, but this is rather special to the group $R(3)$ and some other groups, and also plays no role in making Eq. (17.65) an acceptable GPB). In verifying the Jacobi identity let us temporarily write ϵ^j_{mn} instead of ϵ_{mnj}; in general, the structure constants of a Lie algebra must be written c^j_{mn}, distinguishing the subscripts (in which alone we have antisymmetry always) from the superscript. If f,g, and h are three functions of $\boldsymbol{S}$, use of Eq. (17.65) twice gives:

$$\begin{aligned}\{\{f,g\},h\} =&\epsilon^j_{mn}S_j\frac{\partial}{\partial S_m}\left(\epsilon^k_{rs}S_k\frac{\partial f}{\partial S_r}\frac{\partial g}{\partial S_s}\right)\frac{\partial h}{\partial S_n} = \epsilon^m_{rs}\epsilon^j_{mn}S_j\frac{\partial f}{\partial S_r}\frac{\partial g}{\partial S_s}\frac{\partial h}{\partial S_n}\\ &+\epsilon^j_{mn}S_j\epsilon^k_{rs}S_k\frac{\partial^2 f}{\partial S_m\partial S_r}\frac{\partial g}{\partial S_s}\frac{\partial h}{\partial S_n}+\epsilon^j_{mn}S_j\epsilon^k_{rs}S_k\frac{\partial^2 g}{\partial S_m\partial S_s}\frac{\partial h}{\partial S_n}\frac{\partial f}{\partial S_r}.\end{aligned} \tag{17.66}$$

We must now take the cyclic sum of this expression, with respect to cyclic permutations of f, g,and h, and see if it vanishes. As far as the first of the three terms is concerned, a cyclic sum on f, g, and h is equivalent to one on r, s, and n, and this sum vanishes by itself because the ϵ's obey the Jacobi identity for structure constants of a Lie algebra:

$$\epsilon^m_{rs}\epsilon^j_{mn}+\epsilon^m_{sn}\epsilon^j_{mr}+\epsilon^m_{nr}\epsilon^j_{ms}=0\,. \tag{17.67}$$

We have to consider next the cyclic sums of the second and third terms in Eq. (17.66). Here we make the allowed interchanges $f\to h\to g\to f$ (because it is cyclic) and $r\to n\to s\to m\to r$ (since they are dummy indices) in the third term in Eq. (17.66), and then we find that the cyclic sums of the second and third terms cancel one another:

$$\begin{aligned}&\sum_{\text{cyclic}}\left(\epsilon^j_{mn}S_j\epsilon^k_{rs}S_k\frac{\partial^2 f}{\partial S_m\partial S_r}\frac{\partial g}{\partial S_s}\frac{\partial h}{\partial S_n}+\epsilon^j_{mn}S_j\epsilon^k_{rs}S_k\frac{\partial^2 g}{\partial S_m\partial S_s}\frac{\partial h}{\partial S_n}\frac{\partial f}{\partial S_r}\right)\\ &\quad=\sum_{\text{cyclic}}\left(\epsilon^j_{mn}S_j\epsilon^k_{rs}S_k\frac{\partial^2 f}{\partial S_m\partial S_r}\frac{\partial g}{\partial S_s}\frac{\partial h}{\partial S_n}+\epsilon^j_{rs}S_j\epsilon^k_{nm}S_k\frac{\partial^2 f}{\partial S_r\partial S_m}\frac{\partial g}{\partial S_s}\frac{\partial h}{\partial S_n}\right)\\ &\quad=\sum_{\text{cyclic}}\left(\epsilon^j_{mn}+\epsilon^j_{nm}\right)S_j\epsilon^k_{rs}S_k\frac{\partial^2 f}{\partial S_m\partial S_r}\frac{\partial g}{\partial S_s}\frac{\partial h}{\partial S_n}=0\,.\end{aligned} \tag{17.68}$$

All that is needed here is the antisymmetry property of the ϵ's. Thus we have verified that the natural properties of the ϵ's, which result from their being the structure constants for a Lie algebra, ensure that the definition of Eq. (17.65) is an acceptable GPB.

Taking Eq. (17.63) as the basic or elementary GPB's can also be described in this way: we have converted the abstract Lie brackets in the Lie algebra of $R(3)$ into a GPB by means of the *definitions* of Eqs. (17.63) and (17.65). We have one classical real variable S_k (or more properly, $-S_k$) to correspond to each basic vector l_k in the Lie algebra of $R(3)$, and by definition we have a true realization of this Lie algebra provided by the basic relations of Eq. (17.63). Although the notion "functions of the l_k" is undefined, we treat the S_k as ordinary classical real variables for which the usual processes of formation of functions, and of associative and commutative multiplication of these functions, are defined in the normal classical sense. In addition to these usual processes, we have introduced the bracket operation (Eq. (17.65)), converting the set of these functions into a concrete infinite-dimensional Lie algebra. (Recall the discussion on pages 223-227 in Chapter 14). The reader will appreciate that this method of setting up a GPB modeled on a Lie algebra (of some group) will go through for *any* Lie algebra. In checking the algebraic properties of the GPB, we need only those properties that are common to all sets of structure constants. In general, for an n-parameter Lie algebra, we would introduce n classical real variables and classical functions of these, and suitably generalize Eqs. (17.63) and (17.65) to define elementary and general GPB's. An example using the group $E(3)$ appears in the next chapter.

The physical interpretation of the S_k now depends on the particular pure spin system involved; some examples are given in the sequel. In general, we must give up the interpretation that the S_k are "the components, along some body fixed axes, of the total angular momentum of a rotating rigid body." But in most cases one general and reasonable requirement on the S_k may be imposed: the phase space descriptions given in two Cartesian frames related to one another by a proper rotation shall be connected by a generalized canonical transformation generated by an appropriate linear combination of the S_k; that is, the S_k shall be the generators of the generalized canonical transformations corresponding to rotations of the spatial coordinate system. This provides the justification for calling these pure-spin systems. To a limited extent, that is as long as one restricts oneself to those dynamical quantities that are functions of the S_k alone, and as long as one only asks for the time dependence of the S_k in the course of the motion, one may regard a freely rotating rigid body obeying the equations of motion (Eq. (17.62)) as a classical pure-spin system. As we have already explained, this is a somewhat unphysical example; besides, in this case the generalized canonical transformations generated by the S_k correspond to rotations relative to the body-fixed axes and not to rotations of the spatial coordinate system!

Lest the reader assume the contrary, we must point out one important departure from the general theory of classical realizations of relativity groups developed in the previous chapter that is involved when one talks of non-trivial spin systems. This is that a pure-spin system is not necessarily one whose description is completely invariant under all rotations of the spatial coordinate system, so that we are not dealing here with a classical (generalized) canonical realization of the group $R(3) \otimes T_1$, T_1 being the group of time translations. In a canonical realization of $R(3) \otimes T_1$, the generator H of the subgroup T_1 would necessarily have vanishing GPB's with the generators S_k of $R(3)$, but because the latter generators are the only independent dynamical variables for a spin system, the equations of motion would become trivial and would state that each S_k is a constant of motion. Therefore, as soon as one considers a nontrivial Hamiltonian and nontrivial equations of motion, one is not working with a (generalized) canonical realization of $R(3) \otimes T_1$. (The possibility that the GPB $\{S_k, H\}$ might be a nonvanishing neutral element d_k is easily disposed of; the GPB $\{S_j, \{S_k, H\}\}$ must vanish in any case, and we can apply the Jacobi identity to this to show that d_k=0). When we examine the solutions of the Hamiltonian equations of motion for specific spin systems, we describe in more detail the senses in which the S_k implement rotations of the spatial coordinate system and in which a nontrivial spin-system is not rotationally invariant.

Invariants and Casimir Invariants

We now discuss some general properties of realizations of Lie algebras. These properties will help us understand better the nature of the GPB (Eq. (17.65)) and of the associated generalized canonical transformations. These considerations will also be relevant in later chapters.

Let us begin with the case of a true PB realization of a Lie algebra L corresponding to the Lie group G. (For our present purposes it could equally well be a realization involving a GPB or a Dirac bracket.) To each basic element e_j in L we have one phase-space function $\lambda_j(\omega)$, and these functions obey:

$$\{\lambda_j(\omega), \lambda_k(\omega)\} = c^l_{jk}\lambda_l(\omega)\,. \tag{17.69}$$

We can now look for phase-space functions that are invariant under the canonical transformations generated by the $\lambda_j(\omega)$ corresponding to elements of G. If $\Phi(\omega)$ is such a function, there are two ways in which we could express this property; one is a global statement:

$$\exp(\widetilde{u^j\lambda_j(\omega)})\Phi(\omega) = \Phi(\omega)\,, \tag{17.70}$$

and another is the equivalent infinitesimal statement:

$$\{\lambda_j(\omega), \Phi(\omega)\} = 0\,. \tag{17.71}$$

Each such function $\Phi(\omega)$ is called an invariant of the canonical realization of G under consideration. The importance of invariants of a canonical realization is this: in their action on the underlying phase space, the canonical transformations realizing the group G leave each hypersurface $\Phi(\omega) = C$ invariant, $\Phi(\omega)$ being an invariant and C any constant. That is, hypersurfaces of constant Φ are mapped onto themselves, the value of the constant being maintained. If we have several independent invariants $\Phi(\omega), \Phi'(\omega), \ldots$ we have correspondingly more information on the canonical transformations: each hypersurface $\Phi(\omega) = C, \Phi'(\omega) = C', \ldots$, for each given set of constants $C, C', \ldots$ is mapped onto itself. Thus, even without exhibiting the canonical transformations in detail, knowledge of invariants of the realization tells us something about the nature of the point transformations involved. There is one obvious way in which the number of independent invariants of a canonical realization of G is limited: assuming that the realization is nontrivial, this number cannot exceed the dimensionality of the phase space minus one!

One might think that the invariants associated with different canonical realizations of one and the same group G have little to do with each other, but that is not so. In fact, one can see that if an invariant in a particular realization can be expressed as a function of the generators $\lambda_j(\omega)$ of that realization, then there is a pretty good chance that the same function of the generators in other realizations will turn out to be an invariant in those other realizations. Although this is not always true, one would imagine that some such thing is true, because if Φ in Eq. (17.71) were a function of the λ's alone, the uniform appearance of the basic PB relations in Eq. (17.69) in all realizations shows that on working out the PB in Eq. (17.71) we get an equation that has the same form in all realizations. Invariants of this type are called *Casimir invariants.* More precisely, a function $\mathcal{C}(\lambda_k)$ on phase space, which can be written explicitly as a function of the generators λ_k alone and which is an invariant in *all realizations of the group* G (produced by a true PB realization of L, of course), is called a *Casimir invariant for* L *as well as for* G. We make the further simplification that only Casimir invariants that can be written as *polynomials of finite order* in the generators are considered. We can translate the condition that a Casimir invariant be an invariant in all realizations into this form: if $\mathcal{C}(\lambda)$ is a polynomial in the λ_k and is an invariant, then Eq. (17.71), which now appears as:

$$\{\lambda_j, \mathcal{C}(\lambda)\} \equiv c^l_{jk}\frac{\partial \mathcal{C}(\lambda)}{\partial \lambda_k}\lambda_l = 0, j = 1, 2, \ldots, \tag{17.72}$$

should be an identity in the λ's, no account being taken of any functional relations that may hold among the λ's because such relations are always specific to particular realizations. Now Eq. (17.72) shows that when we evaluate the PB of $\mathcal{C}(\lambda)$ with λ_j, all those terms in $\mathcal{C}$ that are homogeneous of a given degree in the λ's give rise to a set of terms on the right-hand side that are homogeneous and of precisely the same degree; these terms must vanish by themselves because the λ's are to be treated as independent variables. It follows that it is enough to consider special Casimir invariants, each being homogeneous of a given degree in the λ's. A Casimir invariant $\mathcal{C}^{(N)}(\lambda)$ of degree N can always be written in this way:

$$\mathcal{C}^{(N)}(\lambda) = C^{jk\ldots q}\underbrace{\lambda_j\lambda_k\ldots\lambda_q}_{\leftarrow \text{ N factors } \rightarrow}. \tag{17.73}$$

We assume without loss of generality that the coefficients $C^{jk\ldots q}$, which are numbers, are completely symmetric in their indices. Then Eq. (17.72) becomes:

$$\begin{aligned}\frac{\partial \mathcal{C}^{(N)}}{\partial \lambda_v} &= C^{vk\ldots q}\lambda_k\ldots\lambda_q + C^{jv\ldots q}\lambda_j\ldots\lambda_q + \ldots + C^{jk\ldots v}\lambda_j\lambda_k\ldots \\ &= NC^{vk\ldots q}\lambda_k\ldots\lambda_g,\ c^j_{uv}C^{vk\ldots q}\lambda_j\lambda_k\ldots\lambda_q = 0\,. \end{aligned} \tag{17.74}$$

Because this last is to be an identity in the λ's, we have only to symmetrize the coefficients to conclude:

$$c^j_{uv}C^{vk\ldots q} + c^k_{uv}C^{jv\ldots q} + \ldots + c^q_{uv}C^{jk\ldots v} = 0\,. \tag{17.75}$$

Thus all possible Casimir invariants of degree N arise by finding all possible sets of symmetric coefficients $C^{jk\ldots q}$ satisfying Eq. (17.75). This proves what must have become obvious by now: the Casimir invariants for true PB realizations of a Lie algebra L are completely determined by the structure constants, or more abstractly, the structure of L. General theorems assure us that for a given L there is only a finite number of algebraically independent Casimir invariants. Although in principle Eq. (17.75) gives us a way of finding all of them, in practice and in all the cases we shall deal with, we will essentially be able to guess the forms of the independent Casimir invariants.

Now that we have understood the properties of invariants and Casimir invariants, we can reexpress them from a more formal point of view. We will distinguish non-Casimir invariants from Casimir invariants by continuing to write $\Phi(\omega), \Phi'(\omega), \ldots$, for the former. Suppose we are given a canonical realization of G on some phase space, with the associated realization of L being true, as heretofore. We can then consider the set of regular canonical transformations that have the property of *commuting* with the canonical transformations realizing G. It is clear that these form a group. Let us write $\mathcal{G}_c$ for this group. We use a curly $\mathcal{G}_c$, and not G_c, to remind the reader that the group $\mathcal{G}_c$ is not determined by G alone, but rather by the given canonical realization of G. We consider only the identity component of $\mathcal{G}_c$ and further restrict ourselves to those elements that lie on one-parameter subgroups of canonical transformations. Such a regular canonical transformation belonging to $\mathcal{G}_c$ can be written $\exp(\widetilde{a(\omega)})$ for some generating function $a(\omega)$, and it obeys:

$$\exp(\widetilde{a(\omega)})\exp(\widetilde{u^j\lambda_j(\omega)}) = \exp(\widetilde{u^j\lambda_j(\omega)})\exp(\widetilde{a(\omega)}) \tag{17.76}$$

for all $u \in L$. If we take the infinitesimal form of this relation, we get the condition

$$\{\lambda_j(\omega), a(\omega)\} = d_j\,, \tag{17.77}$$

where the d_j are a set of neutral elements characteristic of the function $a(\omega)$. Conversely, if the function $a(\omega)$ obeys Eq. (17.77) and further generates a regular canonical transformation, then use of the Baker-Campbell-Hausdorff formula leads us back to Eq. (17.76). Thus Eq. (17.77) is a complete characterization for generators of one-parameter subgroups in $\mathcal{G}_c$; it then shows that in general each such generator $a(\omega)$ is *not* an invariant of the original canonical realization of G. Surfaces of constant $a(\omega)$ are *not* mapped each *onto itself*, but are mapped *onto one another*, the value of $a(\omega)$ being shifted under the action of G:

$$\exp(\widetilde{u^j\lambda_j(\omega)})a(\omega) \equiv a(T_{\exp(-u)}\omega) \equiv a(\omega) + u^j d_j\,. \tag{17.78}$$

(Following the notation of Chapter 14, $T_{\exp(u)}$ is the point transformation on phase space corresponding to the element $\exp(u)$ in G). Conversely, under this one-parameter subgroup of $\mathcal{G}_c$, the λ's are also shifted in value:

$$\widetilde{e^{\sigma a(\omega)}}\lambda_j(\omega) \equiv \lambda_j(\widetilde{e^{\sigma a(\omega)}}\omega) = \lambda_j(\omega) - \sigma d_j\,. \tag{17.79}$$

[Actually, the demand that the λ_j as well as each $a(\omega)$ be generators of regular finite canonical transformations restricts considerably the appearance of neutral elements d_j in equations such as (17.77). Thus Eq. (17.78) shows that the values accessible to $a(\omega)$ must be such that shifting an allowed value by any amount expressible as a finite linear combination of the d_j must result in another allowed value. That is, the values accessible to $a(\omega)$ must be all real numbers from $-\infty$ to $+\infty$, if some d_j is nonzero. Equation (17.79) implies analogous properties for those λ_j whose corresponding d_j do not vanish.] The set of all functions $a(\omega)$ obeying Eq. (17.77) and generating regular canonical transformations makes up the Lie algebra of the group $\mathcal{G}_c$. Among all these $a(\omega)$, we can now pick up the subset for which the right-hand side of Eq. (17.77) is in fact zero. We write such elements in the Lie algebra of $\mathcal{G}_c$ as $\mathbf{\Phi}(\omega), \mathbf{\Phi}'(\omega), \ldots$; these are the *invariants* of the canonical realization of G as defined earlier. Note the interesting property that the PB of any two elements in the Lie algebra of $\mathcal{G}_c$ is a $\mathbf{\Phi}(\omega)$, thus an invariant! This follows trivially from Eq. (17.77) and the Jacobi identity. The invariants $\mathbf{\Phi}(\omega), \mathbf{\Phi}'(\omega), \ldots$ form a subalgebra, in fact an invariant subalgebra, in the Lie algebra of $\mathcal{G}_c$, for, using again the Jacobi identity,

$$\{\lambda_j, \mathbf{\Phi}\} = 0, \{\lambda_j, \mathbf{\Phi}'\} = 0 \Longrightarrow \{\lambda_j, \{\mathbf{\Phi}, \mathbf{\Phi}'\}\} = 0\,, \tag{17.80}$$

proving that the invariants form a subalgebra. Using the same identity again,

$$\{\lambda_j, a\} = d_j, \{\lambda_j, \mathbf{\Phi}\} = 0 \Longrightarrow \{\lambda_j, \{a, \mathbf{\Phi}\}\} = -\{\mathbf{\Phi}, \{\lambda_j, a\}\} = 0\,, \tag{17.81}$$

The invariants therefore generate an invariant subgroup $\mathcal{G}_c^{(I)}$ in $\mathcal{G}_c$: this is an elegant characterization of them. As we have seen, their values are preserved by the transformations realizing G. The Casimir invariants are but one step away. Generally, the group $\mathcal{G}_c^{(I)}$ is not Abelian, because the PB of two invariants is not necessarily a neutral element. However, the Casimir invariants $\mathcal{C}(\lambda)$, which form a subset of the $\mathbf{\Phi}$'s, also form a subalgebra of the $\mathbf{\Phi}$'s, and an invariant Abelian one at that! These properties are evident from the defining properties of the $\mathcal{C}$'s and the $\mathbf{\Phi}$'s:

$$\{\mathcal{C}(\lambda), \mathcal{C}'(\lambda)\} = 0, \{\mathcal{C}(\lambda), \mathbf{\Phi}(\omega)\} = 0\,. \tag{17.82}$$

Correspondingly, the Casimir invariants are the generators of an Abelian invariant subgroup of $\mathcal{G}_c^{(I)}$ (but this Abelian subgroup need not be an invariant one when embedded in $\mathcal{G}_c$!) We have, then, this picture: a canonical realization of G determines the group $\mathcal{G}_c$ of all those canonical transformations that commute with it. The group $\mathcal{G}_c$ is generated by functions $a(\omega)$ obeying Eq. (17.77). The invariants of the realization of G, called $\mathbf{\Phi}, \mathbf{\Phi}', \ldots$ obey Eq. (17.71) and generate an invariant subgroup $\mathcal{G}_c^{(I)}$ in $\mathcal{G}_c$. The Casimir invariants $\mathcal{C}(\lambda)$ are a subset of the $\mathbf{\Phi}$'s and generate an Abelian invariant subgroup of $\mathcal{G}_c^{(I)}$. It is this last subgroup that is common to all realizations of G. This description is complete except to the extent that there may be invariants or Casimir invariants that do not generate regular canonical transformations; such invariant quantities cannot be obtained from the group $\mathcal{G}_c$ as a starting point, and are not covered by this discussion.

Going back to the definition of an invariant $\mathbf{\Phi}(\omega)$ of a canonical realization of a Lie group G, it is clear that $\mathbf{\Phi}$ has a vanishing PB with any function of the generators $\lambda_j(\omega)$. This means that the surfaces of constant $\mathbf{\Phi}$, in phase space, are mapped each onto itself not only by those canonical transformations that correspond to elements of G but also by every regular canonical transformation generated by any function of the $\lambda_j(\omega)$. These latter, for nonlinear functions of the λ_j, do not correspond to elements of G at all. Thus an invariant $\mathbf{\Phi}(\omega)$ of a canonical realization of G is actually invariant under a somewhat larger set of canonical transformations than those that realize G.

For the case of the group $R(3)$, we have just one independent Casimir invariant, the "squared length of the generator $\boldsymbol{J}(\omega)$". The $J_j(\omega)$ obey Eq. (17.2) (with d_{jk}=0), and this invariant is written $J^2(\omega)$:

$$\begin{aligned} J^2(\omega) &= J_j(\omega)J_j(\omega)\,; \\ \{J_k, J^2\} &= 2J_j\{J_k, J_j\} = 2\epsilon_{kjl}J_jJ_l = 0 \end{aligned} \tag{17.83}$$

This is the one invariant common to all realizations of $R(3)$. In the single particle realization of $R(3)$ given by Eq. (17.6), we can find examples of non-Casimir invariants. We have three of them, $q^2 = q_jq_j, p^2 = p_jp_j$ and $\boldsymbol{q}\cdot\boldsymbol{p} = q_jp_j$. That these are invariants is obvious from the finite transformation properties exhibited in Eq. (17.9). But there are only two independent non-Casimir invariants, because the combination $q^2p^2 - (\boldsymbol{q}\cdot\boldsymbol{p})^2$ coincides with J^2. For a pure spin system, because the generators S_j form a complete set of dynamical variables, there are no invariants other than Casimir invariants, and there is only one of these, to wit, $S^2 = S_jS_j$. Again, for a pure spin system, every generalized canonical transformation is generated by *some* function of the basic coordinates S_j (assuming it lies on a one-parameter subgroup), and thus according to the comments of the previous paragraph, the Casimir invariant S^2 is invariant under all generalized canonical transformations! These two properties of pure spin systems, namely that there are no invariants other than Casimir invariants and that this is an invariant under all generalized canonical transformations, obviously extend to the GPB structure that can be set up based on any Lie algebra L.

Up to now we discussed the invariants of a canonical realization of a Lie group G based on a true PB realization of its Lie algebra L. Let us now relax this restriction and allow for nontrivial realizations up to neutrals of L. One can easily check that most of the earlier discussion goes through with no changes: the definition and characteristic properties of invariant functions $\Phi(\omega)$ are the same as before. Similarly, the group $\mathcal{G}_c$, the properties of its generators, and the fact that the invariant functions $\Phi(\omega)$ generate an invariant subgroup $\mathcal{G}_c^{(I)}$ of $\mathcal{G}_c$, are all unaltered. All these things are independent of any neutral elements that may be present in the realization of L. The differences show up when we consider Casimir invariants: in comparison to the case of a true realization of L, we now generally lose some Casimir invariants. That is, some of the polynomial functions of the λ_j that were Casimir invariants in the true realizations of L cease to be invariant when neutral elements are included in the basic PB relations (Eq. (17.69)). On the other hand, every polynomial in the λ_j that is an invariant for a realization up to neutrals of L will survive as a Casimir invariant of L when the neutral elements go to zero, unless it vanishes identically in this limit or its dependence on the neutral elements is singular. When in place of Eq. (17.69) we have the more general equations:

$$\{\lambda_j(\omega), \lambda_k(\omega)\} = c^l_{jk}\lambda_l(\omega) + d_{jk}\,, \tag{17.84}$$

where we assume that the neutral elements d_{jk} have been reduced to as few independent ones as possible, we clearly cannot insist that each Casimir invariant be homogeneous of a fixed degree in the $\lambda_j(\omega)$, but we can restore such a requirement if we also count the independent neutral elements as contributing to the degree of homogeneity of the invariant. That is, in this more general situation, it suffices to consider Casimir invariants that are homogeneous polynomials of given degree in the λ_j and the independent neutral elements combined. We come across examples of these things when we discuss the Galilei group. In a wider sense, the values given to the independent neutral elements are also invariants of the realization of G and characterize the realization as a whole, but they do not help in breaking up the phase space into invariant hypersurfaces. For the group $R(3)$, of course, none of these problems arises, because there is no nontrivial realization up to neutrals of its Lie algebra; the same is the case for the groups $E(3), O(3,1)$ and $\mathcal{P}$.

The concept of the invariants of canonical realizations leads to that of irreducible realizations. A classical canonical realization of a Lie group G is said to be an *irreducible realization* if every invariant $\Phi(\omega)$ is a number, or more generally, a neutral element. In other words, in such realizations there are no nontrivial invariants, and the phase space does not split up into invariant hypersurfaces. If one is using a singular GPB, what is meant is that the splitting up of the phase space into invariant hypersurfaces is carried no further than that implied by the existence of functions on phase space having vanishing GPB's with all functions. As particular cases of invariants, each Casimir invariant, too, reduces to a number, or neutral element, in an irreducible realization. For such realizations of G, one can immediately prove that the group $\mathcal{G}_c$ consisting of all regular canonical transformations commuting with those realizing G must be Abelian. If $a(\omega)$ and $a'(\omega)$ are generators of two one-parameter subgroups in $\mathcal{G}_c$, we know that $\{a(\omega), a'(\omega)\}$ must be an invariant; hence it must be a neutral element. Use of the Baker-Campbell-Hausdorff formula then shows that $a(\omega)$ and $a'(\omega)$ generate canonical transformations that commute with one another, making $\mathcal{G}_c$ Abelian. In general, one cannot say that $\mathcal{G}_c$ itself is trivial and consists of the unit element alone, but the subgroup $\mathcal{G}_c^{(I)}$ with invariants as generators is certainly trivial. In practice, a sufficient condition for the irreducibility of a canonical realization of G is this: for any two points P, P' in phase space there should be some element in G whose representative canonical transformation carries P to P' (this is easily modified in case we have a singular GPB). If this is so, there can be no invariant hypersurfaces in phase space, and thus no nontrivial invariants either. The single-particle realization of $R(3)$ given by Eq. (17.6) is certainly reducible, as there are three nontrivial invariants. On the other hand, a pure-spin system gives us an irreducible canonical realization of $R(3)$; the only independent invariant is neutral and generates the identity canonical transformation.

The existence and significance of invariants and Casimir invariants extends to *representations* of a Lie group G via linear transformations in linear vector spaces and is by no means restricted to classical canonical realizations of G. We cannot go into the details here, but we can mention the salient points. In Chapter 14, on page 215 we mention that the *generators* of a linear matrix representation of G consist of a set of matrices X_j, one for each basic vector e_j in L, which form a realization (or representation) of L with commutators for Lie brackets:

$$[X_j, X_k] \equiv X_j X_k - X_k X_j = c^l_{jk} X_l \,. \tag{17.85}$$

This is Eq. (14.33). The matrix $D(\exp(u))$ that represents the element $\exp(u)$ of G is now the ordinary exponential of the corresponding linear combination of the generator matrices:

$$\exp(u) \in G \to D(\exp(u)) = \exp(u^j X_j) \,. \tag{17.86}$$

In this kind of realization of G and L, an *invariant* is a matrix Q that commutes with the representation matrices $D(\exp(u))$:

$$Q\exp(u^j X_j) = \exp(u^j X_j) Q \,; \tag{17.87}$$

in infinitesimal form the Lie brackets of Q with the generators X_j must vanish:

$$[X_j, Q]_- = 0 \,. \tag{17.88}$$

Casimir invariants are now called *Casimir operators*: they are polynomials in the X's that obey Eq.(17.88) in all matrix realizations of L and G. It is an easy and interesting exercise (best left to the reader!) to show that Casimir invariants for true PB realizations of L and Casimir operators for linear matrix representations of L are related in this way: if the polynomial in the canonical generators λ_j

$$\begin{aligned} \mathcal{C}(\lambda) =& C^{jk\ldots q} \lambda_j(\omega) \lambda_k(\omega) \ldots \lambda_q(\omega) \\ & \longleftarrow N \text{ factors} \longrightarrow \end{aligned} \tag{17.89}$$

is an invariant in the former case, *and if the coefficients are symmetric in their superscripts*, then the matrix,

$$\begin{aligned} Q(X) =& C^{jk\ldots q} X_j X_k \ldots X_q \\ & \longleftarrow N \text{ factors} \longrightarrow \end{aligned} \tag{17.90}$$

is a Casimir operator for the latter case, and vice versa. The analogue of the statement that hypersurfaces of constant values for the invariants $\mathbf{\Phi}(\omega)$ of a canonical realization of G are mapped onto themselves under the canonical transformations realizing G is this: given a linear matrix representation of G on some vector space, the largest subspace on which an invariant matrix Q reduces to a given multiple of the unit matrix is mapped onto itself by the linear transformations representing elements of G. Thus, as in the classical canonical case, each invariant matrix helps to "reduce" the matrix representation of G. There is an appropriate generalization of this statement to take account of several invariant matrices $Q, Q' \ldots$ at the same time; it is a little more complicated because these invariant matrices may not commute pairwise. Of course, the Casimir operators form a subset of the invariant matrices such that they commute with each of the latter, and thus with each other.

Rotational Behavior And Examples of Pure-Spin Systems

After this digression, we now return to the discussion of pure spin systems. We first dispose of the question of the natural ranges for the basic dynamical variables S_j, j=1,2,3. From the preceding section we have learned that every generalized canonical transformation on the S_j preserves the value of the Casimir invariant $S^2 = S_j S_j$; thus it follows that each such transformation acts as a mapping of surfaces of spheres onto themselves in the three-dimensional space of the S_j, preserving the radii and the basic GPB's. Therefore the phase space for a pure spin system (more accurately, generalized phase space) consists of some set of surfaces of spheres in three-dimensional space, all centered on the origin. For example, it could be the entire three-dimensional space, with each S_j varying independently from $-\infty$ to $+\infty$, or it could be only the surface of some sphere, with S^2 having some given numerical value characteristic of the system and with only two of the three S_j varying independently. Various intermediate possibilities are easily visualized. To preserve the symmetry of the formulae, even if S^2 is constrained to have some given value, we continue to use Eqs.(17.63) and (17.65) in computing GPB's, and thus treat all three S_j as independent variables for partial differentiation. We can be sure that as long as such partial differentiations are done only in order to compute GPB's, there is no loss of consistency in giving S^2 its value at the end of all calculations.

Knowing the form of the phase space as well as of a generalized canonical transformation for a pure-spin system, we should say something about the GPB (Eq.(17.65)) in the light of the comments made in Chapter 9, pages 117-118. Suppose instead of using the S_j as the basic variables we were to use the Casimir invariant S^2 and two others independent of it and express each function of the S_j as a function of these. Then the dependence on S^2 is not subject to partial differentiation in computing GPB's, and all that is relevant is the dependence on the other two variables. Let us choose S_3 as one of these two variables. Then, in agreement with the statements made on page 117, it is possible to choose the other one so that the GPB assumes the same appearance as the *ordinary* PB in one pair of variables. Thus with S^2, S_3 and $\phi = \tan^{-1}(S_1/S_2)$ as the three independent quantities, one has:

$$\{f(\boldsymbol{S}), g(\boldsymbol{S})\} = \frac{\partial f}{\partial \phi}\frac{\partial g}{\partial S_3} - \frac{\partial f}{\partial S_3}\frac{\partial g}{\partial \phi}, \tag{17.91}$$

where, of course, we imagine $f(\boldsymbol{S})$ and $g(\boldsymbol{S})$ being rewritten in terms of S^2, S_3, and ϕ. But this way of reexpressing the GPB based on the Lie algebra of $R(3)$ is not well motivated for two reasons: first, it obscures the three-dimensional rotational symmetry inherent in the GPB, and second, it creates the wrong impression that ϕ, like S_3, generates regular canonical transformations. That ϕ is not the generator of a one-parameter group of regular canonical transformations is obvious from the remark that formally we have

$$e^{\widetilde{\alpha\phi}} S_3 = S_3 + \alpha; \tag{17.92}$$

whatever value α may have, assuming it is not zero, we can exhibit points on the sphere of radius S^2 that cannot be mapped into other points of this sphere under Eq. (17.92). Thus as stated on page 117, although the local existence of the standard PB form is guaranteed for any GPB, this standard form can be quite misleading with respect to the global properties of the GPB.

The manner in which the S_j act as generators of rotations of the spatial coordinate system is quite clear. For a given pure-spin system, in a definite coordinate frame S the three S_j are interpreted as the three Cartesian components of some vectorial quantity associated with the system. Most commonly, it is an angular momentum or a magnetic dipole moment, for example. If $S' = \exp(\alpha_j l_j)S$ is another Cartesian frame rotated with respect to S, then the variables S'_j representing the components of the same vectorial quantity with respect to S' that the S_j represent in S are given in the following way (remember the minus sign in Eq.(17.63)!):

$$S' = \exp(\boldsymbol{\alpha}\cdot\boldsymbol{l})S : S'_j = \exp(\widetilde{\alpha_k S_k})S_j = A_{jk}(\boldsymbol{\alpha})S_k\,. \tag{17.93}$$

Not surprisingly, the S'_j are obtained from the S_j by a proper orthogonal rotation: an example of a generalized canonical transformation. Now let us consider the equations of motion. Write $H(\boldsymbol{S})$ for the Hamiltonian that governs the time development of the dynamical variables S_j used in the coordinate system S. (We assume for simplicity that H has no explicit time dependence.) Then the equations of motion are:

$$\dot{S}_j = -\{H(\boldsymbol{S}), S_j\}, \tag{17.94}$$

and the formal solution is:

$$S_j(t) = \exp(\widetilde{-tH(\boldsymbol{S})})S_j, S_j(0) = S_j\,. \tag{17.95}$$

Now, in the rotated frame $S' = \exp(\boldsymbol{\alpha}\cdot\boldsymbol{l})S$, the quantities S'_j have the same physical meanings that the S_j possess in S; this is true for all t, and thus the connection (Eq.(17.93)) is also valid for all t. Therefore the time development of the S_j in S determines the time development of the S'_j in S'. Because the passage from S_j to S'_j is a generalized canonical transformation, GPB's may be evaluated with respect to the S'_j as well as with respect to the S_j (see Chapter 9). The equations of motion for the S'_j are obtained from Eq. (17.94):

$$\begin{aligned} \dot{S}'_j &= A_{jk}(\boldsymbol{\alpha})\dot{S}_k = -\{H(\boldsymbol{S}), A_{jk}(\boldsymbol{\alpha})S_k\} = -\{H(\boldsymbol{S}), S'_j\} \\ &= -\{H'(\boldsymbol{S}'), S'_j\},\ H'(\boldsymbol{S}') = H(\boldsymbol{S}). \end{aligned} \tag{17.96}$$

The Hamiltonian H' to be used in the new coordinate system is obtained from H by substituting for $\boldsymbol{S}$ in terms of $\boldsymbol{S}'$ in the latter; in general, H and H' do not have the same functional forms; that is, they are different functions of their arguments. Equation (17.95) goes over into:

$$S'_j(t) = \exp(-t\widetilde{H'(\boldsymbol{S}')})S'_j, S'_j(0) = S'_j. \tag{17.97}$$

It is now clear that unless the functional forms H' and H are the same, the pure-spin system is not rotationally invariant. Equation (17.95), which is a formal solution to the equations of motion in S, expresses the S_j at time t in terms of their values at time zero and involves certain functional forms; on the other hand, Eq. (17.97) involves quite different functional forms for the expression of the S'_j at time t in terms of *their* initial values. Therefore the values that the S_j take at various times, starting with certain initial values, are not the same as the values that the S'_j take at the same times if they, too, had started off with the same initial values. This is the meaning of the statement that the dynamics of a nontrivial pure-spin system is not rotationally invariant. (There is, of course, the possibility that H' and H are the same functions, save for a neutral element, in which case the solutions to the equations of motion are the same in all frames, but that merely involves shifting the zero of the energy scale and is uninteresting.) For rotational invariance, H must be a rotationally invariant function of $\boldsymbol{S}$, which makes it a function of S^2 alone!

As our first example of equations of motion of the form in Eq. (17.94), we go back to the Hamiltonian for a freely rotating rigid body expressed in terms of spin variables alone, keeping in mind, of course, the limitations in viewing this system as a genuine pure-spin system. This Hamiltonian is quadratic in the S_j and is given in Eq. (17.55). It is now convenient to assume that the inertia tensor is diagonal and has positive-definite diagonal elements $I_j, j{=}1, 2, 3$. Thus we have the Hamiltonian:

$$H = \frac{1}{2}\left(\frac{S_1^2}{I_1} + \frac{S_2^2}{I_2} + \frac{S_3^2}{I_3}\right), \tag{17.98}$$

and the equations of motion:

$$\begin{aligned} \dot{S}_1 &= \left(\frac{1}{I_3} - \frac{1}{I_2}\right) S_2 S_3, \dot{S}_2 = \left(\frac{1}{I_1} - \frac{1}{I_3}\right) S_3 S_1, \\ \dot{S}_3 &= \left(\frac{1}{I_2} - \frac{1}{I_1}\right) S_1 S_2. \end{aligned} \tag{17.99}$$

Each S_j varies over the range $-\infty$ to ∞ independently, corresponding to their being components of an angular momentum. These equations can be completely integrated in terms of elliptic functions. What helps is that we are guaranteed the existence of two constants of the motion, one being the Casimir invariant S^2, and the other the Hamiltonian itself. For given values of S^2 and H, we are able to reduce Eq. (17.99) to a differential equation for only one component, S_2 say. Let us denote the constant values of H and S^2 by $\frac{1}{2}A$ and B, respectively. These cannot be assigned arbitrarily. For convenience, let us order the diagonal elements of the inertia tensor in this way: $I_1 \geq I_2 \geq I_3 > 0$. Then for a given value of B, A is constrained by:

$$\frac{B}{I_1} \leq A \leq \frac{B}{I_3}, \tag{17.100}$$

or alternatively for a given energy $\frac{1}{2}A, B$ must obey:

$$I_3 A \leq B \leq I_1 A. \tag{17.101}$$

The representative point $S_j(t)$ in phase space is constrained to lie always on the "angular momentum sphere", which has radius $\sqrt{B}$, and on the "energy ellipsoid", which has semi-axes $(\sqrt{AI_1}, \sqrt{AI_2}, \sqrt{AI_3})$; the above inequalities make these two conditions compatible. If either $AI_1 < B$ or $AI_3 > B$, these two surfaces have no points in common, the sphere completely enclosing the ellipsoid in the former case and conversely in the latter. For $AI_1 \geq B > AI_2$, the two surfaces intersect in two closed continuous curves that do not touch one another but are situated symmetrically on the two sides of the $S_2 - S_3$ plane; for $AI_2 > B \geq AI_3$, they again intersect in two closed continuous disjoint curves that are now situated symmetrically on the two sides of the $S_1 - S_2$ plane. The transition occurs at $AI_2 = B$, when the two curves touch at the two points $(0, \pm\sqrt{B}, 0)$. We can solve for S_1 and S_3 and express them in terms of S_2, A, and B; we get:

$$\begin{aligned} S_1^2\left(\frac{1}{I_1} - \frac{1}{I_3}\right) &= S_2^2\left(\frac{1}{I_3} - \frac{1}{I_2}\right) - \left(\frac{B}{I_3} - A\right), \\ S_3^2\left(\frac{1}{I_1} - \frac{1}{I_3}\right) &= S_2^2\left(\frac{1}{I_2} - \frac{1}{I_1}\right) - \left(A - \frac{B}{I_1}\right). \end{aligned} \tag{17.102}$$

Combined with Eq. (17.99), we get the equation of motion for S_2:

$$\begin{aligned} \dot{S}_2 &= \pm\sqrt{\beta\delta}\left[\left(\frac{\alpha}{\beta} - S_2^2\right)\left(\frac{\gamma}{\delta} - S_2^2\right)\right]^{\frac{1}{2}}; \\ \alpha &= A - \frac{B}{I_1} \geq 0, \beta = \frac{1}{I_2} - \frac{1}{I_1} \geq 0, \\ \gamma &= \frac{B}{I_3} - A \geq 0,\ \delta = \frac{1}{I_3} - \frac{1}{I_2} \geq 0. \end{aligned} \tag{17.103}$$

We could have written similar equations for S_1 or S_3, but for S_2 we get the most symmetric expressions, because the principal moments of inertia have been ordered in the sequence $I_1 \geq I_2 \geq I_3$. The zeros of $\dot{S}_2$, which signal the "turning points" in the motion with respect to S_2, occur at either $S_2^2 = \alpha/\beta$ or $S_2^2 = \gamma/\delta$, but which one is realized depends on the relative magnitudes of A and B. In general, we have:

$$\frac{\alpha}{\beta} - \frac{\gamma}{\delta} = \frac{\alpha\delta - \beta\gamma}{\beta\delta} = \frac{1}{\beta\delta}\left(\frac{1}{I_3} - \frac{1}{I_1}\right)\left(A - \frac{B}{I_2}\right). \tag{17.104}$$

Therefore, we have two cases: (1) $AI_2 < B$ with $\alpha/\beta < \gamma/\delta$ and (2) $AI_2 > B$ with $\alpha/\beta > \gamma/\delta$. In case (1), the constant energy and constant angular momentum surfaces intersect in two curves on the two sides of the $S_2 - S_3$ plane, and the point $S_j(t)$ travels on one of them. One sees geometrically that $\dot{S}_2$ vanishes when S_3 also vanishes, and, solving the energy and angular momentum equations, we see that $S_2^2 = \alpha/\beta$ at these points. Thus in this case, S_2^2 always obeys $S_2^2 \leq \alpha/\beta$, and of the two factors in the radical on the right-hand side of the equation for $\dot{S}_2$ in Eq. (17.103), only the first one ever vanishes, the second one being always nonzero. In case (2) the locus of $S_j(t)$ is one of the two curves situated on the two sides of the $S_1 - S_2$ plane and in which the energy and angular momentum surfaces intersect; $\dot{S}_2$ vanishes when S_1 does, and $S_2^2 = \gamma/\delta$ at these points. Now the first factor on the right-hand side of Eq. (17.103) stays positive definite, and the second produces the zeros of $\dot{S}_2$. In either case, S_2^2 remains less than the lesser of α/β and γ/δ. The solution of the equation of motion (Eq. (17.103)), can be written this way:

$$\pm(t - t_0) = \frac{1}{\sqrt{\beta\delta}} \int_0^{S_2} dS_2' \left[\left(\frac{\alpha}{\beta} - S_2'^2\right)\left(\frac{\gamma}{\delta} - S_2'^2\right)\right]^{-\frac{1}{2}}. \tag{17.105}$$

t_0 is the time at which S_2 takes the value zero. One says that t is an "elliptic integral" of S_2, or conversely, that S_2 is an "elliptic function" of t. Considered as analytic functions of their arguments, elliptic functions are doubly periodic functions. We are restricted to real values of S_2 as well as of t. We see that we have two different kinds of periodic motion in the two cases considered above; the periodicities are produced by the vanishing of one or other of the factors in the integrand in Eq. (17.105). At a time, that is for given A and B, only one of the two possible periodicities of S_2 considered as an elliptic analytic function of a (complex) argument t shows up, because t must be real. In case (1), with the zeros of $\dot{S}_2$ occurring at $S_2 = \pm\sqrt{\alpha/\beta}$, the point $S_j(t)$ traces out one of the curves mentioned earlier, again and again, with the time period given by

$$\tau_1 = 2\int_{-\sqrt{\alpha/\beta}}^{+\sqrt{\alpha/\beta}} dx \left[\left(\frac{\alpha}{\beta} - x^2\right)\left(\frac{\gamma}{\delta} - x^2\right)\right]^{-\frac{1}{2}}, \quad \frac{\gamma}{\delta} > \frac{\alpha}{\beta}. \tag{17.106}$$

Similarly, in case (2) the zeros of $\dot{S}_2$ occur at $S_2 = \pm\sqrt{\gamma/\delta}$, and the time period for the motion along one of the two other curves is:

$$\tau_2 = 2\int_{-\sqrt{\gamma/\delta}}^{+\sqrt{\gamma/\delta}} dx \left[\left(\frac{\alpha}{\beta} - x^2\right)\left(\frac{\gamma}{\delta} - x^2\right)\right]^{-\frac{1}{2}}, \quad \frac{\alpha}{\beta} > \frac{\gamma}{\delta}. \tag{17.107}$$

The transitional case, $AI_2 = B$, does not lead to periodic motion. In that case, $\alpha/\beta = \gamma/\delta$, and Eq. (17.103) bccomes:

$$\dot{S}_2 = \pm\sqrt{\beta\delta}\left(\frac{\alpha}{\beta} - S_2^2\right) = \pm\sqrt{\beta\delta}\,(B - S_2^2)\,. \tag{17.108}$$

This can be solved immediately, and we obtain:

$$\frac{\sqrt{B} + S_2(t)}{\sqrt{B} - S_2(t)} = \frac{\sqrt{B} + S_2(t_0)}{\sqrt{B} - S_2(t_0)}\exp(\pm 2\sqrt{B\beta\delta}(t - t_0))\,. \tag{17.109}$$

The sign in the exponent is to be determined knowing the sign of $\dot{S}_2$ at $t = t_0$. If $\dot{S}_2(t_0) > 0$, we take the plus sign in the exponent; thus $S_2(t) \to \sqrt{B}$ as $t \to +\infty$; if $\dot{S}_2(t_0) <0$, we take the minus sign in the exponent, giving $S_2(t) \to -\sqrt{B}$ for $t \to +\infty$. These results are eminently reasonable.

Our second example of a classical pure-spin system is defined by a Hamiltonian linear in the dynamical variables:

$$H(\boldsymbol{S}) = \mu\boldsymbol{B}\cdot\boldsymbol{S} \tag{17.110}$$

Here $\boldsymbol{B}$ is some given vector with components B_j in a fixed coordinate system S. This Hamiltonian can be interpreted as describing the interaction of a particle having a magnetic moment vector $\mu\boldsymbol{S}$ with an external magnetic field $\boldsymbol{B}$. The magnitude of the magnetic moment will be μ if we define the phase space to be the surface $|\boldsymbol{S}| = 1$. This is an "honest" pure-spin system; it is evident that the S_j do generate rotations and also that the system is not rotationally invariant because we have the external field $\boldsymbol{B}$. The equations of motion are:

$$\dot{S}_j = -\{\mu B_k S_k, S_j\} = \mu B_k \epsilon_{kjm} S_m = \mu(\boldsymbol{S}\times\boldsymbol{B})_j\,. \tag{17.111}$$

Of course, S^2 is conserved and so is the component $\boldsymbol{S}\cdot\boldsymbol{B}$ of $\boldsymbol{S}$ along $\boldsymbol{B}$, because this is essentially the Hamiltonian. The projection of $\boldsymbol{S}$ perpendicular to $\boldsymbol{B}$, $(1/B^2)\boldsymbol{B}\times(\boldsymbol{S}\times\boldsymbol{B})$, rotates with a fixed angular velocity $\mu|\boldsymbol{B}|$ about the axis $\boldsymbol{B}$. One says that the magnetic moment vector $\mu\boldsymbol{S}$ *precesses* about the direction of the external field $\boldsymbol{B}$ while preserving its magnitude and its projection along $\boldsymbol{B}$; this is called *Larmor precession.*

The third and last example of a pure-spin system that we consider has a general inhomogeneous quadratic Hamiltonian:

$$H(\boldsymbol{S}) = \frac{1}{2}Q_{jk}S_jS_k + B_jS_j\,. \tag{17.112}$$

For the special case B_j=0 and with the replacement $Q \to I^{-1}$, this goes over to the expression for the kinetic energy of a rotating rigid body, but we prefer not to interpret this term in this way. We would like to interpret the S_j as the spatial components of some vector and thus look upon the quadratic terms also as *interaction* terms. Concerning the quadratic part, Q_{jk} must be symmetric in j and k. We may also assume that it is traceless, because that involves dropping a multiple of S^2 from the Hamiltonian, and this will not affect the equations of motion. We can then rewrite Eq. (17.112) in this way:

$$H(\boldsymbol{S}) = \frac{1}{2}Q_{jk}(S_j S_k - \frac{1}{3}\delta_{jk}S^2) + B_j S_j \, . \tag{17.113}$$

The first part has the form of an (electric) quadrupole interaction, as is felt, for example, by a charged particle (ion) at a crystal lattice site; the second part again has the form of a (magnetic) dipole interaction. Hence the most general quadratic Hamiltonian for a classical spin system can be interpreted as an ion with a magnetic moment situated at a crystalline site and in an external magnetic field. By taking advantage of the conservation of H and S^2, the solution of the equations of motion can be reduced to quadratures.

A general feature of the equations of motion for pure-spin systems is now evident: we always have two simple integrals of the motion, namely the energy and the "total angular momentum" S^2, which allows us to reduce the equations of motion to first order, uncoupled, ordinary differential equations that can be solved immediately. The energy integral always exists in the Hamiltonian formalism, and the variable S^2 is a "kinematic invariant" for pure-spin systems, because it is the Casimir invariant for $R(3)$. Thus to obtain the time dependence of the dynamical variables, all we have to do is solve explicitly for $\dot{S}_2$, say in terms of S_2, A, and B, where $H(\boldsymbol{S}) = \frac{1}{2}A$ and $S^2 = B$. Whether this is easy to do, and the integral computable, depends entirely on the functional form of $H(\boldsymbol{S})$. There is one situation in which one can make a definite statement, and that is when $H(\boldsymbol{S})$ is an *algebraic* function of S_j. (The form of B causes no problems, of course.) In that case, we have the following remarkable result:

The equations of motion of a pure-spin system in classical mechanics, with an algebraic function for the Hamiltonian, can be completely integrated in terms of Abelian functions.

[Abelian functions are defined thus: if *f(z)* is an algebraic function of z and if we define x by:

$$x = \int_{z_0}^{z} dz' f(z') \, ,$$

then x is said to be an *Abelian integral* of z, and z is called an *Abelian function* of x. Note that as in the case of elliptic functions, the upper limit in the integral is considered as the dependent variable and the value of the integral as the independent one. From our knowledge of elementary integrals, it is clear that Abelian integrals include algebraic functions, inverse trigonometric functions, and logarithmic functions.]

Linear Representations of $R(3)$-Vectors and Tensors

Our discussion of the group $R(3)$ given in Chapter 15 is based on the way in which the three Cartesian coordinates of a point P in space change when the spatial coordinate system is rotated. This transformation law for the coordinates is the basic and defining linear representation of the group $R(3)$ by means of 3×3 matrices, and it describes the rotational behavior of vectors. If S' and S

are two Cartesian coordinate systems in space having the same origin O and related by the rotation $\exp(\alpha_j l_j)$, then a vectorial quantity V has components V_j in S and V'_j in S', and these are related thus:

$$\begin{aligned} S' &= \exp(\boldsymbol{\alpha}\cdot\boldsymbol{l})S : V'_j = A_{jk}(\boldsymbol{\alpha})V_k\,, \\ A_{jk}(\boldsymbol{\alpha}) &= \delta_{jk}\cos\alpha + \alpha_j\alpha_k\frac{1-\cos\alpha}{\alpha^2} - \epsilon_{jkm}\alpha_m\frac{\sin\alpha}{\alpha}\,. \end{aligned} \tag{17.114}$$

If a point P has coordinates x_j in S and x'_j in S', the relation between the x and x' is a particular case of the above, with x_k written for V_k, and x'_j for V'_j. In the Hamiltonian formulation of mechanics, a dynamical variable that is a vector consists of three functions $V_j(\omega)$; and according to Eq. (17.12), in the Heisenberg picture, these functions change in going from S to S' in this fashion:

$$V'_j(\omega) = \exp(\widetilde{-\boldsymbol{\alpha}\cdot\boldsymbol{J}(\omega)})V_j(\omega) = A_{jk}(\boldsymbol{\alpha})V_k(\omega)\,. \tag{17.115}$$

The $V'_j(\omega)$ represent in S' what the $V_j(\omega)$ do in S. But the basic PB relations in Eq. (17.11) between $J_j(\omega)$ and $V_k(\omega)$ are valid, whatever be the picture of dynamics used; they merely tell us that the $V_k(\omega)$ are the components of a *vector*. The form of Eq. (17.115) is given by the conventions of Chapter 16.

To specialize the remarks made in Chapter 14 as well as in this one, concerning a general Lie group G: a linear matrix representation of $R(3)$ arises from a matrix representation of its Lie algebra, with commutators for Lie brackets, and conversely. [However, in order to ensure that a representation of the Lie algebra will yield a representation of the group $R(3)$, extra global conditions must be imposed on the former; compare this with the remarks made on page 215] A matrix representation of the Lie algebra of $R(3)$ consists of three matrices S_j, j=1,2,3, obeying Eq. (17.85) for this case:

$$[S_j, S_k]_- = \epsilon_{jkm}S_m\,. \tag{17.116}$$

(There is little fear of confusion arising from the use of the letters S_j here to denote matrices, whereas previously they were used in connection with rigid body rotations and pure-spin systems, because we avoid discussing both topics simultaneously.) If Eq. (17.116) holds, the matrices representing elements of $R(3)$, written $D(\boldsymbol{\alpha})$, are obtained thus:

$$\exp(\boldsymbol{\alpha}\cdot\boldsymbol{l}) \to D(\boldsymbol{\alpha}) = \exp(\boldsymbol{\alpha}\cdot\boldsymbol{S})\,. \tag{17.117}$$

The proper commutation rules for the S_j guarantee the proper multiplication rules for the D's. [The global condition on the S_j turns out to be this: if $|\boldsymbol{\alpha}| = 2\pi$ so that $\exp(\boldsymbol{\alpha}\cdot\boldsymbol{l})$ is a rotation of angle 2π about $\boldsymbol{\alpha}$ and thus is the same as the identity, $D(\boldsymbol{\alpha})$ should be the unit matrix.] Of course, these statements have not been proved in this book, but the relationships involved are quite similar to those that we have proved in connection with realizations of Lie groups via point transformations or canonical transformations. In those cases we saw that we had to find a set of first-order linear differential operators forming a realization of the Lie algebra via commutators. Given such a realization, the point transformations realizing the group were given by exponentials of these differential operators; for a canonical realization the differential operators had to have a special form. For matrix representations, the analogous statements are contained in Eqs. (17.85) and (17.86), or for $R(3)$ in Eqs. (17.116) and (17.117). It is worth analyzing the defining matrix representation of $R(3)$ from this point of view. Comparing Eqs. (17.114) and (17.117) for small α_j [and writing $A(\boldsymbol{\alpha})$ in place of $D(\boldsymbol{\alpha})$] and recalling the definition of the matrix $C(\boldsymbol{\alpha})$ in Eq. (17.21), we find:

$$A(\boldsymbol{\alpha}) \simeq 1 + C(\boldsymbol{\alpha}) \simeq 1 + \boldsymbol{\alpha}\cdot\boldsymbol{S},\; C(\boldsymbol{\alpha}) = \boldsymbol{\alpha}\cdot\boldsymbol{S};$$

$$S_1 = \begin{pmatrix} 0 & 0 & 0 \\ 0 & 0 & -1 \\ 0 & 1 & 0 \end{pmatrix}, S_2 = \begin{pmatrix} 0 & 0 & 1 \\ 0 & 0 & 0 \\ -1 & 0 & 0 \end{pmatrix}, S_3 = \begin{pmatrix} 0 & -1 & 0 \\ 1 & 0 & 0 \\ 0 & 0 & 0 \end{pmatrix} \tag{17.118}$$

These are the matrices realizing the Lie algebra of $R(3)$ in the defining representation;by direct substitution one can verify that the commutation rules in Eq. (17.116) are obeyed. The reason why the matrix $\boldsymbol{\alpha}\cdot\boldsymbol{S}$ has turned out to be the matrix $C(\boldsymbol{\alpha})$ constructed from the structure constants according to the general theory of Chapter 13 is simple: the defining representation of $R(3)$ is also its adjoint representation (This is shown in Chapter 15.) The matrices S_j have been found from the form of $A(\boldsymbol{\alpha})$ for infinitesimal α_j. With the aid of the properties of the matrix $C(\boldsymbol{\alpha})$, contained in Eqs. (17.23), (17.24), and (17.25), we can verify that $A(\boldsymbol{\alpha})$ is indeed the exponential of $C(\boldsymbol{\alpha})$, in agreement with the general statement in Eq. (17.117):

$$\begin{aligned}
\exp(C(\boldsymbol{\alpha})) &= 1 + \sum_{n=1}^{\infty} \frac{[C(\boldsymbol{\alpha})]^{2n}}{(2n)!} + \sum_{n=0}^{\infty} \frac{[C(\boldsymbol{\alpha})]^{2n+1}}{(2n+1)!} \\
&= 1 + P(\boldsymbol{\alpha}) \sum_{n=1}^{\infty} \frac{(-1)^n \alpha^{2n}}{(2n)!} \\
&+ C(\boldsymbol{\alpha}) \sum_{n=0}^{\infty} \frac{(-1)^n \alpha^{2n}}{(2n+1)!} = 1 + (\cos\alpha - 1)P(\boldsymbol{\alpha}) + \frac{\sin\alpha}{\alpha} C(\boldsymbol{\alpha}); \\
\{\exp(C(\boldsymbol{\alpha}))\}_{kj} &= \delta_{kj}\cos\alpha + \frac{\alpha_k\alpha_j}{\alpha^2}(1-\cos\alpha) - \epsilon_{kjm}\alpha_m \frac{\sin\alpha}{\alpha} = A_{kj}(\boldsymbol{\alpha}).
\end{aligned} \tag{17.119}$$

Thus we have satisfied ourselves that the general statements we have made regarding linear matrix representations of $R(3)$ are true in the defining representation.

The set of all matrix representations of $R(3)$ is obtained by finding all possible solutions to the commutation relations of Eq. (17.116) and applying the global condition on them. A general solution obtained in this way may be *reducible*; that is, in the linear vector space on which the matrices $S_j, D(\boldsymbol{\alpha})$ act, there may be subspaces that are carried into themselves under the action of these matrices. A subspace that goes into itself under the S_j will clearly do so under the $D(\boldsymbol{\alpha})$ as well. Such a subspace is called an *invariant subspace* of the original representation space. If the representation space has no invariant subspaces (nontrivial ones, that is), one calls the representation an *irreducible representation* of $R(3)$. For this group, it is a fact that every matrix representation is either irreducible or is a direct sum of irreducible representations. Thus the most general representation of $R(3)$ is obtained by putting together, in a direct sum, any collection of irreducible representations; equally well, the most general solution to Eq. (17.116) is a direct sum of irreducible solutions. Another basic fact concerning $R(3)$ is that every matrix representation of it is equivalent to a unitary representation; that is, one may always assume that the matrices $D(\boldsymbol{\alpha})$ are unitary matrices. From Eq. (17.117) it follows that the matrices S_j may be assumed to be *antihermitian.* For these reasons, it is sufficient to look for and determine all possible sets of three *irreducible, antihermitian* matrices S_j obeying Eq. (17.116), then impose the global condition on each such set to pick out the "allowed" ones. (We repeat the meaning of irreducibility: given the three matrices S_j acting on some vector space, there should be no proper subspace that is mapped into itself under the action of each of the S_j). This analysis is a standard one and is to be found in all texts on quantum mechanics, and one finds that once the global conditions have been imposed, one has the following picture: there is an irreducible, antihermitian set S_j obeying Eq. (17.116) and generating an irreducible matrix representation of $R(3)$, *in every odd dimension.* That is, there is an irreducible, antihermitian set of 1×1 matrices obeying Eq. (17.116), another of 3×3 matrices obeying Eq. (17.116), a third of 5×5 matrices, and so on in $7, 9, \cdots$ dimensions. Using these, one has by exponentiation a unitary irreducible matrix representation of $R(3)$ by 1×1 matrices, another by 3×3 matrices, a third by 5×5 matrices, and so on. This exhausts the irreducible matrix representations of $R(3)$. The one-dimensional representation is the trivial one in which each S_j is zero, and each $D(\boldsymbol{\alpha})$ is the number 1. The 3×3 matrices that solve Eq. (17.116) are the S_j of the defining, adjoint representation and are exhibited in Eq. (17.118). To say that the corresponding representation of $R(3)$ is irreducible is to say that there is no nonzero vector in three-dimensional space whose components are preserved in numerical value under all rotations. Another useful property for $R(3)$ is this: all these irreducible matrix representations, and thus all matrix representations, can be written in such a way that the representation matrices $D(\boldsymbol{\alpha})$ for all $\boldsymbol{\alpha}$ are *real*; combined with the fact that they are also unitary, this means that the $D(\boldsymbol{\alpha})$ are always real orthogonal matrices (or rather, can always be chosen so). When this is so, the S_j, besides being antihermitian, are also real, and thus antisymmetric; for example, this is the case in Eq. (17.118). However, these are not the forms in which the $D(\boldsymbol{\alpha})$ and S_j are expressed in most quantum mechanical treatments.

We do not intend to give a complete derivation, with proofs, of all these interesting properties. Consequently, instead of approaching the problem from the infinitesimal point of view, which is that of finding all possible (globally acceptable) solutions to Eq. (17.116), we shall give a description of geometrical entities called tensors, which form generalizations of vectors in three-dimensional space and whose behaviors under rotations are also generalizations of the behavior of vectors. We shall then indicate how, by considering special kinds of tensors, one can arrive at the various irreducible odd-dimensional matrix representations of $R(3)$.

We know that a physical quantity is called a vector V if it takes three numbers V_j to define it completely in a Cartesian coordinate system S and if on changing to a rotated coordinate system S' these numbers change linearly according to Eq. (17.114). Let $V^{(1)}$ and $V^{(2)}$ be two vectors. Out of their components we can form nine independent products $V_j^{(1)}V_k^{(2)}$, j and k independently taking the values 1, 2, and 3; these nine products also transform linearly under a change of coordinate system. A physical quantity is called a tensor T of the second rank if it requires nine numbers $T_{jk}, 1 \leq j, k \leq 3$ to describe it fully in each coordinate system S and if on switching to a rotated coordinate system $S' = aS$, these numbers transform just like the products $V_j^{(1)}V_k^{(2)}$ of components of vectors. That is, the relation between the "components" T'_{jk} of T in S' and T_{jk} in S must be:

$$S' = aS : T'_{jk} = A_{jl}A_{km}T_{lm}\,. \tag{17.120}$$

[The matrix A corresponds to the rotation a.] A tensor transforms like, but need not be equal to, products of components of vectors. The transformation law for vectors gives rise, by definition, to a representation of $R(3)$. Because each index on a second-rank tensor T transforms by itself just as though it were an index on a vector, the transformation law for these tensors also gives rise to a representation of $R(3)$, in fact, a 9×9 matrix representation. This is easy to verify. Let $S'' = bS' = baS$ be a third coordinate system, and let T''_{jk} be the components of T in it. Then, applying Eq. (17.120) twice, we get:

$$S'' = baS : T''_{jk} = B_{jl}B_{km}T'_{lm} = B_{jl}B_{km}A_{lp}A_{mq}T_{pq} = (BA)_{jl}(BA)_{km}T_{lm}\,, \tag{17.121}$$

demonstrating that we do have a representation of $R(3)$ here. This representation is nine dimensional because a general tensor T of the second rank has nine independent components.

We can immediately generalize these definitions to describe a tensor of arbitrary rank. A physical quantity is called a tensor of rank N if in each coordinate system S it requires 3^N numbers or "components" $T_{jk\ldots q}$, $1 \leq j, k, \ldots q \leq 3$, to specify it fully, and if these components change thus on a rotation of the coordinate system:

$$S' = aS : T'_{jk\ldots q} = A_{jl}A_{km}\ldots A_{qv}T_{lm\ldots v}\,. \tag{17.122}$$

Here, the number of indices on T is N, accounting for 3^N components. As in the case N=2, each index on T transforms by itself as though it were an index on a vector. One can check again that one has a representation of $R(3)$; it is of dimension 3^N, corresponding to the number of independent components of T. A vector is the same thing as a tensor of rank one.

Tensors of a given rank can be multiplied by numbers or scalars (that is, quantities that do not change under rotations), or we can form linear combinations of tensors of the same rank with scalars for coefficients, with the result always being a tensor of the same rank. Thus tensors of a given rank N form a linear space of dimension 3^N. These facts are obvious from the transformation law in Eq. (17.122), and because of them we say that tensors of given rank give us a representation of $R(3)$. Another operation with tensors that leads again to tensors is "outer multiplication": if we have two tensors $T^{(1)}$ and $T^{(2)}$ of ranks N and M, respectively, then by considering all possible products of components of $T^{(1)}$ by components of $T^{(2)}$ we generate $3^N \times 3^M = 3^{N+M}$ quantities that form the components of a tensor T of rank $N+M$. The verification of this is very simple; it simply depends on the fact already mentioned that indices on tensors transform independently.

The representations of $R(3)$ that have been obtained in this way are called the *tensor products* of the defining vector representation with itself. Thus the 3^N dimensional representation provided by the space of tensors of rank N is the N-fold tensor product of the vector representation with itself. These representations of $R(3)$ are reducible, if $N \geq 2$. In the linear space of all tensors of rank N, we can find subsets of tensors with special properties such that (1) these special properties are preserved under multiplication of such a tensor by a scalar as well as under the formation of linear combinations of such tensors with scalars for coefficients, and (2) these properties are preserved under transitions from one Cartesian coordinate system to another. It follows that tensors with these special properties form an *invariant subspace* of the 3^N dimensional space of all tensors of rank N and also that these properties are intrinsic to these tensors. One simple way to generate properties that are intrinsic to certain tensors, and thus help to find invariant subspaces, is to notice that in the basic transformation law (Eq. (17.122)), each index on a tensor transforms exactly like any other index; all of them transform by the common matrix A. This means that if the components $T_{jk\ldots q}$ of T in the coordinate system S have any simple symmetry properties with respect to permutations of the indices $j, k, \ldots q$ (thereby reducing the number of linearly independent components of T), these simple symmetry properties will be valid for the components $T'_{jk\ldots q}$ of T in the rotated coordinate system S'. For example, for a second-rank tensor T_{jk}, there are two possible simple symmetry properties: one is that T may be symmetric under interchange of j and k, and another is that T may be antisymmetric under this interchange. If the former is true, we say that T is a symmetric tensor; in the latter case we call T an antisymmetric tensor. A general second-rank tensor T_{jk} will be neither symmetric nor antisymmetric. However, the content of the statement that there are but two simple symmetry properties for second-rank tensors is just this: any tensor T can be written as the sum of a tensor $T^{(S)}$ that is symmetric and a tensor $T^{(A)}$ that is antisymmetric in a unique way. In fact, in a given frame S, we define the components of $T^{(S)}$ and $T^{(A)}$ in this way:

$$T^{(S)}_{jk} = \frac{1}{2}(T_{jk} + T_{kj}), T^{(A)}_{jk} = \frac{1}{2}(T_{jk} - T_{kj}), \tag{17.123}$$

and then we have:

$$T_{jk} = T^{(S)}_{jk} + T^{(A)}_{jk}. \tag{17.124}$$

What must be done now is to check that $T^{(S)}$ and $T^{(A)}$ are indeed tensors, but this is trivial:

$$\begin{aligned} T'^{(S)}_{jk} &= \frac{1}{2}(T'_{jk} + T'_{kj}) = \frac{1}{2}(A_{jl}A_{km}T_{lm} + A_{km}A_{jl}T_{ml}) \\ &= \frac{1}{2}A_{jl}A_{km}(T_{lm} + T_{ml}) = A_{jl}A_{km}T^{(S)}_{lm}\,, \end{aligned} \tag{17.125}$$

and similarly for $T^{(A)}$. We see that each tensor T can be expressed as the sum of its symmetric part and its antisymmetric part, with each part transforming by itself and preserving its special property of being symmetric or antisymmetric. Thus the full nine-dimensional space of all second-rank tensors splits up into two invariant subspaces. The larger one contains all symmetric tensors. Because any such tensor is easily seen to have six independent components, this is a six dimensional subspace. The smaller one consists of all-antisymmetric tensors and is three dimensional. We shall see shortly that the six-dimensional subspace can be split up further into a five-dimensional subspace and a one-dimensional one, and these are invariant as well as irreducible under $R(3)$.

For a tensor whose rank is greater than or equal to 3, there are more than just two possible elementary symmetry properties in the indices. The simplest possibilities are that the components $T_{jk\ldots q}$ *may be completely symmetric* under all permutations of the indices $j, k, \ldots q$; or *completely antisymmetric*, which means that the components change sign under any odd permutation and are unchanged under any even permutation. (For complete antisymmetry, N cannot exceed 3). In addition to these two extremes, other kinds of symmetry are possible, and these are referred to as *intermediate symmetry.* We do not describe these in detail, but do note the following: a general tensor of rank N can be expressed in a unique way as a sum of several parts, each part being a tensor possessing a definite kind of symmetry property. This property may be complete symmetry, complete antisymmetry, or some intermediate symmetry. Each part transforms by itself under rotations and retains the symmetry type to which it belongs. In this way, the full 3^N dimensional space of all tensors of rank N is split up into various invariant subspaces; each subspace consists of all tensors of rank N and of a specific symmetry type. We now concentrate our attention on the subspace of tensors of rank N, which are completely symmetric in their indices. How many independent components does such a tensor have? Two components $T_{jk\ldots q}$ and $T_{j'k'\ldots q'}$ are equal if the indices $j', k', \ldots q'$ are the same as the indices $j, k, \ldots q$, possibly written in some other order. This means that the value one appears as many times in the set $j', k', \ldots q'$ as it does in the set $j, k, \ldots q$ and similarly for the values two and three. The value of a component of T is dependent only on how many of the indices are ones, and how many are twos and how many are threes, independent of their locations. A simple counting procedure then tells us that such a tensor T has just $\frac{1}{2}(N+1)(N+2)$ independent components - far fewer than the 3^N independent components of a general Nth-rank tensor! Therefore, the fully symmetric Nth-rank tensors give us a linear matrix representation of $R(3)$ of dimensionality $\frac{1}{2}(N+1)(N+2)$. But these are still not irreducible! Symmetry properties under permutations of indices are not the only special properties of tensors that are preserved under rotations and help in the finding of invariant subspaces of special tensors; the other relevant special property arises from the possibility of"taking the trace". This has its origin in the fact that the matrix A that appears in the general transformation law (Eq. (17.122)) is always an orthogonal matrix and thus obeys:

$$A_{jk}A_{jl} = \delta_{kl}\,. \tag{17.126}$$

This allows us to take a general Nth rank tensor $T^{(N)}$ and form from it a tensor $T^{(N-2)}$ of rank $N-2$; this can be done in many ways. $T^{(N)}$ need not be a symmetric tensor of rank N. If, for example, we "take the trace" with respect to the first two indices of $T^{(N)}$, then we define $T^{(N-2)}$ this way:

$$T^{(N-2)}_{jk\cdots} = T^{(N)}_{mmjk\cdots} \tag{17.127}$$

Using Eqs. 122 and 126, we can check that $T^{(N-2)}$ is a tensor:

$$\begin{aligned} T'^{(N-2)}_{jk\ldots} &= T'^{(N)}_{mmjk\cdots} = A_{mp}A_{mq}A_{jl}A_{kn}\ldots T^{(N)}_{pqln\ldots} \\ &= A_{jl}A_{kn}\ldots T^{(N)}_{ppln\cdots} = A_{jl}A_{kn}\ldots T^{(N-2)}_{ln\cdots}\,. \end{aligned} \tag{17.128}$$

Other tensors of rank $N-2$ could be obtained from $T^{(N)}$ by taking the trace with respect to other pairs of indices. By taking the trace with respect to two pairs of indices in $T^{(N)}$, we would obtain a tensor of rank N-4, and so on down the line. For a symmetric tensor of rank N, it does not matter with respect to which pair of indices we take the trace; we will always end up with the same tensor of rank $N-2$, and this, too, will be symmetric. Thus a symmetric tensor T^N of rank N gives rise to a symmetric tensor T^{N-2} of rank $N-2$ by taking the trace once; taking the trace twice in T^N, or once in T^{N-2}, we get a symmetric tensor T^{N-4} of rank $N-4$, and so on. The special intrinsic property of tensors that we can now introduce is that of being *traceless*: a symmetric tensor T^N of rank N is said to be *traceless* if the symmetric tensor T^{N-2} of rank $N-2$ obtained by taking the trace of (any) two indices in T^N vanishes identically. In that case, the lower-rank tensors T^{N-4}, obtained by taking the trace twice or more in T^N will also vanish. (If T is not symmetric, it may be traceless in some pairs of indices but not in others.) Thus a symmetric traceless tensor of rank N obeys the two conditions:

$$T_{jk\cdots q} = T_{j'k'\cdots q'}, j'k'\cdots q' = P(jk\cdots q); \quad T_{mmj\ldots} = 0\,. \tag{17.129}$$

The P denotes any permutation. Just as symmetry is preserved under rotations, so is tracelessness. In fact, this is obvious because the tensor T^{N-2} obtained by taking the trace once transforms in a linear homogeneous fashion (like a tensor, that is !), and thus if its components vanish in one coordinate system, they vanish in all coordinate systems. Therefore the traceless symmetric tensors of rank N form an invariant subspace in the space of symmetric tensors of rank N and give us a representation of $R(3)$ all by themselves. *These are all the irreducible matrix representations of* $R(3)$. Symmetry restricts the number of independent components of T to the number $\frac{1}{2}(N+1)(N+2)$. The tracelessness condition written in the second half of Eq. (17.129) amounts to as many linear relations among the components of T as there are independent components for a symmetric tensor of rank $N-2$, namely $\frac{1}{2}(N-1)N$. Therefore a symmetric traceless tensor of rank N has $\frac{1}{2}[(N+1)(N+2)-N(N-1)]$ independent components, that is, $(2N+1)$ independent components, and this is the dimensionality of the matrix representation of $R(3)$ given by such tensors. By considering symmetric traceless tensors of ranks 0,1,2,3,$\cdots$, we obtain the 1,3,5,7,$\cdots$ dimensional matrix representations of $R(3)$, that is, the complete set of irreducible representations of $R(3)$. Each way of choosing the $(2N+1)$ independent components of a symmetric traceless tensor of rank N leads to a specific form for the $(2N+1)$-dimensional irreducible representation of $R(3)$, with specific matrices S_j and $D(\boldsymbol{\alpha})$. Because the transformation law (Eq. (17.122)), as well as the "subsidiary conditions" (Eq. (17.129)) involve real equations, we can see that the irreducible representations of $R(3)$ can certainly be expressed in real form.

In an obvious sense, the subspace of traceless symmetric tensors of rank N is the largest invariant subspace in the space of symmetric tensors of rank N. In addition to this, the latter space contains as invariant subspaces the traceless symmetric tensors of ranks $N-2, N-4, \cdots$, 1 or 0. Thus the $\frac{1}{2}(N+1)(N+2)$ dimensional representation of $R(3)$ given by symmetric tensors of rank N splits up completely into the $(2N+1)$, $(2N-3)$, $(2N-7), \cdots$, 3 or 1 dimensional irreducible representations of $R(3)$. One verifies that the dimensionalities add up properly. For the case N=2, for example, the six independent components of a symmetric T_{jk} can be separated into the five components of its traceless part and the trace; the former transforms irreducibly according to the five-dimensional representation of $R(3)$, and the latter is a scalar. This breakup is exhibited by:

$$T_{jk} = (T_{jk} - \frac{1}{3}\delta_{jk}T_{mm}) + \frac{1}{3}\delta_{jk}T_{mm}\,. \tag{17.130}$$

Thus we can see that all irreducible matrix realizations of $R(3)$ are gotten by taking the tensor product of the defining representation with itself all possible numbers of times, then "symmetrizing the product" and "removing the trace". We can deal with these representations by considering tensors transforming according to Eq. (17.122) and subject to the subsidiary conditions in Eq. (17.129), or we can explicitly make use of these conditions to put the representation into a form in which only $(2N+1)$-dimensional matrices appear. In order to achieve the latter, the following must be done: out of the Cartesian components $T_{jk\cdots q}$ of a general, symmetric, traceless tensor of rank N (also called an Nth-rank irreducible tensor), we must form precisely $(2N+1)$ linearly independent combinations T_r, r taking $(2N+1)$ distinct values. These linear combinations T_r must be such that by using their definitions in terms of the $T_{jk\cdots q}$ as well as the conditions in Eq.(17.129), we must be able to express each $T_{jk\cdots q}$ as a linear combination of the T_r. If this is so, then the T_r can be treated as completely independent; assigning any set of values to these independently leads to one unique, irreducible tensor of rank N, and vice versa. (This is just like choosing generalized coordinates in Lagrangian mechanics: the idea is to take explicit account of all constraints, which reduce the number of independent variables, but once the constraints are satisfied, one deals with truly independent variables, their number being less than the number of originally constrained ones.) Knowing how to pass from the $T_{jk\cdots q}$ to the T_r and conversely, one can use Eq. (17.122) to determine the transformation law for the latter; it will have the form:

$$S' = \exp(\boldsymbol{\alpha}\cdot\boldsymbol{l})S : T'_r = D_{rs}(\boldsymbol{\alpha})T_s\,. \tag{17.131}$$

The matrices $D_{rs}(\boldsymbol{\alpha})$ are $(2N+1)$-dimensional matrices, forming the corresponding irreducible matrix representation of $R(3)$. Clearly, they are linear combinations of the products appearing in Eq. (17.122), namely of $A_{jl}(\boldsymbol{\alpha})A_{km}(\boldsymbol{\alpha})\cdots$ (N factors), with the forms of the combinations determined by the formulae that express $T_{jk\cdots q}$ in terms of T_r, and vice versa. For example, for an irreducible tensor of rank two, a convenient choice of the five T_r turns out to be the following:

$$T_{23}, T_{31}T_{12}; \frac{1}{2}(T_{11}-T_{22})\,,\ -\frac{\sqrt{3}}{2}T_{33}\,. \tag{17.132}$$

We leave it to the reader to check that the symmetry and tracelessness of T_{jk} allow one to express each Cartesian component T_{jk} in a unique way as a linear combination of these five quantities. It turns out that this choice of five independent T_r's leads to the matrices $||D_{sr}(\boldsymbol{\alpha})||$ being 5×5 real orthogonal matrices. For any irreducible representation, the generator matrices $(S_j)_{sr}$ are obtained by examining the $D(\boldsymbol{\alpha})$ for small $\boldsymbol{\alpha}$, because Eq. (17.117) is valid in any case:

$$(S_j)_{sr} = \frac{\partial}{\partial\alpha_j}D_{sr}(\boldsymbol{\alpha})\bigg|_{\alpha_j=0}\,. \tag{17.133}$$

We conclude this description of the matrix representations of $R(3)$ by giving the equations that characterize tensors in Hamiltonian mechanics. A Cartesian tensor of rank N consists of a set of 3^N phase-space functions, $T_{jk\cdots q}(\omega)$, (N indices here !), obeying:

$$\exp(\widetilde{-\boldsymbol{\alpha}\cdot\boldsymbol{J}}(\omega))T_{jk\cdots q}(\omega) = A_{jl}(\boldsymbol{\alpha})A_{km}(\boldsymbol{\alpha})\cdots A_{qv}(\boldsymbol{\alpha})T_{lm\cdots v}(\omega)\,. \tag{17.134}$$

This generalizes Eq. (17.115) valid for vectors; it may in the first instance be derived by using the Heisenberg picture, but then we realize that it has a picture independent significance. The infinitesimal form of Eq. (17.134) looks like this:

$$\{J_m(\omega), T_{jk\cdots q}(\omega)\} = \epsilon_{mjn}T_{nk\cdots q}(\omega) + \epsilon_{mkn}T_{jn\cdots q}(\omega) + \cdots + \epsilon_{mqn}T_{jk\cdots n}(\omega)\,. \tag{17.135}$$

For an irreducible tensor of rank N, we may continue to describe the rotational properties in this way, and then impose the extra relations (Eq. (17.129)); more simply, we can work with the $(2N+1)$ independent combinations $T_r(\omega)$. In this case, we say that a physical quantity $T(\omega)$ is an irreducible tensor of rank N if it requires $(2N+1)$ phase-space functions $T_r(\omega)$ to describe it fully in each coordinate system S, and if (in the Heisenberg picture), the corresponding functions in a rotated system S' are obtained thus:

$$S' = \exp(\boldsymbol{\alpha}\cdot\boldsymbol{l})S : T'_r(\omega) = \exp(\widetilde{-\boldsymbol{\alpha}\cdot\boldsymbol{J}}(\omega))T_r(\omega) = D_{rs}(\boldsymbol{\alpha})T_s(\omega)\,. \tag{17.136}$$

The corresponding infinitesimal form involves the $(2N+1)$ dimensional generator matrices for the D's:

$$\{J_j(\omega), T_r(\omega)\} = -(S_j)_{rs}T_s(\omega)\,. \tag{17.137}$$

Spinors and the Relation Between $SU(2)$ and $R(3)$

In addition to the irreducible, antihermitian solutions to the commutation rules in Eq. (17.116) that exist in every odd dimension and give rise to representations of $R(3)$, it turns out that irreducible, antihermitian solutions exist in every even dimension as well; however, these fail to obey the global condition and thus do not give us representations of $R(3)$. In each even-dimensional solution, it turns out that the matrix $\exp(\boldsymbol{\alpha}\cdot\boldsymbol{S})$ for $|\boldsymbol{\alpha}| = 2\pi$, is the negative of the unit matrix, rather than the unit matrix! Thus the situation is this: the commutation rules in Eq. (17.116) possess irreducible, antihermitian solutions in *every* dimension, but only every other one qualifies to produce a matrix representation of $R(3)$. But the fact that the commutation rules are obeyed implies that even for the $2, 4, 6, \cdots$ dimensional cases, *for small enough* $|\boldsymbol{\alpha}|$, the matrices $D(\boldsymbol{\alpha})$ obey the multiplication laws of $R(3)$, and it is only globally that they fail to do so. We say that in these cases we have only a local representation of $R(3)$ and not a representation of $R(3)$. What the matrices $D(\boldsymbol{\alpha})$ do give us in these cases are matrix representations of the group $SU(2)$, this being the group of all two-dimensional complex unitary matrices with unit determinant. This is the universal covering group of $R(3)$, and its Lie algebra has the same structure as that of $R(3)$. We now describe briefly the connection between $SU(2)$ and $R(3)$ and their representations.

A reasonable starting point is to exhibit the two-dimensional solution to the commutation rules in Eq. (17.116). This is:

$$S_1 = \frac{-i}{2}\begin{pmatrix}0 & 1\\ 1 & 0\end{pmatrix}, S_2 = \frac{-i}{2}\begin{pmatrix}0 & -i\\ i & 0\end{pmatrix}, S_3 = \frac{-i}{2}\begin{pmatrix}1 & 0\\ 0 & -1\end{pmatrix}. \tag{17.138}$$

Apart from the factors $-i/2$, these matrices are called the *Pauli matrices* σ_j, $j = 1, 2, 3$:

$$\sigma_1 = \begin{pmatrix}0 & 1\\ 1 & 0\end{pmatrix}, \sigma_2 = \begin{pmatrix}0 & -i\\ i & 0\end{pmatrix}, \sigma_3 = \begin{pmatrix}1 & 0\\ 0 & -1\end{pmatrix}. \tag{17.139}$$

Taking account of the factor of proportionality between the S's and the σ's, the commutation rules of the σ's are:

$$\sigma_j\sigma_k - \sigma_k\sigma_j = 2i\epsilon_{jkm}\sigma_m\,. \tag{17.140}$$

Now these σ's have many important properties; we shall be content with brief statements of them, because it would take us too far to motivate all of them.

Each of the σ's is hermitian and traceless, and in fact any 2×2 hermitian traceless matrix can be expressed in a unique way as a real linear combination of the σ's. Along with the 2×2 unit matrix, the σ's form a *complete set* of 2×2 matrices: any 2×2 matrix is obtainable in a unique way as a (complex) linear combination of these four matrices. In addition to Eq. (17.140), the products of the σ's have the following properties: the square of each of them is the unit matrix, and two different σ's multiplied in two different orders give answers differing in sign. These can be summarized thus:

$$\sigma_j \sigma_k + \sigma_k \sigma_j = 2\delta_{jk} \, . \tag{17.141}$$

As already mentioned, the fact that the 2×2 S_j do form a matrix realization of the Lie algebra of $R(3)$ implies that for small enough $|\boldsymbol{\alpha}|$, the exponential matrices do obey the multiplication properties of the group $R(3)$. Let us write for any $\boldsymbol{\alpha}$:

$$a(\boldsymbol{\alpha}) = \exp(\boldsymbol{\alpha} \cdot \boldsymbol{S}) = \exp\left(\frac{-i}{2}\, \boldsymbol{\alpha} \cdot \boldsymbol{\sigma}\right) . \tag{17.142}$$

Then the statement is that if $|\boldsymbol{\alpha}|$ and $|\boldsymbol{\beta}|$ are sufficiently small, we have:

$$a(\boldsymbol{\alpha})a(\boldsymbol{\beta}) = a(\boldsymbol{f}(\boldsymbol{\alpha}, \boldsymbol{\beta})), \exp(\boldsymbol{\alpha} \cdot \boldsymbol{S}) \exp(\boldsymbol{\beta} \cdot \boldsymbol{S}) = \exp(\boldsymbol{f}(\boldsymbol{\alpha}, \boldsymbol{\beta}) \cdot \boldsymbol{S}) \, . \tag{17.143}$$

The functions $f_j(\boldsymbol{\alpha}, \boldsymbol{\beta})$ are the composition functions for $R(3)$. Now consider the product $a(\boldsymbol{\alpha})a(\boldsymbol{\beta})a(-\boldsymbol{\alpha})$, for which we write:

$$\begin{aligned} a(\boldsymbol{\alpha})a(\boldsymbol{\beta})a(-\boldsymbol{\alpha}) &\equiv a(\boldsymbol{\alpha})a(\boldsymbol{\beta})a(\boldsymbol{\alpha})^{-1} = a(\boldsymbol{\gamma}) \, , \\ \gamma_j &= f_j(\boldsymbol{f}(\boldsymbol{\alpha}, \boldsymbol{\beta}), -\boldsymbol{\alpha}) \, ; \end{aligned} \tag{17.144}$$

$\boldsymbol{\alpha}$ and $\boldsymbol{\beta}$ being understood to be sufficiently small, we can use the theory of the adjoint representation of $R(3)$ and conclude that $\boldsymbol{\gamma}$ is the image of $\boldsymbol{\beta}$, under the action of the transformation corresponding to $\exp(\boldsymbol{\alpha} \cdot \boldsymbol{l})$ in this representation. Thus we have:

$$\gamma_j \equiv f_j(\boldsymbol{f}(\boldsymbol{\alpha}, \boldsymbol{\beta}), -\boldsymbol{\alpha}) = A_{jk}(\boldsymbol{\alpha})\beta_k \, , \tag{17.145}$$

and now by expanding both sides of Eq. (17.144) in β and comparing the linear terms we find:

$$a(\boldsymbol{\alpha})S_j a^{-1}(\boldsymbol{\alpha}) = A_{kj}(\boldsymbol{\alpha})S_k, \; a(\boldsymbol{\alpha})\sigma_j a^{-1}(\boldsymbol{\alpha}) = A_{kj}(\boldsymbol{\alpha})\sigma_k \, . \tag{17.146}$$

These equations express the essentials of the relationship between $SU(2)$ and $R(3)$, as we shall see. Now we have discussed these relations under the assumption that $|\boldsymbol{\alpha}|$ is sufiiciently small. But with the properties of the σ's that we have listed, an explicit calculation shows that Eq. (17.146) holds for all $\boldsymbol{\alpha}$. Equation (17.141) tells us that if $\hat{\boldsymbol{\alpha}}$ is a unit vector, then $(\hat{\boldsymbol{\alpha}} \cdot \boldsymbol{\sigma})^2$ is the unit matrix. Consequently, the matrix $a(\boldsymbol{\alpha})$ can be computed by expanding the exponential and gives, for all $\boldsymbol{\alpha}$,

$$\begin{aligned} a(\boldsymbol{\alpha}) &= \exp(-i\boldsymbol{\alpha} \cdot \boldsymbol{\sigma}/2) = \exp(-i\alpha\hat{\boldsymbol{\alpha}} \cdot \boldsymbol{\sigma}/2) \\ &= \cos\frac{\alpha}{2} - i\hat{\boldsymbol{\alpha}} \cdot \boldsymbol{\sigma} \sin\frac{\alpha}{2} \, . \end{aligned} \tag{17.147}$$

A short calculation combining this result with Eqs. (17.140) and (17.141) establishes Eq. (17.146).

The appearance of "half angles" in Eq. (17.147) explains why the matrices $a(\boldsymbol{\alpha})$ fail to obey the global condition for representations of $R(3)$. In fact, if we set $\alpha = 2\pi$, we see that $a(\boldsymbol{\alpha})$ is the negative of the 2×2 unit matrix (but note that this is true for all directions of $\boldsymbol{\alpha}$!). More generally, increasing the magnitude of $\boldsymbol{\alpha}$ by 2π while leaving its direction unaltered changes the sign of $a(\boldsymbol{\alpha})$, whereas it leaves $A(\boldsymbol{\alpha})$ invariant:

$$a\left(\frac{\alpha + 2\pi}{\alpha}\,\boldsymbol{\alpha}\right) = -a(\boldsymbol{\alpha}), A\left(\frac{\alpha + 2\pi}{\alpha}\,\boldsymbol{\alpha}\right) = A(\boldsymbol{\alpha})\,. \tag{17.148}$$

One has to increase the magnitude of $\boldsymbol{\alpha}$ by 4π before one "comes back" to $a(\boldsymbol{\alpha})$; one has then,

$$a\left(\frac{\alpha + 4\pi}{\alpha}\,\boldsymbol{\alpha}\right) = a(\boldsymbol{\alpha}), a^{-1}(\boldsymbol{\alpha}) = a(-\boldsymbol{\alpha}) = a\left(\frac{4\pi - \alpha}{\alpha}\,\boldsymbol{\alpha}\right)\,. \tag{17.149}$$

How can we characterize the matrices $a(\boldsymbol{\alpha})$ that we get by considering all values of $\boldsymbol{\alpha}$? Because the S_j are *antihermitian and traceless*, their exponentials $a(\boldsymbol{\alpha})$ have the corresponding properties of being *unitary* and *unimodular*. We can check these directly, using Eq. (17.147). However, these two properties define the set of matrices $a(\boldsymbol{\alpha})$ completely; that is, *any* 2×2 unitary unimodular matrix is an $a(\boldsymbol{\alpha})$ for some $\boldsymbol{\alpha}$! Let us demonstrate this: by the completeness property, any 2×2 matrix U can be expanded as:

$$U = a_0 - ia_j\sigma_j = a_0 - i\boldsymbol{a} \cdot \boldsymbol{\sigma}\,. \tag{17.150}$$

We choose the parameters a_0, a_j this way because they will turn out to be real. The unitarity of $U, U^+U = \mathbb{I}$, gives:

$$\begin{aligned}(a_0^* + ia_j^*\sigma_j)(a_0 - ia_k\sigma_k) =& a_0^*a_0 + i(a_0a_m^* - a_0^*a_m)\sigma_m \\ &+ a_j^*a_k(\delta_{jk} + i\epsilon_{mjk}\sigma_m) = \mathbb{I}, \\ a_0^*a_0 + a_j^*a_j = 1,\ \epsilon_{mjk}a_j^*a_k =& -a_0a_m^* + a_0^*a_m\,.\end{aligned} \tag{17.151}$$

It follows immediately from the last line that the real three vector $i\epsilon_{mjk}a_j^*a_k$ vanishes identically. Therefore, a_j^* and a_j are proportional: $a_j^* = \lambda a_j$. Putting this back into the last line of Eq. (17.151) we get $a_0^* = \lambda a_0$ as well; then the second line in Eq. (17.151) gives $\lambda(a_0^2 + a_ja_j) = 1$. Now we impose the unimodular condition on U:

$$\det U = \begin{vmatrix} a_0 - ia_3 & -ia_1 - a_2 \\ -ia_1 + a_2 & a_0 + ia_3 \end{vmatrix} = a_0^2 + a_ja_j = 1\,. \tag{17.152}$$

Consequently, $\lambda = 1$. Thus every 2×2 unitary unimodular U can be written in the form of Eq. (17.150) with real a_0, a_j, and these are restricted only by Eq. (17.152). However, this proves the assertion: because $-1 \leq a_0 \leq 1$, we can determine an α uniquely in the range $0 \leq \alpha \leq 2\pi$ for which $a_0 = \cos(\alpha/2)$; then we set $\hat{\boldsymbol{\alpha}} = +\hat{\boldsymbol{a}}$. This makes U equal to $a(\boldsymbol{\alpha})$. *The set of all these* 2×2 *unitary unimodular matrices* $a(\boldsymbol{\alpha})$ *forms the defining representation of the group* $SU(2)$; S means "special", or unimodular, and U "unitary". All the matrices of the defining representation of $SU(2)$ are obtained by choosing all possible directions for $\boldsymbol{\alpha}$ and restricting α to $0 \leq \alpha \leq 2\pi$. There is a vital difference between $SU(2)$ and $R(3)$ that shows up in this way: in the three-dimensional space of the α_j, each point within or on the sphere of radius π represents an element of $R(3)$; however, the two points at the ends of a diameter correspond to the same element, and otherwise distinct points represent distinct elements. In particular, points at the ends of different diameters represent different elements. For $SU(2)$, we must consider all points within or on a sphere of radius 2π to get all elements; whereas distinct points in the interior of the sphere represent distinct elements, *all the points on the surface represent one and the same element of* $SU(2)$. This is because $a(\boldsymbol{\alpha})$ becomes the negative of the unit matrix when $\alpha = 2\pi$, independent of $\hat{\boldsymbol{\alpha}}$.

We can now describe in precise terms the relationship existing between $SU(2)$ and $R(3)$. First, they both have the same structure for their Lie algebras. This is obvious because we took a matrix representation of the Lie algebra of $R(3)$ and by exponentiation we arrived at the *defining* matrix representation of $SU(2)$, and the structure of the Lie algebra of any group can be inferred once one is given any faithful representation of that group ("faithful in some neighborhood of the identity" is sufficient). Second, there is a mapping from $SU(2)$ to $R(3)$ that associates one element of $R(3)$ with each element of $SU(2)$ in such a way that the group operations are preserved. This is easily proved using, for instance, Eq. (17.146). This mapping associates the matrix $A(\boldsymbol{\alpha})$ from the defining representation of $R(3)$ with the matrix $a(\boldsymbol{\alpha})$ from the defining representation of $SU(2)$. That is, with the α's as parameters for both groups, the mapping preserves the values of the parameters. If $A(\boldsymbol{\alpha}), A(\boldsymbol{\beta})$ correspond to $a(\boldsymbol{\alpha}), a(\boldsymbol{\beta})$, respectively, then to the element $a(\boldsymbol{\beta})a(\boldsymbol{\alpha})$ in $SU(2)$ corresponds the element $A(\boldsymbol{\beta})A(\boldsymbol{\alpha})$ of $R(3)$; we see this thus:

$$\begin{aligned} a(\boldsymbol{\beta})a(\boldsymbol{\alpha})\sigma_j a^{-1}(\boldsymbol{\alpha})a^{-1}(\boldsymbol{\beta}) &= a(\boldsymbol{\beta})A_{mj}(\boldsymbol{\alpha})\sigma_m a^{-1}(\boldsymbol{\beta}) = A_{km}(\boldsymbol{\beta})A_{mj}(\boldsymbol{\alpha})\sigma_k\,, \\ \{a(\boldsymbol{\beta})a(\boldsymbol{\alpha})\}\sigma_j\{a(\boldsymbol{\beta})a(\boldsymbol{\alpha})\}^{-1} &= \{A(\boldsymbol{\beta})A(\boldsymbol{\alpha})\}_{kj}\sigma_k\,. \end{aligned} \tag{17.153}$$

Such a mapping from one group to another is called a *homomorphism*. We say that we have a homomorphism from G_1 to G_2 if to each element a in G_1 there corresponds a unique element $\phi(a)$ in G_2 such that $\phi(ba) = \phi(b)\phi(a)$. In a homomorphism, the function (mapping) ϕ must be defined for all possible values of its argument a taken from G_1, but the set of elements $\phi(a)$ in G_2 need not exhaust G_2. Further, although $\phi(a)$ is uniquely determined for a given argument a, several elements $a, b, \ldots$ in G_1 may get mapped onto the same element in G_2: it may happen that $\phi(a) = \phi(b)$ even with $a \neq b$. Because ϕ preserves the group operations, it is trivial to check that every realization and every representation of G_2 (assuming $\phi(G_1) = G_2$) forms, respectively a realization and a representation of G_1 as well: we allow $a \in G_1$ to be realized by the same transformation that corresponds to $\phi(a)$. [If $\phi(G_1) \neq G_2$, it will necessarily be a subgroup of G_2, and then we restrict ourselves to realizations of this subgroup.] However, a realization of G_1, in general, is not one for G_2 because there are in general several elements in G_1 that are mapped to the same element in G_2. In the present case we see that we have a homomorphism from $SU(2)$ to $R(3)$, with the whole of $R(3)$ appearing as the image of $SU(2)$. Every realization or representation for $R(3)$ is one for $SU(2)$ as well. However, the homomorphism is two to one: there are two elements in $SU(2)$ corresponding to each element in $R(3)$. In the defining representations, both $a(\boldsymbol{\alpha})$ and $-a(\boldsymbol{\alpha})$, which is the same as $a[(\alpha + 2\pi)\boldsymbol{\alpha}/\alpha]$, are mapped into $A(\boldsymbol{\alpha})$. This two to one relationship is also clear from the ranges of the parameters α_j for the two groups and the fact that the values of these parameters are preserved under the mapping. Thus not every realization or representation of $SU(2)$ will be one for $R(3)$. Because the Lie algebras are the same for the two groups, we realize that the defining representation for $R(3)$ is also the adjoint representation for $SU(2)$. All the odd-dimensional irreducible representations that we found for $R(3)$ will be irreducible representations for $SU(2)$ as well, but they will not be faithful "in the large" because the two distinct elements of $SU(2)$, corresponding to the matrices $a(\boldsymbol{\alpha})$ and $-a(\boldsymbol{\alpha}) = a[(\alpha + 2\pi)\boldsymbol{\alpha}/\alpha]$ in the defining representation that are mapped into the same element $A(\boldsymbol{\alpha})$ of $R(3)$, will be represented by the same linear transformation.

The faithful irreducible representations of $SU(2)$ are the ones generated by the even-dimensional irreducible solutions to the commutation relations Eq. (17.116); in these the matrix corresponding to the element $a[(\alpha+2\pi)\boldsymbol{\alpha}/\alpha]$ differs in sign from the one corresponding to $a(\boldsymbol{\alpha})$. Thus these are not representations of $R(3)$ at all. The simplest such representation is the defining two-dimensional one; all the others can be obtained from this one by taking the tensor product an odd number of times. A quantity ζ with two complex components ζ_κ, κ=1,2 is called a "spinor" if under an element $a(\boldsymbol{\alpha})$ in $SU(2)$ it transforms linearly via the 2×2 matrix $a(\boldsymbol{\alpha})$:

$$\zeta_\kappa \to \zeta'_\kappa = a_{\kappa\lambda}(\boldsymbol{\alpha})\zeta_\lambda, \zeta' = a(\boldsymbol{\alpha})\zeta \,. \tag{17.154}$$

The fundamental property of spinors is that we can form bilinear expressions in terms of them that transform just like vectors in three-dimensional space do under rotations. Thus, if $\zeta^{(1)}$ and $\zeta^{(2)}$ are any two spinors and if we construct the three quantities $\zeta^{(2)+}\sigma_j\zeta^{(1)}$, then on subjecting $\zeta^{(1)}$ and $\zeta^{(2)}$ to the transformation Eq. (17.154), these products transform linearly via the matrix $A(\boldsymbol{\alpha})$, because of Eq. (17.146):

$$\begin{aligned}(\zeta^{(2)+}\sigma_j\zeta^{(1)})' &= \zeta^{(2)\prime+}\sigma_j\zeta^{(1)\prime} = \zeta^{(2)+}a^{-1}(\boldsymbol{\alpha})\sigma_j a(\boldsymbol{\alpha})\zeta^{(1)} \\ &= A_{jk}(\boldsymbol{\alpha})(\zeta^{(2)+}\sigma_k\zeta^{(1)}) \,.\end{aligned} \tag{17.155}$$

Because it takes two spinors to get something transforming like a vector, spinors are sometimes called "half-vectors." Now the appearance of the complex conjugates of the components of $\zeta^{(2)}$ in Eq. (17.155) is not essential. When the components ζ_κ of ζ are transformed according to Eq. (17.154), the complex conjugates ζ_κ^* transform in *almost the same way.* This is because, for any matrix U in $SU(2)$, one can show from Eq. (17.150), the reality of a_0, $\boldsymbol{a}$, and the forms of the σ_j that:

$$U^* = C^{-1}UC, C = i\sigma_2 = \begin{pmatrix} 0 & 1 \\ -1 & 0 \end{pmatrix} . \qquad (17.156)$$

Therefore, the object $C\zeta^*$ that has ζ_2^* and $-\zeta_1^*$ for its components transforms exactly like ζ itself. Thus with $\zeta^{(1)}$ and $\zeta^{(2)}$ being spinors, we have that the three expressions $\zeta^{(2)\sim}C\sigma_j\zeta^{(1)}$ behave as a vector, but this only means that out of the products $\zeta_\kappa^{(1)}\zeta_{\kappa'}^{(2)}$ we are able to form three linear combinations having the transformation law of a vector. Now it is easy to check that the three matrices $C\sigma_j$ are all symmetric, which means that the expressions $\zeta^{(2)\sim}C\sigma_j\zeta^{(1)}$ are symmetric in the two spinors, or that only the combinations $(\zeta_\kappa^{(1)}\zeta_{\kappa'}^{(2)} + \zeta_{\kappa'}^{(1)}\zeta_\kappa^{(2)})$ are used in constructing a vectorial quantity. This remark leads to another way by which we can arrive at the vectorial representations of $SU(2)$ and $R(3)$. Generalizing Eq. (17.154), we say that the 2^p complex quantities $\zeta_{\kappa\kappa'\ldots}$. with p indices form a spinor of rank p if they transform thus under $SU(2)$:

$$\zeta_{\kappa\kappa'\ldots} \to \zeta'_{\kappa\kappa'\ldots} = a_{\kappa\lambda}(\boldsymbol{\alpha})a_{\kappa'\lambda'}(\boldsymbol{\alpha})\cdots\zeta_{\lambda\lambda'\ldots} \qquad (17.157)$$

Once again we see that completely symmetric spinors of rank p transform all by themselves and form an invariant subspace in the space of all spinors of rank p. Now each index on a spinor takes the values 1, 2; thus a pth-rank completely symmetric spinor has exactly $p{+}1$ independent components. Such spinors therefore give us a $(p{+}1)$-dimensional representation of $SU(2)$. Here p can take all integer values from 0 to ∞. It turns out that these symmetric spinor representations of $SU(2)$ are already irreducible-there are no traces to be removed. Now if p is an even integer, the corresponding representation of $SU(2)$ is odd dimensional. But in these cases, the element in $SU(2)$ corresponding to the matrix $-\mathbb{1}$ is represented by the identity transformation, because there is an even number of a's in Eq. (17.157). More generally, these representations do not distinguish between the elements $a(\boldsymbol{\alpha})$ and $a[(\alpha + 2\pi)\boldsymbol{\alpha}/\alpha]$ in $SU(2)$, and thus they form representations of $R(3)$ as well. *The representation of* $SU(2)$ *given by symmetric spinors of even rank,* $p = 2N$, *is the same as the representation of* $R(3)$ *given by traceless symmetric tensors of rank* N *and is of dimension* $(2N + 1)$, *which is odd.* On the other hand, odd-rank spinors change sign under the element $-\mathbb{1}$ in $SU(2)$; thus their transformation law does distinguish the elements $a(\boldsymbol{\alpha})$ and $a[(\alpha + 2\pi)\boldsymbol{\alpha}/\alpha]$. *The representation of* $SU(2)$ *given by symmetric spinors of odd rank,* $p = 2N + 1$, *is a faithful representation of this group; it is of dimension* $(2N + 2)$, *which is even, and it does not form a representation of* $R(3)$. However, one frequently calls the even-dimensional irreducible representations of $SU(2)$ "double-valued representations of $R(3)$", reserving for the true representations of $R(3)$ the name "single-valued representations of $R(3)$".

This list of representations of $SU(2)$ is a complete one; there are no other irreducible ones. Just like the odd-dimensional ones, the faithful even dimensional irreducible representations of $SU(2)$ can be put into unitary form, but it is impossible to write them in such a way that all the representation matrices become real.

Rotational Properties of Fields — Many-Component Fields

In Chapter 11 we discussed the role of the angular momentum dynamical variables in implementing the effects of rotations on fields. There were two limitations: we discussed only infinitesimal rotations, and we also considered scalar, or single component, fields. We now briefly indicate how these can be extended; in considering multicomponent fields, use is made of what we have learned about the matrix representations of $R(3)$.

We suppress the time dependence of the fields in this discussion. A scalar or single-component field is a single function defined on Euclidean space. (For simplicity we assume that we are dealing with a real field.) A given "state of the field" is described by giving the value of the field at each point in space; in a Cartesian coordinate system S this amounts to specifying a function $\phi(\boldsymbol{x}), x_j$ being the coordinates assigned in S to a general point P. If the same state of the field is viewed from a rotated coordinate system $S', S' = \exp(\alpha_j l_j)S$, one will find a new function of the coordinates because although the value of the field at any point P has not changed (the "state" being the same) the coordinates assigned to P have. If P has coordinates x'_j in S', the new function ϕ' is determined by $\phi'(x'_j) = \phi(x_j) =$ value of field at P.

Concentrating on the change in functional form, we can write:

$$S' = \exp(\boldsymbol{\alpha} \cdot \boldsymbol{l})S : \phi'(x_j) = \phi(A_{kj}(\boldsymbol{\alpha})x_k) \,. \tag{17.158}$$

This law is valid for all states of the field and is the transformation law for the dynamical variable ϕ. Let $\pi(\boldsymbol{x})$ be the "momentum field" canonically conjugate to $\phi(\boldsymbol{x})$ so that the basic PB's are:

$$\{\phi(\boldsymbol{x}), \phi(\boldsymbol{y})\} = \{\pi(\boldsymbol{x}), \pi(\boldsymbol{y})\} = 0, \{\phi(\boldsymbol{x}), \pi(\boldsymbol{y})\} = \delta^{(3)}(\boldsymbol{x} - \boldsymbol{y}) \,. \tag{17.159}$$

As we have seen in Chapter 11, we can define the angular momentum of the field as:

$$J_j = -\int d^3x \, \pi(\boldsymbol{x})\epsilon_{jkm}x_k \frac{\partial}{\partial x_m}\phi(\boldsymbol{x}) \,. \tag{17.160}$$

For an infinitesimal rotation the change in $\phi(\boldsymbol{x})$ given by Eq. (17.158), viewed as a canonical transformation, and the corresponding change in $\pi(\boldsymbol{x})$, is generated by the J_j. We want to show that this is so for a finite rotation as well; namely, we wish to prove:

$$\begin{aligned} S' = \exp(\boldsymbol{\alpha} \cdot \boldsymbol{l})S : \phi'(x_j) &\equiv \phi(A_{kj}(\boldsymbol{\alpha})x_k) \\ &= \exp(\widetilde{-\boldsymbol{\alpha} \cdot \boldsymbol{J}})\phi(x_j) \,. \end{aligned} \tag{17.161}$$

Before we prove Eq. (17.161), we may mention that using Eq. (17.159), the reader can easily verify that the J_j obey the PB relations appropriate to $R(3)$.

Turning to Eq. (17.161), we can first notice that by combining Eqs. (17.159) and (17.160) we have (as in Chapter 11),

$$\begin{aligned}\widetilde{\boldsymbol{\alpha}\cdot\boldsymbol{J}}\phi(\boldsymbol{x}) &= \alpha_j\{J_j,\phi(\boldsymbol{x})\} = \alpha_j\epsilon_{jkm}x_k\frac{\partial}{\partial x_m}\phi(\boldsymbol{x})\\ &= \boldsymbol{\alpha}\cdot\boldsymbol{x}\times\nabla\phi(\boldsymbol{x})\,.\end{aligned}\qquad(17.162)$$

On the other hand, we can re-express the transformed argument $A_{kj}(\boldsymbol{\alpha})x_k$ in Eq. (17.158) in this way: viewing x_k as the three components of a column vector $\boldsymbol{x}$, and because the matrix $A(\boldsymbol{\alpha})$ is the exponential of the matrix $C(\boldsymbol{\alpha})$,

$$\begin{aligned}\{C(\boldsymbol{\alpha})\boldsymbol{x}\}_j =& C_{jk}(\boldsymbol{\alpha})x_k = -\epsilon_{mjk}\alpha_m x_k = \epsilon_{mkn}\alpha_m x_k\frac{\partial}{\partial x_n}x_j\\ =&\{\boldsymbol{\alpha}\cdot\boldsymbol{x}\times\nabla\boldsymbol{x}\}_j\,;\\ \{A^{\sim}(\boldsymbol{\alpha})\boldsymbol{x}\}_j =&\{A(-\boldsymbol{\alpha})\boldsymbol{x}\}_j = \left\{\sum_{n=0}^{\infty}\frac{1}{n!}[C(-\boldsymbol{\alpha})]^n\boldsymbol{x}\right\}_j\\ =&\left\{\sum_{n=0}^{\infty}\frac{1}{n!}[-\boldsymbol{\alpha}\cdot\boldsymbol{x}\times\nabla]^n\boldsymbol{x}\right\}_j = \exp(-\boldsymbol{\alpha}\cdot\boldsymbol{x}\times\nabla)x_j\,.\end{aligned}\qquad(17.163)$$

Therefore for the right-hand side of Eq. (17.158) we can write:

$$\phi(A(-\boldsymbol{\alpha})\boldsymbol{x}) = \phi(\exp(-\boldsymbol{\alpha}\cdot\boldsymbol{x}\times\nabla)\boldsymbol{x})\,.\qquad(17.164)$$

Now because the operator $\boldsymbol{\alpha}\cdot\boldsymbol{x}\times\nabla$ is a differential operator of the *first order*, the reader can easily construct an argument (similar to many that were used in Chapters 13 and 14) to show that the exponential in Eq. (17.164) can be made to act on ϕ itself:

$$\phi(A(-\boldsymbol{\alpha})\boldsymbol{x}) = \phi(\exp(-\boldsymbol{\alpha}\cdot\boldsymbol{x}\times\nabla)\boldsymbol{x}) = \exp(-\boldsymbol{\alpha}\cdot\boldsymbol{x}\times\nabla)\phi(\boldsymbol{x})\,.\qquad(17.165)$$

But now, keeping in mind the fact that the processes of taking the PB and partial differentiation with respect to the x_j are interchangeable, we can substitute $\boldsymbol{\alpha}\cdot\tilde{\boldsymbol{J}}$ for $\boldsymbol{\alpha}\cdot\boldsymbol{x}\times\nabla$ in the exponent in Eq. (17.165), as long as it acts on $\phi(\boldsymbol{x})$. For example, for the square of the operator $\boldsymbol{\alpha}\cdot\boldsymbol{x}\times\nabla$, we have:

$$\begin{aligned}(\boldsymbol{\alpha}\cdot\boldsymbol{x}\times\nabla)^2\phi(\boldsymbol{x}) &= \boldsymbol{\alpha}\cdot\boldsymbol{x}\times\nabla\widetilde{\boldsymbol{\alpha}\cdot\boldsymbol{J}}\phi(\boldsymbol{x}) = \widetilde{\boldsymbol{\alpha}\cdot\boldsymbol{J}}\,\boldsymbol{\alpha}\cdot\boldsymbol{x}\times\nabla\phi(\boldsymbol{x})\\ &= (\widetilde{\boldsymbol{\alpha}\cdot\boldsymbol{J}^2})\phi(\boldsymbol{x})\,.\end{aligned}\qquad(17.166)$$

In this way, we see that Eq. (17.161) is indeed true. The analogous relation for the conjugate field π is proved along the same lines.

We consider next the description of fields with several components. These are needed, for example, to describe the electromagnetic field. Here we have the electric and magnetic fields, each behaving as a vector under rotations. A particular state of the electric field, for example, is described by giving the electric vector at each point in space, and in a Cartesian coordinate system S, the components of the vector at a point P with coordinates x_j will be $E_j(\boldsymbol{x})$. The coordinate system S assigns coordinates $\boldsymbol{x}$ to each point P, and at the same time its axes provide directions along which the electric vector at each point may be resolved. If we switch to a rotated frame $S' = \exp(\boldsymbol{\alpha} \cdot \boldsymbol{l})S$, not only do the coordinates of P change to x'_j; in addition the directions along which the electric vector at P are to be resolved into components will have changed. Thus the same "state of the field" is described in S' by functions $E'_j(\boldsymbol{x}')$ that are obtained thus:

$$S' = \exp(\boldsymbol{\alpha} \cdot \boldsymbol{l})S : E'_j(\boldsymbol{x}') = A_{jk}(\boldsymbol{\alpha})E_k(\boldsymbol{x}) . \tag{17.167}$$

If we write the components of E as a column vector and adopt a similar notation for the argument $\boldsymbol{x}$ (as in Eq. (17.163)), we can exhibit the change in the functions $E(\boldsymbol{x})$ as:

$$S' = \exp(\boldsymbol{\alpha} \cdot \boldsymbol{l})S : E'(\boldsymbol{x}) = A(\boldsymbol{\alpha})E(A(-\boldsymbol{\alpha})\boldsymbol{x}) . \tag{17.168}$$

In comparison with the transformation law for a scalar field as given in Eq. (17.158), we see that the argument $\boldsymbol{x}$ is affected in the same way in both cases, but now, in addition, the components of the vector field undergo a linear transformation. We can generalize Eq. (17.168) to a multicomponent field ϕ that transforms according to any linear matrix representation $D(\boldsymbol{\alpha})$ of $R(3)$, not necessarily just a vector. Here again, the components $\phi_r(\boldsymbol{x})$ of ϕ at a point P with coordinates $\boldsymbol{x}$ in S are determined only with respect to the directions of the axes in S. The analogue to Eq. (17.168) is:

$$\begin{aligned} S' = \exp(\boldsymbol{\alpha} \cdot \boldsymbol{l})S : \phi'(\boldsymbol{x}) &= D(\boldsymbol{\alpha})\phi(A(-\boldsymbol{\alpha})\boldsymbol{x}) , \\ \phi'_r(\boldsymbol{x}) &= D_{rs}(\boldsymbol{\alpha})\phi_s(A(-\boldsymbol{\alpha})\boldsymbol{x}) . \end{aligned} \tag{17.169}$$

We have written ϕ first as a column vector, then, for clarity, have also exhibited explicitly the behavior of the components. We must now determine the forms of the angular momenta such that they generate these transformations. Let $(S_j)_{rs}$ be the matrices that generate the representation matrices $D(\boldsymbol{\alpha})$. Then $D(\boldsymbol{\alpha})$ in Eq. (17.169) can be expressed as an exponential in the S_j, whereas the argument $A(-\boldsymbol{\alpha})\boldsymbol{x}$ can be treated in the same way as in the case of the scalar field, in Eqs. (17.163), (17.164), (17.165). Thus we can put Eq. (17.169) into this form:

$$S' = \exp(\boldsymbol{\alpha} \cdot \boldsymbol{l})S : \phi'(\boldsymbol{x}) = \exp(\alpha_j S_j)\exp(-\boldsymbol{\alpha} \cdot \boldsymbol{x} \times \boldsymbol{\nabla})\phi(\boldsymbol{x}) . \tag{17.170}$$

Now only the S_j affect the indices $r, s \ldots$ labeling the components of ϕ, whereas the expression at $\boldsymbol{\alpha} \cdot \boldsymbol{x} \times \boldsymbol{\nabla}$ has no effect at all on them; we can regard the latter as being a multiple of the unit matrix, δ_{rs}, as far as the space of the components is concerned. This allows us to add the two exponents in Eq. (17.170) to form a single one, and we get the simple expression:

$$\phi'(\boldsymbol{x}) = \exp\left\{-\alpha_j \left(\epsilon_{jkm} x_k \frac{\partial}{\partial x_m} - S_j\right)\right\} \phi(\boldsymbol{x}) . \tag{17.171}$$

Following now the same kinds of arguments as in the scalar case, we see that we can generate this transformation in Hamiltonian mechanics as a canonical transformation produced by the angular momenta J_j if we can choose the latter to have these PB relations with $\phi(\boldsymbol{x})$:

$$\begin{aligned}\widetilde{\boldsymbol{\alpha}\cdot\boldsymbol{J}}\phi(\boldsymbol{x}) \equiv \alpha_j\{J_j,\phi(\boldsymbol{x})\} =&\alpha_j\left(\epsilon_{jkm}x_k\frac{\partial}{\partial x_m} - S_j\right)\phi(\boldsymbol{x})\\ \{J_j,\phi_r(\boldsymbol{x})\} =&\epsilon_{jkm}x_k\frac{\partial}{\partial x_m}\phi_r(\boldsymbol{x}) - (S_j)_{rs}\phi_s(\boldsymbol{x})\,. \end{aligned} \tag{17.172}$$

In that case, we can certainly write:

$$S' = \exp(\boldsymbol{\alpha}\cdot\boldsymbol{l})S : \phi'(\boldsymbol{x}) = \exp(\widetilde{-\boldsymbol{\alpha}\cdot\boldsymbol{J}})\phi(\boldsymbol{x})\,. \tag{17.173}$$

We assume that to each $\phi_r(\boldsymbol{x})$ there is a canonically conjugate momentum field $\pi_r(\boldsymbol{x})$ so that the basic PB's are:

$$\{\phi_r(\boldsymbol{x}),\phi_s(\boldsymbol{y})\} = \{\pi_r(\boldsymbol{x}),\pi_s(\boldsymbol{y})\} = 0, \{\phi_r(\boldsymbol{x}),\pi_s(\boldsymbol{y})\} = \delta_{rs}\delta^{(3)}(\boldsymbol{x}-\boldsymbol{y})\,. \tag{17.174}$$

Then it is fairly easy to construct a set of quantities J_j fulfilling Eq. (17.172):

$$J_j = \int d^3x\left\{\pi_r(\boldsymbol{x})(S_j)_{rs}\phi_s(\boldsymbol{x}) - \pi_r(\boldsymbol{x})\epsilon_{jkm}x_k\frac{\partial}{\partial x_m}\phi_r(\boldsymbol{x})\right\}\,. \tag{17.175}$$

To the natural generalization of Eq. (17.160) we have added an extra term to produce the terms in Eq. (17.172) involving the "spin-matrices" S_j, and now it is trivial to check Eq. (17.172). Thus Eq. (17.173) holds as well The corresponding canonical transformation on the $\pi_r(\boldsymbol{x})$ will be such as to preserve the basic PB's, Eq. (17.174); this already implies that the indices on π will not transform in the same way as they do on ϕ, but in such a way as to be consistent with the Kronecker delta δ_{rs} in the PB between ϕ_r and π_s. With J_j we have the bracket relation:

$$\widetilde{\boldsymbol{\alpha}\cdot\boldsymbol{J}}\pi_r(\boldsymbol{x}) = \boldsymbol{\alpha}\cdot\boldsymbol{x}\times\boldsymbol{\nabla}\pi_r(\boldsymbol{x}) + \alpha_j(S_j)_{sr}\pi_s(\boldsymbol{x})\,,$$

or in matrix notation:

$$\widetilde{\boldsymbol{\alpha}\cdot\boldsymbol{J}}\pi(\boldsymbol{x}) = \{\boldsymbol{\alpha}\cdot\boldsymbol{J},\pi(\boldsymbol{x})\} = (\boldsymbol{\alpha}\cdot\boldsymbol{x}\times\boldsymbol{\nabla} + \alpha_j S_j^{\sim})\pi(\boldsymbol{x})\,. \tag{17.176}$$

Instead of the matrix S_j that appears in Eq. (17.172), we have the matrix $-S_j^{\sim}$; if the S_j obey the basic commutation relations corresponding to the Lie algebra of $R(3)$, then so do the $-S_j^{\sim}$. (The tilde on the matrix S_j denotes the transposed matrix.) The matrix representation of $R(3)$ produced by the latter is said to be *contragredient* to the one produced by the former. Using Eq. (17.176), we can write down the form of the finite canonical transformation on the π's:

$$\begin{aligned}\exp(\widetilde{-\boldsymbol{\alpha}\cdot\boldsymbol{J}})\pi(\boldsymbol{x}) &= \exp(-\alpha_j S_j^{\sim})\exp(-\boldsymbol{\alpha}\cdot\boldsymbol{x}\times\boldsymbol{\nabla})\pi(\boldsymbol{x})\\ &= D^{\sim}(\boldsymbol{\alpha})^{-1}\pi(A(-\boldsymbol{\alpha})\boldsymbol{x})\,. \end{aligned} \tag{17.177}$$

Naturally, in place of the matrix $D(\boldsymbol{\alpha})$ that appeared in the transformation law for $\phi(\boldsymbol{x})$ in Eq. (17.169), there now appears the matrix $D^{\sim}(\boldsymbol{\alpha})^{-1}$: this is the matrix that represents the element $\exp(\boldsymbol{\alpha}\cdot\boldsymbol{l})$ in the representation contragredient to the one in which this element was represented by $D(\boldsymbol{\alpha})$.($D^{\sim}$ is the transpose of D.) The reader can check that with Eq. (17.177), the basic PB's are preserved.

The two terms that appear in the expression (Eq. (17.175)) for the angular momenta J_j are called the "intrinsic" or "spin" part of the angular momentum carried by the field and the "orbital" part, respectively. The former is responsible for the shuffling of the components under a rotation, the latter for the change in the spatial argument $\boldsymbol{x}$. For a scalar field, we naturally have only the "orbital" part. The reader can verify that the spin parts and the orbital parts separately obey the basic PB relations for $R(3)$ and the PB's between them vanish, and thus that the total J_j also obey the basic $R(3)$ PB relations. That is, we have the breakup:

$$\begin{aligned} J_j &= L_j + \mathcal{S}_j \\ L_j &= -\int d^3x\, \pi_r(\boldsymbol{x})\epsilon_{jkm}x_k \frac{\partial}{\partial x_m}\phi_r(\boldsymbol{x})\,, \\ \mathcal{S}_j &= \int d^3x\, \pi_r(\boldsymbol{x})(S_j)_{rs}\phi_s(\boldsymbol{x})\,, \end{aligned} \tag{17.178}$$

and these obey:

$$\{L_j, L_k\} = \epsilon_{jkm}L_m, \quad \{\mathcal{S}_j, \mathcal{S}_k\} = \epsilon_{jkm}\mathcal{S}_m, \quad \{L_j, \mathcal{S}_k\} = 0\,, \tag{17.179}$$

leading to

$$\{J_j, J_k\} = \epsilon_{jkm}J_m\,. \tag{17.180}$$

We could, if we wish, call the integrands appearing in the expressions for L_j and $\mathcal{S}_j$ in Eq. (17.178) the "densities" of orbital and spin angular momentum, respectively, but as already noted in Chapter 11, such identifications are ambiguous to the extent that one could add divergences to these densities, which would then not change L_j and $\mathcal{S}_j$ at all.

Chapter 18

The Three-Dimensional Euclidean Group

Our study of the classical canonical realizations of the group $E(3)$ is a short one. We begin with the basic Lie brackets for the Lie algebra of $E(3)$. There are six basic elements, l_j and d_j, j=1,2,3, the former generating rotations and the latter translations, and they obey:

$$[l_j, l_k] = \epsilon_{jkm} l_m, \ [l_j, d_k] = \epsilon_{jkm} d_m, \ [d_j, d_k] = 0 . \tag{18.1}$$

A general element of $E(3)$ takes the form $\exp(\boldsymbol{a}\cdot\boldsymbol{d})\exp(\boldsymbol{\alpha}\cdot\boldsymbol{l})$ and the composition law is given in Eq. (15.49). In a PB realization up to neutrals of this Lie algebra, we have the correspondence $l_j \to J_j(\omega), d_j \to P_j(\omega)$, and we can assume that the $J_j(\omega)$ give a true PB realization of the $R(3)$ Lie algebra. Thus neutral elements possibly occur in these relations:

$$\{J_j(\omega),\ P_k(\omega)\} = \epsilon_{jkm} P_m(\omega) + d_{jk},\ \{P_j(\omega), P_k(\omega)\} = d'_{jk} . \tag{18.2}$$

However, we now show that d_{jk} must be antisymmetric in j and k and can be transformed away by a redefinition of the P's, then that d'_{jk}, which certainly is antisymmetric in j and k, must vanish. To prove the first statement, we invoke the Jacobi identity for PB's involving two J's and one P. We can write:

$$\tilde{J}_j P_k = \epsilon_{jkm} P_m + d_{jk} , \tag{18.3}$$

and at the same time we note that the Jacobi identity with two J's and any other function is equivalent to the following statement:

$$\tilde{J}_j \tilde{J}_k - \tilde{J}_k \tilde{J}_j = \epsilon_{jkm} \tilde{J}_m,\ \tilde{J}_m = \epsilon_{jkm} \tilde{J}_j \tilde{J}_k . \tag{18.4}$$

Let us now apply both sides of the second equality in Eq. (18.4) to P_l. We use Eq. (18.3) to get:

$$\begin{aligned}\epsilon_{mln} P_n + d_{ml} = \epsilon_{mjk} \tilde{J}_j \epsilon_{kln} P_n =& (\delta_{ml}\delta_{jn} - \delta_{mn}\delta_{jl})(\epsilon_{jnp} P_p + d_{jn}) \\ =& \epsilon_{mlp} P_p + \delta_{ml} d_{jj} - d_{lm} .\end{aligned} \tag{18.5}$$

Thus we get the equality:

$$d_{ml} + d_{lm} = \delta_{ml} d_{jj}\,; \tag{18.6}$$

summing on both sides on l after setting $m = l, d_{jj}$ is seen to vanish. Consequently, d_{jk} is antisymmetric in j and k, and we could rewrite it as $\epsilon_{jkm} d_m$. But with the redefined $P'_m(\omega) = P_m(\omega) + d_m$, it is obvious that there are no neutral elements in the PB relations between J_j and P'_k; we might as well have assumed to start with that $d_{jk} = 0$ in Eq. (18.2). Thus the structure of the first two sets of Lie brackets in Eq. (18.1) is such that in any classical realization of them along with other brackets, we can assume the absence of neutral elements in the corresponding two sets of PB relations. Turning now to the neutral elements d'_{jk} in Eq. (18.2), we first give an infinitesimal-type argument to show that they must vanish, and then another argument based on finite rotations. We have, in any case,

$$\{J_k, \{P_j, P_l\}\} = 0\,, \tag{18.7}$$

and applying the Jacobi identity we get:

$$\{P_l, \{J_k, P_j\}\} + \{P_j, \{P_l, J_k\}\} = \epsilon_{kjm} d'_{lm} + \epsilon_{lkm} d'_{jm} = 0\,. \tag{18.8}$$

Now, because d'_{jk} is antisymmetrie in j and k, we can write $d'_{jk} = \epsilon_{jkm} d'_m$, and with Eq. (18.8) this gives:

$$\begin{aligned} \epsilon_{kjm}\epsilon_{lmn} d'_n + \epsilon_{lkm}\epsilon_{jmn} d'_n &= (\delta_{kn}\delta_{jl} - \delta_{kl}\delta_{jn} + \delta_{ln}\delta_{kj} - \delta_{lj}\delta_{kn}) d'_n \\ &= \delta_{jk} d'_l - \delta_{kl} d'_j = 0\,. \end{aligned} \tag{18.9}$$

Setting $k = l$ in this last step, we find that $d'_j - 3d'_j = 0$ implies the vanishing of d'_j and of d'_{jk}. Thus we have proved that any PB realization up to neutrals of the Lie algebra Eq. (18.1) is equivalent to a true realization, and we have, with no loss of generality,

$$\{J_j, J_k\} = \epsilon_{jkm} J_m,\ \{J_j, P_k\} = \epsilon_{jkm} P_m,\ \{P_j, P_k\} = 0\,. \tag{18.10}$$

As with $R(3)$, we draw two conclusions: (1) when examining a PB realization up to neutrals of the Lie algebras of the Galilei or Poincaré groups, we can assume that there are no neutrals in the PB's corresponding to the $E(3)$ subalgebra spanned by J_j and P_j; (2) in trying to eliminate neutral elements that may be present in other Lie brackets, no attempt must be made to redefine either the J's or the P's.

Now let us give the global argument to show that d'_{jk} must vanish. Under a finite rotation generated by the J's, the P's transform as a three-vector. The definition of d'_{jk} and their neutrality then imply that they form the components of an *invariant* second-rank tensor:

$$\begin{aligned} d'_{jk} = \widetilde{e^{\boldsymbol{\alpha}\cdot\boldsymbol{J}}} d'_{jk} = \widetilde{e^{\boldsymbol{\alpha}\cdot\boldsymbol{J}}}\{P_j, P_k\} =& \{\widetilde{e^{\boldsymbol{\alpha}\cdot\boldsymbol{J}}} P_j, \widetilde{e^{\boldsymbol{\alpha}\cdot\boldsymbol{J}}} P_k\} = A_{mj}(\boldsymbol{\alpha}) A_{nk}(\boldsymbol{\alpha}) \{P_m, P_n\}\,, \\ A_{mj}(\boldsymbol{\alpha}) A_{nk}(\boldsymbol{\alpha}) d'_{mn} =& d'_{jk}\,. \end{aligned} \tag{18.11}$$

However, the only second-rank tensor whose components preserve their numerical values under all rotations is the unit tensor δ_{jk}, and that is *symmetric*; there is no nontrivial *antisymmetric* invariant second-rank tensor. Hence d'_{jk} must vanish.

There are two Casimir invariants for the $E(3)$ Lie algebra: one is the "square of the magnitude of the momentum," $P^2 = P_k P_k$, and the other is "the component of angular momentum along the momentum," $\boldsymbol{J} \cdot \boldsymbol{P} = J_k P_k$. As scalars under rotations, both of them have vanishing PB's with the J's. As for P^2, any two functions of the P_k alone have vanishing PB's with one another, and thus it certainly is a Casimir invariant. The proof for $\boldsymbol{J} \cdot \boldsymbol{P}$ follows:

$$\{P_j, J_k P_k\} = \{P_j, J_k\} P_k = \epsilon_{jkm} P_m P_k = 0\,. \tag{18.12}$$

Therefore, in any canonical realization of $E(3)$, the surfaces in phase space with constant values of P^2 and $\boldsymbol{J} \cdot \boldsymbol{P}$ are mapped onto themselves, the values of these invariants being preserved. Note that the Casimir invariant of $R(3)$, J^2, is not a Casimir invariant for $E(3)$; values of J^2 are not preserved under a general canonical transformation corresponding to an element of $E(3)$. It is the transformations generated by the P's that alter J^2. On the generators themselves, the action is as follows:

$$e^{\widetilde{\boldsymbol{a}\cdot\boldsymbol{P}}} P_j = P_j\,, \quad e^{\widetilde{\boldsymbol{a}\cdot\boldsymbol{P}}} J_j = J_j + \epsilon_{kjm} a_k P_m\,. \tag{18.13}$$

Thus the most general element $\exp(\boldsymbol{a} \cdot \boldsymbol{d}) \exp(\boldsymbol{\alpha} \cdot \boldsymbol{l})$ is represented thus:

$$\begin{aligned} e^{\widetilde{\boldsymbol{a}\cdot\boldsymbol{P}}} e^{\widetilde{\boldsymbol{\alpha}\cdot\boldsymbol{J}}} P_j &= A_{mj}(\boldsymbol{\alpha}) P_m\,, \\ e^{\widetilde{\boldsymbol{a}\cdot\boldsymbol{P}}} e^{\widetilde{\boldsymbol{\alpha}\cdot\boldsymbol{J}}} J_j &= A_{mj}(\boldsymbol{\alpha})(J_m + (\boldsymbol{P} \times \boldsymbol{a})_m)\,. \end{aligned} \tag{18.14}$$

This change in $\boldsymbol{J}$ is just what one expects when the point about which the angular momentum is being measured is shifted.

For a nontrivial realization of the $R(3)$ Lie algebra we noticed that no numerical linear combination of the generators J_j can vanish identically. The situation is different in the case of $E(3)$, owing to differences in the structure of the basic Lie brackets. We know that $E(3)$ is the semidirect product of $R(3)$ by T_3, the three-dimensional translation group. A look at the basic PB relations Eq. (18.10) immediately shows that one could have realizations in which each P_j vanishes identically, without this implying that the J_j, too, vanish identically. Such realizations of $E(3)$ are not quite trivial; the term trivial is reserved for the realization in which *all* the generators J_j, P_j vanish identically. We call realizations of $E(3)$ in which P_j=0, but $J_j \neq 0$, "nonfaithful" realizations. These are just realizations of $R(3)$, the Abelian invariant subgroup T_3 being realized trivially. In these nonfaithful realizations of $E(3)$, the $R(3)$ Casimir invariant $J^2 = J_j J_j$ becomes an $E(3)$ invariant. (We do not call it a Casimir invariant for $E(3)$ because, though it is a polynomial in the $E(3)$ generators, we know that it is not an invariant in all realizations of $E(3)$.) In sum, every $R(3)$ realization gives us a nonfaithful $E(3)$ realization, but we do not cite these in this chapter as examples of $E(3)$ realizations any further (except in connection with "direct products," which are discussed later in this chapter).

A true PB realization of the $E(3)$ Lie algebra is provided in terms of a triplet of canonically conjugate pairs q_j, p_j if we choose:

$$J_j(\omega) = \epsilon_{jkl} q_k p_l, \; P_j(\omega) = p_j \,. \tag{18.15}$$

The phase space is six dimensional, with each q and each p going from $-\infty$ to $+\infty$. In this realization the Casimir invariant $\boldsymbol{J} \cdot \boldsymbol{P}$ vanishes, whereas $P^2 = p^2$ takes on all positive values. Under a general $E(3)$ transformation realizing the element $\exp(\boldsymbol{a} \cdot \boldsymbol{d}) \exp(\boldsymbol{\alpha} \cdot \boldsymbol{l})$, the point q_j, p_j in phase space is mapped onto the point q_j', p_j' as follows:

$$q_j' = e^{\widetilde{\boldsymbol{a}\cdot\boldsymbol{P}}} e^{\widetilde{\boldsymbol{\alpha}\cdot\boldsymbol{J}}} q_j = A_{kj}(\boldsymbol{\alpha})(q_k - a_k), \; p_j' = e^{\widetilde{\boldsymbol{a}\cdot\boldsymbol{P}}} e^{\widetilde{\boldsymbol{\alpha}\cdot\boldsymbol{J}}} P_j = A_{kj}(\boldsymbol{\alpha}) p_k \,. \tag{18.16}$$

We have used here the basic PB's between the q's and p's, and also the vectorial transformation properties of q_j and p_j. Surfaces of spheres in the p space are preserved under Eq. (18.16), corresponding to P^2 being a Casimir invariant, but apart from this, there are no other invariants for this realization of $E(3)$. This is a consequence of the easily verified fact that if q_j, p_j and q_j', p_j' are any two points in phase space, and $p_j', p_j' = p_j p_j$, then one can always find parameters $\boldsymbol{a}$ and $\boldsymbol{\alpha}$ on such that q_j, p_j is carried into q_j', p_j' under the canonical transformation Eq. (18.16).

The above realization of $E(3)$ uses just the phase space variables needed to describe a point particle in classical mechanics and can be called a single particle realization of $E(3)$. We can generalize this realization to get new ones in which the Casimir invariant $\boldsymbol{J} \cdot \boldsymbol{P}$ is not identically zero if we add on a new "spin" degree of freedom. That is, we consider an enlarged phase space with q_j, p_j and S_j as coordinates, the S_j obeying basic PB's corresponding to the Lie algebra of $R(3)$ and having vanishing PB's with q_j and p_j:

$$\{S_j, S_k\} = \epsilon_{jkm} S_m, \; \{S_j, q_k\} = \{S_j, p_k\} = 0 \,. \tag{18.17}$$

We have here an interesting combination with some variables subject to the "ordinary" PB's, and others to generalized ones. The pure spin invariant S^2 is an "absolute" invariant, and no canonical or generalized canonical transformation changes it. If we wish, we can assign it a fixed value. Then we get a true realization of the $E(3)$ Lie algebra if we choose:

$$J_j(\omega) = \epsilon_{jkm} q_k p_m + S_j, \; P_j = (\omega) = p_j \,. \tag{18.18}$$

Both Casimir invariants P^2 and $\boldsymbol{J} \cdot \boldsymbol{P}$ are nonzero; they have the values p^2 and $\boldsymbol{S} \cdot \boldsymbol{p}$ respectively, and are only restricted by $(\boldsymbol{S} \cdot \boldsymbol{p})^2 \leq S^2 p^2$. In the new phase space, Eq. (18.16) is still valid; in addition we have:

$$S_j' = e^{\widetilde{\boldsymbol{a}\cdot\boldsymbol{P}}} e^{\widetilde{\boldsymbol{\alpha}\cdot\boldsymbol{J}}} S_j = A_{kj}(\boldsymbol{\alpha}) S_k \,. \tag{18.19}$$

Surfaces of constant p^2 and $\boldsymbol{p}\cdot\boldsymbol{S}$ in the new phase space are preserved, and once again one checks that there are no invariants besides these Casimir invariants.

In the above realizations, one could have made the identification $P_j(\omega) = q_j$ with $J_j(\omega)$ unchanged, and the basic $E(3)$ brackets would still be valid. But this realization of $E(3)$ would be an "unphysical" one if the transformations of $E(3)$ do represent changes of frame in three-dimensional space, whereas q_j and p_j do represent position and momentum.

The method by which we generalized the "single-particle" realization of $E(3)$ given in Eq. (18.15) to get the realization of Eq. (18.18) in which the Casimir invariant $\boldsymbol{J}\cdot\boldsymbol{P}$ is not identically zero, seems to be a "painless" one. We have two realizations of $E(3)$ to start with, one built up with q_j and p_j, the other a non-faithful one with $J_j \to S_j, P_j \to 0$. The Casimir invariant $\boldsymbol{J}\cdot\boldsymbol{P}$ vanishes in each, although for different reasons. These are independent $E(3)$ realizations in the sense that they operate in distinct phase spaces, and in combining them the generators of the one have zero brackets with those of the other by definition (cf. Eq. (18.17)). The generators given in Eq. (18.18) are simply the sums of corresponding generators in these two realizations. This is an example of the general process of taking the *direct product* of two independent canonical realizations of a group; because we meet it here for the first time and because we use it later again, we now describe this process in some detail.

Let π and π' be two (generalized) phase spaces of dimensions N, N' with coordinates ω^μ, ζ^ρ, respectively. The direct product $\pi\times\pi'$ of π and π' is a space of dimension $N+N'$, points of it being labeled by the $N+N'$ coordinates (ω^μ, ζ^ρ). The ranges of the $\omega^\mu(\zeta^\rho)$ are as in $\pi(\pi')$. Functions on $\pi\times\pi'$ are functions $\Phi(\omega,\zeta)$ of $N+N'$ variables. We make $\pi\times\pi'$ a generalized phase space by defining the GPB in it thus:

$$\{\Phi(\omega,\zeta),\Psi(\omega,\zeta)\} = \frac{\partial\Phi}{\partial\omega^\mu}\frac{\partial\Psi}{\partial\omega^\nu}\{\omega^\mu,\omega^\nu\} + \frac{\partial\Phi}{\partial\zeta^\rho}\frac{\partial\Psi}{\partial\zeta^\sigma}\{\zeta^\rho,\zeta^\sigma\}\,. \tag{18.20}$$

$\{\omega^\mu,\omega^\nu\}$ are the fundamental brackets in π, expressible as functions of ω's alone, analogously for $\{\zeta^\rho,\zeta^\sigma\}$. Taken together, they form the fundamental brackets in $\pi\times\pi'$, with the brackets $\{\omega^\mu,\zeta^\rho\}$ being set equal to zero *by definition.* This GPB in $\pi\times\pi'$ can be rewritten as:

$$\{\Phi,\Psi\}_{\pi\times\pi'} = \{\Phi,\Psi\}_\pi + \{\Phi,\Psi\}_{\pi'}\,; \tag{18.21}$$

the first term on the right is the "bracket in π" of Φ and Ψ considered as functions on π with the ζ^ρ treated as parameters not subject to partial differentiations; the second term is the analogous "bracket in π'" of Φ with Ψ. By *adding* these two terms we get the bracket in $\pi\times\pi'$. This new bracket obeys the Jacobi identity. If we take the cyclic sum on Φ, Ψ, and X of $\{\{\Phi,\Psi\}_{\pi\times\pi'},X\}_{\pi\times\pi'}$, we get four groups of terms. The first, of the form $\{\{\Phi,\Psi\}_\pi,X\}_\pi$, involves brackets in π alone, and such terms add up to zero by themselves. In these terms, the ζ^ρ are treated like parameters, and their presence is irrelevant. The fourth set of terms consists of $\{\{\Phi,\Psi\}_{\pi'},X\}_{\pi'}$, and two others, which again add up to zero by themselves. Typical examples of the second and third groups of terms are

$\{\{\Phi,\Psi\}_{\pi},X\}_{\pi'}$ and $\{\{\Phi,\Psi\}_{\pi'},X\}_{\pi}$. Using the independence of the ω's and the ζ's on $\pi\times\pi'$, so that the two processes of taking the bracket in π and in π' can be applied in any order, one checks that the cyclic sums of these terms exactly cancel one another, establishing the Jacobi identity in $\pi\times\pi'$.

Let us now be given two regular canonical transformations, one in π and one in π', each with a corresponding generating function. By their combined action we get a particular type of regular canonical transformation on $\pi\times\pi'$, one in which the ω's transform by themselves and the ζ's by themselves. In detail, let the transformation in π be written $\omega\to\omega'=T\omega$, with the associated linear mapping of functions on π being given by a generating function $a(\omega)$ thus:

$$\phi(\omega)\to e^{\widetilde{a(\omega)}}\phi(\omega)=\phi(T^{-1}\omega)\,. \tag{18.22}$$

On π' we have a similar equation with ω, T and $a(\omega)$ replaced by ζ, U and $b(\zeta)$, respectively. Then we define in $\pi\times\pi'$ the transformation $(\omega,\zeta)\to(T\omega,U\zeta)$; on functions on $\pi\times\pi'$ we have the action:

$$\Phi(\omega,\zeta)\to e^{\widetilde{a(\omega)}}e^{\widetilde{b(\zeta)}}\Phi(\omega,\zeta)=\Phi(T^{-1}\omega,U^{-1}\zeta)\,. \tag{18.23}$$

These two operators $e^{\widetilde{a(\omega)}}$ and $e^{\widetilde{b(\zeta)}}$ can be applied in either order on a function of ω and ζ because one affects ω alone and the other, ζ alone. It is now easy to see that this mapping preserves brackets in $\pi\times\pi'$. We have to convince ourselves that

$$e^{\widetilde{a(\omega)}}e^{\widetilde{b(\zeta)}}\{\Phi(\omega,\zeta),\Psi(\omega,\zeta)\}=\{e^{\widetilde{a(\omega)}}e^{\widetilde{b(\zeta)}}\Phi(\omega,\zeta),e^{\widetilde{a(\omega)}}e^{\widetilde{b(\zeta)}}\Psi(\omega,\zeta)\}\,. \tag{18.24}$$

But referring to Eq. (18.21), in the first term we certainly have:

$$e^{\widetilde{a(\omega)}}\{\Phi(\omega,\zeta),\Psi(\omega,\zeta)\}_{\pi}=\{e^{\widetilde{a(\omega)}}\Phi(\omega,\zeta),e^{\widetilde{a(\omega)}}\Psi(\omega,\zeta)\}_{\pi} \tag{18.25}$$

from the invariance of brackets in π under canonical transformations in π. Then, because no differentiations with respect to the ζ^{ρ} are involved in this term and because acting on any function of ζ the operator $e^{\widetilde{b(\zeta)}}$ merely changes the argument of that function to $U^{-1}\zeta$, we get:

$$e^{\widetilde{b(\zeta)}}\{e^{\widetilde{a(\omega)}}\Phi(\omega,\zeta),\ e^{\widetilde{a(\omega)}}\Psi(\omega,\zeta)\}_{\pi}=\{e^{\widetilde{a(\omega)}}\ e^{\widetilde{b(\zeta)}}\Phi(\omega,\zeta),\ e^{\widetilde{a(\omega)}}e^{\widetilde{b(\zeta)}}\Psi(\omega,\zeta)\}_{\pi}\,. \tag{18.26}$$

We prove Eq. (18.24) by applying a similar argument to the second term in Eq. (18.21). In the space $\pi\times\pi'$, this is a rather special canonical transformation because its generator splits into a function of ω plus another of ζ. The exponents in Eq. (18.23) can be added because they act on different objects, and we can write

$$e^{\widetilde{a(\omega)}}e^{\widetilde{b(\zeta)}}=e^{\widetilde{a(\omega)+b(\zeta)}}\,. \tag{18.27}$$

Now we can define direct products of canonical realizations of a Lie group G. Let π and π' carry canonical realizations of $G,\omega\to T_{\exp(u)}\omega$ and $\zeta\to U_{\exp(u)}\zeta$, with generators $\lambda_k(\omega)$ in the first case and $\lambda'_k(\zeta)$ in the second. These will give us two PB realizations of the Lie algebra of G, with possibly different sets of neutral elements d_{jk}, d'_{jk} in the two cases. The direct product canonical realization of G on $\pi \times \pi'$ is defined by combining the actions in π and π', of the canonical transformations realizing G. On functions $\Phi(\omega, \zeta)$, and for any element $\exp(u) \in G$, we have:

$$\begin{aligned}\Phi(\omega,\zeta) \to [R_{\exp(u)}\Phi](\omega,\zeta) &= \Phi(T_{\exp(-u)}\omega, U_{\exp(-u)}\zeta)\\ &= \exp(\widetilde{u\cdot\lambda(\omega)})\exp(\widetilde{u\cdot\lambda'(\zeta)})\Phi(\omega,\zeta)\,. \end{aligned} \tag{18.28}$$

This is a canonical realization of G: it is canonical by the remarks of the previous paragraph, and it is a realization because the individual transformations on ω and on ζ form realizations and because the one set commutes with the other. Once again we can add the exponents in Eq. (18.28) and then we find the important result: the generators of the direct product of two independent canonical realizations of a Lie group G are the sums, $\lambda_j(\omega) + \lambda'_j(\zeta)$, of the individual generators $\lambda_j(\omega)$ and $\lambda'_j(\zeta)$. That these sums do yield a realization of the Lie algebra of G on $\pi \times \pi'$ is easy to see. From Eq. (18.20) it follows that the brackets $\{\lambda_j(\omega), \lambda'_k(\zeta)\}$ vanish identically, and thus;

$$\begin{aligned}\{\lambda_j(\omega)+\lambda'_j(\zeta), \lambda_k(\omega)+\lambda'_k(\zeta)\} =&\{\lambda_j(\omega),\lambda_k(\omega)\} + \{\lambda'_j(\zeta), \lambda'_k(\omega)\}\\ =&c^l_{jk}(\lambda_l(\omega)+\lambda'_l(\zeta)) + d_{jk} + d'_{jk}\,. \end{aligned} \tag{18.29}$$

We see, incidentally, that the neutral elements of the individual realizations add up to give those present in the direct product realization.

The reader will appreciate that in order to produce a new realization from two given ones in this way, the really important thing is that the latter *commute*. Therefore one can consider another possibility of the following type. Suppose on some phase space π we have *two* canonical realizations of $G, \omega \to T_{\exp(u)}\omega$ and $\omega \to T'_{\exp(u)}\omega$ such that *every* transformation of one realization commutes with *every* one of the other. Once again it is trivial to see that the product transformations $\omega \to T_{\exp(u)}T'_{\exp(u)}\omega$, which are certainly canonical, do form a realization of G. In this case we have two sets of generators $\lambda_j(\omega)$ and $\lambda'_j(\omega)$ for the two realizations; each set yields a PB realization up to neutrals of the Lie algebra of G (with possibly different sets of neutral elements in the two cases). However, the fact that the transformations $e^{\widetilde{u\cdot\lambda(\omega)}}$ and $e^{\widetilde{v\cdot\lambda'(\omega)}}$ commute for all u and v implies that the brackets $\{\lambda_j(\omega), \lambda'_k(\omega)\}$ are all neutral elements. Therefore, in this case the generators of the product realization are again $\lambda_j(\omega)+\lambda'_j(\omega)$, but the neutral elements associated with the direct product are the sums of those present in the individual realizations, together with neutral elements of the form $\{\lambda_j, \lambda'_k\} - \{\lambda_k, \lambda'_j\}$.

The realization of $E(3)$ generated by the choice of Eq. (18.18) for J_j and P_j can be described now as being the direct product of the single particle realization of Eq. (18.15) and the non faithful realization given by the pure spin realization of $R(3)$. This is why the two sets of generators were just added together. We see more examples of such things in the next chapter when we discuss the Galilei group.

Another true PB realization of $E(3)$, one that we set up in Chapter 11, involves a single component field $\phi(\boldsymbol{x}, t)$ and its canonically conjugate momentum $\pi(\boldsymbol{x}, t)$. The expressions for J_j and P_j, given in Chapter 11, are:

$$\begin{aligned} J_j &= -\int d^3x\, \pi(\boldsymbol{x}, t)\epsilon_{jkm}x_k \frac{\partial}{\partial x_m}\phi(\boldsymbol{x}, t)\,, \\ P_j &= -\int d^3x\, \pi(\boldsymbol{x}, t)\frac{\partial}{\partial x_j}\,\phi(\boldsymbol{x}, t)\,. \end{aligned} \tag{18.30}$$

Using the basic PB's between ϕ and π, the validity of Eq. (18.10) may be checked. Now let us verify that the effects of the finite transformations generated by J and P on ϕ and π are what one expects. If S is a Cartesian frame in three-dimensional space and $S' = \exp(\boldsymbol{a}\cdot\boldsymbol{d})\exp(\boldsymbol{\alpha}\cdot\boldsymbol{l})S$ another one, then (in the Heisenberg picture) the canonical transformation relating dynamical variables in S to those in S' is $e^{\widetilde{-\boldsymbol{\alpha}\cdot\boldsymbol{J}}}e^{\widetilde{-\boldsymbol{a}\cdot\boldsymbol{P}}}$ (See, for instance, page 287). The coordinates x'_j, x_j of a point P measured relative to S', S respectively, are related thus:

$$S' = \exp(\boldsymbol{a}\cdot\boldsymbol{d})\exp(\boldsymbol{\alpha}\cdot\boldsymbol{l})S:\ x'_j = A_{jk}(\boldsymbol{\alpha})x_k + a_j\,. \tag{18.31}$$

For a scalar field we have the transformation law $\phi'(x') = \phi(x)$, or

$$S \to S' : \phi(\boldsymbol{x}) \to \phi'(\boldsymbol{x}) = \phi(A_{kj}(\boldsymbol{\alpha})(x_k - a_k))\,. \tag{18.32}$$

We want to show that this is correctly generated. First, using Eq. (18.30) we have:

$$\begin{aligned} \widetilde{\boldsymbol{a}\cdot\boldsymbol{P}}\phi(\boldsymbol{x}) &= a_j\{P_j, \phi(\boldsymbol{x})\} = a_j\frac{\partial\phi(\boldsymbol{x})}{\partial x_j} = \boldsymbol{a}\cdot\boldsymbol{\nabla}\phi(\boldsymbol{x})\,, \\ \phi(\boldsymbol{x}-\boldsymbol{a}) &= e^{\widetilde{-\boldsymbol{a}\cdot\boldsymbol{P}}}\phi(\boldsymbol{x})\,, \end{aligned} \tag{18.33}$$

using Taylor's theorem. The effect of $e^{\widetilde{-\boldsymbol{\alpha}\cdot\boldsymbol{J}}}$ on $\phi(\boldsymbol{x})$ is given in Eq. (17.161). All in all, we have:

$$e^{\widetilde{-\boldsymbol{\alpha}\cdot\boldsymbol{J}}}e^{\widetilde{-\boldsymbol{a}\cdot\boldsymbol{P}}}\phi(\boldsymbol{x}) = e^{\widetilde{-\boldsymbol{\alpha}\cdot\boldsymbol{J}}}\phi(\boldsymbol{x}-\boldsymbol{a}) = \phi(A_{kj}(\boldsymbol{\alpha})(x_k - a_k))\,, \tag{18.34}$$

and comparing with Eq. (18.32) we see that we do have

$$\phi'(\boldsymbol{x}) = e^{\widetilde{-\boldsymbol{\alpha}\cdot\boldsymbol{J}}}e^{\widetilde{-\boldsymbol{a}\cdot\boldsymbol{P}}}\phi(\boldsymbol{x})\,. \tag{18.35}$$

An identical relation holds for $\pi(\boldsymbol{x})$ as well. If we have a multicomponent field $\phi_r(\boldsymbol{x})$ and conjugates $\pi_r(\boldsymbol{x})$, under rotations they transform the way we described in the last chapter, and under translations only the argument $\boldsymbol{x}$ is changed. The transformation laws for $\phi_r(\boldsymbol{x})$ and $\pi_r(\boldsymbol{x})$ are:

$$\begin{aligned} S' =& \exp(\boldsymbol{a}\cdot\boldsymbol{d})\exp(\boldsymbol{\alpha}\cdot\boldsymbol{l})S:\ \phi'_r(\boldsymbol{x}) = D_{rs}(\boldsymbol{\alpha})\phi_s(A_{kj}(\boldsymbol{\alpha})(x_k - a_k))\,, \\ \pi'_r(\boldsymbol{x}) =& (D^{\sim}(\boldsymbol{\alpha})^{-1})_{rs}\pi_s(A_{kj}(\boldsymbol{\alpha})(x_k - a_k))\,. \end{aligned} \tag{18.36}$$

The only change in the generators is that J_j acquires a spin term:

$$\begin{aligned} J_j &= \int d^3x \{\pi_r(\boldsymbol{x})(S_j)_{rs}\phi_s(\boldsymbol{x}) - \pi_r(\boldsymbol{x})\epsilon_{jkm}x_k \frac{\partial}{\partial x_m}\phi_r(\boldsymbol{x})\}\,, \\ P_j &= -\int d^3x\ \pi_r(\boldsymbol{x})\frac{\partial}{\partial x_j}\phi_r(\boldsymbol{x})\,. \end{aligned} \tag{18.37}$$

We leave it to the reader to check that the following relations obtain:

$$\begin{aligned} e^{\widetilde{-\boldsymbol{\alpha}\cdot\boldsymbol{J}}}e^{\widetilde{-\boldsymbol{a}\cdot\boldsymbol{P}}}\phi(\boldsymbol{x}) &= D(\boldsymbol{\alpha})\phi(A_{kj}(\boldsymbol{\alpha})(x_k - a_k)) = \phi'(\boldsymbol{x})\,, \\ e^{\widetilde{-\boldsymbol{\alpha}\cdot\boldsymbol{J}}}e^{\widetilde{-\boldsymbol{a}\cdot\boldsymbol{P}}}\pi(\boldsymbol{x}) &= D^{\sim}(\boldsymbol{\alpha})^{-1}\pi(A_{kj}(\boldsymbol{\alpha})(x_k - a_k)) = \pi'(\boldsymbol{x})\,. \end{aligned} \tag{18.38}$$

Rigid Body Rotation in a Uniform Gravitational Field

A very interesting application of the $E(3)$ structure occurs in the description of the equations of motion of a rigid body rotating under the influence of a uniform gravitational field. This is in the spirit of the demonstration of the last chapter that for torque-free rotation, the rigid body equations of motion can be set up very simply in terms of a GPB based on the $R(3)$ Lie algebra. Here we use a GPB based on the $E(3)$ algebra. In this application, we use only the mathematical structure of $E(3)$ and do not at all relate it to changes of frame in three-dimensional space.

The kinematics of rigid body motion has been set up in the previous chapter, and we use the same symbols and notation here. First, we derive the Lagrangian equations of motion along the usual lines, then later show how to derive them from a GPB. The expression for the kinetic energy is given in Eq. (17.38); we have seen that it leads to:

$$\frac{d}{dt}\frac{\partial T}{\partial \dot{\alpha}_s} - \frac{\partial T}{\partial \alpha_s} = (\dot{S}_r + \epsilon_{rkj}\frac{\partial T}{\partial S_k}S_j)\hat{\xi}_{rs}(\boldsymbol{\alpha})\,. \tag{18.39}$$

Equating this to zero led to the torque-free equations of motion. Now suppose the rigid body is placed in a uniform gravitational field; let the field act in the direction $-\boldsymbol{n}, \boldsymbol{n}$ being a unit vector and n_k the components of $\boldsymbol{n}$ in the fixed inertial frame S. Then the total Lagrangian is T minus the potential energy V. If the total mass of the rigid body is M, the intensity of the field g, and ξ_j the coordinates of the center of gravity of the body measured in the body fixed axes, then the expression for V is

$$V(\boldsymbol{\alpha}) = Mgn_kA_{kj}(\boldsymbol{\alpha})\xi_j\,. \tag{18.40}$$

In working out the Lagrangian equations of motion, we need an expression for the partial derivatives of the matrix elements $A_{kj}(\boldsymbol{\alpha})$ with respect to the α's. This is easily obtained from Eq. (17.18) if we substitute in it the expression for the angular velocities ω_r in terms of the true generalized velocities $\dot{\alpha}_s$, and then identify the coefficients of $\dot{\alpha}_s$ on both sides. Thus we get:

$$A_{kj,s}(\boldsymbol{\alpha}) \equiv \frac{\partial A_{kj}(\boldsymbol{\alpha})}{\partial \alpha_s} = -A_{kl}(\boldsymbol{\alpha})\epsilon_{ljm}\hat{\xi}_{ms}(\boldsymbol{\alpha})\,. \tag{18.41}$$

Thus with the Lagrangian

$$L(\boldsymbol{\alpha}, \dot{\boldsymbol{\alpha}}) = T(\boldsymbol{\alpha}, \dot{\boldsymbol{\alpha}}) - V(\boldsymbol{\alpha}), \tag{18.42}$$

the equations of motion are:

$$(\dot{S}_r + \epsilon_{rkj}\frac{\partial T}{\partial S_k}S_j)\hat{\xi}_{rs}(\boldsymbol{\alpha}) = -\frac{\partial V}{\partial \alpha_s} = Mgn_k\xi_j A_{kl}(\boldsymbol{\alpha})\epsilon_{ljr}\hat{\xi}_{rs}(\boldsymbol{\alpha}). \tag{18.43}$$

Because the matrix $\hat{\Xi}(\boldsymbol{\alpha})$ is nonsingular, we see that these equations of motion are equivalent to the following:

$$\dot{S}_r = \epsilon_{krj}S_j\frac{\partial T}{\partial S_k} + Mgn_k\xi_j A_{kl}(\boldsymbol{\alpha})\epsilon_{ljr}. \tag{18.44}$$

These have turned out to involve other things besides the S's! Thus let us define three quantities $P_j(\boldsymbol{\alpha})$ as:

$$P_j(\boldsymbol{\alpha}) = Mgn_k A_{kj}(\boldsymbol{\alpha}). \tag{18.45}$$

Then the equations of motion express each $\dot{S}_r$ as a function of the S's and these P's. However, the time derivatives of the P's, too, can be expressed in terms of the S's and P's alone! Once again we use Eq. (17.18):

$$\dot{P}_j = Mgn_k\dot{A}_{kj} = -Mgn_k A_{kl}(\boldsymbol{\alpha})\epsilon_{ljm}\omega_m = -\epsilon_{ljm}P_l I^{-1}_{mn}S_n. \tag{18.46}$$

Actually, this is a kinematical statement if we view the α's and $\dot{\alpha}$'s as the primitive quantities, because the accelerations $\ddot{\alpha}_j$ do not appear at all. Thus we have the following system of equations that express the Hamiltonian and the time derivatives of S_j as well as P_j, all in terms of the S's and P's alone:

$$H = T + V = \frac{1}{2}I^{-1}_{mn}S_m S_n + \xi_m P_m; \tag{18.47a}$$

$$\dot{S}_j = -\epsilon_{jkm}S_m\frac{\partial T(\boldsymbol{S})}{\partial S_k} + \epsilon_{jkm}\xi_m P_k; \tag{18.47b}$$

$$\dot{P}_j = -\epsilon_{ljm}P_l I^{-1}_{mn}S_n. \tag{18.47c}$$

These equations lead to the expectation that the PB's among the S's and P's can *all* be expressed in terms of these variables alone, even though they may be evaluated by starting from the basic PB's among the α_j and their canonical conjugates β_j. For the S_j we know this statement to be true; it is contained in Eq. (17.60). It is also clear that since each P_j is a function of the α's alone, the PB's $\{P_j, P_k\}$ vanish. Let us then compute $\{S_j, P_k\}$. We must use Eqs. (17.56) and (17.57), the definition Eq. (18.45) of $P_k(\boldsymbol{\alpha})$ and Eq. (18.41):

$$\begin{aligned}\{S_j, P_k\} &= \{\hat{\eta}_{mj}(\boldsymbol{\alpha})\beta_m, Mgn_l A_{lk}(\boldsymbol{\alpha})\} = -Mgn_l\hat{\eta}_{mj}(\boldsymbol{\alpha})A_{lk,m}(\boldsymbol{\alpha}) \\ &= Mgn_l\hat{\eta}_{mj}(\boldsymbol{\alpha})A_{lr}(\boldsymbol{\alpha})\epsilon_{rks}\hat{\xi}_{sm}(\boldsymbol{\alpha}) = Mgn_l A_{lr}(\boldsymbol{\alpha})\epsilon_{rkj}, \\ \{S_j, P_k\} &= -\epsilon_{jkm}P_m. \end{aligned} \tag{18.48}$$

Thus actually the S_j and P_j give a true PB realization of the $E(3)$ Lie algebra,

$$\{S_j, S_k\} = -\epsilon_{jkm} S_m, \ \{S_j, P_k\} = -\epsilon_{jkm} P_m, \ \{P_j, P_k\} = 0\,. \tag{18.49}$$

The Eqs. (18.47b) and (18.47c) can certainly be put into the standard Hamiltonian form with the righthand sides of the equations of motion being interpreted as the usual PB's in the α_j and β_j:

$$\dot{S}_j = \{S_j, \boldsymbol{H}(\boldsymbol{S}, \boldsymbol{P})\}, \ \dot{P}_j = \{P_j, \boldsymbol{H}(\boldsymbol{S}, \boldsymbol{P})\}\,. \tag{18.50}$$

But, as in the case of torque-free rotation of a rigid body, here again we only need to use the brackets among the S_j and P_j alone, and the derivation property, to compute the brackets appearing in Eq. (18.50), and we need not invoke the forms of these variables in terms of the original α_j and β_j.

We can repeat the procedure of the preceding chapter and define a pure Euclidean system thus: it is a classical dynamical system all of whose dynamical variables can be expressed in terms of six basic quantities S_j, P_j, between which we postulate a set of basic or elementary GPB's exactly as in Eq. (18.49). The general equation of motion states that the time derivative of any function of $\boldsymbol{S}$ and $\boldsymbol{P}$ equals the GPB of that function with some Hamiltonian $\boldsymbol{H}(\boldsymbol{S}, \boldsymbol{P})$, and the GPB of any two functions is developed from the basic ones, Eq. (18.49), using the derivation property:

$$\{f, g\} = -\epsilon_{jkm} S_m \frac{\partial f}{\partial S_j}\frac{\partial g}{\partial S_k} - \epsilon_{jkm} P_m \left(\frac{\partial f}{\partial S_j}\frac{\partial g}{\partial P_k} - \frac{\partial f}{\partial P_k}\frac{\partial g}{\partial S_j}\right)\,. \tag{18.51}$$

As explained in the preceding chapter, the validity of the Jacobi identity is assured.

The rigid body rotating in a uniform gravitational field is not a pure Euclidean system, just as a freely rotating rigid body is not a pure spin system, but in each case the equations of motion assumed very suggestive forms. The three quantities P_k amount to only two independent functions of the three α_j. It so happens that the particular set of equations Eq. (18.47) can be put into GPB form, and to the extent that one deals only with the variables S_j and P_j, one can base the discussion on the GPB Eq. (18.51). The value of the Casimir invariant P^2 is $M^2 g^2$; it is fixed by the constants in the problem and is not something whose value can be specified as an adjustable initial condition. But the second Casimir invariant $\boldsymbol{S} \cdot \boldsymbol{P}$ has the value $MgJ_j n_j$, J_j being the components along the space axes, of the total angular momentum of the body. If we remove a factor Mg, this is just the component of the total angular momentum of the body in the direction of the gravitational field and this is naturally conserved. Any initial value can be assigned to this invariant, and it will be preserved during the motion. Based on Eq. (18.14) we can say that the "phase space" is a five dimensional space : each point is labeled by the six coordinates S_j, P_j with each S_j varying independently from $-\infty$ to $+\infty$ and with P_j restricted to lie on the sphere of radius Mg. Any generalized canonical transformation, not just the one generated by the Hamiltonian, preserves the value of the "kinematic invariant" $\boldsymbol{S} \cdot \boldsymbol{P}$.

If we were to try to solve the equations of motion Eq. (18.47b) and (18.47c), we could take advantage of the existence of the three constants of the motion $H(\boldsymbol{S}, \boldsymbol{P}), P^2$ and $\boldsymbol{S} \cdot \boldsymbol{P}$. However, in contrast to the problem dealt with in the preceding chapter, this still leaves three unknowns, and the problem does not reduce to quadratures. Some simplifications occur in the case of a *symmetrical* rigid body. We may assume with no loss of generality that principal axes have been chosen; thus the inertia tensor I_{mn} is diagonal. We say that the body is *symmetrical* if then $I_1 = I_2$ (say) and at the same time the center of gravity lies on the third axis in the body-fixed frame. In that case, $\xi_1 = \xi_2 = 0$, only ξ_3 survives, and the Hamiltonian is:

$$H(\boldsymbol{S}, \boldsymbol{P}) = \frac{1}{2}\left(\frac{S_1^2 + S_2^2}{I_1} + \frac{S_3^2}{I_3}\right) + \xi_3 P_3 \equiv \frac{1}{2I}\left(S_1^2 + S_2^2\right) + \frac{1}{2I'} S_3^2 + \xi P_3 \,. \tag{18.52}$$

It is evident that we have gained an extra constant of motion, namely S_3. Thus for a symmetrical rigid body, also called a symmetrical top, the generalized canonical transformation generated by the Hamiltonian preserves the values of $S_3, \boldsymbol{S} \cdot \boldsymbol{P}$, and $H(\boldsymbol{S}, \boldsymbol{P})$. There are, therefore, only two unknown functions of time to be solved for. We could choose these to be the spherical polar angles of $\boldsymbol{P}$, $|\boldsymbol{P}|$ being, of course, fixed at the value Mg. One then finds that the equations of motion reduce to a coupled system of differential equations solvable in terms of elliptic functions, but we forego a detailed discussion of this development here.

Chapter 19

The Galilei Group

The Galilei group is the first of the two groups of space-time transformations to be discussed in this book. For brevity we refer to it as $\mathcal{G}$. This group expresses the geometrical invariance properties of the equations of motion of a non-relativistic classical dynamical system when the system is isolated from external influences. This is so to the extent that these equations of motion are reasonable generalizations of the Newtonian equations of motion for multi particle systems experiencing interparticle forces. If the dynamics of the system can be conveyed via an action principle, so that a Lagrangian and therefore a Hamiltonian description is possible, then the system is described by a classical canonical realization of $\mathcal{G}$.

We recall that an element in the ten-parameter Lie group $\mathcal{G}$ can be written as (A, v_j, b, a_j) : A denotes a 3×3 orthogonal rotation matrix, $\boldsymbol{v}$ is the relative velocity vector between two inertial frames; b a time displacement and $\boldsymbol{a}$ a spatial translation. The composition law is:

$$(A', v'_j, b', a'_j)(A, v_j, b, a_j) = (A'A, v'_j + A'_{jk}v_k, b' + b, a'_j + bv'_j + A'_{jk}a_k)\,. \tag{19.1}$$

Expressed in terms of the ten Lie algebra basis vectors l_j, d_j, g_j and h, the element (A, v_j, b, a_j) is $\exp(\boldsymbol{a}\cdot\boldsymbol{d})\exp(bh)\exp(\boldsymbol{v}\cdot\boldsymbol{g})\exp(\boldsymbol{\alpha}\cdot\boldsymbol{l})$, α_j being the canonical coordinates of A. The basic Lie brackets are:

$$[l_j, l_k] = \epsilon_{jkm}l_m,\ [l_j, g_k] = \epsilon_{jkm}g_m,\ [g_j, g_k] = 0\,; \tag{19.2a}$$

$$[l_j, d_k] = \epsilon_{jkm}d_m,\quad [d_j, d_k] = 0\,; \tag{19.2b}$$

$$[l_j, h] = [d_j, h] = 0,\ [g_j, h] = d_j\,; \tag{19.2c}$$

$$[g_j, d_k] = 0\,. \tag{19.2d}$$

(The arrangement of brackets here differs slightly from that in Eq. (15.73)). In a canonical realization of $\mathcal{G}$, we have the generators $J_j(\omega), G_j(\omega), P_j(\omega)$ and $H(\omega)$ corresponding to l_j, g_j, d_j and h, respectively. The l_j and d_j generate the physical $E(3)$ subgroup of $\mathcal{G}$; l_j and g_j also generate a subgroup having the $E(3)$

structure, and we used the notation $E^{(v)}(3)$ for it in Chapter 15. Our study of $R(3)$ and $E(3)$ in the preceding two chapters tells us that in the PB relations corresponding to Eqs. (19.2a) and (19.2b), we may assume there are no neutral elements present. Once the neutral elements in the bracket relations of the $R(3)$ subalgebra have been removed, the removal of such elements in the $E(3)$ subalgebra needs redefinitions of the $P_j(\omega)$ alone, and in the $E^{(v)}(3)$ subalgebra needs redefinitions of the $G_j(\omega)$ alone. Thus we have in all cases the following PB relations:

$$\begin{aligned}\{J_j, J_k\} = \epsilon_{jkm} J_m,\ \{J_j, G_k\} = \epsilon_{jkm} G_m,\ \{G_j, G_k\} = 0 \quad ; \\ \{J_j, P_k\} = \epsilon_{jkm} P_m,\ \{P_j, P_k\} = 0\,. \end{aligned} \tag{19.3}$$

Next we prove the absence of neutral elements in the PB relations realizing Eq. (19.2c). It is most transparent to use a global argument using the properties of $R(3)$. Consider the PB $\{J_j, H\}$. Suppose it is equal to a neutral element δ_j. Then we have

$$\{J_j, H\} = \delta_j,\ \exp(\widetilde{\alpha_j J_j})H = H + \alpha_j \delta_j\,. \tag{19.4}$$

Using the behavior of J_j under rotations, we then deduce that δ_j must be a three-dimensional vector whose components are unchanged under all rotations:

$$\begin{aligned}\delta_j = \exp(\widetilde{\boldsymbol{\alpha}\cdot\boldsymbol{J}})\delta_j = \exp(\widetilde{\boldsymbol{\alpha}\cdot\boldsymbol{J}})\{J_j, H\} = \{\exp(\widetilde{\boldsymbol{\alpha}\cdot\boldsymbol{J}})J_j, H + \boldsymbol{\alpha}\cdot\boldsymbol{\delta}\} \\ = A_{kj}(\boldsymbol{\alpha})\{J_k, H\}, \qquad A_{kj}(\boldsymbol{\alpha})\delta_k = \delta_j\,. \end{aligned} \tag{19.5}$$

Because there is no nonzero vector with this property, δ_j=0. Like J_j, G_j and P_j also transform as vectors under rotations generated by the J's. Therefore arguments exactly like the above show that in all the PB relations corresponding to Eq. (19.2c) there are no neutral elements; thus one has in all cases:

$$\{J_j, H\} = \{P_j, H\} = 0,\ \{G_j, H\} = P_j\,. \tag{19.6}$$

The situation is that if in a canonical realization of $\mathcal{G}$ we adjust the generators J_j, G_j, and P_j in such a way that Eq. (19.3) is valid (and this is always possible), then Eq. (19.6) will also be valid. Once this has been done, none of the quantities J_j, G_j, and P_j may be altered by the addition of neutral elements. However, $H(\omega)$ could be so altered because the relations of Eqs. (19.3) and (19.6) are not changed in this process. Now what remains is the set of relations Eq. (19.2d). For the PB's $\{G_j, P_k\}$, we must start with the general form

$$\{G_j, P_k\} = M_{jk} \tag{19.7}$$

where the M_{jk} are a set of neutral elements. We see immediately that if the M_{jk} are not identically zero, they cannot be transformed away! The only generator we can possibly alter is $H(\omega)$, but that makes no appearance in Eq. (19.7). On the other hand, there are restrictions on the M_{jk} because of the Jacobi identity, and these will limit the possibilities. We must take the PB of both sides of Eq.

(19.7) with the J's, G's, P's, and H in turn, apply the Jacobi identity to the left-hand side in each case, and see if we get conditions on the M_{jk}. For the J's, we again adopt the global method and get:

$$\begin{aligned} M_{jk} &= \exp\widetilde{(\boldsymbol{\alpha}\cdot\boldsymbol{J})}M_{jk} = \exp\widetilde{(\boldsymbol{\alpha}\cdot\boldsymbol{J})}\{G_j, P_k\} = \{\exp\widetilde{(\boldsymbol{\alpha}\cdot\boldsymbol{J})}G_j, \exp\widetilde{(\boldsymbol{\alpha}\cdot\boldsymbol{J})}P_k\}\,, \\ A_{mj}(\boldsymbol{\alpha})A_{nk}(\boldsymbol{\alpha})M_{mn} &= M_{jk}\,. \end{aligned} \quad (19.8)$$

Thus M_{jk} must be the components of a second-rank Cartesian tensor with the property that its components are unchanged in value after a rotation. Apart from a factor, the only such tensor is the symmetric Kronecker symbol δ_{jk}. It follows that the rotational properties of G_j and P_j restrict the neutral elements to the form:

$$M_{jk} = M\delta_{jk}\,. \quad (19.9)$$

There can be only one independent neutral element M. If we next take the PB of both sides of Eq. (19.7) with a G_m or a P_m or H, and apply the Jacobi identity to the left-hand side and make use of Eqs. (19.6) and (19.7), we find in all cases that both sides vanish identically; thus no new restrictions are generated on the neutral element M. We can conclude that the PB relations:

$$\{G_j, P_k\} = \delta_{jk}M \quad (19.10)$$

with M neutral are consistent with the structure of $\mathcal{G}$, and such an M cannot be transformed away. We have thus proved that the generators of a general canonical realization of $\mathcal{G}$ obey the basic PB relations Eqs. (19.3), (19.6), and (19.10). They involve one neutral element M. If M is zero, we have a true PB realization of the Lie algebra of $\mathcal{G}$; otherwise we have a PB realization up to a neutral element; it is of nontrivial type and is not equivalent to a true realization. Needless to say, the neutrality of M guarantees that the finite canonical transformations built up from the generators J, G, P, and H will in all cases obey the composition law of $\mathcal{G}$. We have also proved that with the standard forms in Eqs. (19.3), (19.6), and (19.10) for the basic PB relations, a canonical realization of $\mathcal{G}$ determines the generators J_j, G_j, and P_j and the neutral element M uniquely, but the generator H is determined only up to an additive neutral element. Thus in a Galilean-invariant theory, with H representing the total energy of the system, the zero of the energy scale can be chosen arbitrarily.

It is pointed out in Chapter 14 (page 228) that a PB realization, with nontrivial neutral elements, of the Lie algebra of a Lie group G can be looked upon as a true PB realization of the Lie algebra of an extended group G'. The increase in the number of parameters in going from G to G' is merely the number of independent neutral elements present. Consequently, the $M \neq 0$ realizations of the Lie algebra of $\mathcal{G}$ can be reinterpreted as true PB realizations of an eleven-parameter group $\mathcal{G}'$ (more precisely, of the Lie algebra of $\mathcal{G}'$). We make M the representative of the eleventh basis element in the Lie algebra of $\mathcal{G}'$. In detail, the Lie algebra of $\mathcal{G}'$ is spanned by the eleven elements l_j, d_j, g_j, h, and μ. The basic Lie bracket relations among these consist of those given in Eqs. (19.2a), (19.2b), (19.2c) together with the following involving μ:

$$[l_j, \mu] = [d_j, \mu] = [g_j, \mu] = 0, \ [h, \mu] = 0 ; \tag{19.11a}$$

$$[g_j, d_k] = \delta_{jk}\mu . \tag{19.11b}$$

By construction, the one-dimensional subspace of the Lie algebra of $\mathcal{G}'$ consisting of all multiples of $\dot{\mu}$ constitutes an Abelian invariant subalgebra in the eleven-parameter Lie algebra; correspondingly, the one-parameter subgroup of $\mathcal{G}'$ having μ for its tangent vector forms an Abelian invariant subgroup of $\mathcal{G}'$. The relation between $\mathcal{G}'$ and $\mathcal{G}$ is that the latter is isomorphic to the factor group of $\mathcal{G}'$ with respect to this Abelian invariant subgroup. If we write $\exp(\theta\mu)$ for elements on the one-parameter subgroup generated by μ, a general element in $\mathcal{G}'$ has parameters (A, v_j, b, a_j, θ) and is given by the product $\exp(\theta\mu)\exp(\boldsymbol{a}\cdot\boldsymbol{d})\exp(bh)\exp(\boldsymbol{v}\cdot\boldsymbol{g})\exp(\boldsymbol{\alpha}\cdot\boldsymbol{l})$; θ lies in the range $-\infty < \theta < \infty$, and the remaining parameters have the same ranges as in the case of $\mathcal{G}$. Concerning the composition law for $\mathcal{G}'$, the parameters $\boldsymbol{\alpha}, \boldsymbol{v}, b$ and $\boldsymbol{a}$ combine exactly as they do in the case of $\mathcal{G}$; whereas the law for θ reflects the difference between Eqs. (19.11b) and (19.2d). Use of the Baker-Campbell-Hausdorff identity easily leads to:

$$\begin{aligned}&(A',v_j', b', a_j', \theta')(A, v_j, b, a_j, \theta)\\&\quad = (A'A, v_j' + A_{jk}'v_k, b' + b, a_j' + bv_j' + A_{jk}'a_k, \theta' + \theta + A_{jk}'v_j'a_k) .\end{aligned} \tag{19.12}$$

We are only interested in true PB realizations of the Lie algebra of $\mathcal{G}'$. (This Lie algebra itself can have nontrivial neutral elements present in a PB realization; specifically the relation $[h, \mu]$=0 could be realized with a nonzero neutral element on the right!) In all such realizations, and in the associated canonical realizations of $\mathcal{G}'$, μ is represented by a quantity that is an invariant of the realization because it has vanishing PB's with all the generators of $\mathcal{G}'$. In fact, the representative of μ is a Casimir invariant for all true PB realizations of the Lie algebra of $\mathcal{G}'$. But according to our work up to now, we have to specialize further before we arrive at canonical realizations of $\mathcal{G}$: we must pick those true realizations of the Lie algebra of $\mathcal{G}'$ in which the representative of μ is a neutral element. It is only these that generate canonical transformations having the same composition law as the elements of $\mathcal{G}$. The point to emphasize is that it is not sufficient for the representative of μ in a true realization of the Lie algebra of $\mathcal{G}'$ to have vanishing PB with all the generators of $\mathcal{G}'$; it must have vanishing PB with all functions on phase space, for otherwise we will not be dealing with a canonical realization of $\mathcal{G}$ at all! These conclusions follow from our work in Chapters 14 and 16. Later on in this chapter we see that in some cases the conditions we have been imposing up to now must be modified. Then we will be able to utilize true realizations of the Lie algebra of $\mathcal{G}'$ in which μ is not realized as a neutral element; this will be compensated for by a suitable modification in the interpretation of the formalism.

For the realizations of $\mathcal{G}$ in which M=0, we do not have to deal with the extended group $\mathcal{G}'$ at all. Alternatively, we can say that true realizations of the

Lie algebra of $\mathcal{G}'$ in which μ is represented by zero (and this is allowed) are true realizations of the Lie algebra of $\mathcal{G}$.

Realizations of $\mathcal{G}$ with M=0

Within the framework of quantum mechanics, representations of $\mathcal{G}$ generated by true representations of its Lie algebra (i.e., not involving any neutral elements in the commutation relations) have been constructed and discussed for the first time by Inönü and Wigner. With some obvious and minor modifications, their treatment can be adapted to the classical case as well. What results is a classification of the various possible types of canonical realizations of $\mathcal{G}$ generated by true realizations of its Lie algebra via PB's. There is, however, one difference to be made clear. In the quantum mechanical case it is reasonable to directly examine and restrict oneself to the *irreducible* true representations of $\mathcal{G}$, and to classify them; in fact, this is what has been done by Inönü and Wigner. Although an analogous procedure could be followed in the classical case, too, we prefer to deal with general canonical realizations of $\mathcal{G}$ (with $M = 0$) rather than restrict ourselves to irreducible canonical realizations alone. This may make the classification of canonical realizations appear more involved than the classification of irreducible quantum mechanical representations, but that is understandable.

The possibility of having various distinct types of true realizations of the Lie algebra of $\mathcal{G}$ arises from the fact that various subsets of the generators could vanish without implying that all the generators vanish. This is similar to the case of $E(3)$ where we found that any realization of $R(3)$ is at the same time a non-faithful one for $E(3)$. In order to survey the possibilities, it is helpful to first determine the action on the generators themselves, of the canonical transformations corresponding to elements of $\mathcal{G}$ in a canonical realization of $\mathcal{G}$. This is just the adjoint action. The generators are $J_j(\omega), G_j(\omega), P_j(\omega)$ and $H(\omega)$; they obey the PB relations Eqs. (19.3) and (19.6) and

$$\{G_j, P_k\} = 0\,. \tag{19.13}$$

With the help of these relations we can draw up Table 19.1. In each row we list the images of the generators standing in the first row under the canonical transformations appearing in the first column.

Table 19.1: ***Transformations of Galilean Generators***

Transformations	Generators			
	J_j	G_j	P_j	H
$\exp(\widetilde{-\boldsymbol{\alpha}\cdot\boldsymbol{J}})$	$A_{jk}(\boldsymbol{\alpha})J_k$	$A_{jk}(\boldsymbol{\alpha})G_k$	$A_{jk}(\boldsymbol{\alpha})P_k$	H
$\exp(\widetilde{-\boldsymbol{v}\cdot\boldsymbol{G}})$	$J_j - (\boldsymbol{G}\times\boldsymbol{v})_j$	G_j	P_j	$H - \boldsymbol{v}\cdot\boldsymbol{P}$
$\exp(\widetilde{-bH})$	J_j	$G_j + bP_j$	P_j	H
$\exp(\widetilde{-\boldsymbol{a}\cdot\boldsymbol{P}})$	$J_j - (\boldsymbol{P}\times\boldsymbol{a})_j$	G_j	P_j	H

With these results, one can quite easily discover the Casimir invariants for $\mathcal{G}$. (We avoid continuous repetition of the fact that we are now dealing with M=0 realizations.) Any Casimir invariant must certainly be a scalar under three-dimensional rotations. A glance at the column headed by P_j tells us that the magnitude of $\boldsymbol{P}$ is preserved by all transformations of $\mathcal{G}$; this leads to one Casimir invariant, $C_1 = P^2$. Although $\boldsymbol{G}$, like $\boldsymbol{P}$, is a vector whose components have zero brackets with one another, under the canonical transformation $e^{\widetilde{-bH}}$ it acquires a component parallel to $\boldsymbol{P}$, and thus its magnitude is not invariant under $\mathcal{G}$. However, we can see that the vector $\boldsymbol{G} \times \boldsymbol{P}$ is unchanged by the canonical transformations generated by $\boldsymbol{G}, \boldsymbol{P}$ and H; thus its magnitude is invariant under $\mathcal{G}$. This leads to the second Casimir invariant, $C_2 = (\boldsymbol{G} \times \boldsymbol{P})^2$. By studying Table 19.1 the reader can satisfy himself that no invariant can be constructed out of the generators of $\mathcal{G}$, which is algebraically independent of C_1 and C_2. Consequently, we can say that C_1 and C_2 form a complete set of independent Casimir invariants for $\mathcal{G}$.

We now turn to the classification of realizations of $\mathcal{G}$ into various kinds. We deal at first with general canonical realizations, not assuming that they are irreducible. The classification depends on which of the generators of $\mathcal{G}$ happen to vanish identically. By examining those PB's in Eqs. (19.3), (19.6), and (19.13) that involve the P_j, we can see that if in a particular realization the P_j *do not* vanish identically, then in that realization *none* of the other generators can vanish identically either. On the other hand, if the P_j *do* vanish identically, that does not impose any constraints on the other generators beyond requiring the brackets $\{G_j, H\}$ to vanish. This happens because there is no basic bracket relation that involves the P_j only on the left-hand side and not on the right.

Let us now examine the first possibility, namely that the P_j do not vanish identically. This property of these realizations can be conveyed equally well by saying that in them the Casimir invariant C_1 does not vanish identically. We may now divide these realizations into two classes based on the properties of the second Casimir invariant C_2. We shall say that a canonical realization of $\mathcal{G}$ belongs to *class A* if neither C_1 nor C_2 vanishes identically, and that it belongs to *class B* if C_1 does not vanish identically but C_2 does. Table 19.1 assures us that it is possible, in principle, to have realizations in which the vectors $\boldsymbol{G}$ and $\boldsymbol{P}$ are parallel, because such a property, if present, is maintained under all transformations of $\mathcal{G}$. For realizations of class A, the Casimir invariants C_1 and C_2 are the only invariants that are common to all such realizations and can be formed out of the generators. For realizations of class B, we lose the invariant C_2 because it vanishes, but then we see that since $\boldsymbol{G}$ and $\boldsymbol{P}$ are parallel, the quantity $\boldsymbol{J}\cdot\boldsymbol{P}$ becomes an invariant. This is easy to see from Table 19.1. Thus in class B the "loss" of the Casimir invariant C_2 is compensated for by the "gain" of the invariant $\boldsymbol{J} \cdot \boldsymbol{P}$. However, we do not call the latter a Casimir invariant because it is specific to class B alone.

Let us now consider the realizations of $\mathcal{G}$ in which P_j=0 identically. It is clear that these are realizations of the group $E^{(v)}(3) \otimes T_1$, the Euclidean group $E^{(v)}(3)$ being generated by the (possibly non-vanishing) quantities J_j, G_j, and the one-dimensional translation group T_1 by H. Now we "lose" both Casimir invariants C_1 and C_2 because they vanish, but the generator H becomes an invariant for all these realizations. The separation of these realizations into various classes is based upon the way in which $E^{(v)}(3)$ is realized, namely, whether it is realized faithfully, nonfaithfully, or trivially; in each case the possible additional invariants can be written down from our study of $E(3)$ in the previous chapter. A canonical realization of $\mathcal{G}$ in which $C_1 = C_2=0$ so that $P_j=0$ but in which neither J_j nor G_j vanish identically is said to belong to *class* C. These are the cases in which $E^{(v)}(3)$ is realized faithfully; conversely, any class C realization of $\mathcal{G}$ arises by combining a faithful realization of $E^{(v)}(3)$ with any realization of T_1, after ensuring that the former canonical transformations commute with the latter. The invariants common to all these realizations of $\mathcal{G}$ are the Casimir invariants of $E^{(v)}(3), G^2$ and $\boldsymbol{J} \cdot \boldsymbol{G}$, along with H. Once again, these are not Casimir invariants for $\mathcal{G}$; the one that certainly does not vanish identically is G^2. If $E^{(v)}(3)$ is realized nonfaithfully, but also nontrivially, that means that the G_j vanish identically, but the J_j do not; the realization of $E^{(v)}(3)$ collapses to a faithful one of $R(3)$ and that of $\mathcal{G}$ to a realization of $R(3)\otimes T_1$. A canonical realization of $\mathcal{G}$ in which $C_1 = C_2 = P_j = G_j=0$ identically, but in which the J_j do not vanish identically, is said to belong to *class* D. Any such realization of $\mathcal{G}$ arises by combining a faithful, nontrivial, realization of $R(3)$ with any realization of T_1, after ensuring that the former canonical transformations commute with the latter. The invariants common to all these realizations of $\mathcal{G}$ are J^2, the Casimir invariant of $R(3)$, and H; the first does not vanish identically. Finally, we have the *class* E realizations of $\mathcal{G}$, which are simply realizations of T_1 generated by H; now all the generators J_j, G_j, and P_j vanish identically, and the only invariant common to all these realizations is H itself. The trivial realization of $\mathcal{G}$ in which all the generators give rise to the identity canonical transformation is a particular case of class E, with H any neutral element. In Table 19.2 we summarize the distinguishing properties of the five classes of canonical realizations of $\mathcal{G}$, along with their characteristic invariants.

Table 19.2: *Classification of Realizations of $\mathcal{G}$ with $M = 0$*

Class	Defining Properties	Characteristic Invariants
A	No generator vanishes identically, and $\boldsymbol{G}$ is not always parallel to $\boldsymbol{P}$	C_1, C_2, neither vanishing identically
B	No generator vanishes identically, but $\boldsymbol{G}$ is parallel to $\boldsymbol{P}$	$C_2 = 0$; $C_1, \boldsymbol{J} \cdot \boldsymbol{P}$; C_1 does not vanish identically
C	$P_j = 0$; none of J_j and G_j vanishes identically	$C_1 = C_2 = 0$; $H, G^2, \boldsymbol{J} \cdot \boldsymbol{G}$; G^2 does not vanish identically
D	$P_j = G_j = 0$; J_j does not vanish identically	$C_1 = C_2 = 0$; H, J^2; J^2 does not vanish identically
E	$P_j = G_j = J_j = 0$	$C_1 = C_2 = 0$; H

As we stated above, the classification of canonical realizations of $\mathcal{G}$ that we have given is meaningful independent of whether one has an irreducible realization or not. In general, a realization of class A, for example, is reducible; neither of the Casimir invariants C_1 and C_2 vanishes identically; each assumes various possible (nonnegative) values. However, there may be invariant hypersurfaces in phase space on which C_2 alone, or both C_1 and C_2, happen to vanish. Nevertheless, we characterize the realization in its entirety when we classify it as being of class A. Similar comments apply to the other classes.

It is when we specialize to the irreducible canonical realizations of $\mathcal{G}$ that we obtain results that are exact analogues to the results Inönü and Wigner obtained in the context of quantum mechanics. To stick to the nomenclature introduced by Inönü and Wigner, we say that an irreducible canonical realization of $\mathcal{G}$ that falls into a definite *class* is of a definite *type*. Thus an irreducible canonical realization that is of class A is said to be a *type-I irreducible realization*. Similarly, irreducible realizations belonging to classes B, C, D, and E are said to be *types II, III, IV and V irreducible realizations*, respectively. The characteristic invariants listed in Table 19.2 reduce, in an irreducible realization, to numbers or neutral elements. This permits us to make more definite statements concerning some of them. For example, a type-I irreducible realization belongs to class A, implying that neither C_1 nor C_2 vanishes identically, but both these are numbers, and thus we conclude that they must both be strictly greater than zero. For a type-II irreducible realization we conclude that $C_1 > 0, C_2 = 0$, and $\boldsymbol{J} \cdot \boldsymbol{P}$ is a number that may or may not vanish. In types III, IV, and V, H is always a number because it is an invariant. Type-III irreducible realizations have a nonzero positive numerical value for the invariant G^2, and similarly type IV for J^2. Type V reduces to the trivial realization of $\mathcal{G}$ because the only possibly nonvanishing generator H is constrained to be neutral, and is therefore ignored. We summarize the important properties of irreducible canonical realizations of types I, II, III and IV in Table 19.3.

Table 19.3: *Irreducible realizations of $\mathcal{G}$ with $M = 0$*

Type	Class	Nature of Invariants
I	A	$C_1 > 0$, $C_2 > 0$
II	B	$C_1 > C_2 = 0$; $\boldsymbol{J} \cdot \boldsymbol{P}$ = some number
III	C	$C_1 = C_2 = 0$; $G^2 > 0$; $\boldsymbol{J} \cdot \boldsymbol{G}$ and H are numbers, as is G^2
IV	D	$C_1 = C_2 = 0$; $J^2 > 0$; J^2 and H are both numbers

We now add a few comments on the relationship between classes and types of realizations of $\mathcal{G}$. For irreducible realizations there is no problem, because the class determines the type, and conversely. (The notion of type is, in any case, defined only for irreducible realizations.) For a general realization, however, the situation is more subtle. Take, for example, a realization of class A. It generally possesses several nontrivial invariants; included among these are C_1 and C_2. We know that the phase space "splits up" into many (continuously many) invariant hypersurfaces of lower dimension than the phase space; each hypersurface is specified by assigning an acceptable value to each of the independent invariants. It may happen that C_2 vanishes on one of these hypersurfaces. Assuming that $C_1 > 0$ on this hypersurface, it is clear that as far as the action of $\mathcal{G}$ on the points of this hypersurface is concerned, $\boldsymbol{J}\cdot\boldsymbol{P}$ becomes an additional invariant. Similarly, there may be an invariant hypersurface on which C_1, and therefore also C_2, vanishes identically. An invariant hypersurface that cannot be "split up" into lower dimensional invariant hypersurfaces offers us, intuitively, the basis for an irreducible realization of $\mathcal{G}$ "embedded" in the original realization of class A. Thus we see that a general realization of class A may be said to "contain" several (in fact continuously many) irreducible realizations, and among these all types could turn up. Admittedly, we have left these statements in a rather imprecise form; we shall not bother to improve on them here.

Let us now construct examples of canonical realizations of $\mathcal{G}$ generated by true PB realizations of its Lie algebra. Because we have already discussed realizations of $R(3)$ and $E(3)$ in the preceding two chapters, there is no point in discussing now realizations of $\mathcal{G}$ belonging to classes C, D, and E. Only realizations in classes A and B involve something new. We can synthesize a realization of class A, in terms of q's and p's, in this way. To construct the two independent vectors P_j and G_j, with all the PB's among them vanishing and with nontrivial expressions for C_1 and C_2, it is natural to work with two Cartesian vectors $q_j^{(1)}$ and $q_j^{(2)}$ and their canonical conjugates $p_j^{(1)}$ and $p_j^{(2)}$. That is, we have a twelve-dimensional phase space with six q's and six p's, each varying from $-\infty$ to $+\infty$, obeying the basic PB relations:

$$\{q_j^{(\alpha)},q_k^{(\beta)}\}=\{p_j^{(\alpha)},p_k^{(\beta)}\}=0,\qquad \{q_j^{(\alpha)},p_k^{(\beta)}\}=\delta_{jk}\delta_{\alpha\beta}\,;$$
$$j,k=1,2,3;\qquad \alpha,\beta=1,2\,. \tag{19.14}$$

We can then take G_j to be $q_j^{(2)}$, say, and P_j to be $p_j^{(1)}$; if the J_j are taken to be the sums of the "single-particle-like" contributions from $\boldsymbol{q}^{(1)},\boldsymbol{p}^{(1)}$ and $\boldsymbol{q}^{(2)},\boldsymbol{p}^{(2)}$, then all the PB relations among J_j, G_j and P_j will be obeyed. Now H must be a scalar under rotations, have vanishing bracket with $P_j=p_j^{(1)}$, and obey $\{G_j,H\}=P_j$. The simplest choice for H is $\boldsymbol{p}^{(1)}\cdot\boldsymbol{p}^{(2)}$, leading to the complete set:

$$\begin{aligned}\boldsymbol{J}&=\boldsymbol{q}^{(1)}\times\boldsymbol{p}^{(1)}+\boldsymbol{q}^{(2)}\times\boldsymbol{p}^{(2)}\,,\\ \boldsymbol{G}&=\boldsymbol{q}^{(2)};\qquad \boldsymbol{P}=\boldsymbol{p}^{(1)};\qquad H=\boldsymbol{p}^{(1)}\cdot\boldsymbol{p}^{(2)}\,.\end{aligned} \tag{19.15}$$

Of course, H could be modified by adding any neutral element, or any invariant, to it. The Casimir invariants are both nontrivial:

$$C_1=\boldsymbol{p}^{(1)}\cdot\boldsymbol{p}^{(1)},\ C_2=(\boldsymbol{q}^{(2)}\times\boldsymbol{p}^{(1)})^2\,. \tag{19.16}$$

It turns out that the realization Eq. (19.15) possesses no nontrivial invariants that are independent of C_1 and C_2. This is made clear by working out the action of $\mathcal{G}$ on the phase space. We express the results in Table 19.4, as in Table 19.1, except that the entries in the first row are now the phase-space coordinates.

Table 19.4: ***Structure of Canonical Class A Realization***

	$q_j^{(1)}$	$q_j^{(2)}$	$p_j^{(1)}$	$p_j^{(2)}$
$\exp(\widetilde{-\boldsymbol{\alpha}\cdot\boldsymbol{J}})$	$A_{jk}(\boldsymbol{\alpha})q_k^{(1)}$	$A_{jk}(\boldsymbol{\alpha})q_k^{(2)}$	$A_{jk}(\boldsymbol{\alpha})p_k^{(1)}$	$A_{jk}(\boldsymbol{\alpha})p_k^{(2)}$
$\exp(\widetilde{-\boldsymbol{v}\cdot\boldsymbol{G}})$	$q_j^{(1)}$	$q_j^{(2)}$	$p_j^{(1)}$	$p_j^{(2)} - v_j$
$\exp(\widetilde{-bH})$	$q_j^{(1)} + bp_j^{(2)}$	$q_j^{(2)} + bp_j^{(1)}$	$p_j^{(1)}$	$p_j^{(2)}$
$\exp(\widetilde{-\boldsymbol{a}\cdot\boldsymbol{P}})$	$q_j^{(1)} + a_j$	$q_j^{(2)}$	$p_j^{(1)}$	$p_j^{(2)}$

By examining these transformations one can satisfy oneself that in general, given any two points in phase space at which C_1 and C_2 have the same values, a suitable element of $\mathcal{G}$ can be found whose representative canonical transformation takes one point into the other.

In the same spirit, that is, using a set of q's and p's as the basic phase-space variables, we can build up a canonical realization of $\mathcal{G}$ belonging to class B. Now the vector $\boldsymbol{G}$ becomes proportional to the vector $\boldsymbol{P}$. This factor of proportionality is clearly a scalar under rotations; all possible values from $-\infty$ to $+\infty$ must be accessible to it, as is clear from the effect of the canonical transformation $e^{\widetilde{-bH}}$ on $\boldsymbol{G}$ (see page 367). Accordingly, we consider an eight-dimensional phase space with four q's and four p's, each varying from $-\infty$ to $+\infty$; three of the q's, written q_j, form a vector under rotations, as do their canonical conjugates p_j, whereas the remaining variables, written q_0 and p_0, are both scalars. The only nonvanishing basic PB's are

$$\{q_j, p_k\} = \delta_{jk}, \ \{q_0, p_0\} = 1 . \tag{19.17}$$

The generators of $\mathcal{G}$ may be chosen to be

$$\boldsymbol{J} = \boldsymbol{q} \times \boldsymbol{p}, \qquad \boldsymbol{G} = q_0\boldsymbol{p}, \qquad \boldsymbol{P} = \boldsymbol{p}, \qquad H = p_0 \ , \tag{19.18}$$

and one can easily verify that Eqs. (19.3), (19.6) and (19.13) are obeyed. The Casimir invariant $C_1 = \boldsymbol{p}\cdot\boldsymbol{p}$ is nontrivial, whereas C_2=0. For this particular realization of class B, the invariant $\boldsymbol{J}\cdot\boldsymbol{P}$ also happens to vanish. The action of $\mathcal{G}$ on the phase space is shown in Table 19.5. On examining the entries there, one sees immediately that $C_1 = \boldsymbol{p}\cdot\boldsymbol{p}$ is the only nontrivial invariant of this class-B realization: given any two points in phase space at which C_1 has the same values, we can find an element of $\mathcal{G}$ whose representative canonical transformation maps one point into the other.

Table: 19.5: *Structure of Canonical Class B Realization*

	q_j	p_j	q_0	p_0
$\exp(-\boldsymbol{\alpha}\cdot\boldsymbol{J})$	$A_{jk}(\boldsymbol{\alpha})q_k$	$A_{jk}(\boldsymbol{\alpha})p_k$	q_0	p_0
$\exp(-\boldsymbol{v}\cdot\boldsymbol{G})$	$q_j+q_0v_j$	p_j	q_0	$p_0-\boldsymbol{v}\cdot\boldsymbol{p}$
$\exp(-bH)$	q_j	p_j	q_0+b	p_0
$\exp(-\boldsymbol{a}\cdot\boldsymbol{P})$	q_j+a_j	p_j	q_0	p_0

The inclusion of spin in these realizations is a straightforward matter. We have the additional variables $S_j, j=1,2,3$ whose basic PB's correspond to the Lie relations of the $R(3)$ Lie algebra; the brackets between the S_j, and the q's and p's vanish, and the extended phase space is defined in the obvious way. (The "kinematic" invariant S^2 may be assigned a numerical value.) Then, the suitably generalized realization of class A is given by the choice:

$$\boldsymbol{J}=\boldsymbol{q}^{(1)}\times\boldsymbol{p}^{(1)}+\boldsymbol{q}^{(2)}\times\boldsymbol{p}^{(2)}+\boldsymbol{S}\,,\qquad \boldsymbol{G}=\boldsymbol{q}^{(2)}\,;$$
$$\boldsymbol{P}=\boldsymbol{p}^{(1)},\ H=\boldsymbol{p}^{(1)}\cdot\boldsymbol{p}^{(2)}\,. \tag{19.19}$$

The action of $\mathcal{G}$ on the extended phase space is given in Table 19.4, together with the statement that under the rotations generated by J_j, the S_j transform as a vector. Of course, S_j does not change under the transformations generated by $\boldsymbol{G},\boldsymbol{P}$, and H. This realization of $\mathcal{G}$ belonging to class A is simply the direct product of the one given by the generators Eq. (19.15) and the non-faithful class D type-IV realization with the generators $\boldsymbol{J}\to\boldsymbol{S},\boldsymbol{G}\to 0$, $\boldsymbol{P}\to 0$, $H\to 0$. We see that the sums of the generators of these two realizations appear in Eq. (19.19). With the inclusion of spin, we have new nontrivial invariants besides the Casimir invariants C_1 and C_2; these are easily constructed from the knowledge that $\boldsymbol{P}=\boldsymbol{p}^{(1)}$ and $\boldsymbol{G}\times\boldsymbol{P}=\boldsymbol{q}^{(2)}\times\boldsymbol{p}^{(1)}$ are, like $\boldsymbol{S}$, invariant under the transformations generated by $\boldsymbol{G},\boldsymbol{P}$, and H, and behave as vectors under rotations, and thus we may choose them to be $\boldsymbol{p}^{(1)}\cdot\boldsymbol{S}$, $\boldsymbol{p}^{(1)}\times\boldsymbol{q}^{(2)}\cdot\boldsymbol{S}$, and $\boldsymbol{p}^{(1)}\times(\boldsymbol{p}^{(1)}\times\boldsymbol{q}^{(2)})\cdot\boldsymbol{S}$. These invariants cannot be written as functions of the generators alone. They are not all "new" because there is an algebraic relation among them, C_1, C_2, and the neutral element S^2:

$$\frac{(\boldsymbol{p}^{(1)}\cdot\boldsymbol{S})^2}{C_1}+\frac{(\boldsymbol{p}^{(1)}\times\boldsymbol{q}^{(2)}\cdot\boldsymbol{S})^2}{C_2}+\frac{(\boldsymbol{p}^{(1)}\times(\boldsymbol{p}^{(1)}\times\boldsymbol{q}^{(2)})\cdot\boldsymbol{S})^2}{C_1C_2}=S^2\,. \tag{19.20}$$

Insight into the nature of these invariants is gained by working with the quantities whose squares appear in Eq. (19.20):

$$\alpha=\frac{\boldsymbol{p}^{(1)}\cdot\boldsymbol{S}}{\sqrt{C_1}},\qquad \beta=\frac{\boldsymbol{p}^{(1)}\times\boldsymbol{q}^{(2)}\cdot\boldsymbol{S}}{\sqrt{C_2}},\qquad \gamma=\frac{\boldsymbol{p}^{(1)}\times(\boldsymbol{p}^{(1)}\times\boldsymbol{q}^{(2)})\cdot\boldsymbol{S}}{\sqrt{C_1C_2}}\,. \tag{19.21}$$

These are, of course, also invariants, being functions of invariants. They obey the relations of the $R(3)$ Lie algebra:

$$\{\alpha,\beta\}=\gamma,\ \{\beta,\gamma\}=\alpha,\ \{\gamma,\alpha\}=\beta\,. \tag{19.22}$$

What this shows is the following: in the terminology introduced in Chapter 17 (page 319), the group $\mathcal{G}_C^{(I)}$ of canonical transformations that commute with those arising in the realization of $\mathcal{G}$ and are generated by the invariants, is a non-Abelian one. And at least among the infinitesimal elements of $\mathcal{G}_C^{(I)}$ we have a family of transformations generated by α, β and γ and having the $R(3)$ group structure.

An analogous inclusion of spin in the class B realization of Eq. (19.18) leads to another class B realization:

$$\boldsymbol{J} = \boldsymbol{q} \times \boldsymbol{p} + \boldsymbol{S};\ \boldsymbol{G} = q_0 \boldsymbol{p}; \qquad \boldsymbol{P} = \boldsymbol{p}; \qquad \boldsymbol{H} = p_0 . \tag{19.23}$$

The direct-product structure is again evident. Now, both the invariants generally known to exist in class B are nontrivial: $C_1 = \boldsymbol{p} \cdot \boldsymbol{p}$ and $\boldsymbol{J} \cdot \boldsymbol{P} = \boldsymbol{S} \cdot \boldsymbol{p}$. Apart from these there are no independent invariants. Again, $\boldsymbol{S}$ is unchanged by the canonical transformations generated by $\boldsymbol{G}, \boldsymbol{P}$ and H, and is a vector under rotations.

The examples discussed up to now have all been reducible realizations of $\mathcal{G}$. The simplest way to generate irreducible realizations is to convert the Lie algebra relations of $\mathcal{G}$ into the basic GPB's among a set of classical real variables equal in number to the number of nonvanishing generators of $\mathcal{G}$; this latter number is determined by the type of irreducible realization one is interested in. Of course we set each Casimir or other invariant equal to a number, which may be zero in some cases (depending on the type of realization); this procedure clearly is not in conflict with the values assigned to the basic GPB's. All this is quite analogous to the theory of pure-spin systems encountered in Chapter 17 or of pure-Euclidean systems in Chapter 18; the details are left to the reader.

All these canonical realizations of $\mathcal{G}$ that are generated from true PB realizations of its Lie algebra are "unphysical" and hard to interpret from a mechanical point of view. The reason is very simple: as Table 19.1 shows, in all such realizations of $\mathcal{G}$ the dynamical variables P_j do not change under the canonical transformations generated by the G_j. Now these canonical transformations cause just those changes in dynamical variables that should result from a transition from one inertial frame to another moving with a uniform velocity relative to the first one; if the P_j are to be interpreted as the total linear momenta of some material system, they *should* change under this change of inertial frame. We do note, however, that the momentum transfers in elastic collisions of two or more material particles have precisely this behavior of not changing under change of inertial frame except for rotations. To obtain canonical realizations of $\mathcal{G}$ that are "physical" in the sense of P_j changing when one changes to a moving frame, we must examine the PB realizations up to neutrals, that is, the $M \neq 0$ realizations, of the Lie algebra of $\mathcal{G}$. This we now proceed to do. We find that such realizations are appropriate to a classical Hamiltonian description of single particles with or without spin, many-particle systems with or without interactions, as well as field systems, as long as the relativity group is the Galilei group. We also find a novel way of interpreting the "unphysical" class A realization of $\mathcal{G}$ constructed in Eq. (19.15) as a direct product of two "physical" realizations of

$\mathcal{G}$, thus establishing contact with the remark above about momentum transfers in elastic collisions.

Realizations of G with $M \neq 0$: General Analysis

With the neutral element M different from zero, the basic PB relations among the generators of $\mathcal{G}$ are given in Eqs. (19.3), (19.6), and (19.10). For convenience, we collect them together here:

$$\begin{aligned} \{J_j, J_k\} &= \epsilon_{jkm} J_m, & \{J_j, G_k\} &= \epsilon_{jkm} G_m, & \{G_j, G_k\} &= 0\,; \\ \{J_j, P_k\} &= \epsilon_{jkm} P_m, & \{P_j, P_k\} &= 0\,; \\ \{J_j, H\} = \{P_j, H\} &= 0, & \{G_j, H\} &= P_j\,; \\ \{G_j, P_k\} &= \delta_{jk} M; & M &\neq 0\,. \end{aligned} \tag{19.24}$$

We first exhibit certain general features of all canonical realizations of $\mathcal{G}$ involving a nonzero neutral M, and later discuss examples of such realizations. The first thing we notice when $M \neq 0$ is that *none* of the generators J_j, G_j, P_j and H can vanish identically; for the quantities G_j and P_j this is evident from the last line in Eq. (19.24); the previous line then implies the same for H; the nonvanishing of G_j or of P_j requires the same of J_j. Thus, in contrast to the M=0 case where various subsets of generators *could* vanish leading to various classes of realizations of $\mathcal{G}$, *from this point of view* there is just one kind of $M \neq 0$ realization. (Of course, one will differentiate between various $M \neq 0$ realizations from *other* points of view!) Any dynamical system described by such a realization of $\mathcal{G}$ has a nontrivial set of linear momenta, a nontrivial energy, nontrivial angular momenta, as well as nontrivial mechanical moments G_j, characterizing it. We soon see from examples that, in general, M may be interpreted as the *total mass* and the $G_j(\omega)$ represent the Cartesian components of the position vector of the center of mass of the relevant dynamical system (multiplied by the mass M).

Analogously to Table 19.1, let us compute the adjoint action corresponding to the basic PB's Eq. (19.24), namely the effects of the canonical transformations realizing $\mathcal{G}$ on the generators themselves. We obtain easily the set of results in Table 19.6. We can put this table to use in several different ways. The Hamiltonian H being the generator of time translations, the third line in the table giving the action of the canonical transformation $\exp(-bH)$ tells us that J_j, P_j and H are always constants of motion for any Galilean invariant Hamiltonian system.

Table 19.6: ***Transformations of Galilean Generators for*** $M \neq 0$.

	J_j	G_j	P_j	H
$\exp(-\boldsymbol{\alpha}\cdot\boldsymbol{J})$	$A_{jk}(\boldsymbol{\alpha})J_k$	$A_{jk}(\boldsymbol{\alpha})G_k$	$A_{jk}(\boldsymbol{\alpha})P_k$	H
$\exp(-\boldsymbol{v}\cdot\boldsymbol{G})$	$J_j - (\boldsymbol{G}\times\boldsymbol{v})_j$	G_j	$P_j - Mv_j$	$H - \boldsymbol{P}\cdot\boldsymbol{v} + \frac{1}{2}Mv^2$
$\exp(-bH)$	J_j	$G_j + bP_j$	P_j	H
$\exp(-\boldsymbol{a}\cdot\boldsymbol{P})$	$J_j - (\boldsymbol{P}\times\boldsymbol{a})_j$	$G_j + Ma_j$	P_j	H

This is not something new to us; these conservation laws are just those of angular momentum, linear momentum, and energy, respectively, and they are consequences of the invariance of the description of the physical system under rotations, translations in space, and translations in time, respectively. Conversely, under the canonical transformations generated by J_j, P_j, and H, corresponding to these changes of frame, H itself is unchanged; this is clear from the last column of the table. From the way in which G_j changes under the canonical transformation $\exp(\widetilde{-bH})$, we see that *the three-vector* $\boldsymbol{G} \times \boldsymbol{P}$ *is also a constant of motion.* The interpretation of this vector emerges shortly. We consider next the forms of the Casimir invariants for such realizations of $\mathcal{G}$. The Casimir invariant $C_1 = P^2$ of the M=0 realizations is no longer an invariant, and we must see if we can generalize it to get one. The only transformations that alter P^2 are those generated by G_j; under $\exp(\widetilde{-\boldsymbol{v} \cdot \boldsymbol{G}})$, P^2 goes into $P^2 - 2M\boldsymbol{P} \cdot \boldsymbol{v} + M^2 v^2$. But we notice that these are also the only transformations under which H is altered, and inspection shows that in the combination $P^2 - 2MH$ all the $\boldsymbol{v}$-dependent terms cancel out. This is a simple generalization of the previous Casimir invariant C_1 and provides us with the first of two Casimir invariants for the present situation:

$$C_1' = P^2 - 2MH \,. \tag{19.25}$$

We have chosen to express C_1' in such a form that in the limit M=0, it reduces to C_1. Notice that we have here a nice illustration of the remarks made on page 321. Whereas C_1 is a homogeneous function of the generators of $\mathcal{G}$ of degree two, C_1' is also such a function provided we count the neutral element M as contributing to the degree of homogeneity of the term - $2MH$, just like H, making this a term of degree two. Eq. (19.25) also shows that for any Galilean invariant system, with $M \neq 0$ of course, we can relate total energy, total momentum, and mass in this way:

$$H = \frac{P^2}{2M} - \frac{C_1'}{2M} = \frac{P^2}{2M} + \text{Galilean invariant terms.} \tag{19.26}$$

The term $P^2/2M$ is just like the Newtonian expression for the kinetic energy of a single particle with mass M and momentum $\boldsymbol{P}$; but of course M and $\boldsymbol{P}$ stand for the total mass and total momentum of the physical system and do not have any "single particle" interpretation in general. The Galilean-invariant terms appearing in Eq. (19.26) generally describe the internal interactions among the parts of the system; they are nontrivial phase-space functions in general, and the important point is that they are simply related to the Casimir invariant C_1'.

We next generalize the Casimir invariant $C_2 = (\boldsymbol{G} \times \boldsymbol{P})^2$ of the M=0 realizations to the second Casimir invariant C_2' for the $M \neq 0$ case. Whereas previously the vector $\boldsymbol{G} \times \boldsymbol{P}$ was an invariant under all transformations of $\mathcal{G}$ save rotations, now it is altered under a space displacement as well:

$$\exp(\widetilde{-\boldsymbol{a} \cdot \boldsymbol{P}})\boldsymbol{G} \times \boldsymbol{P} = \boldsymbol{G} \times \boldsymbol{P} + M\boldsymbol{a} \times \boldsymbol{P} \,. \tag{19.27}$$

However, under this same transformation $\boldsymbol{J}$ also changes, going into $\boldsymbol{J}+\boldsymbol{a}\times\boldsymbol{P}$; thus the combination $M\boldsymbol{J}-\boldsymbol{G}\times\boldsymbol{P}$, which is certainly a vector, is unchanged under a space translation. It is also unaltered by the canonical transformation $\exp(\widetilde{-bH})$, being a function of the constants of motion $\boldsymbol{J}, \boldsymbol{G}\times\boldsymbol{P}$ (and M). But what about velocity transformations generated by $\boldsymbol{G}$? We find that here, too, it is invariant, because of a cancellation of terms:

$$\begin{aligned} \exp(\widetilde{-\boldsymbol{v}\cdot\boldsymbol{G}})(M\boldsymbol{J}-\boldsymbol{G}\times\boldsymbol{P}) &= M(\boldsymbol{J}-\boldsymbol{G}\times\boldsymbol{v})-\boldsymbol{G}\times(\boldsymbol{P}-M\boldsymbol{v}) \\ &= M\boldsymbol{J}-\boldsymbol{G}\times\boldsymbol{P}. \end{aligned} \tag{19.28}$$

This, then, is the correct generalization of $-\boldsymbol{G}\times\boldsymbol{P}$ to the $M\neq 0$ case, and its magnitude gives us the second Casimir invariant:

$$C_2' = (M\boldsymbol{J}-\boldsymbol{G}\times\boldsymbol{P})^2. \tag{19.29}$$

The C_1' and C_2' are, then, a complete set of independent Casimir invariants, and in the limit M=0 they reduce to C_1 and C_2. Note again that with the power of M being counted, C_2' is an invariant homogeneous of degree four in the generators of $\mathcal{G}$.

Let us examine now the physical interpretation of the two quantities $\boldsymbol{G}$ and $\boldsymbol{G}\times\boldsymbol{P}$. As for $\boldsymbol{G}$, we see from Table 19.6 that it transforms as a vector under rotations; with respect to displacements of the spatial coordinate system it is also displaced by the amount $M\boldsymbol{a}$. That is to say, the vector $\boldsymbol{G}/M$ behaves both under rotations and space translations in just the way we would expect a vector characterizing *position* to behave. This is reinforced by the fact that under the canonical transformation $\exp(\widetilde{-\boldsymbol{v}\cdot\boldsymbol{G}})$ corresponding to the transition to a uniformly moving inertial frame, $\boldsymbol{G}/M$ does not change. [Recall from the discussion of the Heisenberg picture in Chapter 16 that the canonical transformation $\exp(\widetilde{-\boldsymbol{v}\cdot\boldsymbol{G}})$ takes a dynamical variable at time zero in an inertial frame S to the same dynamical variable at time zero in the inertial frame $S'=\exp(\boldsymbol{v}\cdot\boldsymbol{g})S$ which moves with constant velocity $\boldsymbol{v}$ relative to S, whereas the time variables used in S' and S are the same.] What is $\boldsymbol{G}/M$ the position of? This is answered by the way $\boldsymbol{G}$ changes in time. If in an inertial frame S the vector $\boldsymbol{G}/M$ represents the dynamical variable "position of *" at time zero, then at time t in S the "position of *" is represented by

$$e^{\widetilde{-tH}}\boldsymbol{G}/M = \boldsymbol{G}/M + t\boldsymbol{P}/M. \tag{19.30}$$

Because $\boldsymbol{P}$ is known to be a constant of motion, being the total linear momentum, and the mass M is an unchanging neutral element, it is clear that for each state of motion, the point * moves in a straight line with a uniform velocity $\boldsymbol{P}/M$. We therefore may identify the point * with the *center of mass* of the physical system in general, and thus the vector $\boldsymbol{G}/M$ with *the position vector of the center of mass* of the system. In particular cases we may find that an alternative interpretation for $\boldsymbol{G}$ exists, but in all cases the result embodied in Eq. (19.30) is valid: for each state of motion of the Galilean invariant system, the time dependence of the dynamical variable $\boldsymbol{G}/M$ is linear. Except for $\boldsymbol{G}$, then, all the generators of $\mathcal{G}$ are constants of motion.

Let us now write $\boldsymbol{Q}$ for the center of mass position vector $\boldsymbol{G}/M$. Then the last line of Eq. (19.24) takes the form:

$$\{Q_j, P_k\} = \delta_{jk} \,. \tag{19.31}$$

The PB's among the P's by themselves, or among the Q's by themselves, vanish. Thus the PB relationships holding among the Q's and P's are just like the basic ones among the q's and p's of a single particle. However, Q_j and P_j do not have a single-particle meaning in general. The basic point is that in a canonical realization of $\mathcal{G}$ with nonzero neutral element M, the bracket relations of the Lie algebra of $\mathcal{G}$ guarantee the existence of a three-dimensional vector $\boldsymbol{Q}$ whose components are canonical conjugates to the three components of the total linear momentum $\boldsymbol{P}$ of the system. In turn this tells us something very general about the set of values accessible to the dynamical variables $P_j(\omega)$. Because M is neutral, the variables $Q_j(\omega)$ act as generators of *regular* canonical transformations, and because for any three-vector $\boldsymbol{a}$

$$\exp(\widetilde{\boldsymbol{a} \cdot \boldsymbol{Q}})\boldsymbol{P} = \boldsymbol{P} + \boldsymbol{a} \,, \tag{19.32}$$

the accessible values for $\boldsymbol{P}$ must be all possible vectors in three-dimensional space. (Of course, this fact is already evident from the behavior of $\boldsymbol{P}$ under the regular canonical transformation $\exp(\widetilde{-\boldsymbol{v} \cdot \boldsymbol{G}})$ as written on page 375.) This property of $\boldsymbol{P}$ is common to all canonical realizations of $\mathcal{G}$ (with $M \neq 0$). The point of this statement is that an analogous property does not hold in the case of realizations of the Poincaré group.

It is clear from Eq. (19.31) and the vanishing of $\{Q_j, Q_k\}$ and $\{P_j, P_k\}$ that the three quantities

$$L_j = \epsilon_{jkm} Q_k P_m \tag{19.33}$$

have PB's with each other corresponding to the $R(3)$ Lie algebra:

$$\{L_j, L_k\} = \epsilon_{jkm} L_m \,. \tag{19.34}$$

These L_j are the components of the angular momentum of the center of mass of the physical system. Because $\boldsymbol{Q}$ and $\boldsymbol{P}$ transform as vectors under the realization of $R(3)$ generated by $\boldsymbol{J}$, the same is true of $\boldsymbol{L}$, and thus we have:

$$\{J_j, L_k\} = \epsilon_{jkm} L_m \,. \tag{19.35}$$

However, our experience with the angular momentum variables for a single particle tells us that between L_j and Q_k, P_k we must have:

$$\{L_j, Q_k\} = \epsilon_{jkm} Q_m, \ \{L_j, P_k\} = \epsilon_{jkm} P_m \,. \tag{19.36}$$

(See Eq. (17.8).) In addition, H turns out to have vanishing brackets with L_j:

$$\{L_j, H\} = \frac{1}{M}\epsilon_{jkm}\{G_k P_m, H\} = \frac{1}{M}\epsilon_{jkm}P_k P_m = 0\,. \tag{19.37}$$

The canonical transformations generated by the L_j, although they do yield a realization of $R(3)$, *do not* correspond to rotations of the spatial coordinate system; *those* transformations by definition are generated by the J_j. What has happened, however, is that by virtue of Eqs. (19.34), (19.36), and (19.37), if in all the PB relations Eq. (19.24) we replace J_j by L_j, these relations remain valid. In other words, given the canonical realization of $\mathcal{G}$ generated by the set of quantities $(J_j, G_j, P_j, H; M)$, we are able to set up another canonical realization of $\mathcal{G}$ with the generators $(L_j, G_j, P_j, H; M)$. Both realizations act on the same phase space, and given the first one, we are able to build the generators of the second. The first is the physical realization of $\mathcal{G}$ reflecting changes of inertial frame, the second is a "mathematical" realization. Except for the $R(3)$ generators, all the other generators are the same in the two realizations; in particular, the neutral element M is the same. What can be said of the differences $(J_j - L_j)$? Writing I_j for these, we know from a subtraction of Eq. (19.34) from Eq. (19.35) that

$$\{L_k, I_j\} = 0\,. \tag{19.38}$$

At the same time, among themselves the I_j obey the $R(3)$ Lie algebra bracket rules:

$$\{I_j, I_k\} = \{J_j, I_k\} - \{L_j, I_k\} = \{J_j, I_k\} = \epsilon_{jkm}I_m\,. \tag{19.39}$$

In the first step, we used Eq. (19.38), and in the next we used the fact that both $\boldsymbol{J}$ and $\boldsymbol{L}$ transform as vectors under the canonical transformations generated by $\boldsymbol{J}$ so that $\boldsymbol{I} = \boldsymbol{J} - \boldsymbol{L}$ has the same behavior. Now each of the quantities $\boldsymbol{G}, \boldsymbol{P}$, and H has the same transformation properties, both with respect to $\boldsymbol{J}$ and with respect to $\boldsymbol{L}$; therefore, they all have vanishing brackets with $\boldsymbol{I}$:

$$\{I_j, G_k\} = \{I_j, P_k\} = 0,\ \{I_j, H\} = 0\,. \tag{19.40}$$

Consequently, *all* the generators $(L_j, G_j, P_j, H; M)$ of the "mathematical" realization of $\mathcal{G}$ have vanishing brackets with the I_j; among themselves the I_j obey the bracket relations of the $R(3)$ Lie algebra. We can set up a non-faithful realization of $\mathcal{G}$ with vanishing neutral element, by the assignment $l_j \to I_j, g_j \to 0, d_j \to 0, h \to 0$. This is a realization of class D. The canonical transformations arising in the "mathematical" realization of $\mathcal{G}$ commute with those arising in the class D "true" realization of $\mathcal{G}$. But the generators $(J_j, G_j, P_j, H; M)$ of the original physical realization of $\mathcal{G}$ are simply the sums of the corresponding generators $(L_j, G_j, P_j, H; M)$ and (I_j,0,0,0;0) in the mathematical $M \neq 0$ realization and the class D "true" realization. Consequently, according to the discussion given on pages 355-357, every canonical realization of $\mathcal{G}$ with nonvanishing neutral element appears in a well-defined way as the *direct product* of a mathematical realization in which the neutral element is the same as in the given realization, and a "true" class D realization of $\mathcal{G}$. Both these latter realizations operate on the same phase space as the originally given one, and their direct product can be taken because they commute. Let us make these statements explicit. Denote the (physical) canonical transformation corresponding to the change of frame $S \to S' = g^{-1}S$, where g is any element in $\mathcal{G}$, by $R(g; \boldsymbol{J}, \boldsymbol{G}, \boldsymbol{P}, H; M)$. This notation makes clear the fact that R has a definite dependence on g and is constructed with the help of the generators $\boldsymbol{J}, \boldsymbol{G}, \boldsymbol{P}, H$, and that the neutral element in the realization is M. Then the statement is:

$$R(g; \boldsymbol{J}, \boldsymbol{G}, \boldsymbol{P}, H; M) = R(g; \boldsymbol{L}, \boldsymbol{G}, \boldsymbol{P}, H; M) R(g; \boldsymbol{I}, \mathbf{0}, \mathbf{0}, 0; 0) \tag{19.41}$$

where for any $g, g' \in \mathcal{G}$ we have the two realization properties

$$\begin{aligned} R(g'; \boldsymbol{L}, \boldsymbol{G}, \boldsymbol{P}, H; M) R(g; \boldsymbol{L}, \boldsymbol{G}, \boldsymbol{P}, H; M) &= R(g'g; \boldsymbol{L}, \boldsymbol{G}, \boldsymbol{P}, H; M)\,, \\ R(g'; \boldsymbol{I}, \mathbf{0}, \mathbf{0}, 0; 0) R(g; \boldsymbol{I}, \mathbf{0}, \mathbf{0}, 0; 0) &= R(g'g; \boldsymbol{I}, \mathbf{0}, \mathbf{0}, 0; 0)\,. \end{aligned} \tag{19.42}$$

and the commuting property:

$$\begin{aligned} &R(g'; \boldsymbol{L}, \boldsymbol{G}, \boldsymbol{P}, H; M) R(g; \boldsymbol{I}, \mathbf{0}, \mathbf{0}, 0; 0) \\ &\qquad = R(g; \boldsymbol{I}, \mathbf{0}, \mathbf{0}, 0; 0) R(g'; \boldsymbol{L}, \boldsymbol{G}, \boldsymbol{P}, H; M)\,. \end{aligned} \tag{19.43}$$

The basic property of the mathematical realization is that in it the $R(3)$ generators L_j are known functions of the other generators G_j, P_j, and the mass M. Whereas the L_j have the interpretation of being the components of the angular momentum of the center of mass, (in other words, the angular momentum of a fictitious "particle" of mass M located at and moving with the center of mass), the I_j can be called the "internal angular momentum" of the system. According to Eq. (19.40), this internal angular momentum is the same in any two inertial frames that are connected by a uniform velocity transformation or a spatial displacement, and in each inertial frame the components I_j are constants of motion. Thus the total angular momenta J_j which are constants of motion in any Galilean invariant theory can be expressed as sums of two parts, L_j and I_j, each of which is separately a constant of motion. The only changes of inertial frame that do cause changes in the I_j are those involving spatial rotations; the corresponding canonical transformations are generated by the J's, and under them the I_j transform as the components of a vector. If one looks for a simple reason why we have been able to split up the canonical transformations realizing $\mathcal{G}$ in this way and express the total angular momenta as sums of kinematically and dynamically independent parts, it is that except for those basic PB relations in Eq. (19.24) that correspond to the $R(3)$ subalgebra, *in no other basic bracket relations do the J_j appear on the right-hand side.* The situation is different in the case of the Poincaré group, where the Lie brackets $[k_j, k_l]$ are equal to $-\epsilon_{jlm} l_m$; in that case no analogous splitting up of the J's is, in general, possible. (For the basic Lie brackets for $\mathcal{P}$ see Eq. (15.132).)

The second Casimir invariant C_2' for $\mathcal{G}$ is given in Eq. (19.29). We now see that it is essentially the square of the magnitude of the internal angular momentum,

$$C_2' = M^2 I^2 \,, \tag{19.44}$$

and we see in a new light why C_2' is a Casimir invariant.

This concludes our general analysis of realizations of $\mathcal{G}$ with nonvanishing neutral element M. The only remark of a general nature still to be made is that from the point of view of physical interpretation, we would require the neutral element M to be a *positive quantity*; this is only to accord with our experience with the Newtonian description of point particles, for which mass is an intrinsically positive attribute. But as far as realizations of $\mathcal{G}$ are concerned, from the purely mathematical point of view there is no reason to exclude negative values of M.

Single-Particle Realizations of $\mathcal{G}$

The first example of a "physical" realization of $\mathcal{G}$ corresponds to a single, free, nonrelativistic particle whose only degrees of freedom are position and momentum. The phase space for such a system is six dimensional, there being three q's and three conjugate p's, each running from $-\infty$ to $+\infty$ and obeying the usual basic PB relations. A realization of the Lie algebra of the Euclidean group $E(3)$ in terms of these variables is presented in Eq. (18.15). We now seek additional functions of q_j and p_j to provide the generators for velocity transformations, G_j, and time translations, H. The construction is simple and involves the parameter m:

$$\boldsymbol{J} = \boldsymbol{q} \times \boldsymbol{p},\ \boldsymbol{G} = m\boldsymbol{q},\ \boldsymbol{P} = \boldsymbol{p},\ H = \frac{p^2}{2m} + U,\ M = m\,. \tag{19.45}$$

Here U is a neutral element: we have included it explicitly in H because we know from the general discussion on pages 364 and 365 that in a canonical realization of $\mathcal{G}$ the Hamiltonian is always undetermined to this extent. We can easily verify that the PB relations Eq. (19.24) hold. The form we have chosen for $\boldsymbol{G}$ results from the general observation that $\boldsymbol{G}/M$ is the center of mass position vector and the fact that for a single massive particle the position of the particle is the same as of the center of mass. The form for the Hamiltonian is essentially determined by Eq. (19.26).

This "single-particle" realization of $\mathcal{G}$ is irreducible. The two Casimir invariants C_1' and C_2' both reduce to numbers:

$$\begin{aligned} C_1' &= P^2 - 2MH = p^2 - 2m\left(\frac{p^2}{2m} + U\right) = -2mU\,, \\ C_2' &= (M\boldsymbol{J} - \boldsymbol{G} \times \boldsymbol{P})^2 = (m\boldsymbol{q} \times \boldsymbol{p} - m\boldsymbol{q} \times \boldsymbol{p})^2 = 0\,. \end{aligned} \tag{19.46}$$

Thus the "internal angular momentum" I_j vanishes in this case. The action of $\mathcal{G}$ on the phase space coordinates $\boldsymbol{q}$ and $\boldsymbol{p}$ is already determined by the entries in Table 19.6 because $\boldsymbol{p}$ coincides with $\boldsymbol{P}$ and $\boldsymbol{q}$ with $\boldsymbol{G}/M$; we have:

$$\exp(\widetilde{-\boldsymbol{\alpha}\cdot\boldsymbol{J}})q_j = A_{jk}(\boldsymbol{\alpha})q_k\,,\qquad \exp(\widetilde{-\boldsymbol{\alpha}\cdot\boldsymbol{J}})p_j = A_{jk}(\boldsymbol{\alpha})p_k\,; \tag{19.47a}$$

$$\exp(\widetilde{-\boldsymbol{v}\cdot\boldsymbol{G}})q_j = q_j\,,\qquad \exp(\widetilde{-\boldsymbol{v}\cdot\boldsymbol{G}})p_j = p_j - mv_j\,; \tag{19.47b}$$

$$\exp(\widetilde{-bH})q_j = q_j + b\frac{p_j}{m}\,,\qquad \exp(\widetilde{-bH})p_j = p_j\,; \tag{19.47c}$$

$$\exp(\widetilde{-\boldsymbol{a}\cdot\boldsymbol{P}})q_j = q_j + a_j\,,\qquad \exp(\widetilde{-\boldsymbol{a}\cdot\boldsymbol{P}})p_j = p_j\,. \tag{19.47d}$$

These equations prove the irreducibility of the realization; given any two points in phase space, we can find an element of $\mathcal{G}$ whose representative canonical transformation maps the one into the other. The physical meaning of Eq. (19.47c) concerning the action of $e^{\widetilde{-bH}}$ is clear: these are the fundamental laws of Newton on the uniform rectilinear motion of a single particle not acted on by any force and observed from an inertial reference frame. Similarly Eqs. (19.47a) and (19.47d) exhibit the elementary facts concerning the behavior of position and momentum under the Euclidean group, and Eq. (19.47b) follows from the usual identification of momentum with mass times velocity. (That $\boldsymbol{q}$ must not change under the velocity transformation generated by $\boldsymbol{G}$ is explained on page 377.) Thus the kinematic as well as the dynamic laws for the description of a single, free, nonrelativistic point particle are recovered from an irreducible canonical realization of $\mathcal{G}$ with a nonzero neutral element (and vanishing internal angular momentum) We can now alter our point of view and say that *a free nonrelativistic point particle in classical mechanics is nothing but an irreducible canonical realization of $\mathcal{G}$ with a positive neutral element* m! If the internal angular momenta I_j vanish, then we have a structureless particle, a complete description of which can be given with the q's and p's alone. The advantage of adopting this point of view, in which we take the invariance group $\mathcal{G}$ as fundamental and identify a particular kind of irreducible realization of it with the notion of point particle that is basic to Newtonian mechanics, is that it gives an appealing basis for the use of three q's and three p's for the description of a particle and for the postulation of the basic PB relations among these variables. An exactly analogous situation obtains in nonrelativistic quantum mechanics as well.

The irreducible single-particle realization of $\mathcal{G}$ described in the two previous paragraphs is not the only type of irreducible realization of $\mathcal{G}$, because in it the internal angular momenta vanish. We obtain the most general irreducible canonical realization of $\mathcal{G}$ if we take the direct product of the preceding irreducible realization with a "true" class D realization of $\mathcal{G}$ provided by a pure-spin system:

$$l_j \to S_j,\; g_j \to 0,\; d_j \to 0,\; h \to 0,\; M \to 0\,. \tag{19.48}$$

The S_j obey the basic PB's corresponding to the $R(3)$ Lie algebra and have vanishing brackets with q_j and p_j. We fix S^2 to be any number (≥ 0). The full phase space is now eight dimensional, and the generators of $\mathcal{G}$ are:

$$\boldsymbol{J} = \boldsymbol{q}\times\boldsymbol{p} + \boldsymbol{S},\; \boldsymbol{G} = m\boldsymbol{q},\; \boldsymbol{P} = \boldsymbol{p},\; H = \frac{p^2}{2m} + U,\; M = m\,. \tag{19.49}$$

The interesting fact is that this realization of $\mathcal{G}$, obtained as the direct product of two irreducible realizations, is again irreducible. Both Casimir invariants C_1' and C_2' are (possibly) nonzero, and both are numbers:

$$C_1' = -2mU,\ C_2' = (M\boldsymbol{J} - \boldsymbol{G} \times \boldsymbol{P})^2 = m^2\boldsymbol{S}^2\,. \tag{19.50}$$

The irreducibility of this realization follows from Eq. (19.47) and the observation that the S_j transform as a vector under rotations but are otherwise unchanged by transformations of $\mathcal{G}$. We can say that such an irreducible realization of $\mathcal{G}$ is a classical nonrelativistic free particle with mass m, magnitude of internal angular momentum $\sqrt{S^2}$ and "internal energy" U. Indeed, the internal angular momenta I_j reduce in this case to the S_j, and in this case the variables S_j can be said to describe the "intrinsic spin" of the particle. Thus the q's and p's do not form a complete set of variables for a single particle in general, but the spin variables must be included. We have here a generalization of the Newtonian concept of a point particle, suggested by our viewing the group $\mathcal{G}$ as the basic object for the definition of a (non-relativistic) particle.

Can we prove that with the construction of Eq. (19.49) we have obtained all possible irreducible canonical realizations of $\mathcal{G}$ with a nonzero neutral element? Indeed we can, and fairly easily, too. In proving the irreducibility of some of the canonical realizations that we have constructed for some groups up to now, we made use of the sufficient condition for irreducibility that was stated on page 321: given any two points in phase space, if we can find an element of the group concerned whose representative canonical transformation maps one point into the other, the realization is irreducible. Now we develop an alternative condition on the generators of a canonical realization of a Lie group, which is necessary as well as sufficient for the realization to be irreducible, and thus is completely equivalent to the definition of irreducibility given previously on page 321. Let us first recall the previously stated definition of irreducibility. A canonical realization of a Lie group G, generated by a set of phase-space functions $X_j(\omega), j = 1, \ldots, n$, is irreducible if and only if

$$\{X_j(\omega), \phi(\omega)\} = 0,\ j = 1, \ldots, n, \Longrightarrow \{f(\omega), \phi(\omega)\} = 0, \text{all} f(\omega)\,. \tag{19.51}$$

This definition is applicable whether the $X_j(\omega)$ give a true PB realization or one up to neutrals of the Lie algebra of G, whether the PB is nonsingular or singular, ordinary or generalized. The foregoing criterion for irreducibility means that with a nonsingular bracket, any invariant of the realization must be just a number; with a singular bracket, it must be a neutral element having vanishing brackets with all phase-space functions We now prove that if Eq. (19.51) holds, then the generators $X_j(\omega)$ enjoy the following property: (1) in the case of a nonsingular bracket, a subset of the set of generators $X_j(\omega)$ can be found, having as many members as the dimension of phase space, such that the functions $X_j(\omega)$ in this subset form an independent set of phase space functions; (2) in the case of a singular bracket, a subset of the $X_j(\omega)$ can be found which, together with a complete independent set of neutral elements, forms a set of independent phase-space functions, the total number in this set being equal to the dimension of phase space. Thus irreducibility implies that, taken together with a complete set of neutral elements, there are enough independent variables among the $X_j(\omega)$ to form a coordinate system in phase space. (The converse of this statement is trivial to establish.) Because we know precisely what we wish to prove, it suffices to give the details for the general case of a singular bracket. We draw upon the discussion of singular GPB's given in Chapter 9 see pages 114 ff. In place of the variables z^μ used there, we now write ω^μ. Let the GPB be specified by the set of functions $\eta^{\mu\nu}(\omega)$ and let $\theta^a(\omega), a = 1, \ldots, m$, be a complete independent set of neutral elements. We can express Eq. (19.51) as a property of the partial differential operators $\tilde{X}_j(\omega)$:

$$\begin{aligned}\tilde{X}_j(\omega)\phi(\omega) = 0,\ j = 1, 2, \ldots n \Longrightarrow \phi = \phi(\theta^a(\omega))\,,\\ \tilde{X}_j(\omega) \equiv \eta^{\mu\nu}(\omega)\frac{\partial X_j(\omega)}{\partial\omega^\mu}\frac{\partial}{\partial\omega^\nu}\,.\end{aligned} \tag{19.52}$$

In other words, there are precisely m independent functions ϕ obeying the system of partial differential equations appearing in Eq. (19.52); we could take them to be the θ^a. On the other hand, this system of equations is already complete, because the commutator of $\tilde{X}_j$ and $\tilde{X}_k$ is a linear combination of the $\tilde{X}$'s, with the structure constants of G as coefficients. We can now apply the theorem concerning such systems of equations, quoted on page 115, to fix the number of independent operators $\tilde{X}_j(\omega)$. If the dimension of the phase space, that is, the number of ω's, is k, then because we know there are m independent solutions to Eq. (19.52), there must be precisely $(k - m)$ independent operators in the set of operators $\tilde{X}_j(\omega)$. It is now an easy matter for the reader to verify that the corresponding set of $(k - m)$ X's together with the $m\,\theta$'s, do give us a set of k independent phase-space functions, which is the desired result.

Let us now use these results in the case of the Galilei group $\mathcal{G}$ to show that the construction of Eq. (19.49) yields the most general irreducible realization of $\mathcal{G}$ with nonzero neutral M. Suppose an irreducible canonical realization of $\mathcal{G}$ is given. Both in the evaluation of GPB's and in the action of generalized canonical transformations, the neutral elements $\theta^a(\omega)$ will be unaffected, and thus we may imagine that they are assigned some numerical values; we need not speak of them any further. The Casimir invariants C_1' and C_2' are both numbers. Given the generators $\boldsymbol{J}, \boldsymbol{G}, \boldsymbol{P}, H$, and the mass M for the realization of $\mathcal{G}$, and following the general analysis contained on pages 375 ff, we can introduce the set of variables:

$$\boldsymbol{Q} = \frac{\boldsymbol{G}}{M},\ \boldsymbol{P},\ \boldsymbol{I} = \boldsymbol{J} - \boldsymbol{Q} \times \boldsymbol{P}\,. \tag{19.53}$$

We know that these variables obey the following bracket relations because of the structure of the Lie algebra of $\mathcal{G}$:

$$\begin{aligned}\{Q_j(\omega), Q_k(\omega)\} = \{P_j(\omega), P_k(\omega)\} = \{Q_j(\omega), I_k(\omega)\} = \{P_j(\omega), I_k(\omega)\} = 0\,;\\ \{Q_j(\omega), P_k(\omega)\} = \delta_{jk};\ \{I_j(\omega), I_k(\omega)\} = \epsilon_{jkm}I_m(\omega)\,.\end{aligned} \tag{19.54}$$

From these relations, or equally well from Table 19.6 and the properties of I_j, we know that the Q's, P's, and I's are capable of being independently varied subject only to the condition that the magnitude $|\boldsymbol{I}|$ is fixed at a value determined by C_2'. This is because we can exhibit three (generalized) canonical transformations, each of which alters only one of $\boldsymbol{Q}, \boldsymbol{P}$ and $\boldsymbol{I}$:

$$\begin{aligned} \exp(\widetilde{-\boldsymbol{a}\cdot\boldsymbol{P}}) &: \quad \boldsymbol{Q}\to\boldsymbol{Q}-\boldsymbol{a},\ \boldsymbol{P}\to\boldsymbol{P}, \boldsymbol{I}\to\boldsymbol{I}\,; \\ \exp(\widetilde{-\boldsymbol{v}\cdot\boldsymbol{G}}) &: \quad \boldsymbol{Q}\to\boldsymbol{Q},\ \boldsymbol{P}\to\boldsymbol{P}-M\boldsymbol{v},\ \boldsymbol{I}\to\boldsymbol{I}\,; \\ \exp(\widetilde{-\boldsymbol{\alpha}\cdot\boldsymbol{I}}) &: \quad \boldsymbol{Q}\to\boldsymbol{Q},\ \boldsymbol{P}\to\boldsymbol{P},\ I_j\to A_{jk}(\boldsymbol{\alpha})I_k\,. \end{aligned} \tag{19.55}$$

Therefore $\boldsymbol{Q}, \boldsymbol{P}$, and $\boldsymbol{I}$ must be independent phase-space functions except for the restriction that $\boldsymbol{I}$ always lies on a sphere of radius $\sqrt{C_2'/M^2}$; Q_j and P_j independently can assume all possible real values from $-\infty$ to $+\infty$. Let us now make use of the knowledge that because the given realization of $\mathcal{G}$ is irreducible, it should be possible to choose a subset of the generators that will all be independent and will be sufficient in number to be used as coordinates in phase space (the neutral elements θ^a are treated as being "frozen" at some permitted values). But then the Q_j, P_j and two of the I_j can always be chosen among the set of new coordinates, because they are independent variables and also are specified functions of the generators of $\mathcal{G}$. For the sake of symmetry, it is better to work with the nine quantities Q_j, P_j and I_j, j=1,2,3, bearing in mind the constraint on $|\boldsymbol{I}|$. Now our general analysis on pages 375 ff shows that these must already give us a sufficient number of variables to form a coordinate system in phase-space, because with C_1' and M being numbers, the remaining generators $\boldsymbol{J}$ and H are expressible as functions of $\boldsymbol{Q}, \boldsymbol{P}$ and $\boldsymbol{I}$, and thus cannot be independent of the latter:

$$\boldsymbol{J} = \boldsymbol{Q}\times\boldsymbol{P}+\boldsymbol{I},\ H = \frac{P^2}{2M} - \frac{C_1'}{2M}\,. \tag{19.56}$$

(cf Eq. (19.26) and the definition of $\boldsymbol{I}$). We see that with the correspondence $\boldsymbol{Q}\to\boldsymbol{q}, \boldsymbol{P}\to\boldsymbol{p}, M\to m, \boldsymbol{I}\to\boldsymbol{S}$, the most general irreducible canonical realization of $\mathcal{G}$ with nonzero M and nonzero C_2' has generators of just the form given in Eq. (19.49), the internal energy U being equal to $-C_1'/2M$, with the familiar bracket relations being valid because of Eq. (19.54). In case C_2' vanishes, there is no need to introduce the variables I_j at all (they vanish), and the irreducible realization of $\mathcal{G}$ reduces to the form Eq. (19.45) corresponding to a spinless, nonrelativistic, free particle. This completes the proof that the constructions of Eqs. (19.45) and (19.49) exhaust all the irreducible canonical realizations of $\mathcal{G}$ with nonzero neutral M.

Many-Particle Systems—More Realizations of $\mathcal{G}$

A canonical realization of $\mathcal{G}$ that describes two massive but free particles is obtained if we construct the direct product of two single-particle irreducible realizations of $\mathcal{G}$, corresponding to masses m_1 and m_2, say. (For simplicity we leave out spin.) As we explained in Chapter 18 (see pages 355 to 357), the generators of the direct product of two independent canonical realizations of a

Lie group G are the sums of the generators of the individual realizations. In particular, the neutral elements present in the PB relations in the individual realizations must also be added appropriately to give those present in the product realization. For the case of $\mathcal{G}$, this simply means that the "total mass" associated with a direct product of two realizations is the sum of the individual masses. This is merely the additivity law for masses familiar from Newtonian mechanics. The generators for the two-particle system then are:

$$\begin{aligned} &\boldsymbol{J} = \boldsymbol{q}_1 \times \boldsymbol{p}_1 + \boldsymbol{q}_2 \times \boldsymbol{p}_2\,, \quad \boldsymbol{G} = m_1\boldsymbol{q}_1 + m_2\boldsymbol{q}_2 \quad , \ \boldsymbol{P} = \boldsymbol{p}_1 + \boldsymbol{p}_2\,, \\ &H = \frac{p_1^2}{2m_1} + \frac{p_2^2}{2m_2} + U_1 + U_2\,; \qquad M = m_1 + m_2 \qquad : \ m_1, m_2 > 0\,. \end{aligned} \tag{19.57}$$

The U_1 and U_2 are the individual internal energies, and the q's and p's obey the normal PB relations; thus the three components of $\boldsymbol{p}_1$ are conjugate to the corresponding components of $\boldsymbol{q}_1, \boldsymbol{p}_2$ to those of $\boldsymbol{q}_2$, whereas the variables with subscript 1 have zero bracket with those with subscript 2.

This realization of $\mathcal{G}$ describes two free *noninteracting* nonrelativistic particles. It is not an irreducible realization; both Casimir invariants are nontrivial functions of the q's and p's. An interesting feature of this realization is that *it has nonvanishing internal angular momentum.* This is just the difference between the total angular momentum $\boldsymbol{J}$ and the center of mass angular momentum $(1/M)\boldsymbol{G} \times \boldsymbol{P}$, and is physically the *relative* angular momentum of the two-particle system. Therefore, although in the case of an irreducible single-particle realization of $\mathcal{G}$ the internal angular momentum $\boldsymbol{I}$ is naturally interpreted as the intrinsic spin $\boldsymbol{S}$ of the particle, which cannot be constructed in any natural sense as the vector product of some position-type vector and its conjugate momentum, in the present case the internal angular momentum does have the "$\boldsymbol{q} \times \boldsymbol{p}$" type structure and meaning. This serves to remind us that the internal angular momentum contained in any Galilean invariant system (as long as C_2' is nonzero) can neither be always interpreted as "intrinsic spin" nor always as "orbital"; depending on the particular system, it could be one or the other or a combination of both. We have learned in our general analysis that any canonical realization of $\mathcal{G}$ appears as the direct product of two commuting canonical realizations of $\mathcal{G}$, the first one having zero internal angular momentum, and mass equal to that of the given realization, the second a "true" class D realization of $\mathcal{G}$ which is just an $R(3)$ realization generated by the I_j. For the realization corresponding to two free massive particles, given by Eq. (19.57), the transition to the familiar center of mass and relative variables exhibits explicitly this splitting of the generators. Let us write $\boldsymbol{Q}^c, \boldsymbol{P}^c$ for the variables describing the center of mass motion, $\boldsymbol{q}, \boldsymbol{p}$ for those describing the relative motion. These are defined in such a way that they obey the standard PB rules, given that the individual q's and p's do:

$$\boldsymbol{Q}^c = \frac{1}{M}(m_1\boldsymbol{q}_1 + m_2\boldsymbol{q}_2),\ \boldsymbol{P}^c = \boldsymbol{p}_1 + \boldsymbol{p}_2\,;$$
$$\boldsymbol{q} = \boldsymbol{q}_1 - \boldsymbol{q}_2,\ \boldsymbol{p} = \frac{1}{M}(m_2\boldsymbol{p}_1 - m_1\boldsymbol{p}_2)\,;$$
$$\{Q_j^c, Q_k^c\} = \{P_j^c, P_k^c\} = \{Q_j^c, q_k\} = \{Q_j^c, p_k\} = \{P_j^c, q_k\} = \{P_j^c, p_k\} = 0\,,$$
$$\{Q_j^c, P_k^c\} = \{q_j, p_k\} = \delta_{jk}\,. \tag{19.58}$$

(Note that the change of variables $\boldsymbol{q}_1, \boldsymbol{p}_1, \boldsymbol{q}_2, \boldsymbol{p}_2 \rightarrow \boldsymbol{Q}^c, \boldsymbol{P}^c, \boldsymbol{q}, \boldsymbol{p}$ is a canonical transformation.) In these variables, the generators of Eq.(19.57) of the two particle system become:

$$\boldsymbol{J} = \boldsymbol{Q}^c \times \boldsymbol{P}^c + \boldsymbol{q} \times \boldsymbol{p},\ \boldsymbol{G} = M\boldsymbol{Q}^c,\ \boldsymbol{P} = \boldsymbol{P}^c\,,$$
$$H = \frac{\boldsymbol{P}^{c^2}}{2M} + \frac{1}{2M}\left(2M(U_1 + U_2) + \frac{M^2}{m_1 m_2}\boldsymbol{p}^2\right);\ M = m_1 + m_2\,. \tag{19.59}$$

The Casimir invariants and the internal angular momenta are immediately identified:

$$C_1 = -\,2M(U_1 + U_2) - \frac{M^2}{m_1 m_2}\boldsymbol{p}^2\,,$$
$$C_2' = M^2(\boldsymbol{q} \times \boldsymbol{p})^2;\ \boldsymbol{I} = \boldsymbol{q} \times \boldsymbol{p}\,. \tag{19.60}$$

One sees explicitly that $\boldsymbol{I}$ is simply the relative orbital angular momentum. It has vanishing PB's with $\boldsymbol{G}, \boldsymbol{P}, H$, and the center of mass angular momentum, as it should.

Before we proceed to the generalization to more than two particles and the inclusion of interactions, let us fulfill the promise made on page 374. We want to show how the "true" class A realization of $\mathcal{G}$ given by the generators of Eq. (19.15) can be obtained from the direct product of two "physical" realizations of $\mathcal{G}$, each having a nonzero neutral element M. In the two particle realization of $\mathcal{G}$ that we have been discussing, we assume that the individual masses m_1 and m_2 are both greater than zero; this accords with our experience in nature. The mass of any interesting physical system must be positive. However, this is an extra requirement on the canonical realizations of $\mathcal{G}$ in order that they may have a reasonable physical interpretation. The structure of $\mathcal{G}$ does not demand that M be positive, and we can consider canonical realizations of $\mathcal{G}$ in which the neutral element is negative; there is no mathematical inconsistency. Further, the rule that the neutral elements in individual realizations must be added when we take the direct product of these realizations is also valid whatever these neutral elements may be.

This immediately gives us the clue to the construction of a "true" realization of $\mathcal{G}$ having zero total mass from the direct product of two realizations of $\mathcal{G}$ each of which has a nonzero neutral element: we merely choose these two neutral elements to be equal in magnitude but opposed in sign! Thus we take two "single-particle" type realizations like the one given in Eq. (19.45), one having mass m, the other "mass" $-m$, and add the generators to get the direct-product generators. (We ignore the internal energies U.) The resulting generators will look exactly like those in Eq. (19.57), but with m_1 and m_2 replaced by m and $-m$, respectively (and the U_i omitted):

$$\begin{aligned} &\boldsymbol{J} = \boldsymbol{q}_1 \times \boldsymbol{p}_1 + \boldsymbol{q}_2 \times \boldsymbol{p}_2,\ \boldsymbol{G} = m(\boldsymbol{q}_1 - \boldsymbol{q}_2),\ \boldsymbol{P} = \boldsymbol{p}_1 + \boldsymbol{p}_2\,, \\ &H = \frac{1}{2m}(\boldsymbol{p}_1^2 - \boldsymbol{p}_2^2);\ M = m - m = 0\,. \end{aligned} \tag{19.61}$$

This is a "true" canonical realization of $\mathcal{G}$, but this is none other than the class A realization of Eq. (19.15) expressed in another form! If we introduce the new variables:

$$\begin{aligned} \boldsymbol{q}^{(1)} &= \frac{1}{2}(\boldsymbol{q}_1 + \boldsymbol{q}_2),\ \boldsymbol{p}^{(1)} = \boldsymbol{p}_1 + \boldsymbol{p}_2\,, \\ \boldsymbol{q}^{(2)} &= m(\boldsymbol{q}_1 - \boldsymbol{q}_2),\ \boldsymbol{p}^{(2)} = \frac{1}{2m}(\boldsymbol{p}_1 - \boldsymbol{p}_2) \end{aligned} \tag{19.62}$$

that obey the standard PB relations, then write the generators in Eq. (19.61) in terms of them, we obtain precisely the expressions synthesized in Eq. (19.15). Comparing the new variables introduced here with the center of mass and relative variables introduced in Eq. (19.58), in the physical case of positive masses m_1 and m_2, we could think of these changes of variables as being analogous to one another.

Let us now return to the "physical" realizations of $\mathcal{G}$. The example of two free particles is easily generalized to n free, noninteracting particles. Incorporating a slight change in notation, we let an index r, taking the values $1, 2, \cdots n$, enumerate the n particles, and write $m^{(r)}, \boldsymbol{q}^{(r)}$ and $\boldsymbol{p}^{(r)}$ for the mass, Cartesian position, and conjugate Cartesian momentum, of the rth particle. Then the generators of $\mathcal{G}$ that correspond to the direct product of n single-particle realizations are:

$$\begin{aligned} &\boldsymbol{J} = \sum_{r=1}^{n} \boldsymbol{q}^{(r)} \times \boldsymbol{p}^{(r)},\ \boldsymbol{G} = \sum_{r=1}^{n} m^{(r)}\boldsymbol{q}^{(r)},\ \boldsymbol{P} = \sum_{r=1}^{n} \boldsymbol{p}^{(r)}\,, \\ &H = \sum_{r=1}^{n}\left(\frac{\boldsymbol{p}^{(r)^2}}{2m^{(r)}} + U^{(r)}\right),\ M = \sum_{r=1}^{n} m^{(r)}\,. \end{aligned} \tag{19.63}$$

The fundamental brackets are, of course,

$$\{q_j^{(r)}, q_k^{(s)}\} = \{p_j^{(r)}, p_k^{(s)}\} = 0,\ \{q_j^{(r)}, p_k^{(s)}\} = \delta_{rs}\delta_{jk}\,. \tag{19.64}$$

For simplicity, we do not consider particles with spin, but this modification is a straightforward matter.

Now let us consider a many-particle system *with interactions.* From elementary mechanics we are familiar with such systems, and in all such systems the energy or Hamiltonian contains additional "interaction terms" apart from the sum of the single-particle kinetic energies. These extra terms signal the presence of forces among the particles. On the other hand, the expressions for the linear and angular momenta of an interacting system are generally taken to be the same as for the free system, and there are no additional interaction terms in these dynamical quantities. But we know that the (total) linear and angular momenta act as generators for spatial displacements and rotations, respectively, whereas the Hamiltonian is the generator of time displacements. Thus we have two classes of generators, one class not containing interaction terms, the other containing such terms. We are thus led to a natural classification of the generators of $\mathcal{G}$ into two classes. The class of generators that have the same composition in terms of the Hamiltonian variables $\boldsymbol{q}^{(r)}, \boldsymbol{p}^{(r)}$ (and $\boldsymbol{S}^{(r)}$ if present) for both the free and the interacting systems are called "kinematic" generators; whereas those whose composition changes by interaction are called "dynamic" generators. (Note that the distinction between "free" and "interacting" systems involved here is a "legal" definition; that is, it is beyond dynamics, but is based on convention, or common sense!). It is clear that by a suitable choice of basic elements in the Lie algebra of $\mathcal{G}$, the subset of kinematic generators will, in fact, form a subalgebra in the Lie algebra, and therefore the corresponding subgroup of $\mathcal{G}$ is singled out as being realized in the same fashion for the free and the interacting systems. The previously stated convention for the structure of the generators $\boldsymbol{J}$ and $\boldsymbol{P}$ amounts to saying that these are included in the set of kinematic generators; therefore the Euclidean subgroup $E(3)$ of $\mathcal{G}$ is usually assumed to be realized in the same way for the free and the interacting systems. (More specifically, by convention it is assumed that in an inertial frame $\boldsymbol{J}$ and $\boldsymbol{P}$ are kinematic generators.) The Hamiltonian H is, by definition, dynamic. The question then arises whether the generators for moving frames, $\boldsymbol{G}$, are kinematic or dynamic. The structure of $\mathcal{G}$ is such that it allows both possibilities, and in this respect the situation is very different from the case of the Poincaré group $\mathcal{P}$. (In that case we see that if $\boldsymbol{J}$ and $\boldsymbol{P}$ are kinematic and H is dynamic, then the Lie algebra relations for $\mathcal{P}$ force the generators for moving frames, $\boldsymbol{K}$, to be dynamic.) Let us indicate with the superscript $^{(0)}$ the free-particle expressions given in Eq. (19.63) for the generators of $\mathcal{G}$. Then for an interacting system of n particles, the generators of $\mathcal{G}$ can be written as:

$$\begin{aligned} &\boldsymbol{J} = \boldsymbol{J}^{(0)},\ \boldsymbol{P} = \boldsymbol{P}^{(0)},\ M = \sum_{r=1}^{n} m^{(r)}\,, \\ &\boldsymbol{G} = \boldsymbol{G}^{(0)} + \boldsymbol{W},\ H = H^{(0)} + V\,. \end{aligned} \tag{19.65}$$

If $\boldsymbol{W}$ is nonzero, then $\boldsymbol{G}$ too is a dynamic generator, like H. $\boldsymbol{W}$ and V are functions of all the $\boldsymbol{q}^{(r)}$ and $\boldsymbol{p}^{(r)}$. Some of the bracket relations in the Lie algebra of $\mathcal{G}$, namely those corresponding to the $E(3)$ subgroup, are automatically satisfied. The remaining ones give restrictions on the "interaction terms" $\boldsymbol{W}$ and V:

$$\begin{aligned} \{J_j, G_k\} &= \epsilon_{jkm} G_m \Longrightarrow \{J_j^{(0)}, W_k\} = \epsilon_{jkm} W_m\,; \\ \{G_j, P_k\} &= \delta_{jk} M \Longrightarrow \{P_j^{(0)}, W_k\} = 0\,; \end{aligned} \tag{19.66a}$$

$$\{J_j, H\} = 0 \Longrightarrow \{J_j^{(0)}, V\} = 0\,;$$
$$\{P_j, H\} = 0 \Longrightarrow \{P_j^{(0)}, V\} = 0\,; \qquad (19.66b)$$
$$\{G_j, G_k\} = 0 \Longrightarrow \{G_j^{(0)}, W_k\} + \{W_j, G_k^{(0)}\} + \{W_j, W_k\} = 0\,;$$
$$\{G_j, H\} = P_j \Longrightarrow \{G_j^{(0)}, V\} + \{W_j, H^{(0)}\} + \{W_j, V\} = 0\,. \qquad (19.66c)$$

Equations (19.66a) and (19.66b) have a simple interpretation, precisely because $E(3)$ is being realized kinematically. The former tell us that the $W_j(\boldsymbol{q}^{(r)}, \boldsymbol{p}^{(r)})$ form the components of a three-dimensional vector under rotations of the spatial coordinate system and are invariant under space translations. The latter tell us that V is an $E(3)$ invariant, being unchanged under spatial rotations as well as spatial translations. The remaining Eqs. (19.66c) are non-linear in the interaction terms W_j, V and do not, in general, seem to have a simple interpretation. The real difficulty in constructing a realization of $\mathcal{G}$ with dynamic generators $\boldsymbol{G}$ and H lies in solving these nonlinear equations. But all these problems of nonlinearity disappear if G_j is chosen to be kinematic as well, and thus equal to $G_j^{(0)}$. Then, of the ten generators of $\mathcal{G}$, only H has an interaction term V and the constraints on V are simply Eq. (19.66b) and a degenerate form of Eq. (19.66c):

$$\{J_j^{(0)}, V\} = 0,\ \{P_j^{(0)}, V\} = 0\,; \qquad (19.67a)$$
$$\{G_j^{(0)}, V\} \equiv \sum_r m^{(r)}\{q_j^{(r)}, V\} \equiv \sum_r \frac{\partial V}{\partial(p_j^{(r)}/m^{(r)})} = 0\,. \qquad (19.67b)$$

Let us call $\boldsymbol{p}^{(r)}/m^{(r)}$ the "Hamiltonian velocity" of the rth particle. The first set of the Hamiltonian equations of motion shows that this Hamiltonian velocity differs in general from the kinetic velocity $\dot{\boldsymbol{q}}^{(r)}$, which is the time rate of change of position; the two are the same only if the interaction V is $\boldsymbol{p}^{(r)}$-independent. In any case, Eq. (19.67b) says that V is unchanged if all the Hamiltonian velocities are altered by the same amount; hence V can only depend on the differences of the Hamiltonian velocities of pairs of particles.

The previous condition that V be an $E(3)$ invariant has already restricted its dependence on the $\boldsymbol{q}^{(r)}$ to a dependence on the relative coordinates $\boldsymbol{q}^{(r)} - \boldsymbol{q}^{(s)}$, since the transformation $\boldsymbol{q}^{(r)} \to \boldsymbol{q}^{(r)} + \boldsymbol{a}, \boldsymbol{p}^{(r)} \to \boldsymbol{p}^{(r)}$, for any fixed numerical three-vector $\boldsymbol{a}$, belongs to $E(3)$. Thus by choosing V to be any rotationally invariant function of the relative coordinates and the relative Hamiltonian velocities and by taking $\boldsymbol{J}, \boldsymbol{P}$, and $\boldsymbol{G}$ to be all kinematic, we can have a Galilean invariant interaction between particles. We see in Chapter 21 that when the extra requirement of manifest covariance is imposed, this same class of interactions is again permitted. (The spin can be included with no essential changes in principle). In particular, a velocity-independent potential V depending only on the relative coordinates, is acceptable. This takes us back to the elementary types of forces encountered in Newtonian mechanics.

These results on the possible forms of a Galilean-invariant interacting many-particle system have been obtained within the Hamiltonian formalism. The basic assumption is that in each inertial frame the equations of motion could be expressed in Hamiltonian form so that changes of inertial frame are implemented in the mathematics by canonical transformations. The choice of basic variables and the form of the fundamental PB's are both determined by the corresponding noninteracting system. It is instructive at this point to go back to the Lagrangian description and see briefly how Galilean invariance expresses itself. For simplicity we consider only the case in which the interaction V depends on the $\boldsymbol{q}^{(r)}$ alone; then the Hamiltonian velocity and the kinetic velocity coincide for each particle, and the passage to the Lagrangian is immediate. (Otherwise the relation between these two velocities depends on V and solving for the p's in terms of the q's is not easy). We have in this case the system of equations:

$$\begin{aligned} H &= \sum_r \frac{\boldsymbol{p}^{(r)^2}}{2m^{(r)}} + V(\boldsymbol{q}^{(r)} - \boldsymbol{q}^{(s)}) \,, \\ \boldsymbol{p}^{(r)} &= m^{(r)} \dot{\boldsymbol{q}}^{(r)} \,; \\ L(q, \dot{q}) &\equiv \sum_r \boldsymbol{p}^{(r)} \cdot \dot{\boldsymbol{q}}^{(r)} - H(p, q) \\ &\equiv \frac{1}{2} \sum_r m^{(r)} \dot{\boldsymbol{q}}^{(r)2} - V(\boldsymbol{q}^{(r)} - \boldsymbol{q}^{(s)}) \,. \end{aligned} \tag{19.68}$$

The internal energies have been omitted for simplicity. Being the generator of time translations within the Galilei group $\mathcal{G}$, the Hamiltonian H is found to have vanishing PB's with both P_j and J_j, thus rendering these two objects constants of motion. As we have already seen in Chapter 11, these properties of rotational and translational invariance of the Hamiltonian imply and are equivalent to similar invariance properties for the Lagrangian. The more interesting point is the behavior of L when we make a genuine velocity transformation of the form:

$$\boldsymbol{q}^{(r)}(t) \rightarrow \boldsymbol{q}^{(r)'}(t) = \boldsymbol{q}^{(r)}(t) + \boldsymbol{v}t, \; t \rightarrow t' = t \,. \tag{19.69}$$

For the present discussion, instead of looking upon this transformation as reflecting the passage from one inertial frame to another moving uniformly with respect to the first, let us look upon it as a transformation that, when applied to one solution of the equations of motion, gives rise to another solution of the same equations of motion. (The former is the passive and the latter the active interpretation; in any event, the invariance of the equations of motion under Eq. (19.69) is obvious). Given any trajectory $\boldsymbol{q}^{(r)}(t)$ in configuration space, whether or not the equations of motion are satisfied on it, the transformation Eq. (19.69) yields a new trajectory $\boldsymbol{q}^{(r)'}(t)$. Call the former C, the latter C'. For corresponding instants on C and C', the Lagrangian does not have the same values; that is, it is not invariant under the transformation Eq. (19.69); but the change in L is a total time derivative:

$$\begin{aligned} L(\boldsymbol{q}'(t),\dot{\boldsymbol{q}}'(t)) &= \frac{1}{2}\sum_r m^{(r)}(\dot{\boldsymbol{q}}^{(r)}(t)+\boldsymbol{v})^2 - V(\boldsymbol{q}^{(r)}-\boldsymbol{q}^{(s)}) \\ &= L(\boldsymbol{q}(t),\dot{\boldsymbol{q}}(t)) + \frac{d}{dt}F(\boldsymbol{q}(t),t)\,, \\ F(\boldsymbol{q}(t),t) &= \boldsymbol{v}\cdot\sum_r m^{(r)}\boldsymbol{q}^{(r)} + \frac{t}{2}v^2\sum_r m^{(r)}\,. \end{aligned} \tag{19.70}$$

Consequently, if we choose two instants of time, t_1 and t_2, and compute the values of the action between these two times on C and C', we will find:

$$\Phi[C'] = \Phi[C] + [F(\boldsymbol{q}(t),t)]_{t_1}^{t_2}\,; \tag{19.71}$$

in other words, the difference between the two actions consists of end-point contributions only. This immediately explains why the equations of motion are invariant under the velocity transformation Eq. (19.69) even though the Lagrangian is not: the crucial point is that the change in L is a total time derivative, which produces only end-point contributions in the difference of the two action integrals. Any infinitesimal variation of the trajectory C specifies an infinitesimal variation of C' through Eq. (19.69), and conversely, because the transformation is reversible, all variations of C' can be gotten this way. Consequently we see from Eq. (19.71) that if the equations of motion are obeyed on C, so that a small variation of C gives rise to only end-point contributions in the change $\Delta\Phi[C]$ of $\Phi[C]$, the same is true of $\Delta\Phi[C']$ under any small change in C', and the equations of motion must be obeyed on C'. This illustrates in a practical example the discussion in Chapter 4, page 29.

This discussion serves to emphasize the fact that whereas the Newtonian equations of motion for a many-particle system may be explicitly invariant under all the transformations of $\mathcal{G}$, if these equations are expressed through the action principle with a Lagrangian L not all these invariances of the equations of motion are achieved in the same way. For the transformations of the Euclidean subgroup $E(3)$, we do have the invariance of L, leading to P_j and J_j being constants of motion. However, for velocity transformations, the invariance of the equations of motion is achieved, not by the invariance of L, but by the change in L being a total time derivative, that is, by L being quasi-invariant.

Realization of $\mathcal{G}$ with a Single-Component Field — a New Principle

We now turn to some interesting examples of realizations of $\mathcal{G}$ constructed in terms of fields. We first build up a realization using a single-component or scalar field and later generalize it to a many-component field. After these mathematical examples, we show how the equations of classical fluid dynamics can also be put into Hamiltonian form and how they, too, yield a canonical realization of $\mathcal{G}$. In all these cases the mass M is a nonvanishing quantity.

Let us begin by recalling the form of the realization of $E(3)$ given by a single-component field, as given in the preceding chapter. The expressions for the generators of $E(3)$ are contained in Eq. (18.30), and are repeated here:

$$\begin{aligned} J_j &= -\int d^3x\,\pi(\boldsymbol{x})\epsilon_{jkm}x_k\frac{\partial}{\partial x_m}\phi(\boldsymbol{x})\,,\\ P_j &= -\int d^3x\,\pi(\boldsymbol{x})\frac{\partial}{\partial x_j}\phi(\boldsymbol{x})\,. \end{aligned} \tag{19.72}$$

The ϕ is the field and π its canonical conjugate. It is not a difficult matter to extend this to get a set of quantities obeying the bracket relations of the Lie algebra of the eleven-parameter group $\mathcal{G}'$, the group that is the "central extension" of $\mathcal{G}$ (see pages 365, 366). What we need are additional dynamical variables G_j, H, and M to stand for the elements g_j, h, and μ, respectively in the Lie algebra of $\mathcal{G}'$ such that along with J_j and P_j given above the Lie relations of Eqs. (19.2) and (19.11) are realized via PB's. The following construction fits the bill:

$$\begin{aligned} G_j &= m\int d^3x\,\pi(\boldsymbol{x})\,x_j\phi(\boldsymbol{x}),\ H=\frac{1}{2m}\int d^3x\,\pi(\boldsymbol{x})\nabla^2\phi(\boldsymbol{x})\,,\\ M &= m\int d^3x\,\pi(\boldsymbol{x})\phi(\boldsymbol{x})\,. \end{aligned} \tag{19.73}$$

Here m is a fixed (positive) constant, and with Eqs. (19.72) and (19.73) we have a true PB realization of the Lie algebra of $\mathcal{G}'$.

But the reader will immediately recognize that what we have been led to by the above construction *is not a canonical realization of* $\mathcal{G}$! Indeed, on the basis of our considerations up to now, we have seen that not every true PB realization of the Lie algebra of $\mathcal{G}'$ yields a canonical realization of $\mathcal{G}$, but only those in which the element μ is represented by a neutral element, that is by an element that generates the identity canonical transformation (see the discussion on pages 365, 366). Now, the quantity M in Eq. (19.73) is certainly not neutral; later on we exhibit the form of the canonical transformation it generates. Therefore, to make use of the foregoing construction and interpret it as describing a Galilean-invariant dynamical system, some of our basic assumptions must be modified. To see where such a modification must be introduced, let us go back to the equations that show that a Galilean invariant dynamical system is described by a *canonical realization of* $\mathcal{G}$, which in turn imply that we must look for PB realizations up to neutrals of the Lie algebra of $\mathcal{G}$. These are Eqs. (16.15) and (16.25); we repeat the latter here, with the difference that the dynamical variable A on which the various canonical transformations act is not dropped on the two sides of the equation:

$$\begin{aligned} &R(v^j, X_j(S,0;\omega))R(u^j, X_j(S,0;\omega))A(S,0;\omega)\\ &\qquad=R(w^j, X_j(S,0;\omega))A(S,0;\omega)\\ &R(u^j, X_j(S,0;\omega))\equiv\exp(u\cdot\widetilde{X(S,0;\omega)}),\ \exp(v)\exp(u)=\exp(w)\,. \end{aligned} \tag{19.74}$$

To remind the reader of the meanings of the quantities appearing here, the u's, v's, and w's are canonical coordinates of the first kind for elements of the relativity group $\mathcal{G}$ under consideration; the X 's are a set of phase space functions, one for each basic vector in the Lie algebra of $\mathcal{G}$; each R is a canonical transformation depending on an element of $\mathcal{G}$ and built up in the manner indicated with the help of the X's; A is a general dynamical variable; S is a reference frame (observer); and the 0 denotes the zero of time in S. Now Eq. (19.74) by itself does not imply that the R's constitute a canonical realization of $\mathcal{G}$; such a conclusion can be reached only if we make the assumption that *every phase-space function is an observable.* (In any case, it is clear that Eq. (19.74) need be true only for those dynamical variables A that are observables). If this assumption is made, then in particular every one of the phase-space coordinates ω^μ is an observable; one can then "cancel" the $A(S,0;\omega)$ on the two sides of Eq. (19.74), and conclude that the composition law for the canonical transformations R follows that of the group $\mathcal{G}$,

$$\begin{aligned}\exp(v)\exp(u) = \exp(w) \quad &\Longrightarrow \quad R(v^j, X_j(S,0;\omega))R(u^j, X_j(S,0;\omega)) \\ &= \quad R(w^j, X_j(S,0;\omega)), \qquad (19.75)\end{aligned}$$

and then, on the basis of our work in Chapter 14, we find that the X_j give a PB realization *up to neutrals* of the Lie algebra of $\mathcal{G}$.

It is at this stage of the argument that we can introduce a modification. Suppose that Eq. (19.74) need be true only for a certain, specially chosen set of phase-space functions to be named *observables*, and that specifically not every phase-space function need be an observable. Then the possibility of "cancelation" of $A(S,0;\omega)$ on the two sides of Eq. (19.74) no longer exists, and thus we lose Eq. (19.75) as well. What conclusion we can draw concerning the composition law for the canonical transformations $R(u^j, X_j)$ now depends on the nature of the rule that selects observables. In the present context, where we are dealing with the Galilei group $\mathcal{G}$ as the relativity group and where the X_j are the dynamical variables $\boldsymbol{J}, \boldsymbol{P}, \boldsymbol{G}$, and H of Eqs. (19.72) and (19.73), let us introduce this rule: *a dynamical variable A is an observable if and only its PB with M vanishes.* In other words, observables are those phase-space functions that are unchanged by the one-parameter group of canonical transformations generated by the non-neutral quantity M given in Eq. (19.73). Equipped with this rule, the field realization of $\mathcal{G}'$ generated by the eleven quantities in Eqs. (19.72) and (19.73) can be interpreted as a model of a physical system possessing invariance under $\mathcal{G}$. We now verify this in detail. Evidently, what is involved is the way in which space translations and pure velocity transformations are realized. In the abstract group $\mathcal{G}$, these elements commute:

$$\exp(\boldsymbol{v}\cdot\boldsymbol{g})\exp(\boldsymbol{a}\cdot\boldsymbol{d}) = exp(\boldsymbol{a}\cdot\boldsymbol{d})\exp(\boldsymbol{v}\cdot\boldsymbol{g}) \ . \qquad (19.76)$$

In the canonical realizations of $\mathcal{G}$ thus far discussed, with M truly neutral, the representative canonical transformations also did commute:

$$\exp(\widetilde{\boldsymbol{v}\cdot\boldsymbol{G}})\exp(\widetilde{\boldsymbol{a}\cdot\boldsymbol{P}}) = \exp(\widetilde{\boldsymbol{a}\cdot\boldsymbol{P}})\exp(\widetilde{\boldsymbol{v}\cdot\boldsymbol{G}}) \text{ if } M \text{ neutral}. \qquad (19.77)$$

But now, with M nonneutral, the canonical transformations generated by G_j and P_j do not commute. In fact, using the Baker-Campbell-Hausdorff identity, and the bracket relations

$$\{G_j, G_k\} = \{P_j, P_k\} = 0,\ \{G_j, P_k\} = \delta_{jk} M,\ \{G_j, M\} = \{P_j, M\} = 0\,, \tag{19.78}$$

we obtain:

$$\exp(\widetilde{\boldsymbol{v}\cdot\boldsymbol{G}})\exp(\widetilde{\boldsymbol{a}\cdot\boldsymbol{P}}) = \exp(\widetilde{\boldsymbol{a}\cdot\boldsymbol{P}})\exp(\widetilde{\boldsymbol{v}\cdot\boldsymbol{G}})\exp(\widetilde{\boldsymbol{a}\cdot\boldsymbol{v}M})\,. \tag{19.79}$$

This is why these canonical transformations do not constitute a canonical realization of $\mathcal{G}$; we have here a canonical realization of $\mathcal{G}'$ that, because μ is not realized as a neutral, is not, in effect, only a realization of $\mathcal{G}$. We have compensated for this by a rule that selects observables. We no longer require that the canonical transformations generated by $\boldsymbol{J}, \boldsymbol{P}, \boldsymbol{G}$, and H have the multiplication law of $\mathcal{G}$, but are satisfied if in their effect on observables they follow the multiplication law of $\mathcal{G}$. The composition law to be checked involves $\exp(\widetilde{\boldsymbol{v}\cdot\boldsymbol{G}})$ and $\exp(\widetilde{\boldsymbol{a}\cdot\boldsymbol{P}})$; we get, then, with the use of Eq. (19.79),

$$\exp(\widetilde{\boldsymbol{v}\cdot\boldsymbol{G}})\exp(\widetilde{\boldsymbol{a}\cdot\boldsymbol{P}})A = \exp(\widetilde{\boldsymbol{a}\cdot\boldsymbol{P}})\exp(\widetilde{\boldsymbol{v}\cdot\boldsymbol{G}})A \quad \text{if} \quad \widetilde{M}A \equiv \{M, A\} = 0\,. \tag{19.80}$$

Thus these canonical transformations do commute when acting on observables. Quite generally, if $g = (A, v_j, b, a_j)$ is any element of $\mathcal{G}$, it is represented by the canonical transformation

$$R(g) = \exp(\widetilde{\boldsymbol{a}\cdot\boldsymbol{P}})\exp(\widetilde{bH})\exp(\widetilde{\boldsymbol{v}\cdot\boldsymbol{G}})\exp(\widetilde{\boldsymbol{\alpha}\cdot\boldsymbol{J}})\,, \tag{19.81}$$

$\boldsymbol{\alpha}$ denoting the axis and angle of the rotation A; we have

$$R(g')R(g)A = R(g'g)A \quad \text{if } \widetilde{M}A = 0\,. \tag{19.82}$$

The canonical transformation $R(g')R(g)$ generally differs from $R(g'g)$, although for some particular choices of g' and g they may be equal. What is true for all g and g' is that the product $R^{-1}(g'g)R(g')R(g)$ is a canonical transformation belonging to the one-parameter subgroup of transformations generated by M.

Rules of the type introduced here, which characterize certain dynamical variables as observables and others as nonobservables, are quite familiar in quantum mechanics. They go by the name of "superselection rules", and we may use the same name here. In the case of nonrelativistic dynamical systems whose space-time symmetry is described by $\mathcal{G}$, whenever the mass M is not a neutral element, we recover the physical interpretation by imposing a superselection rule; this is a feature common to both classical and quantum mechanics. The mass superselection rule should be called the "Bargmann superselection rule." In the foregoing discussion we obtained this superselection rule by an examination of Eq. (19.74), which was derived in Chapter 16 using the Heisenberg picture of dynamics. Now we must recognize that ultimately the quantities that must possess a physically acceptable behavior under the transformations of the Galilei group are the average values of observables in physical states. What this means is that instead of using a superselection rule for determining observables, we could equally well limit the possible physical states to be those corresponding to phase space distribution functions $\rho(\omega)$ that are invariant under the canonical transformations generated by M. The final effect can be seen to be the same. From one point of view, however, it is desirable to impose a condition on the observables; that is,if it were imposed on the states, we might face some problems in ensuring that an allowed density function be normalized properly to unity.

Let us now return to an examination of some properties of the realization of $\mathcal{G}$ given by the generators in Eqs. (19.72) and (19.73). Because M has vanishing brackets with all the other generators, all the generators are observables. This is gratifying, because we would certainly want the momenta, angular momenta, and energy to be counted among the observables of the system. On the other hand, neither the field $\phi(\boldsymbol{x}, t)$ nor its conjugate $\pi(\boldsymbol{x}, t)$ is observable, because both have nonzero PB's with M! In fact, the canonical transformation generated by M is very simple to work out. We find:

$$\begin{aligned}\widetilde{M}\phi(\boldsymbol{x}) &\equiv \{M, \phi(\boldsymbol{x})\} = m \int d^3y \{\pi(\boldsymbol{y})\phi(\boldsymbol{y}), \phi(\boldsymbol{x})\} = -m\phi(\boldsymbol{x});\\ \widetilde{M}\pi(\boldsymbol{x}) &= m\pi(\boldsymbol{x});\ e^{\lambda\tilde{M}}\phi(\boldsymbol{x}) = e^{-m\lambda}\phi(\boldsymbol{x}),\ e^{\lambda\tilde{M}}\pi(\boldsymbol{x}) = e^{m\lambda}\pi(\boldsymbol{x}). \end{aligned} \tag{19.83}$$

Thus a functional of $\phi(\boldsymbol{x})$ and $\pi(\boldsymbol{x})$ is an observable if, roughly speaking, it is expressible as a sum of terms in each of which there are as many factors ϕ as there are factors π. More precisely, in order to be an observable, a functional $A[\phi, \pi]$ must obey

$$A[\lambda\phi, \lambda^{-1}\pi] = A[\phi, \pi] \text{ for all } \lambda > 0. \tag{19.84}$$

Because each of the generators $\boldsymbol{J}, \boldsymbol{P}, \boldsymbol{G}, H$, as well as M itself involves one ϕ and one π, they are all observables. Next we consider the canonical transformations generated by $\boldsymbol{J}, \boldsymbol{G}, \boldsymbol{P}$, and H. We compute their effects on the variables ϕ and π, although these are not observables, because the transformation laws for ϕ and π determine those for any observable. The form of the $E(3)$ transformations is known from our work in Chapter 18; we have:

$$\begin{aligned}\exp(\widetilde{\boldsymbol{\alpha} \cdot \boldsymbol{J}})\phi(x_j) &= \phi(A_{jk}(\boldsymbol{\alpha})x_k); \qquad \exp(\widetilde{\boldsymbol{\alpha} \cdot \boldsymbol{J}})\pi(x_j) = \pi(A_{jk}(\boldsymbol{\alpha})x_k);\\ \exp(\widetilde{\boldsymbol{a} \cdot \boldsymbol{P}})\phi(\boldsymbol{x}) &= \phi(\boldsymbol{x} + \boldsymbol{a}), \qquad \exp(\widetilde{\boldsymbol{a} \cdot \boldsymbol{P}})\pi(\boldsymbol{x}) = \pi(\boldsymbol{x} + \boldsymbol{a}). \end{aligned} \tag{19.85}$$

The transformations generated by G_j are also quite simple. We have:

$$\begin{aligned}\widetilde{G}_j\phi(\boldsymbol{x}) &= -mx_j\phi(\boldsymbol{x}), \qquad \widetilde{G}_j\pi(\boldsymbol{x}) = mx_j\pi(\boldsymbol{x}),\\ \exp(\widetilde{\boldsymbol{v} \cdot \boldsymbol{G}})\phi(\boldsymbol{x}) &= \exp(-m\boldsymbol{v} \cdot \boldsymbol{x})\phi(\boldsymbol{x}), \qquad \exp(\widetilde{\boldsymbol{v} \cdot \boldsymbol{G}})\pi(\boldsymbol{x}) = \exp(m\boldsymbol{v} \cdot \boldsymbol{x})\pi(\boldsymbol{x}). \end{aligned} \tag{19.86}$$

(At this point the reader can convince himself of the validity of Eq. (19.79), and hence of the noncommutativity of space translations and velocity transformations when acting on ϕ and π, by suitably combining Eqs. (19.83), (19.85), and

(19.86).) In contrast to the results given in Eqs. (19.85) and (19.86), the time translation generated by H involves a nonlocal operation, because the result of applying H to $\phi(\boldsymbol{x})$ or $\pi(\boldsymbol{x})$ is represented by a *second*-order differential operator acting on the argument $\boldsymbol{x}$. Formally, we have:

$$\begin{aligned} \widetilde{H}\phi(\boldsymbol{x}) &= -\frac{1}{2m}\nabla^2\phi(\boldsymbol{x})\,; \\ \phi(\boldsymbol{x}, b) \equiv \exp(-b\widetilde{H})\phi(\boldsymbol{x}) &= \exp[(b/2m)\nabla^2]\phi(\boldsymbol{x})\,; \\ \widetilde{H}\pi(\boldsymbol{x}) &= \frac{1}{2m}\nabla^2\pi(\boldsymbol{x})\,; \\ \pi(\boldsymbol{x}, \boldsymbol{b}) \equiv \exp(-b\widetilde{H})\pi(\boldsymbol{x}) &= \exp[(-b/2m)\nabla^2]\pi(\boldsymbol{x})\,. \end{aligned} \tag{19.87}$$

For the sake of the interested reader we would like to point out the connection of the Galilean field theory discussed up to now with an operator realization of $\mathcal{G}$ that arises in the quantum mechanical description of a free spinless nonrelativistic particle. This arises upon noticing that each one of the eleven generators given in Eqs. (19.72) and (19.73) can be written in the form:

$$A[\phi, \pi] = \int d^3x\, \pi(\boldsymbol{x})a\left(\boldsymbol{x}, \frac{\partial}{\partial \boldsymbol{x}}\right)\phi(\boldsymbol{x})\,, \tag{19.88}$$

where $a(\boldsymbol{x}, (\partial/\partial\boldsymbol{x}))$ is a local partial differential operator (or in some cases only a function of $\boldsymbol{x}$) corresponding to the particular generator A. Thus we have the correspondence:

$$\begin{aligned} A &= J_j \to a = -(\boldsymbol{x} \times \boldsymbol{\nabla})_j\,; \\ A &= P_j \to a = -\nabla_j\,; \\ A &= G_j \to a = m x_j\,; \\ A &= H \to a = \frac{1}{2m}\nabla^2\,; \\ A &= M \to a = m\,. \end{aligned} \tag{19.89}$$

Now the fact that the A's give a PB realization of the Lie algebra of $\mathcal{G}'$ is linked to the fact that the a's give an operator realization, via commutators, of the same Lie algebra. If A and B are two generators having the form of Eq. (19.88), their PB can be computed thus:

$$\begin{aligned} \{A, B\} &= \int d^3x\, d^3y \left\{\pi(\boldsymbol{x})a\left(\boldsymbol{x}, \frac{\partial}{\partial\boldsymbol{x}}\right)\phi(\boldsymbol{x}), \pi(\boldsymbol{y})b\left(\boldsymbol{y}, \frac{\partial}{\partial\boldsymbol{y}}\right)\phi(\boldsymbol{y})\right\} \\ &= \int d^3x\, d^3y \left[\pi(\boldsymbol{x})\left(b\left(\boldsymbol{y}, \frac{\partial}{\partial\boldsymbol{y}}\right)\phi(\boldsymbol{y})\right) a\left(\boldsymbol{x}, \frac{\partial}{\partial\boldsymbol{x}}\right)\delta^{(3)}(\boldsymbol{x}-\boldsymbol{y})\right. \\ &\quad \left. - \left(a\left(\boldsymbol{x}, \frac{\partial}{\partial\boldsymbol{x}}\right)\phi(\boldsymbol{x})\right)\pi(\boldsymbol{y})b\left(\boldsymbol{y}, \frac{\partial}{\partial\boldsymbol{y}}\right)\delta^{(3)}(\boldsymbol{x}-\boldsymbol{y})\right] \end{aligned}$$

$$
\begin{aligned}
&= \int d^3x\,\pi(\boldsymbol{x})a\left(\boldsymbol{x},\frac{\partial}{\partial \boldsymbol{x}}\right)\left(\int d^3y\delta^{(3)}(\boldsymbol{x}-\boldsymbol{y})b\left(\boldsymbol{y},\frac{\partial}{\partial \boldsymbol{y}}\right)\phi(\boldsymbol{y})\right)\\
&\quad -\int d^3y\,\pi(\boldsymbol{y})b\left(\boldsymbol{y},\frac{\partial}{\partial \boldsymbol{y}}\right)\left(\int d^3x\delta^{(3)}(\boldsymbol{x}-\boldsymbol{y})a\left(\boldsymbol{x},\frac{\partial}{\partial \boldsymbol{x}}\right)\phi(\boldsymbol{x})\right)\\
&= \int d^3x\,\pi(\boldsymbol{x})\left[a\left(\boldsymbol{x},\frac{\partial}{\partial \boldsymbol{x}}\right)b\left(\boldsymbol{x},\frac{\partial}{\partial \boldsymbol{x}}\right)-b\left(\boldsymbol{x},\frac{\partial}{\partial \boldsymbol{x}}\right)a\left(\boldsymbol{x},\frac{\partial}{\partial \boldsymbol{x}}\right)\right]\phi(\boldsymbol{x}),\\
\{A,B\} &= \int d^3x\,\pi(\boldsymbol{x})[a,b]_-\phi(\boldsymbol{x}). \qquad (19.90)
\end{aligned}
$$

Thus the PB of two expressions of the form in Eq. (19.88) is another expression of the same form, with the commutator of the individual local differential operators appearing as the local differential operator in the final result. It can be checked easily that the various a's listed in Eq. (19.89) obey the same commutation rules as the Lie algebra of $\mathcal{G}'$, namely:

$$
\begin{aligned}
[-(\boldsymbol{x}\times\boldsymbol{\nabla})_j,-(\boldsymbol{x}\times\boldsymbol{\nabla})_k]_- &= \epsilon_{jkm}(-(\boldsymbol{x}\times\boldsymbol{\nabla})_m);\\
[-(\boldsymbol{x}\times\boldsymbol{\nabla})_j,-\boldsymbol{\nabla}_k]_- &= \epsilon_{jkm}(-\boldsymbol{\nabla}_m);[\boldsymbol{\nabla}_j,\boldsymbol{\nabla}_k]_-=0;\\
[-(\boldsymbol{x}\times\boldsymbol{\nabla})_j,mx_k]_- &= \epsilon_{jkm}mx_m;[mx_j,mx_k]_-=0;\\
\left[-(\boldsymbol{x}\times\boldsymbol{\nabla})_j,\frac{\boldsymbol{\nabla}^2}{2m}\right]_- &= \left[-\boldsymbol{\nabla}_j,\frac{\boldsymbol{\nabla}^2}{2m}\right]_-=0;\\
\left[mx_j,\frac{\boldsymbol{\nabla}^2}{2m}\right]_- &= -\boldsymbol{\nabla}_j,[mx_j,-\boldsymbol{\nabla}_k]_-=m\delta_{jk}. \qquad (19.91)
\end{aligned}
$$

Combining these commutation rules with the general formula of Eq. (19.90), we see in a new light why it is that the classical expressions of Eqs. (19.72) and (19.73) have PB's with one another following the Lie algebra of $\mathcal{G}'$. These eleven operators $a(\boldsymbol{x},(\partial/\partial\boldsymbol{x}))$ furnish an *irreducible* operator representation of the Lie algebra of $\mathcal{G}'$; apart from occasional factors of i, these are precisely the operators for generating a unitary representation of $\mathcal{G}$ to describe a spinless nonrelativistic free particle in quantum mechanics. Although in the operator representation the mass m is an invariant and a trivial dynamical variable, the corresponding quantity M in the classical realization is a nontrivial dynamical variable. It is, finally, clear that such a straightforward correspondence between a free classical Galilean field and a free quantum mechanical particle is possible only as long as the generators for the field realization are taken to be bilinear in the field.

Extension to Multicomponent Fields, Local Interactions

There are two ways in which we can extend the single-component free Galilean field to a multicomponent field. If we examine the corresponding extension in the case of the Euclidean group $E(3)$, we see that the essential change occurred only in the generators of the $R(3)$ subgroup, whereas the translation generators remained essentially unchanged in form. This is evident from the structure of the $E(3)$ generators given in Eq. (18.37). Stated in terms of the finite canonical transformations realizing $E(3)$, the effect of translations is always simply a change of the spatial coordinates that occur as arguments of the field variables, whether we have a single-component or a multicomponent field; whereas a rotation not only changes the argument of a multicomponent field but also shuffles the components. This difference between the J_j and the P_j on extension from a single to a multicomponent field realization of $E(3)$ is linked to the semidirect product structure, $R(3) \times T_3$, of $E(3)$. In the case of $\mathcal{G}$, there are two interesting ways in which we can exhibit it as a semidirect product, and at the outset one might think that this leads to two essentially different ways of dealing with multi-component fields. We describe both these cases here, but then we indicate how by means of a suitable canonical transformation one form can be made equivalent to the other. The first alternative is to realize that the elements g_j, d_j, and h span an *invariant subalgebra* in the Lie algebra of $\mathcal{G}$, and that $\mathcal{G}$ is the semidirect product of $R(3)$ with the invariant subgroup generated by g_j, d_j, and h. Thus we have the possibility of a multicomponent field $\phi_r(\boldsymbol{x}, t)$, with canonical conjugate $\pi_r(\boldsymbol{x}, t)$, with which we can set up a canonical realization of $\mathcal{G}$ with only the angular momenta J_j containing essentially new terms. The several components of ϕ are just associated with a matrix representation of $R(3)$. If the generator matrices of this representation of $R(3)$ are written as S_j, then the $E(3)$ generators given in Eq. (18.37) can be extended to give the generators of $\mathcal{G}$ thus:

$$\begin{aligned} J_j &= \int d^3x\, \pi_r(\boldsymbol{x})\{(S_j)_{rs} - \delta_{rs}(\boldsymbol{x} \times \boldsymbol{\nabla})_j\}\phi_s(\boldsymbol{x})\,, \\ P_j &= -\int d^3x\, \pi_r(\boldsymbol{x})\frac{\partial}{\partial x_j}\phi_r(\boldsymbol{x})\,, \\ G_j &= m\int d^3x\, \pi_r(\boldsymbol{x})x_j\phi_r(\boldsymbol{x}),\ H = \frac{1}{2m}\int d^3x\, \pi_r(\boldsymbol{x})\boldsymbol{\nabla}^2\phi_r(\boldsymbol{x})\,; \\ M &= m\int d^3x\, \pi_r(\boldsymbol{x})\phi_r(\boldsymbol{x})\,. \end{aligned} \tag{19.92}$$

(The mass superselection rule must be invoked again!) Using the characteristic commutation rules obeyed by the matrices S_j and the standard PB's among the ϕ's and π's, the reader can check that these generators obey the expected PB relations. The canonical transformations generated by P_j, G_j, H, and M are essentially unchanged because they do not act on the indices on ϕ and π. They are given by the obvious extensions of the transformations exhibited in Eqs. (19.83), (19.85), (19.86), and (19.87). The transformations generated by J_j are given as in the $E(3)$ case by Eq. (18.38). As for the single-component field, here again the mass M is an invariant of the realization of $\mathcal{G}$. In addition, it can be verified that the square of the spin contribution to the angular momentum,

$$\mathcal{S}^2 = \sum_j \left\{\int d^3x\, \pi_r(\boldsymbol{x})(S_j)_{rs}\phi_s(\boldsymbol{x})\right\}^2 \tag{19.93}$$

is an invariant of the realization.

The second possible way to introduce a multicomponent field is to use the breakup $\mathcal{G} = E^{(v)}(3) \times T_4; E^{(v)}(3)$ is generated by l_j and g_j, and T_4 is the invariant Abelian subgroup of space-time translations generated by d_j and h. Now the many components of the field are to be associated with a linear matrix-representation of $E^{(v)}(3)$, and among the generators of $\mathcal{G}$ both J_j and G_j contain essentially new terms as compared with the case of a single-component field. The group $E^{(v)}(3)$ has the same structure as the Euclidean group $E(3)$, and a matrix representation of it involves six matrices that act as the generators of the representation; they obey commutation relations corresponding to the Lie algebra relations of the Euclidean group. If we write S_j and T_j for these six matrices, the former generating rotations and the latter "translations" in $E^{(v)}(3)$, the characteristic commutation rules are:

$$[S_j, S_k]_- = \epsilon_{jkm} S_m, \ [S_j, T_k]_- = \epsilon_{jkm} T_m, \ [T_j, T_k]_- = 0 \,. \tag{19.94}$$

It would take us too far afield to describe in detail the kinds of solutions that exist for these commutation relations. It is sufficient for us to remark that many of the interesting solutions involve *infinite-dimensional* matrices S_j and T_j; this is so even for the interesting irreducible solutions to Eq. (19.94). Given such a solution, the matrices representing the elements $\exp(\boldsymbol{\alpha}\cdot\boldsymbol{l})$ and $\exp(\boldsymbol{v}\cdot\boldsymbol{g})$ in $E^{(v)}(3)$ are, respectively, $\exp(\alpha_j S_j)$ and $\exp(v_j T_j)$, and these, too, are infinite dimensional. If $\alpha, \beta, \ldots$ label the rows and columns of this matrix representation of $E^{(v)}(3)$, we can introduce an infinite component field $\phi_\alpha(\boldsymbol{x})$ with conjugate field $\pi_\alpha(\boldsymbol{x})$ subject to the basic PB relations:

$$\begin{aligned}
\{\phi_\alpha(\boldsymbol{x}), \phi_\beta(\boldsymbol{y})\} &= \{\pi_\alpha(\boldsymbol{x}), \pi_\beta(\boldsymbol{y})\} = 0 \,, \\
\{\phi_\alpha(\boldsymbol{x}), \pi_\beta(\boldsymbol{y})\} &= \delta_{\alpha\beta}\delta^{(3)}(\boldsymbol{x}-\boldsymbol{y}); \ \alpha, \beta = 1, 2, \ldots, \infty
\end{aligned} \tag{19.95}$$

The generators for this infinite component free-field realization of $\mathcal{G}$ are:

$$\begin{aligned}
J_j &= \int d^3x \, \pi_\alpha(\boldsymbol{x})\{(S_j)_{\alpha\beta} - \delta_{\alpha\beta}(\boldsymbol{x}\times\boldsymbol{\nabla})_j\}\phi_\beta(\boldsymbol{x}) \,, \\
G_j &= \int d^3x \, \pi_\alpha(\boldsymbol{x})\{(T_j)_{\alpha\beta} + \delta_{\alpha\beta} m x_j\}\phi_\beta(\boldsymbol{x}) \,, \\
P_j &= -\int d^3x \, \pi_\alpha(\boldsymbol{x})\frac{\partial}{\partial x_j}\phi_\alpha(\boldsymbol{x}), \ H = \frac{1}{2m}\int d^3x \, \pi_\alpha(\boldsymbol{x})\nabla^2\phi_\alpha(\boldsymbol{x}) \,, \\
M &= m\int d^3x \, \pi_\alpha(\boldsymbol{x})\phi_\alpha(\boldsymbol{x}) \,.
\end{aligned} \tag{19.96}$$

One has to use both Eqs. (19.94) and (19.95) to check that the proper PB relations hold. The canonical transformations generated by J_j and G_j, which are the interesting ones, can be checked to be as follows:

$$\begin{aligned}
\exp(\widetilde{\boldsymbol{\alpha}\cdot\boldsymbol{J}})\phi_\alpha(\boldsymbol{x}) &= \{\exp(-\boldsymbol{\alpha}\cdot\boldsymbol{S})\}_{\alpha\beta}\phi_\beta(A_{jk}(\boldsymbol{\alpha})x_k) \,, \\
\exp(\widetilde{\boldsymbol{\alpha}\cdot\boldsymbol{J}})\pi_\alpha(\boldsymbol{x}) &= \{\exp(\boldsymbol{\alpha}\cdot\boldsymbol{S})\}_{\beta\alpha}\pi_\beta(A_{jk}(\boldsymbol{\alpha})x_k) \,; \\
\exp(\widetilde{\boldsymbol{v}\cdot\boldsymbol{G}})\phi_\alpha(\boldsymbol{x}) &= \{\exp(-\boldsymbol{v}\cdot\boldsymbol{T})\}_{\alpha\beta}\exp(-m\boldsymbol{v}\cdot\boldsymbol{x})\phi_\beta(\boldsymbol{x}) \,, \\
\exp(\widetilde{\boldsymbol{v}\cdot\boldsymbol{G}})\pi_\alpha(\boldsymbol{x}) &= \{\exp(\boldsymbol{v}\cdot\boldsymbol{T})\}_{\beta\alpha}\exp(m\boldsymbol{v}\cdot\boldsymbol{x})\pi_\beta(\boldsymbol{x}) \,.
\end{aligned} \tag{19.97}$$

For this realization of $\mathcal{G}$, in addition to the invariant M, we have two more invariants corresponding to the Casimir invariants of $E^{(v)}(3)$. These are:

$$\begin{aligned}
\mathcal{J}^2 &= \sum_j \left\{ \int d^3x\, \pi_\alpha(\boldsymbol{x})(T_j)_{\alpha\beta}\phi_\beta(\boldsymbol{x}) \right\}^2 , \\
\mathcal{J}\cdot\mathcal{S} &= \sum_j \left\{ \int d^3x\, \pi_\alpha(\boldsymbol{x})(T_j)_{\alpha\beta}\phi_\beta(\boldsymbol{x}) \right\} \left\{ \int d^3x\, \pi_{\alpha'}(\boldsymbol{x})(S_j)_{\alpha'\beta'}\phi_{\beta'}(\boldsymbol{x}) \right\} .
\end{aligned} \tag{19.98}$$

The $\mathcal{S}^2$ is no longer an invariant.

However, it now turns out that by means of a canonical transformation, the generators in Eq. (19.96) can be made to assume the same forms as the generators in Eq. (19.92); in other words, the extra terms in G_j in Eq. (19.96) can be transformed away! If we define a dynamical variable X as follows:

$$X = \frac{1}{m} \int d^3x\ \pi_\alpha(\boldsymbol{x})(T_j)_{\alpha\beta} \frac{\partial}{\partial x_j} \phi_\beta(\boldsymbol{x}) , \tag{19.99}$$

then after a certain amount of routine algebra one finds, with J_j, G_j, P_j, H, and M being the generators given in Eq. (19.96),

$$\begin{aligned}
\exp(-\tilde{X})(J_j, P_j, H, M) &= (J_j, P_j, H, M); \\
\exp(-\tilde{X})G_j &= \int d^3x\ \pi_\alpha(\boldsymbol{x}) m x_j \phi_\alpha(\boldsymbol{x}) .
\end{aligned} \tag{19.100}$$

Therefore the infinite-component free-field realization of $\mathcal{G}$, constructed with the help of an infinite-dimensional matrix solution to Eq. (19.94), yields nothing essentially new. Because *any* solution to the commutation rules obeyed by the S_j can always be chosen to be a direct sum of (irreducible) finite-dimensional solutions, the infinite-component free field is canonically equivalent to a direct sum (or collection) of an infinite number of finite component fields, each part in the summand being associated with a particular matrix representation of $R(3)$!

The generators exhibited in Eqs. (19.92) and (19.96) are of the general form of Eq. (19.88), with the difference that the various local differential operators $a(\boldsymbol{x}, (\partial/\partial\boldsymbol{x}))$ are at the same time matrices in the space of the representation of $R(3)$ or $E^{(v)}(3)$ as the case may be. In the case of Eq. (19.92) we have the association

$$\begin{gathered}
J_j \to S_j - (\boldsymbol{x}\times\nabla)_j,\ P_j \to -\nabla_j,\ G_j \to m x_j , \\
H \to \tfrac{1}{2m}\nabla^2,\ M \to m ,
\end{gathered} \tag{19.101}$$

and in the case of Eq. (19.96) we have:

$$\begin{gathered}
J_j \to S_j - (\boldsymbol{x}\times\nabla)_j,\ P_j \to -\nabla_j,\ G_j \to T_j + m x_j \\
H \to \tfrac{1}{2m}\nabla^2,\ M \to m .
\end{gathered} \tag{19.102}$$

In both cases, these matrix cum differential operators obey commutation rules corresponding to $\mathcal{G}'$, and apart from some factors of i, they act as generators of quantum mechanical operator representations of $\mathcal{G}$. With Eq. (19.101) is associated the quantum mechanical description of a free non-relativistic particle with mass m and nonvanishing spin. The canonical equivalence of the generators in Eq. (19.96) to a direct sum of generators of the type in Eq. (19.92) has a counterpart in quantum mechanics. By an operator unitary transformation, the generators of the operator representation given in Eq. (19.102) become direct sums of generators of the type given in Eq. (19.101)!

We now discuss briefly the possibility of having an interacting Galilean field. As in the many-particle situation, it is conventional to require that J_j and P_j do not alter their forms on the inclusion of interactions, that is, they are kinematic, whereas H and/or $\boldsymbol{G}$ is dynamic. But with both dynamic, we again end up with the nonlinear relations that are hard to solve. Thus we restrict ourselves in the field case also to G_j being kinematic. (We assume the expression for M is also unchanged by interaction.) To make matters as simple as possible, we further assume that the interaction term V in H is the integral of some local density v which is a function of $\phi_\alpha(\boldsymbol{x}), \pi_\alpha(\boldsymbol{x})$, and their spatial gradients. With this assumption for V it is necessary to work with the general case of an infinite-component Galilean field, because the transformation $\exp(-\tilde{X})$ used in the foregoing equivalence proof is a nonlocal transformation in space. [This happens because in the definition of X in Eq. (19.99) the gradient $(\partial/\partial x_j)$ appears.] To specialize to a finite component field, we just set the T_j equal to zero. The Lie relations of $\mathcal{G}'$ now require that V have vanishing PB's with $J_j^{(0)}, P_j^{(0)}, G_j^{(0)}$ and M, the superscript $^{(0)}$ indicating that these are kinematic generators as defined in Eq. (19.96). If we write

$$V[\phi_\alpha, \pi_\alpha] = \int d^3x\ v(\boldsymbol{x}),\ \ v(\boldsymbol{x}) \equiv v(\phi_\alpha(\boldsymbol{x}), \pi_\alpha(\boldsymbol{x}); \partial_j\phi_\alpha, \partial_j\pi_\alpha; \cdots), \quad (19.103)$$

then we must demand that the PB's of $v(\boldsymbol{x})$ with the densities of any of the kinematic generators $J_j^{(0)}, P_j^{(0)}, G_j^{(0)}$ and M either vanish or are expressible as gradients which vanish on integration (dropping surface terms at infinity). The conditions that follow from the vanishing of $\{J_j^{(0)}, V\}$ and $\{P_j^{(0)}, V\}$ are not very hard to satisfy: they require that V be a translational and rotational invariant. The vanishing of $\{M, V\}$ implies that v be a sum of terms each consisting of an *equal* number of ϕ and π factors, or more generally that V be unchanged under the change of scale $\phi_\alpha \to \lambda\phi_\alpha, \pi_\alpha \to \lambda^{-1}\pi_\alpha, \lambda > 0$. This means, for example, that cubic interaction terms are not allowed. The vanishing of $\{G_j^{(0)}, V\}$ imposes special restrictions on the kinds of derivative terms that can be present in v. The simplest interactions are the nonderivative ones: we take polynomials, each term in which is a product of an equal number of ϕ's and π's, constructed in such a way that they are invariant under the *index* transformations

$$\begin{aligned}&\phi_\alpha(\boldsymbol{x}) \to \{\exp(-\boldsymbol{\alpha}\cdot\boldsymbol{S})\}_{\alpha\beta}\phi_\beta(\boldsymbol{x})\,, \pi_\alpha(\boldsymbol{x}) \to \{\exp(\boldsymbol{\alpha}\cdot\boldsymbol{S})\}_{\beta\alpha}\pi_\beta(\boldsymbol{x})\,;\\ &\phi_\alpha(\boldsymbol{x}) \to \{\exp(-\boldsymbol{v}\cdot\boldsymbol{T})\}_{\alpha\beta}\phi_\beta(\mathrm{x})\,, \pi_\alpha(\boldsymbol{x}) \to \{\exp(\boldsymbol{v}\cdot\boldsymbol{T})\}_{\beta\alpha}\pi_\beta(\boldsymbol{x})\,. \end{aligned} \quad (19.104)$$

Such terms give rise to "scalar" interactions. Other interactions are possible; for example, a term like $[\boldsymbol{\nabla}(\phi_\alpha(\boldsymbol{x})\pi_\alpha(\boldsymbol{x}))]^2$ is an acceptable v, but the characterization and classification of all these would take us too far afield.

For an interacting system, the equations of motion are nonlinear, and the canonical time evolution $\exp(-\widetilde{bH})$ cannot be given in closed form. But J_j, G_j, and P_j being all kinematic, the behavior under the other transformations of $\mathcal{G}$ is, of course, unchanged.

Lagrangian for the Free Galilean Field.

As in the many-particle case, it is interesting to rephrase some of our conclusions regarding field realizations of $\mathcal{G}$ in the Lagrangian framework. One difference between the two situations, however, is that in the case of fields the interaction V must depend on the momentum $\pi(\boldsymbol{x})$ because of the mass superselection rule, whereas potentials depending on the q's alone are permitted in the particle case. In order to be able to pass to the Lagrangian we therefore consider only free Galilean fields. But then it suffices to consider finite-component fields alone, the components transforming according to some (odd-dimensional irreducible real) matrix representation of $R(3)$; the generators for $\mathcal{G}$ will then be given by Eq. (19.92). In this connection it is necessary to remind the reader of one important property of the matrices S_j that appear in the expression for J_j, their *antisymmetry*. This follows from the fact that every matrix representation of $R(3)$ can be assumed to be given by real orthogonal matrices (of some odd dimension if irreducible), implying that the generator matrices may be taken real antisymmetric (page 332).

Let us then take the generators given in Eq. (19.92) and see how to pass to the Lagrangian. Now in terms of the variables ϕ, π, the Hamiltonian has the form $H \sim \pi\phi$; this means that the first set of the Hamiltonian equations of motion, $\dot{\phi} = \{\phi, H\}$, do not involve π at all, and thus cannot be solved to express π in terms of ϕ and $\dot{\phi}$! In terms of the ϕ, π variables, H is singular; we must make a canonical transformation to new variables, in terms of which H is nonsingular. (From Chapter 8, pages 78-79. we know that this can always be achieved.) This can be done as follows:

$$\begin{aligned}\xi_r(\boldsymbol{x}) =& \frac{1}{\sqrt{2}}(\pi_r(\boldsymbol{x}) + \phi_r(\boldsymbol{x})),\ \eta_r(\boldsymbol{x}) = \frac{1}{\sqrt{2}}(\pi_r(\boldsymbol{x}) - \phi_r(\boldsymbol{x}));\\ \{\xi_r(\boldsymbol{x}), \xi_s(\boldsymbol{y})\} =& \{\eta_r(\boldsymbol{x}), \eta_s(\boldsymbol{y})\} = 0;\ \{\xi_r(\boldsymbol{x}), \eta_s(\boldsymbol{y})\} = \delta_{rs}\delta^{(3)}(\boldsymbol{x} - \boldsymbol{y}). \quad (19.105)\end{aligned}$$

The ξ's are the new q's, the η's the conjugate p's. We can now rewrite all the generators of $\mathcal{G}$ as functionals of ξ and η. Dropping surface terms at (spatial) infinity whenever necessary and using the antisymmetry of the matrices S_j, we arrive at these expressions:

$$\begin{aligned}J_j &= \int d^3x\ \eta_r(\boldsymbol{x})\{(S_j)_{rs} - (\boldsymbol{x} \times \boldsymbol{\nabla})_j\delta_{rs}\}\xi_s(\boldsymbol{x}),\\ P_j &= -\int d^3x\ \eta_r(\boldsymbol{x})\frac{\partial}{\partial x_j}\xi_r(\boldsymbol{x}),\end{aligned}$$

$$\begin{aligned} G_j &= \frac{m}{2}\int d^3x\; x_j(\xi_r(\boldsymbol{x})\xi_r(\boldsymbol{x})) - -\eta_r(\boldsymbol{x})\eta_r(\boldsymbol{x}))\,, \\ H &= \frac{1}{4m}\int d^3x\{\xi_r(\boldsymbol{x})\nabla^2\xi_r(\boldsymbol{x}) - \eta_r(\boldsymbol{x})\nabla^2\eta_r(\boldsymbol{x})\} \\ &= \frac{1}{4m}\int d^3x\{(\boldsymbol{\nabla}\eta_r(\boldsymbol{x}))^2 - (\boldsymbol{\nabla}\xi_r(\boldsymbol{x}))^2\}\,, \\ M &= \frac{m}{2}\int d^3x(\xi_r(\boldsymbol{x})\xi_r(\boldsymbol{x}) - \eta_r(\boldsymbol{x})\eta_r(\boldsymbol{x}))\,. \end{aligned} \tag{19.106}$$

The densities occurring here differ from those occurring in Eq. (19.92) by gradients. We see that the ξ's and η's make contributions of opposing sign both to the energy and to the total mass.

We can now solve for the η's in terms of the ξ's and $\dot{\xi}$'s. The full set of Hamiltonian equations of motion is

$$\dot{\xi}(\boldsymbol{x},t) = -\frac{1}{2m}\nabla^2\eta_r(\boldsymbol{x},t)\,, \tag{19.107a}$$

$$\dot{\eta}_r(\boldsymbol{x},t) = -\frac{1}{2m}\nabla^2\xi_r(\boldsymbol{x},t)\,. \tag{19.107b}$$

We can invert Eq. (19.107a) using the function $D(\boldsymbol{x})$:

$$\begin{aligned} D(\boldsymbol{x}) &= \frac{-1}{(2\pi)^3}\int \frac{\exp(i\boldsymbol{k}\cdot\boldsymbol{x})}{k^2}d^3k = -\frac{1}{4\pi}\frac{1}{|\boldsymbol{x}|}\,; \\ \nabla^2 D(\boldsymbol{x}) &= \delta^{(3)}(\boldsymbol{x})\,. \end{aligned} \tag{19.108}$$

We get:

$$\eta_r(\boldsymbol{x},t) = -2m\int d^3y D(\boldsymbol{x}-\boldsymbol{y})\dot{\xi}_r(\boldsymbol{y},t)\,. \tag{19.109}$$

The Lagrangian can then be set up according to the usual rules as a functional of ξ and $\dot{\xi}$ and turns out to be:

$$\begin{aligned} L[\xi,\dot{\xi}] &\equiv \int d^3x\;\eta_r(\boldsymbol{x},t)\dot{\xi}_r(\boldsymbol{x},t) - H \\ &= -m\int\!\!\int d^3x d^3y\;\dot{\xi}_r(\boldsymbol{x},t)D(\boldsymbol{x}-\boldsymbol{y})\dot{\xi}_r(\boldsymbol{y},t) + \frac{1}{4m}\int d^3x(\boldsymbol{\nabla}\xi_r(\boldsymbol{x},t))^2\,. \end{aligned} \tag{19.110}$$

This is the Lagrangian for the free finite component Galilean field. It is quadratic in the velocities $\dot{\xi}_r$ and also in the gradients of the coordinates ξ_r. It is interesting that the Lagrangian formulation is not local even though the Hamiltonian is local in the field variables; the cause of this is the nonlocal relation between the velocities and the momenta, as follows from Eq. (19.109). (To be precise, the velocity is a local function of the momentum, but not conversely.) The

Lagrangian equation of motion is, of course, equivalent to Eq. (19.107); it appears as:

$$\frac{\partial}{\partial t}\left(-2m\int d^3y D(\boldsymbol{x}-\boldsymbol{y})\dot{\xi}_r(\boldsymbol{y},t)\right) = -\frac{1}{2m}\nabla^2\xi_r(\boldsymbol{x},t)\,,$$

that is,

$$\int d^3y D(\boldsymbol{x}-\boldsymbol{y})\ddot{\xi}_r(\boldsymbol{y},t) = \frac{1}{4m^2}\nabla^2\xi_r(\boldsymbol{x},t)\,. \tag{19.111}$$

Using the definition of $D(\boldsymbol{x})$, this can be converted into a local form:

$$\left[\frac{\partial^2}{\partial t^2}-\frac{1}{4m^2}(\nabla^2)^2\right]\xi_r(\boldsymbol{x},t) \equiv \left(\frac{\partial}{\partial t}+\frac{\nabla^2}{2m}\right)\left(\frac{\partial}{\partial t}-\frac{\nabla^2}{2m}\right)\xi_r(\boldsymbol{x},t) = 0\,. \tag{19.112}$$

Note that the first term in L, involving the velocities $\dot{\xi}$, is positive definite; this can be seen using the form of the Fourier transform of $D(\boldsymbol{x})$, and the reality of the field variables $\xi_r(\boldsymbol{x},t)$.

The invariance properties of the Hamiltonian under the Euclidean group go over into invariance properties of the Lagrangian as well. The corresponding transformations on the ξ's are defined thus:

$$\begin{aligned} R(3) &\;:\; \xi_r(\boldsymbol{x},t) \to D_{rs}(\boldsymbol{\alpha})\xi_s(A_{kj}(\boldsymbol{\alpha})x_k,t)\,; \\ T_3 &\;:\; \xi_r(\boldsymbol{x},t) \to \xi_r(\boldsymbol{x}+\boldsymbol{a},t)\,. \end{aligned} \tag{19.113}$$

These invariances of L lead to the conservation of angular and linear momenta. More interesting is the action of a velocity transformation in $\mathcal{G}$. The transformation law for $\xi_r(\boldsymbol{x},t)$ is not immediately evident, but can be inferred from the Hamiltonian version. There, for a general time t, the canonical transformation that implements a velocity transformation is actually independent of t. (See Chapter 16, page 286; this is a property of the Heisenberg picture.) Using the fields ϕ_r, π_r, we have:

$$\phi_r'(\boldsymbol{x},t) = \exp(\widetilde{\boldsymbol{v}\cdot\boldsymbol{G}})\phi_r(\boldsymbol{x},t) = \exp(\widetilde{\boldsymbol{v}\cdot\boldsymbol{G}})\exp(\widetilde{-tH})\phi_r(\boldsymbol{x})\,. \tag{19.114}$$

We can invert the sequence of canonical transformations here this way:

$$\exp(\widetilde{\boldsymbol{v}\cdot\boldsymbol{G}})\exp(\widetilde{-tH}) = \exp(\widetilde{-tH})\exp(\widetilde{\boldsymbol{v}\cdot\boldsymbol{G}})\exp(-t\boldsymbol{v}\cdot\widetilde{\boldsymbol{P}})\,. \tag{19.115}$$

This can be derived from the Baker-Campbell-Hausdorff formula, the PB relations of $\mathcal{G}'$, and the fact that $\tilde{M}$ commutes with $\tilde{P}_j, \tilde{G}_j$, and $\tilde{H}$. We now use the known action of these canonical transformations (see Eqs. (19.85) and (19.86)), and thus evaluate the right-hand side of Eq. (19.114):

$$\begin{aligned} \phi_r'(\boldsymbol{x},t) &= \exp(\widetilde{-tH})\exp(\widetilde{\boldsymbol{v}\cdot\boldsymbol{G}})\exp(\widetilde{-t\boldsymbol{v}\cdot\boldsymbol{P}})\phi_r(\boldsymbol{x}) \\ &= \exp(\widetilde{-tH})\exp(\widetilde{\boldsymbol{v}\cdot\boldsymbol{G}})\phi_r(\boldsymbol{x}-\boldsymbol{v}t) \\ &= \exp(\widetilde{-tH})\exp[-m\boldsymbol{v}\cdot(\boldsymbol{x}-\boldsymbol{v}t)]\phi_r(\boldsymbol{x}-\boldsymbol{v}t)\,, \\ \phi_r'(\boldsymbol{x},t) &= \exp[-m\boldsymbol{v}\cdot(\boldsymbol{x}-\boldsymbol{v}t)]\phi_r(\mathbf{x}-\boldsymbol{v}t,t)\,. \end{aligned} \tag{19.116}$$

In a similar way we get the transformation law for $\pi_r(\mathbf{x},t)$:

$$\pi'_r(\boldsymbol{x},t) = \exp(\widetilde{\boldsymbol{v}\cdot\boldsymbol{G}})\pi_r(\boldsymbol{x},t) = \exp[m\boldsymbol{v}\cdot(\boldsymbol{x}-\boldsymbol{v}t)]\pi_r(\boldsymbol{x}-\boldsymbol{v}t,t)\,. \tag{19.117}$$

We must now transcribe these canonical transformation laws for ϕ and π into a transformation law for the Lagrangian variable $\xi_r(\boldsymbol{x},t)$; the result is expressible entirely in terms of ξ and $\dot{\xi}$ and is:

$$\begin{aligned}\xi_r(x,t) \to \xi'_r(\boldsymbol{x},t) &= \cosh(m\boldsymbol{v}\cdot(\boldsymbol{x}-\boldsymbol{v}t))\xi_r(\boldsymbol{x}-\boldsymbol{v}t,t)\\ &\quad - 2m\sinh(m\boldsymbol{v}\cdot(\boldsymbol{x}-\boldsymbol{v}t))\int d^3yD(\boldsymbol{x}-\boldsymbol{y}-\boldsymbol{v}t)\dot{\xi}_r(\boldsymbol{y},t)\,.\end{aligned} \tag{19.118}$$

Within the Lagrangian framework, this then is the transformation law for the q's of the theory under a finite velocity transformation belonging to $\mathcal{G}$. To discuss the invariance of the Lagrangian equations of motion, we need only the infinitesimal form of Eq. (19.118). For ξ and its velocity, this is:

$$\begin{aligned}\xi_r(\boldsymbol{x},t) \to \xi'_r(\boldsymbol{x},t) &= \xi_r(\boldsymbol{x},t) + \delta\xi_r(\boldsymbol{x},t)\,,\\ \delta\xi_r(\boldsymbol{x},t) &= -t\boldsymbol{v}\cdot\nabla\xi_r(\boldsymbol{x},t) - 2m^2\boldsymbol{v}\cdot\boldsymbol{x}\int d^3yD(\boldsymbol{x}-\boldsymbol{y})\dot{\xi}_r(\boldsymbol{y},t)\,;\\ \delta\dot{\xi}_r(\boldsymbol{x},t) &= -\boldsymbol{v}\cdot\nabla\xi_r(\boldsymbol{x},t) - t\boldsymbol{v}\cdot\nabla\dot{\xi}_r(\boldsymbol{x},t)\\ &\quad -2m^2\boldsymbol{v}\cdot\boldsymbol{x}\int d^3yD(\boldsymbol{x}-\boldsymbol{y})\ddot{\xi}_r(\boldsymbol{y},t)\,.\end{aligned} \tag{19.119}$$

The Lagrangian is not invariant under this transformation. But the change in it can be expressed as a total time derivative, thus it is quasi-invariant, and that accounts in the standard way for the invariance of the equations of motion under this transformation. By discarding terms at (spatial) infinity, we find:

$$\begin{aligned}\delta L[\xi,\dot{\xi}] &\equiv L[\xi+\delta\xi,\dot{\xi}+\delta\dot{\xi}] - L[\xi,\dot{\xi}] = \frac{d}{dt}\boldsymbol{v}\cdot\boldsymbol{F}[\xi,\dot{\xi}]\,,\\ \boldsymbol{F}[\xi,\dot{\xi}] &= \frac{m}{2}\int d^3x\;\boldsymbol{x}\xi_r(\boldsymbol{x},t)\xi_r(\boldsymbol{x},t)\\ &\quad + 2m^2\int\!\!\int\!\!\int d^3xd^3yd^3z\;\dot{\xi}_r(\boldsymbol{x},t)D(\boldsymbol{x}-\boldsymbol{y})\boldsymbol{y}D(\boldsymbol{y}-\boldsymbol{z})\dot{\xi}_r(\boldsymbol{z},t)\,.\end{aligned} \tag{19.120}$$

Therefore, if C is any trajectory in configuration space and if C' is obtained from C by the infinitesimal transformation Eq. (19.119) (in which the only infinitesimal quantities are the components of $\boldsymbol{v}$), then the action functionals for C and C' obey:

$$\Phi[C'] = \Phi[C] + \boldsymbol{v}\cdot\boldsymbol{F}[\xi(\boldsymbol{x},t),\dot{\xi}(\boldsymbol{x},t)]|_{t_1}^{t_2}\,. \tag{19.121}$$

Therefore as in the case of a many-particle system, the equations of motion are valid on C' if they are valid on C. Combining Eq. (19.121) with the alternative and general expression for the difference $\Phi[C'] - \Phi[C]$, we check that this invariance property of the equations of motion is not associated with a conservation law for some explicitly time-independent quantity, but leads back to the known fact that the generator $\boldsymbol{G}$ has a linear time dependence. [The linear t-dependence comes in from the explicit linear time dependence in $\delta\xi_r(\boldsymbol{x},t)$ in Eq. (19.119).]

We conclude this discussion of the Lagrangian for the free Galilean field by considering the conservation law for the mass. Analogous to Eq. (19.118), we can work out the canonical transformation generated by M and express it entirely in Lagrangian variables. If the parameter of the transformation is λ, the finite form is:

$$\xi_r(\boldsymbol{x},t) \to \xi_r'(\boldsymbol{x},t) = \cosh(m\lambda)\xi_r(\boldsymbol{x},t) - 2m\sinh(m\lambda)\int d^3y D(\boldsymbol{x}-\boldsymbol{y})\dot{\xi}_r(\boldsymbol{y},t)\,, \tag{19.122}$$

whereas for $|\delta\lambda| \ll 1$, we have:

$$\begin{aligned} \delta\xi_r(\boldsymbol{x},t) &= -2m^2\delta\lambda\int d^3y D(\boldsymbol{x}-\boldsymbol{y})\dot{\xi}_r(\boldsymbol{y},t)\,, \\ \delta\dot{\xi}_r(\boldsymbol{x},t) &= -2m^2\delta\lambda\int d^3y D(\boldsymbol{x}-\boldsymbol{y})\ddot{\xi}_r(\boldsymbol{y},t)\,. \end{aligned} \tag{19.123}$$

Again the Lagrangian is changed by the addition of a total derivative:

$$\begin{aligned} \delta L[\xi,\dot{\xi}] = -\,\delta\lambda\frac{d}{dt}\Big[m\int\!\!\int\!\!\int d^3x d^3y d^3z\,\xi_r(\boldsymbol{x},t)D(\boldsymbol{x}-\boldsymbol{y})D(\boldsymbol{y}-\boldsymbol{z})\dot{\xi}_r(\boldsymbol{z},t) \\ +\frac{1}{4m}\int d^3x\,\xi_r(\boldsymbol{x},t)\xi_r(\boldsymbol{x},t)\Big]\,. \end{aligned} \tag{19.124}$$

Thus once again the equations of motion are invariant under the transformations Eqs. (19.122) and (19.123). But this time, because neither $\delta\xi_r(\boldsymbol{x},t)$ nor the expression whose time derivative appears on the right-hand side of Eq. (19.124) has any *explicit* time dependence, this invariance property of the equations of motion is associated with a conservation law, and the conserved quantity is the total mass M expressed in Lagrangian variables! (The details of checking this statement are left to the reader.)

Standard Versus Nonstandard States for Galilean Fields

We must remark on the fact that in all our discussion of Galilean fields, none of the eleven generators J_j, P_j, G_j, H, and M is of the "finite polynomial" type, (For this notion, see Chapter 10.) By their very definition each of them involves all the degrees of freedom of the field system. Therefore in general only the PB's of the *densities* of these generators with one another are well-defined; the PB's of the integrated densities may or may not exist. There are two alternatives open to us. The first alternative is to consider only those configurations of the field for which these eleven generators exist. By virtue of the physical interpretation of H as the generator of time translations, if this requirement is imposed at one instant of time it is automatically true at all times. We call these, as in Chapter 11, the "standard configurations". By definition, then, the standard configurations of the free Galilean field furnish a (real) realization of the Galilei group (use being made of the mass superselection rule to select observables). We know that this realization is reducible, because of the existence of invariants like the total mass (which is finite for these configurations!). Subsets of standard configurations of the field, all with one value of M, are invariant under the transformations realizing $\mathcal{G}$.

The second alternative is to consider arbitrary, in general nonstandard, configurations of the classical field, but then some or all of the eleven generators may not exist. In this case, the evaluation of the PB's of the eleven generators with the local field quantities ϕ, π is purely formal, because the former do not exist. Nevertheless, we can look upon the transformations of the type in Eq. (19.86) say, as well-defined correspondences that preserve the fundamental PB relations. Taking just the case of the spin-zero field, we have, for instance, the correspondences

$$\begin{aligned}\phi(\boldsymbol{x}), \pi(\boldsymbol{x}) &\to \phi(\boldsymbol{x}+\boldsymbol{a}), \pi(\boldsymbol{x}+\boldsymbol{a})\,;\\ &\to \phi(A_{jk}(\boldsymbol{\alpha})x_k), \pi(A_{jk}(\boldsymbol{\alpha})x_k)\,;\\ &\to \exp(-m\boldsymbol{v}\cdot\boldsymbol{x})\phi(\boldsymbol{x}), \exp(m\boldsymbol{v}\cdot\boldsymbol{x})\pi(\boldsymbol{x})\,;\\ &\to \exp[(t/2m)\nabla^2]\phi(\boldsymbol{x}), \exp[(-t/2m)\nabla^2]\pi(\boldsymbol{x})\,. \end{aligned} \tag{19.125}$$

These are all automorphisms of the fundamental field variables, which preserve the PB relations, whether or not we have the standard configurations. We have, in all cases, an eleven-parameter group of automorphisms that furnishes an explicit realization of the extended group $\mathcal{G}'$. The infinitesimal forms of these automorphisms are derivations on the ϕ's and π's. In the standard configurations, these automorphisms are inner automorphisms, the derivations are inner derivations, and each derivation can be written as a well-defined tilded operator ($\tilde{J}, \tilde{P}$ etc.) acting on the ϕ's and π's. In the nonstandard configurations, we do not have the J's, P's, and so on defined, but the automorphisms are still valid as are the derivations. They now become, however, outer automorphisms and outer derivations. (For automorphisms and derivations of a Lie algebra see Chapter 15, pages 246-247.)

Fluid Mechanics as a Galilean Field Theory

The examples of Galilean invariant field theories discussed up to now may strike the reader as being purely mathematical examples, not referring to any recognizable physical situations. This is especially so because the expression for the total mass in all these examples was not, and could not be restricted to be, positive definite. Partly to remedy this situation, we give now a discussion of the basic equations of fluid dynamics and show how they lead to a field realization of the Galilei group. Although these examples are more physical than the previous ones, they are still a little bit unrealistic in the sense that we only discuss a fluid flowing unrestrictedly in all of space, unlimited by obstacles like cylinders and spheres, and more importantly, is not subject to any externally applied forces like, say, the gravitational force. If either of these things were to be included, we would lose Galilean invariance! We also must ignore the presence of viscous forces in favor of getting a realization of $\mathcal{G}$. Thus the only force involved is that due to the isotropic hydrostatic pressure, which is taken to be some function of the density of the fluid; the density itself is a basic variable, meaning that we are considering the case of a compressible fluid.

The equations of motion for fluid dynamics are derived on the basis of elementary mechanics as follows. Let $\rho(\boldsymbol{x}, t)$ and $\boldsymbol{u}(\boldsymbol{x}, t)$ be the mass density of the fluid and the velocity of the fluid at time t at the point $\boldsymbol{x}$ in space. It is important to understand that $\rho(\boldsymbol{x}, t)$ is not the density of a fixed portion of the fluid, but is the density of that part of the fluid that happens to be in the vicinity of the point $\boldsymbol{x}$ at the time t. Similar remarks apply to the meaning of $\boldsymbol{u}(\boldsymbol{x}, t)$. If we consider a small portion of the fluid contained in a small element of volume δv, this element of volume will undergo changes as this portion of the fluid moves around from place to place, but by definition the total mass of fluid contained in this variable volume remains the same. If this volume of fluid is located around the point $\boldsymbol{x}$ at time t, its momentum at this time will be $\rho(\boldsymbol{x}, t)\, \delta v_t\, \boldsymbol{u}(\boldsymbol{x}, t)$. After a small interval of time $\triangle t$, the location of the fluid is $\boldsymbol{x} + \triangle t\, \boldsymbol{u}(\boldsymbol{x}, t)$; its volume will have changed to some $\delta v_{t+\triangle t}$, but in such a way that $\rho(\boldsymbol{x}, t)\delta v_t$ is equal to the new density, $\rho(\boldsymbol{x}+\boldsymbol{u}(\boldsymbol{x}, t)\triangle t,\ t+\triangle t)$, times $\delta v_{t+\triangle t}$. Therefore the momentum of this same portion of the fluid at the time $t + \triangle t$ is $\rho(\boldsymbol{x}, t)\, \delta v_t\, \boldsymbol{u}(\boldsymbol{x} + \boldsymbol{u}(\boldsymbol{x}, t)\triangle t,\ t + \triangle t)$. We can now compute the acceleration; it is:

$$\underset{\triangle t \to 0}{lt.}\ \frac{1}{\triangle t}\rho(\boldsymbol{x}, t)\delta v_t[\boldsymbol{u}(\boldsymbol{x} + \boldsymbol{u}(\boldsymbol{x}, t)\triangle t, t + \triangle t) - \boldsymbol{u}(\boldsymbol{x}, t)]$$
$$= \rho(\boldsymbol{x}, t)\delta v_t \left[\frac{\partial \boldsymbol{u}(\boldsymbol{x}, t)}{\partial t} + \boldsymbol{u}(\boldsymbol{x}, t) \cdot \boldsymbol{\nabla} \boldsymbol{u}(\boldsymbol{x}, t)\right]. \tag{19.126}$$

The coefficient of $\rho\delta v$ is, then, the acceleration per unit mass of that part of the fluid that is at $\boldsymbol{x}$ at time t; it contains two parts, a "temporal" change $\partial \boldsymbol{u}/\partial t$, and a "convective" change $\boldsymbol{u} \cdot \boldsymbol{\nabla} \boldsymbol{u}$ arising from the motion of the fluid and the basic variables $\rho, \boldsymbol{u}$ not referring to fixed portions of the fluid. If now $\boldsymbol{F}(\boldsymbol{x}, t)$ is the force acting per unit volume of the fluid, as a function of $\boldsymbol{x}$ and time, Newton's equation of motion applied to the moving element of fluid δv_t reads:

$$\rho(\boldsymbol{x}, t)\delta v_t \left[\frac{\partial \boldsymbol{u}(\boldsymbol{x}, t)}{\partial t} + \boldsymbol{u}(\boldsymbol{x}, t) \cdot \boldsymbol{\nabla} \boldsymbol{u}(\boldsymbol{x}, t)\right] = \boldsymbol{F}(\boldsymbol{x}, t)\delta v_t\,. \tag{19.127}$$

The only force we are considering is that induced by the pressure $p(\boldsymbol{x}, t)$ in terms of which $\boldsymbol{F} = -\boldsymbol{\nabla} p$. We are thus led to the Euler equations for a fluid:

$$\frac{\partial \boldsymbol{u}}{\partial t} + \boldsymbol{u} \cdot \boldsymbol{\nabla} \boldsymbol{u} = -\frac{1}{\rho}\boldsymbol{\nabla} p\,. \tag{19.128}$$

The equation of state expresses $p(\boldsymbol{x}, t)$ as a local function of $\rho(\boldsymbol{x}, t)$, $p = p(\rho)$.

We take advantage of the foregoing derivation to note the following. Let $\psi(\boldsymbol{x}, t)$ denote *any* local property of the fluid. Then the rate of change in time of this property as we move with the fluid and consider a fixed portion of the fluid is always given by:

$$\frac{D\psi}{Dt} \equiv \frac{\partial \psi}{\partial t} + \boldsymbol{u} \cdot \boldsymbol{\nabla} \psi \,. \tag{19.129}$$

To the equation of motion Eq. (19.128) must be added the *equation of continuity* to get the full set of equations describing fluid flow. This is obtained in the standard way by noticing that if we consider a small volume δv_0 *fixed in space (not* moving with the fluid), then the changes in the density ρ at points inside δv_0 must be directly related to the inflow and outflow of fluid through the surface bounding δv_0. This leads to:

$$\frac{\partial \rho(\boldsymbol{x}, t)}{\partial t} + \boldsymbol{\nabla} \cdot (\rho \boldsymbol{u}) = 0 \,. \tag{19.130}$$

Notice that this can be rewritten in the form:

$$\frac{D\rho}{Dt} + \rho \boldsymbol{\nabla} \cdot \boldsymbol{u} = 0 \,,$$

that is,

$$\frac{D}{Dt} \ln \rho = -\boldsymbol{\nabla} \cdot \boldsymbol{u} \,, \tag{19.131}$$

which helps us to see that if ρ is nowhere negative at one instant of time, then it preserves this property for all times. It also follows from Eq. (19.130) that the total mass,

$$M = \int d^3x \; \rho(\boldsymbol{x}, t) \,, \tag{19.132}$$

is a constant of motion.

We now want to show how Eqs. (19.128) and (19.130) can be derived from a Lagrangian, how one can then pass to a Hamiltonian and a set of basic PB's, and also to set up the full set of generators needed to give a canonical realization of $\mathcal{G}$. The expression for the Hamiltonian density, or energy density, is expected on physical grounds to be of the form

$$\mathcal{H}(\boldsymbol{x}, t) = \frac{1}{2} \rho \boldsymbol{u} \cdot \boldsymbol{u} + \rho W(\rho) \,, \tag{19.133}$$

where $W(\rho)$ is the "potential energy" per unit mass which is related to the pressure $p(\rho)$ in a manner to be specified. Similarly, on physical grounds we would expect the densities of linear and angular momentum to be

$$P_j(\boldsymbol{x}, t) = \rho u_j, \, J_j(\boldsymbol{x}, t) = \epsilon_{jkm} x_k \rho u_m \,, \tag{19.134}$$

and the linear and angular momenta to be spatial integrals of these. We must now consider separately the special case of *irrotational* fluid flow for which $\nabla \times \boldsymbol{u} = 0$ for all time, and the general case of flow with *non-vanishing* vorticity $\boldsymbol{\Omega}(\boldsymbol{x}, t) = \nabla \times \boldsymbol{u}(\boldsymbol{x}, t)$, because the basic field variables to be introduced in the two cases are not the same. (Clearly there are more degrees of freedom for general flow as compared to irrotational flow.)

Irrotational Fluid Flow

For irrotational motion, the velocity $\boldsymbol{u}$ is the gradient of a scalar velocity potential $-\phi(\boldsymbol{x}, t)$,

$$\boldsymbol{u} = -\nabla\phi\,. \tag{19.135}$$

We can use this to simplify Eq. (19.128) so that it becomes a *scalar* equation. Because the pressure p is some function of ρ, the right-hand side of Eq. (19.128) can be taken to be the (negative of the) gradient of some function $f(\rho)$. Using simple vector algebra, we can put Eq. (19.128) into this form:

$$\nabla(-\dot{\phi} + \frac{1}{2}(\nabla\phi)^2 + f(\rho)) = 0, \quad \nabla f = \frac{1}{\rho}\nabla p\,. \tag{19.136}$$

Considering only those configurations of the system in which the relevant field quantities vanish at spatial infinity, we may drop the ∇ operator in the equation here containing ϕ; thus the basic equations of motion become both scalar equations and are:

$$\frac{\partial\phi}{\partial t} - \frac{1}{2}(\nabla\phi)^2 - f(\rho) = 0\,, \tag{19.137a}$$

$$\frac{\partial\rho}{\partial t} - \nabla\cdot(\rho\nabla\phi) = 0\,. \tag{19.137b}$$

These are the equations to be obtained from a suitable Lagrangian. We can see right away that because these equations are first-order differential equations in time, the Lagrangian must be linear in the velocities $\dot{\rho}$ and $\dot{\phi}$. Therefore, the theory described on pages 129-130 in Chapter 9 can be used. For such Lagrangians, the corresponding Hamiltonian is simply the negative of the velocity independent term in the Lagrangian; thus from Eq. (19.133) we know what these terms in the Lagrangian should look like. We expect that the Lagrangian density has the following form:

$$\mathcal{L}(\boldsymbol{x}, t) = -\frac{1}{2}\rho(\nabla\phi)^2 - \rho W(\rho) + \text{terms linear in } \dot{\rho} \text{ and } \dot{\phi}\,. \tag{19.138}$$

Comparing this with Eq. (19.137a), we see that the equation of motion for ϕ will be gotten correctly if we choose the Lagrangian to be

$$\begin{aligned} L &= \int d^3x\; \mathcal{L}(\boldsymbol{x}, t)\,, \\ \mathcal{L}(\boldsymbol{x}, t) &= \rho\dot{\phi} - \frac{1}{2}\,\rho(\nabla\phi)^2 - \rho W(\rho)\,, \end{aligned} \tag{19.139}$$

with the potential energy per unit mass, $W(\rho)$, determined thus:

$$\frac{d}{d\rho}(\rho W(\rho)) = f(\rho), \frac{df}{d\rho} = \frac{1}{\rho}\frac{dp}{d\rho}. \tag{19.140}$$

In fact, Eq. (19.137a) is simply the Euler-Lagrange variational equation corresponding to the coordinate ρ,because L as chosen here is not dependent on $\dot{\rho}$ and the left-hand side of Eq. (19.137a) is the partial functional derivative of L with respect to $\rho(\boldsymbol{x}, t)$. But with the choice of Eq. (19.139), we get the equation of motion for ρ correctly as well, by considering the variational equation for the coordinate ϕ. We find:

$$\begin{aligned} \frac{\delta L}{\delta\dot{\phi}(\boldsymbol{x},t)} &= \rho(\boldsymbol{x},t), \ \frac{\delta L}{\delta\phi(\boldsymbol{x},t)} = \boldsymbol{\nabla}\cdot(\rho\boldsymbol{\nabla}\phi); \\ \frac{d}{dt}\frac{\delta L}{\delta\dot{\phi}(\boldsymbol{x},t)} - \frac{\delta L}{\delta\phi(\boldsymbol{x},t)} &= 0 \Rightarrow \dot{\rho} - \boldsymbol{\nabla}\cdot(\rho\boldsymbol{\nabla}\phi) = 0. \end{aligned} \tag{19.141}$$

Therefore, this Lagrangian gives correctly both the equation of motion and the equation of continuity for the case of irrotational fluid flow, the generalized coordinates being the fields ρ and ϕ.

We now apply the theory of Chapter 9 to this Lagrangian to get the form of the basic PB's. We enumerate the continuously infinite set of coordinates by $\rho(\boldsymbol{x})$, $\phi(\boldsymbol{x}), \boldsymbol{x}$ running over all space. Then L has the form:

$$\begin{aligned} L[\rho,\phi;\dot{\rho},\dot{\phi}] &= \int d^3x(A_{\rho(\boldsymbol{x})}\dot{\rho}(\boldsymbol{x}) + A_{\phi(\boldsymbol{x})}\dot{\phi}(\boldsymbol{x})) - V[\rho,\phi], \\ A_{\rho(\boldsymbol{x})} = 0, &\quad A_{\phi(\boldsymbol{x})} = \rho(\boldsymbol{x}); \\ V[\rho,\phi] &= \int d^3x\ \rho(\boldsymbol{x},t)[\frac{1}{2}(\boldsymbol{\nabla}\phi(\boldsymbol{x},t))^2 + W(\rho)]. \end{aligned} \tag{19.142}$$

The antisymmetric matrix $\eta_{\cdot,\cdot}$ defined as the "curl" of the "vector" $A_\cdot$ has components

$$\begin{aligned} \eta_{\rho(\boldsymbol{x}),\rho(\boldsymbol{y})} &\equiv \frac{\delta A_{\rho(\boldsymbol{y})}}{\delta\rho(\boldsymbol{x})} - \frac{\delta A_{\rho(\boldsymbol{x})}}{\delta\rho(\boldsymbol{y})} = 0, \\ \eta_{\rho(\boldsymbol{x}),\phi(\boldsymbol{y})} &\equiv \frac{\delta A_{\phi(\boldsymbol{y})}}{\delta\rho(\boldsymbol{x})} - \frac{\delta A_{\rho(\boldsymbol{x})}}{\delta\phi(\boldsymbol{y})} = \delta^{(3)}(\boldsymbol{x}-\boldsymbol{y}), \\ \eta_{\phi(\boldsymbol{x}),\rho(\boldsymbol{y})} &= -\eta_{\rho(\boldsymbol{y}),\phi(\boldsymbol{x})} = -\delta^{(3)}(\boldsymbol{x}-\boldsymbol{y}), \\ \eta_{\phi(\boldsymbol{x}),\phi(\boldsymbol{y})} &\equiv \frac{\delta A_{\phi(\boldsymbol{y})}}{\delta\phi(\boldsymbol{x})} - \frac{\delta A_{\phi(\boldsymbol{x})}}{\delta\phi(\boldsymbol{y})} = 0. \end{aligned} \tag{19.143}$$

(All the derivatives appearing here are functional derivatives.) The only nonvanishing components of the inverse matrix $\eta^{\cdot\cdot}$ are

$$\eta^{\rho(\boldsymbol{x}),\phi(\boldsymbol{y})} \equiv -\eta^{\phi(\boldsymbol{y}),\rho(\boldsymbol{x})} = -\delta^{(3)}(\boldsymbol{x}-\boldsymbol{y}); \tag{19.144}$$

and this leads to the basic bracket relations among the $\rho(\boldsymbol{x})$ and $\phi(\boldsymbol{y})$:

$$\{\rho(\boldsymbol{x}), \rho(\boldsymbol{y})\} = \{\phi(\boldsymbol{x}), \phi(\boldsymbol{y})\} = 0, \qquad \{\phi(\boldsymbol{x}), \rho(\boldsymbol{y})\} = \delta^{(3)}(\boldsymbol{x} - \boldsymbol{y}). \tag{19.145}$$

From the sign of the nonvanishing PB, we may think of the density ρ as a "p" and of the velocity potential ϕ as its canonically conjugate "q". The Hamiltonian coincides with $V[\rho, \phi]$ because the terms linear in the velocities cancel out in passing from L to H; thus we have:

$$H[\rho, \phi] = \int d^3x\ \rho(\boldsymbol{x})[\frac{1}{2}(\boldsymbol{\nabla}\phi(\boldsymbol{x}))^2 + W(\rho(\boldsymbol{x}))]. \tag{19.146}$$

(At this point, in accordance with the Hamiltonian formalism, we have dropped the explicit t dependence of the field variables.) Using Eqs. (19.145) and (19.146), the reader can check that the Hamiltonian equations of motion coincide with the Lagrangian ones, and one can write:

$$\begin{aligned} \dot{\phi} &= \{\phi, H\}, \qquad \dot{\rho} = \{\rho, H\}; \\ \phi(\boldsymbol{x}, t) &= \exp(\widetilde{-tH})\phi(\boldsymbol{x}), \rho(\boldsymbol{x}, t) = \exp(\widetilde{-tH})\rho(\boldsymbol{x}). \end{aligned} \tag{19.147}$$

Now we must construct the other generators of $\mathcal{G}$, namely J_j, P_j, and G_j, and recognize M. The canonical transformations generated by J_j and P_j must be:

$$\begin{aligned} \exp(\widetilde{\boldsymbol{\alpha} \cdot \boldsymbol{J}}) \begin{pmatrix} \rho(\boldsymbol{x}) \\ \phi(\boldsymbol{x}) \end{pmatrix} &= \begin{pmatrix} \rho(A_{jk}(\boldsymbol{\alpha})x_k) \\ \phi(A_{jk}(\boldsymbol{\alpha})x_k) \end{pmatrix}, \\ \exp(\widetilde{\boldsymbol{a} \cdot \boldsymbol{P}}) \begin{pmatrix} \rho(\boldsymbol{x}) \\ \phi(\boldsymbol{x}) \end{pmatrix} &= \begin{pmatrix} \rho(\boldsymbol{x} + \boldsymbol{a}) \\ \phi(\boldsymbol{x} + \boldsymbol{a}) \end{pmatrix}; \end{aligned} \tag{19.148}$$

from our experience with the case of free Galilean fields we know how to construct J_j and P_j in order that this be so. We must have:

$$J_j = -\int d^3x\ \rho(\boldsymbol{x})(\boldsymbol{x} \times \boldsymbol{\nabla})_j\phi(\boldsymbol{x}), \qquad P_j = -\int d^3x\ \rho(\boldsymbol{x})\frac{\partial}{\partial x_j}\phi(\boldsymbol{x}). \tag{19.149}$$

But the densities appearing here coincide exactly with the ones expected on physical grounds, as given in Eq. (19.134)! As for the G_j, we can work out the expected forms of the canonical transformations generated by them. If S and $S' = \exp(\boldsymbol{v} \cdot \boldsymbol{g})S$ are two inertial frames in relative motion, then at the common zero of time in the two frames we expect the two densities $\rho'(\boldsymbol{x})$ and $\rho(\boldsymbol{x})$ to be equal, whereas the velocities $\boldsymbol{u}'(\boldsymbol{x})$ and $\boldsymbol{u}(\boldsymbol{x})$ must fulfill $\boldsymbol{u}'(\boldsymbol{x}) = \boldsymbol{u}(\boldsymbol{x}) - \boldsymbol{v}$. Therefore the effect of the canonical transformation $\exp(\widetilde{-\boldsymbol{v} \cdot \boldsymbol{G}})$ must be:

$$\exp(\widetilde{-\boldsymbol{v} \cdot \boldsymbol{G}})\rho(\boldsymbol{x}) = \rho(\boldsymbol{x}), \qquad \exp(\widetilde{-\boldsymbol{v} \cdot \boldsymbol{G}})\phi(\boldsymbol{x}) = \phi(\boldsymbol{x}) + \boldsymbol{v} \cdot \boldsymbol{x}, \tag{19.150}$$

which requirement suggests the form:

$$G_j = \int d^3x\ x_j\rho(\boldsymbol{x}). \tag{19.151}$$

Thus G_j is the moment of the mass distribution, or, apart from a factor M, it is the position of the center of mass of the fluid. With the generators H, J_j, P_j, and G_j defined as in Eqs. (19.146), (19.149), and (19.151), one can verify that all the Lie algebra relations for $\mathcal{G}'$ are satisfied. In particular, one can check that one has

$$\{G_j, H\} = P_j, \qquad \{G_j, P_k\} = \delta_{jk} M, \tag{19.152}$$

with the mass M as given in Eq. (19.132).

We have thus succeeded in exhibiting irrotational fluid flow as a Galilean invariant field theory. One must, of course, impose the mass superselection rule and restrict the observables to be just those functionals of ρ and ϕ that have vanishing PB's with M. Although such a superselection rule might have appeared rather forced in the free Galilean field theories constructed earlier, in the present case it makes very good physical sense. The canonical transformation generated by M is the following:

$$\exp(\widetilde{\lambda M})\rho(\boldsymbol{x}) = \rho(\boldsymbol{x}), \exp(\widetilde{\lambda M})\phi(\boldsymbol{x}) = \phi(\boldsymbol{x}) - \lambda \,. \tag{19.153}$$

That is, it leaves ρ invariant and shifts ϕ by a constant. Only those variables that are left unchanged by this shift in ϕ are observable. But such a restriction is eminently reasonable, because the original observables for the system are just the density and the fluid velocity, $\rho(\boldsymbol{x})$ and $\boldsymbol{u}(\boldsymbol{x})$. The latter determines $\phi(\boldsymbol{x})$ only up to an additive constant! Of course, all the generators of $\mathcal{G}$ are observable; this is also evident from the fact that they can all be explicitly expressed in terms of $\rho(\boldsymbol{x})$ and $\boldsymbol{\nabla}\phi(\boldsymbol{x})$. Regarding the nature of the variable $\phi(\boldsymbol{x})$, the following comment must be made. We have already noted that the equations of motion ensure that if $\rho(\boldsymbol{x})$ is nowhere negative at one instant of time, then it remains so for all time. More generally, the canonical transformations generated by all the generators of $\mathcal{G}$ preserve such a property for $\rho(\boldsymbol{x})$ (See Eqs. (19.148), (19.150), and (19.153)). On the other hand, the field $\phi(\mathbf{x})$ produces shifts in $\rho(\boldsymbol{x})$ because it is canonically conjugate to $\rho(\boldsymbol{x})$. For any function $a(\boldsymbol{x})$, we have formally:

$$\exp\left(-\widetilde{\int d^3x\, a(\boldsymbol{x})\phi(\boldsymbol{x})}\right)\rho(\boldsymbol{x}) = \rho(\boldsymbol{x}) - a(\boldsymbol{x})\,. \tag{19.154}$$

Therefore, if the canonical transformation generated by $\int d^3x\; a(\boldsymbol{x})\phi(\boldsymbol{x})$ is admitted to be a regular canonical transformation, it can convert a physically reasonable configuration in which $\rho(\boldsymbol{x}) \geqq 0$ into one in which $\rho(\boldsymbol{x}) < 0$ for some $\boldsymbol{x}$. It is therefore better to define the allowed, physically acceptable configurations of the field to be just those in which $\rho(\boldsymbol{x}) \geqq 0$, all $\boldsymbol{x}$, and declare that the variables $\phi(\boldsymbol{x})$ *do not* generate regular canonical transformations. This, of course, does not preclude the existence of variables depending on $\phi(\boldsymbol{x})$ which *do* generate regular canonical transformations. Examples are H, J_j, and P_j. We can, of course, regard $\rho(\boldsymbol{x})$ itself as a generator of regular canonical transformations.

The invariance of the Hamiltonian under the transformations realizing $E(3)$ corresponds exactly to the invariance of the Lagrangian under these transformations:

$$\begin{aligned} \rho(\boldsymbol{x},t) &\rightarrow \rho(\boldsymbol{x}+\boldsymbol{a},t), \quad \phi(\boldsymbol{x},t) \rightarrow \phi(\boldsymbol{x}+\boldsymbol{a},t)\,; \\ \rho(\boldsymbol{x},t) &\rightarrow \rho(A_{jk}(\boldsymbol{\alpha})x_k,t), \; \phi(\boldsymbol{x},t) \rightarrow \phi(A_{jk}(\boldsymbol{\alpha})x_k,t)\,. \end{aligned} \tag{19.155}$$

Let us work out the behavior of the Lagrangian under the transformations generated by the G_j. The transformations themselves, for arbitrary time t, are:

$$\begin{aligned} \exp(\widetilde{\boldsymbol{v}\cdot\boldsymbol{G}})\phi(\boldsymbol{x},t) &= \exp(\widetilde{\boldsymbol{v}\cdot\boldsymbol{G}})\exp(\widetilde{-tH})\phi(\boldsymbol{x}) \\ &= \exp(\widetilde{-tH})\exp(\widetilde{\boldsymbol{v}\cdot\boldsymbol{G}})\exp(\widetilde{-t\boldsymbol{v}\cdot\boldsymbol{P}})\phi(\boldsymbol{x}) \\ &= \exp(\widetilde{-tH})\exp(\widetilde{\boldsymbol{v}\cdot\boldsymbol{G}})\phi(\boldsymbol{x}-\boldsymbol{v}t) \\ &= \exp(\widetilde{-tH})[\phi(\boldsymbol{x}-\boldsymbol{v}t)-\boldsymbol{v}\cdot(\boldsymbol{x}-\boldsymbol{v}t)] \\ &= \phi(\boldsymbol{x}-\boldsymbol{v}t,t)-\boldsymbol{v}\cdot(\boldsymbol{x}-\boldsymbol{v}t)\,; \\ \exp(\widetilde{\boldsymbol{v}\cdot\boldsymbol{G}})\rho(\boldsymbol{x},t) &= \rho(\boldsymbol{x}-\boldsymbol{v}t,t)\,. \end{aligned} \tag{19.156}$$

For infinitesimal $\boldsymbol{v}$, we must consider the behavior of the Lagrangian under the change:

$$\begin{aligned} \rho(\boldsymbol{x},t) \rightarrow \rho'(\boldsymbol{x},t) &= \rho(\boldsymbol{x}-\boldsymbol{v}t,t) \cong \rho(\boldsymbol{x},t) - t\boldsymbol{v}\cdot\boldsymbol{\nabla}\rho(\boldsymbol{x},t)\,, \\ \phi(\boldsymbol{x},t) \rightarrow \phi'(\boldsymbol{x},t) &\cong \phi(\boldsymbol{x}-\boldsymbol{v}t,t) - \boldsymbol{v}\cdot\boldsymbol{x} = \phi(\boldsymbol{x},t) - t\boldsymbol{v}\cdot\boldsymbol{\nabla}\phi(\boldsymbol{x},t) - \boldsymbol{v}\cdot\boldsymbol{x}\,, \\ \dot\phi(\boldsymbol{x},t) \rightarrow \dot\phi'(\boldsymbol{x},t) &\cong \dot\phi(\boldsymbol{x}-\boldsymbol{v}t,t) - \boldsymbol{v}\cdot\boldsymbol{\nabla}\phi(\boldsymbol{x},t)\,. \end{aligned} \tag{19.157}$$

Interestingly enough, we find that L is invariant under this infinitesimal variation in ρ and ϕ, for:

$$\begin{aligned} L[\rho',\phi'] &= \int d^3x[\rho(\boldsymbol{x}-\boldsymbol{v}t,t)(\dot\phi(\boldsymbol{x}-\boldsymbol{v}t,t)-\boldsymbol{v}\cdot\boldsymbol{\nabla}\phi(\boldsymbol{x},t)) \\ &- \frac{1}{2}\rho(\boldsymbol{x}-\boldsymbol{v}t,t)(\boldsymbol{\nabla}\phi(\boldsymbol{x}-\boldsymbol{v}t,t)-\boldsymbol{v})^2 \\ &- \rho(\boldsymbol{x}-\boldsymbol{v}t,t)W(\rho(\boldsymbol{x}-\boldsymbol{v}t,t))] \\ &\cong L[\rho,\phi] - \int d^3x\ \rho(\boldsymbol{x},t)\boldsymbol{v}\cdot\boldsymbol{\nabla}\phi(\boldsymbol{x},t) + \int d^3x\ \rho(\boldsymbol{x},t)\boldsymbol{v}\cdot\boldsymbol{\nabla}\phi(\boldsymbol{x},t)\,, \\ \delta L[\rho,\phi] &= 0\,. \end{aligned} \tag{19.158}$$

Therefore the values of the actions $\Phi[C], \Phi[C']$ computed along the trajectories $\rho(\boldsymbol{x},t), \phi(\boldsymbol{x},t)$ and $\rho'(\boldsymbol{x},t), \phi'(\boldsymbol{x},t)$ between the same times t_1, t_2 are the same. Thus we understand why the equations of motion are Galilean invariant; we also see that even though the Lagrangian is invariant, there are no constants of motion because the variations $\delta\rho(\boldsymbol{x},t)$ and $\delta\phi(\boldsymbol{x},t)$ are explicitly time dependent. In fact, putting Eqs. (19.157) and (19.158) into the standard expression for the difference $\Phi[C'] - \Phi[C]$, we immediately recover the linear time dependence of G_j. The behavior of L under the transformation Eq. (19.153) is simpler to discuss. It is obviously invariant, so is the action $\Phi[C]$, and from the action principle we get back the conservation of M.

General Fluid Flow

We now develop a Lagrangian treatment of the case of fluid flow with nonvanishing vorticity $\boldsymbol{\Omega}(\boldsymbol{x}t)$,

$$\boldsymbol{\Omega}(\boldsymbol{x},t) = \boldsymbol{\nabla} \times \boldsymbol{u}(\boldsymbol{x},t)\,, \tag{19.159}$$

and also exhibit it as a Galilean-invariant field theory. In the absence of viscous forces, vorticity can be neither created nor destroyed, and this accounts for the clean separation in the discussion of the two cases of vanishing and nonvanishing vorticity. Once again, in order to put the basic Eqs. (19.128) and (19.130) into Lagrangian form, we need to introduce a set of scalar velocity potentials and rewrite Eq. (19.128) as a set of scalar equations. Because these potentials may be somewhat unfamiliar and are needed for the Lagrangian treatment as well as for the subsequent passage to the Hamiltonian, we devote some space to a description of them.

The usual method of expressing a general vector field in terms of auxiliary "potentials", familiar from electromagnetism, is to write it as the gradient of a scalar potential plus the curl of a vector potential; the ambiguity in the latter could be settled by requiring it to be divergenceless. Here we shall use an alternative method of introducing potentials, invented by Clebsch, in which all the potentials are scalar quantities. We write $\phi(\boldsymbol{x},t)$, $\alpha(\boldsymbol{x},t)$, $\beta(\boldsymbol{x},t)$ for them (we certainly need three because $\boldsymbol{u}(\boldsymbol{x},t)$ has three components) and construct $\boldsymbol{u}$ from them thus:

$$\boldsymbol{u}(\boldsymbol{x},t) = -\boldsymbol{\nabla}\phi(\boldsymbol{x},t) - \alpha(\boldsymbol{x},t)\boldsymbol{\nabla}\beta(\boldsymbol{x},t)\,. \tag{19.160}$$

Clearly, a set of potentials determines $\boldsymbol{u}$ uniquely but not conversely. Thus there is a gauge-type ambiguity. We can describe this ambiguity precisely, as well as get a feeling for these "Clebsch potentials", in the following way. It is evident that the potentials α and β must be chosen so as to produce the right vorticity in the form:

$$\boldsymbol{\Omega}(\boldsymbol{x},t) = -(\boldsymbol{\nabla}\alpha(\boldsymbol{x},t)) \times (\boldsymbol{\nabla}\beta(\boldsymbol{x},t))\,. \tag{19.161}$$

In fact, if α and β do obey Eq. (19.161), we can always choose $\phi(\boldsymbol{x},t)$ so that Eq. (19.160) holds. Now what are the constraints on α and β? Given the vorticity $\boldsymbol{\Omega}$ (which must obey $\boldsymbol{\nabla}\cdot\boldsymbol{\Omega} = 0$), how can we find a pair α, β to satisfy Eq. (19.161)? We can see that α and β must satisfy the equations

$$\boldsymbol{\Omega}(\boldsymbol{x},t)\cdot\boldsymbol{\nabla}\alpha(\boldsymbol{x},t) = \boldsymbol{\Omega}(\boldsymbol{x},t)\cdot\boldsymbol{\nabla}\beta(\boldsymbol{x},t) = 0\,. \tag{19.162}$$

Now if we consider the partial differential equation:

$$\boldsymbol{\Omega}(\boldsymbol{x},t)\cdot\boldsymbol{\nabla}\psi(\boldsymbol{x},t) = 0 \tag{19.163}$$

for an unknown ψ, (and ignore the time variable for the present), we can certainly construct *two* independent solutions. The proof of this statement is left to the reader. But any *three* solutions of Eq. (19.163) must be functionally dependent; that is, we cannot have *three* independent solutions to this equation. For, if ψ_j, $j = 1, 2, 3$, are any three solutions, then Eq. (19.163) implies a linear relation among the rows of the Jacobian matrix

$$\left\| \frac{\partial \psi_j}{\partial x_k} \right\| ;$$

the determinant of this matrix must then vanish, and there must be a functional relation among ψ_1, ψ_2, and ψ_3. Conversely, ψ_1, ψ_2 and ψ_3 are solutions to Eq. (19.163) and ψ_1 and ψ_2 are independent of one another, then ψ_3 can be written as a function of ψ_1 and ψ_2. Let us now choose two independent solutions ψ_1, ψ_2 of Eq.(19.163). Because space is three dimensional and $\boldsymbol{\Omega}$ is orthogonal to both $\boldsymbol{\nabla}\psi_1$ and $\boldsymbol{\nabla}\psi_2$, we must have:

$$\boldsymbol{\Omega}(\boldsymbol{x}, t) = P(\boldsymbol{x}, t)(\boldsymbol{\nabla}\psi_1(\boldsymbol{x}, t)) \times (\boldsymbol{\nabla}\psi_2(\boldsymbol{x}, t)), \tag{19.164}$$

$P(\boldsymbol{x}, t)$ being some scalar. Now we impose the condition that $\boldsymbol{\Omega}$ itself is the curl of $\boldsymbol{u}$:

$$\boldsymbol{\nabla} \cdot \boldsymbol{\Omega}(\boldsymbol{x}, t) = (\boldsymbol{\nabla}\psi_1) \times (\boldsymbol{\nabla}\psi_2) \cdot \boldsymbol{\nabla} P = 0 . \tag{19.165}$$

and taking P times this equation and using Eq. (19.164), it follows that P is itself a solution to Eq. (19.163), like ψ_1 and ψ_2. We conclude that P is some function of ψ_1 and ψ_2; at the risk of a slight abuse of notation, we write $P = P(\psi_1; \psi_2)$. It is now easy to find an α and a β to satisfy Eq. (19.161); we can, for example, choose α to be the indefinite integral of P with respect to ψ_1, at constant ψ_2, and β to be the negative of ψ_2. Now suppose α, β and α', β' are two pairs of potentials producing the same vorticity. As long as $\boldsymbol{\Omega}$ is nonvanishing, α and β must be functionally independent, and α' and β' must obey Eq. (19.163), just as α and β do. Thus α' and β' must be expressible as functions of α and β. The equality of the two vorticities then gives:

$$\begin{aligned} \boldsymbol{\nabla}\alpha' \times \boldsymbol{\nabla}\beta' &= \left(\frac{\partial \alpha'}{\partial \alpha} \frac{\partial \beta'}{\partial \beta} - \frac{\partial \alpha'}{\partial \beta} \frac{\partial \beta'}{\partial \alpha} \right) \boldsymbol{\nabla}\alpha \times \boldsymbol{\nabla}\beta = \boldsymbol{\nabla}\alpha \times \boldsymbol{\nabla}\beta , \\ & \frac{\partial \alpha'}{\partial \alpha} \frac{\partial \beta'}{\partial \beta} - \frac{\partial \alpha'}{\partial \beta} \frac{\partial \beta'}{\partial \alpha} = 1 . \end{aligned} \tag{19.166}$$

This equation specifies completely the arbitrariness in the pair α, β: any two allowed pairs $(\alpha, \beta), (\alpha', \beta')$ are such that the members of the latter are expressible as functions of the members of the former, such that Eq. (19.166) holds. (This is simply the set of all canonical transformations in one pair of canonical variables, because the left-hand side of the last line in Eq. (19.166) is the standard PB expression in one q and one p!)

Thus we see that any given vorticity $\boldsymbol{\Omega}$ can be built up in the form of Eq. (19.161), and then ϕ chosen so that Eq. (19.160) is true. That is, the representation of $\boldsymbol{u}$ in terms of the three scalar potentials is always possible though not unambiguous. Now this discussion of the possibility as well as the ambiguity in the choice of the potentials ϕ, α, β refers to one instant of time. If we consider the question of development in time, it is clear that the ambiguity in the choice of α and β at each instant of time will prevent us from being able to compute $\dot{\alpha}$ and $\dot{\beta}$ given $\dot{\boldsymbol{u}}$! Even if at time t some definite choice of the potentials has been made, this will not prevent the ambiguity from reappearing at a subsequent instant of time, $t + \delta t$, in spite of the fact that the change in $\boldsymbol{u}$ in this small interval of time may be known. Therefore from the equation of motion Eq. (19.128) for $\boldsymbol{u}(\boldsymbol{x}, t)$, we are not able to obtain separate equations of motion for ϕ, α and β, unless some assumptions are made in resolving the ambiguity in the choice of the potentials at each instant of time. *We must choose some equations motion for α and β, consistent with Eqs. (19.128) and (19.130), and then from Eq. (19.128) we can deduce an equation of motion for ϕ.* If we simply substitute Eq. (19.160) in Eq. (19.128), we obtain:

$$\boldsymbol{\nabla}(-\dot{\phi}(\boldsymbol{x}, t) + f(\rho)) = \dot{\alpha}(\boldsymbol{x}, t)\boldsymbol{\nabla}\beta(\boldsymbol{x}, t) + \alpha(\boldsymbol{x}, t)\boldsymbol{\nabla}\dot{\beta}(\boldsymbol{x}, t) - \boldsymbol{u} \cdot \boldsymbol{\nabla}\boldsymbol{u}(\boldsymbol{x}, t) . \tag{19.167}$$

We must now examine carefully what possible assumptions we can make for $\dot{\alpha}$ and $\dot{\beta}$. Now the constraints on $\alpha(\boldsymbol{x}, t)$ and $\beta(\boldsymbol{x}, t)$ are two: (1) for each time t they must be solutions of Eq. (19.163), and (2) at each time t they must be such solutions of Eq. (19.163) that they produce the right vorticity via Eq. (19.161). Let us first impose requirement (1) and then (2). From the equation of motion Eq. (19.128) for $\boldsymbol{u}(x, t)$, we can derive the following one for $\boldsymbol{\Omega}(\boldsymbol{x}, t)$:

$$\dot{\boldsymbol{\Omega}}(\boldsymbol{x}, t) + \boldsymbol{\nabla} \times (\boldsymbol{u} \cdot \boldsymbol{\nabla}\boldsymbol{u}(\boldsymbol{x}, t)) = 0 . \tag{19.168}$$

(This equation for $\boldsymbol{\Omega}$ is valid whenever the right-hand side of Eq. (19.128) is the gradient of a scalar.) The velocity $\boldsymbol{u}(\boldsymbol{x}, t)$ at one instant of time is adequate to compute the vorticity at a succeeding instant of time. If now $\psi(\boldsymbol{x}, t)$ is a solution of Eq. (19.163) for each t, then knowing the equation for $\dot{\boldsymbol{\Omega}}(\boldsymbol{x}, t)$ will give a corresponding condition on $\dot{\psi}(\boldsymbol{x}, t)$. We simply differentiate Eq. (19.163) with respect to time and use Eq. (19.168) in the term involving $\dot{\boldsymbol{\Omega}}(\boldsymbol{x}, t)$. After a bit of routine algebra, we find that Eqs. (19.163) and (19.168) together imply the following:

$$\boldsymbol{\Omega}(\boldsymbol{x}, t) \cdot \boldsymbol{\nabla}\frac{D\psi(\boldsymbol{x}, t)}{Dt} = 0 . \tag{19.169}$$

[$(D\psi/Dt)$ is defined in Eq. (19.129) and is the rate of change of ψ as we move with the fluid.] Therefore, the manner in which $\boldsymbol{\Omega}$ changes in time is such that if ψ is a solution of Eq. (19.163), then so is $(D\psi/Dt)$. The first constraint on the time development of the potentials α and β is then this: $(D\alpha/Dt)$ and $(D\beta/Dt)$, like α and β, must be solutions to Eq. (19.163). But because α and β are two independent solutions at time t, this means that $(D\alpha/Dt)$ should be expressible as some function of α, β and t, and $(D\beta/Dt)$ as some other such function; all in all,

$$\boldsymbol{\Omega} \cdot \boldsymbol{\nabla}\alpha = \boldsymbol{\Omega} \cdot \boldsymbol{\nabla}\beta = 0, \text{ all } t \Rightarrow \frac{D\alpha}{Dt} = F(\alpha, \beta, t), \quad \frac{D\beta}{Dt} = G(\alpha, \beta, t) . \tag{19.170}$$

Next, we must see what restrictions there are on F and G, coming from the second requirement on α and β, namely that $\boldsymbol{\Omega}$ be constructed at all times in terms of α and β as in Eq. (19.161). We get this condition by differentiating both sides of Eq. (19.161) with respect to time, then substituting Eqs. (19.168) and (19.170) into it. The resulting equation is:

$$\begin{aligned} -\dot{\boldsymbol{\Omega}}(\boldsymbol{x},t) =& \boldsymbol{\nabla} \times (\boldsymbol{u} \cdot \boldsymbol{\nabla} \boldsymbol{u}(\boldsymbol{x},t)) \\ =& (\boldsymbol{\nabla}(F - \boldsymbol{u} \cdot \boldsymbol{\nabla}\alpha)) \times \boldsymbol{\nabla}\beta + \boldsymbol{\nabla}\alpha \times \boldsymbol{\nabla}(G - \boldsymbol{u} \cdot \boldsymbol{\nabla}\beta) . \end{aligned} \tag{19.171}$$

If one brings the terms involving F and G to one side, this has the form:

$$\begin{aligned} \left(\frac{\partial F}{\partial \alpha} + \frac{\partial G}{\partial \beta}\right) \boldsymbol{\nabla}\alpha \times \boldsymbol{\nabla}\beta =& \boldsymbol{\nabla} \times (\boldsymbol{u} \cdot \boldsymbol{\nabla}\boldsymbol{u}) + \boldsymbol{\nabla}(\boldsymbol{u} \cdot \boldsymbol{\nabla}\alpha) \times \boldsymbol{\nabla}\beta + \boldsymbol{\nabla}\alpha \times \boldsymbol{\nabla}(\boldsymbol{u} \cdot \boldsymbol{\nabla}\beta) \\ \equiv& \boldsymbol{\nabla} \times \left[\boldsymbol{u} \cdot \boldsymbol{\nabla}\boldsymbol{u} + (\boldsymbol{u} \cdot \boldsymbol{\nabla}\alpha)\boldsymbol{\nabla}\beta - (\boldsymbol{u} \cdot \boldsymbol{\nabla}\beta)\boldsymbol{\nabla}\alpha\right] . \end{aligned} \tag{19.172}$$

A short calculation shows that the right-hand side vanishes identically, because the vector inside the square brackets turns out to be the gradient of $\frac{1}{2}u^2$. Therefore, the only restriction on F and G is:

$$\frac{\partial F(\alpha,\beta,t)}{\partial \alpha} + \frac{\partial G(\alpha,\beta,t)}{\partial \beta} = 0 . \tag{19.173}$$

The most general equations of motion that may be assumed for α and β are those given in Eq. (19.170), with F and G restricted by Eq.(19.173) alone. Any allowed choice for F and G then permits us to simplify the right-hand side of Eq. (19.167) by substituting in it for $\dot{\alpha}$ and $\dot{\beta}$, and the resulting expression appears as the gradient of some scalar. On comparing the two sides we would deduce an equation of motion for ϕ. Although the *most general* possibility is of this nature, the *most simple* choice we can make is to set both F and G equal to zero! This certainly obeys Eq. (19.173). We therefore assume that α and β obey the equations of motion:

$$\frac{\partial \alpha(\boldsymbol{x},t)}{\partial t} + \boldsymbol{u} \cdot \boldsymbol{\nabla}\alpha(\boldsymbol{x},t) = 0 , \tag{19.174a}$$

$$\frac{\partial \beta(\boldsymbol{x},t)}{\partial t} + \boldsymbol{u} \cdot \boldsymbol{\nabla}\beta(\boldsymbol{x},t) = 0 . \tag{19.174b}$$

With this, the remaining arbitrariness in the choice of the potentials α, β is of the following form: in place of the pair (α, β), we can use the pair (α', β'), where α' and β' are any two functions of α and β *not explicitly dependent on time,* which obey Eq. (19.166). The validity of Eq. (19.174) for α and β then implies exactly similar equations for α' and β'. The variables $\rho(\boldsymbol{x},t)$ and $\boldsymbol{u}(\boldsymbol{x},t)$, which are the basic observables of the system, are unaltered by such changes in the choice of the Clebsch potentials. The change in ϕ consequent on a change in α and β is easy enough to compute, but we omit the details.

The equation of motion for ϕ emerges from using Eq. (19.174) on the right-hand side of Eq. (19.167) to substitute for $\dot{\alpha}$ and $\dot{\beta}$, rearranging terms so that it all appears as the gradient of some scalar. Restricting ourselves to configurations of the system in which all relevant field quantities vanish at spatial infinity, we may equate the two scalars whose gradients are supposed to be equal, and in this way we get:

$$\frac{\partial\phi(\boldsymbol{x},t)}{\partial t} = f(\rho) + \frac{1}{2}[(\boldsymbol{\nabla}\phi)^2 - \alpha^2(\boldsymbol{\nabla}\beta)^2]\,. \tag{19.175}$$

Our rewriting of the basic equations of motion, Eqs. (19.128) and (19.130), as a set of equations for the Clebsch potentials is now done. The full set consists of Eqs. (19.130), (19.174) and (19.175), and it is this set (or rather an easily derived equivalent set—see the following) that we will cast into Lagrangian form. Before we do so, let us notice that by combining the equation of continuity for ρ, Eq. (19.130), with the postulated equations of motion for α and β in Eq. (19.174), we get continuity equations for $\rho\alpha$ as well as $\rho\beta$, and thus derive that the spatial integrals of these quantities, too, are constants of motion:

$$\begin{gathered}\frac{\partial}{\partial t}(\rho\alpha) + \boldsymbol{\nabla}\cdot(\rho\alpha\boldsymbol{u}) = \frac{\partial}{\partial t}(\rho\beta) + \boldsymbol{\nabla}\cdot(\rho\beta\boldsymbol{u}) = 0\,,\\ \int d^3x\ \rho(\boldsymbol{x},t)\alpha(\boldsymbol{x},t) = \text{constant}, \quad \int d^3x\ \rho(\boldsymbol{x},t)\beta(\boldsymbol{x},t) = \text{constant}\,.\end{gathered} \tag{19.176}$$

But such constants of motion must be capable of immediate generalization, because we know that there is a lot of freedom in the choice of α and β; for example, α could be replaced by any time-independent function of α and β, $F(\alpha,\beta)$, with a related replacement for β, and what is true for α must be true for $F(\alpha,\beta)$. This is indeed so; from the equations of motion for ρ, α and β, not only do we get the constants of motion exhibited in Eq. (19.176), but more generally, for any time-independent function $F(\alpha,\beta)$, we have:

$$\begin{aligned}\frac{\partial}{\partial t}(\rho F(\alpha,\beta)) + \boldsymbol{\nabla}\cdot(\rho F(\alpha,\beta)\boldsymbol{u}) &= 0\,,\\ \int d^3x\ \rho(\boldsymbol{x},t)F(\alpha(\boldsymbol{x},t),\beta(\boldsymbol{x},t)) &= \text{constant}\,.\end{aligned} \tag{19.177}$$

The existence of this uncountably large family of constants of motion corresponds precisely to the freedom in the choice of α and β. However, all these constants of motion are "useless" in the sense that the only choice of $F(\alpha,\beta)$ that will allow us to express the constant of motion in terms of the observables ρ and $\boldsymbol{u}$ is F=constant, and then the constant of motion is just the total mass!

Let us now proceed to the construction of a suitable Lagrangian. We again need one which is at most linear in the velocities. The energy density is expected to be:

$$\mathcal{H}(\boldsymbol{x},t) = \frac{1}{2}\rho u^2 + \rho W(\rho) = \frac{1}{2}\rho(\boldsymbol{\nabla}\phi)^2 + \rho\alpha\boldsymbol{\nabla}\phi\cdot\boldsymbol{\nabla}\beta + \frac{1}{2}\rho\alpha^2(\boldsymbol{\nabla}\beta)^2 + \rho W(\rho)\,, \tag{19.178}$$

with $W(\rho)$ being related to the pressure $p(\rho)$ in the same way as in the irrotational case. Thus we look for a Lagrangian density whose velocity independent term is $-\mathcal{H}(\boldsymbol{x},t)$. Because the pressure term $\rho W(\rho)$ contributes only in the equation of motion for ϕ, it would be convenient to arrange things so that the Euler-Lagrange variational equation corresponding to a variation in ρ "alone" *is* the equation of motion for ϕ. This suggests that we replace the set of independent variables $\rho, \phi, \alpha, \beta$ by a new set. On comparing the terms in the energy density $\mathcal{H}(\boldsymbol{x},t)$ with the terms appearing on the right-hand side of Eq. (19.175), it is clear that on a variation in ρ "alone" we would like only $\rho(\boldsymbol{\nabla}\phi)^2$ and $\rho\alpha^2(\boldsymbol{\nabla}\beta)^2$ to contribute, and $\rho\alpha\boldsymbol{\nabla}\phi\cdot\boldsymbol{\nabla}\beta$ not to contribute at all. Thus we switch over to a new set of independent variables $\rho, \phi, \sigma, \beta$ where $\sigma = \rho\alpha$, and set up the Lagrangian density

$$\begin{aligned}\mathcal{L}(\boldsymbol{x},t) &= \rho(\boldsymbol{x},t)\dot{\phi}(\boldsymbol{x},t) - \frac{1}{2}\rho(\boldsymbol{\nabla}\phi)^2 - \sigma\boldsymbol{\nabla}\phi\cdot\boldsymbol{\nabla}\beta \\ &\quad -\frac{1}{2}\frac{\sigma^2}{\rho}(\boldsymbol{\nabla}\beta)^2 - \rho W(\rho) + \ldots \end{aligned} \tag{19.179}$$

The omitted terms here must all be linear homogeneous in the velocities. We now see that with these omitted terms *independent of ρ and $\dot{\rho}$*, the Euler-Lagrange variational equation with respect to ρ already yields the equation of motion for ϕ, for then:

$$\frac{\delta L}{\delta\dot{\rho}(\mathbf{x},t)} = 0, \quad \frac{\delta L}{\delta\rho(\boldsymbol{x},t)} = \dot{\phi}(\boldsymbol{x},t) - \frac{1}{2}(\boldsymbol{\nabla}\phi)^2 + \frac{1}{2}\frac{\sigma^2}{\rho^2}(\boldsymbol{\nabla}\beta)^2 - f(\rho), \tag{19.180}$$

and the vanishing of $(\delta L/\delta\rho)$ is precisely the equation of motion for ϕ. We can now see that the value of using the new set of independent variables ρ, ϕ, σ and β is that the terms represented by dots in Eq. (19.179) may be assumed not to contain ρ and $\dot{\rho}$. [There is no harm in using ρ, ϕ, α and β as the basic variables, but in that case we would not have secured the ϕ equation so easily. In particular because the term $\rho\alpha\boldsymbol{\nabla}\phi\cdot\boldsymbol{\nabla}\beta$ would have made an unwanted contribution to $(\delta L/\delta\rho)$, we would have had to arrange for another contribution to cancel this one either directly or by virtue of the other Euler-Lagrange equations.] With the terms explicitly written down in Eq. (19.179), we next find that the Euler-Lagrange equation corresponding to ϕ is already the equation of continuity for ρ, and there is no need for the omitted terms to depend on ϕ and $\dot{\phi}$ either! In that case, we have:

$$\begin{aligned}\frac{\delta L}{\delta\dot{\phi}(\boldsymbol{x},t)} &= \rho(\boldsymbol{x},t), \\ \frac{\delta L}{\delta\phi(\boldsymbol{x},t)} &= \boldsymbol{\nabla}\cdot(\rho\boldsymbol{\nabla}\phi) + \boldsymbol{\nabla}\cdot(\sigma\boldsymbol{\nabla}\beta) = -\boldsymbol{\nabla}\cdot(\rho\boldsymbol{u}), \\ \frac{\partial}{\partial t}\frac{\delta L}{\delta\dot{\phi}} - \frac{\delta L}{\delta\phi} &= 0 \to \frac{\partial\rho}{\partial t} + \boldsymbol{\nabla}\cdot(\rho\boldsymbol{u}) = 0. \end{aligned} \tag{19.181}$$

Therefore, the terms to be added to Eq. (19.179) must depend on $\sigma, \dot{\sigma}, \beta, \dot{\beta}$ alone, be linear homogeneous in $\dot{\sigma}, \dot{\beta}$ and be such as to yield the proper equations of motion for σ and β. All this can be arranged by just adding the term $\sigma\dot{\beta}$ to

Eq. (19.179). The full Lagrangian density, and the complete set of equations of motion in the new variables, appears thus:

$$\mathcal{L}(\boldsymbol{x},t) = \rho(\boldsymbol{x},t)\dot{\phi}(\boldsymbol{x},t) + \sigma(\boldsymbol{x},t)\dot{\beta}(\boldsymbol{x},t) - \frac{1}{2}\rho\left[\boldsymbol{\nabla}\phi + \frac{\sigma}{\rho}\boldsymbol{\nabla}\beta\right]^2 - \rho W(\rho)\,; \tag{19.182a}$$

$$\boldsymbol{u}(\boldsymbol{x},t) \equiv -\boldsymbol{\nabla}\phi - \frac{\sigma}{\rho}\boldsymbol{\nabla}\beta \text{ (by definition)}\,; \tag{19.182b}$$

$$\rho \text{ equation : } \dot{\phi} = f(\rho) + \frac{1}{2}\left[(\boldsymbol{\nabla}\phi)^2 - \frac{\sigma^2}{\rho^2}(\boldsymbol{\nabla}\beta)^2\right]\,; \tag{19.182c}$$

$$\phi \text{ equation : } \dot{\rho} + \boldsymbol{\nabla}\cdot(\rho\boldsymbol{u}) = 0\,; \tag{19.182d}$$

$$\sigma \text{ equation : } \dot{\beta} + \boldsymbol{u}\cdot\boldsymbol{\nabla}\beta = 0\,; \tag{19.182e}$$

$$\beta \text{ equation : } \dot{\sigma} + \boldsymbol{\nabla}\cdot(\sigma\boldsymbol{u}) = 0\,. \tag{19.182f}$$

By "ρ equation" we mean the Euler-Lagrange equation

$$\left(\frac{d}{dt}\right)\frac{\delta L}{\delta\dot{\rho}} - \frac{\delta L}{\delta\rho} = 0, \quad \text{and so on}\,.$$

With this Lagrangian, we have achieved our aim of expressing the equations of motion for general fluid flow via an action principle. Before proceeding to the Hamiltonian, let us briefly see how we recover the family of conservation laws Eq. (19.177) from the Lagrangian formulation. The form of an infinitesimal change in the potentials ϕ, α, and β that amounts to an allowed redefinition of these potentials leaving ρ and $\boldsymbol{u}$ fixed is given by the equations:

$$\begin{aligned}
\delta\alpha \equiv &\quad \alpha' - \alpha = \delta\epsilon\tfrac{\partial F(\alpha,\beta)}{\partial\beta}\,,\\
\delta\beta \equiv &\quad \beta' - \beta = -\delta\epsilon\tfrac{\partial F(\alpha,\beta)}{\partial\alpha}\,,\\
\delta\phi \equiv &\quad \phi' - \phi = \delta\epsilon\left(\alpha\tfrac{\partial F}{\partial\alpha} - F\right)\,;
\end{aligned} \tag{19.183}$$

$\delta\epsilon$ is a small parameter, F any (time-independent) function of α and β, and these equations are obtained by using Eq. (19.166) and the invariance of the velocity $\boldsymbol{u}$. In terms of the variables ρ, ϕ, σ and β we consider then the variation of L under the variation:

$$\begin{aligned}
\delta\rho = 0,\ \ \delta\phi &= \delta\epsilon\left(\alpha\frac{\partial F}{\partial\alpha} - F\right),\ \ \delta\sigma = \delta\epsilon\rho\frac{\partial F}{\partial\beta}\,,\\
\delta\beta &= -\delta\epsilon\frac{\partial F}{\partial\alpha}\,.
\end{aligned} \tag{19.184}$$

(There is no harm in using α in these equations and looking upon F as an explicit function of α and β, although the independent Lagrangian coordinates are taken to be ρ, ϕ, σ, and β.) We find:

$$\begin{aligned} \delta L &= \int d^3x\{\rho\delta\dot{\phi} + \delta\sigma\dot{\beta} + \sigma\delta\dot{\beta}\} \\ &= \int d^3x\ \delta\epsilon\left\{\rho\left(\dot{\alpha}\frac{\partial F}{\partial\alpha} + \alpha\dot{\alpha}\frac{\partial^2 F}{\partial\alpha^2} + \alpha\dot{\beta}\frac{\partial^2 F}{\partial\alpha\partial\beta} - \dot{\alpha}\frac{\partial F}{\partial\alpha} - \dot{\beta}\frac{\partial F}{\partial\beta}\right)\right. \\ &\quad \left. + \rho\dot{\beta}\frac{\partial F}{\partial\beta} - \rho\alpha\left(\dot{\alpha}\frac{\partial^2 F}{\partial\alpha^2} + \dot{\beta}\frac{\partial^2 F}{\partial\alpha\partial\beta}\right)\right\} = 0\,. \end{aligned} \tag{19.185}$$

Therefore, the Lagrangian is invariant under the variation in Eq. (19.184). Because these variations are not explicitly time dependent, the action principle yields a constant of motion that is also not explicitly time dependent. Because L depends only on the velocities $\dot{\phi}$ and $\dot{\beta}$ and not on $\dot{\rho}$ and $\dot{\sigma}$, this constant of motion is

$$\begin{aligned} \int d^3x\left\{\frac{\delta L}{\delta\dot{\phi}(\boldsymbol{x},t)}\delta\phi(\boldsymbol{x},t) + \frac{\delta L}{\delta\dot{\beta}(\boldsymbol{x},t)}\delta\beta(\boldsymbol{x},t)\right\} \\ = \delta\epsilon\int d^3x\left\{\rho\left(\alpha\frac{\partial F}{\partial\alpha} - F\right) - \sigma\frac{\partial F}{\partial\alpha}\right\} \\ = -\,\delta\epsilon\int d^3x\ \rho F(\alpha,\beta)\,. \end{aligned} \tag{19.186}$$

This is precisely the conservation law Eq. (19.177).

To develop the PB's and the Hamiltonian, we write the Lagrangian as:

$$\begin{aligned} L = \textstyle\int d^3x(A_{\rho(\boldsymbol{x})})\dot{\rho}(\boldsymbol{x}) + A_{\phi(\boldsymbol{x})}\dot{\phi}(\boldsymbol{x}) + A_{\sigma(\boldsymbol{x})}\dot{\sigma}(\boldsymbol{x}) + A_{\beta(\boldsymbol{x})}\dot{\beta}(\boldsymbol{x}) - V[\rho,\phi,\sigma,\beta]\,, \\ A_{\rho(\boldsymbol{x})} = 0,\ A_{\phi(\boldsymbol{x})} = \rho(\boldsymbol{x}),\ A_{\sigma(\boldsymbol{x})} = 0,\ A_{\beta(\boldsymbol{x})} = \sigma(\boldsymbol{x})\,; \\ V[\rho,\phi,\sigma,\beta] = \textstyle\int d^3x\left(\frac{1}{2}\rho\left(\boldsymbol{\nabla}\phi + \frac{\sigma}{\rho}\boldsymbol{\nabla}\beta\right)^2 + \rho W(\rho)\right) \end{aligned} \tag{19.187}$$

The only nonvanishing components of the "curl" $\eta_{\cdot,\cdot}$ of A are:

$$\begin{aligned} \eta_{\rho(\boldsymbol{x}),\phi(\boldsymbol{y})} &\equiv \frac{\delta A_{\phi(\boldsymbol{y})}}{\delta\rho(\boldsymbol{x})} - \frac{\delta A_{\rho(\boldsymbol{x})}}{\delta\phi(\boldsymbol{y})} = \delta^{(3)}(\boldsymbol{x}-\boldsymbol{y}) = -\eta_{\phi(\boldsymbol{y}),\rho(\boldsymbol{x})}\,, \\ \eta_{\sigma(\boldsymbol{x}),\beta(\boldsymbol{y})} &\equiv \frac{\delta A_{\beta(\boldsymbol{y})}}{\delta\sigma(\boldsymbol{x})} - \frac{\delta A_{\sigma(\boldsymbol{x})}}{\delta\beta(\boldsymbol{y})} = \delta^{(3)}(\boldsymbol{x}-\boldsymbol{y}) = -\eta_{\beta(\boldsymbol{y}),\sigma(\boldsymbol{x})}\,; \end{aligned} \tag{19.188}$$

and the only nonvanishing components of the inverse $\eta^{\cdot,\cdot}$ are:

$$\begin{aligned} \eta^{\rho(\boldsymbol{x}),\phi(\boldsymbol{y})} &= -\eta^{\phi(\boldsymbol{y}),\rho(\boldsymbol{x})} = -\delta^{(3)}(\boldsymbol{x}-\boldsymbol{y})\,, \\ \eta^{\sigma(\boldsymbol{x}),\beta(\boldsymbol{y})} &= -\eta^{\beta(\boldsymbol{y}),\sigma(\boldsymbol{x})} = -\delta^{(3)}(\boldsymbol{x}-\boldsymbol{y})\,. \end{aligned} \tag{19.189}$$

From the theory of Lagrangians linear in the velocities, we set up the basic PB relations:

$$\begin{aligned} \{\phi(\boldsymbol{x}),\rho(\boldsymbol{y})\} = \{\beta(\boldsymbol{x}),\sigma(\boldsymbol{y})\} = \delta^{(3)}(\boldsymbol{x}-\boldsymbol{y})\,, \\ \{\rho,\rho\} = \{\rho,\sigma\} = \{\rho,\beta\} = \{\phi,\phi\} = \{\phi,\sigma\} = \{\phi,\beta\} = \{\sigma,\sigma\} = \{\beta,\beta\} = 0\,. \end{aligned} \tag{19.190}$$

The Hamiltonian H is the same as V:

$$H[\rho,\phi,\sigma,\beta] = \int d^3x\rho(\boldsymbol{x})\left[\frac{1}{2}\left(\boldsymbol{\nabla}\phi(\boldsymbol{x}) + \frac{\sigma(\boldsymbol{x})}{\rho(\boldsymbol{x})}\boldsymbol{\nabla}\beta(\boldsymbol{x})\right)^2 + W(\rho(\boldsymbol{x}))\right] . \quad (19.191)$$

The reader can verify by using Eq. (19.190) that the Hamiltonian equations of motion reproduce Eqs. (19.182c-19.182f). To construct the linear and angular momenta, we note that Eqs. (19.148) must be valid again, and that exactly similar equations must be true in the case of $\sigma(\boldsymbol{x})$ and $\beta(\boldsymbol{x})$ as well. The natural choice for J_j and P_j is, then:

$$\begin{aligned} J_j &= -\int d^3x(\rho(\boldsymbol{x})(\boldsymbol{x}\times\boldsymbol{\nabla})_j\phi(\boldsymbol{x}) + \sigma(\boldsymbol{x})(\boldsymbol{x}\times\boldsymbol{\nabla})_j\beta(\boldsymbol{x}))\,, \\ P_j &= -\int d^3x\left(\rho(\boldsymbol{x})\frac{\partial}{\partial x_j}\phi(\boldsymbol{x}) + \sigma(\boldsymbol{x})\frac{\partial}{\partial x_j}\beta(\boldsymbol{x})\right) . \end{aligned} \quad (19.192)$$

Then the action of the Euclidean group $E(3)$ is just as expected. We notice that the densities of the angular and linear momenta are $\boldsymbol{x}\times\rho(\boldsymbol{x})\boldsymbol{u}(\boldsymbol{x})$ and $\rho(\boldsymbol{x})\boldsymbol{u}(\boldsymbol{x})$, respectively; this is physically reasonable, and further these densities are unaffected by the ambiguities in the Clebsch potentials. Arguments similar to those used in the irrotational case lead us again to expect the canonical transformations generated by the G_j to obey Eq. (19.150) and to leave the variables $\sigma(\boldsymbol{x}), \beta(\boldsymbol{x})$ unchanged, so that G_j should be:

$$G_j = \int d^3x\; x_j\rho(\boldsymbol{x})\,. \quad (19.193)$$

All the PB relations corresponding to the Lie algebra of $\mathcal{G}$ can now be checked to be valid, the mass M being the spatial integral of $\rho(\boldsymbol{x})$. The mass *superselection rule* states the *unobservability* of changing $\phi(\mathbf{x})$ by addition of a constant. More generally, on physical grounds we should treat every conservation law of the form of Eq. (19.177) as a superselection rule, then restrict observables to those functionals of $\rho, \phi, \sigma, \beta$ that are invariant under the transformation Eq. (19.184). The invariance of L under the variation Eq. (19.184) now appears as the invariance of H under the canonical transformation generated by any variable of the form $\int d^3x\,\rho(\boldsymbol{x})F(\sigma(\boldsymbol{x})/\rho(\boldsymbol{x}), \beta(\boldsymbol{x}))$:

$$\left\{H, \int d^3x\; \rho(\boldsymbol{x})F\left(\frac{\sigma(\boldsymbol{x})}{\rho(\boldsymbol{x})}, \beta(\boldsymbol{x})\right)\right\} = 0\,. \quad (19.194)$$

With respect to the transformations of $\mathcal{G}$, the behavior of L is the same as in the irrotational case. It is explicitly unchanged under the action of $E(3)$. The reader can verify that L is also invariant under the action of an infinitesimal velocity transformation in which the fields change as follows:

$$\begin{aligned} \delta\rho(\boldsymbol{x},t) &= -t\boldsymbol{v}\cdot\boldsymbol{\nabla}\rho(\boldsymbol{x},t), \qquad & \delta\phi(\boldsymbol{x},t) &= -t\boldsymbol{v}\cdot\boldsymbol{\nabla}\phi(\boldsymbol{x},t) - \boldsymbol{v}\cdot\boldsymbol{x}\,, \\ \delta\sigma(\boldsymbol{x},t) &= -t\boldsymbol{v}\cdot\boldsymbol{\nabla}\sigma(\boldsymbol{x},t), \qquad & \delta\beta(\boldsymbol{x},t) &= -t\boldsymbol{v}\cdot\boldsymbol{\nabla}\beta(\boldsymbol{x},t)\,. \end{aligned} \quad (19.195)$$

The explicit time dependences here give back, via the action principle, the linear time dependence of the dynamical variables G_j. The mass conservation law also comes out in the same manner as in the irrotational case, on account of the invariance of L under the change $\phi \to \phi$+constant (special case of Eq. (19.184)!). This completes our discussion of the equations of fluid flow as constituting a Galilean invariant field theory, except for the following remarks. The system with *nonzero vorticity* has the form of a *two-field theory*, one component associated with ϕ, ρ and the other with β, σ. This is of interest in connection with the "two-fluid" theory of the hydrodynamics of helium at very low temperatures. Note that the component (β, σ) *does not contribute to the mass*!

In concluding this chapter on the Galilei group, it is instructive to compare the behaviors of the Lagrangians in the various types of dynamical systems we have considered, under some of the infinitesimal transformations of $\mathcal{G}$. We have considered many-particle interacting systems, systems of multicomponent free Galilean fields, and fluid flow of two types. In all cases, the Lagrangians are invariant under the action of the Euclidean group $E(3)$; because the transformations of $E(3)$ are always not explicitly dependent on time, the action principle gives the conservation laws of J_j and P_j. But the description of G_j and the mass M differs from case to case. Of course, in all cases, under an infinitesimal velocity transformation the Lagrangian is either invariant or quasi-invariant so that the equations of motion are always Galilean invariant. The same is true for the effect of an infinitesimal transformation generated by M. Schematically, if we write $\delta L = (d/dt)\delta F$, we have the set of results in Table 19.7. Here, $(\partial/\partial t)(\ldots) \neq 0$ means that $(\ldots)$ has an explicit time dependence; $(\partial/\partial t)(\ldots) = 0$ means that it does not; and q stands generally for the basic Lagrangian variables of the system. Thus, for the systems discussed by us, the situation with respect to infinitesimal velocity transformations is that δF is sometimes zero and sometimes nonzero, but it never has an explicit time dependence; in all cases the explicit linear time dependence of the δq produces, via the action principle, a linear time dependence for the G_j. As for the mass, the associated transformation has always a δF and a δq not explicitly time dependent, leading always to a useful conservation law (except that it is trivial in the many-particle case!).

Table 19.7

System	Infinitesimal transformation generated by G_j	Infinitesimal transformation generated by M
Many-particle system	$\delta F \neq 0$, $\frac{\partial}{\partial t}\delta F = 0$ $\frac{\partial}{\partial t}(\delta q) \neq 0$.	-
Free Galilean fields	$\delta F \neq 0$, $\frac{\partial}{\partial t}\delta F = 0$ $\frac{\partial}{\partial t}(\delta q) \neq 0$	$\delta F \neq 0$, $\frac{\partial}{\partial t}\delta F = 0$ $\frac{\partial}{\partial t}(\delta q) = 0$
Fluid mechanics	$\delta F = 0$ $\frac{\partial}{\partial t}(\delta q) \neq 0$	$\delta F = 0$ $\frac{\partial}{\partial t}(\delta q) = 0$

Chapter 20

The Poincaré Group

This chapter is devoted to the study of some aspects and examples of classical canonical realizations of the Poincaré group, or as it is also called, the inhomogeneous Lorentz group. For brevity, we write $\mathcal{P}$ for this group; for the homogeneous Lorentz group we use the abbreviation HLG or sometimes $SO(3, 1)$ as we did in Chapter 15. In passing over from the Galilei group, which describes the invariance properties of the equations of motion of Newtonian mechanics, to $\mathcal{P}$, what is retained is the idea that there is a privileged class of reference frames in nature, the inertial frames, in which the laws of particle mechanics are especially simple. In particular a particle "not acted on by any forces" follows a straight-line trajectory in any one of these frames. As in the case of $\mathcal{G}$, two inertial frames in special relativity theory must be related to one another by a space-time translation, or by a spatial rotation, or by the fact that one of them moves with a uniform velocity relative to the other, or by a combination of all these. What distinguishes $\mathcal{P}$ from $\mathcal{G}$ is the way in which we relate the spacetime coordinates assigned to an event by two observers in two different inertial frames. In the case of $\mathcal{G}$ this relation is the one that is "reasonable" and "obvious" from the stand point of Newtonian mechanics (and in fact from common, everyday experience). In the case of $\mathcal{P}$ we have, on the other hand, the Lorentz transformation formulae discussed in detail in Chapter 15. These different formulae for the way the space-time coordinates of events change in going from one inertial frame to another give different structures to the group of all possible changes of inertial frame, leading to $\mathcal{G}$ in the one case and $\mathcal{P}$ in the other. The transformation formulae in the case of $\mathcal{P}$ contain the velocity of light in vacuum, c, in an essential manner; one can, in a physically meaningful way, think of $\mathcal{G}$ as the limiting form of $\mathcal{P}$ as $c \to \infty$. (Of course, c is a fixed real number in any given system of units and never really goes to infinity; what is meant is that in specific situations if certain characteristic velocities are neglected in comparison with c, in a consistent way, then $\mathcal{P}$ is essentially replaced by $\mathcal{G}$). For the most part, we assume that our system of units is such that c can be set equal to unity, so that the writing becomes easier.

Let us begin by recalling from Chapter 15 the form of the Lorentz transformation equations. If the space-time origins of two inertial frames S, S' coincide, then we have here an element of the HLG. The coordinates x^μ, x'^μ assigned by S, S', respectively, to a given event, are linearly related thus:

$$x'^\mu = \Lambda^\mu{}_\nu(\boldsymbol{\zeta}, \boldsymbol{\alpha})x^\nu\,. \tag{20.1}$$

The $\boldsymbol{\alpha}$ are the canonical coordinates of a rotation, $\boldsymbol{\zeta}$ the parameters of a pure velocity transformation, and the matrix $||\Lambda^\mu{}_\nu||$ is given in detail in Eq. (15.85). With this relation between x' and x, we write $S' = \exp(\boldsymbol{\zeta}\cdot\boldsymbol{k})\exp(\boldsymbol{\alpha}\cdot\boldsymbol{l})S$, and quite generally any element of the HLG, as long as it is in the identity component, can be uniquely written as the product $\exp(\boldsymbol{\zeta}\cdot\boldsymbol{k})\exp(\boldsymbol{\alpha}\cdot\boldsymbol{l})$. The product of two elements of the HLG corresponds to the product of the corresponding matrices $||\Lambda||$. The basic Lie bracket relations among the k_j and l_j can be exhibited in the split form:

$$[l_j, l_m] = \epsilon_{jmn}l_n,\ [l_j, k_m] = \epsilon_{jmn}k_n,\ [k_j, k_m] = -\epsilon_{jmn}l_n\,, \tag{20.2}$$

emphasizing the structure with respect to the subgroup of rotations of three-dimensional space, generated by the l_j. We could alternatively write them in covariant four-dimensional notation as:

$$\begin{aligned}[l_{\mu\nu}, l_{\sigma\rho}] = g_{\mu\sigma}l_{\nu\rho} - g_{\nu\sigma}l_{\mu\rho} + g_{\mu\rho}l_{\sigma\nu} - g_{\nu\rho}l_{\sigma\mu},\ \mu, \nu, \ldots = 0, 1, 2, 3\,;\\ l_{\mu\nu} = -l_{\nu\mu}, l_{mn} = \epsilon_{mnj}l_j, l_{0m} = k_m\,.\end{aligned} \tag{20.3}$$

For a general pair of inertial frames S, S', with noncoincident space-time origins, the relation of Eq. (20.1) is replaced by:

$$x'^\mu = \Lambda^\mu{}_\nu(\boldsymbol{\zeta}, \boldsymbol{\alpha})x^\nu + a^\mu\,. \tag{20.4}$$

Clearly, a^μ are the space-time coordinates assigned by S' to the spacetime origin of S. With this relation between x' and x, we write $S' = \exp(a^\mu d_u)\exp(\boldsymbol{\zeta}\cdot\boldsymbol{k})\exp(\boldsymbol{\alpha}\cdot\boldsymbol{l})S$, with $d_0 = -h, d_j$ the additional basic elements in the Lie algebra of $\mathcal{P}$. We have here a general element of $\mathcal{P}$, denoted either as $\exp(a^\mu d_\mu)$ $\exp(\boldsymbol{\zeta}\cdot\boldsymbol{k})\exp(\boldsymbol{\alpha}\cdot\boldsymbol{l})$ or the pair (Λ, a^μ) consisting of a 4×4 Lorentz transformation matrix Λ and a space-time translation a^μ. The latter notation is convenient for expressing the composition law in $\mathcal{P}$:

$$(\Lambda', a'^\mu)(\Lambda, a^\mu) = (\Lambda'\Lambda, a'^\mu + +\Lambda'^\mu{}_\nu a^\nu)\,. \tag{20.5}$$

The group $\mathcal{P}$ is the semidirect product of the HLG by the Abelian invariant subgroup T_4 of space-time translations. The basic Lie relations for $\mathcal{P}$ consist of Eq. (20.2) together with the following:

$$\begin{aligned}[l_j, d_m] = \epsilon_{jmn}d_n,\ [l_j, h] = 0;\ [k_j, d_m] = \delta_{jm}h,\ [k_j, h] = d_j\,;\\ [d_j, d_m] = 0,\ [d_j, h] = 0\,.\end{aligned} \tag{20.6}$$

In four-dimensional form, we can write Eq. (20.6) in the form:

$$[l_{\mu\nu}, d_\rho] = g_{\mu\rho}d_\nu - g_{\nu\rho}d_\mu,\ [d_\mu, d_\nu] = 0\,, \tag{20.7}$$

and then these together with Eq. (20.3) constitute the full set of basic Lie brackets for $\mathcal{P}$.

A classical dynamical system for which a Lagrangian, and thus a Hamiltonian, description exists and whose laws are invariant under $\mathcal{P}$ is described by a classical canonical realization of $\mathcal{P}$ on the phase-space of the system. For such a realization of $\mathcal{P}$, we need ten phase-space functions $J_j(\omega), K_j(\omega), P_j(\omega), H(\omega)$ to represent the Lie algebra elements l_j, k_j, d_j, and h, respectively, and they must give us a PB realization up to neutrals of the relations in Eqs. (20.2) and (20.6). However, we now show that these generators can always be adjusted in such a way that there are no neutral elements at all, so that we can always assume that a canonical realization of $\mathcal{P}$ is generated by a true PB realization of its Lie algebra. We first prove this for the HLG, and then for $\mathcal{P}$. We can start by assuming that $J_j(\omega)$ and $K_j(\omega)$ have been chosen in such a way that there are no neutral elements in the PB's between two J's or a J and a K; the argument needed is the same as in the case of $E(3)$ and is independent of the bracket between two K's. Thus we have:

$$\{J_j, J_k\} = \epsilon_{jkm} J_m, \ \{J_j, K_k\} = \epsilon_{jkm} K_m \,, \tag{20.8}$$

and as a consequence, for a finite rotation,

$$\widetilde{e^{\boldsymbol{\alpha}\cdot \boldsymbol{J}}} K_j = A_{kj}(\boldsymbol{\alpha}) K_k \,. \tag{20.9}$$

If we now suppose that the bracket $\{K_j, K_k\}$ is equal to $-\epsilon_{jkm} J_m$ together with a neutral element d_{jk}, it follows by applying the finite rotation transformation $\widetilde{e^{\boldsymbol{\alpha}\cdot \boldsymbol{J}}}$ that d_{jk} is an antisymmetric second-rank tensor with respect to $R(3)$, whose components are unchanged by any rotation. Because there are no nontrivial tensors of this type, we can conclude that the d_{jk} vanish, so that if Eq. (20.8) holds good, then so will the following:

$$\{K_j, K_k\} = -\epsilon_{jkm} J_m \,. \tag{20.10}$$

This proves that all neutral elements in the Lie relations corresponding to the HLG can be transformed away. Let us next consider the PB's involving the translation generators $P_j(\omega), H(\omega)$. Arguments used in the cases of $E(3)$ and $\mathcal{G}$ show again that with a suitable choice of $P_j(\omega)$ we have:

$$\begin{aligned} \{J_j, P_k\} = \epsilon_{jkm} P_m, \qquad & \{J_j, H\} = 0; \\ \{P_j, P_k\} = 0, \qquad & \{P_j, H\} = 0 \,. \end{aligned} \tag{20.11}$$

In achieving this, only the $P_j(\omega)$, and not $H(\omega)$, may have to be redefined. We are now left with the two sets of brackets $\{K_j, P_k\}$ and $\{K_j, H\}$. If the latter were equal to P_j plus some neutral elements c_j, then once again application of the rotation $\widetilde{e^{\boldsymbol{\alpha}\cdot \boldsymbol{J}}}$ leads to the conclusion that c_j is a numerically invariant three-dimensional vector, which must then vanish because there are no such vectors. Therefore, with the previously adjusted choices of J_j, K_j, and P_j, but with H

still undetermined to the extent that one could add a neutral element to it, we have secured

$$\{K_j, H\} = P_j \,. \tag{20.12}$$

Let us now write, for the remaining brackets,

$$\{K_j, P_k\} = \delta_{jk} H + d'_{jk} \,, \tag{20.13}$$

d'_{jk} being neutrals. Apply again $\widetilde{e^{\boldsymbol{\alpha}\cdot\boldsymbol{J}}}$ to both sides to conclude that d'_{jk} is a numerically invariant second-rank tensor. But now we have no restriction such as antisymmetry in the indices, and the most general possibility for d'_{jk} is that it is δ_{jk} times a neutral element d'. We can now remove the arbitrariness in H by combining the terms on the right-hand side in Eq. (20.13), and rewriting $H + d'$ as the new H. With H so redefined, we have:

$$\{K_j, P_k\} = \delta_{jk} H \,, \tag{20.14}$$

and we have proved that all neutral elements in the Lie algebra of $\mathcal{P}$ can be removed. At the same time, we see that when this is done, there is no remaining arbitrariness in the choice of the generators. Thus in contrast to the case of $\mathcal{G}$, in a Poincaré invariant theory there is no arbitrariness in the *zero point of the energy*. More generally, a canonical realization of $\mathcal{P}$ possesses a uniquely determined set of generators that give a true PB realization of the Lie algebra of $\mathcal{P}$.

It is interesting that while there are no PB realizations of the Lie algebra of $\mathcal{P}$ involving neutral elements in a nontrivial way, the group $\mathcal{G}$ obtained as a *limiting form* of $\mathcal{P}$ does have such realizations. The reason why this happens is essentially this: the two processes of transforming away any neutral elements that may be artificially introduced in a realization of the Lie algebra of $\mathcal{P}$, and of taking the limit $c \to \infty$ to arrive at a realization of the Lie algebra of $\mathcal{G}$, do not commute; the final result can depend on the order of these two operations. In particular, a trivial neutral element in the Lie algebra of $\mathcal{P}$ can turn into a nontrivial one in the Lie algebra of $\mathcal{G}$ after the limit $c \to \infty$ has been taken. We illustrate this later on by showing how a single-particle realization of $\mathcal{P}$ goes over into a single particle realization of $\mathcal{G}$ with finite mass.

General Properties of Realizations of $\mathcal{P}$

We have proved that we may assume that the generators of a canonical realization of $\mathcal{P}$ obey the set of PB relations given in Eqs. (20.8), (20.10), (20.11), (20.12), and (20.14). Whereas in the case of $\mathcal{G}$ we could consider various classes of realizations (with M=0) in which various subsets of the generators vanished identically, the situation with $\mathcal{P}$ is in this sense more like the $M \neq 0$ realizations of $\mathcal{G}$, and there are limitations on the possibility of some generators of $\mathcal{P}$ vanishing identically. At the level of the HLG, the PB relations Eq. (20.8) and (20.10) show that if the J's are nonvanishing, then the K's must also be nonvanishing; in other words, the only alternatives are to have either all the J's and K's vanishing identically, leading to the trivial realization of the HLG as well as of $\mathcal{P}$, or to have none of the J_j and K_j vanishing identically. Similarly, turning to the realizations of $\mathcal{P}$ and assuming that we have nontrivial J's and K's, Eqs. (20.12) and (20.14) show that either all the P's and H vanish identically, or none of them does. Therefore, leaving aside the totally trivial realization of $\mathcal{P}$, we have only two possibilities: either we have a faithful realization of the Lie algebra of $\mathcal{P}$ with none of the ten generators vanishing identically, or we have a realization in which $P_j = H = 0$. Clearly, the latter correspond to having a canonical realization of the HLG alone, the translations T_4 in the semidirect product $\mathcal{P} = HLG \times T_4$ being realized trivially. We do not discuss these types of realizations of $\mathcal{P}$ very much, limiting ourselves to computing the (Casimir) invariants for them.

The behaviors of the generators themselves under the canonical transformations corresponding to elements of $\mathcal{P}$, in a canonical realization of $\mathcal{P}$, are obtained in part from the results contained in Chapter 15. We refer here to Eqs. (15.111) and (15.127). Just as for the abstract Lie algebra of $\mathcal{P}$, we combine the generators $J_j(\omega)$ and $K_j(\omega)$ into a two-index quantity $J_{\mu\nu}(\omega)$, and the translation generators $H(\omega)$ and $P_j(\omega)$ into a single-index quantity $P_\mu(\omega)$:

$$J_{mn}(\omega) = \epsilon_{mnj}J_j(\omega),\ J_{0m}(\omega) = K_m(\omega),\ J_{\mu\nu}(\omega) = -J_{\nu\mu}(\omega)\,; \\ P_0(\omega) = P^0(\omega) = H(\omega)\,. \tag{20.15}$$

Then these obey the PB relations corresponding to Eqs. (20.3) and (20.7):

$$\begin{aligned}\{J_{\mu\nu}(\omega), J_{\sigma\rho}(\omega)\} &= g_{\mu\sigma}J_{\nu\rho}(\omega) - g_{\nu\sigma}J_{\mu\rho}(\omega) + g_{\mu\rho}J_{\sigma\nu}(\omega) - g_{\nu\rho}J_{\sigma\mu}(\omega)\,,\\ \{J_{\mu\nu}(\omega), P_\rho(\omega)\} &= g_{\mu\rho}P_\nu(\omega) - g_{\nu\rho}P_\mu(\omega), \{P_\mu(\omega), P_\nu(\omega)\} = 0\,.\end{aligned} \tag{20.16}$$

Let us consider now an element $\Lambda = \exp(-\frac{1}{2}\lambda^{\mu\nu}l_{\mu\nu})$ in the HLG, with corresponding Lorentz-transformation matrix $\|\Lambda(\lambda)^\rho{}_\sigma\|$. [The determination of the matrix elements $\Lambda(\lambda)^\rho{}_\sigma$ in terms of the canonical coordinates $\lambda^{\mu\nu}$ is described later when we consider linear matrix representations of the HLG.] Then from the quoted equations of Chapter 15 we obtain:

$$\begin{aligned}\exp(-\frac{1}{2}\lambda^{\mu\nu}\widetilde{J_{\mu\nu}(\omega)})J_{\alpha\beta}(\omega) &= \Lambda(\lambda)^\rho{}_\alpha\Lambda(\lambda)^\sigma{}_\beta J_{\rho\sigma}(\omega)\,,\\ \exp(-\frac{1}{2}\lambda^{\mu\nu}\widetilde{J_{\mu\nu}(\omega)})P_\alpha(\omega) &= \Lambda(\lambda)^\rho{}_\alpha P_\rho(\omega)\,.\end{aligned} \tag{20.17}$$

Thus, under the HLG, each index on $J_{..}(\omega)$ and the one on $P_.(\omega)$ transforms by itself as though it were an index on a four vector. $J_{\mu\nu}(\omega)$ forms an antisymmetric tensor of the second rank, $P_\mu(\omega)$ a four vector. In the particular cases of pure rotations or pure Lorentz transformations, $\Lambda = \exp(\boldsymbol{\alpha}\cdot\boldsymbol{l})$ or $\exp(\boldsymbol{\zeta}\cdot\boldsymbol{k})$, the corresponding matrices $||\Lambda^\rho{}_\sigma||$ are the matrices $\Lambda(\mathbf{0},\boldsymbol{\alpha}), \Lambda(\boldsymbol{\zeta},\mathbf{0})$, respectively, given explicitly in Eq. (15.85). Under a space-time translation, we obtain with the use of Eq. (20.16),

$$
\begin{aligned}
e^{\widetilde{a^\mu P_\mu(\omega)}} J_{\alpha\beta}(\omega) &= J_{\alpha\beta}(\omega) - a^\mu\{J_{\alpha\beta}(\omega), P_\mu(\omega)\} \\
&= J_{\alpha\beta}(\omega) - a_\alpha P_\beta(\omega) + a_\beta P_\alpha(\omega)\,; \\
e^{\widetilde{a^\mu P_\mu(\omega)}} P_\alpha(\omega) &= P_\alpha(\omega)\,.
\end{aligned}
\tag{20.18}
$$

Thus with Eqs. (20.17) and (20.18) we have the behaviors of the generators under the finite canonical transformations realizing $\mathcal{P}$. However, although these equations have an elegant form and contain the required information in a concise manner, from the point of view of physical interpretation it is better to return to the three-dimensional form in which we discuss the transformations generated by $\boldsymbol{J}, \boldsymbol{K}, \boldsymbol{P}$, and H individually. This is similar to the property of the basic PB relations as given in Eq. (20.16) that they exhibit the geometric form, but the dynamical features, in which the time direction has a preferred significance, tend to be obscured by too much emphasis on the four-dimensional geometry. The $J_j(\omega)$ retain, of course, the physical meaning of the three-dimensional total angular momentum of the physical system, whereas $P_j(\omega)$ and $H(\omega)$ are the total three-dimensional linear momentum and energy, respectively. Either from the relevant PB relations or the integrated form given in Eq. (20.18), we see that these seven quantities are all constants of motion; conversely the Hamiltonian is the same in any two inertial frames connected by an element of $E(3) \otimes T_1$. On the other hand, the generators of pure velocity transformations, $K_j(\omega)$, are not constants of motion. For them we have from Eq. (20.18):

$$
e^{\widetilde{-tH(\omega)}} K_j(\omega) = K_j(\omega) + tP_j(\omega)\,. \tag{20.19}
$$

We discuss the meaning of the K's in a moment, but let us now note once again the *dual roles* played by the functions $J_j(\omega), K_j(\omega), P_j(\omega)$, and $H(\omega)$ of being the *generators* of canonical transformations to different inertial frames, at the same time representing important *physical quantities*. For example, the bracket relations between the J's and the P's may be interpreted either as stating that the linear momentum transforms like a vector under a rotation (i.e. the change in $P_j(\omega)$ in an infinitesimal rotation is perpendicular to itself and to the axis of the rotation), *or* as stating that under a displacement the angular momentum changes by a quantity proportional to the normal component of momentum. Similarly the bracket relation Eq. (20.12) between $K_j(\omega)$ and $H(\omega)$ may be taken to mean that the energy changes on transforming to a moving frame by a quantity proportional to the component of linear momentum along the direction of the relative velocity. Equally well it may be taken to mean that the $K_j(\omega)$ are not constants of motion but change linearly with respect to time by a quantity proportional to the linear momentum.

The $K_j(\omega)$ correspond to the variables G_j of the preceding chapter. We saw there that in a realization of $\mathcal{G}$ with nonvanishing mass $M, G_j/M$ had the interpretation of the position coordinates of the center of mass of the system. An analogous interpretation is sometimes possible for K_j. Under spatial rotations the $K_j(\omega)$ do transform as the three components of a vector. If one makes a spatial translation, one obtains from Eq. (20.18):

$$e^{\widetilde{\boldsymbol{a}\cdot\boldsymbol{P}(\omega)}} K_j(\omega) = K_j(\omega) + a_j P_0(\omega) = K_j(\omega) - a_j H(\omega)\,. \tag{20.20}$$

This means that if K_j/H were a well-defined triplet of dynamical variables, it defines a vector whose behavior under the operations of the Euclidean group is just like the behavior of something characterizing position; and taking account of the fact that K_j/H has a linear dependence on time, as is shown by Eq. (20.19), we could interpret the K_j as being the "moment of energy" (energy times the position coordinates of the center of energy of the system). This hypothetical point in the system moves with the velocity P_j/H. (The nonrelativistic *center of mass* is here replaced by the *center of energy*). But there is a certain amount of freedom here, because we could add to K_j/H any three-dimensional vector that is invariant under both space and time translations, such as P_j, and the result would still be a position-type quantity having a linear time dependence. Sometimes it is K_j itself, and sometimes it is a vector modified in this way, that has the meaning of the moment of the energy. We should now note two differences as compared to the case of $\mathcal{G}$. The first is that as long as the mass M was a nonzero constant, division of G_j by it was a straightforward operation, and it always led to sensible position-type variables $Q_j = G_j/M$. However, in the case of $\mathcal{P}$ we need to consider division of K_j by H, and if the latter is not known to be nonsingular we would have states of the system in which the position of the center of energy is undefined. Thus we may always retain the interpretation of a relativistic moment of energy for K_j (or for K_j plus something suitable), and in case $H(\omega)$ is positive definite we can build up something like position out of K_j/H. In many physically important realizations of $\mathcal{P}$, we can indeed verify that H is nonsingular. But even so, there is a second difference compared to $\mathcal{G}$. The basic PB's for the generators of $\mathcal{G}$ inform us that the G_j/M have vanishing brackets with one another, and combining this with the values of their brackets with P_k, we are able to interpret the G_j/M as position-type variables *canonically conjugate to the* P_k. This in turn leads to the possibility of splitting the total angular momentum into two kinematically independent parts, one being the center-of-mass angular momentum and the other internal, with the property that the latter is unchanged by space as well as time translations. In order to achieve something similar in a realization of $\mathcal{P}$, assuming that $H(\omega)$ is positive definite, more work has to be done. We do have, from Eq. (20.14), the relations:

$$\left\{\frac{K_j}{H}, P_k\right\} = \delta_{jk}\,, \tag{20.21}$$

but in general the bracket $\{K_j/H, K_k/H\}$ does not vanish, and thus we cannot always interpret the K_j/H as canonical conjugates to the P_k. These variables must be modified to obtain such canonical conjugates, and this is related to a splitting of the total angular momentum into independent parts, one part being invariant again under space and time translations. We shall see how this happens after we discuss the construction of Casimir invariants for the HLG and $\mathcal{P}$.

The independent generators for a canonical realization of the HLG are six in number, being the $J_j(\omega)$ and the $K_j(\omega)$, and they obey the PB relations in Eqs. (20.8) and (20.10). Any Casimir invariant must in particular be a rotational invariant. There are three such that can be formed out of the J's and K's, namely J_jJ_j, J_jK_j, and K_jK_j. Of these one finds easily that the second is already a Casimir invariant, because its PB with the K's vanishes:

$$\{K_k, J_jK_j\} = \epsilon_{kjm}K_mK_j - J_j\epsilon_{kjm}J_m = 0\,. \tag{20.22}$$

On the other hand, neither J_jJ_j nor K_jK_j has a vanishing bracket with the K_k, but the difference does and thus is another Casimir invariant:

$$\begin{aligned}\{K_k, K_jK_j - J_jJ_j\} =& 2K_j\{K_k, K_j\} - 2J_j\{K_k, J_j\}\\ =& -2K_j\epsilon_{kjm}J_m - 2J_j\epsilon_{kjm}K_m = 0\,.\end{aligned} \tag{20.23}$$

Therefore, the two independent Casimir invariants for a canonical realization of the HLG are J_jK_j and $K_jK_j - J_jJ_j$, and there are no others. If one remembers the tensor transformation law Eq. (20.17) for the generators of the HLG and writes the Casimir invariants in four-dimensional form, one sees more immediately why these are invariants; one has:

$$K_jK_j - J_jJ_j = -\frac{1}{2}J^{\mu\nu}J_{\mu\nu}, \quad J_jK_j = -\frac{1}{8}\epsilon_{\mu\nu\lambda\rho}J^{\mu\nu}J^{\lambda\rho}\,. \tag{20.24}$$

Here, $\epsilon_{\mu\nu\lambda\rho}$ is the completely antisymmetric four-dimensional Levi-Civita symbol, with values determined by the definition $\epsilon_{0123} = +1$.

Turning to the case of $\mathcal{P}$, it is preferable to use the four-dimensional structure from the start in the construction of Casimir invariants. Because there are two Casimir invariants in the $M \neq 0$ realizations of $\mathcal{G}$, and because we expect $\mathcal{G}$ to be the limit of $\mathcal{P}$ as $c \to \infty$ in suitable situations, we must expect to find two Casimir invariants for $\mathcal{P}$ as well. One of them is easily constructed in terms of the energy-momentum four-vector $P_\mu(\omega)$. These transform linearly under a transformation of the HLG, as shown in Eq. (20.17), and are invariant under space-time translations. Therefore, a function of $P_\mu(\omega)$ which is unchanged under the HLG automatically is a Casimir invariant for $\mathcal{P}$ at the same time. Such a function is the four-dimensional scalar product of P_μ with itself: all four-dimensional scalar products are preserved by the matrices $||\Lambda^\rho{}_\sigma||$ of homogeneous Lorentz transformations. This Casimir invariant for $\mathcal{P}$ is called the square of the total mass, and we define it precisely as

$$M^2 = -P^\mu(\omega)P_\mu(\omega) = (H(\omega))^2 - P_j(\omega)P_j(\omega)\,. \tag{20.25}$$

Though we have written M^2 here, it is only M^2 that is a real quantity; M is also real only when M^2 is nonnegative. Toward building up a second Casimir invariant for $\mathcal{P}$, we define now a combination of the generators $J_{\mu\nu}$ and P_μ that transforms as a four-vector under the HLG:

$$W_\mu(\omega) = \frac{1}{2}\epsilon_{\mu\nu\lambda\rho}P^\nu(\omega)J^{\lambda\rho}(\omega)\,. \tag{20.26}$$

This vectorial transformation property of $W_\mu(\omega)$ is evident from the way it is constructed, the properties of $\epsilon_{\mu\nu\lambda\rho}$, and Eq. (20.17). It is called the Pauli-Lubanski four-vector. The point of constructing a four-vector in this particular way is that, like the $P_\mu(\omega)$, the $W_\mu(\omega)$ are invariant under space-time translations. This is caused by the antisymmetry of the ϵ tensor:

$$\begin{aligned}\{P_\mu, W_\sigma\} &= \frac{1}{2}\epsilon_{\sigma\nu\lambda\rho}P^\nu\{P_\mu, J^{\lambda\rho}\} \\ &= -\frac{1}{2}\epsilon_{\sigma\nu\lambda\rho}P^\nu(\delta^\lambda_\mu P^\rho - \delta^\rho_\mu P^\lambda) = 0\,. \end{aligned} \tag{20.27}$$

One could have combined $J_{\mu\nu}$ and P_λ to form a four-vector in another way, namely, $P^\lambda J_{\lambda\mu}$, but unlike the W_μ this would not have been invariant under space-time translations. Having secured this property for $W_\mu(\omega)$, we then see that the four-dimensional scalar product of W_μ with itself is invariant under the HLG and is therefore a Casimir invariant for $\mathcal{P}$ as well. This is the second Casimir invariant for $\mathcal{P}$; we define it more precisely as

$$W^2(\omega) = W^\mu(\omega)W_\mu(\omega)\,. \tag{20.28}$$

Once again it is not implied that W itself is always real. Thus the independent Casimir invariants for $\mathcal{P}$ are M^2 and W^2. These generalize the two Casimir invariants $C_1' = P_jP_j - 2MH$ and $C_2' = (M\boldsymbol{J} - \boldsymbol{G}\times\boldsymbol{P})^2$ that we found for the Galilei group in the $M \neq 0$ realizations. To see this connection, one has to put in the appropriate powers of c everywhere. In the non-relativistic limit, M goes into the mass M that appears as a neutral element in the realization of $\mathcal{G}$. (We see this "happen" later on when we describe the single-particle case as an example). At the same time the Hamiltonian for the realization of $\mathcal{G}, H_1$ say, is to be computed by subtracting Mc^2 from H and developing the rest in inverse powers of c and keeping just the c-independent terms:

$$\begin{aligned} H_1 &= \sqrt{M^2c^4 + P_jP_jc^2} - Mc^2 \\ &= Mc^2\left(1 + \frac{1}{2}\frac{P_jP_j}{M^2c^2} + \ldots\right) - Mc^2 \cong \frac{P_jP_j}{2M}\,. \end{aligned} \tag{20.29}$$

To say next that $C_1' = P_jP_j - 2MH_1$ is a Casimir invariant for $\mathcal{G}$ is to recognize the fact that in any realization of G the Hamiltonian is defined only up to an additive neutral element. The situation with regard to W_μ is as follows. Let us write out its components explicitly:

$$W_0 = \boldsymbol{J}\cdot\boldsymbol{P},\ \boldsymbol{W} = -H\boldsymbol{J} + \boldsymbol{K}\times\boldsymbol{P}\,. \tag{20.30}$$

The W_0 is the helicity and remains the helicity in the nonrelativistic limit. The space part, $\boldsymbol{W}$, can be seen to be the analogue of the internal angular momentum vector $\boldsymbol{J} - (1/M)\boldsymbol{G} \times \boldsymbol{P}$ defined in realizations of $\mathcal{G}$. In the limit $c \to \infty$, $\boldsymbol{W}$ dominates W_0, and $\boldsymbol{W}/c^2$ approaches a finite limit. Thus one can see that W^2 is the relativistic extension of C_2'. (In these arguments, we must assume that M^2 is positive definite.)

On the basis of the two Casimir invariants of $\mathcal{P}$, we can get an idea of the different possible kinds of canonical realizations of $\mathcal{P}$. In an irreducible realization, both M^2 and W^2 are some numbers. We separate the different realizations from one another on the basis of the value of M^2. This value tells us what type of energy-momentum four-vector $P^\mu(\omega)$ we are dealing with. As we know from the kinematics of Lorentz transformations, the set of all four-vectors breaks up into several distinct types that are invariant under the action of Lorentz transformations. Writing p^μ for a general numerical four-vector, the simplest possibility is that it is the null vector, p^μ=0; this by itself is invariant under all Lorentz transformations. Apart from this we have the three distinct possibilities that p^μ may be timelike, lightlike, or spacelike. These correspond to the "squared length" $p^2 = p^\mu p_\mu = |\boldsymbol{p}|^2 - (p^0)^2$ being negative, zero, or positive, respectively. In the timelike and lightlike cases, we can split up the four-vectors further according to the sign of the time component p^0. If $p^0 > 0 (< 0)$ we have positive (negative) timelike and lightlike four-vectors, and each such family is invariant under all Lorentz transformations in the identity component of the HLG. If we consider a positive timelike or positive lightlike four-vector p^μ, then its time component is always bounded below by $\sqrt{-p^2}$ when we subject p^μ to all possible Lorentz transformations. In all cases other than these, that is if p^μ is negative timelike, negative lightlike, or spacelike, there is no lower bound on p^0. Given such a p^μ, in a suitable reference frame we can make p^0 take on negative values arbitrarily large in magnitude. In an irreducible canonical realization of $\mathcal{P}$, any one of these possibilities exists for $P_\mu(\omega)$. If P_μ=0 identically, we just have a realization of the HLG. If P_μ is positive timelike (lightlike), then the energy $H(\omega)$ exceeds (equals) the magnitude of the space part $|\boldsymbol{P}|$ and thus is bounded below. If P_μ is of any other type, it involves an energy component that is unbounded below, which is unphysical. Therefore, only those irreducible canonical realizations of $\mathcal{P}$ in which $M^2 \geq 0$, and $P^0 > 0$, are physical in this sense. As for the possible values of the other invariant W^2, we must deal with the fact that W_μ and P_μ are "orthogonal" to one another:

$$W^\mu P_\mu = 0\,. \tag{20.31}$$

From the kinematics of four-vectors, we then conclude: if P^μ is timelike, then W^μ must be spacelike or null and thus $W^2 \geq 0$; if P^μ is lightlike, then W^μ must be spacelike, lightlike or null, thus again $W^2 \geq 0$; and finally if P^μ is spacelike, then no definite statement can be made on the nature of W^μ and thus on the sign of W^2. We thus have a classification of irreducible canonical realizations of $\mathcal{P}$ according to the value of M^2 (and, if relevant, the sign of P^0), and the possible values of W^2. A general canonical realization of $\mathcal{P}$ can be thought of as being obtained by "stacking" together several, and even continuously many, irreducible realizations. In such a realization, either or both of M^2 and W^2 can be nontrivial phase-space functions, and both positive and negative values of

M^2 may occur in the same realization of $\mathcal{P}$. But in most physical examples of realizations of $\mathcal{P}$ one expects that the energy $P^0(\omega)$ should be bounded below; if this happens, it can only be because the four-vector P^μ is positive timelike or positive light-like in the entire realization, so that $M^2(\omega) \geq 0$ as well. In any case, in the more physical examples we discuss, the Hamiltonian $H(\omega) = P^0(\omega)$ is explicitly positive definite, and this is compatible with a positive timelike or lightlike P^μ alone. It is also clear that only with $M^2 > 0$ is there a possibility of going to a nonrelativistic limit. In particular, there is no such limit if P^μ is lightlike, and certainly not if it is spacelike. For all these reasons, unless the contrary is explicitly stated, we always consider only those classical canonical realizations of $\mathcal{P}$ in which $M^2(\omega) > 0, P^0(\omega) > 0$; in such a case, we also have $W^2(\omega) \leq 0$. Thus the energy momentum four-vector is positive timelike, and the Pauli-Lubanski vector is spacelike or null. (The exceptions are some of the field realizations of $\mathcal{P}$, those on pages 472-476).

With the restriction to such realizations of $\mathcal{P}$, we can now give a general analysis similar in spirit to the one given in the preceding chapter. There we found it convenient to recognize first from the basic PB relations of $\mathcal{G}$ the existence of canonical conjugates to the linear momenta P_j, and we used these to split the angular momentum into independent parts; in the present case it is easier to follow the converse procedure - at any rate the basic PB realizations for $\mathcal{P}$ do not immediately present us with canonical conjugates to the P_j! On the other hand, the translational invariance of the W_μ and the form of the components exhibited in Eq. (20.30) both suggest that this vector describes an internal part of the total angular momentum. Although the brackets $\{W_\mu, P_\nu\}$ vanish, the $\{W_\mu, W_\nu\}$ do not; it is easiest to evaluate them using the known bracket relations between $J_{\lambda\mu}$ and W_ν that characterize the latter as a four vector:

$$\begin{aligned} \{W_\mu, W_\nu\} &= \frac{1}{2}\epsilon_{\mu\lambda\rho\sigma}P^\lambda\{J^{\rho\sigma}, W_\nu\} = \frac{1}{2}\epsilon_{\mu\lambda\rho\sigma}P^\lambda(\delta^\rho_\nu W^\sigma - \delta^\sigma_\nu W^\rho)\,, \\ \{W_\mu, W_\nu\} &= -\epsilon_{\mu\nu\lambda\sigma}P^\lambda W^\sigma\,. \end{aligned} \tag{20.32}$$

Now, because of the orthogonality of W^μ and P_μ, we can regard W_0 as given in terms of the space components W_j and P_μ:

$$W_0 = -\frac{\boldsymbol{W}\cdot\boldsymbol{P}}{H}\,. \tag{20.33}$$

Specializing Eq. (20.32) to the space components, we have

$$\{W_j, W_k\} = \epsilon_{jkl}(-HW_l - W_0P_l) = \epsilon_{jkl}\left(-HW_l + \boldsymbol{W}\cdot\boldsymbol{P}\frac{P_l}{H}\right),$$

or temporarily introducing the vector $V_j = -W_j/H$ in place of W_j, we have the PB relations:

$$\begin{aligned} \{V_j, V_k\} &= \epsilon_{jkl}V_l - \frac{\boldsymbol{V}\cdot\boldsymbol{P}}{H^2}\epsilon_{jkl}P_l\,, \\ \{V_j, P_k\} &= \{V_j, H\} = 0\,. \end{aligned} \tag{20.34}$$

If the term involving $\boldsymbol{V}\cdot\boldsymbol{P}$ had not been there, we might have identified V_j with an internal angular momentum. Let us try adding a term $\alpha\boldsymbol{V}\cdot\boldsymbol{P}P_j$ to V_j, α being a function of the P's alone, and see if we get an angular momentum:

$$\begin{aligned}
\{V_j+\alpha\boldsymbol{V}\cdot\boldsymbol{P}P_j, V_k{+}\alpha\boldsymbol{V}\cdot\boldsymbol{P}P_k\} &= \epsilon_{jkl}V_l - \frac{\boldsymbol{V}\cdot\boldsymbol{P}}{H^2}\epsilon_{jkl}P_l \\
&+\alpha(\{V_j,\boldsymbol{V}\cdot\boldsymbol{P}\}P_k - \{V_k,\boldsymbol{V}\cdot\boldsymbol{P}\}P_j) \\
&=\epsilon_{jkl}\left(V_l - \frac{\boldsymbol{V}\cdot\boldsymbol{P}}{H^2}P_l\right) + \alpha(\epsilon_{jml}P_mV_lP_k - \epsilon_{kml}P_mV_lP_j) \\
&=\epsilon_{jkl}\left(V_l - \frac{\boldsymbol{V}\cdot\boldsymbol{P}}{H^2}P_l\right) + \alpha((\boldsymbol{P}\times\boldsymbol{V})_jP_k - (\boldsymbol{P}\times\boldsymbol{V})_kP_j) \\
&=\epsilon_{jkl}\left(V_l - \frac{\boldsymbol{V}\cdot\boldsymbol{P}}{H^2}P_l + \alpha\{(\boldsymbol{P}\times\boldsymbol{V})\times\boldsymbol{P}\}_l\right), \\
\{V_j+\alpha\boldsymbol{V}\cdot\boldsymbol{P}P_j, V_k{+}\alpha\boldsymbol{V}\cdot\boldsymbol{P}P_k\} &= \epsilon_{jkl}((1+\alpha\boldsymbol{P}^2)V_l - \left(\alpha+\frac{1}{H^2}\right)\boldsymbol{V}\cdot\boldsymbol{P}P_l)
\end{aligned} \tag{20.35}$$

It is clear that we get an angular momentum if we allow for a scale factor, β say, which again is a function of P_μ alone. We try to solve for α and β from the equations:

$$1+\alpha\boldsymbol{P}\cdot\boldsymbol{P} = \beta\,, \; -\left(\alpha+\frac{1}{H^2}\right) = \alpha\beta\,. \tag{20.36}$$

There are two possible solutions, and we choose the one that does not involve any singular denominators. Writing M for the positive square root of the Casimir invariant $M^2(\omega)$, we have

$$\alpha = -\frac{1}{H(H+M)},\; \beta = \frac{M}{H}; \tag{20.37}$$

and the vector that has the angular momentum structure for its PB's with itself is:

$$S_j = \beta^{-1}(V_j+\alpha\boldsymbol{V}\cdot\boldsymbol{P}P_j) = -\frac{W_j}{M} + \frac{\boldsymbol{W}\cdot\boldsymbol{P}P_j}{MH(M+H)}\,. \tag{20.38}$$

This is the internal angular momentum of the system described by the given canonical realization of $\mathcal{P}$. Its algebraic properties are given by:

$$\{S_j, S_k\} = \epsilon_{jkl}S_l,\; \{J_j, S_k\} = \epsilon_{jkl}S_l,\; \{S_j, P_\mu\} = 0\,. \tag{20.39}$$

As expected, this internal angular momentum is unchanged by spatial translations and is a constant of motion. It is possible to rewrite Eq. (20.38) in this way:

$$S_j = -\frac{W_j}{H} + \frac{[\boldsymbol{P}\times(\boldsymbol{P}\times\boldsymbol{W})]_j}{MH(M+H)}\,. \tag{20.40}$$

From Eqs. (20.38) and (20.40), we derive two useful relations:

$$\boldsymbol{W} \times \boldsymbol{P} = -M\boldsymbol{S} \times \boldsymbol{P},\ \boldsymbol{W} \cdot \boldsymbol{P} = -H\boldsymbol{S} \cdot \boldsymbol{P}\,. \tag{20.41}$$

Thus far we have succeeded in expressing the S_j in terms of P_μ and W_j, and thus essentially in terms of the generators of $\mathcal{P}$ and the Casimir invariant M^2. We can now turn things around to see if, for example, $\boldsymbol{J}$ can be expressed as $\boldsymbol{S}$ plus something else. Equation (20.40) tells us how to do this: using the definition of $\boldsymbol{W}$ in the first term on the right-hand side, we find:

$$\boldsymbol{J} = \boldsymbol{S} + \frac{\boldsymbol{K} \times \boldsymbol{P}}{H} - \frac{\boldsymbol{P} \times (\boldsymbol{P} \times \boldsymbol{W})}{MH(M+H)} = \boldsymbol{S} + \left(\frac{\boldsymbol{K}}{H} + \frac{\boldsymbol{P} \times \boldsymbol{W}}{MH(M+H)}\right) \times \boldsymbol{P}\,. \tag{20.42}$$

This decomposition of $\boldsymbol{J}$ is very much like the decomposition achieved in analyzing realizations of $\mathcal{G}$: $\boldsymbol{J}$ consists of an internal part $\boldsymbol{S}$ and another part that is the vector product of some three-dimensional vector with $\boldsymbol{P}$. This suggests that this new vector introduced here may, in fact, be canonically conjugate to $\boldsymbol{P}$. Let us write $\boldsymbol{Q}$ for it, so that we have:

$$Q_j = \frac{K_j}{H} + \frac{(\boldsymbol{P} \times \boldsymbol{W})_j}{MH(M+H)}\,; \tag{20.43a}$$

$$J_j = S_j + (\boldsymbol{Q} \times \boldsymbol{P})_j\,. \tag{20.43b}$$

Just as Eq. (20.40) constitutes the definition of S_j as a function of the generators of $\mathcal{P}$, Eq. (20.43a) constitutes a similar definition of Q_j. We know that the PB $\{Q_j, P_k\}$ has the expected value δ_{jk} and we wish to find out whether the bracket $\{Q_j, Q_k\}$ vanishes. To check this, we need to use, in addition to the basic PB relations of $\mathcal{P}$, the following derived ones:

$$\begin{aligned} \left\{\frac{K_j}{H}, \frac{K_k}{H}\right\} &= \frac{\epsilon_{jkl} W_l}{H^3};\ \{K_j, W_0\} = -W_j,\ \{K_j, W_k\} = -\delta_{jk} W_0\,; \\ \{K_j, (\boldsymbol{P} \times \boldsymbol{W})_k\} &= -\,\epsilon_{jkl}(HW_l + W_0 P_l)\,; \\ \{(\boldsymbol{P} \times \boldsymbol{W})_j, (\boldsymbol{P} \times \boldsymbol{W})_k\} &= \epsilon_{jkl} M^2 W_0 P_l\,. \end{aligned} \tag{20.44}$$

With the help of these, and doing some routine calculations, we discover

$$\{Q_j, Q_k\} = 0\,. \tag{20.45}$$

We have succeeded on both counts. Given any canonical realization of $\mathcal{P}$ in which $M^2(\omega) > 0, P^0(\omega) > 0$, we have constructed functions S_j and Q_j of the generators of $\mathcal{P}$ such that the former is a translation-invariant internal angular momentum, and the latter are true canonical conjugates to the $P_j(\omega)$. That the PB's among the three components of $\boldsymbol{Q} \times \boldsymbol{P}$ have the familiar angular momentum pattern is evident from the first line in Eq. (20.39) and the breakup of J_j exhibited in Eq. (20.43b); now it becomes even more immediately obvious. It is the $\boldsymbol{Q} \times \boldsymbol{P}$ piece in $\boldsymbol{J}$ that is changed by the required amount under a spatial displacement generated by P_j. The kinematic independence of S_j and P_j

follows from the second line of Eq. (20.39). That of S_j and Q_j is assured by the remark that Q_j is a vector under space rotations, which fixes the values of $\{J_j, Q_k\}$, and the obvious fact that these brackets have the same values as do $\{(\boldsymbol{Q} \times \boldsymbol{P})_j, Q_k\}$; thus we have, for the full set of variables S_j, Q_j, and P_j,

$$\begin{aligned} \{S_j, S_k\} = \epsilon_{jkl} S_l,\ \{Q_j, P_k\} = \delta_{jk}\,; \\ \{S_j, Q_k\} = \{S_j, P_k\} = \{Q_j, Q_k\} = \{P_j, P_k\} = 0\,. \end{aligned} \tag{20.46}$$

We stress once more that all these relations, which are just like the ones involved in the kinematical description of a particle with spin, follow from the definitions of S_j and Q_j in terms of, and the basic PB relations among, the generators of $\mathcal{P}$.

It is interesting to note that the generators $\boldsymbol{K}$ of pure Lorentz transformations can also be expressed explicitly as a function of $\boldsymbol{Q}, \boldsymbol{P}, \boldsymbol{S}$, and the mass invariant M. Such an expression for $\boldsymbol{J}$ is the one given in Eq. (20.43b). Combining Eqs. (20.43a) and (20.41) we immediately get:

$$K_j = HQ_j + \frac{(\boldsymbol{P} \times \boldsymbol{S})_j}{(M+H)}\,. \tag{20.47}$$

This expression, together with Eq. (20.43b), can be interpreted in this way: instead of setting up a realization of $\mathcal{P}$ by giving the ten independent quantities J_j, K_j, P_j, H obeying the proper PB relations, we can equally well set up the ten independent quantities Q_j, P_j, S_j and M; we require the PB relations Eq. (20.46) to be valid and also require the vanishing of the brackets of M with the Q's, P's, and S's. In terms of these, we now define J_j and K_j by means of Eqs. (20.43b) and (20.47), and H by

$$H(\omega) = [M^2 + P_j P_j]^{1/2}\,. \tag{20.48}$$

Then all the PB relations for $\mathcal{P}$ will be obeyed. Although this tells us what the most general (physical) realization of $\mathcal{P}$ looks like, (we will see that it is like a "particle" with variable mass and spin), it is not necessarily useful from a practical point of view. It is interesting to see what form the Casimir invariant W^2 has in these variables. Combining Eqs. (20.33), (20.38), and (20.41), we have:

$$\begin{aligned} W_0 =& \boldsymbol{S} \cdot \boldsymbol{P},\ \boldsymbol{W} = -M\boldsymbol{S} - \frac{\boldsymbol{S} \cdot \boldsymbol{P}\boldsymbol{P}}{M+H}\,; \\ W^2 =& M^2 S^2 + (\boldsymbol{S} \cdot \boldsymbol{P})^2 \left(\frac{2M}{M+H} + \frac{H-M}{M+H} - 1 \right) = M^2 S^2\,. \end{aligned} \tag{20.49}$$

Thus this Casimir invariant is essentially the squared magnitude of the internal angular momentum. The similarity of this result to the result concerning the Casimir invariant C_2' for $\mathcal{G}$ is evident (see Eq. (19.44)).

From this general point of view, both the similarities and the differences between a (physical) realization of $\mathcal{P}$ and a realization of $\mathcal{G}$ with nonzero mass

M are now clear. In both cases we need the spatial momenta P_j, their conjugates Q_j, and the internal angular momenta S_j kinematically independent of the P's and Q's. In addition, for $\mathcal{P}$ we need the Casimir invariant M^2 and for $\mathcal{G}$ the invariant C_1' [which essentially determines the "internal energy" $U(\omega)$ in the Hamiltonian, see Eqs. (19.25) and (19.26)]. The differences lie in the forms of $H(\omega)$ and $K_j(\omega)$ for a realization of $\mathcal{P}$, involving Eqs. (20.47) and (20.48), and in the forms of $H(\omega)$ and $G_j(\omega)$ in a realization of $\mathcal{G}$, given by Eq. (19.26) and the definition $\boldsymbol{G} = M\boldsymbol{Q}$. Although the generators of $E(3)$ are built up in the same way in both cases, the energy and the velocity transformation generators are not. In particular, in the case of $\mathcal{P}$ the entire realization does not split in any obvious way into a direct product of two commuting realizations, as it did in the case of $\mathcal{G}$. The reason is that the K_j have an essential dependence on the internal angular momenta (see Eq. (20.47)); this dependence is necessary because the bracket $\{K_j, K_k\}$ must reproduce the internal angular momentum present in J_j.

This concludes our general analysis of canonical realizations of $\mathcal{P}$. We emphasize again that we have considered only "physical" realizations in which the four-vector $P^\mu(\omega)$ is positive timelike. If it were allowed to be lightlike or spacelike, then the internal variables are no longer of the nature of an angular momentum describable in terms of the $R(3)$ subgroup of the HLG, but involve other characteristic subgroups of the HLG. At the same time, there are no well-defined canonical conjugates to the $P_j(\omega)$ either, at least none with several desirable properties! But we do not explore these problems here. We go on now to a consideration of examples of canonical realizations of $\mathcal{P}$.

Single-Particle Realization of $\mathcal{P}$: Spinless Case

The first example we take up describes a single relativistic free particle with no internal angular momentum. We consider an irreducible canonical realization of $\mathcal{P}$ in which the energy-momentum is positive timelike, and the internal angular momentum variables S_j vanish. The Casimir invariants will reduce to numbers; $M^2(\omega)$ becomes a positive number which we write as m^2, and W^2 vanishes. From our general analysis given above, we see that such an irreducible realization of $\mathcal{P}$ can be formulated in terms of a triplet of canonical degrees of freedom, q_j, p_j, j=1,2,3, obeying the standard fundamental PB relations. By specializing formulae given in the general analysis, we get the generators of $\mathcal{P}$ as functions of q_j, p_j thus:

$$\begin{aligned} J_j &= \epsilon_{jkl} q_k p_l, \; P_j = p_j, K_j = (m^2 + \boldsymbol{p}\cdot\boldsymbol{p})^{1/2} q_j\,, \\ H &= (m^2 + \boldsymbol{p}\cdot\boldsymbol{p})^{1/2} \equiv E(\boldsymbol{p})\,. \end{aligned} \tag{20.50}$$

Though there is no need to do so, the reader can verify that with this construction all the PB relations for $\mathcal{P}$ are obeyed. The Pauli-Lubanski vector W_μ is also explicitly seen to vanish. Let us now see how the canonical transformations realizing $\mathcal{P}$ act on the phase-space coordinates q_j, p_j. (The phase space is six dimensional with each q and each p running over the entire real line). The action of the canonical transformations realizing $E(3)$ is familiar enough by now and needs no repetition. We only need to examine the transformations $\widetilde{e^{-tH}}$ and $\widetilde{e^{-\boldsymbol{\zeta}\cdot\boldsymbol{K}}}$. The former describes the time development. The p_j are known to be constants of motion. As for the q_j, only the term linear in t survives, because $\{H, q_j\}$ is a function of the p's alone, causing all multiple PB's to vanish. Thus we have:

$$\begin{aligned}\widetilde{e^{-tH}}p_j \equiv& p_j(t) = p_j\,;\\ \widetilde{e^{-tH}}q_j \equiv& q_j(t) = q_j - t\{H, q_j\} = q_j + t\frac{\partial E(\boldsymbol{p})}{\partial p_j} = q_j + \frac{p_j}{E}t\,.\end{aligned} \tag{20.51}$$

The equations of motion are thereby solved, and they describe uniform motion in a straight line. This is as it should be for a free particle. That the particle is relativistic is manifested by the dependence of energy, $E(\mathbf{p})$, and velocity, $p_j/E(\mathbf{p})$, on momentum. Next let us consider the canonical transformation $\widetilde{e^{-\boldsymbol{\zeta}\cdot\boldsymbol{K}}}$. This corresponds to the transition from an inertial frame S to another, $S' = \exp(\boldsymbol{\zeta}\cdot\boldsymbol{k})S$. The behavior of the p_j is known because they form part of the generators of $\mathcal{P}$. From Eq. (20.17) we have:

$$\begin{aligned}\widetilde{e^{-\boldsymbol{\zeta}\cdot\boldsymbol{K}}}p_j =& \Lambda(-\boldsymbol{\zeta},\mathbf{0})^{\mu}{}_{j}p_\mu = \left(\delta_{kj} + \zeta_k\zeta_j\frac{\cosh\zeta - 1}{\zeta^2}\right)p_k + \zeta_j\frac{\sinh\zeta}{\zeta}p_0\\ =& p_j + \zeta_j\boldsymbol{\zeta}\cdot\boldsymbol{p}\frac{\cosh\zeta - 1}{\zeta^2} - \zeta_j\frac{\sinh\zeta}{\zeta}E\,;\\ \widetilde{e^{-\boldsymbol{\zeta}\cdot\boldsymbol{K}}}E(\boldsymbol{p}) \equiv& -\widetilde{e^{-\boldsymbol{\zeta}\cdot\boldsymbol{K}}}p_0 = -\Lambda(-\boldsymbol{\zeta},\mathbf{0})^{\mu}{}_{0}p_\mu = -\cosh\zeta\, p_0 - \zeta_j\frac{\sinh\zeta}{\zeta}p_j\\ =& E\cosh\zeta - \boldsymbol{\zeta}\cdot\boldsymbol{p}\frac{\sinh\zeta}{\zeta} = E(\widetilde{e^{-\boldsymbol{\zeta}\cdot\boldsymbol{K}}}\boldsymbol{p})\,.\end{aligned} \tag{20.52}$$

These are the correct relativistic transformation laws for energy and momentum. For example, if $\boldsymbol{\zeta}$ were chosen to be in the positive x-direction, the frame S' moves relative to S in the positive x-direction, and the space-time coordinates in the two are related by:

$$x' = x\cosh\zeta - t\sinh\zeta,\; t' = t\cosh\zeta - x\sinh\zeta,\; y' = y,\; z' = z;$$

in such a case, Eq. (20.52) yields:

$$\begin{aligned}p_x' &= p_x\cosh\zeta - E\sinh\zeta,\; E' = E\cosh\zeta - p_x\sinh\zeta\,,\\ p_y' &= p_y,\; p_z' = p_z\,.\end{aligned}$$

The behavior of the q_j under this canonical transformation is slightly more complicated. We first derive its transformation law and then explain its form. It is simplest to recognize that in the present situation the q_j can be expressed in terms of the generators of $\mathcal{P}$ as K_j/H and then use the known transformation laws for K_j and H. For K_j we may use Eq. (20.17) again, or even more simply take over Eq. (15.113) (after changing $\boldsymbol{\zeta}$ to $-\boldsymbol{\zeta}$ there):

$$\widetilde{e^{-\boldsymbol{\zeta}\cdot\boldsymbol{K}}}K_j = \left(\delta_{jm}\cosh\zeta - \zeta_j\zeta_m\frac{\cosh\zeta - 1}{\zeta^2}\right)K_m - \epsilon_{jmn}\zeta_m\frac{\sinh\zeta}{\zeta}J_n\,. \tag{20.53}$$

Therefore we get:

$$\begin{aligned}
e^{\widetilde{-\boldsymbol{\zeta}\cdot\boldsymbol{K}}}\boldsymbol{q} =& e^{\widetilde{-\boldsymbol{\zeta}\cdot\boldsymbol{K}}}\left(\frac{\boldsymbol{K}}{E}\right) = \frac{e^{\widetilde{-\boldsymbol{\zeta}\cdot\boldsymbol{K}}}\boldsymbol{K}}{e^{\widetilde{-\boldsymbol{\zeta}\cdot\boldsymbol{K}}}E} \\
=& \frac{\boldsymbol{K}\cosh\zeta - \boldsymbol{\zeta}\boldsymbol{\zeta}\cdot\boldsymbol{K}\frac{\cosh\zeta-1}{\zeta^2} - \frac{\sinh\zeta}{\zeta}\boldsymbol{\zeta}\times\boldsymbol{J}}{E'} \\
=& \frac{1}{E'}\left(E\boldsymbol{q}\cosh\zeta - \boldsymbol{\zeta}\boldsymbol{\zeta}\cdot\boldsymbol{q}E\frac{\cosh\zeta-1}{\zeta^2} - \frac{\sinh\zeta}{\zeta}\boldsymbol{\zeta}\times(\boldsymbol{q}\times\boldsymbol{p})\right) \\
=& \frac{1}{E'}\left(\boldsymbol{q}\left(E\cosh\zeta - \boldsymbol{\zeta}\cdot\boldsymbol{p}\frac{\sinh\zeta}{\zeta}\right) - \boldsymbol{\zeta}\boldsymbol{\zeta}\cdot\boldsymbol{q}E\frac{\cosh\zeta-1}{\zeta^2} + \boldsymbol{\zeta}\cdot\boldsymbol{q}\boldsymbol{p}\frac{\sinh\zeta}{\zeta}\right), \\
e^{\widetilde{-\boldsymbol{\zeta}\cdot\boldsymbol{K}}}\boldsymbol{q} =& \boldsymbol{q} + \frac{\boldsymbol{\zeta}\cdot\boldsymbol{q}}{E'}\left(\frac{\sinh\zeta}{\zeta}\boldsymbol{p} - E\frac{\cosh\zeta-1}{\zeta^2}\boldsymbol{\zeta}\right). \qquad (20.54)
\end{aligned}$$

There are two ways in which we can make this transformation law for $\boldsymbol{q}$ appear reasonable: one is a general argument based on the use of the Hamiltonian formalism and its meaning, the other is a geometrical argument that happens to be valid in the present case. The first argument is this. The point is that the variables $p_j, E(\boldsymbol{p})$ transform in the familiar four-vector form under Lorentz transformations, whereas the behavior of the q_j is intrinsically more complicated. This is because we have always treated the time t and the q_j on different footings; in the Hamiltonian formalism the canonical transformation $e^{\widetilde{-\boldsymbol{\zeta}\cdot\boldsymbol{K}}}$ carries us from one notion of simultaneity (valid in one frame) directly to a new notion of simultaneity (valid in a different frame) without introducing separate time variables t, t'. Thus if q_j represents the position of the particle at time zero in the frame S, then $e^{\widetilde{-\boldsymbol{\zeta}\cdot\boldsymbol{K}}}q_j$ represents the position *at time zero in the frame* $S' = \exp(\boldsymbol{\zeta}\cdot\boldsymbol{k})S$. On the other hand, the choice of Hamiltonian makes the particle a free one so that the momenta are constant in time. Hence for these variables the change in the meaning of simultaneity is irrelevant, and their transformation law together with E is the normal one familiar from elementary kinematics of special relativity. Thus we can understand that the operation $e^{\widetilde{-\boldsymbol{\zeta}\cdot\boldsymbol{K}}}$ has a relatively complicated action on the q_j. The second argument we give now (and which, as we say, happens to be valid in the present context) explains in detail the right-hand side of Eq. (20.54). For this purpose, consider the description in S of the space-time trajectory traced out by the particle in a particular state of motion. Any possible state of motion is completely determined by giving the values of the momenta p_j that the particle has, then the position q_j occupied by the particle at time zero, all as seen in S. The p_j remain fixed in time, whereas the trajectory of the particle in space-time is determined by knowing the position $q_j(t)$ for any time t, as given in Eq. (20.51). A general point P on this trajectory can be labeled by t and is assigned the space-time coordinates $P \to (t, q_j(t))$ in S. Now suppose we view this P as an "event," and determine the space-time coordinates assigned to it in $S' = \exp(\boldsymbol{\zeta}\cdot\boldsymbol{k})S$ by applying the Lorentz-transformation formula to $(t, q_j(t))$. We should write $(t', q'_j(t'))$ for the space-time coordinates of P assigned in S'; these are given thus:

$$\begin{aligned} t' &= t\cosh\zeta - \boldsymbol{\zeta}\cdot\boldsymbol{q}(t)\frac{\sinh\zeta}{\zeta}, \\ q'_j(t') &= q_j(t) + \zeta_j\boldsymbol{\zeta}\cdot\boldsymbol{q}(t)\frac{\cosh\zeta - 1}{\zeta^2} - \zeta_j\frac{\sinh\zeta}{\zeta}t. \end{aligned} \tag{20.55}$$

Now this state of motion, viewed from S', is again characterized by a set of (constant) values p'_j for the momenta, these being obtained from p_j and E via Eq. (20.52) and by the values $q'_j \equiv q'_j(0)$ of the position coordinates of the particle *at time zero in* S'. We want now to relate $q'_j(0)$ directly to $q_j(0)$, that is, the position at time zero in S' to the position at time zero in S, because that is what the canonical transformation $e^{-\widetilde{\boldsymbol{\zeta}\cdot\boldsymbol{K}}}$ is designed to accomplish. Let P_0 be the point on the space-time trajectory at which the time in S' is zero, that is, for which $t' = 0$. What coordinates does S assign to P_0? We get these by setting $t' = 0$ in the first of Eq. (20.55), and then using the known dependence of $q_j(t)$ on t:

$$\begin{aligned} t' =&0 : t\cosh\zeta = \boldsymbol{\zeta}\cdot\boldsymbol{q}(t)\frac{\sinh\zeta}{\zeta} = \frac{\sinh\zeta}{\zeta}\left(\boldsymbol{\zeta}\cdot\boldsymbol{q} + \frac{t\boldsymbol{\zeta}\cdot\boldsymbol{p}}{E}\right), \\ t =& \left(\boldsymbol{\zeta}\cdot\boldsymbol{q}\frac{\sinh\zeta}{\zeta}\right)\left(\cosh\zeta - \boldsymbol{\zeta}\cdot\boldsymbol{p}\frac{\sinh\zeta}{E\zeta}\right)^{-1} = \frac{E}{E'}\boldsymbol{\zeta}\cdot\boldsymbol{q}\frac{\sinh\zeta}{\zeta}; \\ q_j(t) =& q_j + \frac{p_j}{E'}\boldsymbol{\zeta}\cdot\boldsymbol{q}\frac{\sinh\zeta}{\zeta}. \end{aligned} \tag{20.56}$$

We can now use these specific values of $q_j(t)$ and t on the right-hand side in Eq. (20.55), after setting $t' = 0$ on the left, and we then obtain the desired connection between q'_j and q_j:

$$\begin{aligned} q'_j &= q_j + \frac{p_j}{E'}\boldsymbol{\zeta}\cdot\boldsymbol{q}\frac{\sinh\zeta}{\zeta} + \zeta_j\left(1 + \frac{\boldsymbol{\zeta}\cdot\boldsymbol{p}}{E'}\frac{\sinh\zeta}{\zeta}\right)\boldsymbol{\zeta}\cdot\boldsymbol{q}\frac{\cosh\zeta - 1}{\zeta^2} \\ &- \zeta_j\frac{E}{E'}\boldsymbol{\zeta}\cdot\boldsymbol{q}\left(\frac{\sinh\zeta}{\zeta}\right)^2 \\ &= q_j + \frac{\boldsymbol{\zeta}\cdot\boldsymbol{q}}{E'}\left(\frac{\sinh\zeta}{\zeta}p_j + E\cosh\zeta\frac{\cosh\zeta - 1}{\zeta^2}\zeta_j - E\frac{\sinh^2\zeta}{\zeta^2}\zeta_j\right) \\ &= q_j + \frac{\boldsymbol{\zeta}\cdot\boldsymbol{q}}{E'}\left(\frac{\sinh\zeta}{\zeta}p_j - E\frac{\cosh\zeta - 1}{\zeta^2}\zeta_j\right). \end{aligned} \tag{20.57}$$

This is precisely Eq. (20.54); in a sense, the canonical transformation $e^{-\widetilde{\boldsymbol{\zeta}\cdot\boldsymbol{K}}}$ is "doing what it should do." But if we reflect a moment, it becomes clear that an extra assumption, going beyond the Hamiltonian framework, is involved here. The variables t and $q_j(t)$ are essentially different in character; the former is a parameter, the latter are dynamical variables. Yet we are demanding that for each state of motion of the particle, the four variables $(t, q_j(t))$ that are the space-time coordinates of points on the space-time trajectory must transform according to the Lorentz transformation formulae. In other words, recalling the manner in which the Lorentz group is defined in Chapter 15, each point on the trajectory of the particle constitutes an "event" and $(t, q_j(t))$ are the space-time coordinates of it. Although this is a reasonable assumption to make, it is independent of the assumptions thus far made. The latter amount to the requirement that for a relativistically invariant Hamiltonian system, every inertial observer must be capable of describing the system in the Hamiltonian form and that the descriptions of any two observers must be mathematically "equivalent". In classical canonical mechanics, "equivalence" means "equivalence by means of a canonical transformation", and this then leads to the existence of a canonical realization of $\mathcal{P}$. The extra assumption now being made is that, given a canonical realization of $\mathcal{P}$ and having decided on its physical interpretation, certain dynamical variables in the theory are asked to transform in a specific way; this notion goes by the name of "manifest covariance". Extra relations are thereby imposed on the theory, because the known canonical transformation properties determined by the realization of $\mathcal{P}$ must be made to agree with certain other transformation properties given "from the outside." What we have seen is that for a free spinless relativistic particle these two transformation laws agree. As we have mentioned in the preceding chapter, such considerations are relevant in the case of the Galilei group as well; we discuss these matters more fully in the following chapter.

We now wish to discuss this spinless free-particle realization of $\mathcal{P}$ from two other points of view. The first is to see how, in the limit $c \to \infty$, it can yield a realization of the Lie algebra of $\mathcal{G}$ with a nontrivial neutral element. For this purpose, we must rewrite the expressions for the generators of $\mathcal{P}$ putting in appropriate powers of c wherever they should be present; the generators J_j and P_j remain unchanged, but K_j and H become:

$$K_j = (m^2c^4 + \boldsymbol{p}\cdot\boldsymbol{p}c^2)^{1/2}q_j, \; H = (m^2c^4 + \boldsymbol{p}\cdot\boldsymbol{p}c^2)^{1/2}. \tag{20.58}$$

This introduces extra powers of c on the right-hand sides of some of the basic PB relations; those so affected are

$$\{K_j, K_l\} = -c^2\epsilon_{jlm}J_m\,; \; \{K_j, H\} = c^2 P_j\,. \tag{20.59}$$

The remaining PB relations retain their previous forms. Now although J_j and P_j, being independent of c, are well defined in the limit $c \to \infty$, both K_j and H become infinite. In order that we may extract expressions with a finite limiting form, we must scale K_j and subtract the rest energy mc^2 from H. Thus we must work with:

$$\begin{aligned} K'_j &= \frac{1}{c^2}K_j = \left(m^2 + \frac{\boldsymbol{p}\cdot\boldsymbol{p}}{c^2}\right)^{1/2} q_j \underset{c\to\infty}{\longrightarrow} mq_j\,, \\ H' &= H - mc^2 = (m^2c^4 + \boldsymbol{p}\cdot\boldsymbol{p}c^2)^{1/2} - mc^2 \underset{c\to\infty}{\longrightarrow} \frac{\boldsymbol{p}\cdot\boldsymbol{p}}{2m}. \end{aligned} \tag{20.60}$$

Let us now write the full set of basic PB's for $\mathcal{P}$ in terms of the generators J_j, P_j, K'_j and H'; we divide them into two groups depending on the appearance of c explicitly on the right-hand side:

$$\begin{aligned}\{J_j, J_k\} =&\epsilon_{jkl}J_l,\ \{J_j, P_k\} = \epsilon_{jkl}P_l,\ \{J_j, K'_k\} = \epsilon_{jkl}K'_l\,,\\ \{J_j, H'\} =&0,\ \{P_j, P_k\} = 0,\ \{P_j, H'\} = 0,\ \{K'_j, H'\} = P_j\,; \qquad (20.61a)\\ \{K'_j, K'_k\} =&\frac{-1}{c^2}\epsilon_{jkl}J_l,\ \{K'_j, P_k\} = \frac{1}{c^2}\delta_{jk}H' + m\delta_{jk}\,. \qquad (20.61b)\end{aligned}$$

We see that in the last relation a neutral element m has explicitly appeared. As long as c is finite, so that we are dealing with $\mathcal{P}$, this is a trivial neutral element; it can always be recombined with H' in the form $H' + mc^2$, because the term $(1/c^2)\delta_{jk}H'$ is also present on the right-hand side of the PB relation for $\{K'_j, P_k\}$. We then recover the true realization of the Lie algebra of $\mathcal{P}$. (In fact this was the way we were able to eliminate all neutral elements in the case of $\mathcal{P}$!) But suppose we leave the PB relations of $\mathcal{P}$ in the foregoing form, and take the limit $c \to \infty$ everywhere. Then both K'_j and H' tend to finite limits, given in Eq. (20.60); but at the same time the right-hand side of the last relation in Eq. (20.61b) reduces to just $m\delta_{jk}$, and now we have no other generator with which to combine m and thus transform it away! A neutral element introduced "by hand" for finite c, which is therefore trivial, has become a nontrivial one after the limit $c \to \infty$ has been taken. The limits for K'_j and H' are of course the generators G_j and the Hamiltonian, respectively, corresponding to the single-particle spinless realization of $\mathcal{G}$ (see Eq. (19.45)). Whereas the relations in Eq. (20.61a) go over into the corresponding ones for $\mathcal{G}$, the last two relations in Eq. (20.61b) give rise to $\{G_j, G_k\}$=0 and $\{G_j, P_k\} = m\delta_{jk}$, respectively. This, then, is the way in which the neutral element in the realization of $\mathcal{G}$ emerges. We can also see why the Hamiltonian in the case of $\mathcal{G}$ remains indeterminate up to an additive constant. It is that instead of replacing H by H' in order to get a finite nonrelativistic limit, we could equally well have replaced it by $H'' = H - mc^2 + U$, where U is any finite constant. Then the limiting form for H'' would be $p^2/2m + U$. But this has no effect on the limiting forms of the PB relations Eq. (20.61); the only place where U appears is on the right-hand side of the second part of Eq. (20.61b), and there it appears in the form U/c^2 which disappears when $c \to \infty$!

Finally, we discuss briefly this realization of $\mathcal{P}$ from the Lagrangian viewpoint. (We revert to units in which c=1.) The Lagrangian is easily obtained:

$$\begin{aligned}\dot{q}_j &= \frac{\partial H}{\partial p_j} = p_j(m^2 + p_k p_k)^{-1/2} \Rightarrow p_j = m\dot{q}_j(1 - \dot{q}_k\dot{q}_k)^{-1/2}\,;\\ L(q, \dot{q}) &= p_j\dot{q}_j - H = -m(1 - \dot{q}_j\dot{q}_j)^{1/2}\,. \qquad (20.62)\end{aligned}$$

The Lagrangian L is explicitly invariant under spatial translations and rotations and has no explicit time dependence, just like H; these properties lead to the conservation laws of momentum, angular momentum, and energy. More interesting is the behavior under pure Lorentz transformations; we expect to recover the property that the equations of motion show invariance under these transformations, but that there is no associated explicitly time independent constant of motion.

Configuration space is three-dimensional Euclidean space with the q_j as Cartesian coordinates. A kinematically conceivable trajectory in configuration space is specified by giving the q's as some functions of time, $q_j(t)$, subject to the restriction that always the magnitude of the velocity, $(\dot{q}_j\dot{q}_j)^{1/2}$, does not exceed or equal unity. Viewed as a trajectory in space-time, the points on it are labeled by the four coordinates $(t, q_j(t))$. Adopting the active point of view, a pure Lorentz transformation acts on this trajectory to give another kinematically allowed one. It is enough to consider an infinitesimal transformation with parameters $\delta\zeta_j$. The space-time coordinates of the points on the new space-time trajectory are $(t', q'_j(t'))$ given by:

$$t' = t - \delta\zeta_j q_j(t),\; q'_j(t') = q_j(t) - t\delta\zeta_j\,. \tag{20.63}$$

That is, on the new trajectory, the particle is at $q'_j(t')$ at time t'. Reverting to configuration space, the new trajectory is given by the three functions $q'_j(t)$ as follows:

$$\begin{aligned} q'_j(t) &\approx q_j(t + \delta\zeta_k q_k(t)) - t\delta\zeta_j \approx q_j(t) + \delta q_j(t)\,, \\ \delta q_j(t) &= \dot{q}_j(t)\delta\zeta_k q_k(t) - t\delta\zeta_j\,. \end{aligned} \tag{20.64}$$

We now compare the values of L at points on the two trajectories corresponding to the same time and compute the difference:

$$\begin{aligned} \delta L \equiv L(\dot{q}'_j(t)) - L(\dot{q}_j(t)) &= \frac{\partial L}{\partial \dot{q}_k}\delta\dot{q}_k(t) \\ &= m\dot{q}_k(1 - \dot{q}_j\dot{q}_j)^{-1/2}(\ddot{q}_k\delta\zeta_l q_l + \dot{q}_k\delta\zeta_l\dot{q}_l - \delta\zeta_k) \\ &= m\delta\zeta_l q_l(1 - \dot{q}_j\dot{q}_j)^{-1/2}\dot{q}_k\ddot{q}_k - m\delta\zeta_l\dot{q}_l(1 - \dot{q}_j\dot{q}_j)^{1/2} \\ &= \frac{d}{dt}\delta F(q, \dot{q})\,, \\ \delta F(q, \dot{q}) &= -m(1 - \dot{q}_j\dot{q}_j)^{1/2}\delta\zeta_l q_l\,. \end{aligned} \tag{20.65}$$

The change in L is a perfect differential with respect to time, thus L is quasi-invariant under this transformation. Therefore, from the action principle we conclude that if on the original trajectory the equations of motion are obeyed, they will also be obeyed on the Lorentz transformed trajectory; in other words, the equations of motion are invariant under the transformations Eqs. (20.63) and (20.64). On the other hand, because $\delta q_j(t)$ is explicitly time dependent, we do not get a conservation law from the action principle. What is conserved is the time-dependent quantity obtained by computing the change in the action functional in the two independent ways; this quantity is:

$$\begin{aligned}\frac{\partial L}{\partial \dot{q}_j}\delta q_j - \delta F = & \frac{m\dot{q}_j}{(1-\dot{q}_k\dot{q}_k)^{1/2}}(\dot{q}_j\delta\zeta_l q_l - t\delta\zeta_j) + m(1-\dot{q}_k\dot{q}_k)^{1/2}\delta\zeta_l q_l \\ = & \ \delta\zeta_j\left(\frac{m}{(1-\dot{q}_k\dot{q}_k)^{1/2}}q_j - \frac{t\dot{q}_j m}{(1-\dot{q}_k\dot{q}_k)^{1/2}}\right) = \delta\zeta_j(Eq_j - tp_j)\,.\end{aligned} \tag{20.66}$$

The constancy of this expression amounts to the same thing as the solution of the equations of motion for the generators K_j in the Hamiltonian language.

Single Particle with Spin

The realization of $\mathcal{P}$ studied above is not the most general irreducible realization in which the energy-momentum four-vector P^μ is positive timelike. This is obtained by the inclusion of spin in the previous realization. Now both Casimir invariants will be nonzero; M^2 will still be the square of the mass, m^2, while W^2 is the square of the spin times m^2. In this realization of $\mathcal{P}$ the internal angular momentum and the spin of the particle coincide. The basic phase space variables are q_j, p_j, and S_j, obeying the PB relations Eq. (20.46) (with the replacements $Q \to q, P \to p$); each q and each p goes independently over the real line, and the vector $\mathbf{S}$ is restricted to lie on a sphere of radius $(W^2/M^2)^{1/2}$. From our general analysis we know what the generators of $\mathcal{P}$ look like; they are:

$$\begin{aligned}J_j &= \epsilon_{jkl}q_k p_l + S_j,\ P_j = p_j,\ H = E(\mathbf{p}) = (m^2 + p_j p_j)^{1/2}\,, \\ K_j &= (m^2 + p_k p_k)^{1/2} q_j + \frac{\epsilon_{jkl}p_k S_l}{m + (m^2 + p_m p_m)^{1/2}}\,.\end{aligned} \tag{20.67}$$

That this yields the most general irreducible canonical realization of $\mathcal{P}$ in which P^μ is positive timelike is proved by an argument similar to the one used in the preceding chapter. We know that in any irreducible realization, we can choose a subset of the generators, or more generally form a set of functions of the generators, that will be independent and suffice to form a set of coordinates in phase space. From our general analysis we know that in any realization of $\mathcal{P}$ with timelike P^μ we can introduce the set of phase-space functions Q_j, P_j, and S_j, and from the bracket relations among them we infer that they are capable of being independently varied and thus form a set of independent phase-space functions. (The only restriction is that S^2 cannot be varied by any of the generators of $\mathcal{P}$.) On the other hand, we are able to express all the generators of $\mathcal{P}$ in terms of Q_j, P_j and S_j and the Casimir invariants. Putting these two facts together yields the desired result. The irreducibility of the realization generated by the expressions in Eq. (20.67) (and in case S_j=0 the expressions in Eq. (20.50)) is also evident by now.

As we have already remarked, the inclusion of spin in the realization of $\mathcal{P}$ is slightly more complicated than in the case of $\mathcal{G}$. This is because, once the term S_j has been added on to $(\boldsymbol{q}\times\boldsymbol{p})_j$ in the definition of the angular momenta, there must necessarily be additional terms in K_j, too; whereas in the case of $\mathcal{G}$ the generators G_j did not have to be altered. For this same reason, whereas the realization of $\mathcal{G}$ with spin included appeared immediately as the direct product of a spinless realization and a nonfaithful mass-zero realization, no such breakup is possible in the present case, at least as long as one works with generators that are real phase-space functions.

The spin-dependent terms in the generators of $\mathcal{P}$ have no effect on the Lorentz-transformation properties of the momentum variables p_j, and Eq. (20.52) remains valid. The solution of the equations of motion for the positions q_j is also unchanged, and Eq. (20.51) can be used. Like p_j, the spin components S_j are also constants of motion. The new features to be examined involve the behavior of the S_j under the elements of $\mathcal{P}$, and the altered behavior of the q_j. Let us look at the former first. Under $E(3)$, the properties of the spin are standard: unaltered by translations, vector under rotations. To discover the effect of a pure Lorentz transformation, we fall back upon the known transformation laws for the Pauli-Lubanski vector W_μ and p_μ and the relations connecting these two vectors with S_j. Rewriting Eqs. (20.38) and (20.49) for the present case of an irreducible realization with mass m, we have:

$$\begin{aligned} W_0 &= \boldsymbol{S}\cdot\boldsymbol{p}, \boldsymbol{W} = -m\boldsymbol{S} - \frac{\boldsymbol{p}\boldsymbol{p}\cdot\boldsymbol{S}}{(m+E)}, \quad E = (m^2 + \boldsymbol{p}\cdot\boldsymbol{p})^{1/2}; \\ \boldsymbol{S} &= -\frac{1}{m}\left[\boldsymbol{W} + \frac{W_0\boldsymbol{p}}{m+E}\right]. \end{aligned} \tag{20.68}$$

The effects of the canonical transformation $e^{\widetilde{-\boldsymbol{\zeta}\cdot\boldsymbol{K}}}$ on p_μ are given in Eq. (20.52), and exactly similar equations hold for W_μ; we write these here again:

$$\begin{aligned} e^{\widetilde{-\boldsymbol{\zeta}\cdot\boldsymbol{K}}}p_j &= \Lambda_{jk}p_k - \zeta_j E\frac{\sinh\zeta}{\zeta}, \quad e^{\widetilde{-\boldsymbol{\zeta}\cdot\boldsymbol{K}}}E = E' = E\cosh\zeta - \zeta_j p_j\frac{\sinh\zeta}{\zeta}; \\ e^{\widetilde{-\boldsymbol{\zeta}\cdot\boldsymbol{K}}}W_j &= \Lambda_{jk}W_k + \zeta_j W_0\frac{\sinh\zeta}{\zeta}, \quad e^{\widetilde{-\boldsymbol{\zeta}\cdot\boldsymbol{K}}}W_0 = W_0\cosh\zeta + \zeta_j W_j\frac{\sinh\zeta}{\zeta}; \\ \Lambda_{jk} &= \delta_{jk} + \zeta_j\zeta_k\frac{\cosh\zeta - 1}{\zeta^2}. \end{aligned} \tag{20.69}$$

We can now compute the effect of a Lorentz transformation on S_j. We get:

$$\begin{aligned} e^{\widetilde{-\boldsymbol{\zeta}\cdot\boldsymbol{K}}}S_j &= -\frac{1}{m}\left[e^{\widetilde{-\boldsymbol{\zeta}\cdot\boldsymbol{K}}}W_j + \frac{(e^{\widetilde{-\boldsymbol{\zeta}\cdot\boldsymbol{K}}}W_0)(e^{\widetilde{-\boldsymbol{\zeta}\cdot\boldsymbol{K}}}p_j)}{(m+E')}\right] \\ &= -\frac{1}{m}\left[\Lambda_{jk}W_k + \zeta_j W_0\frac{\sinh\zeta}{\zeta}\right. \\ &\quad \left. + \frac{\left(W_0\cosh\zeta + \zeta_k W_k\frac{\sinh\zeta}{\zeta}\right)\left(\Lambda_{jl}p_l - \zeta_j E\frac{\sinh\zeta}{\zeta}\right)}{(m+E')}\right]. \end{aligned} \tag{20.70}$$

This last expression involves the W_μ linearly and is otherwise a function depending on p_j and ζ_j, which means that on substituting for W_μ in terms of S_j

and p_j from Eq. (20.68) we will obtain an expression linear in the S's, with coefficients depending on $\boldsymbol{p}$ and $\boldsymbol{\zeta}$. Thus we write:

$$\widetilde{e^{-\boldsymbol{\zeta}\cdot\boldsymbol{K}}} S_j = R_{kj}(\boldsymbol{\zeta},\boldsymbol{p})S_k\,. \tag{20.71}$$

(Actually, that the Lorentz-transformed spin components must be momentum-dependent linear combinations of the original spin components can already be seen from the form of K_j.) Simplifying the right-hand side of Eq. (20.70) to get the coefficients $R_{kj}(\boldsymbol{\zeta},\boldsymbol{p})$ involves tedious algebra justified only by the form of the final result. In the first step one can collect together the terms of the five different forms that occur, namely, $\delta_{jk}, \zeta_k\zeta_j, p_kp_j, \zeta_kp_j$, and $p_k\zeta_j$, and separately simplify the five different coefficients. After this, if these terms are put together again, it turns out that the matrix $||R_{kj}||$ can be written in terms of a certain simpler matrix. We recall from our study of the group $R(3)$ in Chapter 17 that it is possible to associate a three-dimensional antisymmetric matrix $C(\boldsymbol{\alpha})$ with every real three-dimensional vector $\boldsymbol{\alpha}$ in a linear fashion— the definition is in Eq. (17.21). What is involved now is the matrix $C(\boldsymbol{\zeta}\times\mathbf{p})$ corresponding to the triplet of real numbers $\boldsymbol{\zeta}\times\boldsymbol{p}$. We find that the matrix $||R_{kj}||$ can be written this way:

$$\begin{aligned} R(\boldsymbol{\zeta},\boldsymbol{p}) &= 1\!\!1 + a(\boldsymbol{\zeta},\boldsymbol{p})C(\boldsymbol{\zeta}\times\boldsymbol{p}) + b(\boldsymbol{\zeta},\boldsymbol{p})C^2(\boldsymbol{\zeta}\times\boldsymbol{p})\,,\\ a(\boldsymbol{\zeta},\boldsymbol{p}) &= (m+E')^{-1}\left\{\frac{-\sinh\zeta}{\zeta} + \frac{(\cosh\zeta-1)}{\zeta^2}\frac{\boldsymbol{\zeta}\cdot\boldsymbol{p}}{E+m}\right\},\\ b(\boldsymbol{\zeta},\boldsymbol{p}) &= (m+E')^{-1}\left\{\frac{\cosh\zeta-1}{\zeta^2}\frac{1}{E+m}\right\}. \end{aligned} \tag{20.72}$$

With this, we know how the spin components S_j transform under the most general element of $\mathcal{P}$. We would have been surprised if the matrix $||R||$ did not correspond to some rotation, namely, if it were not an orthogonal matrix. A direct computation shows that it is indeed orthogonal. We know that the square of the matrix $C(\boldsymbol{\zeta}\times\boldsymbol{p})$ is essentially a projection matrix (see Eq. (17.23)):

$$C^2(\boldsymbol{\zeta}\times\boldsymbol{p}) = -|\boldsymbol{\zeta}\times\boldsymbol{p}|^2P(\boldsymbol{\zeta}\times\boldsymbol{p})\,; \tag{20.73}$$

for the products involving $C(\boldsymbol{\zeta}\times\boldsymbol{p})$ and $P(\boldsymbol{\zeta}\times\boldsymbol{p})$ we can use Eq. (17.24). Thus we find, omitting the arguments $\boldsymbol{\zeta},\boldsymbol{p}$, $\boldsymbol{\zeta}\times\boldsymbol{p}$:

$$\begin{aligned} R^{\sim}R &= (1\!\!1 - aC - b|\boldsymbol{\zeta}\times\boldsymbol{p}|^2P)(1 + aC - b|\boldsymbol{\zeta}\times\boldsymbol{p}|^2P)\\ &= (1\!\!1 - b|\boldsymbol{\zeta}\times\boldsymbol{p}|^2P)^2 - a^2C^2\\ &= 1\!\!1 + (\boldsymbol{\zeta}\times\boldsymbol{p})^2(a^2 + b^2|\boldsymbol{\zeta}\times\boldsymbol{p}|^2 - 2b)P\,. \end{aligned} \tag{20.74}$$

Putting in the values of a and b, the coefficient of P here is seen to vanish; thus the orthogonality of R follows. This shows that under any element of the HLG, the spin variables S_j undergo some three-dimensional rotation corresponding to some element of $R(3)$. In analogy with the situation in quantum mechanics, we call this the "Wigner rotation." If the element of the HLG happens to be in

$R(3)$, the Wigner rotation coincides with this element of $R(3)$. If the element of the HLG happens to be a pure Lorentz transformation, the Wigner rotation becomes a function of momentum in addition to being a function of the Lorentz transformation and is described by the orthogonal matrix $R(\boldsymbol{\zeta}, \boldsymbol{p})$. The point to stress is that the S_j always experience a rotation under a change of inertial frame.

Turning to the behavior of the position variables q_j, although the effect of elements of $E(3)$ is the expected one, spin complications arise when we consider pure Lorentz transformations. Specifically, we can see that because the spin-dependent terms in K_j are also momentum dependent, the effect of the canonical transformation $e^{\widetilde{-\boldsymbol{\zeta}\cdot\boldsymbol{K}}}$ on q_j will not agree with the change in q_j computed by treating $(t, q_j(t))$ as the space-time coordinates of events on the particle trajectory and applying the Lorentz transformation formulae to them. The latter calculation yields Eq. (20.57) once again, in which no reference to spin is made, whereas the former calculation yields the right-hand side of Eq. (20.54) or (20.57) together with terms which are momentum-dependent and linear in the spin variables. Thus the conditions for manifest covariance (to be derived in the next chapter) will not be obeyed in the case of a free particle with spin. Nevertheless, let us complete the task of exhibiting the canonical transformations realizing $\mathcal{P}$ by computing the Lorentz transform of the q_j. For what appears on the right-hand side of Eqs. (20.54), or (20.57), let us write $\phi_j(\boldsymbol{q}, \boldsymbol{p}, \boldsymbol{\zeta})$:

$$\phi_j(\boldsymbol{q}, \boldsymbol{p}, \boldsymbol{\zeta}) = q_j + \frac{\boldsymbol{\zeta}\cdot\boldsymbol{q}}{E'}\left(\frac{\sinh\zeta}{\zeta}p_j - E\frac{\cosh\zeta - 1}{\zeta^2}\zeta_j\right). \tag{20.75}$$

The Lorentz transform of q_j consists of ϕ_j and a spin-dependent piece; we can write:

$$e^{\widetilde{-\boldsymbol{\zeta}\cdot\boldsymbol{K}}} q_j = \phi_j(\boldsymbol{q}, \boldsymbol{p}, \boldsymbol{\zeta}) + A_{kj}(\boldsymbol{\zeta}, \boldsymbol{p}) S_k\,, \tag{20.76}$$

the coefficients $A_{kj}(\boldsymbol{\zeta}, \boldsymbol{p})$ being the quantities to be determined. To see what these coefficients may look like, let us first consider an infinitesimal Lorentz transformation; in that case, the A's are given essentially by the PB of q_j with the spin dependent part in $\boldsymbol{K}$:

$$\begin{aligned} |\zeta_j| \ll 1: \; A_{kj}(\boldsymbol{\zeta}, \boldsymbol{p}) S_k \;&\approx\; \left\{q_j, \frac{\boldsymbol{\zeta}\cdot\boldsymbol{p}\times\boldsymbol{S}}{E+m}\right\} = \epsilon_{lmk}\zeta_l S_k \left\{q_j, \frac{p_m}{E+m}\right\}, \\ A_{kj}(\boldsymbol{\zeta}, \boldsymbol{p}) \;&\approx\; -\epsilon_{kjl}\frac{\zeta_l}{E+m} + \frac{(\boldsymbol{p}\times\boldsymbol{\zeta})_k p_j}{E(E+m)^2}\,. \end{aligned} \tag{20.77}$$

More generally, for a finite Lorentz transformation, we can use Eq. (20.43a) (specialized to the present case), together with Eqs. (20.53) and (20.69):

$$\begin{aligned}
e^{\widetilde{-\boldsymbol{\zeta}\cdot\boldsymbol{K}}}q_j \equiv & e^{\widetilde{-\boldsymbol{\zeta}\cdot\boldsymbol{K}}}\left(\frac{K_j}{E}+\frac{\epsilon_{jkl}p_kW_l}{mE(E+m)}\right) \\
= & \frac{1}{E'}\left(K_j\cosh\zeta-\zeta_j\boldsymbol{\zeta}\cdot\boldsymbol{K}\frac{\cosh\zeta-1}{\zeta^2}-\epsilon_{jmn}\zeta_mJ_n\frac{\sinh\zeta}{\zeta}\right) \\
& +\frac{\epsilon_{jkl}}{mE'(E'+m)}\left(\Lambda_{km}p_m-\zeta_kE\frac{\sinh\zeta}{\zeta}\right)\left(\Lambda_{ln}W_n+\zeta_lW_0\frac{\sinh\zeta}{\zeta}\right).
\end{aligned} \tag{20.78}$$

If we now put in the forms of $\boldsymbol{J},\boldsymbol{K}$, and W_μ, given in Eqs. (20.67) and (20.68), the terms independent of spin lead to $\phi_j(\boldsymbol{q},\boldsymbol{p},\boldsymbol{\zeta})$; the spin-dependent terms are seen to be in fact linear in the spin variables, and they will yield $A_{kj}(\boldsymbol{\zeta},\boldsymbol{p})$. There is some ambiguity in simplifying these coefficients to a neat form; if we try to make them appear as much like Eq. (20.77) as possible, we get:

$$\begin{aligned}
A_{kj}(\boldsymbol{\zeta},\boldsymbol{p}) = & \frac{\sinh\zeta}{\zeta(E'+m)}\left(-\epsilon_{kjl}\zeta_l+\frac{(\boldsymbol{p}\times\boldsymbol{\zeta})_kp_j}{E(E+m)}\right) \\
& +\frac{(\cosh\zeta-1)}{\zeta^2(E'+m)(E+m)}\left(\zeta^2\epsilon_{kjl}p_l+(1-E/E')(\boldsymbol{p}\times\boldsymbol{\zeta})_k\zeta_j\right).
\end{aligned} \tag{20.79}$$

With this, the calculation of the right-hand side of Eq. (20.76) is completed; for small ζ_j we immediately recover Eq. (20.77).

We conclude this description of a spinning particle by pointing out its relationship to a general canonical realization of $\mathcal{P}$ with timelike energy momenta (and positive energy!). In fact, our earlier analysis shows that any such realization can be made to look like that of a spinning particle, except that the "mass" $M(\omega)$ and the "square of the spin" $W^2(\omega)/M^2(\omega)$ are both generally variable. The variables Q_j, S_j introduced in Eqs. (20.43a) and (20.38), respectively, are in many ways just like the q_j and S_j of the present discussion. In particular the Lorentz transformation properties of the general variables Q_j, S_j are given by Eqs (20.76) and (20.71), respectively, with appropriate replacements. It goes without saying, however, that in the general situation there will be dynamical variables other than just functions of Q_j, P_j and S_j, and also that S_j may be something other than "spin."

Many-Particle Systems

The simplest system in this category consists of a system of two free particles. The corresponding realization of $\mathcal{P}$ is the direct product of two irreducible realizations, each characterized by a value of the mass. (For simplicity, we omit the spin variables.) The generators for the product are, as usual, sums of the individual generators. If we write $\boldsymbol{q}^{(1)},\boldsymbol{p}^{(1)}$ for the variables for the first particle, $\boldsymbol{q}^{(2)},\boldsymbol{p}^{(2)}$ for the second particle, and $m^{(1)}$ and $m^{(2)}$ denote the masses, the generators of $\mathcal{P}$ are:

$$\begin{aligned} \boldsymbol{J} &= \boldsymbol{q}^{(1)} \times \boldsymbol{p}^{(1)} + \boldsymbol{q}^{(2)} \times \boldsymbol{p}^{(2)}, & \boldsymbol{P} &= \boldsymbol{p}^{(1)} + \boldsymbol{p}^{(2)}, \\ \boldsymbol{K} &= E^{(1)}\boldsymbol{q}^{(1)} + E^{(2)}\boldsymbol{q}^{(2)}, & H &= E^{(1)} + E^{(2)}; \\ E^{(1)} &= (m^{(1)2} + \boldsymbol{p}^{(1)} \cdot \boldsymbol{p}^{(1)})^{1/2}, & E^{(2)} &= (m^{(2)2} + \boldsymbol{p}^{(2)} \cdot \boldsymbol{p}^{(2)})^{1/2}. \end{aligned} \tag{20.80}$$

The canonical transformations realizing $\mathcal{P}$ are easily obtained by taking direct products of the canonical transformations that were exhibited in the single-particle case.

This realization of $\mathcal{P}$ corresponds to two free noninteracting relativistic particles. It is a reducible realization of $\mathcal{P}$ with both Casimir invariants M^2 and W^2 being nontrivial quantities. From elementary relativistic kinematics we know that the range of values for M^2, which could be written in the form

$$\begin{aligned} M^2 = (E^{(1)} + E^{(2)})^2 - (\boldsymbol{p}^{(1)} + \boldsymbol{p}^{(2)})^2 &= (m^{(1)})^2 + (m^{(2)})^2 \\ &+ 2(E^{(1)}E^{(2)} - \boldsymbol{p}^{(1)} \cdot \boldsymbol{p}^{(2)}), \end{aligned} \tag{20.81}$$

is $(m^{(1)} + m^{(2)})^2 \leq M^2 < \infty$. As for the invariant W^2, we first write down the components of W:

$$\begin{aligned} W_0 &= (\boldsymbol{q}^{(1)} - \boldsymbol{q}^{(2)}) \times \boldsymbol{p}^{(1)} \cdot \boldsymbol{p}^{(2)}, \\ \boldsymbol{W} &= (\boldsymbol{q}^{(1)} - \boldsymbol{q}^{(2)}) \times (E^{(1)}\boldsymbol{p}^{(2)} - E^{(2)}\boldsymbol{p}^{(1)}). \end{aligned} \tag{20.82}$$

Thus the expression for W^2 is

$$\begin{aligned} W^2 =& (\boldsymbol{q}^{(1)} - \boldsymbol{q}^{(2)})^2 (E^{(1)}\boldsymbol{p}^{(2)} - E^{(2)}\boldsymbol{p}^{(1)})^2 \\ &- [(\boldsymbol{q}^{(1)} - \boldsymbol{q}^{(2)}) \cdot (E^{(1)}\boldsymbol{p}^{(2)} - E^{(2)}\boldsymbol{p}^{(1)})]^2 - [(\boldsymbol{q}^{(1)} - \boldsymbol{q}^{(2)}) \cdot \boldsymbol{p}^{(1)} \times \boldsymbol{p}^{(2)}]^2. \end{aligned} \tag{20.83}$$

In any case we know that because W_μ is a spacelike four-vector, W^2 cannot become negative. From the explicit expressions given here we can easily visualize configurations in the phase space of the two particle system for which W^2 vanishes as well as others for which it has an arbitrarily large positive value. Thus, the range of W^2 is $0 \leq W^2 < \infty$. Even though we are dealing with spinless particles, the combined system can have a non-vanishing internal angular momentum, essentially a relative orbital angular momentum; this is similar to the Galilean case. But now the actual form of the S_j is quite different: there is no longer a simple separation into center of energy and relative variables such that the S_j are built up from the latter in the "$\mathbf{q} \times \mathbf{p}$" manner. If we use Eqs. (20.38) and (20.82) we get:

$$\begin{aligned} \boldsymbol{S} = \frac{1}{M}\Bigg[&(\boldsymbol{q}^{(1)} - \boldsymbol{q}^{(2)}) \times (E^{(2)}\boldsymbol{p}^{(1)} - E^{(1)}\boldsymbol{p}^{(2)}) \\ &- (\boldsymbol{p}^{(1)} + \boldsymbol{p}^{(2)}) \frac{(\boldsymbol{q}^{(1)} - \boldsymbol{q}^{(2)}) \cdot \boldsymbol{p}^{(1)} \times \boldsymbol{p}^{(2)}}{M + E^{(1)} + E^{(2)}} \Bigg]. \end{aligned} \tag{20.84}$$

That the dependence on the q's should be only via the difference $(\boldsymbol{q}^{(1)} - \boldsymbol{q}^{(2)})$ is already evident from the fact that the S_j are invariant under space translations, so that the PB's $\{S_j, P_k\}$ vanish. In the case of $\mathcal{G}$, we could go further and limit the dependence of the internal angular momenta on the p's too, because they are known to be invariant under Galilean velocity transformations too. The generators G_j are functions of the q's alone (and in case they are chosen to be kinematic generators this is true in the interacting case as well), and thus we could conclude that the internal angular momenta depended on the p's only via the relative (Hamiltonian) velocities

$$\left(\frac{\boldsymbol{p}^{(r)}}{m^{(r)}} - \frac{\boldsymbol{p}^{(s)}}{m^{(s)}}\right).$$

This is explicitly borne out by Eqs. (19.58) and (19.59). But the situation is different now on two counts: in the first place the K_j are not functions of the q's alone, and in the second place the S's are *not* invariant under a pure Lorentz transformation! They experience the Wigner rotation. Therefore, we have no reason to expect a simple dependence of the S_j on the p's, say via some characteristic combination, and there is none. In the frame in which the total momentum is zero, the center of energy or center of momentum frame, the S_j do appear as the components of the vector product of relative position and relative velocity (apart from factors), but this is special to this frame.

Similar differences from the Galilean situation show up when we work out the canonical conjugates to the P_j. We use Eqs. (20.43a) and (20.82) to obtain:

$$\begin{aligned} \boldsymbol{Q} =& \frac{1}{E^{(1)} + E^{(2)}} \Bigg[E^{(1)}\boldsymbol{q}^{(1)} + E^{(2)}\boldsymbol{q}^{(2)} \\ &+ \frac{(\boldsymbol{p}^{(1)} + \boldsymbol{p}^{(2)}) \times \{(\boldsymbol{q}^{(1)} - \boldsymbol{q}^{(2)}) \times (E^{(1)}\boldsymbol{p}^{(2)} - E^{(2)}\boldsymbol{p}^{(1)})\}}{M(M + E^{(1)} + E^{(2)})} \Bigg]. \end{aligned} \tag{20.85}$$

This differs, as expected, from $\boldsymbol{K}/H$. However, in the nonrelativistic limit, which must be taken after putting in appropriate powers of c everywhere, the second term in the square bracket appearing with a denominator does not contribute, and $\boldsymbol{Q}$ goes over into $\boldsymbol{G}/M$. From the examples we have so far seen, we can discern several possible relationships between $\boldsymbol{Q}, \boldsymbol{K}/H$, and the center of energy for a relativistic system. For a single spinless free particle, all three things coincide. For a single free particle with spin, the first and the third coincide, but both differ from $\boldsymbol{K}/H$. For a system of two spinless free particles, $\boldsymbol{K}/H$ is the center of energy, but this differs from $\boldsymbol{Q}$. If we had considered two free particles with at least one possessing spin, all three objects would have been different!

The generalization to describe a system of n free particles is straightforward. Still omitting spin, the generators are:

$$\begin{aligned} \boldsymbol{J} &= \sum_{r=1}^{n} \boldsymbol{q}^{(r)} \times \boldsymbol{p}^{(r)}, \ \boldsymbol{P} = \sum_{r=1}^{n} \boldsymbol{p}^{(r)}, \ \boldsymbol{K} = \sum_{r=1}^{n} E^{(r)}\boldsymbol{q}^{(r)}, \ H = \sum_{r=1}^{n} E^{(r)}; \\ E^{(r)} &= ({m^{(r)}}^2 + \boldsymbol{p}^{(r)} \cdot \boldsymbol{p}^{(r)})^{1/2}. \end{aligned} \tag{20.86}$$

(The q's and p's obey standard PB rules, as in the case of Eq. (20.80).) It is again a consequence of relativistic kinematics that the range for M is from

$$\sum_{r=1}^{n} m^{(r)}$$

to infinity; W^2 goes from zero to infinity. In the same spirit as in the preceding chapter, we can now try to introduce interactions, but the situation is much more complex for the Poincaré group than for the Galilei group. One immediately recognizes that provided J_j and P_j are kinematic generators (and this is generally assumed), it is impossible to choose K_j to be a kinematic generator because this would violate the bracket relation:

$$\{K_j, P_k\} = \delta_{jk} H \,. \tag{20.14}$$

Let us indicate the "noninteracting" generators in Eq. (20.86) with the superscript (0); then we can write:

$$J_j = J_j^{(0)}, P_j = P_j^{(0)}; K_j = K_j^{(0)} + W_j, H = H^{(0)} + V\,, \tag{20.87}$$

and the W_j cannot vanish if V does not. The PB relations for the $E(3)$ subalgebra are automatically satisfied; the remaining ones give conditions on the "interaction terms" W_j and V:

$$\begin{aligned}
\{J_j, K_k\} &= \epsilon_{jkm} K_m \to \{J_j^{(0)}, W_k\} = \epsilon_{jkm} W_m\,;\\
\{J_j, H\} &= 0 \to \{J_j^{(0)}, V\} = 0;\ \{P_j, H\} = 0 \to \{P_j^{(0)}, V\} = 0\,; && (20.88a)\\
\{K_j, P_k\} &= \delta_{jk} H \to \{W_j, P_k^{(0)}\} = \delta_{jk} V\,; && (20.88b)\\
\{K_j, K_k\} &= -\epsilon_{jkm} J_m \to \{K_j^{(0)}, W_k\} + \{W_j, K_k^{(0)}\} + \{W_j, W_k\} = 0\,;\\
\{K_j, H\} &= P_j \to \{K_j^{(0)}, V\} + \{W_j, H^{(0)}\} + \{W_j, V\} = 0\,. && (20.88c)
\end{aligned}$$

Subset (a) of these equations tells us that W_k and V form a vector and a scalar, respectively, under three-dimensional spatial rotations, whereas V is invariant under space translations. Equation (20.88b), which still involves the interaction terms linearly, denies this property for the W_j. Whereas V can depend only on the relative coordinates $\boldsymbol{q}^{(r)} - \boldsymbol{q}^{(s)}$ (and the p's), the W_j *must* depend on the q's in some other fashion: this is very different from the case of the Galilei group! Conditions (20.88c) are nonlinear in the unknowns and are more complicated; the successful construction of a realization of $\mathcal{P}$ by an interacting many-particle system involves finding nontrivial solutions to these differential equations. Such solutions are not obvious, and there seems to be no simple method of characterizing this system. More serious is the fact that if we now impose the conditions for manifest covariance (this is done in the next chapter), the entire formalism turns out to be incapable of describing anything but noninteracting particles! No such thing happens in the case of $\mathcal{G}$, and the difference in structures of $\mathcal{G}$ and $\mathcal{P}$ accounts for this.

Linear Representations of SO(3,1)-Vectors and Tensors

We digress at this point to give a brief description of the linear matrix representations of the group $SO(3,\,1)$, in the same spirit as the discussion given in Chapter 17 for the group $R(3)$. For the most part we restrict ourselves to the finite-dimensional representations.

Our study of the group $SO(3,1)$ in Chapter 15 was based on the defining four-dimensional matrix representation describing the properties of four-vectors. This representation, set up on physical grounds, has the following appearance: if S and S' are two inertial frames in special relativity, such that $S' = \exp(\boldsymbol{\zeta}\cdot\boldsymbol{k})\exp(\boldsymbol{\alpha}\cdot\boldsymbol{l})S$, then a four-vector V has components V^μ in S and V'^μ in S' related by

$$S' = \exp(\boldsymbol{\zeta}\cdot\boldsymbol{k})\exp(\boldsymbol{\alpha}\cdot\boldsymbol{l})S:\; V'^\mu = \Lambda^\mu{}_\nu(\boldsymbol{\zeta},\boldsymbol{\alpha})V^\nu\,. \tag{20.89}$$

To the element $\exp(\boldsymbol{\zeta}\cdot\boldsymbol{k})\exp(\boldsymbol{\alpha}\cdot\boldsymbol{l})$ in $SO(3,1)$ corresponds the matrix $||\Lambda^\mu{}_\nu(\boldsymbol{\zeta},\boldsymbol{\alpha})||$, and multiplication of these matrices yields the group composition law for $SO(3,\,1)$. [The index "to the left" is the row index, the one "to the right" is the column index; the elements $\Lambda^\mu{}_\nu(\boldsymbol{\zeta},\boldsymbol{\alpha})$ are given in Eq. (15.85)]. If we use canonical coordinates, a general element of the group is written as $\exp(\frac{1}{2}\lambda^{\mu\nu}l_{\mu\nu})$; the corresponding matrix $||\Lambda^\mu{}_\nu(\lambda)||$ is, at least in principle, determined subsequently.

In a general linear matrix representation of SO(3,1), of dimension d say, there is an assignment of a nonsingular $d\times d$ matrix $D(\boldsymbol{\zeta},\boldsymbol{\alpha})$ to each element $\exp(\boldsymbol{\zeta}\cdot\boldsymbol{k})\exp(\boldsymbol{\alpha}\cdot\boldsymbol{l})$ in $SO(3,1)$ such that the group laws are preserved: the product of the matrices corresponding to two elements is the matrix corresponding to the product of the elements. Specializing the remarks made in Chapters 14 and 15, every such matrix representation of $SO(3,1)$ arises from a matrix representation of its Lie algebra, with commutators for Lie brackets, and conversely. [However, in order to ensure that a matrix representation of the Lie algebra will yield a representation of the group $SO(3,1)$, a global condition must be imposed on the former; this is exactly analogous to the $R(3)$ situation and is developed below.] A matrix representation of the Lie algebra of $SO(3,1)$ consists of six matrices M_j, N_j, j=1,2,3, obeying the commutation rules

$$\begin{aligned}
[M_j, M_k]_- &= \epsilon_{jkm}M_m,\; [M_j,N_k]_- = \epsilon_{jkm}N_m\,,\\
[N_j, N_k]_- &= -\epsilon_{jkm}M_m\,.
\end{aligned} \tag{20.90}$$

We can also write these relations in covariant form thus:

$$\begin{aligned}
M_{jk} &= \epsilon_{jkm}M_m, M_{0j} = -M_{j0} = N_j\,;\\
[M_{\mu\nu}, M_{\sigma\rho}]_- &= g_{\mu\sigma}M_{\nu\rho} - g_{\nu\sigma}M_{\mu\rho} + g_{\mu\rho}M_{\sigma\nu} - g_{\nu\rho}M_{\sigma\mu}\,.
\end{aligned} \tag{20.91}$$

If these commutation rules are satisfied, the matrices representing elements of $SO(3,\,1)$ are obtained thus:

$$\exp(\boldsymbol{\zeta}\cdot\boldsymbol{k})\exp(\boldsymbol{\alpha}\cdot\boldsymbol{l}) \to D(\boldsymbol{\zeta},\boldsymbol{\alpha}) = e^{\boldsymbol{\zeta}\cdot\boldsymbol{N}}e^{\boldsymbol{\alpha}\cdot\boldsymbol{M}}$$

or

$$\exp(\frac{1}{2}\lambda^{\mu\nu}l_{\nu\mu}) \to D(\lambda^{\mu\nu}) = e^{(1/2)\lambda^{\mu\nu}M_{\nu\mu}}\,. \tag{20.92}$$

The proper multiplication rules for the D's are guaranteed by the proper commutation rules for the M's and N's, except that the following global condition must be imposed on a solution of Eq. (20.90): if $\boldsymbol{\zeta}$=0 and $|\boldsymbol{\alpha}| = 2\pi$ so that the element considered is a spatial rotation of amount 2π about some axis, which is the same as the identity element of $SO(3, 1)$, then $D(\mathbf{0}, \boldsymbol{\alpha}) = e^{\boldsymbol{\alpha}\cdot\boldsymbol{M}}$ must be the unit matrix. There is no global condition, however, involving elements $\exp(\boldsymbol{\zeta}\cdot\boldsymbol{k})$, that is, involving the matrices N_j.

Before turning to the properties of general solutions of Eqs. (20.90) and (20.91), let us see what the matrices $M_{\mu\nu}$ look like in the case of the defining representation of $SO(3,1)$. Here the information we need is contained on page 265, Chapter 15. It is enough to consider elements close to the identity. Abstractly, such an element has the alternative forms $\exp(\delta\zeta_j k_j + \delta\alpha_j l_j)$, or $\exp(-\frac{1}{2}\delta\lambda^{\mu\nu} l_{\mu\nu})$; to pass from one form to the other we use Eqs. (15.106) and (15.108). On the other hand, the defining representation of $SO(3,1)$ assigns the matrix $||\delta^\mu{}_\nu + \delta\lambda^\mu{}_\nu||$ to this element. Therefore, because of Eq. (20.92), the generator matrices $(M_{\mu\nu})^\alpha{}_\beta$ in the defining representation must be such that

$$(1 - \frac{1}{2}\delta\lambda^{\mu\nu} M_{\mu\nu})^\alpha{}_\beta = \delta^\alpha_\beta + \delta\lambda^\alpha{}_\beta\,;$$

this suffices to determine the generator matrices, and they are:

$$(M_{\mu\nu})^\alpha{}_\beta = \delta^\alpha_\nu g_{\mu\beta} - \delta^\alpha_\mu g_{\nu\beta}\,. \tag{20.93}$$

Here, α is the row index (the index to the left), and β is the column index. One can easily verify that this specific set of 4×4 matrices constitutes a solution to the commutation rules in Eq. (20.91). At least in principle one can now compute explicitly the matrix $||\Lambda^\mu{}_\nu||$, determined by the exponential $\exp(-\frac{1}{2}\lambda^{\mu\nu} M_{\mu\nu})$, the M 's being the above 4×4 matrices; this $||\Lambda^\mu{}_\nu||$ can then be recast in the form $\Lambda(\boldsymbol{\zeta}, \boldsymbol{\alpha})$, which would yield the connection between canonical coordinates for $SO(3,1)$ and the coordinates ζ_j, α_j. However, we do not carry out this computation here.

A general finite-dimensional matrix representation of $SO(3,1)$ turns out always to be a direct sum of irreducible representations; this is similar to the $R(3)$ situation and can be traced to the close relationship that exists between finite dimensional representations of $SO(3,1)$ and those of $SO(4)$. Thus we can confine ourselves to the irreducibles. In contrast to $R(3)$, however, it is impossible to arrange things so that a nontrivial finite dimensional representation of $SO(3,1)$ be unitary; it will always be essentially nonunitary. The reason is that whereas $R(3)$ is a compact Lie group, $SO(3,1)$ is noncompact. [There do exist nontrivial unitary representations of $SO(3,1)$, but then they must be infinite dimensional!] On the other hand, if we take a finite dimensional nonunitary representation of $SO(3,1)$ and restrict it to elements of the subgroup $R(3)$, we obtain a finite dimensional representation of $R(3)$ that can always be put into unitary form. That is to say, denoting the representation matrices by $D(\boldsymbol{\zeta}, \boldsymbol{\alpha})$, the representation of $SO(3,1)$ can be set up in such a way that the matrices $D(\mathbf{0}, \boldsymbol{\alpha})$ are unitary, but the matrices $D(\boldsymbol{\zeta}, \mathbf{0})$ are always nonunitary. More specifically, it turns out that the generators M_j can always be chosen to be *antihermitian*, whereas the N_j can be chosen *hermitian*. This makes the matrices $D(\boldsymbol{\zeta}, \mathbf{0})$ corresponding to pure Lorentz transformations into hermitian matrices, and the matrix $D(\boldsymbol{\zeta}, \boldsymbol{\alpha})$ representing a general element of $SO(3,1)$ appears, in a suitable basis, as the product of a (positive-definite) hermitian matrix and a unitary matrix, $D(\boldsymbol{\zeta}, \boldsymbol{\alpha}) = D(\boldsymbol{\zeta}, \mathbf{0})D(\mathbf{0}, \boldsymbol{\alpha})$.

Luckily, the problem of finding the finite-dimensional irreducible solutions to Eq. (20.90) can be reduced to the problem of finding solutions to the commutation rules corresponding to the $R(3)$ Lie algebra, and we know what the latter look like. We see this as follows: in terms of M_j and N_j define two triplets of matrices $S_j^{(1)}$ and $S_j^{(2)}$ thus:

$$S_j^{(1)} = \frac{1}{2}(M_j - iN_j), \; S_j^{(2)} = \frac{1}{2}(M_j + iN_j). \tag{20.94}$$

[What we are doing here is "illegal" in this sense: the abstract Lie algebra of $SO(3,1)$ is a *real* linear vector space, and in a matrix representation of it each vector in the Lie algebra is represented by a real linear combination of the basic matrices M_j and N_j. Therefore, $S_j^{(1)}$ and $S_j^{(2)}$ do not represent vectors in the Lie algebra of $SO(3,1)$, but that need not stop us from using their properties if it helps in finding all solutions to Eq. (20.90)!] The *M*'s and *N*'s can be recovered from the *S*'s. Then the basic commutation relations in Eq. (20.90) lead to the following ones:

$$\begin{aligned} [S_j^{(1)}, S_k^{(1)}]_- &= \epsilon_{jkm} S_m^{(1)}, \; [S_j^{(2)}, S_k^{(2)}]_- = \epsilon_{jkm} S_m^{(2)}, \\ [S_j^{(1)}, S_k^{(2)}]_- &= 0. \end{aligned} \tag{20.95}$$

We see that we need essentially two independent (i.e., commuting) solutions to the commutation relations of the $SU(2)$ Lie algebra! The snag is that we are not assured in advance that the matrices $S_j^{(1)}$ and $S_j^{(2)}$ are antihermitian. Nevertheless it is a fact (which we do not establish here) that any finite-dimensional solution to the commutation relations of the $SU(2)$ Lie algebra can actually be used to construct a representation of $SU(2)$; this representation can be assumed to be unitary, which means that any finite-dimensional matrix solution to the $SU(2)$ Lie algebra commutation rules can be assumed to be by antihermitian matrices. Thus we may assume that $S_j^{(1)}$ and $S_j^{(2)}$ are all antihermitian; the possible choices correspond exactly to the matrices that give rise to unitary representations of $SU(2)$ in various dimensions. These we have described in Chapter 17. Up to unitary equivalence, there is just one set of irreducible antihermitian matrices S_j obeying the $SU(2)$ commutation rules in each dimension. Generally, these distinct possibilities are labeled by their "spin-value" rather than their dimension, the latter being twice the former plus one. Thus the solutions in 1,2, 3,.... dimensions correspond to spin $0, \frac{1}{2}, 1 \ldots$. A general spin value is $n/2$, n being some integer. The most general irreducible solution to Eq. (20.95) can then be synthesized in this way: we take two vector spaces $V^{(n)}$ and $V^{(m)}$ of dimensions $(n+1)$ and $(m+1)$ respectively, and form their direct product, a space of dimension $(n+1)(m+1)$. We then choose $S_j^{(1)}$ to be the $(n+1)$-dimensional irreducible solution to the $SU(2)$ commutation rules, operating in $V^{(n)}$, times the unit operator in $V^{(m)}$; and $S_j^{(2)}$ to be the unit operator in $V^{(n)}$ times the $(m+1)$-dimensional irreducible solution to the $SU(2)$ commutation rules operating in $V^{(m)}$. The $S^{(1)}$'s commute with the $S^{(2)}$'s because they operate on independent spaces. This irreducible solution to Eq. (20.95) is usually written as $(n/2, m/2)$; by allowing n and m to independently run over the values 0,1,2,3..., we get all possible irreducible infinite-dimensional solutions to Eq. (20.95) and thus to Eq. (20.90). $[(n/2, m/2)$ and $(n'/2, m'/2)$ are inequivalent unless $n = n', m = m']$. Now we must determine under what conditions the global conditions are satisfied so that we do obtain a representation of $SO(3,1)$. Going back to $SU(2)$ and $R(3)$, we recall that only the odd-dimensional irreducible representations of the former are also representations of the latter. This is because in the $(n+1)$-dimensional representation of $SU(2)$ (irreducible, of course), the matrix $e^{\boldsymbol{\alpha}\cdot\boldsymbol{S}}$ is $(-1)^n$ times the unit matrix, when $|\boldsymbol{\alpha}| = 2\pi$. Now the representation labeled $(n/2,\ m/2)$ of the commutation relations in Eqs. (20.90) or (20.95) must be such that for $|\boldsymbol{\alpha}| = 2\pi, e^{\boldsymbol{\alpha}\cdot\boldsymbol{M}}$ reduces to the unit matrix, if we wish the M's and N's to give a representation of $SO(3,1)$. Because $M_j = S_j^{(1)} + S_j^{(2)}$ and because the $S^{(1)}$'s commute with the $S^{(2)}$s, it follows that in general $e^{\boldsymbol{\alpha}\cdot\boldsymbol{M}} = (-1)^{n+m}$ when $|\boldsymbol{\alpha}| = 2\pi$. Therefore only those solutions $(n/2,\ m/2)$ to the commutation relations in Eqs. (20.90) or (20.95) can be exponentiated to give a representation of $SO(3,1)$, for which $(n+m)$ is an *even* integer — equivalently, $(n+m)/2$ is an integer. If $(n+m)$ is an *odd* integer, the matrices M_j and N_j lead to a representation of the group $SL(2,C)$, the universal covering group of $SO(3,1)$ and not of $SO(3,1)$; we describe these later on. To summarize, the irreducible finite-dimensional matrix representations of $SO(3,1)$ can be labeled in the form $(n/2, m/2)$ where n and m are both even integers or both odd integers. The dimensionality of the representation is $(n+1)(m+1)$. It may be of interest to know what representation of $R(3)$ is obtained if we take the irreducible representation $(n/2, m/2)$ of $SO(3,1)$ and restrict it to $R(3)$. We generally obtain a reducible representation of $R(3)$; the irreducible ones that occur are those of dimension $(n+m+1), (n+m-1), (n+m-3), \cdots (|n-m|+1)$, once each. All of these are of odd dimension, as they must be; the associated "spin-values" are $(n/2+m/2), (n/2+m/2-1), (n/2+m/2-2) \cdots (|n-m|/2)$. Whereas in the case of $R(3)$ there is, up to equivalence, just one irreducible representation in each odd dimension, for $SO(3,\ 1)$ there can be two or more irreducible inequivalent representations of the same dimension. For example, if $n \neq m, (n/2, m/2)$ and $(m/2,\ n/2)$ are inequivalent.

In the case of $R(3)$ we know that all the irreducible representations of this group could be arrived at by considering the transformation properties of special kinds of Cartesian tensors. In particular it is enough to consider symmetric traceless tensors of all possible ranks (rank = number of indices) in order to obtain each irreducible representation once. We now describe briefly how each of the irreducible representations of $SO(3,1)$ can also be obtained by using tensors of special types. However, we now need to go beyond the completely symmetric tensors in order to get all the representations.

To begin with, let us generalize Eq. (20.89) so as to define a tensor with respect to $SO(3,1)$. A physical quantity T is said to be a tensor of rank N if it requires 4^N numbers $T^{\mu\nu\cdots\sigma}, 0 \leq \mu, \nu, \cdots \sigma \leq 3$ (N indices), to specify it completely in each inertial frame S, and if on changing to another frame S' these numbers change linearly as follows:

$$S' = \Lambda S : \; T'^{\mu\nu\ldots\sigma} = \Lambda^\mu{}_{\mu'} \Lambda^\nu{}_{\nu'} \cdots \Lambda^\sigma{}_{\sigma'} \, T^{\mu'\nu'\ldots\sigma'} \\ \leftarrow N - \text{factors} \rightarrow . \tag{20.96}$$

$T^{\mu\nu\ldots\sigma}$ are the components of T as measured in $S, T'^{\mu\nu\ldots\sigma}$ are its components as measured in S'. Each index on a tensor behaves as though it were an index on a four-vector and changes accordingly under the transition $S \rightarrow S' = \Lambda S$. We verify immediately that if we have three inertial frames $S, S' = \Lambda S$, and $S'' = \Lambda' S' = (\Lambda'\Lambda)S$, the two possible ways of relating the components of T in S'' to its components in S give the same result. We can form linear combinations of several tensors, all of the same rank, with scalars for coefficients, the result again is a tensor of the same rank. Thus the set of all (real) tensors of rank N forms a linear space of dimension 4^N, and in this space we obtain via Eq. (20.96) a linear matrix representation of $SO(3,1)$. Another operation with tensors is, again, "outer multiplication": given two tensors $T^{(1)}$ and $T^{(2)}$ of ranks N_1 and N_2, respectively, by considering all possible products of components of $T^{(1)}$ by components of $T^{(2)}$ we generate $4^{N_1+N_2}$ quantities that form the components of a tensor T of rank $(N_1 + N_2)$. This is easily verified.

The 4^N-dimensional representation of $SO(3,1)$ provided by the set of all (real) tensors of rank N is called the N-fold tensor product of the defining four-dimensional representation with itself. For $N \geq 2$, this is a reducible representation. Generalizing what was said in the case of $R(3)$, we can find subsets of tensors that transform among themselves by imposing special kinds of symmetry with respect to permutations of the indices and by asking that certain traces with respect to pairs of indices vanish. Both these kinds of properties are preserved under a transformation such as Eq. (20.96), the first because all indices on a tensor transform in the same way, the second because the matrices $||\Lambda^\mu{}_\nu||$ are pseudo-orthogonal:

$$\Lambda^\mu{}_\nu \Lambda_\mu{}^{\nu'} = \delta_\nu^{\nu'} . \tag{20.97}$$

The simplest kind of irreducible tensor is a completely symmetric traceless tensor of rank N. The components of such a tensor obey:

$$T^{\mu\nu\ldots\sigma} = T^{\mu'\nu'\ldots\sigma'}, \; \mu'\nu'\ldots\sigma' = P(\mu\nu\ldots\sigma), \; g_{\mu\nu}T^{\mu\nu\ldots\sigma} = 0 . \tag{20.98}$$

The symbol P denotes any permutation. Tensors of this type form an invariant subspace in the 4^N-dimensional space of all Nth-rank tensors and thus give us a representation of $SO(3,1)$ all by themselves. In fact, they give us an irreducible representation. Which one? Symmetry restricts the number of independent components of T to $\frac{1}{6}(N+1)(N+2)(N+3)$. Tracelessness amounts to as many linear relations on the components of T as the number of independent components for an (N-2)-th rank symmetric tensor, which is $\frac{1}{6}(N-1)N(N+1)$. Thus T has $\frac{1}{6}(N+1)((N+2)(N+3)-N(N-1)) = (N+1)^2$ independent components. This number coincides with the dimension of the representation $(N/2, N/2)$ of $SO(3,1)$, and it is, in fact, true that one does obtain this representation in this way; Because the conditions imposed in Eq. (20.98), as well as the transformation law (Eq. (20.96)), are all real in character, it should not be hard to believe that the irreducible representations $(N/2, N/2)$ of $SO(3,1)$ can be given in real form.

We see that by considering the simplest possible types of traceless tensors, namely fully symmetric ones, we are led to a special set of representations of $SO(3,1)$. To obtain the other representations we must consider traceless tensors of more complicated symmetry types. A certain amount of antisymmetry in the indices must be allowed. For simplicity let us consider first a second-rank antisymmetric tensor $T^{\mu\nu} = -T^{\nu\mu}$; we can now make use of the existence of the invariant Levi-Civita symbol $\epsilon^{\lambda\mu\nu\rho}$ normalized to $\epsilon_{0123} = +1$ and ask if T can possibly possess any simple property on contraction with $\epsilon^{\cdots}$: Specifically, we can ask for what values of η an equation of this type is permissible:

$$\epsilon^{\lambda\mu}{}_{\nu\rho}T^{\nu\rho} = \eta T^{\lambda\mu}\,. \tag{20.99}$$

Now a product of two ϵ's can always be rewritten in terms of the metric tensor g in this way:

$$\epsilon^{\lambda\mu\nu\rho}\epsilon_{\lambda'\mu'\nu'\rho'} = -g^{\lambda}_{\lambda'}\cdot g^{\mu}_{\mu'}\cdot g^{\nu}_{\nu'}\cdot g^{\rho}_{\rho'} + \cdots \tag{20.100}$$

The dots here stand for the 23 remaining terms that make the right-hand side completely antisymmetric in $\lambda\mu\nu\rho$ (and simultaneously in $\lambda'\mu'\nu'\rho'$). Combining Eqs. (20.99) and (20.100) we find that $\eta^2 = -4, \eta = \pm 2i$. An anti-symmetric tensor that obeys Eq. (20.99) will do so for one of these two values of η, and its components cannot all be real. It is said to be self-dual or anti-self-dual according as $\eta = 2i$ or $-2i$ respectively. A general second-rank anti-symmetric tensor will not obey Eq. (20.99), but it can be decomposed into two parts, one obeying Eq. (20.99) with $\eta = -2i$ and the other with $\eta = +2i$. If T does obey Eq. (20.99), then this denotes a property that is preserved under the action of SO(3,1): this is because $\epsilon^{\cdots\cdot}$ is an invariant tensor. Generalizing, we can say that if a tensor of arbitrary rank is antisymmetric in a particular pair of indices and if it obeys Eq. (20.99) with respect to that pair of indices, this denotes a special SO(3,1)-invariant property of that tensor. We can add this idea of "ϵ-parity" to the earlier ones of symmetry and tracelessness in determining special tensors that give rise to the irreducible representations $(n/2, m/2)$ of $SO(3,1)$ with $n \neq m$. Specifically, in case $n > m$, one has to proceed as follows: consider all those nth-rank tensors T whose components we write as:

$$T^{\lambda_1\cdots\lambda_r\nu_1\ldots\nu_s}{}_{\mu_1\cdots\mu_r}\,,\; r = \frac{n-m}{2}, s = m$$

(all the indices are upper ones!) and which enjoy the following properties: (1) complete symmetry in the indices $\nu_1 \ldots, \nu_s$; (2) complete symmetry in the index pairs $\binom{\lambda_1}{\mu_1}, \ldots \binom{\lambda_r}{\mu_r}$, that is under the exchange of any two index pairs $\binom{\lambda_j}{\mu_j}$ and $\binom{\lambda_k}{\mu_k}$; (3) antisymmetry under the interchange of any λ and the associated μ below it; (4) ϵ parity =-$2i$ with respect to each index pair $\binom{\lambda}{\mu}$; (5) vanishing of all traces. Properties (1) to (4) can be schematically indicated thus:

$$T^{\cdots\nu_j\cdots\nu_k\cdots}_{\cdots} = T^{\cdots\nu_k\cdots\nu_j\cdots}_{\cdots}\,; \tag{20.101a}$$

$$T^{\cdots\lambda_j\cdots\lambda_k\cdots}_{\cdots\mu_j\cdots\mu_k\cdots} = T^{\cdots\lambda_k\cdots\lambda_j\cdots}_{\cdots\mu_k\cdots j\cdots}\,; \tag{20.101b}$$

$$T^{\cdots\lambda\cdots}_{\cdots\mu\cdots} = -\,T^{\cdots\mu\cdots}_{\cdots\lambda\cdots}\,; \tag{20.101c}$$

$$\epsilon_{\lambda\mu}{}^{\nu\rho}\, T^{\cdots\lambda\cdots}_{\cdots\mu\cdots} = -\,2i\, T^{\cdots\nu\cdots}_{\cdots\rho\cdots} \tag{20.101d}$$

(The dots denote indices that occupy the same positions on both sides of the equation.) All these are intrinsic properties, preserved under $SO(3{,}1)$; it then turns out that tensors with these properties are irreducible and in fact give rise to the representation $(n/2, m/2)$ of $SO(3,1)$. From the form of the condition Eq. (20.101d) it follows that the components of such tensors cannot all be real, showing that if $n > m$ the representation $(n/2, m/2)$ is essentially complex and cannot be put into real form while preserving the dimension.

To obtain tensors corresponding to the representation $(n/2, m/2)$ with $m > n$, all we need to do is to reverse the ϵ parity. We consider all mth-rank tensors T whose components we write as

$$T^{\lambda_1\cdots\lambda_r\nu_1\cdots\nu_s}_{\mu_1\cdots\mu_r}\,, r = \frac{m-n}{2}, s = n$$

(again, all indices are upper ones.) We impose complete tracelessness, and Eqs. (20.101a)-(20.101c) again, and in place of Eq. (20.101d) we require:

$$\epsilon_{\lambda\mu}{}^{\nu\rho}\, T^{\cdots\lambda\cdots}_{\cdots\mu\cdots} = +2i\, T^{\cdots\nu\cdots\cdots}_{\cdots\rho\cdots} \tag{20.101d'}$$

The set of all these tensors gives us the irreducible representation $(n/2, m/2)$ of $SO(3,1)$ if $m > n$. This again is an essentially complex representation of $SO(3,1)$. But since the only difference between Eqs. (20.101d) and (20.101d′) is that i has been replaced by $-i$, we can deduce the following: if we take all those irreducible tensors that provide us with the representation $(n/2, m/2)$ for some $n > m$ and consider their complex conjugates, the latter now obey the conditions that define irreducible tensors for the representation $(m/2, n/2)$! (Recall that all the conditions of symmetry, antisymmetry, and tracelessness are real conditions.) Therefore, for $n \neq m$, neither of the representations $(n/2, m/2)$, $(m/2, n/2)$ can be made real, but one is essentially the complex conjugate of the other.

Thus we see that every irreducible matrix representation of $SO(3,1)$ can be generated by considering traceless tensors of appropriate rank, symmetry type and ϵ parity. An arbitrary tensor will not of course belong to any one representation of $SO(3,1)$, but it can be systematically decomposed into irreducible parts and finally expressed as a linear combination of several tensors, each belonging to one irreducible representation of $SO(3,1)$. By way of illustration, consider a general second-rank tensor $T^{\mu\nu}$. We can as a first step split it into a symmetric part and an antisymmetric part:

$$T^{\mu\nu} = S^{\mu\nu} + A^{\mu\nu}, S^{\mu\nu} = \frac{1}{2}(T^{\mu\nu} + T^{\nu\mu}), A^{\mu\nu} = \frac{1}{2}(T^{\mu\nu} - T^{\nu\mu}). \tag{20.102}$$

We can next split off from $S^{\mu\nu}$ its trace:

$$S^{\mu\nu} = S^{(1)\mu\nu} + \frac{1}{4}g^{\mu\nu}S, S^{(1)\mu\nu} = S^{\mu\nu} - \frac{1}{4}g^{\mu\nu}S^{\lambda}{}_{\lambda}, S = S^{\lambda}{}_{\lambda}. \tag{20.103}$$

The tensor $S^{(1)}$ is now irreducible, and provided it does not vanish, it constitutes the part in T belonging to the representation (1,1); similarly, if S does not vanish, it is the part in T belonging to the identity representation (0, 0). We next split A into parts with different ϵ parities:

$$\begin{aligned} A^{\mu\nu} &= A^{(1)\mu\nu} + A^{(2)\mu\nu}, \\ A^{(1)\mu\nu} &= \frac{1}{2}\left(A^{\mu\nu} + \frac{i}{2}\epsilon^{\mu\nu}{}_{\sigma\rho}A^{\sigma\rho}\right), A^{(2)\mu\nu} = \frac{1}{2}\left(A^{\mu\nu} - \frac{i}{2}\epsilon^{\mu\nu}{}_{\sigma\rho}A^{\sigma\rho}\right); \\ \epsilon^{\mu\nu}{}_{\sigma\rho}A^{(1)\sigma\rho} &= -2iA^{(1)\mu\nu}, \epsilon^{\mu\nu}{}_{\sigma\rho}A^{(2)\sigma\rho} = +2iA^{(2)\mu\nu}. \end{aligned} \tag{20.104}$$

The tensors $A^{(1)}$ and $A^{(2)}$ are irreducible, and they are the parts in T belonging to the representations (1, 0) and (0, 1), respectively. All in all, the decomposition of a general T takes the form:

$$T^{\mu\nu} = S^{(1)\mu\nu} + \frac{1}{4}g^{\mu\nu}S + A^{(1)\mu\nu} + A^{(2)\mu\nu}, \tag{20.105}$$

indicating the presence of parts belonging to (1,1), (0,0), (1,0), and (0, 1) representations, respectively. [The dimensionalities add up, too: 16=9+1+3+3]. As a special case, if T were antisymmetric to begin with, then the representations (1,1) and (0,0) do not appear at all, and we are left with (1,0),(0,1) alone. This happens for the generators $M_{\mu\nu}$! Using Eqs. (20.104) and (20.91), one sees that the part belonging to the representation (1,0) is precisely the triplet $S_j^{(1)}$ defined in Eq. (20.94), whereas the $S_j^{(2)}$ belong to the representation (0,1).

In dealing with the representations of $SO(3,1)$ in practice, one has, as with $R(3)$, two possible choices. A tensor T belonging to the representation $(n/2, m/2)$ could be supplied with the proper number of four-vector indices and made to transform according to Eq. (20.96), but then one must carry along all the linear relations existing among these Cartesian components $T^{\lambda\mu\cdots}$. An alternative procedure is to take account of all these subsidiary conditions explicitly and work with precisely $(n+1)(m+1)$ linearly independent linear combinations T_α of the $T^{\mu\nu\cdots}$. The definitions of the T_α, together with all the relations among the $T^{\mu\nu\cdots}$ (such as Eq. (20.101)) must allow us to express each $T^{\mu\nu\cdots}$ as a linear combination of the independent T_α's. Assigning a set of values to the components T_α independently will determine a unique array of quantities $T^{\mu\nu\cdots}$ obeying all the subsidiary conditions, and vice versa. One can now rewrite the transformation law Eq. (20.96) in terms of the T_α's; this involves the $(n+1)(m+1)$-dimensional representation matrices $D(\boldsymbol{\zeta},\boldsymbol{\alpha})$ of Eq. (20.92):

$$S' = \exp(\boldsymbol{\zeta}\cdot\boldsymbol{k})\exp(\boldsymbol{\alpha}\cdot\boldsymbol{l})S:\ T'_\alpha = D_{\alpha\beta}(\boldsymbol{\zeta},\boldsymbol{\alpha})T_\beta\,. \tag{20.106}$$

The T_α are the components of T as measured in S, T'_α as measured in S'. There is no unique way of choosing the T_α; this reflects the freedom in the choice of basis in which to express the matrices $D(\boldsymbol{\zeta},\boldsymbol{\alpha})$.

Let us conclude this section on the finite-dimensional representations of $SO(3,1)$ by describing tensors in the Hamiltonian language. An Nth-rank Cartesian tensor T is a set of 4^N phase-space functions $T^{\mu\nu\ldots\sigma}(\omega)$, behaving under the canonical transformations realizing $SO(3,1)$ as follows:

$$e^{\widetilde{-\boldsymbol{\alpha}\cdot\boldsymbol{J}}}e^{\widetilde{-\boldsymbol{\zeta}\cdot\boldsymbol{K}}}T^{\mu\nu\ldots\sigma}(\omega) = \underbrace{\Lambda^\mu{}_{\mu'}(\boldsymbol{\zeta},\boldsymbol{\alpha})\Lambda^\nu{}_{\nu'}(\boldsymbol{\zeta},\boldsymbol{\alpha})\cdots\Lambda^\sigma{}_{\sigma'}(\boldsymbol{\zeta},\boldsymbol{\alpha})}_{\longleftarrow\ N\text{ factors}\ \longrightarrow}T^{\mu'\nu'\ldots\sigma'}(\omega)\,. \tag{20.107}$$

[Note the sequence of canonical transformations on the left; this can be derived by considering the frame change $S\to S' = \exp(\boldsymbol{\zeta}\cdot\boldsymbol{k})\exp(\boldsymbol{\alpha}\cdot\boldsymbol{l})S$ and using the Heisenberg picture.] If we use canonical coordinates, the same equation reads:

$$\exp(\frac{1}{2}\widetilde{\lambda^{\rho\tau}J_{\rho\tau}})T^{\mu\nu\ldots\sigma}(\omega) = \underbrace{\Lambda^\mu{}_{\mu'}(\lambda)\Lambda^\nu{}_{\nu'}(\lambda)\ldots\Lambda^\sigma{}_{\sigma'}(\lambda)}_{\longleftarrow\ N\text{ factors}\ \longrightarrow}T^{\mu'\nu'\ldots\sigma'}(\omega)\,. \tag{20.108}$$

[The matrix $||\Lambda^\mu{}_\nu(\lambda)||$ is, of course, the exponential of the matrix $-\frac{1}{2}\lambda^{\mu\nu}M_{\mu\nu}$, the M's being the 4×4 matrices given in Eq. (20.93).] The infinitesimal form of Eq. (20.108) looks like this:

$$\begin{aligned}\{J^{\rho\tau}, T^{\mu\nu\ldots\sigma}\} =& g^{\rho\mu}T^{\tau\nu\ldots\sigma} - g^{\tau\mu}T^{\rho\nu\ldots\sigma} + g^{\rho\nu}T^{\mu\tau\ldots\sigma}\\ &- g^{\tau\nu}T^{\mu\rho\ldots\sigma} + \ldots + g^{\rho\sigma}T^{\mu\nu\ldots\tau} - g^{\tau\sigma}T^{\mu\nu\ldots\rho}\,.\end{aligned} \tag{20.109}$$

This system of PB relations with the generators $J^{\alpha\beta}(\omega)$ is also adequate to say that the $T^{\mu\nu\ldots\sigma}(\omega)$ are the components of an Nth-rank tensor. For an irreducible tensor $T_\alpha(\omega)$, Eqs. (20.107), (20.108), and (20.109) must be replaced by the following (we give both the split forms in which rotations and pure Lorentz transformations are separated and the covariant form):

$$\begin{aligned} e^{\widetilde{-\boldsymbol{\alpha}\cdot\boldsymbol{J}}}e^{\widetilde{-\boldsymbol{\zeta}\cdot\boldsymbol{K}}}T_\alpha(\omega) &= D_{\alpha\beta}(\boldsymbol{\zeta},\boldsymbol{\alpha})T_\beta(\omega) \equiv [e^{\boldsymbol{\zeta}\cdot\boldsymbol{N}}e^{\boldsymbol{\alpha}\cdot\boldsymbol{M}}]_{\alpha\beta}T_\beta(\omega)\,; \\ \exp(\frac{1}{2}\lambda^{\mu\nu}\widetilde{J_{\mu\nu}})T_\alpha(\omega) &= D_{\alpha\beta}(\lambda)T_\beta(\omega) \equiv [\exp(-\frac{1}{2}\lambda^{\mu\nu}M_{\mu\nu})]_{\alpha\beta}T_\beta(\omega)\,; \\ \{J_j, T_\alpha\} &= -(M_j)_{\alpha\beta}T_\beta, \{K_j, T_\alpha\} = -(N_j)_{\alpha\beta}T_\beta\,; \\ \{J_{\mu\nu}, T_\alpha\} &= -(M_{\mu\nu})_{\alpha\beta}T_\beta\,. \end{aligned} \tag{20.110}$$

[Indices α, β in Eq. (20.110) are not four-vector indices; they label the rows and columns of the representation of $SO(3,1)$ to which T belongs.]

Spinors And The Relation Between $SL(2,C)$ and $SO(3,1)$

The group $SL(2,C)$ is the group formed by all two-dimensional complex matrices with determinant equal to unity. The relationship between this group and $SO(3,1)$ is quite similar to that between $SU(2)$ and $R(3)$; there is a two-to-one homomorphism, which is a two-to-one mapping that preserves the group operations, from $SL(2,C)$ to $SO(3,1)$. The Lie algebras for the two groups have the same structure, and every representation (or realization) of $SO(3,1)$ is also one for $SL(2,C)$, but not vice versa.

The simplest way to study this connection between $SO(3,1)$ and $SL(2,C)$ is to look at the two-dimensional matrix solutions to the $SO(3,1)$ commutation relations. We shall begin with the one labeled $\left(\frac{1}{2},0\right)$, and by convention this leads to the defining representation of $SL(2,C)$. Referring to Eqs. (20.94) and (20.95), this representation $\left(\frac{1}{2},0\right)$ arises by taking the $S_j^{(1)}$ to be the generators of the defining representation of $SU(2)$ and setting the $S_j^{(2)}$ equal to zero. [In the representation $\left(0,\frac{1}{2}\right)$, $S_j^{(1)}$ and $S_j^{(2)}$ are switched.] Thus M_j and N_j are given by:

$$\begin{aligned} &\left(\tfrac{1}{2},0\right) : M_j = -(i/2)\sigma_j, N_j = \left(\tfrac{1}{2}\right)\sigma_j\,; \\ \sigma_1 = \begin{pmatrix} 0 & 1 \\ 1 & 0 \end{pmatrix}, \ &\sigma_2 = \begin{pmatrix} 0 & -i \\ i & 0 \end{pmatrix}, \sigma_3 = \begin{pmatrix} 1 & 0 \\ 0 & -1 \end{pmatrix}. \end{aligned} \tag{20.111}$$

Here, then, is a nontrivial solution to Eq. (20.90). But for $|\boldsymbol{\alpha}| = 2\pi, e^{\boldsymbol{\alpha}\cdot\boldsymbol{M}}$ is the negative of the 2×2 unit matrix. Thus the global condition for a representation of $SO(3,1)$ is not obeyed; it is the same reason why the defining representation of $SU(2)$ is not an $R(3)$ representation. Nevertheless we can define, for all $(\boldsymbol{\zeta},\boldsymbol{\alpha})$, the exponentiated 2×2 matrices:

$$a(\boldsymbol{\zeta},\boldsymbol{\alpha}) = e^{\boldsymbol{\zeta}\cdot\boldsymbol{N}}e^{\boldsymbol{\alpha}\cdot\boldsymbol{M}} = e^{\boldsymbol{\zeta}\cdot\boldsymbol{\sigma}/2}e^{-i\boldsymbol{\alpha}\cdot\boldsymbol{\sigma}/2}\,. \tag{20.112}$$

Even though we do not have a representation of $SO(3,1)$ here, the fact that the $SO(3,1)$ commutation relations are obeyed means that for small enough values of ζ_j and α_j the matrices $a(\boldsymbol{\zeta},\boldsymbol{\alpha})$ do obey the multiplication rules of $SO(3,1)$. Thus, if $(\boldsymbol{\zeta},\boldsymbol{\alpha}), (\boldsymbol{\zeta}',\boldsymbol{\alpha}')$ are two elements of $SO(3,1)$ sufficiently close to the identity, and $(\boldsymbol{\zeta}',\boldsymbol{\alpha}')(\boldsymbol{\zeta},\boldsymbol{\alpha}) = (\boldsymbol{\zeta}'',\boldsymbol{\alpha}'')$ is the product element in $SO(3,1)$ also close to the identity, then we have $a(\boldsymbol{\zeta}',\boldsymbol{\alpha}')a(\boldsymbol{\zeta},\boldsymbol{\alpha}) = a(\boldsymbol{\zeta}'',\boldsymbol{\alpha}'')$. It is only globally that we do not have a representation of $SO(3,1)$. The validity of the $SO(3,1)$ multiplication law in a neighborhood of the identity allows, as before, the derivation of the transformation law of the generators $M_{\mu\nu}$ according to the adjoint representation, under conjugation by the matrices $a(\boldsymbol{\zeta},\boldsymbol{\alpha})$:

$$\begin{aligned} a(\boldsymbol{\zeta},\boldsymbol{\alpha})M_{\mu\nu}a(\boldsymbol{\zeta},\boldsymbol{\alpha})^{-1} &= \Lambda^{\rho}{}_{\mu}(\boldsymbol{\zeta},\boldsymbol{\alpha})\Lambda^{\sigma}{}_{\nu}(\boldsymbol{\zeta},\boldsymbol{\alpha})M_{\rho\sigma}\,, \\ a(\boldsymbol{\zeta},\boldsymbol{\alpha})^{-1}M_{\mu\nu}a(\boldsymbol{\zeta},\boldsymbol{\alpha}) &= \Lambda_{\mu}{}^{\rho}(\boldsymbol{\zeta},\boldsymbol{\alpha})\Lambda_{\nu}{}^{\sigma}(\boldsymbol{\zeta},\boldsymbol{\alpha})M_{\rho\sigma}\,. \end{aligned} \tag{20.113}$$

These are analogous to Eq. (17.146), and follow from general theory. However, although such an equation in the $SU(2)$ case led to a simple understanding of the $SU(2)-R(3)$ relationship, such is not the case here; we need a different property of the matrices $a(\boldsymbol{\zeta},\boldsymbol{\alpha})$, and we develop this next.

Let us first examine the matrices $a(\boldsymbol{\zeta},\boldsymbol{\alpha})$ in more detail. Had we been dealing with $SO(3,1)$, the ranges of the ζ's and α's would have been as follows: each component ζ_j of $\boldsymbol{\zeta}$ could assume any real value from $-\infty$ to $+\infty$ independently, whereas $\boldsymbol{\alpha}$ could have any direction but $|\boldsymbol{\alpha}| \leq \pi$. This is because the α_j denote an element of $R(3)$. Thus $(\boldsymbol{\zeta},\boldsymbol{\alpha})$ and $(\boldsymbol{\zeta},(|\boldsymbol{\alpha}|+2\pi)\boldsymbol{\alpha}/|\boldsymbol{\alpha}|)$ stand for one and the same element from $SO(3,1)$. There is no such cyclic condition on the ζ_j: if $\boldsymbol{\zeta}$ and $\boldsymbol{\zeta}'$ are two different vectors, the corresponding pure Lorentz transformations $\exp(\boldsymbol{\zeta}\cdot\boldsymbol{k})$ and $\exp(\boldsymbol{\zeta}'\cdot\boldsymbol{k})$ are different. Knowing now the relationship between $SU(2)$ and $R(3)$, and more specifically the difference in the ranges of α_j in the two cases, we can see that all the matrices $a(\boldsymbol{\zeta},\boldsymbol{\alpha})$ for all possible parameter values are already obtained if we restrict these parameters in the following way. As in $SO(3,1)$, we permit each ζ_j to vary independently from $-\infty$ to $+\infty$, whereas for $\boldsymbol{\alpha}$ we allow all directions but impose $|\boldsymbol{\alpha}| \leq 2\pi$. In fact, it is clear from properties of the defining representation of $SU(2)$ that the matrices $a(\boldsymbol{\zeta},\boldsymbol{\alpha})$ obey:

$$\begin{aligned} a\left(\boldsymbol{\zeta},\frac{|\boldsymbol{\alpha}|+2\pi}{|\boldsymbol{\alpha}|}\,\boldsymbol{\alpha}\right) &= -a(\boldsymbol{\zeta},\boldsymbol{\alpha})\,, \\ a(\boldsymbol{\zeta},\boldsymbol{\alpha})\Big|_{|\boldsymbol{\alpha}|=2\pi} &= -a(\boldsymbol{\zeta},\boldsymbol{0}) \text{ independent of } \hat{\boldsymbol{\alpha}}\,. \end{aligned} \tag{20.114}$$

In dealing with these 2×2 matrices, we therefore define the ranges of ζ_j and α_j in this way. Can we now characterize the set of all matrices $a(\boldsymbol{\zeta},\boldsymbol{\alpha})$ in a simple way? It is clear that because the σ_j are traceless $a(\boldsymbol{\zeta},\boldsymbol{\alpha})$ is a complex 2×2 matrix with unit determinant. But as $\boldsymbol{\zeta}$ and $\boldsymbol{\alpha}$ vary in the prescribed domains, we do get in this way all 2×2 complex matrices with unit determinant! Equation (20.112) indicates why this is so: it expresses $a(\boldsymbol{\zeta},\boldsymbol{\alpha})$ as the product of a *positive-definite hermitian* matrix $e^{\boldsymbol{\zeta}\cdot\boldsymbol{\sigma}/2}$ and a *unitary* matrix $e^{-i\boldsymbol{\alpha}\cdot\boldsymbol{\sigma}/2}$, each factor being unimodular. Because we give all possible values to $\boldsymbol{\zeta}$, the first factor goes over all possible positive-definite hermitian 2×2 matrices with unit determinant, because any such matrix is the exponential of some traceless hermitian 2×2 matrix, and for these the σ_j form a basis. The second factor varies over all elements of the defining representation of $SU(2)$, that is, all possible unitary unimodular 2×2 matrices. On the other hand, by means of the polar decomposition, *any* 2×2 complex unimodular matrix can be expressed in a unique way as the product of a hermitian positive definite matrix and a unitary matrix, each factor being unimodular. This proves the assertion that what we obtain by constructing the matrices $a(\boldsymbol{\zeta},\boldsymbol{\alpha})$ for $\boldsymbol{\zeta},\boldsymbol{\alpha}$ in the prescribed ranges is the collection of all complex unimodular 2×2 matrices; these provide us with the defining representation of the group $SL(2,C)$, and we have seen that for small values of the ζ's and α's the group composition laws for $SL(2,C)$ and $SO(3,1)$ coincide. Using the uniqueness of the polar decomposition, we can convince ourselves that except in some special cases there is a one-to-one correspondence between the values of the parameters $\boldsymbol{\zeta},\boldsymbol{\alpha}$ (within the prescribed ranges $-\infty < \zeta_j < \infty$, $|\boldsymbol{\alpha}| \leq 2\pi$) and elements of $SL(2,C)$; two matrices $a(\boldsymbol{\zeta},\boldsymbol{\alpha})$ and $a(\boldsymbol{\zeta}',\boldsymbol{\alpha}')$ certainly differ if $\boldsymbol{\zeta} \neq \boldsymbol{\zeta}'$. If $\boldsymbol{\zeta} = \boldsymbol{\zeta}'$, they are equal only if $\boldsymbol{\alpha} = \boldsymbol{\alpha}'$ and $|\boldsymbol{\alpha}| = |\boldsymbol{\alpha}'| < 2\pi$, or if $|\boldsymbol{\alpha}| = |\boldsymbol{\alpha}'| = 2\pi$, in which event the directions of $\boldsymbol{\alpha}$ and $\boldsymbol{\alpha}'$ are irrelevant. Therefore, $(\boldsymbol{\zeta},\boldsymbol{\alpha})$ form a set of coordinates for $SL(2,C)$; a unique set of values is assigned to these coordinates for each element in $SL(2,C)$ except for those that correspond to negative-definite hermitian unimodular matrices, for then $\boldsymbol{\zeta}$ is determinate, $|\boldsymbol{\alpha}| = 2\pi$, but $\hat{\boldsymbol{\alpha}}$ is arbitrary. The same parameters thus appear as coordinates for both $SL(2,C)$ and $SO(3,1)$, except that the range is greater in the former case; we will see that $SL(2,C)$ has "twice as many elements" as $SO(3,1)$.

We recall now that the connection between spinors and vectors in the case of $SU(2)$ is essentially based on the following equation:

$$a(\mathbf{0},\boldsymbol{\alpha})^\dagger \sigma_j a(\mathbf{0},\boldsymbol{\alpha}) = A_{jk}(\boldsymbol{\alpha})\sigma_k\,. \tag{20.115}$$

We have taken advantage of the fact that the matrices $a(\mathbf{0},\boldsymbol{\alpha})$ describe elements of $SU(2)$ and are therefore unitary, so that $a(\mathbf{0},\boldsymbol{\alpha})^{-1}$ coincides with $a(\mathbf{0},\boldsymbol{\alpha})^\dagger$. Thus Eq. (20.115) is really the same as Eq. (17.146). Let us next compute the left-hand side of Eq. (20.115), after replacing, however, $a(\mathbf{0},\boldsymbol{\alpha})$ by $a(\boldsymbol{\zeta},\mathbf{0})$. Using the properties of the Pauli matrices, the result is definitely some linear combination of the σ's and the 2×2 unit matrix:

$$\begin{aligned} a(\boldsymbol{\zeta},\mathbf{0})^\dagger \sigma_j a(\boldsymbol{\zeta},\mathbf{0}) &\equiv a(\boldsymbol{\zeta},\mathbf{0})\sigma_j a(\boldsymbol{\zeta},\mathbf{0}) \\ &= \left(\cosh\frac{\zeta}{2} + \frac{\boldsymbol{\zeta}\cdot\boldsymbol{\sigma}}{\zeta}\sinh\frac{\zeta}{2}\right)\sigma_j\left(\cosh\frac{\zeta}{2} + \frac{\boldsymbol{\zeta}\cdot\boldsymbol{\sigma}}{\zeta}\sinh\frac{\zeta}{2}\right) \\ &= \left(\delta_{jk} + \zeta_j\zeta_k\frac{\cosh\zeta - 1}{\zeta^2}\right)\sigma_k + \zeta_j\frac{\sinh\zeta}{\zeta}\mathbb{1}\,. \end{aligned} \tag{20.116}$$

(We have simply used the multiplication properties of the Pauli matrices.) We see that the coefficients of the Lorentz transformation matrix $\Lambda^\mu{}_\nu(\boldsymbol{\zeta},\mathbf{0})$ have appeared. Let us also compute $a(\boldsymbol{\zeta},\mathbf{0})^\dagger \mathbb{1} a(\boldsymbol{\zeta},\mathbf{0})$:

$$a(\boldsymbol{\zeta},\mathbf{0})^\dagger \mathbb{1} a(\boldsymbol{\zeta},\mathbf{0}) = a(\boldsymbol{\zeta},\mathbf{0})a(\boldsymbol{\zeta},\mathbf{0}) = \cosh\zeta\cdot\mathbb{1} + \frac{\sinh\zeta}{\zeta}\zeta_j\sigma_j\,. \tag{20.117}$$

It is not hard to see that these two equations can be combined into one concise statement if we formally define a quartet of matrices $\sigma_\mu = (\sigma_j, \mathbb{1})$, namely

$\sigma_0 = \mathbb{1}$, and $\sigma^\mu = (\sigma_j, -\mathbb{1})$:

$$\begin{aligned} a(\boldsymbol{\zeta},\mathbf{0})^\dagger \sigma_\mu \sigma(\boldsymbol{\zeta},\mathbf{0}) &= \Lambda_\mu{}^\nu(\boldsymbol{\zeta},\mathbf{0})\sigma_\nu\,, \\ a(\boldsymbol{\zeta},\mathbf{0})^\dagger \sigma^\mu a(\boldsymbol{\zeta},\mathbf{0}) &= \Lambda^\mu{}_\nu(\boldsymbol{\zeta},\mathbf{0})\sigma^\nu\,. \end{aligned} \tag{20.118}$$

Let us combine this with the known effect on the σ^μ of an element from $SU(2)$:

$$a(\mathbf{0},\boldsymbol{\alpha})^\dagger \sigma^\mu a(\mathbf{0},\boldsymbol{\alpha}) = \Lambda^\mu{}_\nu(\mathbf{0},\boldsymbol{\alpha})\sigma^\nu\,. \tag{20.119}$$

[This is simply Eq. (20.115) again, combined with the unitarity of $a(\mathbf{0},\boldsymbol{\alpha})$]. Because $a(\boldsymbol{\zeta},\boldsymbol{\alpha}) = a(\boldsymbol{\zeta},\mathbf{0})a(\mathbf{0},\boldsymbol{\alpha})$ on the one hand, and $\Lambda^\mu{}_\nu(\boldsymbol{\zeta},\boldsymbol{\alpha}) = \Lambda^\mu{}_\rho(\boldsymbol{\zeta},\mathbf{0})$ $\Lambda^\rho{}_\nu(\mathbf{0},\boldsymbol{\alpha})$ on the other, we get, with no restrictions on $\boldsymbol{\zeta}$ and $\boldsymbol{\alpha}$:

$$a(\boldsymbol{\zeta},\boldsymbol{\alpha})^\dagger \sigma^\mu a(\boldsymbol{\zeta},\boldsymbol{\alpha}) = \Lambda^\mu{}_\nu(\boldsymbol{\zeta},\boldsymbol{\alpha})\sigma^\nu\,. \tag{20.120}$$

This is the extension of Eq. (20.115) that we need. It is a rather special property of the $SL(2,C)$ matrices and is not suggested directly by any general arguments of a group-theoretical nature, such as led to Eq. (20.113). Therefore, let us briefly motivate it from another point of view. Every traceless hermitian 2×2 matrix is a unique real linear combination of the σ_j and conversely. If we remove the trace condition, the 2×2 unit matrix must be included; then every hermitian 2×2 matrix is a unique real linear combination of the four matrices σ_μ, and conversely. If x^μ is a quartet of real numbers, which may be thought of as the components of some real space-time four vector, then $X = x^\mu \sigma_\mu$ is some hermitian $2{\times}2$ matrix, and conversely with any X we can associate a four-vector x^μ. Thus it is better to write $X(x)$, and we have a one-one correspondence between real four-vectors x^μ and hermitian $2{\times}2$ matrices $X(x)$. For a general $X(x)$ and any unimodular matrix $a \in SL(2,C)$, $a^\dagger X(x) a$ is another hermitian $2{\times}2$ matrix; by expanding it in terms of the σ_μ, it appears as $X(x')$ for some x'^μ, and x'^μ will be real linear combinations (dependent on a) of the x^ν:

$$\begin{aligned} X(x) &= x^\mu \sigma_\mu;\ a^\dagger X(x) a = X(x') = x'^\mu \sigma_\mu\,, \\ x'^\mu &= L_\nu{}^\mu(a) x^\nu\,. \end{aligned} \tag{20.121}$$

What sort of matrix is $||L_\nu{}^\mu(a)||$? Because a is unimodular, $X(x)$ and $a^\dagger X(x) a$ have the same determinant. But $X(x)$ is a 2×2 matrix, thus $\det X(x)$ must be bilinear in the x's:

$$\begin{aligned} \det X(x) = \det x^\mu \sigma_\mu = \begin{vmatrix} x^0 + x^3 & x^1 - ix^2 \\ x^1 + ix^2 & x^0 - x^3 \end{vmatrix} &= (x^0)^2 - (x^1)^2 - (x^2)^2 - (x^3)^2 \\ = -\, x^\mu x_\mu\,. \end{aligned} \tag{20.122}$$

Det $X(x)$ is simply the negative of the Lorentz square of the vector x^μ, and this is preserved in passing from $X(x)$ to $a^\dagger X(x) a$, that is, from x^μ to x'^μ. The matrix $||L_\nu{}^\mu(a)||$ gives a linear transformation that preserves the indefinite quadratic form of special relativity; thus it is some homogeneous Lorentz transformation! Using Λ in place of L, we can rewrite Eq. (20.121) in this way:

$$a \in SL(2,C):\ a^\dagger \sigma_\mu a = \Lambda_\mu{}^\nu(a)\sigma_\nu,\ \Lambda(a) \in SO(3,1)\,. \tag{20.123}$$

That $\Lambda(a)$ is actually in $SO(3,1)$ follows from the result $\Lambda_0{}^0(a) = Tra^\dagger a > 0$. Therefore we are led to expect that by means of this action of matrices a in the defining representation of $SL(2,C)$ on the σ_μ, we can associate some element of $SO(3,1)$ with each element in $SL(2,C)$.

Going back now to Eq. (20.120), we see that it establishes just this correspondence from $SL(2,C)$ to $SO(3,1)$. In the first place, because that equation is valid for all $\boldsymbol{\zeta}$ and $\boldsymbol{\alpha}$, it follows that all elements of $SO(3,1)$ do turn up in this correspondence as we allow $a(\boldsymbol{\zeta},\boldsymbol{\alpha})$ to vary over all of $SL(2,C)$. In the second place, it shows that if we use the parameters $\boldsymbol{\zeta},\boldsymbol{\alpha}$ for both groups, these parameters are preserved under the correspondence; that is, the element in $SO(3,1)$ that corresponds to $a(\boldsymbol{\zeta},\boldsymbol{\alpha})$ in $SL(2,C)$ is $\Lambda(\boldsymbol{\zeta},\boldsymbol{\alpha})$. Because the parameters go over different ranges in the two groups, the correspondence will not be one-one both ways. Two sets of parameters $(\boldsymbol{\zeta},\boldsymbol{\alpha})$ and $(\boldsymbol{\zeta},(|\boldsymbol{\alpha}|+2\pi)\boldsymbol{\alpha}/|\boldsymbol{\alpha}|)$ lead to different elements in $SL(2,C)$ but to the same one in $SO(3,1)$; the two matrices in $SL(2,C)$ differ but in sign. Thus $a(\boldsymbol{\zeta},\boldsymbol{\alpha})$ and $-a(\boldsymbol{\zeta},\boldsymbol{\alpha})$ both go into the same $\Lambda(\boldsymbol{\zeta},\boldsymbol{\alpha})$ and apart from this no other element in $SL(2,C)$ will be mapped into $\Lambda(\boldsymbol{\zeta},\boldsymbol{\alpha})$. Lastly, the group properties are preserved, namely, products go into products and inverses into inverses, as is easily checked with Eq. (20.123). Thus what we have here is a two-to-one homomorphism from $SL(2,C)$ onto $SO(3,1)$, which is exactly like the $SU(2)-R(3)$ relationship. Incidentally, by now it is clear that $SL(2,C)$ has the same Lie algebra as $SO(3,1)$; after all, the defining matrix representation of the former has been constructed using a matrix solution to the Lie algebra commutation rules of the latter. Because of the homomorphism existing between $SL(2,C)$ and $SO(3,1)$, every representation of the latter is also a representation of the former. But we can go further: $SL(2,C)$ happens to be the universal covering group of $SO(3,1)$, and every finite-dimensional matrix representation of the Lie algebra of $SL(2,C)$ leads to a representation of the entire group. There is now no need to impose extra global conditions. However, we know all the irreducible finite dimensional matrix solutions to the $SL(2,C)$ Lie algebra: they are simply the solutions labeled $(n/2,m/2)$ with $n,m=0,1,2,\cdots$ independently. Thus the finite dimensional irreducible matrix representations of $SL(2,C)$ can be also labeled $(n/2,m/2)$, with dimension $(n+1)(m+1)$. If $(-1)^{n+m}=+1$, it is also a representation of $SO(3,1)$, but if $(-1)^{n+m}=-1$, we have a representation of $SL(2,C)$ alone. The former ones are not faithful representations of $SL(2,C)$ because they do not distinguish the elements $(\boldsymbol{\zeta},\boldsymbol{\alpha})$ and $(\boldsymbol{\zeta},(|\boldsymbol{\alpha}|+2\pi)\boldsymbol{\alpha}/|\boldsymbol{\alpha}|)$; the latter ones do distinguish these elements and are faithful representations.

A quantity ξ with two complex components ξ_κ is called a spinor of type $(\frac{1}{2},0)$ with respect to $SL(2,C)$ if the element $(\boldsymbol{\zeta},\boldsymbol{\alpha})$ in $SL(2,C)$ subjects it to the linear transformation determined by the matrix $a(\boldsymbol{\zeta},\boldsymbol{\alpha})$:

$$\xi_\kappa \to \xi'_\kappa = a_{\kappa\lambda}(\boldsymbol{\zeta},\boldsymbol{\alpha})\xi_\lambda,\ \xi' = a(\boldsymbol{\zeta},\boldsymbol{\alpha})\xi\,. \tag{20.124}$$

The fundamental point is that with two spinors $\xi^{(1)},\xi^{(2)}$ of this type we can construct four quantities, $\xi^{(2)^\dagger}\sigma^\mu\xi^{(1)}$, that behave as a four-vector: this is the consequence of Eq. (20.120). We have:

$$
\begin{aligned}
(\xi^{(2)^\dagger}\sigma^\mu\xi^{(1)})' &\equiv \xi^{(2)'^\dagger}\sigma^\mu\xi^{(1)'} = \xi^{(2)^\dagger}a(\boldsymbol{\zeta},\boldsymbol{\alpha})^\dagger\sigma^\mu a(\boldsymbol{\zeta},\boldsymbol{\alpha})\xi^{(1)} \\
&= \Lambda^\mu{}_\nu(\boldsymbol{\zeta},\boldsymbol{\alpha})(\xi^{(2)^\dagger}\sigma^\nu\xi^{(1)})\,. \qquad (20.125)
\end{aligned}
$$

Thus if $\xi^{(1)}$ and $\xi^{(2)}$ transform as spinors of type $\left(\frac{1}{2},0\right)$, $\xi^{(2)^\dagger}\sigma^\mu\xi^{(1)}$ transform like a four-vector. (This does not imply that every four-vector can be actually constructed out of two spinors in this way!) Thus again spinors are like "half-vectors."

In addition to the law in Eq. (20.124), we can introduce two-component quantities that belong to the representation $\left(0,\frac{1}{2}\right)$. The generators M_j, N_j for this representation of $SL(2,C)$ and the matrix $a'(\boldsymbol{\zeta},\boldsymbol{\alpha})$ that stands for the element that is represented by $a(\boldsymbol{\zeta},\boldsymbol{\alpha})$ in $\left(\frac{1}{2},0\right)$ are given by simply reversing the sign of N_j in Eq. (20.111):

$$
\left(0,\frac{1}{2}\right):M_j = -\frac{i}{2}\sigma_j, N_j = -\frac{1}{2}\sigma_j\,;
$$

$$
a'(\boldsymbol{\zeta},\boldsymbol{\alpha}) = \exp(\boldsymbol{\zeta}\cdot\boldsymbol{N})\exp(\boldsymbol{\alpha}\cdot\boldsymbol{M}) = \exp(-\boldsymbol{\zeta}\cdot\boldsymbol{\sigma}/2)\exp(-i\boldsymbol{\alpha}\cdot\boldsymbol{\sigma}/2) \qquad (20.126)
$$

This matrix $a'(\boldsymbol{\zeta},\boldsymbol{\alpha})$ is, apart from a similarity transformation, simply the complex conjugate of $a(\boldsymbol{\zeta},\boldsymbol{\alpha})$. We see this by using the relation $\sigma_j^* = -C^{-1}\sigma_j C, C = i\sigma_2$:

$$
\begin{aligned}
a(\boldsymbol{\zeta},\boldsymbol{\alpha})^* &= \exp(\boldsymbol{\zeta}\cdot\boldsymbol{\sigma}^*/2)\exp(i\boldsymbol{\alpha}\cdot\boldsymbol{\sigma}^*/2) = C^{-1}\exp(-\boldsymbol{\zeta}\cdot\boldsymbol{\sigma}/2)\exp(-i\boldsymbol{\alpha}\cdot\boldsymbol{\sigma}/2)C\,, \\
a'(\boldsymbol{\zeta},\boldsymbol{\alpha}) &= Ca(\boldsymbol{\zeta},\boldsymbol{\alpha})^*C^{-1}\,. \qquad (20.127)
\end{aligned}
$$

We may therefore define a two-component quantity to belong to the representation $(0,\frac{1}{2})$ if it transforms via the matrix $a'(\boldsymbol{\zeta},\boldsymbol{\alpha})$ or equivalently, making a change of basis, via the matrix $a(\boldsymbol{\zeta},\boldsymbol{\alpha})^*$. It is simplest to choose the latter law. Thus a quantity η with two complex components $\eta_{\dot\kappa}$ is said to be a spinor of type $(0,\frac{1}{2})$ under $SL(2,C)$ if it follows the transformation law:

$$
\eta_{\dot\kappa} \to \eta'_{\dot\kappa} = \{a_{\kappa\lambda}(\boldsymbol{\zeta},\boldsymbol{\alpha})\}^*\eta_{\dot\lambda}, \eta' = a(\boldsymbol{\zeta},\boldsymbol{\alpha})^*\eta\,. \qquad (20.128)
$$

It is conventional to indicate by a dot $\cdot$ an index that transforms by the complex conjugate of the rule for an ordinary undotted index given in Eq. (20.124). There is, of course, no need to put dots above the indices appearing with a. We have, therefore, two types of two-dimensional spinors under $SL(2,C)$, the undotted $(\frac{1}{2},0)$ type and the dotted $(0,\frac{1}{2})$ type. These are distinct nonequivalent types. The vector $\xi^{(2)^\dagger}\sigma^\mu\xi^{(1)}$ constructed from two spinors of type $(\frac{1}{2},0)$ involves the complex conjugate of one of them, that is, something really belonging to the representation $(0,\frac{1}{2})$. This construction of a vector is, therefore, exactly equivalent to the statement that from two spinors, ξ of type $(\frac{1}{2},0)$ and η of type $(0,\frac{1}{2})$, we can form four quantities, $\eta^T\sigma^\mu\xi$, linear in each, that will transform as a four-vector when ξ is subjected to Eq. (20.124) and η to Eq. (20.128). Thus from one quantity belonging to the representation $(\frac{1}{2},0)$ and one to $(0, \frac{1}{2})$, we are able to construct a four-vector that belongs to the representation $(\frac{1}{2}, \frac{1}{2})$! This recipe generalizes, and for $SL(2,C)$ every representation $(n/2, m/2)$ can be obtained by consideration of spinors of higher rank, and more specifically, *symmetric* higher rank spinors. A general spinorial quantity ξ has several indices, some dotted and some not; if there are m of the former and n of the latter, its transformation law is:

$$\xi_{\kappa\kappa'\ldots,\dot{\mu}\dot{\mu}'\ldots} \to \xi'_{\kappa\kappa'\ldots,\dot{\mu}\dot{\mu}'\ldots} = \underbrace{a_{\kappa\lambda}a_{\kappa'\lambda'}\cdots}_{\leftarrow n \text{ factors} \rightarrow} \quad \underbrace{a^*_{\mu\nu}a^*_{\mu'\nu'}\cdots}_{\leftarrow m \text{ factors} \rightarrow} \quad \xi_{\lambda\lambda'\ldots,\dot{\nu}\dot{\nu}'\ldots} \tag{20.129}$$

Such a ξ generally has 2^{n+m} components because each index on ξ takes on two values. But those ξ's that are symmetric in the n undotted indices, and separately also symmetric in the m dotted ones, obviously transform by themselves and have only $(n+1)(m+1)$ independent components, again because each index can take on only two values. (There is, of course, no question of symmetry between a dotted and an undotted index, because they transform differently.) Such symmetric spinors supply us precisely with the matrix representation $(n/2, m/2)$ of $SL(2,C)$. Thus in going from $SU(2)$ to $SL(2,C)$ we remain with completely symmetric spinors only, but there are now two spinor types, (i.e., it suffices to deal with symmetric spinors in constructing all the representations of $SL(2,C)$).

As is to be expected, a representation $(n/2, m/2)$ of $SL(2,C)$ with $(-1)^{n+m} = +1$ given by a symmetric multispinor with n undotted and m dotted indices is the same as the representation of $SO(3,1)$ constructed from irreducible tensors of special type in the way described previously. The space of such spinors is "the same" as the space of such tensors. If $(-1)^{n+m} = -1$, the $SL(2,C)$ representation is provided by a multispinor with an odd total number of spinor indices; thus such a spinor changes sign under the element -1 of $SL(2,C)$. However, one frequently calls these faithful representations of $SL(2,C)$ "double-valued representations of $SO(3,\ 1)$," the true $SO(3,1)$ representations being termed "single-valued."

This concludes all that we wish to say regarding the finite-dimensional linear matrix representations of the groups $SO(3,1)$ and $SL(2,C)$. We comment very briefly on the infinite-dimensional representations. As we stated previously, all the nontrivial finite-dimensional representations of $SL(2,C)$ and $SO(3,1)$ are necessarily nonunitary, and to obtain unitary representations one has to consider infinite-dimensional ones. However, "most" of the infinite-dimensional irreducible representations are again nonunitary, only a "few" are unitary. Let us try to make these statements more precise. Going back to the finite-dimensional representations [of $SL(2,C)$, say], the representation $(n/2, m/2)$ could equally well be specified by giving the pair of numbers $\{(n-m)/2, (n+m)/2\}$. The significance of these numbers is the following: when the irreducible $SL(2,C)$ representation $(n/2, m/2)$ is restricted to $SU(2)$, we obtain several $SU(2)$ representations, namely those of dimensions $(|n-m|+1), (|n-m|+3), \ldots, (n+m+1)$ once each. The "spin values" of these are $|n-m|/2, (|n-m|/2)+1, \ldots (n+m)/2$. Therefore, the pair $\{(n-m)/2, (n+m)/2\}$ tells us the "spectrum of $SU(2)$ representations" appearing in the irreducible representation of $SL(2, C)$: there is a lowest spin value [the "smallest" $SU(2)$ representation present], then a sequence of spin values increasing in unit steps up to a maximum. (If $n \neq m$, $(n/2, m/2)$ and $(m/2, n/2)$ have the same $SU(2)$ spectrum; they are distinguished by the difference in sign of $(n-m)/2$ and $(m-n)/2$). The infinite-dimensional irreducible representations of $SL(2, C)$ are labeled in a somewhat similar fashion, by a pair $\{j_0, \rho\}$, where j_0 takes on one of the values $0, \frac{1}{2}, 1, \ldots$ *and* ρ *is any complex number.* The j_0 is like $|n-m|/2$, and it tells us what is the "smallest" $SU(2)$ representation occurring in the representation $\{j_0, \rho\}$ of $SL(2, C)$. Given j_0, the full set of $SU(2)$ representations that is present consists of those of dimensionalities $(2j_0+1), (2j_0+3), (2j_0+5), \ldots$, ad infinitum, once each; in other words, each spin value $j_0, j_0+1, j_0+2, \ldots \infty$ appears once, and those less than j_0 are absent. There is now no highest spin value: the representation of $SL(2, C)$ is infinite dimensional, and $(n+m)/2$ has been replaced by the complex parameter ρ. In a certain convention for defining ρ, it turns out that the representation $\{j_0, \rho\}$ is unitary only if ρ is pure imaginary, or if it is real and in the range $0 < \rho < 1$. If ρ takes on any other complex or real value, the representation is nonunitary! It is also always infinite dimensional unless ρ happens to be a real number whose magnitude equals j_0 or exceeds j_0 by some positive integer, $|\rho| - j_0 = 0, 1, 2, 3, \ldots$; in such a case, we simply get one of the finite-dimensional representations studied previously. A representation $\{j_0, \rho\}$ of $SL(2, C)$ is also a representation of $SO(3, 1)$ if j_0 is an integer or zero, but not if it is half an odd integer. Alternatively, one calls the former "single-valued" and the latter "double-valued" representations of $SO(3, 1)$.

Realizations of P Using Fields

In our discussion of canonical realizations of the Galilei group using field variables, given in the preceding chapter, we constructed a class of realizations for which the generators were closely related to the generators of unitary operator representations of $\mathcal{G}$ taken from quantum mechanics. We had a correspondence $A[\Phi, \pi] \to a(\boldsymbol{x}, \partial/\partial\boldsymbol{x})$ between the generators A of a canonical realization of $\mathcal{G}$ and local partial differential operators $a(\boldsymbol{x}, \partial/\partial\boldsymbol{x})$ whose commutators with one another gave an operator realization of the Lie algebra of $\mathcal{G}$. We now construct analogous realizations of $\mathcal{P}$. There are, however, two important differences compared with the previous situation. The first is that whereas all the operators $a(\boldsymbol{x}, \partial/\partial\boldsymbol{x})$ are local differential operators in the case of $\mathcal{G}$, we now have to deal with nonlocal operators. Let us indicate briefly why this is so: whereas the classical nonrelativistic energy-momentum relationship $E = p^2/2m$ leads in nonrelativistic quantum mechanics to a local operator $-\nabla^2/2m$ as the representative of the energy, the classical relativistic relation $E = (m^2+p^2)^{1/2}$ goes over in relativistic quantum mechanics to the nonlocal operator $(m^2 - \nabla^2)^{1/2}$. The action of this operator on a function $f(\boldsymbol{x})$ is defined by saying that the Fourier transform $\tilde{f}(\boldsymbol{k})$ of $f(\boldsymbol{x})$ gets multiplied by the factor $(m^2+k^2)^{1/2}$. We can write:

$$(m^2 - \nabla^2)^{1/2} f(\boldsymbol{x}) = (2\pi)^{-3} \int d^3k \int d^3y (m^2 + \boldsymbol{k}^2)^{1/2} \exp(i\boldsymbol{k}.(\boldsymbol{x}-\boldsymbol{y})) f(\boldsymbol{y})$$
$$= \int d^3y D(\boldsymbol{x}-\boldsymbol{y}) f(\boldsymbol{y}),$$
$$D(\boldsymbol{x}-\boldsymbol{y}) = (2\pi)^{-3} \int d^3k (m^2 + \boldsymbol{k}^2)^{1/2} \exp(i\boldsymbol{k}.(\boldsymbol{x}-\boldsymbol{y})) . \qquad (20.130)$$

Although this makes precise the meaning of acting with $(m^2 - \nabla^2)^{1/2}$ on a function $f(\boldsymbol{x})$, in many computations one need not make explicit use of this representation as long as every product of operators and functions is written with the factors in the proper sequence. When in doubt, one can fall back upon Eq. (20.130). The second difference from the case of $\mathcal{G}$ arises from the essential presence of factors of i in certain quantum-mechanical operators. For a classical canonical realization we need real phase-space functions for generators, and it is not obvious that in making use of a quantum mechanical operator representation to get a canonical realization all such factors of i can be eliminated consistently. In the case of $\mathcal{G}$ it turns out that this could be done essentially because the Hamiltonian H does not appear on the right-hand side of any basic PB relation for $\mathcal{G}$, but the situation is different in the present case. To set up a canonical realization of $\mathcal{P}$ based on a spinless field, we still need two pairs of canonically conjugate fields in order that the right PB relations may hold and the generators be real. With these remarks, we now describe these realizations of $\mathcal{P}$.

Let $\phi_\alpha(\boldsymbol{x}), \alpha$=1,2, be two fields with canonical conjugates $\pi_\alpha(\boldsymbol{x})$, subject to the PB relations:

$$\{\phi_\alpha(\boldsymbol{x}), \pi_\beta(\boldsymbol{y})\} = \delta_{\alpha\beta}\delta^{(3)}(\boldsymbol{x}-\boldsymbol{y}), \ \{\phi_\alpha, \phi_\beta\} = \{\pi_\alpha, \pi_\beta\} = 0 . \qquad (20.131)$$

The $E(3)$ generators are obtained by an obvious extension from the single component spinless case:

$$J_j = -\int d^3x \ \pi_\alpha(\boldsymbol{x}) \epsilon_{jkm} x_k \frac{\partial}{\partial x_m} \phi_\alpha(\boldsymbol{x}) ,$$
$$P_j = -\int d^3x \ \pi_\alpha(\boldsymbol{x}) \frac{\partial}{\partial x_j} \phi_\alpha(\boldsymbol{x}), \ \text{(sum on } \alpha!) . \qquad (20.132)$$

To construct K_j and H, it is necessary to introduce a two-dimensional antisymmetric matrix $||\epsilon_{\alpha\beta}||, \epsilon_{12} = -\epsilon_{21} = 1$. Then if we choose:

$$K_j = \frac{1}{2} \int d^3x \ \epsilon_{\alpha\beta} \pi_\alpha(\boldsymbol{x}) ((m^2 - \nabla^2)^{1/2} x_j + x_j (m^2 - \nabla^2)^{1/2}) \phi_\beta(\boldsymbol{x}) ,$$
$$H = \int d^3x \ \epsilon_{\alpha\beta} \pi_\alpha(\boldsymbol{x}) (m^2 - \nabla^2)^{1/2} \phi_\beta(\boldsymbol{x}) , \qquad (20.133)$$

we can show that we do get a realization of $\mathcal{P}$. Note specially the structure of K_j : $(m^2 - \nabla^2)^{1/2} x_j$ and $x_j (m^2 - \nabla^2)^{1/2}$ are not the same thing; an operator like $(m^2 - \nabla^2)^{1/2}$ acts on whatever function of $\boldsymbol{x}$ stands to its right. Let us now show as an example that $\{K_j, P_k\}$ is equal to $\delta_{jk}H$. From Eq. (20.132), we have:

$$\{P_k, \phi_\alpha(\boldsymbol{x})\} = \frac{\partial}{\partial x_k}\phi_\alpha(\boldsymbol{x}), \{P_k, \pi_\alpha(\boldsymbol{x})\} = \frac{\partial}{\partial x_k}\pi_\alpha(\boldsymbol{x}). \tag{20.134}$$

Thus we have:

$$\begin{aligned}
\{K_j, P_k\} &= -\frac{1}{2}\int d^3x\ \epsilon_{\alpha\beta}\{P_k, \pi_\alpha(\boldsymbol{x})((m^2 - \nabla^2)^{1/2}x_j + x_j(m^2 - \nabla^2)^{1/2})\phi_\beta(\boldsymbol{x})\} \\
&= -\frac{1}{2}\int d^3x\ \epsilon_{\alpha\beta}\left(\frac{\partial \pi_\alpha(\boldsymbol{x})}{\partial x_k}((m^2 - \nabla^2)^{1/2}x_j + x_j(m^2 - \nabla^2)^{1/2})\phi_\beta(\boldsymbol{x})\right. \\
&\quad \left. + \pi_\alpha(\boldsymbol{x})((m^2 - \nabla^2)^{1/2}x_j + x_j(m^2 - \nabla^2)^{1/2})\frac{\partial}{\partial x_k}\phi_\beta(\boldsymbol{x})\right) \\
&= -\frac{1}{2}\int d^3x\ \epsilon_{\alpha\beta}\left(\frac{\partial}{\partial x_k}(\pi_\alpha(\boldsymbol{x})((m^2 - \nabla^2)^{1/2}x_j + x_j(m^2 - \nabla^2)^{1/2})\phi_\beta(\mathbf{x}))\right. \\
&\quad \left. - \pi_\alpha(\boldsymbol{x})((m^2 - \nabla^2)^{1/2}\delta_{jk} + \delta_{jk}(m^2 - \nabla)^{1/2})\phi_\beta(\boldsymbol{x})\right) \\
&= \delta_{jk}H\,.
\end{aligned} \tag{20.135}$$

(The calculation is somewhat formal because, as usual, a surface term has been dropped.) In a similar way, we can compute $\{K_j, H\}$:

$$\begin{aligned}
\{K_j, H\} &= \frac{1}{2}\epsilon_{\alpha\beta}\epsilon_{\gamma\delta}\int d^3x d^3x'\{\pi_\alpha(\boldsymbol{x})((m^2 - \nabla^2)^{1/2}x_j + x_j(m^2 - \nabla^2)^{1/2}) \\
&\quad \times\phi_\beta(\boldsymbol{x}), \pi_\gamma(\boldsymbol{x}')(m^2 - \nabla'^2)^{1/2}\phi_\delta(\boldsymbol{x}')\} \\
&= \frac{1}{2}\epsilon_{\alpha\beta}\epsilon_{\gamma\delta}\int d^3x\ d^3x'(\pi_\alpha(\boldsymbol{x})(m^2 - \nabla'^2)^{1/2}\phi_\delta(\boldsymbol{x}') \\
&\quad \times\delta_{\beta\gamma}((m^2 - \nabla^2)^{1/2}x_j + x_j(m^2 - \nabla^2)^{1/2})\delta^{(3)}(\boldsymbol{x} - \boldsymbol{x}') \\
&\quad -\pi_\gamma(\boldsymbol{x}')((m^2 - \nabla^2)^{1/2}x_j + x_j(m^2 - \nabla^2)^{1/2})\phi_\beta(\boldsymbol{x}) \\
&\quad \times\delta_{\alpha\delta}(m^2 - \nabla'^2)^{1/2}\delta^{(3)}(\boldsymbol{x} - \boldsymbol{x}')) \\
&= \frac{1}{2}\epsilon_{\alpha\beta}\epsilon_{\gamma\delta}\int d^3x\ \delta_{\beta\gamma}\pi_\alpha(\boldsymbol{x})((m^2 - \nabla^2)^{1/2}x_j \\
&\quad +x_j(m^2 - \nabla^2)^{1/2})(m^2 - \nabla^2)^{1/2}\phi_\delta(\boldsymbol{x}) \\
&\quad -\frac{1}{2}\epsilon_{\alpha\beta}\epsilon_{\gamma\delta}\int d^3x'\delta_{\alpha\delta}\pi_\gamma(\boldsymbol{x}')(m^2 - \nabla'^2)^{1/2}((m^2 - \nabla'^2)^{1/2}x'_j \\
&\quad +x'_j(m^2 - \nabla'^2)^{1/2})\phi_\beta(\boldsymbol{x}') \\
&= \frac{1}{2}\int d^3x\ \pi_\alpha(\boldsymbol{x})(\epsilon^2)_{\alpha\beta}(x_j(m^2 - \nabla^2) - (m^2 - \nabla^2)x_j)\phi_\beta(\boldsymbol{x}) \\
&= -\frac{1}{2}\int d^3x\ \pi_\alpha(\boldsymbol{x})(\nabla^2 x_j - x_j\nabla^2)\phi_\alpha(\boldsymbol{x}) = P_j\,.
\end{aligned} \tag{20.136}$$

This exercise shows that the presence of the matrix $\epsilon_{\alpha\beta}$, with a square equal to -1, is essential; had it been replaced by the unit matrix in both K_j and H, Eq. (20.135) would still be valid; whereas in place of Eq. (20.136) we would have obtained $\{K_j, H\} = -P_j$! The PB relation $\{K_j, K_k\} = -\epsilon_{jkl}J_l$ can also be checked to be true, and once again the presence of the matrix ϵ in K_j turns out to be essential.

A correspondence analogous to Eq. (19.89) can now be set up. Each generator A constructed above has the form:

$$A[\phi, \pi] = \int d^3x\ \pi(\boldsymbol{x})a\left(\boldsymbol{x}, \frac{\partial}{\partial \boldsymbol{x}}\right)\phi(\boldsymbol{x}) \tag{20.137}$$

where $a(\boldsymbol{x}, \partial/\partial\boldsymbol{x})$ is a 2×2 matrix and at the same time either a local or a nonlocal operator acting on $\phi(\boldsymbol{x})$ which stands to its right. We have:

$$\begin{aligned} A &= J_j \to a = -(\boldsymbol{x}\times\boldsymbol{\nabla})_j, A = P_j \to a = -\nabla_j\,; \\ A &= K_j \to a = \frac{\epsilon}{2}[(m^2-\nabla^2)^{1/2}x_j + x_j(m^2-\nabla^2)^{1/2}]\,, \\ A &= H \to a = \epsilon(m^2-\nabla^2)^{1/2}\,. \end{aligned} \tag{20.138}$$

The real classical quantities A give a PB realization of the Lie algebra of $\mathcal{P}$ because the a's, via their commutation rules, give an operator representation of this Lie algebra. (In the quantum mechanical representation of $\mathcal{P}$ with Hermitian generators, there occur explicit factors of i in J_j and P_j, but none in K_j and H; to get the right Lie algebra relations, we have had to include the matrix ϵ in K_j and H, while removing i from J_j and P_j to obtain real generators!)

The above construction can be extended to cover multicomponent fields in two ways, exactly as in the case of $\mathcal{G}$. In the first instance we can consider a multicomponent field whose components are related to some (real) matrix representation of $R(3)$. We are led to include new "spin" terms in J_j, having a simple structure. Although in the corresponding situation in the preceding chapter no extra terms were needed in G_j (see Eq. (19.92)), now a change in J_j must be accompanied by a change in K_j! Let the matrices generating the representation of $R(3)$ be written $(S_j)_{rs}$, and let the basic field quantities be $\phi_{\alpha r}(\boldsymbol{x}), \pi_{\beta s}(\boldsymbol{x})$ obeying

$$\{\phi_{\alpha r}(\boldsymbol{x}), \pi_{\beta s}(\boldsymbol{y})\} = \delta_{\alpha\beta}\delta_{rs}\delta^{(3)}(\boldsymbol{x}-\boldsymbol{y})\,. \tag{20.139}$$

Then we get a realization of the Lie algebra of $\mathcal{P}$ if we define:

$$\begin{aligned} J_j &= \int d^3x\ \pi_{\alpha r}(\boldsymbol{x})[(S_j)_{rs} - \delta_{rs}(\boldsymbol{x}\times\boldsymbol{\nabla})_j]\phi_{\alpha s}(\boldsymbol{x})\,, \\ P_j &= -\int d^3x\ \pi_{\alpha r}(\boldsymbol{x})\frac{\partial}{\partial x_j}\phi_{\alpha r}(\boldsymbol{x})\,, \end{aligned}$$

$$\begin{aligned} K_j &= \int d^3x\, \epsilon_{\alpha\beta}\pi_{\alpha r}(\boldsymbol{x}) \left[\frac{1}{2}\delta_{rs}((m^2-\nabla^2)^{1/2}x_j + x_j(m^2-\nabla^2)^{1/2}) \right. \\ &\quad \left. - \epsilon_{jkl}(S_k)_{rs}\frac{\partial/\partial x_l}{m+(m^2-\nabla^2)^{1/2}}\right]\phi_{\beta s(\boldsymbol{x})}\,, \\ H &= \int d^3x\, \epsilon_{\alpha\beta}\pi_{\alpha r}(\boldsymbol{x})(m^2-\nabla^2)^{1/2}\phi_{\beta r}(\boldsymbol{x})\,. \end{aligned} \tag{20.140}$$

The terms induced in K_j by the "spin" terms in J_j are complicated and, again, nonlocal; to define the action of $[m+(m^2-\nabla^2)^{1/2}]^{-1}$ on $\phi_{\beta s}(\boldsymbol{x})$ we use the same method as in the case of the operator $(m^2-\nabla^2)^{1/2}$. We leave it to the reader to check that the PB relations of $\mathcal{P}$ are obeyed. This construction, too, is motivated by a quantum mechanical representation of $\mathcal{P}$, one that corresponds to a particle with spin.

The second kind of multicomponent field arises by considering any real matrix representation of $SO(3,1)$. This involves six real generator matrices, $(S_j)_{rs}$ and $(M_j)_{rs}$, subject to the commutation relations

$$[S_j, S_k]_- = -[M_j, M_k]_- = \epsilon_{jkl}S_l\,,\ [S_j, M_k]_- = \epsilon_{jkl}M_l\,. \tag{20.141}$$

This is Eq. (20.90); we have replaced M_j and N_j by S_j and M_j respectively. [If the representation of $SO(3,1)$ is real orthogonal, hence unitary, the generator matrices will be infinite dimensional and real antisymmetric.] Now we will have extra terms in J_j and K_j, but both of a simple form. With Eq. (20.139) now being valid for the appropriate set of values of the indices r,s, we have:

$$\begin{aligned} J_j &= \int d^3x\, \pi_{\alpha r}(\boldsymbol{x})[(S_j)_{rs} - \delta_{rs}(\boldsymbol{x}\times\boldsymbol{\nabla})_j]\phi_{\alpha s}(\boldsymbol{x}) \\ P_j &= -\int d^3x\, \pi_{\alpha r}(\boldsymbol{x})\frac{\partial}{\partial x_j}\phi_{\alpha r}(\boldsymbol{x})\,, \\ K_j &= \int d^3x\, \pi_{\alpha r}(\boldsymbol{x})\left[\epsilon_{\alpha\beta}\frac{\delta_{rs}}{2}((m^2-\nabla^2)^{1/2}x_j + x_j(m^2-\nabla^2)^{1/2}) \right. \\ &\quad \left. + \delta_{\alpha\beta}(M_j)_{rs}\right]\phi_{\beta s}(\boldsymbol{x})\,, \\ H &= \int d^3x\, \epsilon_{\alpha\beta}\pi_{\alpha r}(\boldsymbol{x})(m^2-\nabla^2)^{1/2}\phi_{\beta r}(\boldsymbol{x})\,. \end{aligned} \tag{20.142}$$

It is a matter of detailed verification to see that the term in K_j involving M_j is accompanied by $\delta_{\alpha\beta}$ rather than $\epsilon_{\alpha\beta}$. This solution for $\mathcal{P}$ is analogous to that given in Eq. (19.96) for $\mathcal{G}$. But as in that case, it can here be shown that this realization of $\mathcal{P}$ is canonically equivalent to one of the form of Eq. (20.140), in which only the matrices S_j appear! However, because the proof involves manipulation with rather complicated quantities, we omit the details.

In all these realizations of $\mathcal{P}$ the canonical transformations generated by J_j and P_j have the customary simple forms; whereas those generated by H and K_j are quite complicated and involve nonlocal operations. But the more important features of these realizations are (1) that to every $\phi_{\alpha r}$ there is by definition a canonically conjugate field $\pi_{\alpha r}$, (2) that H is not positive definite, and (3) the generators K_j and H are not integrals of suitable local densities. For these reasons, such realizations of $\mathcal{P}$ have not received much attention; thus we turn now to the more conventional field realizations of $\mathcal{P}$.

Realizations of $\mathcal{P}$ with Local Generators

The first such realization of $\mathcal{P}$ we describe corresponds to a free scalar field. The basic variable is a field $\phi(\boldsymbol{x}, t)$, which may also be written $\phi(x^\mu)$. Its behavior under the Euclidean group $E(3)$ is as usual. Under a change of inertial frame $S \to S' = \exp(\zeta_j k_j)S$ corresponding to a pure Lorentz transformation, if a given "state of the field" is described in S by a function $\phi(x^\mu)$ having a specific functional form, that same state is described in S' by ϕ' having a different functional form:

$$S' = \exp(\zeta_j k_j)S : \ \phi'(x'^\mu) = \phi(x^\mu), \ x'^\mu = \Lambda^\mu{}_\nu(\boldsymbol{\zeta}, \mathbf{0})x^\nu \,. \tag{20.143}$$

The change in functional form, $\phi(x^\mu) \to \phi'(x^\mu)$, is easily obtained from the above law. Now, the usual treatment proceeds from a Lagrangian that is written as the space integral of a Lagrangian density; by choosing the latter to transform like a scalar field also, one has a four-dimensional invariant form for the action. In this way the Lorentz invariance of the field equation of motion is ensured. We are interested here in starting with the Lagrangian and then determining the generators of the canonical realization of $\mathcal{P}$ that acts in the corresponding phase space.

Accordingly, we consider the Lagrangian

$$\begin{aligned} L(t) = \int d^3x \ \mathcal{L}(\boldsymbol{x}, t), \ \mathcal{L} &= -\frac{1}{2}\frac{\partial\phi}{\partial x^\mu}\frac{\partial\phi}{\partial x_\mu} - \frac{1}{2}m^2\phi^2 \\ &= \frac{1}{2}[\dot{\phi}^2 - (\boldsymbol{\nabla}\phi)^2 - m^2\phi^2] \,. \end{aligned} \tag{20.144}$$

(Recall the metric $g_{00} = -1$). The equation of motion that ensues is called the Klein-Gordon equation:

$$\frac{d}{dt}\frac{\delta L}{\delta\dot{\phi}} \equiv \frac{d}{dt}\dot{\phi} = \frac{\delta L}{\delta\phi} = \nabla^2\phi - m^2\phi, \left(\frac{\partial}{\partial x^\mu}\frac{\partial}{\partial x_\mu} - m^2\right)\phi(x) = 0 \,. \tag{20.145}$$

It is thus "manifestly Lorentz invariant", meaning that if ϕ is a solution of this equation, then a ϕ' obtained from ϕ by a Lorentz transformation acting according to Eq. (20.143) is also a solution. Let us now proceed to put this theory into the canonical form by introducing the canonically conjugate field and then constructing the generators of $\mathcal{P}$. In obtaining the latter we can take recourse to the action principle and the invariance and quasi-invariance properties of L under Lorentz transformations. The canonical conjugate to ϕ is $\pi = \dot{\phi}$, and the Hamiltonian is

$$\begin{aligned} H &= \int d^3x \ \pi\dot{\phi} - L = \int d^3x \ \mathcal{H}(\boldsymbol{x}, t) \,, \\ \mathcal{H}(\boldsymbol{x}, t) &= \frac{1}{2}(\pi^2 + (\boldsymbol{\nabla}\phi)^2 + m^2\phi^2) \,. \end{aligned} \tag{20.146}$$

Thus H is positive definite and is the integral of a local density. The linear and angular momenta have the expected structures. A direct way to derive these constants of motion is to utilize the invariance of L under $E(3)$. Thus, corresponding to a small spatial displacement causing the variation $\delta\phi = -\boldsymbol{a}\cdot\nabla\phi$, $|\boldsymbol{a}| << 1$, L is unchanged, and the action principle gives us the constant of motion:

$$\int d^3x\,\frac{\delta L}{\delta\dot\phi}\delta\dot\phi = -a_j\int d^3x\;\pi(\boldsymbol{x},t)\frac{\partial}{\partial x_j}\phi(\boldsymbol{x},t) = a_jP_j\,. \tag{20.147}$$

That determines P_j. Similarly, rotational invariance of L leads to the expected form of J_j. Turning now to a small pure Lorentz transformation, $(\boldsymbol{x},t)\to(\boldsymbol{x}-\boldsymbol{\zeta}t, t-\boldsymbol{\zeta}\cdot\boldsymbol{x})$ with $|\boldsymbol{\zeta}|\ll 1$, we are led to the variation $\delta\phi$ given by Eq. (20.143):

$$\delta\phi(\boldsymbol{x},t) = \phi(\boldsymbol{x}+\boldsymbol{\zeta}t, t+\boldsymbol{\zeta}\cdot\boldsymbol{x}) - \phi(\boldsymbol{x},t) = \boldsymbol{\zeta}\cdot(t\nabla\phi + \boldsymbol{x}\dot\phi)\,. \tag{20.148}$$

A quick calculation shows that (dropping surface terms) $L(t)$ is not invariant but quasi-invariant, the change δL being a total time derivative:

$$\begin{aligned}\delta L =& \delta\int d^3x\left[\frac{1}{2}\dot\phi^2 + \frac{1}{2}\phi\nabla^2\phi - \frac{1}{2}m^2\phi^2\right]\\ =& \int d^3x\left[\dot\phi\delta\dot\phi + \nabla^2\phi\cdot\delta\phi - m^2\phi\delta\phi\right]\\ =& \int d^3x\Big[\dot\phi(\boldsymbol{\zeta}\cdot\nabla\phi + t\boldsymbol{\zeta}\cdot\nabla\dot\phi + \boldsymbol{\zeta}\cdot\boldsymbol{x}\ddot\phi) + \nabla^2\phi(t\boldsymbol{\zeta}\cdot\nabla\phi + \boldsymbol{\zeta}\cdot\boldsymbol{x}\dot\phi)\\ & -m^2\phi(t\boldsymbol{\zeta}\cdot\nabla\phi + \boldsymbol{\zeta}\cdot\boldsymbol{x}\dot\phi)\Big]\\ =& \int d^3x[\dot\phi\boldsymbol{\zeta}\cdot\nabla\phi + \boldsymbol{\zeta}\cdot\boldsymbol{x}\dot\phi\ddot\phi + \boldsymbol{\zeta}\cdot\boldsymbol{x}\dot\phi\nabla^2\phi - m^2\boldsymbol{\zeta}\cdot\boldsymbol{x}\phi\dot\phi]\\ =& \frac{d}{dt}\int d^3x[-\frac{1}{2}m^2\phi^2\boldsymbol{\zeta}\cdot\boldsymbol{x} + \frac{1}{2}\dot\phi^2\boldsymbol{\zeta}\cdot\boldsymbol{x} - \frac{1}{2}\boldsymbol{\zeta}\cdot\boldsymbol{x}(\nabla\phi)^2]\,.\end{aligned} \tag{20.149}$$

The associated constant of motion is:

$$\begin{aligned}&\int d^3x\;\dot\phi(t\boldsymbol{\zeta}\cdot\nabla\phi + \boldsymbol{\zeta}\cdot\boldsymbol{x}\dot\phi) + \int d^3x[\frac{1}{2}m^2\phi^2 - \frac{1}{2}\dot\phi^2 + \frac{1}{2}(\nabla\phi)^2]\boldsymbol{\zeta}\cdot\boldsymbol{x}\\ &= -t\zeta_jP_j + \zeta_j\frac{1}{2}\int d^3x\;x_j[\dot\phi^2 + (\nabla\phi)^2 + m^2\phi^2]\,.\end{aligned} \tag{20.150}$$

This permits us to identify the coefficient of ζ_j in the second term as the relativistic moment K_j. Now all the generators of $\mathcal{P}$ for this system are obtained. We have the customary basic brackets between $\phi(\boldsymbol{x})$ and $\pi(\boldsymbol{x})$, and then we have:

$$\begin{aligned}J_j &= -\int d^3x\;\pi(\boldsymbol{x})\epsilon_{jkm}x_k\frac{\partial}{\partial x_m}\phi(\boldsymbol{x}),\; P_j = -\int d^3x\;\pi(\boldsymbol{x})\frac{\partial}{\partial x_j}\phi(\boldsymbol{x})\,;\\ K_j &= \int d^3x\;x_j\mathcal{H}(\boldsymbol{x}),\; H = \int d^3x\,\mathcal{H}(\boldsymbol{x})\,,\\ \mathcal{H}(\boldsymbol{x}) &= \frac{1}{2}[\pi^2(\boldsymbol{x}) + (\nabla\phi(\boldsymbol{x}))^2 + m^2\phi^2(\boldsymbol{x})]\,.\end{aligned} \tag{20.151}$$

The reader can satisfy himself that all the PB relations of $\mathcal{P}$ are obeyed. This field realization of $\mathcal{P}$ has now the attractive features that all the generators are integrals of corresponding densities and that H is positive definite; no nonlocality is involved in these expressions.

One might wonder whether generators just like Eq. (20.151) can be built up using a multicomponent field, preserving the positive definiteness of H and making the simplest alterations in J_j and K_j. However,one finds very quickly that this is possible only for *an infinite component field.* Let us demonstrate this. We have in mind some real matrix representation of $SO(3,1)$, with generator matrices $(S_j)_{rs}, (M_j)_{rs}$, and a set of field variables $\phi_r(\boldsymbol{x}), \pi_s(\boldsymbol{x})$ subject to the customary basic bracket rules. (As in Eq. (20.141), here again we use the notation S_j, M_j in place of M_j, N_j appearing in Eq. (20.90)). The desired modification of the generators in Eq. (20.151) of the scalar field case is given by:

$$\begin{aligned} J_j &= \int d^3x\ \pi_r(\boldsymbol{x})[(S_j)_{rs} - \delta_{rs}(\boldsymbol{x}\times\boldsymbol{\nabla})_j]\phi_s(\boldsymbol{x})\,, \\ P_j &= -\int d^3x\ \pi_r(\boldsymbol{x})\frac{\partial}{\partial x_j}\phi_r(\boldsymbol{x})\,, \\ K_j &= \int d^3x[x_j\mathcal{H}(\boldsymbol{x}) + \pi_r(\boldsymbol{x})(M_j)_{rs}\phi_s(\boldsymbol{x})],\ H = \int d^3x\mathcal{H}(\boldsymbol{x})\,, \\ \mathcal{H}(\boldsymbol{x}) &= \frac{1}{2}[\pi_r(\boldsymbol{x})\pi_r(\boldsymbol{x} + \boldsymbol{\nabla}\phi_r(\boldsymbol{x})\cdot\boldsymbol{\nabla}\phi_r(\boldsymbol{x}) + m^2\phi_r(\boldsymbol{x})\phi_r(\boldsymbol{x})]\,. \end{aligned} \tag{20.152}$$

In verifying the relation $\{K_j, H\} = P_j$, our experience with the scalar field case tells us that the required right-hand side is already produced by the bracket of the term in K_j involving $\mathcal{H}(\boldsymbol{x})$, with H; thus the bracket of the M_j term in K_j and H must vanish. That leads to:

$$\begin{aligned} 0 =&(M_j)_{rs}\int d^3x\int d^3x'\{\phi_s(\boldsymbol{x})\pi_r(\boldsymbol{x}), \pi_u(\boldsymbol{x}')\pi_u(\boldsymbol{x}') + \boldsymbol{\nabla}'\phi_u(\boldsymbol{x}')\cdot\boldsymbol{\nabla}'\phi_u(\boldsymbol{x}') \\ &+ m^2\phi_u(\boldsymbol{x}')\phi_u(\boldsymbol{x}')\} \\ =&(M_j)_{rs}2\int\int d^2xd^3x'(\pi_r(\boldsymbol{x})\delta_{su}\pi_u(\boldsymbol{x}')\delta^{(3)}(\boldsymbol{x}-\boldsymbol{x}') \\ &- \phi_s(\boldsymbol{x})\delta_{ru}(\boldsymbol{\nabla}'\phi_u(\boldsymbol{x}')\cdot\boldsymbol{\nabla}'\delta^{(3)}(\boldsymbol{x}-\boldsymbol{x}') + m^2\phi_u(\boldsymbol{x}')\delta^{(3)}(\boldsymbol{x}-\boldsymbol{x}'))) \\ =&2(M_j)_{rs}\int d^3x[\pi_r(\boldsymbol{x})\pi_s(\boldsymbol{x}) - \boldsymbol{\nabla}\phi_r(\boldsymbol{x})\cdot\boldsymbol{\nabla}\phi_s(\boldsymbol{x}) - m^2\phi_r(\boldsymbol{x})\phi_s(\boldsymbol{x})]\,. \end{aligned} \tag{20.153}$$

Because the coefficient of $(M_j)_{rs}$ is symmetric in r and s, the only way that this can vanish identically is for the matrix M_j to be antisymmetric. But in that case, the representation of $SO(3,1)$ involved is not only real, it is by orthogonal matrices. In other words, it is a unitary infinite-dimensional one. Given that it is so, there is then no difficulty in checking that all the remaining PB relations for $\mathcal{P}$ are obeyed by the set of quantities in Eq. (20.152). Using this Hamiltonian, we see that each component $\phi_r(\boldsymbol{x}, t)$ obeys the Klein-Gordon equation by itself.

Therefore, there is no generalization of the scalar field realization of $\mathcal{P}$, given by Eq. (20.151), to a multicomponent field *with a finite number of components*, such that H remains the integral of a local positive definite density and such that *each component of ϕ possesses its own independent canonically conjugate field.* What has to be given up in generalizing to a field with a finite number of components is this last mentioned property: whereas several components ϕ_r may be needed so as to transform in a linear fashion according to some finite dimensional (real) matrix representation of $SO(3,1)$ under the transformations of the Lorentz group, only a subset of these components forms a set of independent dynamical quantities from the Hamiltonian viewpoint, possessing independent canonical conjugates; the remaining components are determined in terms of the previous ones and (possibly) their momentum densities.

We illustrate this situation shortly by describing the case of a vector field, but let us here say a few words about introducing interactions either in the case of a scalar field or the infinite component field. For the non-interacting situation the generators of $\mathcal{P}$ are given by Eqs. (20.151) or (20.152) as the case may be. Now by convention P_j and J_j are unchanged in the presence of interactions. We have already remarked that by virtue of the PB relation $\{K_j, P_k\} = \delta_{jk}H$, it is not possible to have the Hamiltonian contain an interaction term without at the same time K_j also having a corresponding term. Interaction terms of a suitable kind must be added to both H and K_j. But the only such addition that would preserve the manifest covariance of the Lagrangian density, and thus of the equations of motion, is essentially to replace the mass term $m^2\phi^2$ in $\mathcal{L}(\boldsymbol{x},t)$ by $(m^2\phi^2 + V\{\phi\})$ where $V\{\phi\}$ is some local function of ϕ constructed to look like and behave like a scalar field! Such a term is called an interaction Lagrangian. Its inclusion in L does not alter the previous definition of π, and also does not alter the expressions for J_j and P_j. It changes the energy density $\mathcal{H}(\boldsymbol{x})$ appearing in Eqs. (20.151) or (20.152) by having an additional term $V\{\phi\}$, but the expressions for K_j and H in terms of $\mathcal{H}(\boldsymbol{x})$ remain unaffected. A short calculation shows that the addition of this term to $\mathcal{H}(\boldsymbol{x})$ does no harm to the PB relations among the generators, the essential reason being that V has no dependence on $\partial_\mu\phi$.

(A Lagrangian treatment of this system composed of an infinite component field is not of much value, for roughly the same reason that we did not give a Lagrangian treatment of the free particle with spin!)

Realization by Free Massive Vector Field

We describe now the realization of $\mathcal{P}$ corresponding to a field made up of four components transforming like a four-vector under Lorentz transformations. We consider the case of finite "mass"; that is, the equation of motion is simply the Klein-Gordon Eq. (20.145) with a nonzero m^2 for each field component. This example indicates how it happens that not all field components are independent canonical variables, although they are formally needed to state the transformation law under the Lorentz group in a simple way. As mentioned earlier, this is a characteristic property of finite multicomponent fields giving rise to realizations of $\mathcal{P}$, when we work from an action integral having a local invariant density, and insist on having a positive definite Hamiltonian.

Let us first fix the transformation law for a vector field, that is, set up the analogue to Eq. (20.143). A particular "state of the field" as observed in an inertial frame S is given by specifying the values of four space-time functions $\phi^\mu(x)$. In another frame, $S' = \Lambda S$, $\Lambda \in SO(3,1)$, this same physical situation is described by four functions ϕ'^μ having generally different functional dependences on their arguments:

$$S' = \Lambda S : \phi'^\mu(x') = \Lambda^\mu{}_\nu \phi^\nu(x), \ x'^\alpha = \Lambda^\alpha{}_\beta x^\beta . \tag{20.154}$$

(In connection with this equation and Eq. (20.143), it is perhaps better to say that they relate two different descriptions of a given *state of motion* of the field rather than of a given *state*, because the latter term in Hamiltonian mechanics refers to conditions at one instant of time, and we have here in general a change in the meaning of simultaneity.) This equation specifies the change in functional forms, $\phi^\mu(x) \to \phi'^\mu(x)$. It is customary to introduce, along with $\phi_\mu(x)$, its "four-dimensional curl" defined as:

$$G_{\mu\nu}(x) = \frac{\partial \phi_\nu(x)}{\partial x^\mu} - \frac{\partial \phi_\mu(x)}{\partial x^\nu} . \tag{20.155}$$

The equation that we endeavor to obtain from a Lagrangian is:

$$\frac{\partial G_{\mu\nu}(x)}{\partial x_\mu} - m^2 \phi_\nu(x) = 0 . \tag{20.156}$$

Using Eq. (20.155) and the antisymmetry of $G_{\mu\nu}$ in μ and ν (and also the assumption $m^2 \neq 0$) would immediately lead to the Klein-Gordon equation for each $\phi_\nu(x)$ together with a "subsidiary relation" involving the first derivatives of these quantities:

$$\frac{\partial \phi_\nu}{\partial x_\nu} = \frac{1}{m^2} \frac{\partial^2 G_{\mu\nu}}{\partial x_\mu \partial x_\nu} = 0, \ \left(\frac{\partial}{\partial x^\mu} \frac{\partial}{\partial x_\mu} - m^2 \right) \phi_\nu = 0 . \tag{20.157}$$

[The Klein-Gordon equations alone, without the subsidiary relation, would have led to an indefinite expression for the Hamiltonian.]

We adopt here a Lagrangian that is linear in the velocities. With suitable modifications, the theory given in Chapter 9 for the treatment of such Lagrangians is applicable. Starting with the ten field functions $\phi_\nu(x), G_{\mu\nu}(x)$ as independent dynamical variables, it is possible to choose the Lagrangian density in such a way that both the connection between $G_{\mu\nu}$ and ϕ_ν given by Eq. (20.155) and the desired Eq. (20.156) become consequences of the basic Lagrangian equations of motion. Because our basic object is to discover the true independent variables between which the fundamental PB's must be imposed and also to determine the generators of $\mathcal{P}$, we distinguish between the time and space components in our equations. The Lagrangian suitable for our purposes is:

$$L(t) = \int d^3x\, \mathcal{L}(\boldsymbol{x}, t)\,,$$
$$\mathcal{L} = -\frac{1}{2}G^{\mu\nu}\left(\frac{\partial\phi_\nu}{\partial x^\mu} - \frac{\partial\phi_\mu}{\partial x^\nu}\right) + \frac{1}{4}G^{\mu\nu}G_{\mu\nu} - \frac{1}{2}m^2\phi^\nu\phi_\nu\,. \quad (20.158)$$

As in the case of the scalar field, $\mathcal{L}$ is constructed so that it explicitly transforms like a scalar field under Lorentz transformations, given that ϕ_μ and $G_{\mu\nu}$ behave as vector and (antisymmetric) tensor fields, respectively. Thus the "manifest Lorentz covariance" of the equations of motion is built in from the start. Let us now introduce the following notations for the different components of $G_{\mu\nu}$:

$$G_{k0} = \mathcal{E}_k,\; G_{jk} = \epsilon_{jkl}\mathcal{B}_l\,. \quad (20.159)$$

($\mathcal{E}_k$ and $\mathcal{B}_k$ are like electric and magnetic fields.) Then the Lagrangian density is:

$$\mathcal{L} = -\; \mathcal{E}_k\dot{\phi}_k + \mathcal{E}_k\partial_k\phi_0 - \epsilon_{klm}\mathcal{B}_k\partial_l\phi_m + \frac{1}{2}(\mathcal{B}_k\mathcal{B}_k - \mathcal{E}_k\mathcal{E}_k)$$
$$-\; \frac{m^2}{2}(\phi_k\phi_k - \phi_0^2)\,. \quad (20.160)$$

At this point the basic variables are $\mathcal{E}_k(\boldsymbol{x}, t)$, $\mathcal{B}_k(\boldsymbol{x}, t)$, $\phi_k(\boldsymbol{x}, t)$ and $\phi_0(\boldsymbol{x}, t)$. The Lagrangian equations of motion follow:

$$\frac{d}{dt}\frac{\delta L}{\delta\dot{\mathcal{E}}_k} - \frac{\delta L}{\delta\mathcal{E}_k} = 0 \to \dot{\phi}_k = \partial_k\phi_0 - \mathcal{E}_k\,; \quad (20.161a)$$
$$\frac{d}{dt}\frac{\delta L}{\delta\dot{\mathcal{B}}_k} - \frac{\delta L}{\delta\mathcal{B}_k} = 0 \to \mathcal{B}_k = \epsilon_{klm}\partial_l\phi_m\,; \quad (20.161b)$$
$$\frac{d}{dt}\frac{\delta L}{\delta\dot{\phi}_k} - \frac{\delta L}{\delta\phi_k} = 0 \to \dot{\mathcal{E}}_k = m^2\phi_k + \epsilon_{klm}\partial_l\mathcal{B}_m\,; \quad (20.161c)$$
$$\frac{d}{dt}\frac{\delta L}{\delta\dot{\phi}_0} - \frac{\delta L}{\delta\phi_0} = 0 \to \phi_0 = \frac{1}{m^2}\partial_k\mathcal{E}_k\,. \quad (20.161d)$$

Equations (20.161a) and (20.161b) are the same as Eq. (20.155), Eq. (20.161c) and (20.161d) the same as Eq. (20.156), so that we have indeed obtained the desired system of equations from the Lagrangian. Now, neither L nor any of the equations of motion contains the time derivatives of the variables $\mathcal{B}_k(\boldsymbol{x}, t)$ and $\phi_0(\boldsymbol{x}, t)$. Instead, the Lagrangian equations "corresponding" to these, namely Eqs. (20.161b) and (20.161d), turn out to be equations that determine these variables explicitly at each instant of time in terms of the other variables $\mathcal{E}_k, \phi_k$. It is in fact only for $\mathcal{E}_k$ and ϕ_k that we have genuine equations of motion. The treatment of Lagrangians linear in the velocities given in Chapter 9 is not wide enough to cover this kind of situation; there we assumed that the Lagrangian equations of motion determined the time derivative of each generalized coordinate as some expression involving the coordinates. It can now be seen that because Eq. (20.161b) and (20.161d) *are* consequences of the principle of stationary action, we may use them to eliminate the coordinates $\mathcal{B}_k$ and ϕ_0 from the original Lagrangian L, obtaining thereby a new Lagrangian L', which is a functional only of $\mathcal{E}_k$, ϕ_k, $\dot{\mathcal{E}}_k$, and $\dot{\phi}_k$; we may rest assured that the Lagrangian equations of motion following from L' coincide with Eqs. (20.161a) and (20.161c). (We leave a proof of this statement to the reader.) Carrying out this substitution, we obtain:

$$\begin{aligned} L'(t) &= \int d^3x \; \mathcal{L}'(\boldsymbol{x},t), \; \mathcal{L}' = -\mathcal{E}_k\dot{\phi}_k - \mathcal{H}(\boldsymbol{x},t)\,, \\ \mathcal{H}(\boldsymbol{x},t) &= \frac{1}{2}\left[\mathcal{E}_k\mathcal{E}_k + \mathcal{B}_k\mathcal{B}_k + m^2\phi_k\phi_k + \frac{1}{m^2}(\partial_k\mathcal{E}_k)^2\right], \\ \mathcal{B}_k &\equiv \epsilon_{klm}\partial_l\phi_m\,. \end{aligned} \tag{20.162}$$

Note that now we have used $\mathcal{B}_k$ only for ease in writing, as it is a shorthand for the curl of the vector $\boldsymbol{\phi}$. A comparison of Eqs. (20.158) and (20.162) illustrates very well just the point we wish to make. The former is in a form that makes relativistic invariance "manifest", and in order to achieve this we had to use a sufficient number of fields so that the transformation law under the Lorentz group could be stated in a simple way, using the finite-dimensional matrix representations of the Lorentz group corresponding to four-vectors and antisymmetric tensors. But some of the "equations of motion" that followed from this Lagrangian served only to express some of these field components in terms of the others. Eliminating these field components led to the "reduced" Lagrangian L' in Eq. (20.162), involving the independent variables $\mathcal{E}_k$ and ϕ_k, for which this Lagrangian gives genuine equations of motion. The Lagrangian L' is not the integral of a manifestly relativistic scalar density, but the relativistic invariance has not been lost, as we confirm by constructing all the generators of $\mathcal{P}$.

To L' we can apply the theory developed in Chapter 9, which leads to the fundamental PB relations:

$$\{\mathcal{E}_j(\boldsymbol{x}), \phi_k(\boldsymbol{x}')\} = \delta_{jk}\delta^{(3)}(\boldsymbol{x}-\boldsymbol{x}'),\; \{\mathcal{E}_j(\boldsymbol{x}), \mathcal{E}_k(\boldsymbol{x}')\} = \{\phi_j(\boldsymbol{x}), \phi_k(\boldsymbol{x}')\} = 0\,. \tag{20.163}$$

It is clear that the space integral of $\mathcal{H}(\boldsymbol{x},t)$ in Eq. (20.162) is the Hamiltonian of the system, and the reader can satisfy himself that the Hamiltonian equations of motion,

$$\dot{\mathcal{E}}_k = \{\mathcal{E}_k, H\},\; \dot{\phi}_k = \{\phi_k, H\},\; H = \int d^3x\; \mathcal{H}(\boldsymbol{x},t)\,, \tag{20.164}$$

give back Eqs. (20.161a) and (20.161c). What is left is to identify the remaining generators P_j, J_j, and K_j of $\mathcal{P}$. Because we have explicit invariance of L' under $E(3)$, the forms of P_j and J_j are easy to derive. For a small displacement in space by a vector $\boldsymbol{a}, |\boldsymbol{a}| << 1$, we utilize the invariance of L' under the variation $\delta\mathcal{E}_k = \boldsymbol{a}\cdot\nabla\mathcal{E}_k, \delta\phi_k = \boldsymbol{a}\cdot\nabla\phi_k$ to obtain from the action principle the conservation of

$$\int d^3x \frac{\delta L'}{\delta\dot{\phi}_k(\boldsymbol{x},t)}\delta\phi_k(\boldsymbol{x},t) = -a_j\int d^3x\, \mathcal{E}_k(\boldsymbol{x},t)\frac{\partial}{\partial x_j}\phi_k(\boldsymbol{x},t) \equiv -a_jP_j\,. \quad (20.165)$$

That determines P_j. Under an infinitesimal rotation with parameters $\alpha_j, |\boldsymbol{\alpha}| \ll 1$, the variables $\mathcal{E}_k$ and ϕ_k change their arguments and at the same time are transformed as vectors, leading to the variation:

$$\delta\mathcal{E}_k = \epsilon_{jlm}\alpha_j x_l\partial_m\mathcal{E}_k - \epsilon_{kjl}\alpha_j\mathcal{E}_l, \delta\phi_k = \epsilon_{jlm}\alpha_j x_l\partial_m\phi_k - \epsilon_{kjl}\alpha_j\phi_l\,. \quad (20.166)$$

Again L' is invariant; thus we have conservation of

$$\int d^3x\frac{\delta L'}{\delta\dot{\phi}_k(\boldsymbol{x},t)}\delta\phi_k(\boldsymbol{x},t) = -\alpha_jJ_j\,,$$

$$J_j = \int d^3x\,\mathcal{E}_k(\boldsymbol{x},t)[\delta_{kl}(\boldsymbol{x}\times\nabla)_j - (S_j)_{kl}]\phi_l(\boldsymbol{x},t),\ (S_j)_{kl} = -\epsilon_{jkl}\,. \quad (20.167)$$

We have introduced here the three antisymmetric matrices S_j that generate the three-dimensional defining representation of $R(3)$: see Chapter 17, page 331. Thus J_j is determined. Finally, we consider K_j. We expect that under the corresponding variation $\delta\mathcal{E}_k$, $\delta\phi_k$, L' will only be quasi-invariant. Let the parameters of an infinitesimal Lorentz transformation be $\zeta_j, |\boldsymbol{\zeta}| \ll 1$, so that under it $(\boldsymbol{x},t)\to(\boldsymbol{x}-\boldsymbol{\zeta}t, t-\boldsymbol{\zeta}\cdot\boldsymbol{x})$. The transformation law Eq. (20.154) fixes the form of the variation to be considered. Working to first order in ζ_j, we have in the first instance:

$$\begin{aligned}
\delta\phi_0 &\equiv \phi_0'(\boldsymbol{x},t) - \phi_0(\boldsymbol{x},t) = \zeta_k\phi_k(\boldsymbol{x},t) + d_\zeta\phi_0(\boldsymbol{x},t)\,,\\
\delta\phi_k &\equiv \phi_k'(\boldsymbol{x},t) - \phi_k(\boldsymbol{x},t) = \zeta_k\phi_0(\boldsymbol{x},t) + d_\zeta\phi_k(\boldsymbol{x},t)\,,\\
d_\zeta &\equiv t\boldsymbol{\zeta}\cdot\nabla + \boldsymbol{\zeta}\cdot\boldsymbol{x}\frac{\partial}{\partial t}\cdot
\end{aligned} \quad (20.168)$$

All the additional variations $\delta\mathcal{E}_k, \delta\mathcal{B}_k, \delta\dot{\phi}_k, \delta\nabla\cdot\boldsymbol{\mathcal{E}}$ needed in computing $\delta L'$ are easily obtained from here. After substituting for ϕ_0 the full set is:

$$\begin{aligned}
\delta\mathcal{E}_k &= d_\zeta\mathcal{E}_k + (\boldsymbol{\zeta}\times\boldsymbol{\mathcal{B}})_k,\ \delta\phi_k = d_\zeta\phi_k + \frac{\zeta_k}{m^2}\nabla\cdot\boldsymbol{\mathcal{E}}\,,\\
\delta\dot{\phi}_k &= d_\zeta\dot{\phi}_k + \boldsymbol{\zeta}\cdot\nabla\phi_k + \frac{\zeta_k}{m^2}\nabla\cdot\dot{\boldsymbol{\mathcal{E}}}\,,\\
\delta\mathcal{B}_k &= d_\zeta\mathcal{B}_k + (\boldsymbol{\zeta}\times\dot{\boldsymbol{\phi}})_k - \frac{1}{m^2}(\boldsymbol{\zeta}\times\nabla)_k\nabla\cdot\boldsymbol{\mathcal{E}}\,,\\
\delta\nabla\cdot\boldsymbol{\mathcal{E}} &= d_\zeta\nabla\cdot\boldsymbol{\mathcal{E}} + \boldsymbol{\zeta}\cdot\dot{\boldsymbol{\mathcal{E}}} - \boldsymbol{\zeta}\cdot\nabla\times\boldsymbol{\mathcal{B}}\,.
\end{aligned} \quad (20.169)$$

Now the computation of $\delta L'$ is straightforward. After some cancellations and dropping of surface terms, the result is:

$$\delta L' = \frac{d}{dt}\left[\zeta_k\int d^3x\,x_k(-\mathcal{E}_j(\boldsymbol{x},t)\dot{\phi}_j(\boldsymbol{x},t) - \mathcal{H}(\boldsymbol{x},t)) - \frac{\zeta_k}{m^2}\int d^3x\,\mathcal{E}_k\nabla\cdot\boldsymbol{\mathcal{E}}\right]\,. \quad (20.170)$$

This quasi-invariance of L' guarantees that the equations of motion are Lorentz invariant. The associated constant of motion is

$$\int d^3x \, \frac{\delta L'}{\delta \dot{\phi}_k(\boldsymbol{x},t)} \delta\phi_k(\boldsymbol{x},t) + \zeta_k \int d^3x \, x_k(\mathcal{E}_j\dot{\phi}_j + \mathcal{H})$$
$$+ \frac{\zeta_k}{m^2} \int d^3x \, \mathcal{E}_k \boldsymbol{\nabla}\cdot\boldsymbol{\mathcal{E}} = \zeta_j K_j - t\zeta_j P_j \,,$$
$$K_j = \int d^3x \; x_j \mathcal{H}(\boldsymbol{x},t) \,. \tag{20.171}$$

All the generators of $\mathcal{P}$ can now be collected. We drop the argument t in the variables, as we must, and we have the full set:

$$J_j = \int d^3x \, \mathcal{E}_k(\boldsymbol{x})[\delta_{kl}(\boldsymbol{x}\times\boldsymbol{\nabla})_j - (S_j)_{kl}]\phi_l(\boldsymbol{x}) \,,$$
$$P_j = \int d^3x \, \mathcal{E}_k(\boldsymbol{x}) \frac{\partial}{\partial x_j}\phi_k(\boldsymbol{x}) \,,$$
$$K_j = \int d^3x \, x_j \mathcal{H}(\mathbf{x}), \; H = \int d^3x \, \mathcal{H}(\boldsymbol{x}) \,;$$
$$\mathcal{H}(\boldsymbol{x}) = \frac{1}{2}\left[(\boldsymbol{\mathcal{E}}(\boldsymbol{x}))^2 + (\boldsymbol{\nabla}\times\boldsymbol{\phi}(\boldsymbol{x}))^2 + m^2(\boldsymbol{\phi}(\boldsymbol{x}))^2 + \frac{1}{m^2}(\boldsymbol{\nabla}\cdot\boldsymbol{\mathcal{E}}(\boldsymbol{x}))^2\right] . \tag{20.172}$$

By using the basic PB relations between $\mathcal{E}_j$ and ϕ_k, all the PB relations of $\mathcal{P}$ can be verified to be true. [The apparent difference in sign between J_j and P_j above, and in the corresponding expressions in, say, Eq. (20.152), is explained by the fact that here ϕ_k is the "momentum" canonically conjugate to the field $\mathcal{E}_k$, as is clear from Eq. (20.163).] We have here a realization of $\mathcal{P}$ in which each generator is the spatial integral of a suitable local density, and furthermore the energy density $\mathcal{H}$ (and so also the total energy H) is positive definite. As a consequence, the energy-momentum four-vector $P^\mu = (H, P_k)$ is guaranteed to be positive timelike. In this final form only those variables that are elements of canonical pairs, that is, which possess independent canonical conjugates, appear; this makes the structure transparent from the Hamiltonian viewpoint. But the effect of a pure Lorentz transformation on these canonical variables $\mathcal{E}_k(\boldsymbol{x}), \phi_k(\boldsymbol{x})$, if expressed in terms of these variables alone, would certainly look ungainly; it is to make this effect appear simple that one must introduce the additional quantities $\phi_0(\boldsymbol{x}), \mathcal{B}_k(\boldsymbol{x})$ which are not supplied with canonical conjugates of their own!

Classical Electromagnetic Field as a Realization of $\mathcal{P}$.

We conclude this study of examples of realizations of $\mathcal{P}$ by discussing the Hamiltonian formulation of Maxwell's electrodynamics and the realization of $\mathcal{P}$ corresponding to the free electromagnetic field. The electromagnetic and the gravitational fields together constitute the two most important fundamental fields in classical physics, and as is well known, the structure of the field equations of electrodynamics is at the root of the special theory of relativity.

Unfortunately the scope of this book does not permit a detailed study of the classical gravitational field.

The electromagnetic field is, roughly speaking, the limit as $m \to 0$ of the massive vector field described previously. However, this is a singular limit, as evidenced by the presence of terms in the generators of $\mathcal{P}$ in Eq. (20.172) in which m appears in the denominator. Therefore, the problems of determining the true dynamical variables of the system and the setting up of fundamental PB's are special ones, and some care must be exercised. All this is bound up with the invariance property of electrodynamics called gauge invariance - there is no such thing in the case of the massive vector field. In the process of choosing dynamical variables and setting up the Hamiltonian, we allow for the interaction of the electromagnetic field with an externally prescribed electromagnetic current, because this makes it easier to understand what is going on. In determining the generators of $\mathcal{P}$, we shall revert to the free electromagnetic field, setting the external source equal to zero.

Maxwell's equations in the presence of prescribed sources split up, as is well-known, into two sets: one set is source independent, implying general restrictions on possible electric and magnetic fields $\boldsymbol{E}(\boldsymbol{x},t)$, $\boldsymbol{B}(\boldsymbol{x},t)$, whereas the other set involves the sources explicitly. The two sets of equations are:

$$\boldsymbol{\nabla} \times \boldsymbol{E} + \frac{\partial \boldsymbol{B}}{\partial t} = 0,\ \boldsymbol{\nabla} \cdot \boldsymbol{B} = 0\,; \tag{20.173a}$$

$$\boldsymbol{\nabla} \times \boldsymbol{B} - \frac{\partial \boldsymbol{E}}{\partial t} = \boldsymbol{j},\ \boldsymbol{\nabla} \cdot \boldsymbol{E} = \rho\,. \tag{20.173b}$$

Here, $\rho(\boldsymbol{x},t)$ and $\boldsymbol{j}(\boldsymbol{x},t)$ are the externally specified charge and current densities, respectively. Of course, Eq. (20.173b) leads to inconsistencies unless

$$\frac{\partial \rho}{\partial t} + \boldsymbol{\nabla} \cdot \boldsymbol{j} = 0\,, \tag{20.174}$$

thus the external sources must obey this continuity condition. To put these equations into relativistically invariant-looking forms (though with a specified source we do not really have relativistic invariance!), E_k and B_k combine to form a second-rank antisymmetric tensor $F_{\mu\nu}$, (cf Eq. (20.159)), whereas ρ and $\boldsymbol{j}$ combine into a vector j^μ:

$$E_k = F_{k0},\ B_k = \frac{1}{2}\epsilon_{klm}F_{lm},\ F_{\mu\nu} = -F_{\nu\mu};\ j^\mu = (\rho, j_k)\,. \tag{20.175}$$

Then, Eqs. (20.173a) and (20.173b) become, respectively:

$$\partial_\lambda F_{\mu\nu} + \partial_\mu F_{\nu\lambda} + \partial_\nu F_{\lambda\mu} = 0\,, \tag{20.176a}$$

$$\partial_\mu F^{\mu\nu} = -\,j^\nu\,. \tag{20.176b}$$

Equations (20.173a) or (20.176a) can be solved by introducing four potentials A_μ (written as though they form the components of a four-vector) and writing:

$$E_k = \partial_k A_0 - \dot{A}_k,\ B_k = \epsilon_{klm}\partial_l A_m\,;$$

that is,

$$F_{\mu\nu} = \partial_\mu A_\nu - \partial_\nu A_\mu . \tag{20.177}$$

However, from the fact that $F_{\mu\nu}$ transforms as a tensor under Lorentz transformations, we cannot conclude that A_μ is a four-vector! This is because two sets of potentials A_μ and A'_μ, which differ by the four-dimensional gradient of a scalar, $\partial_\mu \chi$, say, lead to the same tensor $F_{\mu\nu}$. The replacement of A_μ by $A'_\mu = A_\mu + \partial_\mu \chi$ is called a gauge transformation. Because the mathematical basis for introducing the A_μ is that thereby one set of the Maxwell equations, namely Eqs. (20.173a) or (20.176a), is automatically obeyed, the most general behavior that must be permitted for A_μ under a Lorentz transformation is to allow it to be first transformed as a four-vector, then subject it to some gauge transformation. The gauge transformation would depend on the Lorentz transformation and on other quantities, too. What complicates the treatment of the Maxwell field in comparison, say, with the vector field with finite mass, is that one must take account of both Lorentz and gauge invariances.

We now show that, as in the previous case, both Eqs. (20.177) and (20.176b) can be made to follow as the Lagrangian equations from a suitably chosen Lagrangian. We choose:

$$\begin{aligned} L(t) &= \int d^3x\, \mathcal{L}(\boldsymbol{x}, t) , \\ \mathcal{L} &= -\frac{1}{2} F^{\mu\nu}(\partial_\mu A_\nu - \partial_\nu A_\mu) + \frac{1}{4} F^{\mu\nu} F_{\mu\nu} + A^\mu j_\mu \\ &= -E_k \dot{A}_k + E_k \partial_k A_0 - \epsilon_{klm} B_k \partial_l A_m + \frac{1}{2}(B_k B_k - E_k E_k) \\ &\quad +(A_k j_k - A_0 j_0) . \end{aligned} \tag{20.178}$$

To begin with, the independent variables are E_k, B_k, A_k, and A_0; the corresponding Euler-Lagrange equations are:

$$\frac{d}{dt}\frac{\delta L}{\delta \dot{E}_k} - \frac{\delta L}{\delta E_k} = 0 \rightarrow \dot{A}_k = \partial_k A_0 - E_k ; \tag{20.179a}$$

$$\frac{d}{dt}\frac{\delta L}{\delta \dot{B}_k} - \frac{\delta L}{\delta B_k} = 0 \rightarrow B_k = \epsilon_{klm} \partial_l A_m ; \tag{20.179b}$$

$$\frac{d}{dt}\frac{\delta L}{\delta \dot{A}_k} - \frac{\delta L}{\delta A_k} = 0 \rightarrow \dot{E}_k = \epsilon_{klm} \partial_l B_m - j_k ; \tag{20.179c}$$

$$\frac{d}{dt}\frac{\delta L}{\delta \dot{A}_0} - \frac{\delta L}{\delta A_0} = 0 \rightarrow \partial_k E_k = -j_0 (= \rho) . \tag{20.179d}$$

We see that Eqs. (20.179a) and (20.179b) coincide with Eq. (20.177), whereas Eqs. (20.179c) and (20.179d) reproduce Eq. (20.176b). However, only the first and third equations look like genuine equations of motion; the other two are like constraints. We can handle Eq. (20.179b) in the same way as in the vector field case: because we have no equation of motion for B_k, and $\dot{B}_k$ does not appear in

L either, we can everywhere substitute for B_k in terms of A_k. This will now be assumed to be done. But the nature of Eq. (20.179d) is quite different from the corresponding Eq. (20.161d) in the vector field case: because we now have no "mass" term in the Lagrangian, we are unable to use Eq. (20.179d) to eliminate A_0 in terms of other quantities. In fact it turns out to be a constraint on the variables E_k, so that even the three components of E_k cannot be treated any more as independent field variables. The field quantity $A_0(\boldsymbol{x}, t)$ neither obeys an equation of motion nor is determined in terms of other field quantities referring to time t. Both these facts as well as the constraint on E_k can be traced to the properties of the Lagrangian L under a gauge transformation. Consider the infinitesimal variation $\delta F_{\mu\nu} = 0$, $\delta A_k = \partial_k \chi$, $\delta A_0 = \dot{\chi}$, where $\chi(\boldsymbol{x}, t)$ is an arbitrary infinitesimal function of space and time; the change in L is:

$$\begin{aligned}\delta L &= \int d^3x\, \delta\mathcal{L}((\boldsymbol{x}, t) = \int d^3x[\delta A_k j_k - \delta A_0 j_0] = \int d^3x[-\dot{\chi} j_0 - \chi \partial_k j_k] \\ &= -\int d^3x \left[\dot{\chi} j_0 + \chi \frac{\partial j_0}{\partial t}\right] = -\frac{d}{dt}\int d^3x \chi(\boldsymbol{x}, t) j_0(\boldsymbol{x}, t)\,. \qquad (20.180)\end{aligned}$$

Thus L is quasi-invariant under this variation. This means two things: on the one hand, given any solution of the Lagrangian equations of motion we can get another solution by adding $\dot{\chi}$ to A_0 and $\partial_k\chi$ to A_k where χ is *arbitrary*, but this clearly implies that the original Lagrangian equations could not have been such that they would permit an unambiguous determination of A_0 (or the longitudinal part of A_k) for later times given their values at an initial time. This explains why A_0 remains an arbitrary function not restricted by the equations of motion in any way. On the other hand, the quasi-invariance of L leads, in principle, to a constant of motion. This constant is:

$$\begin{aligned}\int d^3x \frac{\delta L}{\delta \dot{A}_k}\delta A_k + \int d^3x \chi j_0 &= \int d^3x(-E_k \partial_k \chi + \chi j_0) \\ &= \int d^3x(\partial_k E_k + j_0)\chi\,. \qquad (20.181)\end{aligned}$$

Here we have a quantity that contains an arbitrary function $\chi(\boldsymbol{x}, t)$ but which must nevertheless remain constant in time! As discussed in Chapter 8, the only way the action principle can save itself is to ensure that the coefficient of χ vanishes, either identically or by virtue of the Lagrangian equations; the former is not the case; thus we end up with a constraint on E_k.

Having clarified the relationship of the Euler-Lagrange Eq. (20.179d) to the gauge properties of the Lagrangian, we must recognize the fact that in Eq. (20.179c) we do have an equation of motion for E_k. Thus the question arises: is the constraint on E_k preserved in time? What must be done is to equate the time derivative of Eq. (20.179d) to the divergence of the right-hand side of Eq. (20.179c). If we do this we simply get the continuity equation for the external sources ρ, j_k, and this by assumption is satisfied. Thus the constraint on E_k is preserved in time, and that part of Eq. (20.179c) that refers to this fact need not any longer be treated as an equation of motion but can be dropped in favor of Eq. (20.179d). In other words, part of the information contained in Eq. (20.179c) amounts to taking the time derivative of both sides of Eq. (20.179d), and this part need not be kept any longer. What needs to be done is that we must separate the vector fields E_k and A_k into their lamellar (curl free, or longitudinal) and solenoidal (divergence free, or transverse) parts:

$$\begin{aligned} E_k(\boldsymbol{x},t) &= E_k^T(\boldsymbol{x},t) + \partial_k \mu(\boldsymbol{x},t)\,, \\ A_k(\boldsymbol{x},t) &= A_k^T(\boldsymbol{x},t) + \partial_k \lambda(\boldsymbol{x},t)\,; \\ \partial_k E_k^T &= \partial_k A_k^T = 0\,. \end{aligned} \tag{20.182}$$

This separation of a vector field into two such parts is a nonlocal operation, in the sense that both $E_k^T(\boldsymbol{x},t)$ and $\mu(\boldsymbol{x},t)$ at one point $\boldsymbol{x}$ in space depend on the values of $E_k(\boldsymbol{x}',t)$ at *all* points $\boldsymbol{x}'$. Formally we can write, in the case of the electric field:

$$\begin{aligned} \nabla^2 \mu &= \partial_k E_k\,, \\ \mu(\boldsymbol{x},t) &= \frac{1}{\nabla^2}\partial_k E_k(\boldsymbol{x},t) \equiv -\frac{1}{4\pi}\int d^3x' \frac{\partial_k' E_k(\boldsymbol{x}',t)}{|\boldsymbol{x}-\boldsymbol{x}'|}\,; \\ E_k^T(\boldsymbol{x},t) &= \left(\delta_{kj} - \frac{\partial_k\partial_j}{\nabla^2}\right) E_j(\boldsymbol{x},t) \equiv E_k(\boldsymbol{x},t) + \frac{1}{4\pi}\int d^3x' \frac{\partial_k'\partial_j' E_j(\boldsymbol{x}',t)}{|\boldsymbol{x}-\boldsymbol{x}'|}\,. \end{aligned} \tag{20.183}$$

Similar equations hold for the case of $A_k(\boldsymbol{x},t)$. Then, going back to the Lagrangian equations of motion Eq. (20.179), we interpret them thus: only the transverse part of Eq. (20.179a) is a genuine equation of motion for A_k^T, free from arbitrariness because of gauge transformations, but λ is undetermined to the same extent that A_0 is; in Eq. (20.179b) we can substitute A_m^T in place of A_m and retain this equation as the definition of B_k; only the transverse part of Eq. (20.179c) is to be retained, giving an equation of motion for E_k^T; Eq. (20.179d) determines μ in terms of the external charge density j_0. Thus the genuine dynamical variables are A_k^T and E_k^T, for which we have the equations of motion:

$$\dot{A}_k^T(\boldsymbol{x},t) = -E_k^T(\boldsymbol{x},t),\ \dot{E}_k^T(\boldsymbol{x},t) = -\nabla^2 A_k^T(\boldsymbol{x},t) - j_k^T(\boldsymbol{x},t)\,. \tag{20.184}$$

(Here we have separated off the transverse part j_k^T of the external current as well.) The remaining equations are:

$$B_k = \epsilon_{klm}\partial_l A_m^T,\ \nabla^2\mu = -j_0,\ \dot{\lambda} = A_0 - \mu\,. \tag{20.185}$$

At this point, because A_0 is arbitrary, we set it equal to zero; one calls this procedure "working in the Coulomb gauge." Using Eq. (20.185), which are defining equations, and Eq. (20.182), we can express the original Lagrangian L as a new function L' of the true variables E_k^T, A_k^T, their time derivatives, and the external source. In simplifying the expressions obtained, one can use the fact that an integral over all space of the product of two vector fields, one of which is curl free and the other divergence free, vanishes. One then gets:

$$\begin{aligned}
L'(t) &= \int d^3x\, \mathcal{L}'(\boldsymbol{x},t)\,, \\
\mathcal{L}'(\boldsymbol{x},t) &= -\, E_k^T(\boldsymbol{x},t)\dot{A}_k^T(\boldsymbol{x},t) - \frac{1}{2}(E_k^T(\boldsymbol{x},t)E_k^T(\boldsymbol{x},t) \\
&\quad + \partial_j A_k^T(\boldsymbol{x},t)\partial_j A_k^T(\boldsymbol{x},t)) + A_k^T(\boldsymbol{x},t)j_k^T(\boldsymbol{x},t) \\
&\quad - \frac{1}{2}j_0(\boldsymbol{x},t)\mu(\boldsymbol{x},t) - \frac{\partial}{\partial t}(\lambda(\boldsymbol{x},t)j_0(\boldsymbol{x},t))\,.
\end{aligned} \tag{20.186}$$

The reader can easily see that the equations of motion Eq. (20.184) for the transverse parts E_k^T and A_k^T follow as the Euler-Lagrange equations from this new Lagrangian; as in the case of the vector field, this is expected and should cause no surprise. Now in L' the last two terms may be dropped at least as far as getting the proper equations of motion is concerned, because they do not involve E_k^T and A_k^T at all. The last term is a total time derivative in L', which can always be dropped in a Lagrangian, whereas the next to the last term is the static Coulomb energy of the external source (apart from a sign):

$$-\frac{1}{2}\int j_0(\boldsymbol{x},t)\mu(\boldsymbol{x},t)d^3x = -\frac{1}{8\pi}\int\int d^3x d^3x' \frac{j_0(\boldsymbol{x},t)j_0(\boldsymbol{x}',t)}{|\boldsymbol{x}-\boldsymbol{x}'|}\,. \tag{20.187}$$

We can therefore work with an effective Lagrangian $L^{\text{eff}}(t)$:

$$\begin{aligned}
L^{\text{eff}}(t) &= \int d^3x\, \mathcal{L}^{\text{eff}}(\boldsymbol{x},t)\,, \\
\mathcal{L}^{\text{eff}}(\boldsymbol{x},t) &= -\, E_k^T \dot{A}_k^T - \frac{1}{2}(E_k^T E_k^T + \partial_j A_k^T \partial_j A_k^T) + A_k^T j_k^T\,.
\end{aligned} \tag{20.188}$$

This now is in a form in which the passage to the Hamiltonian is straightforward. The basic PB's must be set up between the transverse fields E_k^T and A_k^T; these brackets differ from Eq. (20.163) in the vector field case in that only the "transverse part" of $\delta^{(3)}(\boldsymbol{x}-\boldsymbol{x}')$ must be used:

$$\begin{aligned}
\{E_j^T(\boldsymbol{x}), A_k^T(\boldsymbol{x}')\} &= \left(\delta_{jk} - \frac{\partial_j\partial_k}{\nabla^2}\right)\delta^{(3)}(\boldsymbol{x}-\boldsymbol{x}') \equiv \delta_{jk}^T(\boldsymbol{x}-\boldsymbol{x}')\,, \\
\{E_j^T(\boldsymbol{x}), E_k^T(\boldsymbol{x}')\} &= \{A_j^T(\boldsymbol{x}), A_k^T(\boldsymbol{x}')\} = 0\,.
\end{aligned} \tag{20.189}$$

One sees the fundamental differences in the Hamiltonian structure of the electromagnetic field (in the Coulomb gauge !) and the vector field discussed earlier. Where we previously had three canonical pairs $\mathcal{E}_j, \phi_k$ associated with each point $\boldsymbol{x}$, we have now just two canonical pairs. No PB's are defined for the longitudinal parts of E_k and A_k, these being determined by the external sources. The Hamiltonian can be read off from Eq. (20.188);

$$\begin{aligned}
H^{\text{eff}} &= \int d^3x\, \mathcal{H}^{\text{eff}}(\boldsymbol{x},t)\,, \\
\mathcal{H}^{\text{eff}} &= \frac{1}{2}(E_k^T E_k^T + \partial_j A_k^T \partial_j A_k^T) - A_k^T j_k^T\,,
\end{aligned} \tag{20.190}$$

and it is easily checked that the Hamiltonian equations of motion reproduce Eq. (20.184).

In the presence of a nonvanishing external source j_k^T the system does not possess relativistic invariance, and thus there is no corresponding realization of $\mathcal{P}$ either. But with j_k^T set equal to zero, we deal with the free electromagnetic field, and then there is a set of generators of $\mathcal{P}$ producing a canonical realization of $\mathcal{P}$. We can obtain these generators by using again the invariance and quasi-invariance properties of the free Lagrangian:

$$L^{\text{free}} = \int d^3x\, \mathcal{L}^{\text{free}}(\boldsymbol{x}, t)\,,$$
$$\mathcal{L}^{\text{free}} = -\, E_k^T \dot{A}_k^T - \mathcal{H}(\boldsymbol{x}, t),\ \ \mathcal{H}(\boldsymbol{x}, t) = \frac{1}{2}(E_k^T E_k^T + \partial_j A_k^T \partial_j A_k^T)\,. \qquad (20.191)$$

Even though we deal here with transverse fields, this is an $E(3)$ invariant condition; thus L^{free} is manifestly invariant under $E(3)$. The computation for P_j and J_j follows the same steps as in the vector field case, remembering always this transversality condition, and we obtain the expressions:

$$\begin{aligned} P_j &= \int d^3x\, E_k^T(\boldsymbol{x}, t)\frac{\partial}{\partial x_j} A_k^T(\boldsymbol{x}, t) = \text{constant}\,, \\ J_j &= \int d^3x\, E_k^T(\boldsymbol{x}, t)[\delta_{kl}(\boldsymbol{x} \times \boldsymbol{\nabla})_j - (S_j)_{kl}]A_l^T(\boldsymbol{x}, t) = \text{constant}\,, \\ (S_j)_{kl} &= -\epsilon_{jkl}\,. \end{aligned} \qquad (20.192)$$

The calculation of K_j is a little more complicated. If under an infinitesimal pure Lorentz transformation, A_μ were treated as a four-vector field, we would obtain the variation:

$$\delta A_0 = d_\zeta A_0 + \zeta_k A_k,\ \ \delta A_k = d_\zeta A_k + \zeta_k A_0\,. \qquad (20.193)$$

But we know that this has to be modified: even if initially A_0=0 and A_k consists purely of its transverse part, after this variation $A_0 \neq 0$ and $A_k \neq A_k^T$. Thus we must then carry out an infinitesimal gauge transformation to restore the original conditions. But we actually only require the variation in A_k^T, and this is unaffected by the subsequent gauge transformation that follows the variation in Eq. (20.193). Thus we may set $A_0 = 0, A_k = A_k^T$ in δA_k in the above, and then project out the transverse part:

$$\begin{aligned} \delta A_k^T &= (d_\zeta A_k^T)^T = (t\boldsymbol{\zeta} \cdot \boldsymbol{\nabla} A_k^T + \boldsymbol{\zeta} \cdot \boldsymbol{x} \dot{A}_k^T)^T \\ &= t\boldsymbol{\zeta} \cdot \boldsymbol{\nabla} A_k^T + (\boldsymbol{\zeta} \cdot \boldsymbol{x} \dot{A}_k^T)^T \\ &= t\boldsymbol{\zeta} \cdot \boldsymbol{\nabla} A_k^T + \left(\delta_{kj} - \frac{\partial_k \partial_j}{\nabla^2}\right)(\boldsymbol{\zeta} \cdot \boldsymbol{x} \dot{A}_j^T) \\ &= d_\zeta A_k^T - \frac{\partial_k}{\nabla^2}\zeta_j \dot{A}_j^T\,. \end{aligned} \qquad (20.194)$$

In a similar way to compute δE_k^T we start with an expression such as appears in Eq. (20.169) for δE_k, then project out the transverse part:

$$\begin{aligned}
\delta E_k^T =&(d_\zeta E_k^T + (\boldsymbol{\zeta}\times\boldsymbol{B})_k)^T = d_\zeta E_k^T - \frac{\partial_k}{\nabla^2}\zeta_j \dot{E}_j^T + (\boldsymbol{\zeta}\times\boldsymbol{B})_k^T\,;\\
(\boldsymbol{\zeta}\times\boldsymbol{B})_k^T =& \left(\delta_{kj} - \frac{\partial_k\partial_j}{\nabla^2}\right)(\boldsymbol{\zeta}\times\boldsymbol{B})_j = \left(\delta_{kj} - \frac{\partial_k\partial_j}{\nabla^2}\right)\epsilon_{jlm}\zeta_l\epsilon_{mrs}\partial_r A_s^T\\
=&\zeta_l(\delta_{jr}\delta_{ls} - \delta_{js}\delta_{lr})\left(\delta_{kj} - \frac{\partial_k\partial_j}{\nabla^2}\right)\partial_r A_s^T = -\boldsymbol{\zeta}\cdot\boldsymbol{\nabla}A_k^T\,;\\
\delta E_k^T =&d_\zeta E_k^T - \frac{\partial_k}{\nabla^2}\zeta_j\dot{E}_j^T - \zeta_j\partial_j A_k^T\,.
\end{aligned} \tag{20.195}$$

Armed with these expressions for the variations in A^T and E^T produced by a pure infinitesimal Lorentz transformation, we can now examine the behavior of L^{free}. In the calculation of δL^{free}, many terms drop out after partial integrations, because of the transverse nature of the two fields, and one finally obtains:

$$\delta L^{\text{free}} = \frac{d}{dt}\int d^3x\; \zeta_j x_j(-E_k^T\dot{A}_k^T - \frac{1}{2}E_k^T E_k^T - \frac{1}{2}\partial_j A_k^T\partial_j A_k^T)\,. \tag{20.196}$$

Combining this quasi-invariance of L^{free} with the action principle, we get:

$$\begin{aligned}
&\int d^3x \frac{\delta L^{\text{free}}}{\delta \dot{A}_k^T(\boldsymbol{x},t)}\delta A_k^T(\boldsymbol{x},t) + \int d^3x\; \boldsymbol{\zeta}\cdot\boldsymbol{x}(E_k^T\dot{A}_k^T + \frac{1}{2}E_k^T E_k^T + \frac{1}{2}\partial_j A_k^T\partial_j A_k^T)\\
&= \int d^3x[-E_k^T(\boldsymbol{x},t)\{d_\zeta A_k^T\} + \boldsymbol{\zeta}\cdot\boldsymbol{x}\{E_k^T\dot{A}_k^T + \frac{1}{2}E_k^T E_k^T + \frac{1}{2}\partial_j A_k^T\partial_j A_k^T\}]\\
&= -t\zeta_j P_j + \zeta_j\int d^3x\, x_j\,\mathcal{H}(\boldsymbol{x},t) = \text{constant}\,.
\end{aligned} \tag{20.197}$$

The relativistic moment K_j is thereby identified. Collecting all the generators of $\mathcal{P}$, we have:

$$\begin{aligned}
J_j &= \int d^3x\, E_k^T(\boldsymbol{x})[\delta_{kl}(\boldsymbol{x}\times\boldsymbol{\nabla})_j - (S_j)_{kl}]A_l^T(\boldsymbol{x})\,,\\
P_j &= \int d^3x\, E_k^T(\boldsymbol{x})\frac{\partial}{\partial x_j}A_k^T(\boldsymbol{x})\,,\\
K_j &= \int d^3x\, x_j\mathcal{H}(\boldsymbol{x}),\; H = \int d^3x\,\mathcal{H}(\boldsymbol{x})\,,\\
\mathcal{H}(\boldsymbol{x}) &= \frac{1}{2}[E_k^T(\boldsymbol{x})E_k^T(\boldsymbol{x}) + \partial_j A_k^T(\boldsymbol{x})\partial_j A_k^T(\boldsymbol{x})]\,.
\end{aligned} \tag{20.198}$$

The interested reader may verify all the PB relations of $\mathcal{P}$. This generates then the canonical realization of $\mathcal{P}$ corresponding to the free electromagnetic field. As in the vector case we have a positive definite energy density; hence the energy-momentum four-vector is positive timelike. The expressions for energy density and momentum density are familiar from classical theory; remembering throughout that $\boldsymbol{E}$ and $\boldsymbol{A}$ are transverse fields, we could rewrite P_j and H as:

$$P_j = \int d^3x(\boldsymbol{E}\times\boldsymbol{B})_j,\; H = \int d^3x\;\frac{1}{2}(\boldsymbol{E}^2 + \boldsymbol{B}^2)\,, \tag{20.199}$$

taking us back to the expressions that appear in the usual treatments of the Maxwell field.

Standard Versus Nonstandard Configurations for Lorentzian Fields

Several times in our treatment of realizations of $\mathcal{P}$ generated by dynamical fields we have remarked on the formal nature of the manipulations involved. We have, whenever convenient, dropped "surface terms" in the process of converting an integral from one form to another. The problem that arises here is exactly similar to that met with in the preceding chapter and concerns the nature of the configurations of the field system. Strictly speaking, each element of $\mathcal{P}$ acts as an automorphism on the basic dynamical variables of the system in a manner that preserves the PB relations among these variables. It is not always the case that these automorphisms are inner ones generated by well-defined dynamical variables. The existence of relativistically invariant equations of motion is not sufficient to guarantee that the field system furnishes a canonical realization of $\mathcal{P}$. Whereas expressions could be written down for the generators of the Poincaré transformations, they are of a formal nature, because none of them are "finite polynomials" in the sense of Chapter 10. In what we have called the "standard configurations," the generators do, in fact,exist. We characterize these subsequently. However, there are always nonstandard configurations also, for which the automorphisms determined by $\mathcal{P}$ are outer ones; these configurations describe a system that does not exhibit the full symmetry of the automorphisms of the classical field.

As an example of these considerations, we take the case of the free spinless field; formally we have the generators given in Eq. (20.151) for this realization of $\mathcal{P}$. If the total energy H is to exist, it is necessary that the quantities $\pi(\boldsymbol{x})$, $\phi(\boldsymbol{x})$, and $\nabla\phi(\boldsymbol{x})$ be each square integrable. By a simple application of Cauchy's inequality, provided H exists, it follows that P_j exists. However, this does not guarantee that J_j or K_j will exist! If we also require that

$$\int d^3x|x_j|\pi^2(\boldsymbol{x}) < \infty, \quad \int d^3x|x_j|\phi^2(\boldsymbol{x}) < \infty, \quad \int d^3x|x_j|(\nabla\phi(\boldsymbol{x})^2) < \infty\,, \tag{20.200}$$

then the existence of J_j and K_j would be guaranteed. Hence a sufficient condition that the free scalar field furnish a realization of $\mathcal{P}$ is that $\pi(\boldsymbol{x}), \phi(\boldsymbol{x})$, and $\nabla\phi(\boldsymbol{x})$ all be bounded at all points and that they fall off at least as fast as $1/|\boldsymbol{x}|^{2+\epsilon}$ as $|\boldsymbol{x}| \to \infty$, (where ϵ is non-negative and nonzero). Unless they behave in this way, some generators may not be properly defined. We could allow the field variables to develop mild singularities that are square integrable in a finite region of space without any serious harm.

It is clear that configurations in which the fields do not fall off fast enough do not furnish a realization of $\mathcal{P}$. In particular, those configurations in which the field excitations (and so the energy density) are "homogeneous in the large" so that a thermodynamic limit can be considered are *not* relativistically invariant.

The set of configurations for which the generator H, say, does not exist is itself reducible into subsets not connected to one another by proper canonical transformations. Thus, for example, the configuration in which $\phi(\boldsymbol{x})$ is uniform all over space is not only disjoint from one in which $\phi(\boldsymbol{x})$ is square integrable but also from one in which $\phi(\boldsymbol{x})$ vanishes in one-half of space but has a more or less uniform nonvanishing value elsewhere. Any number of such disjoint sets of configurations can be displayed. The relativistic scalar field thus describes an infinite number of disjoint sets of configurations, and it is therefore necessary to specify the set of configurations one is interested in. Unless otherwise dictated, one normally refers to those cases where the configuration corresponding to no excitation is included. In this case all the generators exist and the field does furnish a realization of the Poincaré group.

Chapter 21

Manifest Covariance in Hamiltonian Mechanics

The structures of the Galilei and Poincaré groups have been abstracted from the transformation laws for the space and time coordinates of "events," corresponding to the transition from one inertial observer to another. These transformation laws are essentially geometrical in character; thus we may regard the two groups $\mathcal{G}$ and $\mathcal{P}$ as describing two possible space-time geometries. (The geometrical flavor is probably more pronounced in the relativistic case, but both are capable of geometrical interpretation). Because of this, one has the feeling, especially in the relativistic case, that space and time coordinates "ought to be treated on the same footing". The principle of relativity, be it Galilean or Poincaré, as applied to mechanics is usually stated as follows: "All equivalent observers deduce the same laws of physics". Let us summarize very briefly the way in which this is implemented in our discussion in Chapter 16. In Hamiltonian mechanics, the laws of motion are stated in terms of PBs. On the other hand, working for definiteness in the Heisenberg picture, the specific "legal" definitions of dynamical variables in different frames yield *different* (but related) dynamical variables. Hence we look for relations or transformations between the dynamical variables given by legal definitions in one frame and those similarly given in another frame, which preserve PB relations. But these are simply canonical transformations. (This is certainly so for finite numbers of degrees of freedom, but only formally so otherwise.) Thus, given any two inertial observers, there are two sets of dynamical variables identified by the legal definitions in each frame, and there is a canonical transformation connecting the two sets. In effect, not only is the development in time given by the action of a canonical transformation, but so is any other conceivable change of inertial frame. (Here we think of a change in the zero of time as a particular but simple change of frame.) Because the set of all transformations from one observer to an equivalent one has the structure of $\mathcal{G}$ or $\mathcal{P}$, our analysis shows that we must have a realization of the relativity group by means of canonical transformations. Thus we find precisely the condition under which a canonical system has relativistic invariance, that is, for which physical laws deduced by different observers are the same: namely, when we have a canonical realization of the appropriate relativity group.

Note that thus far we have not referred to manifest tensor covariance of the equations of motion. Although a theory may be relativistic, the time parameter may play a special role. This is seen particularly clearly in the case of a system of many particles, where the time parameter is treated essentially differently from the space coordinates of the particles; the latter alone are dynamical variables. More generally, in developing the Hamiltonian formalism to encompass relativistic invariance, we have given an essentially distinguished role to dynamical quantities defined at one instant of time, be they field variables $\phi_r(\boldsymbol{x}, t), \pi_r(\boldsymbol{x}, t)$ or particle positions and canonically conjugate momenta $\boldsymbol{q}^{(r)}(t), \boldsymbol{p}^{(r)}(t)$. In such a treatment, the generators of frame transformations, be they Galilean or Poincaré, are also expressed in terms of dynamical quantities at a prescribed time.

Turning to manifest covariance, there is actually a new concept involved here! Let us consider the mechanics of a many-particle system. We may (if we so insist!) strengthen the relativity postulate by requiring that, interacting or not, the labeled trajectory of a particle in configuration space, that is, the correspondence $t \to \boldsymbol{q}^{(r)}(t)$, must behave like a "world-line"; it must be possible to interpret it as a succession of "events" whose space-time coordinates are $(\boldsymbol{q}^{(r)}(t), t)$ which transform in the geometric manner under frame changes. Such a requirement is quite *reasonable*; after all, the group structures of $\mathcal{G}$ and $\mathcal{P}$ are derived from the transformation laws of space-time coordinates of events, and we wish to identify certain quantities occurring in the formalism as events. However, one must appreciate that *this requirement is distinct from, and supplementary to, the postulate requiring identity of physical laws.* Here the transformation is that of particular "states" or trajectories. Similar considerations apply in the case of dynamical fields. All in all, the original relativity postulate leads, via the group structure of $\mathcal{G}$ or $\mathcal{P}$, only to the existence of and the proper PB relations between the generators of the realization of $\mathcal{G}$ or $\mathcal{P}$. The concept of manifest covariance now requires that under the canonical transformations realizing $\mathcal{G}$ or $\mathcal{P}$, certain dynamical quantities [such as the Cartesian coordinates $\boldsymbol{q}^{(r)}(t)$ of a particle] transform in a "geometric" manner [e.g., the quartet $(t, \boldsymbol{q}^{(r)}(t))$, made up of one parameter and three dynamical variables, constitute an event]. As we will see, this leads to specific PB relations between the generators of $\mathcal{G}$ or $\mathcal{P}$ on the one hand, and the dynamical variables concerned on the other.

Somewhat related to this notion of manifest covariance is the idea of treating dynamical quantities for all values of time "on the same footing." This approach has been employed to a large extent during the past two decades, particularly within the framework of quantum field theory; one treats on the same footing the field components at all times and all space points, and manifest covariance is demanded. Similar ideas could be considered for classical Galilean and Lorentzian fields as well. We now proceed to a brief but illustrative discussion of this point of view by treating a system with a finite number of degrees of freedom. Following this, we consider the consequences of imposing manifest covariance for particle mechanics, first in the nonrelativistic situation described by $\mathcal{G}$ and then in the special relativistic situation corresponding to $\mathcal{P}$. After this, we take up the cases of Galilean and Lorentzian fields.

Geometric Formalism for a Finite Number of Degrees of Freedom

Let us consider a system described by a single canonical pair q, p subject to the PB relation:

$$\{q, p\} = 1\,. \tag{21.1}$$

If $H(q,p)$ (assumed not explicitly time dependent) is the Hamiltonian of the system, we would define $q(t)$, $p(t)$ as the solution to the equations of motion and boundary conditions:

$$\begin{gathered} \frac{dq(t)}{dt} = \frac{\partial H(q(t),p(t))}{\partial p(t)}\,, \qquad \frac{dp(t)}{dt} = -\frac{\partial H(q(t),p(t))}{\partial q(t)}\,, \\ q(0) = q\,, \qquad p(0) = p\,; \end{gathered} \tag{21.2}$$

we then view the time evolution as a mapping from q, p to $q(t)$, $p(t)$. The canonical time evolution is known to be an automorphism because we can combine Eqs. (21.1) and (21.2) to deduce:

$$\{q(t),\; p(t)\} = 1\,. \tag{21.3}$$

Instead of such a discussion, however, we could *postulate* a time-dependent automorphism as the law of time evolution,

$$q \to q(t),\; p \to p(t)\,, \tag{21.4}$$

with $q(t)$, $p(t)$ being prescribed functions of q, p, t such that Eq. (21.3) is satisfied. In the simple case of a nonrelativistic particle $p(t)$ could be identified with $m\dot{q}(t)$ where m is the mass of the particle. In such a case it suffices to specify the single mapping $q \to q(t)$, and one can deal (or hope to deal!) directly with the general PB

$$\{q(t), q(t')\} = D(t,t')\,. \tag{21.5}$$

In general, $D(t,t')$ is a nontrivial dynamical variable, not a pure number; it is subject to the boundary conditions:

$$D(t,t) = 0;\; \frac{\partial}{\partial t'}D(t,t')\bigg|_{t'=t} = -\frac{\partial}{\partial t}D(t,t')\bigg|_{t'=t} = \frac{1}{m}\,. \tag{21.6}$$

The interesting thing to observe is that in the form of Eqs. (21.5) and (21.6), all reference to the interaction part $V(q)$ in the Hamiltonian

$$H(q,p) = \frac{1}{2m}p^2 + V(q) \tag{21.7}$$

has apparently disappeared! But this is not so. Actually, the nature of V is concealed in the explicit form of the relation of $q(t)$ with $q(0)$ and $\dot{q}(0)$, and the existence of a Hamiltonian of the form of Eq. (21.7) also puts in restrictions on the mapping $q \to q(t)$ and the quantity $D(t, t')$.

It is remarkable that although the foregoing equations appear to describe a system with only one degree of freedom, they could actually be used to describe a system with arbitrarily many degrees of freedom. As an example consider a set of N simple harmonic oscillators with distinct frequencies ω_n and time-dependent dynamical variables:

$$\begin{aligned} q_n(t) &= q_n \cos\omega_n t + \frac{p_n}{m\omega_n}\sin\omega_n t\,, \\ p_n(t) &= p_n \cos\omega_n t - q_n m\omega_n \sin\omega_n t\,, \end{aligned} \tag{21.8}$$

and write:

$$q(t) = \frac{1}{\Lambda}\sum_{n=1}^{N} \lambda_n q_n(t)\,. \tag{21.9}$$

Then, assuming the standard PBs among the independent q_n and p_n, we have:

$$D(t, t') = \{q(t), q(t')\} = \frac{1}{\Lambda^2}\sum_{n=1}^{N} \frac{\lambda_n^2}{m\omega_n}\sin\omega_n(t' - t)\,,$$

so that

$$\frac{\partial}{\partial t'}D(t, t')\bigg|_{t'=t} = \frac{1}{m\Lambda^2}\sum_{n=1}^{N}\lambda_n^2\,. \tag{21.10}$$

It is then a trivial matter to choose

$$\Lambda^2 = \sum_{n=1}^{N}\lambda_n^2$$

so that Eq. (21.6) is obeyed. But the time-dependent dynamical variable in Eq. (21.9) describes not one but N degrees of freedom!

Galilean Many-Particle System

Let us now consider the expression of manifest covariance for a many-particle system described by a canonical realization of $\mathcal{G}$. The manner in which such a system may include interactions has been described in Chapter 19 (388 ff). The first part of our discussion relates to the action of $E(3)$; thus this is relevant in the relativistic case as well. The basic canonical variables are the position coordinates $q_j^{(r)}$ and their conjugates $p_j^{(r)}$. The interpretation of the former as the *three Cartesian coordinates* of particle number r requires the following familiar PB relations between the q's and the generators J_j and P_j of space rotations and space translations:

$$\{J_j, q_k^{(r)}\} = \epsilon_{jkl}q_l^{(r)}\,,\ \{P_j, q_k^{(r)}\} = -\delta_{jk}\,. \tag{21.11}$$

Our work in Chapters 17 and 18 assures us that if these "infinitesimal transformation" relations are obeyed, then under any element of $E(3)$, the effects on the q's of a *canonical* transformation and a *geometrical* transformation agree:

$$\widetilde{e^{\boldsymbol{\alpha}\cdot\boldsymbol{J}}} q_k^{(r)} = A_{jk}(\boldsymbol{\alpha}) q_j^{(r)}, \ \widetilde{e^{\boldsymbol{a}\cdot\boldsymbol{P}}} q_k^{(r)} = q_k^{(r)} - a_k \,. \tag{21.12}$$

Thus for the system under consideration, manifest covariance under $E(3)$ amounts to the basic PB's among the generators J_j and P_j, together with the additional relations in Eq. (21.11). Now the question arises as to the transformation properties of the canonical momenta $p_j^{(r)}$. These would be known once one has the connection between the p's and the Cartesian velocities $\dot{q}_j^{(r)}$, because the behavior of the latter is determined by Eq. (21.12). If for each particle the canonical and kinetic momenta coincide, that is, $p_j^{(r)} = m^{(r)}\dot{q}_j^{(r)}$, then we would demand the following additional covariance conditions:

$$\{J_j, p_k^{(r)}\} = \epsilon_{jkl} p_l^{(r)}, \{P_j, p_k^{(r)}\} = 0 \,. \tag{21.13}$$

If we demand both Eqs. (21.11) and (21.13), and of course the $E(3)$ relations among the J's and P's, then we can see very quickly that these generators have the standard kinematic forms whether or not there is interaction. For, defining $J_j^{(0)}$ and $P_j^{(0)}$ in the usual fashion,

$$J_j^{(0)} = \sum_r \epsilon_{jkl} q_k^{(r)} p_l^{(r)}, \ P_j^{(0)} = \sum_r p_j^{(r)}, \tag{21.14}$$

it is clear that Eqs. (21.11) and (21.13) are obeyed with $J_j^{(0)}$, $P_j^{(0)}$ written in place of J_j, P_j; thus the differences $J_j - J_j^{(0)}$, $P_j - P_j^{(0)}$ have vanishing brackets with all the q's and p's; hence they must be pure numbers. However, because both J_j, P_j and $J_j^{(0)}$, $P_j^{(0)}$ are solutions to the $E(3)$ Lie algebra relations, our proof that all neutral elements in these relations can be transformed away now shows that these two sets of generators must be identical.

Suppose that instead of imposing Eq. (21.13), we adopt only Eq. (21.11) as the expression of manifest Euclidean invariance, thus leaving room for a more general connection between canonical and kinetic momenta. This is also more reasonable because the groups $E(3)$, $\mathcal{G}$ and $\mathcal{P}$ all have their bases in transformation laws for space-time coordinates of events, not in laws for velocities or momenta. In that case we cannot conclude that J_j and P_j have the kinematic forms given in Eq. (21.14). But now we can demonstrate that the two sets of $E(3)$ generators J_j, P_j and $J_j^{(0)}$, $P_j^{(0)}$ are canonically equivalent, the equivalence being via a canonical transformation that leaves the $q_j^{(r)}$ invariant and alters only the $p_j^{(r)}$. This would mean that, assuming Eq. (21.11) alone, we can also assume that the $E(3)$ generators are kinematic, the only thing needed to achieve this being a possible change in the canonical momenta, with no change in the interpretation of the $q_j^{(r)}$. To show this, let us first examine the form of the most general canonical transformation $(q_1 \ldots, q_k, p_1, \ldots, p_k) \to (Q_1, \ldots, Q_k, P_1, \ldots, P_k)$

on k canonical pairs, such that $Q_1 = q_1, \ldots, Q_k = q_k$. It is clear that because the q_r and p_r form $2k$ independent phase-space variables, so do the variables Q_r, p_r made up of k old and k new quantities. The transformation can then be described by means of a generating function $F_{(3)}(p, Q)$ according to Eq. (6.49). The identity of the q_r and Q_r leads to the system of equations

$$\begin{aligned} F_{(3)}(p, Q) &= -\sum_r p_r Q_r - \psi(Q); \\ q_s &= -\frac{\partial F_{(3)}}{\partial p_s} = Q_s, \\ P_s &= -\frac{\partial F_{(3)}}{\partial Q_s} = \frac{\partial \psi(Q)}{\partial Q_s} + p_s = p_s + \frac{\partial \psi(q)}{\partial q_s}. \end{aligned} \tag{21.15}$$

This transformation can be rewritten in the compact form:

$$Q_s = e^{\widetilde{\psi(q)}} q_s = q_s, \; P_s = e^{\widetilde{\psi(q)}} p_s = p_s + \frac{\partial \psi}{\partial q_s}, \tag{21.16}$$

and thus we see that the most general transformation of this variety is fully determined by one arbitrary function $\psi(q)$. Conversely, of course, such a transformation determines $\psi(q)$ up to an additive constant.

Let us now suppose we are given two canonical realizations of $E(3)$, one with J_j, P_j as generators, the other with $J_j^{(0)}, P_j^{(0)}$, such that the effects on the $q_j^{(r)}$ coincide. A general element $\exp(\boldsymbol{a} \cdot \boldsymbol{d})\exp(\boldsymbol{\alpha} \cdot \boldsymbol{l})$ in $E(3)$, with coordinates (α_j, a_j), is simply written as g; the corresponding canonical transformations in the two realizations are:

$$R(g) \equiv R(\boldsymbol{\alpha}, \boldsymbol{a}) = e^{\widetilde{\boldsymbol{a}\cdot\boldsymbol{P}}} e^{\widetilde{\boldsymbol{\alpha}\cdot\boldsymbol{J}}}, \; R^{(0)}(g) \equiv R^{(0)}(\boldsymbol{\alpha}, \boldsymbol{a}) = e^{\widetilde{\boldsymbol{a}\cdot\boldsymbol{P}^{(0)}}} e^{\widetilde{\boldsymbol{\alpha}\cdot\boldsymbol{J}^{(0)}}}. \tag{21.17}$$

In proving the canonical equivalence of these two realizations, we can work either with the generators or with the finite transformations; we choose the latter method. According to Eq. (21.12) we have, say:

$$R(g)q_j^{(r)} = A_{kj}(\boldsymbol{\alpha})(q_k^{(r)} - a_k) \equiv g^{-1} q_j^{(r)}, \tag{21.18}$$

and an identical equation is valid with $R^{(0)}(g)$ in place of $R(g)$. Therefore, the canonical transformation $R^{(0)}(g^{-1})R(g)$ is one that leaves each $q_j^{(r)}$ invariant; by the arguments of the previous paragraph, we can then write:

$$R(g) = R^{(0)}(g)\exp(\widetilde{\psi(g; q_j^{(r)})}), \tag{21.19}$$

$\psi(g; q_j^{(r)})$ being determined up to an additive constant that may depend on g. Now both $R(g)$ and $R^{(0)}(g)$ obey the same group composition law; thus if g' and g are any two elements of $E(3)$, we must have:

$$R^{(0)}(g')\exp(\widetilde{\psi(g'; q_j^{(r)})})R^{(0)}(g)\exp(\widetilde{\psi(g; q_j^{(r)})}) = R^{(0)}(g'g)\exp(\widetilde{\psi(g'g; q_j^{(r)})}). \tag{21.20}$$

On the left-hand side we can shift the factor $R^{(0)}(g)$ to the left of the factor involving $\psi(g';q_j^{(r)})$ by using the formulae:

$$\begin{aligned} R^{(0)}(g)f(q_j^{(r)}) &= f(g^{-1}q_j^{(r)}), \\ R^{(0)}(g)\exp(\widetilde{f(q_j^{(r)})}) &= \exp(\widetilde{f(g^{-1}q_j^{(r)})})R^{(0)}(g). \end{aligned} \tag{21.21}$$

These are valid for any g and any function $f(q)$. We then obtain a condition on ψ:

$$\exp(\widetilde{\psi(g';gq_j^{(r)})})\exp(\widetilde{\psi(g;q_j^{(r)})}) = \exp(\widetilde{\psi(g'g;q_j^{(r)})}). \tag{21.22}$$

Now all canonical transformations involved in this equation are generated by functions of the q's alone; thus they commute with one another. This allows us to add the two exponents on the left and identify the sum with the exponent on the right. To save writing, let us indicate equations that are valid apart from the possible presence of functions that are constant in phase space, by placing a dot on the equality sign; then we obtain a functional equation for ψ:

$$\psi(g;q_j^{(r)}) \doteq \psi(g'g;q_j^{(r)}) - \psi(g';gq_j^{(r)}). \tag{21.23}$$

The important point here is that the element g' is quite arbitrary, yet it does not appear on the left-hand side. We exploit this fact to write ψ in terms of a simpler function.

One procedure that suggests itself is that we average the right-hand side of Eq. (21.23) over the group $E(3)$ with respect to g'. Given a sufficiently well-behaved function $f(g)$ defined over $E(3)$, it is possible to define a "volume element" dg such that if g_0 is any fixed element we have the equality:

$$\int_{E(3)} f(g)dg = \int_{E(3)} f(gg_0)dg. \tag{21.24}$$

In effect, keeping g_0 fixed, the Jacobian involved in the change of integration variable $g \to gg_0$ is unity. Using the coordinates (α_j, a_j), up to a positive factor the above integral has the explicit form:

$$\begin{aligned} \int_{E(3)} f(g)dg \equiv & \int_{-\infty}^{\infty}\int_{-\infty}^{\infty}\int_{-\infty}^{\infty} da_1da_2da_3 \frac{1}{2\pi^2}\iiint_{|\boldsymbol{\alpha}|\leq\pi} d\alpha_1 d\alpha_2 d\alpha_3 \\ & \left(\frac{\sin(|\boldsymbol{\alpha}|/2)}{|\boldsymbol{\alpha}|}\right)^2 f(\alpha_j, a_j). \end{aligned} \tag{21.25}$$

This integral converges only if f is integrable over every finite volume and goes to zero sufficiently rapidly as $|\boldsymbol{a}| \to \infty$. In particular, if f=1, the integral diverges; the "total volume" of $E(3)$ is infinite, and it is a non-compact group. On the other-hand, the part of the volume element referring to the α_j is simply the one to be used in averaging a function defined on $R(3)$. Analogously to Eq.(21.24), we have the important property:

$$\begin{aligned}&\frac{1}{2\pi^2}\iiint_{|\boldsymbol{\alpha}|\le\pi} d\alpha_1 d\alpha_2 d\alpha_3 \left(\frac{\sin(|\boldsymbol{\alpha}|/2)}{|\boldsymbol{\alpha}|}\right)^2 f(\exp(\boldsymbol{\alpha}\cdot\boldsymbol{l}))\\ &\quad=\frac{1}{2\pi^2}\iiint_{|\boldsymbol{\alpha}|\le\pi} d\alpha_1 d\alpha_2 d\alpha_3 \left(\frac{\sin(|\boldsymbol{\alpha}|/2)}{|\boldsymbol{\alpha}|}\right)^2 f(\exp(\boldsymbol{\alpha}\cdot\boldsymbol{l})\exp(\boldsymbol{\beta}\cdot\boldsymbol{l}))\,,\end{aligned} \tag{21.26}$$

for a function defined over $R(3)$. One can see that $R(3)$ has a finite total volume which has been normalized to unity by including the factor $(1/2\pi^2)$; this corresponds to $R(3)$ being a compact group. Now letting V denote any finite region in $E(3)$ with volume $[V]$, we can integrate the right-hand side of Eq. (21.23) with respect to g' over V to get:

$$\psi(g;q_j^{(r)})\doteq\frac{1}{[V]}\int_V dg'\psi(g'g;q_j^{(r)})-\frac{1}{[V]}\int_V dg'\psi(g';gq_j^{(r)})\,. \tag{21.27}$$

In the first integral, let us change the variable of integration from g' to $g'g$; the range of integration changes from V to Vg obtained by multiplying the elements in V by g on the right, but because the Jacobian is unity we have $[Vg]=[V]$. Thus Eq. (21.27) becomes:

$$\psi(g;q_j^{(r)})\doteq\frac{1}{[Vg]}\int_{Vg} dg'\psi(g';q_j^{(r)})-\frac{1}{[V]}\int_V dg'\psi(g';gq_j^{(r)})\,. \tag{21.28}$$

If in this equation we are permitted to take the limit $V\to E(3)$, and the limiting values of these integrals do not depend on the manner in which the limit is taken, then we have:

$$\begin{aligned}\psi(g;q_j^{(r)}) &\doteq \phi(q_j^{(r)})-\phi(gq_j^{(r)})\,,\\ \phi(q_j^{(r)}) &= \operatorname*{Lt.}_{V\to E(3)}\frac{1}{[V]}\int_V dg'\psi(g';q_j^{(r)})\,.\end{aligned} \tag{21.29}$$

Thus the functional relation Eq. (21.23) for ψ allows us to express the function ψ depending on one group element and all the q's in terms of a simpler function of the q's alone. We can now use Eq. (21.29) in Eq. (21.19), and also take account of Eq. (21.21) to obtain the canonical equivalence of the two realizations of $E(3)$:

$$\begin{aligned}R(g) &= R^{(0)}(g)\exp(-\widetilde{\phi(gq_j^{(r)})})\exp(\widetilde{\phi(q_j^{(r)})})\\ &= \exp(-\widetilde{\phi(q_j^{(r)})})R^{(0)}(g)\exp(\widetilde{\phi(q_j^{(r)})})\,;\\ J_j &= \exp(-\widetilde{\phi(q_j^{(r)})})J_j^{(0)},\ P_j=\exp(-\widetilde{\phi(q_j^{(r)})})P_j^{(0)}\,.\end{aligned} \tag{21.30}$$

The argument given above fails if the limiting process involved does not lead to a finite answer. If this happens, it is because $E(3)$ is noncompact and the integration region is not closed. We give next a modified procedure, in which the idea of group averaging is used only for the compact $R(3)$ subgroup of $E(3)$. For this reason, the canonical equivalence is established in two steps. Starting again with Eq. (21.23) and taking both g and g' to be elements of $R(3)$, we average the right-hand side over $R(3)$; this gives a finite answer because the integration region is a closed one, and ψ exists and is finite for all values of its arguments. Thus we obtain:

$$\begin{aligned}\psi((\boldsymbol{\alpha},\mathbf{0});q_j^{(r)}) &\doteq \phi_1(q_j^{(r)})-\phi_1(A_{jk}(\boldsymbol{\alpha})q_k^{(r)})\,,\\ \phi_1(q_j^{(r)}) &= \frac{1}{2\pi^2}\iiint_{|\boldsymbol{\alpha}|\leq\pi} d\alpha_1 d\alpha_2 d\alpha_3\left(\frac{\sin(|\boldsymbol{\alpha}|/2)}{|\boldsymbol{\alpha}|}\right)^2\psi((\boldsymbol{\alpha},\mathbf{0});q_j^{(r)})\,.\end{aligned} \tag{21.31}$$

Specializing Eq. (21.19) to elements in $R(3)$ and using arguments like those needed to obtain Eq. (21.30), we establish the equivalence of the $R(3)$ realizations contained in the two $E(3)$ realizations:

$$\begin{aligned}R((\boldsymbol{\alpha},\mathbf{0})) &= \exp(-\widetilde{\phi_1(q_j^{(r)})})R^{(0)}((\boldsymbol{\alpha},\mathbf{0}))\exp(\widetilde{\phi_1(q_j^{(r)})})\,,\\ J_j &= \exp(-\widetilde{\phi_1(q_j^{(r)})})J_j^{(0)}\,.\end{aligned} \tag{21.32}$$

Again, the equivalence is via a canonical transformation not affecting the q's at all. Let us then perform the inverse of this transformation on $R(g)$, and thus consider in place of the originally given realization of $E(3)$ the equivalent one

$$R'(g)=\exp(\widetilde{\phi_1(q_j^{(r)})})R(g)\exp(-\widetilde{\phi_1(q_j^{(r)})})\,, \tag{21.33}$$

and compare *this* with $R^{(0)}(g)$. Between $R'(g)$ and $R^{(0)}(g)$ there again holds an equation of the form of Eq. (21.19) with some function ψ' in place of ψ; ψ' also obeys the functional relation of Eq. (21.23). However, the point is that now $\psi'(g;q_j^{(r)})$ may be taken to vanish if $g\in R(3)$. Thus we have:

$$\begin{aligned}\psi'(g;q_j^{(r)}) &\doteq \psi'(g'g;q_j^{(r)})-\psi'(g';gq_j^{(r)})\,,\\ \psi'((\boldsymbol{\alpha},\mathbf{0});q_j^{(r)}) &= 0\,.\end{aligned} \tag{21.34}$$

Now in the first line of the above, take $g=(\boldsymbol{\alpha},\mathbf{0})\in R(3)$, $g'=(\mathbf{0},\boldsymbol{a})\in T_3$; then we get:

$$\psi'((\boldsymbol{\alpha},\boldsymbol{a});q_j^{(r)})\doteq\psi'((\mathbf{0},\boldsymbol{a});A_{jk}(\boldsymbol{\alpha})q_k^{(r)})\,. \tag{21.35}$$

Therefore it is enough to deal with the values of $\psi'(g;q_j^{(r)})$ when g is a pure spatial translation; the values for other g are determined in terms of these. Writing $\psi'((\mathbf{0},\boldsymbol{a});q_j^{(r)})$ as $\psi'(a_j;q_j^{(r)})$ for simplicity, we must obtain from Eq.

(21.34) all the conditions on this function. Taking both g and g' to be pure translations, we get:

$$\psi'(a_j; q_j^{(r)}) \doteq \psi'(a_j + b_j; q_j^{(r)}) - \psi'(b_j; q_j^{(r)} + a_j)\,. \tag{21.36}$$

There is one other relation that can be gotten from Eq. (21.34), which arises because T_3 is an invariant subgroup of $E(3)$. With g and g' any two elements and using Eq. (21.34) twice, we have:

$$\begin{aligned}\psi'(g'gg'^{-1}; q_j^{(r)}) &\doteq \psi'(g'^{-1}; q_j^{(r)}) + \psi'(g'g; g'^{-1}q_j^{(r)}) \\ &\doteq \psi'(g'^{-1}; q_j^{(r)}) + \psi'(g; g'^{-1}q_j^{(r)}) \\ &\quad + \psi'(g'; gg'^{-1}q_j^{(r)})\,.\end{aligned} \tag{21.37}$$

Now choose $g = \exp(\boldsymbol{a} \cdot \boldsymbol{d}) \in T_3$ and $g' = \exp(\boldsymbol{\alpha} \cdot \boldsymbol{l}) \in R(3)$; then only the second term on the right survives to give

$$\psi'(A_{jk}(\boldsymbol{\alpha})a_k; q_j^{(r)}) \doteq \psi'(a_j; A_{kj}(\boldsymbol{\alpha})q_k^{(r)})\,,$$

which is better written as

$$\psi'(A_{jk}(\boldsymbol{\alpha})a_k; A_{jk}(\boldsymbol{\alpha})q_k^{(r)}) \doteq \psi'(a_j; q_j^{(r)})\,. \tag{21.38}$$

This relation informs us that ψ' is invariant (in the $\doteq$ sense) if all its arguments $a_j, q_j^{(1)}, \ldots$ are treated as three-vectors and subjected to any rotation. Now we use Eq. (21.36) to express ψ' in terms of a function ϕ' depending on one vector argument less than does ψ'; for this, use the freedom in the choice of the vector b_j to make the first two arguments in ψ' on the right equal but opposed in sign, that is, choose $\boldsymbol{a} + \boldsymbol{b} = -\boldsymbol{q}^{(1)}$:

$$\begin{aligned}\psi'(a_j; q_j^{(1)}, q_j^{(2)}, \ldots) \doteq{}& \psi'(-q_j^{(1)}; q_j^{(1)}, q_j^{(2)}, \ldots) \\ &- \psi'(-q_j^{(1)} - a_j;\, q_j^{(1)} + a_j, q_j^{(2)} + a_j \ldots)\,.\end{aligned} \tag{21.39}$$

This shows that if we define a function $\phi'(q_j^{(1)}, \ldots)$ as follows:

$$\phi'(q_j^{(1)}, q_j^{(2)}, \ldots) = \psi'(-q_j^{(1)}; q_j^{(1)}, q_j^{(2)}, \ldots)\,, \tag{21.40}$$

then we have:

$$\psi'(a_j; q_j^{(r)}) \doteq \phi'(q_j^{(r)}) - \phi'(q_j^{(r)} + a_j)\,. \tag{21.41}$$

In addition, from Eq. (21.38) follows the rotational invariance of $\phi'(q_j^{(r)})$:

$$\begin{aligned}\phi'(A_{jk}(\boldsymbol{\alpha})q_k^{(r)}) &= \phi'(q_j^{(r)})\,, \\ R^{(0)}((\boldsymbol{\alpha}, \mathbf{0}))\exp(\widetilde{\phi'(q_j^{(r)})}) &= \exp(\widetilde{\phi'(q_j^{(r)})})R^{(0)}((\boldsymbol{\alpha}, \mathbf{0}))\,.\end{aligned} \tag{21.42}$$

Going back now to Eq. (21.19) with $R'(g)$ and ψ' in place of $R(g)$ and ψ, and then making use of Eqs. (21.32), (21.33), (21.41), and (21.42), we have:

$$\begin{aligned} R'((\boldsymbol{\alpha},\mathbf{0})) &= \exp(-\widetilde{\phi'(q_j^{(r)})})R^{(0)}((\boldsymbol{\alpha},\mathbf{0}))\exp(\widetilde{\phi'(q_j^{(r)})}), \\ R'((\mathbf{0},\boldsymbol{a})) &= \exp(-\widetilde{\phi'(q_j^{(r)})})R^{(0)}((\mathbf{0},\boldsymbol{a}))\exp(\widetilde{\phi'(q_j^{(r)})}), \end{aligned}$$

and thus quite generally,

$$R'(g) = \exp(-\widetilde{\phi'(q_j^{(r)})})R^{(0)}(g)\exp(\widetilde{\phi'(q_j^{(r)})}). \tag{21.43}$$

Once again we have proved the canonical equivalence of the realizations $R(g)$ and $R^{(0)}(g)$ of $E(3)$, given that they have the same effects on the Cartesian position coordinates $q_j^{(0)}$. Combining Eqs. (21.43) with (21.33), the connection between the original generators J_j, P_j and the kinematic ones is:

$$\begin{aligned} J_j &= \exp(-\widetilde{\phi(q_j^{(r)})})J_j^{(0)}, \\ P_j &= \exp(-\widetilde{\phi(q_j^{(r)})})P_j^{(0)}, \\ \phi(q_j^{(r)}) &= \phi_1(q_j^{(r)}) + \phi'(q_j^{(r)}). \end{aligned} \tag{21.44}$$

These arguments justify to some extent the assumption made in several previous chapters that the $E(3)$ generators can be chosen to be kinematic ones; this may be assumed without loss of generality if the manifest covariance conditions in Eq. (21.11) are assumed. It also follows that if the canonical conjugates $p_j^{(r)}$ have been adjusted so that J_j and P_j are kinematic, then they transform as implied by Eq. (21.13) under $E(3)$; even so, they need not be the same as the kinetic momenta.

The remaining covariance conditions on the description of a Galilean invariant multiparticle system stem from transformations to a moving coordinate system. If S and $S' = \exp(\boldsymbol{v}\cdot\boldsymbol{g})S$ are two inertial frames with a uniform relative velocity $\boldsymbol{v}$ and if we demand that the particle positions transform in the geometric manner, then for any time t we must have:

$$S' = \exp(\boldsymbol{v}\cdot\boldsymbol{g})S : q_j^{(r)'}(t) = e^{\widetilde{-\boldsymbol{v}\cdot\boldsymbol{G}}}q_j^{(r)}(t) \equiv q_j^{(r)}(t) - v_j t. \tag{21.45}$$

(We have used again the Heisenberg picture, in which the same generators G_j are to be used in connecting variables at all times t in S and S'). This finite transformation law implies and is implied by the infinitesimal version:

$$\{G_k, q_j^{(r)}(t)\} = \delta_{kj}t. \tag{21.46}$$

Making use of Eq. (21.11) here, this can be simplified to read:

$$\{G_k + tP_k, q_j^{(r)}(t)\} \equiv e^{\widetilde{-tH}}\{G_k, q_j^{(r)}(0)\} = 0;$$

thus it is enough to demand that at time $t = 0$, we have the relation

$$\{G_k, q_j^{(r)}\} = 0\,. \tag{21.47}$$

The complete set of "world-line" conditions for a Galilean multiparticle system consists then of Eqs. (21.11) and (21.47). These are stated in terms of the particle positions at time zero; but because J_j and P_j are time-independent, and because of the known solution to the equation of motion for G_j, the covariance conditions for general t do hold.

Let us now see what restrictions these world-line conditions impose on the interaction terms in the generators of $\mathcal{G}$. In chapter 19 (page 389 ff) we have obtained the conditions resulting from Galilean invariance. If both G_j and H are dynamic, the corresponding interaction terms W_j and V must obey Eq. (19.66); if only H is dynamic, the situation is simpler and only Eq. (19.67) need be obeyed. In either case we assume that J_j and P_j are kinematic; thus the covariance conditions in Eq. (21.11) are obeyed. Take first the case that both G_j and H are dynamic. Then, because $G_j^{(0)}$ is a function of the $q_j^{(r)}$ alone, the world-line condition Eq. (21.47) demands that W_j also be a function of the q's alone:

$$W_j = W_j(q_k^{(r)})\,. \tag{21.48}$$

Putting this back into Eq. (19.66), the conditions for manifest covariance on a Galilean multiparticle system become the following:

$$W_j(q_k^{(r)}) = \text{a vector under } R(3), \text{ invariant under } T_3\,; \tag{21.49a}$$

$$V(p_k^{(r)}, q_k^{(r)}) = \text{an } E(3) \text{ scalar}\,; \tag{21.49b}$$

$$\{G_j^{(0)}, V\} + \{W_j, H^{(0)}\} + \{W_j, V\} = 0\,. \tag{21.49c}$$

Any set of interaction terms W_j, V obeying these conditions is acceptable.

The situation is much simpler if G_j is also taken to be kinematic, for then the world-line condition Eq, (21.47) is automatically obeyed. The only interaction term V must be an $E(3)$ scalar, which depends only on the differences of the "Hamiltonian velocities" $p^{(r)}/m^{(r)} - p^{(s)}/m^{(s)}$. Thus by choosing any rotationally invariant function of the relative coordinates and relative Hamiltonian velocities for V and taking J_j, P_j, and G_j to be kinematic, we obtain a manifestly covariant multiparticle Galilean system with interaction. In sum, the imposition of manifest covariance does not imply any very strong restrictions on the allowed interactions in a Galilean system.

Manifest Covariance in Lorentz-Invariant Particle Mechanics

The situation changes considerably when we switch from Galilean to Poincaré relativity. The increase in complexity is because the time ascribed to an event depends more intimately on the particular observer than it did before. There are two ways in which we may proceed with the description of a many-particle system: we may as hitherto describe each particle by three position coordinates and their canonical momenta, all functions of one common time parameter, and then derive world-line conditions analogous to Eqs. (21.11) and (21.47), but now involving the generators K_j in place of the G_j. These conditions express the requirement that the four quantities $(q_j(t), t)$ transform like the space-time coordinates of an event in the Lorentzian manner. One can examine the consequences of adding these world-line conditions to the already existing PB relations corresponding to the Lie algebra of $\mathcal{P}$. When we do this, we see that no interactions are allowed! The alternative method of obtaining manifest relativistic covariance is to use a parameter formalism: the events constituting the world-line of a particle may be labeled by some new parameter chosen independently of any observer, and then the four space and time coordinates of each point on the world line will be functions of the parameter. These functions change linearly according to the Lorentz-transformation formulae when we change from one inertial frame to another one. We describe the latter formalism first, and then go into the details of the former.

We consider the description of a free particle, and then of a charged particle in a given electromagnetic field as examples of the relativistic parameter formalism. We have already given the relativistic Lagrangian and Hamiltonian description of a free particle in the preceding chapter; each event on the world line of the particle is identified by its time coordinate, and the three spatial coordinates are functions of the time. The latter three quantities are the true dynamical variables. Although this treatment is fully in accord with the requirement of relativistic invariance, it is not manifestly relativistic in appearance; on the other hand, it is suited to the physical interpretation. Now we introduce a parameter θ to label the points on the world line of the particle; this parameter increases monotonically from $-\infty$ to $+\infty$ as the world line is traversed from the infinite past to the infinite future. Given an inertial frame S, a point P on the world line is assigned four space-time coordinates $x^\mu(\theta), \theta$ being the value of the parameter at P and x^μ four functions of θ. In a different inertial frame $S' = \Lambda S, P$ is assigned the coordinates $x'^\mu(\theta)$ given by:

$$S' = \Lambda S: \; x'^\mu(\theta) = \Lambda^\mu{}_\nu x^\nu(\theta)\,, \tag{21.50}$$

Λ being some element of the HLG. The point is that θ does not change in going from S to S'. We regard all four functions $x^\mu(\theta)$ as dynamical quantities; the Lagrangian equations of motion are differential equations with θ as the independent variable. When we go over to the Hamiltonian formalism, it is the "development in θ" rather than the development in time that appears as the gradual unfolding of a canonical transformation; the corresponding generator is a relativistic invariant "Hamiltonian" differing from the energy. Because each infinitesimal stretch of the world line must be a timelike vector pointing to the future, we have the nonholonomic constraints:

$$\frac{dx^\mu}{d\theta}\frac{dx_\mu}{d\theta} \equiv \frac{dx_j}{d\theta}\frac{dx_j}{d\theta} - \left(\frac{dt}{d\theta}\right)^2 \langle 0, \frac{dt}{d\theta}\rangle 0\,. \tag{21.51}$$

One immediately evident feature of this treatment is that we necessarily deal with a constrained system. This is because there is no unique way of choosing the parameter θ; θ can always be replaced by θ' where the latter is any monotonically increasing function of the former, and vice versa. It follows that the Lagrangian equations of motion must reflect this arbitrariness and cannot determine all four "accelerations" $d^2x^\mu/d\theta^2$; the Lagrangian must be singular. The most general solution to the equations of motion must involve (at least) one arbitrary function. This, then, indicates that in the Hamiltonian version we must find (at least) one primary first class constraint, because it is the coefficients of such constraint functions that enter as arbitrary functions (of θ) in the Hamiltonian.

With this description in advance, let us proceed. The free particle Lagrangian $L(\theta)$ to be used can be identified by rewriting the expression for the action, given as a time integral of the Lagrangian of Eq. (20.62) as an integral with respect to θ.

$$\Phi[C] = \int_C dt L(t) = -\, m \int_C dt \left(1 - \frac{dx_j}{dt}\frac{dx_j}{dt}\right)^{\frac{1}{2}} = \int_C d\theta L(\theta)\,,$$
$$L(\theta) = -\, m(-\dot{x}_\mu \dot{x}^\mu)^{\frac{1}{2}}\,. \tag{21.52}$$

The x^μ being functions of θ, the dot now signifies differentiation with respect to θ. In writing out the Lagrangian equations of motion, let us introduce the matrix $||W_{\mu\nu}||$ of second partial derivatives of L with respect to the velocities (see Eq. (8.1)):

$$\frac{\partial L}{\partial \dot{x}^\mu} \equiv m\dot{x}_\mu/(-\dot{x}_\lambda \dot{x}^\lambda)^{\frac{1}{2}}\,,$$
$$W_{\mu\nu} \equiv \frac{\partial^2 L}{\partial \dot{x}^\mu \partial \dot{x}^\nu} = [m/(-\dot{x}_\lambda \dot{x}^\lambda)^{3/2}](\dot{x}_\mu \dot{x}_\nu - g_{\mu\nu}\dot{x}_\rho \dot{x}^\rho)\,. \tag{21.53}$$

Then the equations of motion are:

$$W_{\mu\nu}\ddot{x}^\nu = 0\,. \tag{21.54}$$

The singular nature of L follows from the fact that $||W_{\mu\nu}||$ possesses the null eigenvector $\dot{x}^\nu$; however, because the right-hand side in Eq. (21.54) vanishes, no constraints are generated as a result of this fact, and one is simply unable to solve for all the "accelerations". We can express three of them in terms of the fourth one and the four velocities:

$$\ddot{x}_j = \left(\frac{\ddot{x}_0}{\dot{x}_0}\right)\dot{x}_j\,. \tag{21.55}$$

Choosing $x^0(\theta)$ in any way we like, as long as $\dot{x}^0 > 0$, these equations can be solved to express each x_j as a function of θ; one arbitrary function appears in this solution. The solution is easily obtained; for each value of j, we have from Eq. (21.55):

$$\ddot{x}_j/\dot{x}_j = \ddot{x}^0/\dot{x}^0 \quad \Rightarrow \quad \frac{d}{d\theta}\ln \dot{x}_j = \frac{d}{d\theta}\ln \dot{x}^0 \Rightarrow \dot{x}_j = a_j \dot{x}^0$$
$$\Longrightarrow \quad x_j(\theta) = a_j x^0(\theta) + b_j\,. \tag{21.56}$$

Once the arbitrary (monotonic) function $x^0(\theta)$ is chosen, the dependence of x_j on θ is also determined. But though the structure of this singular Lagrangian system is quite transparent, the situation is so simple that the appearance of an arbitrary function of θ on the right in Eq. (21.56) is somewhat irrelevant; in each inertial frame and for each possible world line of the particle, the space coordinates are already expressed as functions of the time.

Turning to the invariance and quasi-invariance properties of L and conservation laws, if we consider an infinitesimal Poincaré transformation causing the variation

$$\delta x^\mu(\theta) = \delta\lambda^\mu{}_\nu x^\nu(\theta) + \delta a^\mu, \delta\lambda^{\mu\nu} + \delta\lambda^{\nu\mu} = 0\,, \tag{21.57}$$

then L is easily seen to be invariant. Previously, when dealing with the Lagrangian $L(t)$ with time as the parameter, we had only quasi-invariance under pure Lorentz transformations. The invariance of L under Eq. (21.57) gives a set of conservation laws:

$$\begin{aligned} \delta x^\mu(\theta)\frac{\partial L}{\partial \dot{x}^\mu} &= \text{constant} \implies \\ p_\mu \equiv \frac{m\dot{x}_\mu}{(-\dot{x}_\lambda \dot{x}^\lambda)^{\frac{1}{2}}} &= \text{constant}\,, \\ J_{\mu\nu} = \frac{m(x_\mu \dot{x}_\nu - x_\nu \dot{x}_\mu)}{(-\dot{x}_\lambda \dot{x}^\lambda)^{\frac{1}{2}}} &= \text{constant}\,. \end{aligned} \tag{21.58}$$

"Constant" now means "constant in θ"; thus these constants of motion preserve their values as θ varies; in other words, they have the same values at all points of the world line of the particle. In addition, because they are homogeneous functions of degree zero in the velocities, their values for a given state of motion and in a given inertial frame do not depend on the particular parameter θ chosen to describe the trajectory in space-time. In other words, they are invariant under changes in choice of θ. The p_μ is, of course, the energy-momentum four-vector; J_{kl}, the space-space components of $J_{\mu\nu}$, are the angular momenta; and the constancy of J_{0k} gives back the familiar solution to the equations of motion *in time* for the relativistic moment.

The freedom in the choice of the parameter θ manifests itself in the quasi-invariance of L under the infinitesimal variation

$$\delta x^\mu(\theta) = \epsilon \dot{x}^\mu(\theta) f(\theta) \tag{21.59}$$

wherein ϵ is a small parameter and $f(\theta)$ an arbitrary function. This variation results from changing the value of the parameter assigned to a general point P on the trajectory from θ to $\theta - \epsilon f(\theta)$. Under Eq. (21.59), we do find:

$$\delta L = \frac{d}{d\theta}(-m\epsilon f(\theta)(-\dot{x}_\mu \dot{x}^\mu)^{\frac{1}{2}})\,. \tag{21.60}$$

The action principle would yield a constant of motion, however, that would be proportional to the completely arbitrary function $f(\theta)$; the only way out is for this constant of motion to vanish identically, and indeed this is just what happens:

$$\frac{\partial L}{\partial \dot{x}_\mu}\delta x_\mu + m\epsilon f(\theta)(-\dot{x}_\mu \dot{x}^\mu)^{\frac{1}{2}} = \frac{m\dot{x}_\mu}{(-\dot{x}_\lambda \dot{x}^\lambda)^{\frac{1}{2}}}\epsilon \dot{x}^\mu f + m\epsilon(-\dot{x}_\mu \dot{x}^\mu)^{\frac{1}{2}} f = 0\,. \quad (21.61)$$

To pass over to the Hamiltonian form, we must define four canonical momenta, one corresponding to each x^μ; the values of these p_μ already appear in Eq. (21.58):

$$p_\mu = \frac{\partial L}{\partial \dot{x}^\mu} = \frac{m\dot{x}_\mu}{(-\dot{x}_\lambda \dot{x}^\lambda)^{\frac{1}{2}}}\,. \quad (21.62)$$

It is obvious that these are not algebraically independent functions of the velocities; we have one primary constraint connecting the p's alone:

$$\phi(p) \equiv p^\mu p_\mu + m^2\,,\ \phi \approx 0\,. \quad (21.63)$$

This was anticipated. If we now make use of Eq. (21.62) and compute the Hamiltonian $p_\mu \dot{x}^\mu - L$, Euler's theorem on homogeneous functions tells us that this expression vanishes; in other words, when the constraint Eq. (21.63) is obeyed, the Hamiltonian vanishes. According to Chapter 8, we must adopt as the general Hamiltonian a multiple of the constraint function ϕ, the coefficient being to start with an arbitrary function of θ:

$$H = v\phi(p)\,. \quad (21.64)$$

The unknown v stands for the velocity variable which could not be solved for in terms of the p's and x's using Eq. (21.62). We see that it is or can be chosen to be essentially $\dot{x}^0$. We now have the basic brackets:

$$\{x^\mu, p_\nu\} = \delta^\mu_\nu\,,\ \{x^\mu, x^\nu\} = 0, \{p_\mu, p_\nu\} = 0 \quad (21.65)$$

and the Hamiltonian equations of motion

$$\dot{x}^\mu \approx \{x^\mu, H\} \approx vp^\mu\,,\ \dot{p}^\mu \approx \{p^\mu, H\} \approx 0\,; \quad (21.66)$$

finally the primary constraint Eq. (21.63) applies. According to the general theory of Chapter 8, we must see whether any conditions are imposed on v by demanding the primary constraints be maintained for all θ, and also whether any further (secondary) constraints are generated. In the present simple situation, both H and ϕ are functions of p_μ alone; thus neither of these things happens. We just have the one constraint $\phi \approx 0$, and that turns into a first-class constraint.

It may seem a little puzzling at first sight that the Hamiltonian should itself vanish, yet lead to nontrivial equations of motion. The point, of course, is that it only vanishes weakly. The situation is clarified by showing that if we eliminate the arbitrary function $v(\theta)$ from the Hamiltonian equations of motion, we simply recover the Lagrangian Eq. (21.55), as we must:

$$\begin{aligned} \dot{x}^\mu \approx vp^\mu \Rightarrow \ddot{x}^\mu \approx \dot{v}p^\mu \approx \frac{\dot{v}}{v}\dot{x}^\mu \Rightarrow \frac{\dot{v}}{v} &= \frac{\ddot{x}^0}{\dot{x}^0}, \\ \ddot{x}^\mu &= \frac{\ddot{x}^0}{\dot{x}^0}\dot{x}^\mu . \end{aligned} \tag{21.67}$$

(We choose to express v essentially in terms of $\dot{x}^0$, making this the unknown velocity). Thus the presence of the arbitrary coefficient v in H amounts to the same thing as having to specify $x^0(\theta)$ before being able to solve the Lagrangian equations.

We can write the constants of motion Eq. (21.58) as functions of the p's and x's alone:

$$P_\mu = p_\mu, J_{\mu\nu} = x_\mu p_\nu - x_\nu p_\mu . \tag{21.68}$$

Using the PB's Eq. (21.65), we check that P_μ and $J_{\mu\nu}$ obey the Lie algebra relations of $\mathcal{P}$, so that they generate a canonical realization of $\mathcal{P}$ in the eight-dimensional phase space of the p's and x's. However, all these generators have vanishing PB with $\phi(p)$, which ensures that the realization of $\mathcal{P}$ carries the constraint hypersurface into itself so that the constraint $\phi \approx 0$ is compatible with relativistic invariance (this is manifest from the form of ϕ).

As the next example of the use of the parameter formalism, consider the interaction of a charged particle with mass m and charge e with an externally specified electromagnetic field. The basic Lagrangian variables again are the four $x^\mu(\theta)$, whereas the external field is described by a numerical four-vector potential function $A_\mu(x)$. The Lagrangian for the system is:

$$L(x(\theta), \dot{x}(\theta)) = -m(-\dot{x}_\mu\dot{x}^\mu)^{\frac{1}{2}} + eA_\mu(x(\theta))\dot{x}^\mu . \tag{21.69}$$

In the interaction term, A_μ is to be evaluated at the world point of the particle. The Lagrangian equations of motion reproduce the Lorentz force properly:

$$\begin{aligned} p_\mu = \frac{\partial L}{\partial \dot{x}^\mu} &= \frac{m\dot{x}_\mu}{(-\dot{x}_\lambda\dot{x}^\lambda)^{\frac{1}{2}}} + eA_\mu(x(\theta)); \\ \frac{\partial L}{\partial x^\mu} &= e\frac{\partial A_\nu}{\partial x^\mu}\dot{x}^\nu ; \\ \frac{d}{d\theta}\frac{\partial L}{\partial \dot{x}^\mu} &= \frac{\partial L}{\partial x^\mu} \Rightarrow \frac{d}{d\theta}\left(\frac{m\dot{x}_\mu}{(-\dot{x}_\lambda\dot{x}^\lambda)^{\frac{1}{2}}}\right) = e\left(\frac{\partial A_\nu}{\partial x^\mu} - \frac{\partial A_\mu}{\partial x^\nu}\right)\dot{x}^\nu \\ &= eF_{\mu\nu}(x(\theta))\dot{x}^\nu , \end{aligned}$$

that is,

$$W_{\mu\nu}\ddot{x}^\nu = eF_{\mu\nu}(x(\theta))\dot{x}^\nu . \tag{21.70}$$

The matrix $\|W_{\mu\nu}\|$ is the same as appears in Eq. (21.53); there is no change because the interaction term in L is linear in the velocities. Once again, we have a singular Lagrangian system, because $\|W_{\mu\nu}\|$ is singular and not all accelerations can be obtained. But as in the free-particle case, the existence of a null eigenvector for $\|W_{\mu\nu}\|$ does not lead to any new constraints, just because $F_{\mu\nu}$ is an antisymmetric tensor. Therefore, for the sake of a manifestly relativistic appearance, we have a set of four equations of motion in Eq. (21.70). There are actually only three independent equations here, and we may assign the one function $x^0(\theta)$ in any way we please (subject to $\dot{x}^0 > 0$).

The Hamiltonian treatment introduces the momenta p_μ as in Eq. (21.70), and the PB's Eq. (21.65). There is the primary constraint:

$$\phi'(p,x) = (p_\mu - eA_\mu(x))(p^\mu - eA^\mu(x)) + m^2 \approx 0 \tag{21.71}$$

now involving both p and x; because the Lagrangian Eq. (21.69) is again homogeneous of degree one in the velocities, the general Hamiltonian is an arbitrary multiple of ϕ':

$$H(p,x) = v(\theta)\phi'(p,x)\,. \tag{21.72}$$

Because there is just one primary constraint and H is proportional to it, the condition $\phi' \approx 0$ is maintained in θ; thus it becomes a first-class constraint leaving v unconstrained.

Our treatment of these singular Lagrangians arising out of the use of the parameter formalism followed the lines of Chapter 8, in that one coordinate is to be specified in its dependence on the parameter before the Lagrangian equations can be solved. A common choice for the parameter θ is the "proper time" of the particle; let us see how this fits in with the general treatment of singular Lagrangians. This choice for θ amounts to imposing a constraint on the Lagrangian velocity variables $\dot{x}^\mu$,

$$\dot{x}^\mu \dot{x}_\mu \equiv \frac{dx_j}{d\theta}\frac{dx_j}{d\theta} - \left(\frac{dx^0}{d\theta}\right)^2 = -1 \tag{21.73}$$

after the Lagrangian equations of motion such as Eqs. (21.54) or (21.70) have been obtained from the action principle. This is a constraint introduced "from the outside," because it is not a consequence of the action principle. In the terminology of Chapter 8, this is a Lagrangian constraint of type B, because the velocity variables are involved. A single differentiation of Eq. (21.73) leads to one more equation for the accelerations, namely,

$$\dot{x}_\mu \ddot{x}^\mu = 0\,, \tag{21.74}$$

and now we have enough equations to solve for all four accelerations. It can be easily seen that the left-hand sides of Eqs. (21.54) and (21.70) both reduce to $m\ddot{x}_\mu$ on using Eqs. (21.73) and (21.74). This method of making the equations of motion "determinate" is preferable to the earlier one of specifying $x^0(\theta)$, in that it maintains the manifest relativistic invariance throughout. But one must not think that we now have four independent equations for the $\ddot{x}^\mu$; we do not, because Eq. (21.74) must be satisfied identically, and these constraint equations must be carried along.

We have seen how the parameter formalism can be used to maintain explicit relativistic invariance, and also how the Lagrangian and Hamiltonian treatments work out. But this Hamiltonian treatment exists only for a single particle, not for two or more particles! The reason is simple: a straightforward extension of the last few pages would be to introduce one labeling parameter θ for the world-line of each particle; thus we have as many parameters as particles. But then even if the dynamics of the multiparticle system is obtained from an action principle, the expression for the action generally involves integrations over the world-lines of several particles, and in fact the action is not the integral with respect to one unique parameter θ of some function that could be interpreted as the Lagrangian of the system. Going even further, it is possible to set up directly relativistically invariant equations of motion for multiparticle systems that do not have their basis in an action principle at all. The most important example of the former is a formulation of classical electrodynamics in which one does not introduce the electromagnetic field at all but deals directly with interparticle interactions at a distance; the interaction terms in the Action are double-integrals over the world-lines of pairs of charged particles. We will study such theories in the following chapter. Aside from the fact that there are various nontrivial difficulties in setting up a consistent dynamical theory of this kind, we wish only to note here that it does not yield to a canonical Hamiltonian formulation. An example of the second-mentioned variety of manifestly relativistic particle theories with interaction is due to Wigner and van Dam. [E. P. Wigner and H. van Dam. Phys. Rev., **138**, B1576 (1965).] Here one directly writes down equations of motion for particle coordinates, maintaining explicit relativistic invariance; these equations of motion are actually not "local" differential equations (by "local" we mean "local with respect to the independent parameter") but integro-differential systems. Theories of this type too are non-canonical.

World-Line Conditions for Relativistic Particle Mechanics

We revert to the canonical formalism with all dynamical variables functions of one independent time parameter. For a multiparticle system the basic variables are the Cartesian positions $q_j^{(r)}$ and the canonical momenta $p_j^{(r)}$. Manifest covariance for the positions under $E(3)$ is expressed again by Eq. (21.11); thus we assume that the $p_j^{(r)}$ have been chosen so as to put J_j and P_j into kinematic forms. We must develop next the relativistic analogue to the nonrelativistie relation Eq. (21.47) that guarantees the geometric transformation laws in going from one frame to another by a velocity transformation. Let S and S' be two inertial frames connected by a pure Lorentz transformation, $S' = \exp(\boldsymbol{\zeta} \cdot \boldsymbol{k})S$. The canonical transformation $e^{-\widetilde{\boldsymbol{\zeta}\cdot\boldsymbol{K}}}$ acting on some dynamical quantity referring to time t in S produces the corresponding quantity at time t in S'. For an event occurring at a space-time point P, the coordinates x^μ and x'^μ assigned in

S and S' are related by

$$\begin{aligned} x'^{\mu} &= \Lambda^{\mu}{}_{\nu}(\boldsymbol{\zeta},\mathbf{0})x^{\nu}\,; \\ \Lambda^{j}{}_{k} &= \delta_{jk}+\frac{\cosh\zeta-1}{\zeta^2}\zeta_j\zeta_k\,, \\ \Lambda^{j}{}_{0} &= \Lambda^{0}{}_{j}=-\frac{\sinh\zeta}{\zeta}\zeta_j\,, \\ \Lambda^{0}{}_{0} &= \cosh\zeta\,. \end{aligned} \tag{21.75}$$

Let P be a point on the world line of one of the particles, its coordinates in S being $(t, q_j(t))$. (For the moment, the particle number is omitted.) The time assigned to P in S' is computed using the Lorentz transformation formula:

$$t' = t\cosh\zeta - \frac{\sinh\zeta}{\zeta}\zeta_k q_k(t)\,. \tag{21.76}$$

If the dynamical variable describing the particle position in S' is written q'_j, we have two ways of computing the space coordinates of P in S'. One way is to use the Lorentz formulae, which gives:

$$q'_j(t') = q_j(t) + \frac{\cosh\zeta-1}{\zeta^2}\zeta_j\zeta_k q_k(t) - \frac{\sinh\zeta}{\zeta}\zeta_j t\,. \tag{21.77}$$

The other way is to use the canonical transformation formula:

$$q'_j(t') = (e^{\widetilde{-\boldsymbol{\zeta}\cdot\boldsymbol{K}}}q_j(\sigma))_{\sigma=t'=t\cosh\zeta-(\sinh\zeta/\zeta)\zeta_k q_k(t)}\,. \tag{21.78}$$

The equality of these two results gives the finite form of the relativistic world-line conditions:

$$\left(e^{\widetilde{-\boldsymbol{\zeta}\cdot\boldsymbol{K}}}q_j(\sigma)\right)_{\sigma=t\cosh\zeta-(\sinh\zeta/\zeta)\zeta_k q_k(t)} = q_j(t) + \frac{\cosh\zeta-1}{\zeta^2}\zeta_j\zeta_k q_k(t) - \frac{\sinh\zeta}{\zeta}\zeta_j t\,. \tag{21.79}$$

It is well to appreciate the somewhat special nature of what appears on the left-hand side of this condition: we are instructed to first compute the effect of a canonical transformation on the dynamical variable $q_j(\sigma)$ where σ is treated just as a parameter, and *then* substitute for σ a value that involves a dynamical variable. Let us obtain an infinitesimal form of this condition, and then prove that such a form is also sufficient for the validity of Eq. (21.79). To first order in ζ_j, we have $\sigma = t - \zeta_k q_k(t)$; thus the condition becomes:

$$\begin{aligned} (q_j(\sigma))_{\sigma=t-\zeta_k q_k(t)} - (\widetilde{\boldsymbol{\zeta}\cdot\boldsymbol{K}}q_j(\sigma))_{\sigma=t} &= q_j(t)-\zeta_j t\,, \\ -\zeta_k q_k(t)\{q_j(t),H\} - \zeta_k\{K_k,q_j(t)\} &= -\zeta_j t\,, \\ \zeta_k\{K_k+P_k t, q_j(t)\} &= \zeta_k q_k(t)\{H,q_j(t)\}\,, \\ \{e^{\widetilde{-tH}}K_k, e^{\widetilde{-tH}}q_j\} &= e^{\widetilde{-tH}}(q_k\{H,q_j\})\,, \\ \{K_k,q_j\} &= q_k\{H,q_j\}\,. \end{aligned} \tag{21.80}$$

This is the replacement for the Galilean condition Eq. (21.47); we refer to it as the relativistic world-line condition. It is stated in terms of dynamical quantities referring to time zero.

It must be intuitively evident that Eq. (21.80) implies Eq. (21.79), but the explicit proof is somewhat more involved than the corresponding proof in the Galilean case that Eq. (21.47) implies Eq. (21.45). Given the two inertial frames S and $S' = \exp(\boldsymbol{\zeta} \cdot \boldsymbol{k})S$, imagine a one-parameter family of inertial frames $S_\lambda = \exp(\lambda\hat{\boldsymbol{\zeta}} \cdot \boldsymbol{k})S$. For a point P on the particle trajectory, with coordinates $(t, q_j(t))$ in S, imagine that we compute the coordinates in S_λ by a sequence of infinitesimal steps: to go from the coordinates assigned to P in S_λ to those assigned to P in $S_{\lambda+\delta\lambda}$, we use the Lorentz-transformation formula for the time coordinate and the canonical transformation formula for the space coordinates. If the dynamical variables of position in S_λ are $q_j^{(\lambda)}$, P has coordinates $(t(\lambda), q_j^{(\lambda)}(t(\lambda)))$ in S_λ. We get a differential equation for $t(\lambda)$ using Eq. (21.76):

$$t(\lambda + \delta\lambda) = t(\lambda)\cosh\delta\lambda - \sinh\delta\lambda \cdot \hat{\zeta}_k q_k^{(\lambda)}(t(\lambda)),$$

that is,

$$\frac{d}{d\lambda}t(\lambda) = -(\hat{\zeta}_k q_k^{(\lambda)}(\sigma))_{\sigma=t(\lambda)}. \tag{21.81}$$

This is simply the infinitesimal form of a pure Lorentz transformation. We get a differential equation for the space coordinates of P by using the canonical transformation formula. Here we must keep in mind the fact that because the frames S_λ are connected by a one-parameter group of transformations, one and the same generator $\hat{\boldsymbol{\zeta}} \cdot \boldsymbol{K}$ is to be used to connect them to one another. Thus we get:

$$(q_j^{(\lambda+\delta\lambda)}(\sigma))_{\sigma=t(\lambda+\delta\lambda)} = (e^{-\delta\lambda\widetilde{\hat{\boldsymbol{\zeta}}\cdot\boldsymbol{K}}} q_j^{(\lambda)}(\sigma))_{\sigma=t(\lambda+\delta\lambda)}. \tag{21.82}$$

To expand the right-hand side in $\delta\lambda$, we need to know the Hamiltonian to be used in the frame S_λ; this is

$$H_\lambda = e^{-\lambda\widetilde{\hat{\zeta}\cdot\boldsymbol{K}}}H = H\cosh\lambda - \hat{\boldsymbol{\zeta}}\cdot\boldsymbol{P}\sinh\lambda. \tag{21.83}$$

Thus we expand the right-hand side of Eq. (21.82) as:

$$\begin{aligned}(e^{-\delta\lambda\widetilde{\hat{\zeta}\cdot\boldsymbol{K}}} q_j^{(\lambda)}(\sigma))_{\sigma=t(\lambda+\delta\lambda)} &\simeq (q_j^{(\lambda)}(\sigma))_{\sigma=t(\lambda+\delta\lambda)} - \delta\lambda(\hat{\zeta}_k\{K_k, q_j^{(\lambda)}(\sigma)\})_{\sigma=t(\lambda)} \\ &\simeq (q_j^{(\lambda)}(\sigma))_{\sigma=t(\lambda)} - \frac{dt(\lambda)}{d\lambda}\delta\lambda(\{H_\lambda, q_j^{(\lambda)}(\sigma)\})_{\sigma=t(\lambda)} \\ &\quad - \delta\lambda(\hat{\zeta}_k\{K_k, q_j^{(\lambda)}(\sigma)\})_{\sigma=t(\lambda)}.\end{aligned} \tag{21.84}$$

Putting this into Eq. (21.82) we get a differential equation for $q_j^{(\lambda)}$:

$$\frac{d}{d\lambda}\left\{\left(q_j^{(\lambda)}(\sigma)\right)_{\sigma=t(\lambda)}\right\} = -\left(\left\{H_\lambda, q_j^{(\lambda)}(\sigma)\right\}\frac{dt(\lambda)}{d\lambda} + \hat{\zeta}_k\left\{K_k, q_j^{(\lambda)}(\sigma)\right\}\right)_{\sigma=t(\lambda)} \tag{21.85}$$

Therefore, the pair of differential Eqs. (21.81) and (21.85) results from using the geometric formula for t and the canonical formula for q_j. Now suppose the world-line condition is valid in the form of Eq. (21.80). Multiplying both sides of this condition by $\hat{\zeta}_k$ and then applying the canonical transformation $e^{-\lambda\widetilde{\hat{\zeta}\cdot\boldsymbol{K}}}$ to both sides gives:

$$\hat{\zeta}_k\{K_k, q_j^{(\lambda)}(0)\} = \hat{\zeta}_k q_k^{(\lambda)}(0)\{H_\lambda, q_j^{(\lambda)}(0)\}\,. \tag{21.86}$$

Applying next the time translation $e^{\widetilde{-\sigma H_\lambda}}$ in the frame S_λ,

$$\hat{\zeta}_k\{K_k + \sigma P_{\lambda k}, q_j^{(\lambda)}(\sigma)\} = \hat{\zeta}_k q_k^{(\lambda)}(\sigma)\{H_\lambda, q_j^{(\lambda)}(\sigma)\}\,,$$

that is,

$$\hat{\zeta}_k\{K_k, q_j^{(\lambda)}(\sigma)\} - \hat{\zeta}_k q_k^{(\lambda)}(\sigma)\{H_\lambda, q_j^{(\lambda)}(\sigma)\} = \sigma\hat{\zeta}_j\,. \tag{21.87}$$

This is true for any σ. Putting in the value $\sigma = t(\lambda)$, and using both Eqs. (21.81) and (21.87), the right-hand side of Eq. (21.85) simplifies, and we simply have:

$$\frac{d}{d\lambda}\{(q_j^{(\lambda)}(\sigma))_{\sigma=t(\lambda)}\} = -\hat{\zeta}_j t(\lambda)\,. \tag{21.88}$$

The pair of equations, (21.81) and (21.88), can now be solved immediately because the boundary conditions at $\lambda = 0$ are known; namely the coordinates $(t, q_j(t))$ of P in S are known. We get:

$$\begin{aligned} t(\lambda) &= t\cosh\lambda - \sinh\lambda\hat{\zeta}_k q_k(t)\,, \\ (q_j^{(\lambda)}(\sigma))_{\sigma=t(\lambda)} &= q_j(t) + \hat{\zeta}_j\hat{\zeta}_k(\cosh\lambda - 1)q_k(t) - \hat{\zeta}_j\sinh\lambda\cdot t\,. \end{aligned} \tag{21.89}$$

This is simply the result we would have obtained had we used the Lorentz transformation formulae for q_j as well. Therefore, the infinitesimal form of the relativistic world-line condition implies the validity of the finite form of Eq. (21.79).

The generators of $\mathcal{P}$ for the realization that describes a single spinless free particle are given in Eq. (20.50). Whereas covariance conditions under $E(3)$ are obviously obeyed, we can see that the relativistic world-line condition Eq. (21.80) also is valid because K_k coincides with Hq_k. This fact justifies the explanation given in the preceding chapter for the transformation law of position under a velocity transformation, worked out in Eq. (20.54) there. On the other hand, for a free particle with spin, the generator K_k contains besides Hq_k a spin-dependent term *that is also momentum dependent*, thus the world-line condition is violated. For such a particle, therefore, we must give up the geometrical transformation law for position and adopt only the one given by the canonical transformation.

The No-Interaction Theorem

Let us now list the entire set of manifest covariance conditions for a multi-particle system within the special theory of relativity:

$$\{J_j, q_k^{(r)}\} = \epsilon_{jkl} q_l^{(r)}, \ \{P_j, q_k^{(r)}\} = -\delta_{jk} \, ; \tag{21.90a}$$

$$\{K_j, q_k^{(r)}\} = q_j^{(r)} \{H, q_k^{(r)}\} \, . \tag{21.90b}$$

The generators J_j, P_j, K_j and H are functions of $q_k^{(r)}, p_k^{(r)}$ and among themselves obey the PB relations corresponding to the Lie algebra of $\mathcal{P}$. It was proved by Currie, Jordan, and Sudarshan that, for the case of two particles, this combined set of conditions is so restrictive that no interactions are allowed. (D.G. Currie, T.F. Jordan and E.C.G. Sudarshan, *Rev. Mod. Phys.*,**35**, 350 (1963)). They showed that by means of a canonical transformation that leaves the $q_k^{(r)}$ invariant the generators of $\mathcal{P}$ could all be made to assume the free particle forms. This proof was extended to the case of three particles by Cannon and Jordan and to the case of an arbitrary but finite number of particles by Leutwyler. (J. T. Cannon and T. F. Jordan, *J. Math. Phys.* **5**, 299 (1964). H. Leutwyler, Nuovo Cimento, **37**, 556 (1965)). We describe here the proof due to Leutwyler. Because a fair amount of work is involved and we wish to get the main ideas across as clearly as possible, at several points in the argument we introduce plausible and provable assumptions concerning the forms of the various functions involved; we take care to mention explicitly these points, and at the end of the argument we indicate how they may be proved. In addition to the manifest covariance conditions in Eq. (21.90) and, of course, the PB relations among the generators of $\mathcal{P}$, Leutwyler makes essential use of one further assumption, namely that the Hamiltonian of the many-particle system is of standard type; this means that the transition to a Lagrangian is possible in the normal manner, or that the matrix made up of the second derivatives of H, with respect to the p's, is nonsingular:

$$\det \left\| \frac{\partial^2 H}{\partial p_k^{(r)} \partial p_l^{(s)}} \right\| \neq 0 \, . \tag{21.91}$$

One might recall at this point the fact that such a property for H is not a canonically invariant one (see Chapter 8). In any case, we make use of it.

By virtue of Eq. (21.90a) we know that J_j and P_j can be taken as kinematic, so that $E(3)$ acts on $q_j^{(r)}$ and $p_j^{(r)}$ in the obvious way. It is now worthwhile to list in one place those other relations that we need to use a great deal; apart from the statements that H is an $E(3)$ invariant and K_j a vector under spatial rotations, these are:

$$\{K_j, q_k^{(r)}\} = q_j^{(r)} \{H, q_k^{(r)}\} \, , \tag{21.92a}$$

$$\{K_j, P_k\} = \delta_{jk} H \, , \tag{21.92b}$$

$$\{K_j, H\} = P_j \, . \tag{21.92c}$$

We first make use of Eqs. (21.92a) and (21.92b) to deduce as much as we can about the functional forms of K_j and H, and then turn to Eq. (21.92c). The world-line condition in Eq. (21.92a) reads in detail:

$$\frac{\partial}{\partial p_k^{(r)}} K_j = q_j^{(r)} \frac{\partial}{\partial p_k^{(r)}} H \,. \tag{21.93}$$

If we take the derivative of both sides with respect to $p_l^{(s)}$, the left-hand side is then symmetric under simultaneous interchange of r and s, k and l, thus the right must necessarily be symmetric too. This gives:

$$(q_j^{(r)} - q_j^{(s)}) \frac{\partial^2 H}{\partial p_k^{(r)} \partial p_l^{(s)}} = 0, \text{ no sum on } r, s \,. \tag{21.94}$$

For different particle labels, that is, $r \neq s$, it follows that the second derivative of H appearing here must vanish or that H is a sum of terms each of which is a function of the canonical momenta of one particle only (and all the q's); in other words, H must have the form:

$$H = \sum_r h^{(r)}(q, p^{(r)}) \,, \tag{21.95}$$

the rth term depending on all the q's (written collectively as q) and on $p^{(r)}$.

The world-line condition has thus led to some information on the way in which H can depend on the p's. But there is some more information to be obtained from this condition, relating to the dependence of the K_j on the p's. Let us use the form obtained in Eq. (21.95) for H, in Eq. (21.93); we can write the result as:

$$\frac{\partial}{\partial p_k^{(r)}} (K_j - q_j^{(r)} h^{(r)}(q, p^{(r)})) = 0,$$

$$\frac{\partial}{\partial p_k^{(r)}} (K_j - \sum_s q_j^{(s)} h^{(s)}(q, p^{(s)})) = 0 \,,$$

$$K_j = \sum_s q_j^{(s)} h^{(s)}(q, p^{(s)}) + k_j(q) \,. \tag{21.96}$$

We simply used the fact that each h depends on only one triplet of p's, to go from the first to the second line. Therefore, the dependence of H on the p's fixes that of K_j as well, and, in fact, knowing $h^{(s)}$, one needs only to be given the three functions $k_j(q)$ to get K_j. Now there is no more information to be gotten from the world-line conditions (Eq. (21.92a)), because with the foregoing forms for H and K_j those conditions are obeyed. The splitting up of H into several parts is not unique, and the question arises as to whether this can be done in such a way that the invariance of H under $E(3)$ passes over into the invariance under $E(3)$ for each individual term $h^{(r)}(q, p^{(r)})$. This can, indeed, be done, and we assume that the $h^{(r)}(q, p^{(r)})$ have been so chosen. (Call this point A;

it is the first of the points whose proofs are outlined at the end). This, then, means that, in particular, each $h^{(r)}$ is unchanged by spatial translations. If we make use of this fact and insert K_j and H from Eqs. (21.96) and (21.95) into Eq. (21.92b), we see immediately that the functions $k_j(q)$ are also translation invariant. In addition, of course, they behave as the components of a vector under spatial rotations because the K_j do, whereas the $h^{(r)}$ are invariant. With the knowledge, then, that H and K_j have the forms given in Eqs. (21.95) and (21.96), that each $h^{(r)}(q,p^{(r)})$ can be made $E(3)$ invariant and that then $\boldsymbol{k}(q)$ is a translation-invariant vector, we have exhausted all the information contained in Eqs. (21.92a) and (21.92b) and the properties of H and K_j under $E(3)$. From this point on, we use Eq. (21.92c) alone, along with the property that H is standard, as expressed in Eq. (21.91). Our aim is to isolate the dependences of $h^{(r)}(q,p^{(r)})$ and $k_j(q)$ on the q's.

The matrix that appears in Eq. (21.91) is of dimension $3N, N$ being the number of particles; (r,k) acts as a composite row index, (s,l) as a column index. With the split form of Eq. (21.95) for H, we immediately see that this $3N \times 3N$ matrix becomes a block-diagonal matrix with N submatrices on the diagonal, each a 3×3 matrix. The condition in Eq. (21.91) is equivalent to a set of conditions, one for each value of r:

$$\det \left\| \frac{\partial^2 h^{(r)}(q,p^{(r)})}{\partial p_k^{(r)} \partial p_l^{(r)}} \right\| \neq 0 . \tag{21.97}$$

For each r, we have here a 3×3 symmetric nonsingular matrix, k and l labeling the rows and columns; therefore we have a set of inverse matrices, $||h_{jk}^{(r)-1}(q,p^{(r)})||$, also symmetric, satisfying

$$\sum_k h_{jk}^{(r)-1} \frac{\partial^2 h^{(r)}}{\partial p_k^{(r)} \partial p_l^{(r)}} = \delta_{jl} . \tag{21.98}$$

[The superscript r on $h^{(r)}$ and $h_{jk}^{(r)-1}$ indicates that these functions do not depend on the canonical momenta $p^{(s)}$ for $s \neq r$]. We use Eq. (21.98) in the sequel. Let us now write out the condition in Eq. (21.92c) in terms of the $h^{(r)}$ and k_j:

$$\sum_{rs} \{q_j^{(r)} h^{(r)}, h^{(s)}\} + \sum_r \{k_j, h^{(r)}\} = \sum_r p_j^{(r)} ,$$

$$\sum_{rs} q_j^{(r)} \{h^{(r)}, h^{(s)}\} + \sum_{rs} h^{(r)} \{q_j^{(r)}, h^{(s)}\} + \sum_r \{k_j, h^{(r)}\} = \sum_r p_j^{(r)} ,$$

$$\sum_{rsk} (q_j^{(r)} - q_j^{(s)}) \frac{\partial h^{(r)}}{\partial q_k^{(s)}} \frac{\partial h^{(s)}}{\partial p_k^{(s)}} + \frac{1}{2} \sum_r \frac{\partial}{\partial p_j^{(r)}} [(h^{(r)})^2 - \boldsymbol{p}^{(r)} \cdot \boldsymbol{p}^{(r)}]$$

$$+ \sum_{rk} \frac{\partial k_j}{\partial q_k^{(r)}} \frac{\partial h^{(r)}}{\partial p_k^{(r)}} = 0 . \tag{21.99}$$

We use this last equation repeatedly. It represents a very stringent condition on the h's and k's, which must be exploited. We see right away that whereas the second and third groups of terms are sums of functions, each of which depends on the momenta of one particle alone, the first group of terms is not already in this form. We can put this fact to use in this way: letting u and v be two distinct particle labels, $u \neq v$, we differentiate Eq. (21.99) first with respect to $p_m^{(u)}$ and then with respect to $p_n^{(v)}$; clearly the second and third groups of terms drop out in this process, and we obtain the condition:

$$(q_j^{(u)} - q_j^{(v)}) \sum_k \left(\frac{\partial^2 h^{(u)}}{\partial q_k^{(v)} \partial p_m^{(u)}} \frac{\partial^2 h^{(v)}}{\partial p_k^{(v)} \partial p_n^{(v)}} - \frac{\partial^2 h^{(u)}}{\partial p_k^{(u)} \partial p_m^{(u)}} \frac{\partial^2 h^{(v)}}{\partial q_k^{(u)} \partial p_n^{(v)}} \right) = 0 \,, \qquad u \neq v \,.$$

Because the factor $(q_j^{(u)} - q_j^{(v)})$ is nonvanishing for $u \neq v$ and is not involved in any summations, we may infer that for $u \neq v$:

$$\sum_k \frac{\partial^2 h^{(u)}}{\partial p_m^{(u)} \partial q_k^{(v)}} \frac{\partial^2 h^{(v)}}{\partial p_k^{(v)} \partial p_n^{(v)}} = \sum_k \frac{\partial^2 h^{(u)}}{\partial p_m^{(u)} \partial p_k^{(u)}} \frac{\partial^2 h^{(v)}}{\partial q_k^{(u)} \partial p_n^{(v)}} \,. \tag{21.100}$$

Now we can use the existence of the inverse matrices $||h_{jk}^{(u)-1}||$, $||h_{jk}^{(v)-1}||$ to bring all dependence on $h^{(u)}$ to one side, on $h^{(v)}$ to the other; multiplying both sides of Eq. (21.100) by $h_{k'm}^{(u)-1} h_{nk''}^{(v)-1}$ and summing on m and n gives:

$$\sum_m h_{k'm}^{(u)-1} \frac{\partial^2 h^{(u)}}{\partial p_m^{(u)} \partial q_{k''}^{(v)}} = \sum_n h_{k''n}^{(v)-1} \frac{\partial^2 h^{(v)}}{\partial p_n^{(v)} \partial q_{k'}^{(u)}} \,, u \neq v \,. \tag{21.101}$$

But now, apart from being a function of the q's, the left-hand side involves the canonical momentum $p^{(u)}$ alone, the right $p^{(v)}$ alone. It must therefore be the case that on both sides of Eq. (21.101) we have functions of the q's alone. Because the right-hand side of Eq. (21.101) arises from the left by the interchanges $u \leftrightarrow v, k' \leftrightarrow k''$, we have that for $u \neq v$ there exist functions

$$\lambda_{k'k''}^{uv}(q) = \lambda_{k''k'}^{vu}(q) = \sum_m h_{k'm}^{(u)-1} \frac{\partial^2 h^{(u)}}{\partial p_m^{(u)} \partial q_{k''}^{(v)}} \,. \tag{21.102}$$

It can now be shown, by the use of a slightly more involved argument, that the foregoing equations hold for $u = v$ as well; functions $\lambda_{k'k''}^{uu}(q)$ do exist, symmetric in k' and k'', fulfilling Eq. (21.102) for $v = u$. (This is point B.) Taking advantage of this fact, we can write Eq. (21.102), for u and v unrestricted, in the alternative form:

$$\frac{\partial^2 h^{(u)}}{\partial p_m^{(u)} \partial q_{k''}^{(v)}} = \sum_{k'} \frac{\partial^2 h^{(u)}}{\partial p_m^{(u)} \partial p_{k'}^{(u)}} \lambda_{k'k''}^{uv}(q), \ \lambda_{k'k''}^{uv} = \lambda_{k''k'}^{vu} \,. \tag{21.102$'$}$$

Recalling that $h^{(u)}$ depends on $p^{(u)}$ alone (and the q's), we have, more generally:

$$\frac{\partial}{\partial p_m^{(r)}} \left(\frac{\partial h^{(u)}}{\partial q_{k''}^{(v)}} - \sum_{k'} \frac{\partial h^{(u)}}{\partial p_{k'}^{(u)}} \lambda_{k'k''}^{uv}(q) \right) = 0 \,, \qquad \text{all } r \,. \tag{21.103}$$

In fact, if $r \neq u$, this is an identity, and for $r{=}u$ it is simply the earlier equation. We can integrate Eq. (21.103) once and thus derive a statement on the dependence of the $h^{(u)}$ on the q's:

$$\frac{\partial h^{(u)}}{\partial q_k^{(v)}} = \sum_{k'} \frac{\partial h^{(u)}}{\partial p_{k'}^{(u)}} \lambda_{k'k}^{uv}(q) + \mu_k^{vu}(q)\,. \tag{21.104}$$

This is then a consequence of the PB relation $\{K_j, H\} = P_j$ rewritten as in Eq. (21.99); other consequences follow in the sequel. Now it is clear that Eq. (21.104) must lead to integrability conditions on the expressions on the right; thus we get some conditions on the functions $\lambda_{k'k}^{uv}$ and μ_k^{vu}. Taking the derivative of the right-hand side with respect to $q_j^{(r)}$ and demanding symmetry under $r \leftrightarrow v, j \leftrightarrow k$ we get:

$$\begin{aligned}\sum_{k'} \frac{\partial^2 h^{(u)}}{\partial p_{k'}^{(u)} \partial q_j^{(r)}} \lambda_{k'k}^{uv} + \sum_{k'} \frac{\partial h^{(u)}}{\partial p_{k'}^{(u)}} \frac{\partial \lambda_{k'k}^{uv}}{\partial q_j^{(r)}} + \frac{\partial \mu_k^{vu}}{\partial q_j^{(r)}} \\ = \sum_{k'} \frac{\partial^2 h^{(u)}}{\partial p_{k'}^{(u)} \partial q_k^{(v)}} \lambda_{k'j}^{ur} + \sum_{k'} \frac{\partial h^{(u)}}{\partial p_{k'}^{(u)}} \frac{\partial \lambda_{k'j}^{ur}}{\partial q_k^{(v)}} + \frac{\partial \mu_j^{ru}}{\partial q_k^{(v)}}\,.\end{aligned}$$

On using Eq. (20.102′), the first term on the left exactly cancels the first one on the right; thus the condition that Eq. (21.104) be integrable becomes:

$$\sum_{k'} \left(\frac{\partial \lambda_{k'k}^{uv}}{\partial q_j^{(r)}} - \frac{\partial \lambda_{k'j}^{ur}}{\partial q_k^{(v)}} \right) \frac{\partial h^{(u)}}{\partial p_{k'}^{(u)}} = \frac{\partial \mu_j^{ru}}{\partial q_k^{(v)}} - \frac{\partial \mu_k^{vu}}{\partial q_j^{(r)}}\,. \tag{21.105}$$

But now, whereas the λ's and μ's are functions of the q's alone, the derivatives $\partial h^{(u)}/\partial p_{k'}^{(u)}$ involve *essential* dependences on the canonical momenta $p^{(u)}$ – witness the nonsingularity of

$$\left\| \frac{\partial^2 h^{(u)}}{\partial p_j^{(u)} \partial p_k^{(u)}} \right\|\,.$$

Thus Eq. (21.105) can be satisfied only if both sides vanish identically, term by term. We thus get very restrictive conditions on the λ's and μ's:

$$\frac{\partial \lambda_{k'k}^{uv}}{\partial q_j^{(r)}} = \frac{\partial \lambda_{k'j}^{ur}}{\partial q_k^{(v)}}, \quad \frac{\partial \mu_j^{ru}}{\partial q_k^{(v)}} = \frac{\partial \mu_k^{vu}}{\partial q_j^{(r)}}\,; \tag{21.106}$$

these conditions have the same form as ones that state the vanishing of the "curl" of a vector field; hence both the λ's and μ's must have the forms of "gradients" of suitable functions $L_{k'}^{(v)}(q), M^{(u)}(q)$:

$$\lambda_{k'k}^{uv} = \frac{\partial L_{k'}^{(u)}}{\partial q_k^{(v)}}, \quad \mu_k^{vu} = \frac{\partial M^{(u)}}{\partial q_k^{(v)}}\,. \tag{21.107}$$

So far, in developing the consequences of Eq. (21.102), no use has been made of the symmetry of $\lambda^{uv}_{k'k}$ under the interchanges $u \leftrightarrow v, k' \leftrightarrow k$; taking note of it now, in conjunction with Eq. (21.107), we see that the $L^{(u)}_{k'}(q)$ themselves form the components of a "vector" with vanishing "curl":

$$\begin{aligned}\lambda^{uv}_{k'k} = \lambda^{vu}_{kk'} \Rightarrow \frac{\partial L^{(u)}_{k'}}{\partial q^{(v)}_k} &= \frac{\partial L^{(v)}_k}{\partial q^{(u)}_{k'}} \Rightarrow L^{(u)}_k(q) = \frac{\partial L(q)}{\partial q^{(u)}_k}\,,\\ \lambda^{uv}_{kl} &= \frac{\partial^2 L(q)}{\partial q^{(u)}_k \partial q^{(v)}_l}\,.\end{aligned} \tag{21.108}$$

All the λ's are thus given in terms of one function $L(q)$, and all the μ's in terms of the N functions $M^{(u)}(q)$. Now we fix the $E(3)$ transformation properties of $L(q), M^{(u)}(q)$. Because each $h^{(u)}$ is an $E(3)$ invariant, we see from Eq. (21.104) that for fixed u and v, λ^{uv}_{kl} form the components of a second rank Cartesian tensor and μ^{vu}_k those of a Cartesian vector, under spatial rotations; under space translations, both λ's and μ's must be invariant. All these properties would follow automatically if it were possible to choose all the functions $L(q), M^{(u)}(q)$ to be $E(3)$ invariant; one can show that such a choice does exist, thus we adopt it. (This is point C).

Having obtained all this information on the λ's and μ's from the integrability conditions for Eq. (21.104), we can go back to that equation now and integrate it. Using Eqs. (21.107) and (21.108), we can state Eq. (21.104) using PB notation as:

$$\begin{aligned}\{h^{(u)}, p^{(v)}_k\} = -\{h^{(u)}, L^{(v)}_k\} + \{M^{(u)}, p^{(v)}_k\} &= -\,\{h^{(u)}, \{L, p^{(v)}_k\}\} + \{M^{(u)}, p^{(v)}_k\}\,,\\ \{h^{(u)} - M^{(u)}, p^{(v)}_k + \{L, p^{(v)}_k\}\} &=0\,,\\ \{h^{(u)}(q,p^{(u)}) - M^{(u)}(q), e^{\widetilde{L(q)}} p^{(v)}_k\} &=0\,.\end{aligned} \tag{21.109}$$

The form of this result suggests that we switch to a new set of canonical momenta p' obtained from the original ones via the canonical transformation $e^{\widetilde{L(q)}}$. This transformation does not change the particle position variables q at all. With $p' = e^{\tilde{L}}p$, we take the generators of $\mathcal{P}$ given originally as functions of q's and p's and rewrite them in terms of the q and p' variables. Because $L(q)$ could be chosen to be a Euclidean invariant function, the $E(3)$ generators J_j and P_j preserve their functional forms in this process of rewriting; whereas the K_j and H may change theirs because a change in the forms of the functions $h^{(u)}(q,p^{(u)})$ may occur. In any case, neither the PB relations of $\mathcal{P}$ nor the world-line conditions are disturbed by this (canonical) change of variables. Any function of the q's alone, such as $M^{(u)}(q)$ and $k_j(q)$, is left unaltered. In the new variables, if $h^{(u)}(q,p^{(u)}) = h'^{(u)}(q,p'^{(u)})$, Eq. (21.109) takes the form:

$$\begin{aligned}\{h'^{(u)}(q,p'^{(u)}) - M^{(u)}(q), p'^{(v)}_k\} &= 0\,,\\ h'^{(u)}(q,p'^{(u)}) &= M^{(u)}(q) + N^{(u)}(p'^{(u)})\,.\end{aligned} \tag{21.110}$$

Therefore, in the new variables, the q and p' dependences of each $h'^{(u)}$ split up in this way. We may now assume without loss of generality that the original choice of canonical momenta was already such as to give each $h^{(u)}$ the foregoing structure. If this were not initially so, we carry out the canonical transformation $e^{\tilde{L}}$, achieve the desired form, and then omit primes in Eq. (21.110). Although the standardness criteria Eqs. (21.91) and (21.97) for the Hamiltonian are not invariant under arbitrary canonical transformations, it is easily checked that they are preserved under the canonical transformation $e^{\tilde{L}}$; therefore the functions $N^{(r)}(p^{(r)})$ are such that

$$\det\left\|\frac{\partial^2 N^{(r)}(p^{(r)})}{\partial p_j^{(r)}\partial p_k^{(r)}}\right\| \neq 0\,. \tag{21.111}$$

We are now ready to go back to Eq. (21.99) and grapple with it again. If we write it in terms of the $M^{(r)}(q)$ and $N^{(r)}(p^{(r)})$, it takes the form:

$$\begin{aligned}
&\frac{1}{2}\sum_r \frac{\partial}{\partial p_j^{(r)}}[(N^{(r)})^2 - \boldsymbol{p}^{(r)}\cdot\boldsymbol{p}^{(r)}] + \sum_{rk}\frac{\partial N^{(r)}}{\partial p_k^{(r)}}C_{kj}^{(r)} = 0\,,\\
&C_{kj}^{(r)} \equiv M^{(r)}(q)\delta_{kj} + \frac{\partial k_j(q)}{\partial q_k^{(r)}} + \sum_s (q_j^{(s)} - q_j^{(r)})\frac{\partial M^{(s)}(q)}{\partial q_k^{(r)}}\\
&\qquad \equiv \frac{\partial}{\partial q_k^{(r)}}[k_j(q) + \sum_s q_j^{(s)}M^{(s)}(q)] - q_j^{(r)}\frac{\partial}{\partial q_k^{(r)}}\sum_s M^{(s)}(q)\,.
\end{aligned} \tag{21.112}$$

This way of putting it tells us the following: because the first term in the first line is a function of $p^{(r)}$ alone, and in the second term the derivatives $\partial N^{(r)}/\partial p_k^{(r)}$ carry essential dependences on the $p_k^{(r)}$, the coefficients $C_{kj}^{(r)}$ which are in any event functions of the q's alone must actually be numbers. The point is that because of Eq. (21.111) there can be no q dependences in the $C_{kj}^{(r)}$ which somehow cancel out in Eq.(21.112). Further, from the invariances of $h^{(r)}$ and $M^{(r)}$ under $E(3)$ follows that of $N^{(r)}$; hence the covariance of Eq. (21.112) with respect to three-dimensional rotations tells us that the constants $C_{jk}^{(r)}$ form the components of an invariant second-rank Cartesian tensor for each r. The only possible form for these quantities thus is $C_{jk}^{(r)} = C^{(r)}\delta_{jk}$, $C^{(r)}$ a pure number. The equation involving $N^{(r)}$ in Eq. (21.112) becomes:

$$\sum_r \frac{\partial}{\partial p_j^{(r)}}[(N^{(r)} + C^{(r)})^2 - \boldsymbol{p}^{(r)}\cdot\boldsymbol{p}^{(r)}] = 0\,. \tag{21.113}$$

Each term in this sum depends on a single canonical momentum $p^{(r)}$; thus they can all add up to zero only if each of them is constant. But each term also has to be the jth component of a vector in three dimensions, and the only vector with constant components is the zero vector; hence, actually, each term in Eq. (21.113) vanishes by itself! There must then be N constants $m^{(r)}$ such that

$$N^{(r)}(p^{(r)}) = [(m^{(r)})^2 + \boldsymbol{p}^{(r)}\cdot\boldsymbol{p}^{(r)}]^{1/2} - C^{(r)}\,. \tag{21.114}$$

We are close to determining completely the forms of H and K_j. Going back to Eq. (21.112), the equation involving $k_j(q)$ and $M^{(s)}(q)$ is:

$$\frac{\partial}{\partial q_k^{(r)}}\left[k_j(q)+\sum_s q_j^{(s)}M^{(s)}(q)\right] = C^{(r)}\delta_{kj} + q_j^{(r)}\frac{\partial}{\partial q_k^{(r)}}\sum_s M^{(s)}(q)\,,$$

$$\frac{\partial}{\partial q_k^{(r)}}\left[k_j(q)+\sum_s q_j^{(s)}M^{(s)}(q) - C^{(s)}\right] = q_j^{(r)}\frac{\partial}{\partial q_k^{(r)}}\sum_s M^{(s)}(q)\,. \tag{21.115}$$

The right-hand side must obey integrability conditions, and they force it to vanish. It is enough to differentiate the right-hand side with respect to $q_l^{(r)}$ and demand symmetry under $k \leftrightarrow l$:

$$\delta_{jl}\frac{\partial}{\partial q_k^{(r)}}\sum_s M^{(s)} + q_j^{(r)}\frac{\partial^2}{\partial q_k^{(r)}\partial q_l^{(r)}}\sum_s M^{(s)}$$
$$= \delta_{jk}\frac{\partial}{\partial q_l^{(r)}}\sum_s M^{(s)} + q_j^{(r)}\frac{\partial^2}{\partial q_l^{(r)}\partial q_k^{(r)}}\sum_s M^{(s)}\,,$$

that is,

$$\frac{\partial}{\partial q_k^{(r)}}\sum_s M^{(s)}(q) = 0 \Rightarrow \sum_s M^{(s)}(q) = \text{constant}\,. \tag{21.116}$$

Therefore the vector appearing in the square brackets on the left-hand side in the second line in Eq. (21.115) has constant components; that is, it vanishes:

$$k_j(q) + \sum_s q_j^{(s)}M^{(s)}(q) = \sum_s q_j^{(s)}C^{(s)}\,. \tag{21.117}$$

Now both $k_j(q)$ and $M^{(s)}(q)$ are functions that are invariant under spatial translations; if we subject all the $q_j^{(s)}$ to a common translation, then we deduce that:

$$\sum_s M^{(s)}(q) = \sum_s C^{(s)}\,, \tag{21.118}$$

thus fixing the constant that had appeared in Eq. (21.116). The forms of H and K_j now follow by putting together all this information:

$$H = \sum_r h^{(r)}(q, p^{(r)}) = \sum_r [N^{(r)}(p^{(r)}) + M^{(r)}(q)]$$
$$= \sum_r [(m^{(r)})^2 + \boldsymbol{p}^{(r)}\cdot\boldsymbol{p}^{(r)}]^{1/2} + \sum_r M^{(r)}(q) - \sum_r C^{(r)}$$
$$= \sum_r [(m^{(r)})^2 + \boldsymbol{p}^{(r)}\cdot\boldsymbol{p}^{(r)}]^{1/2}\,;$$

$$
\begin{aligned}
K_j &= \sum_r q_j^{(r)} h^{(r)}(q, p^{(r)}) + k_j(q) \\
&= \sum_r q_j^{(r)} [[(m^{(r)})^2 + \boldsymbol{p}^{(r)} \cdot \boldsymbol{p}^{(r)}]^{1/2} - C^{(r)} + M^{(r)}(q)] + k_j(q) \\
&= \sum_r q_j^{(r)} [(m^{(r)})^2 + \boldsymbol{p}^{(r)} \cdot \boldsymbol{p}^{(r)}]^{1/2} .
\end{aligned} \tag{21.119}
$$

Use of some of the PB relations of the Lie algebra of $\mathcal{P}$, and imposition of the manifest covariance conditions for a multiparticle system, have thus led to the result that apart from having to carry out the canonical transformation $e^{\widetilde{L(q)}}$, the generators of $\mathcal{P}$ all have the standard free-particle forms.

What remains is to comment on the points A, B, C that appeared in the foregoing work. Some of these statements are easy to establish, whereas some require more work; we refer the reader to the paper of Leutwyler for full details. [H. Leutwyler, Nuovo Cimento, **37**, 556 (1965).] Here let us briefly indicate the nature of the arguments. As for point A: starting with any breakup of the total Hamiltonian into the sum of terms $h^{(r)}(q, p^{(r)})$, we can utilize the translation invariance of the sum to replace each term by $h^{(r)}(q - q^{(1)}, p^{(r)})$, say. That is, we take the original terms $h^{(r)}(q, p^{(r)})$ and in them replace $q^{(1)}, q^{(2)}, \ldots, q^{(s)}, \ldots$ by $0, q^{(2)} - q^{(1)}, \ldots, q^{(s)} - q^{(1)}, \ldots$ and re-identify the functions $h^{(r)}$ accordingly. But now, each term in the decomposition of H has become translation invariant. Next we utilize the invariance of H under $R(3)$ to replace each term in the sum over $h^{(r)}$ by its average with respect to $R(3)$: this is simply the averaging process described in Eq. (21.26). That is, after having chosen each $h^{(r)}$ to be translation invariant, we next replace it by

$$
h^{(r)}(q, p^{(r)}) \to \frac{1}{2\pi^2} \iiint_{|\boldsymbol{\alpha}| \leq \pi} d\alpha_1 d\alpha_2 d\alpha_3 \left(\frac{\sin(|\boldsymbol{\alpha}|/2)}{|\boldsymbol{\alpha}|} \right)^2 e^{\widetilde{\boldsymbol{\alpha} \cdot \boldsymbol{J}}} h^{(r)}(q, p^{(r)}) . \tag{21.120}
$$

This averaging process leaves H unaltered, but makes each $h^{(r)}$ an $R(3)$ invariant. The replacement Eq. (21.120) does not spoil the previously arranged translation invariance; hence all in all we have achieved $E(3)$ invariance for each $h^{(r)}$. Now we can go to point B, that is, the existence and symmetry properties of functions $\lambda_{kk'}^{uu}(q)$ fullfilling Eq. (21.102) for $v = u$. That $h^{(u)}$ is translation invariant means:

$$
\sum_v \frac{\partial h^{(u)}(q, p^{(u)})}{\partial q_k^{(v)}} = 0 ,
$$

that is,

$$
\frac{\partial h^{(u)}}{\partial q_k^{(u)}} = - \sum_v{}' \frac{\partial h^{(u)}}{\partial q_k^{(v)}}
$$

where the prime indicates that the value u for v must be omitted. Differentiating the above with respect to $p_j^{(u)}$ and using on the right the result contained in Eq.

(21.102) (we need it for $v \neq u$ alone), we get:

$$\begin{aligned}
\frac{\partial^2 h^{(u)}}{\partial p_j^{(u)} \partial q_k^{(u)}} &= -\sum_v{}' \sum_{k'} \frac{\partial^2 h^{(u)}}{\partial p_j^{(u)} \partial p_{k'}^{(u)}} \lambda_{k'k}^{uv}(q) \\
&= \sum_{k'} \frac{\partial^2 h^{(u)}}{\partial p_j^{(u)} \partial p_{k'}^{(u)}} \lambda_{k'k}^{uu}(q) \,, \\
\lambda_{k'k}^{uu}(q) &\equiv -\sum_v{}' \lambda_{k'k}^{uv}(q) \,. \qquad (21.121)
\end{aligned}$$

This establishes the existence of the functions $\lambda_{k'k}^{uu}(q)$, such that Eq. (21.103) is valid for all u and v. But to show that these functions are symmetric in k' and k is harder. One uses the fact that the λ's could be written in terms of functions $L_k^{(u)}$ as in Eq. (21.107) (recall that this did not need the symmetry properties of the λ's). Combining this judiciously with the rotational invariance of $h^{(u)}$ as well as the way in which $\lambda_{k'k}^{uu}$ was defined above, one proves the symmetry relation for $\lambda_{k'k}^{uu}$. Point C, namely that the functions $L(q), M^{(u)}(q)$ can be made $E(3)$ invariant, uses again the $E(3)$ invariance of $h^{(u)}$ and the consequences of this for the λ's and μ's because of Eq. (21.104). This much is already sufficient to take care of the problem for the functions $M^{(u)}(q)$; whereas in the case of $L(q)$ some more work involving an averaging process over $R(3)$ is required.

Summarizing the situation for many-particle systems, in the Galilean case the imposition of the geometric transformation law for particle position variables,

$$q_j(t) \to q_j'(t') = A_{jk}(\boldsymbol{\alpha})q_k(t) - v_j t + a_j, \; t' = t - b \qquad (21.122)$$

under a general change of inertial frame, is compatible with an interacting system described by a Hamiltonian scheme. That is, the foregoing equation of transformation for position, combined with the PB relations among the generators of $\mathcal{G}$, include among their solutions (when understood to apply to a collection of particles) Hamiltonian interacting systems. With a Lorentzian system the situation is that if we impose the law:

$$\begin{aligned}
q_j(t) \to q_j'(t') &= \Lambda_{jk}(\boldsymbol{\zeta}, \boldsymbol{\alpha})q_k(t) + \Lambda_{j0}(\boldsymbol{\zeta}, \boldsymbol{\alpha})t \,, \\
t' &= \Lambda^0{}_j(\boldsymbol{\zeta}, \boldsymbol{\alpha})q_j(t) + \Lambda^0{}_0(\boldsymbol{\zeta}, \boldsymbol{\alpha})t \,, \qquad (21.123)
\end{aligned}$$

we do not have *any* Hamiltonian systems of interacting particles. The world-line form of dynamics, coupled with the assumption that the canonical Hamiltonian formalism applies, rules out the possibility of any interactions. However, there are noncanonical systems that admit the conditions for manifest covariance and include particle interactions. We describe an example in the next chapter and show there that, suitably reinterpreted, a manifestly relativistic many-particle system including interaction can also be described as a canonical Hamiltonian system of particles *and* fields in mutual interaction!

Covariant Galilean Field Theory

Part of what we have to say under this heading has already appeared in Chapter 19. There we constructed realizations of $\mathcal{G}$ using a free scalar field, a free finite-component field, and a free infinite-component field. We also saw that with the simplifying assumption that the generators G_j (as well as J_j, P_j, and the mass M) were kinematic, it was easy to introduce interactions. This assumption is maintained here. For a scalar field, the mass parameter being m, the kinematic generators act on ϕ at time $t = 0$ as follows:

$$\begin{aligned} \{P_j, \phi(\boldsymbol{x})\} &= \frac{\partial \phi(\boldsymbol{x})}{\partial x_j}, \ \{J_j, \phi(\boldsymbol{x})\} = (\boldsymbol{x} \times \boldsymbol{\nabla})_j \phi(\boldsymbol{x}), \\ \{G_j, \phi(\boldsymbol{x})\} &= -m x_j \phi(\boldsymbol{x}). \end{aligned} \tag{21.124}$$

If we replace ϕ by π and change m to $-m$, we get the action on the conjugate momentum field $\pi(\boldsymbol{x})$. To generalize to a general time t, it is sufficient to know that P_j and J_j are constants of motion, whereas $\{G_j, H\} = P_j$. It is not necessary to know the Hamiltonian; the Hamiltonian by definition produces thc equation of motion. We thus have:

$$\begin{aligned} \{P_j, \phi(\boldsymbol{x}, t)\} &= \frac{\partial \phi(\boldsymbol{x}, t)}{\partial x_j}, \ \{J_j, \phi(\boldsymbol{x}, t)\} = (\boldsymbol{x} \times \boldsymbol{\nabla})_j \phi(\boldsymbol{x}, t), \\ \{G_j, \phi(\boldsymbol{x}, t)\} &= -\left(m x_j + t \frac{\partial}{\partial x_j}\right) \phi(\boldsymbol{x}, t); \\ \{H, \phi(\boldsymbol{x}, t)\} &= -\frac{\partial}{\partial t} \phi(\boldsymbol{x}, t). \end{aligned} \tag{21.125}$$

Let us now work out the general formula relating the scalar fields in two different inertial frames. Let S and $S' = gS$ be the two frames, g an element of $\mathcal{G}$ written in detail as $\exp(\boldsymbol{a} \cdot \boldsymbol{d}) \exp(bh) \exp(\boldsymbol{v} \cdot \boldsymbol{g}) \exp(\boldsymbol{\alpha} \cdot \boldsymbol{l})$. The transformation laws for the space-time coordinates of events are

$$x'_j = A_{jk}(\boldsymbol{\alpha}) x_k - v_j t + a_j, \ t' = t - b. \tag{21.126}$$

Using the rules of the Heisenberg picture as developed in Chapter 16, we can relate the field $\phi'(\boldsymbol{x}, t)$ in S' to $\phi(\boldsymbol{x}, t)$ in S as:

$$\phi'(\boldsymbol{x}, t) = \widetilde{e^{-\alpha \cdot \boldsymbol{J}}} \widetilde{e^{-\boldsymbol{v} \cdot \boldsymbol{G}}} \widetilde{e^{-bH}} \widetilde{e^{-\boldsymbol{a} \cdot \boldsymbol{P}}} \phi(\boldsymbol{x}, t). \tag{21.127}$$

The right-hand side is easily computed; apart from the knowledge of the action of $E(3)$, we need to use Eq. (19.116). We get in a straightforward way:

$$\phi'(\boldsymbol{x}, t) = e^{m \boldsymbol{v} \cdot (\boldsymbol{x} - \boldsymbol{a} + (t+b) \boldsymbol{v})} \phi(A_{kj}(\boldsymbol{\alpha})(x_k - a_k + (t+b) v_k, t + b). \tag{21.128}$$

This law looks much simpler if we use $\boldsymbol{x}'$, t' as the arguments of ϕ' and recall the connection in Eq. (21.126) between these arguments and $\boldsymbol{x}$, t:

$$\phi'(\boldsymbol{x}', t') = \exp(m v_j A_{jk}(\boldsymbol{\alpha}) x_k) \phi(\boldsymbol{x}, t). \tag{21.129}$$

The accompanying law for π is:

$$\pi'(\boldsymbol{x}', t') = \exp(-mv_j A_{jk}(\boldsymbol{\alpha})x_k)\pi(\boldsymbol{x}, t)\,. \tag{21.130}$$

We see that at a given space-time point the field in S determines that in S' in a geometric fashion. (Note, however, the occurrence of the mass parameter m here!)

At this point we can abstract the relations in Eqs. (21.129) and (21.130) and define a manifestly covariant Galilean scalar field as the pair $\phi(\boldsymbol{x}, t)$, $\pi(\boldsymbol{x}, t)$ subject to the foregoing transformation law in going from one inertial frame S to another S', with a suitable value of m. No reference to the nature or even presence of interaction appears in these statements. If we choose we could then introduce the fundamental equal-time PB relations:

$$\{\phi(\boldsymbol{x}, t), \pi(\boldsymbol{y}, t)\} = \delta^{(3)}(\boldsymbol{x} - \boldsymbol{y}),\ \{\phi, \phi\} = \{\pi, \pi\} = 0\,. \tag{21.131}$$

It is immediately evident that the geometric transformations Eqs. (21.129) and (21.130) preserve these relations; that is, they act as automorphisms. Our result on the possibility of introducing interactions, derived in Chapter 19, can be stated thus: there do exist Galilean fields transforming in the manifestly covariant manner defined above, which can be described in the canonical Hamiltonian framework and include interaction.

We can, in a straightforward manner, extend Eqs. (21.129) and (21.130) to multi-component fields. In the case of an infinite-component field ϕ_α with both J_j and G_j generating canonical transformations that affect the index α, we have the infinite-dimensional matrices S_j, T_j that generate a linear representation of $E^{(v)}(3)$, and the kinematic generators have the forms given in Eq. (19.96). (H is not the same as given there, but may include interaction.) Then Eqs. (21.129) and (21.130) are, respectively, replaced by:

$$\begin{aligned}\phi'_\alpha(\boldsymbol{x}', t') &= \exp(mv_j A_{jk}(\boldsymbol{\alpha})x_k)\{e^{\boldsymbol{v}\cdot\boldsymbol{T}} e^{\boldsymbol{\alpha}\cdot\boldsymbol{S}}\}_{\alpha\beta}\phi_\beta(\boldsymbol{x}, t)\,,\\ \pi'_\alpha(\boldsymbol{x}', t') &= \exp(-mv_j A_{jk}(\boldsymbol{\alpha})x_k)\{e^{-\boldsymbol{\alpha}\cdot\boldsymbol{S}} e^{-\boldsymbol{v}\cdot\boldsymbol{T}}\}_{\beta\alpha}\pi_\beta(\boldsymbol{x}, t)\,.\end{aligned} \tag{21.132}$$

The index transformations here are such as to make these laws act as automorphisms on the obvious generalization of Eq. (21.131). We get the case of a finite-component field by setting $T_j = 0$.

Finally, for completeness let us record the covariant transformation laws for the fundamental field variables that appear in the discussion of the equations of motion of a nonviscous fluid. We know from our work in Chapter 19 that we have the two cases of irrotational motion and general motion; that in each case a Hamiltonian description of the system exists and gives rise to a realization of $\mathcal{G}$. In the irrotational case we have the two basic fields $\rho(\boldsymbol{x}, t)$ and $\phi(\boldsymbol{x}, t)$: the former is the fluid density, the latter the velocity potential, and they are conjugate variables. For the general change of inertial frame described by Eq. (21.126), our work in Chapter 19 leads to the transformation laws

$$\rho'(\boldsymbol{x}', t') = \rho(\boldsymbol{x}, t),\ \phi'(\boldsymbol{x}', t') = \phi(\boldsymbol{x}, t) + v_j A_{jk}(\boldsymbol{\alpha})x_k\,. \tag{21.133}$$

These laws are perfectly reasonable from a physical point of view, and they act as automorphisms on the equal-time PB relations:

$$\{\phi(\boldsymbol{x},t),\rho(\boldsymbol{y},t)\} = \delta^{(3)}(\boldsymbol{x}-\boldsymbol{y}),\ \{\phi,\phi\} = \{\rho,\rho\} = 0\,. \tag{21.134}$$

In the case of general motion, we have in addition to ϕ, ρ the pair of field functions β, σ (recall that $\phi, \alpha = \sigma/\rho$ and β are the Clebsch potentials for the velocity field); Eq. (21.133) is supplemented by:

$$\beta'(\boldsymbol{x}',t) = \beta(\boldsymbol{x},t),\ \sigma'(\boldsymbol{x}',t') = \sigma(\boldsymbol{x},t)\,, \tag{21.135}$$

whereas to Eq. (21.134) must be added:

$$\begin{aligned}\{\beta(\boldsymbol{x},t),\sigma(\boldsymbol{y},t)\} = \delta^{(3)}(\boldsymbol{x}-\boldsymbol{y}),\ \{\beta,\beta\} =& \{\sigma,\sigma\} = \{\beta,\rho\} = \{\beta,\phi\}\\ =& \{\sigma,\rho\} = \{\sigma,\phi\} = 0\,.\end{aligned} \tag{21.136}$$

These are both physically important nonlinear field theories that fit into the canonical framework and are manifestly covariant in the Galilean sense.

Covariant Lorentzian Field Theory

Manifestly covariant relativistic field theory is a vast subject to which several excellent treatises have been devoted, and it would be impossible for us to give an exhaustive discussion of it here. We must content ourselves with showing how some of our work in the previous chapter can be moulded into a manifestly covariant form, with a discussion of some special features. In the preceding chapter we looked at two basically different kinds of realizations of $\mathcal{P}$ using fields as dynamical variables: in the first kind the generators are not expressible as integrals over space of suitable local densities, in the second kind they are. (By a local density is meant, of course, an expression constructed out of the field variables and a finite number of their spatial derivatives all at one point of space.) The former class of realizations is not discussed at all here. The latter class again is separated into two quite distinct types, one of which involves fields with infinite numbers of components, the other finite multicomponent fields. The distinguishing feature of the examples that we gave of the former is that each component $\phi_r(\boldsymbol{x})$ of the field possesses its own canonical conjugate momentum $\pi_r(\boldsymbol{x})$; this is generally not true of the finite component field. We see from the examples of the massive vector field as well as the electromagnetic field how it happens that not all the components of a finite multicomponent field are independent dynamical variables in the Hamiltonian sense. (The scalar field does not need a separate discussion.) We first show how the generators of $\mathcal{P}$ as well as their action on the basic field can be put into a manifestly covariant form in the case of the (interacting) infinite-component field. Then we discuss the transformation laws for finite-component fields. Finally, we give a treatment of the electro-magnetic field that displays manifest relativistic covariance by using what is called the Lorentz gauge; this is in contrast to the Coulomb gauge treatment of the preceding chapter, which is not manifestly covariant.

The generators of $\mathcal{P}$ for the free infinite-component field are given in Eq. (20.152). If interaction is to be included, we add a term $V\{\phi\}$ to the Hamiltonian density $\mathcal{H}(\boldsymbol{x})$, thereby making both K_j and H dynamic generators. For the moment let us assume we are dealing with the free case. The action of the generators on the field $\phi_r(\boldsymbol{x})$ at time $t = 0$ is given by:

$$\{P_j, \phi_r(\boldsymbol{x})\} = \frac{\partial \phi_r(\boldsymbol{x})}{\partial x_j}, \ \{J_j, \phi_r(\boldsymbol{x})\} = (\boldsymbol{x} \times \boldsymbol{\nabla})_j \phi_r(\boldsymbol{x}) - (S_j)_{rs}\phi_s(\boldsymbol{x}),$$
$$\{K_j, \phi_r(\boldsymbol{x})\} = -x_j \pi_r(\boldsymbol{x}) - (M_j)_{rs}\phi_s(\boldsymbol{x}). \tag{21.137}$$

We can now immediately get the equations that hold at any time t:

$$\{P_j, \phi_r(\boldsymbol{x},t)\} = \frac{\partial \phi_r(\boldsymbol{x},t)}{\partial x_j}, \{J_j, \phi_r(\boldsymbol{x},t)\} = (\boldsymbol{x} \times \boldsymbol{\nabla})_j \phi_r(\boldsymbol{x},t) - (S_j)_{rs}\phi_s(\boldsymbol{x},t),$$
$$\{H, \phi_r(\boldsymbol{x},t)\} \equiv \{P^0, \phi_r(\boldsymbol{x},t)\} \equiv -\frac{\partial \phi_r(\boldsymbol{x},t)}{\partial t} = -\pi_r(\boldsymbol{x},t),$$
$$\{K_j, \phi_r(\boldsymbol{x},t)\} = -\left(t\frac{\partial}{\partial x_j} + x_j\frac{\partial}{\partial t}\right)\phi_r(\boldsymbol{x},t) - (M_j)_{rs}\phi_s(\boldsymbol{x},t). \tag{21.138}$$

We can now put these equations into a manifestly covariant form: the matrices S_j, M_j which are real antisymmetric infinite-dimensional generators of some real unitary representation of $SO(3,1)$ can be combined into a second-rank antisymmetric tensor $S_{\mu\nu}$ in the usual fashion, with $S_{12} = S_3, \ldots$ and $S_{0j} = M_j$; in the same way J_j and K_j are parts of the tensor $J_{\mu\nu}$ and $P_j, -H$ those of P_μ. In this language, we have:

$$\begin{aligned} \{J_{\mu\nu}, \phi_r(x)\} &= (x_\mu \partial_\nu - x_\nu \partial_\mu)\phi_r(x) - (S_{\mu\nu})_{rs}\phi_s(x), \\ \{P_\mu, \phi_r(x)\} &= \partial_\mu \phi_r(x); \end{aligned} \tag{21.139}$$

(x now stands for $\boldsymbol{x}, t$).

In a similar fashion we can try to express the generators themselves in a manifestly covariant form. In Eq. (20.152) we gave them as functions of $\phi_r(\boldsymbol{x})$ and $\pi_r(\boldsymbol{x})$, the fields at time $t = 0$. If we express them in terms of $\phi_r(x)$ and $\pi_r(x) = \dot{\phi}_r(x)$, clearly P_j, J_j, and H remain unchanged in form, whereas K_j is altered by the term $-tP_j$. Hence we can begin with:

$$\begin{aligned} J_{kl} &= \int d^3x\, \dot{\phi}_r(x)[(S_{kl})_{rs} - \delta_{rs}(x_k\partial_l - x_l\partial_k)]\phi_s(x), \\ J_{0j} &= \int d^3x\{\dot{\phi}_r(x)(s_{0j})_{rs}\phi_s(x) + \frac{x_j}{2}(\dot{\phi}_r(x)\dot{\phi}_r(x) + \partial_k\phi_r(x)\partial_k\phi_r(x) \\ &\quad + m^2\phi_r(x)\phi_r(x)) + x^0\dot{\phi}_r(x)\partial_j\phi_r(x)\}; \\ P_j &= -\int d^3x\, \dot{\phi}_r(x)\partial_j\phi_r(x), \\ P_0 &= -\frac{1}{2}\int d^3x(\dot{\phi}_r(x)\dot{\phi}_r(x) + \partial_k\phi_r(x)\partial_k\phi_r(x) + m^2\phi_r(x)\phi_r(x)). \end{aligned} \tag{21.140}$$

Utilizing the form of the Lagrangian density,

$$\mathcal{L}(x) = -\frac{1}{2}\partial_\mu\phi_r(x)\partial^\mu\phi_r(x) - \frac{m^2}{2}\phi_r(x)\phi_r(x)\,, \tag{21.141}$$

it is easy to put the generators into the following forms:

$$\begin{aligned} J_{kl} &= \int d^3x\,\partial_0\phi_r[(S_{kl})_{rs} - \delta_{rs}(x_k\partial_l - x_l\partial_k)]\phi_s\,, \\ J_{0j} &= \int d^3x\,\partial_0\phi_r[(S_{0j})_{rs} - \delta_{rs}(x_0\partial_j - x_j\partial_0)]\phi_s - \int d^3x\,x_j\mathcal{L}(x)\,; \\ P_j &= -\int d^3x\,\partial_0\phi_r\partial_j\phi_r,\ P_0 = -\int d^3x\,\partial_0\phi_r\partial_0\phi_r + \int d^3x\,\mathcal{L}(x)\,. \end{aligned} \tag{21.142}$$

Comparing J_{kl} with J_{0j}, and P_j with P_0, we see that a little more rewriting is necessary to achieve a covariant form. The clue lies in the observation that the integration over all space that is involved here singles out a specific inertial frame, and this is what must be avoided. All space at one instant of time (in a specific frame) can be pictured as a "plane" in four-dimensional space-time, perpendicular to the time axis in that frame. Its most important property is that it is a spacelike plane: if x and y are any two distinct points on it, the vector $x - y$ is spacelike. We must generalize by replacing a spacelike *plane* by a spacelike *surface*: this is such that any two points on it are again relatively spacelike, and the normal to it at every point is timelike. Whereas integration over a spacelike plane uses as measure the spatial volume element d^3x in that frame in which the plane is all space at one time, in the case of a general spacelike surface we have a vectorial volume element $d\sigma_\mu$. This is built up thus: being a three-dimensional surface, the points on the spacelike surface can be parametrized by three parameters u_1, u_2, u_3, and the coordinates of points on the surface can be given as functions of the u's, in the fashion $x^\mu = x^\mu(u)$. Then we have:

$$d\sigma_\lambda = \epsilon_{\lambda\mu\nu\rho}\frac{\partial x^\mu}{\partial u_1}\frac{\partial x^\nu}{\partial u_2}\frac{\partial x^\rho}{\partial u_3}\,du_1du_2du_3\,. \tag{21.143}$$

If we go back to the simple case of a plane, all space at time t, we can take u_j equal to x_j, and then only one component of $d\sigma_\lambda$ survives: $d\sigma_0 = d^3x$, $d\sigma_j = 0$. In general, the coefficient multiplying d^3u in $d\sigma_\lambda$ is a timelike vector parallel at each point of the surface to the normal at that point. With this notation, the expressions for J_{kl} and J_{0j} can be written as:

$$\begin{aligned} J_{kl} &= \int d\sigma^0\partial_0\phi_r[\delta_{rs}(x_k\partial_l - x_l\partial_k) - (S_{kl})_{rs}]\phi_s\,, \\ J_{0j} &= \int d\sigma^0\{\partial_0\phi_r[\delta_{rs}(x_0\partial_j - x_j\partial_0) - (S_{0j})_{rs}]\phi_s + x_j\mathcal{L}(x)\}\,. \end{aligned} \tag{21.144}$$

The integration here is still over all space at the time t. We cannot, of course, just replace this integration region by any arbitrarily chosen space-like surface without checking the properties of the integrand. To be able to effect such a replacement, what we need is an analogue in four dimensions to Gauss's theorem. If the integrands in Eq. (21.144) could be exhibited as the $\lambda = 0$ components of some quantity $(*)_\lambda$ indexed by λ (and other indices) such that the divergence $\partial^\lambda (*)_\lambda$ vanished, then we would be justified in altering the integration region. The following expression naturally suggests itself:

$$\chi_{\lambda,\mu\nu} \equiv \partial_\lambda \phi_r [\delta_{rs}(x_\mu \partial_\nu - x_\nu \partial_\mu) - (S_{\mu\nu})_{rs}]\phi_s - (g_{\lambda\mu} x_\nu - g_{\lambda\nu} x_\mu)\mathcal{L} \,. \quad (21.145)$$

The integrand in the case of J_{kl} is $\chi_{0,kl}$ and in the case of J_{0j} is $\chi_{0,0j}$; using the equations of motion it is not hard to verify that $\chi_{\lambda,\mu\nu}$ obeys:

$$\partial^\lambda \chi_{\lambda,\mu\nu} = 0 \,. \quad (21.146)$$

We are now free to use Gauss's theorem in four dimensions, and discarding terms at spatial infinity write $J_{\mu\nu}$ as integrals of certain densities over an arbitrary spacelike surface in space-time:

$$\begin{aligned} J_{\mu\nu} &= \int d\sigma^\lambda \chi_{\lambda,\mu\nu} \\ &= \int d\sigma^\lambda [\partial_\lambda \phi_r [\delta_{rs}(x_\mu \partial_\nu - x_\nu \partial_\mu) - (S_{\mu\nu})_{rs}]\phi_s - (g_{\lambda\mu} x_\nu - g_{\lambda\nu} x_\mu)\mathcal{L}] \,. \end{aligned} \quad (21.147)$$

For a similar transcription of the expressions for P_j and P_0 in Eq. (21.142), the following construction suggests itself:

$$\theta_{\lambda\mu} = \partial_\lambda \phi_r \partial_\mu \phi_r + g_{\lambda\mu}\mathcal{L} \,; \quad (21.148)$$

use of the equations of motion again shows that $\partial^\lambda \theta_{\lambda\mu} = 0$. This permits us to change the integration region in P_j and P_0 also to an arbitrary spacelike surface and to write:

$$P_\mu = \int d\sigma^\lambda \theta_{\lambda\mu} = \int d\sigma^\lambda (\partial_\lambda \phi_r \partial_\mu \phi_r + g_{\lambda\mu}\mathcal{L}) \,. \quad (21.149)$$

The generators $J_{\mu\nu}$ also could be expressed using $\theta_{\lambda\mu}$ as:

$$J_{\mu\nu} = \int d\sigma^\lambda (x_\mu \theta_{\lambda\nu} - x_\nu \theta_{\lambda\mu} - \partial_\lambda \phi_r (S_{\mu\nu})_{rs} \phi_s) \,. \quad (21.150)$$

With these expressions for the generators of $\mathcal{P}$, and Eq. (21.139) for their action on the field, we have achieved a manifestly covariant form for all the basic equations of the system.

All the foregoing work goes through also in the case of the scalar field: we simply omit the indices r, s, and the matrices $S_{\mu\nu}$. We leave it to the reader to check that Eqs. (21.149), (21.150) with (21.148) are valid also in the presence of the term $V\{\phi\}$ expressing interaction, as long as V is constructed to behave just like a scalar field.

Let us now turn to a brief discussion of finite-component fields. In this case we know that, in general, there is no one-one correspondence between components of ϕ and conjugate momenta. It is usual here to start from the other end and define *directly* the law of transformation of the field components under a Lorentz transformation. One does not assume *to begin with* that one is working within a Hamiltonian framework with generators to implement these transformations, but one chooses the transformation law on "geometrical" grounds. For example, in the case of a vector field $\phi_\mu(x)$, under a change of frame $S \to S' = \Lambda S, \Lambda \in SO(3,1)$, we have the law given in Eq. (20.154):

$$S' = \Lambda S: \ \phi'_\mu(x') = \Lambda_\mu{}^\nu \phi_\nu(x), \ x'^\mu = \Lambda^\mu{}_\nu x^\nu \,. \tag{21.151}$$

For a scalar field, there are no indices on ϕ, and there is simply a change of argument. For fields transforming by some other matrix representation $D_{\alpha\beta}(\Lambda)$ of $SO(3,1)$, we postulate the law

$$\phi'_\alpha(x') = D_{\alpha\beta}(\Lambda)\phi_\beta(x)\,. \tag{21.152}$$

No explicit reference to interactions or even mass (cf. the Galilean case) appears in these transformation laws! If now the canonical Hamiltonian framework is assumed, these transitions $S \to S'$ must also be accomplished by suitable canonical transformations built up using generators $J_{\mu\nu}$. We must demand that the dynamical variable $\phi'_\lambda(x)$ is the canonical transform of $\phi_\lambda(x)$, the space-time arguments x being the same. If we express Λ as

$$\begin{aligned} \Lambda &= \exp(-\frac{1}{2}\lambda^{\mu\nu} l_{\mu\nu}), \ \Lambda^\mu{}_\nu = \Lambda^\mu{}_\nu(\lambda)\,, \\ \Lambda^\mu{}_\nu(\delta\lambda) &\simeq \delta^\mu_\nu + \delta\lambda^\mu{}_\nu\,, \end{aligned} \tag{21.153}$$

then for the vector field we must have:

$$\exp(\widetilde{\frac{1}{2}\lambda^{\mu\nu} J_{\mu\nu}})\phi_\lambda(x^\rho) \equiv \phi'_\lambda(x^\rho) = \Lambda_\lambda{}^\sigma \phi_\sigma(\Lambda_\tau{}^\rho x^\tau)\,,$$

that is,

$$\exp(\widetilde{\frac{1}{2}\lambda^{\mu\nu} J_{\mu\nu}})\phi_\lambda(x^\rho) = \Lambda_\lambda{}^\sigma \phi_\sigma(\Lambda_\tau{}^\rho x^\tau)\,. \tag{21.154}$$

Specializing to the case of an infinitesimal transformation we derive the form of the PB between J and ϕ:

$$\begin{aligned} \{J_{\mu\nu}, \phi_\lambda(x)\} &= (x_\mu \partial_\nu - x_\nu \partial_\mu)\phi_\lambda(x) + (g_{\mu\lambda} g^\sigma_\nu - g_{\nu\lambda} g^\sigma_\mu)\phi_\sigma(x) \\ &= (x_\mu \partial_\nu - x_\nu \partial_\mu)\phi_\lambda(x) - (M_{\mu\nu})_\lambda{}^\sigma \phi_\sigma(x)\,. \end{aligned} \tag{21.155}$$

[The matrices $M_{\mu\nu}$ generate the defining four-dimensional representation of $SO(3,1)$ and are given in Eq. (20.93)]. Similar PB relations hold for any other multicomponent field $\phi_\alpha(x)$ transforming as in Eq. (21.152). Although these considerations do give the action of the generators of $\mathcal{P}$ on the basic fields in a manifestly covariant form, what they conceal is the fact that, in general, there is a separation of the components of the field into dynamically independent and dependent ones, this distinction being essential in setting up the Hamiltonian scheme and the basic PB's.

By a consideration of space-time translations that in a canonical theory are generated by P_μ, we get in analogy to Eqs. (21.154) and (21.155) equations of the form

$$\exp\left(\widetilde{a^\mu P_\mu}\right)\phi_r(x) = \phi_r(x+a),\ \{P_\mu, \phi_r(x)\} = \partial_\mu \phi_r(x)\,, \tag{21.156}$$

valid for each value of the index r. In this approach, such PB relations as Eqs. (21.155) and (21.156) express the manifest covariance of the theory (like the world-line conditions in particle mechanics they postulate the identity of geometrical and canonical transformation laws), and in every consistent theory a choice of basic canonical variables and generators of $\mathcal{P}$ must be made so as to fulfill these demands. The point is that these demands do not rule out interactions.

Manifestly Covariant Treatments of the Electromagnetic Field

The treatment of the electromagnetic field given in the preceding chapter is as follows: We started from an action principle using a Lagrangian linear in the velocities. The Lagrangian equations of motion, which are identical in content to the Maxwell equations, separate into genuine equations of motion and equations of constraint. Using the latter and setting A_0 equal to zero because this variable is left completely free by the Lagrangian equations, we set up an effective Lagrangian involving only those degrees of freedom that obey genuine equations of motion. This effective Lagrangian is then used to pick out the basic pairs of canonically conjugate variables and to state the fundamental PB relations at equal time. With the external current set equal to zero, we use this effective Lagrangian also to construct the generators of $\mathcal{P}$ corresponding to the free electromagnetic field. Although this whole process is physically transparent and certainly relativistic in content, it lacks *manifest* covariance; the "vector" potential does not transform under Lorentz transformations as a local vector field. (Of course, the original Maxwell equations and relativity also do not say that the "vector" potential *must* transform in any particular way, they only fix the properties of the field strengths $F_{\mu\nu}$.) We now treat the electromagnetic field in a manner in which the four-vector transformation property for A_μ is incorporated; that is, the relativistic covariance is made manifest. We have to deal here again with the theory of constrained systems; hence we give two descriptions, one in which manifest covariance is welded to the Lagrangian theory of constrained systems and the other to the Hamiltonian form. We now use Lagrangians not linear in the velocities.

Let us begin with the electromagnetic field in interaction with a specified external source $j_\mu(x)$, so that the Maxwell equations are Eq. (20.176):

$$\partial_\lambda F_{\mu\nu} + \partial_\mu F_{\nu\lambda} + \partial_\nu F_{\lambda\mu} = 0\,, \tag{21.157a}$$

$$\partial_\mu F^{\mu\nu} = -\,j^\nu\,. \tag{21.157b}$$

For consistency, the source j_μ must necessarily be conserved: $\partial^\mu j_\mu=0$. We again solve Eq. (21.157a) by introducing auxiliary potentials $A_\mu(x)$ and setting:

$$F_{\mu\nu} = \partial_\mu A_\nu - \partial_\nu A_\mu \,. \tag{21.158}$$

In contrast to the previous treatment, we do not now try to obtain Eq. (21.158) as one of the Lagrangian equations of motion, but rather regard this as a *definition* of $F_{\mu\nu}$ in terms of the basic dynamical variables A_μ. The Lagrangian is a functional of A_μ and $\partial_\mu A_\nu$, and the object is to derive the field equations:

$$\partial_\mu \partial^\mu A_\nu - \partial_\nu \partial^\mu A_\mu = -j_\nu \tag{21.159}$$

which is the set Eq. (21.157b) of Maxwell's equations. We choose the (quasi-gauge-invariant) Lagrangian as

$$\begin{aligned} L &= \int d^3x\, \mathcal{L}(x)\,, \\ \mathcal{L} &= -\frac{1}{4}(\partial_\mu A_\nu - \partial_\nu A_\mu)(\partial^\mu A^\nu - \partial^\nu A^\mu) + A_\mu j^\mu \\ &= \frac{1}{2}\dot{A}_k\dot{A}_k - \dot{A}_k\partial_k A_0 + \frac{1}{2}\partial_k A_0 \partial_k A_0 - \frac{1}{2}\partial_j A_k(\partial_j A_k - \partial_k A_j) + A_k j_k - A_0 j_0 \,. \end{aligned} \tag{21.160}$$

The Lagrangian equations of motion are easily worked out; we have:

$$\frac{\delta L}{\delta \dot{A}_0} = 0, \quad \frac{\delta L}{\delta A_0} = \partial_k \dot{A}_k - \nabla^2 A_0 - j_0\,; \tag{21.161a}$$

$$\frac{\delta L}{\delta \dot{A}_k} = \dot{A}_k - \partial_k A_0, \quad \frac{\delta L}{\delta A_k} = \nabla^2 A_k - \partial_k \nabla\cdot\boldsymbol{A} + j_k\,; \tag{21.161b}$$

$$\frac{d}{dt}\frac{\delta L}{\delta \dot{A}_k} - \frac{\delta L}{\delta A_k} = 0 \Rightarrow \ddot{A}_k = \nabla^2 A_k + \partial_k(\dot{A}_0 - \nabla\cdot\boldsymbol{A}) + j_k\,; \tag{21.161c}$$

$$\frac{d}{dt}\frac{\delta L}{\delta \dot{A}_0} - \frac{\delta L}{\delta A_0} = 0 \Rightarrow \partial_k \dot{A}_k - \nabla^2 A_0 - j_0 = 0\,. \tag{21.161d}$$

It is easily verified that Eqs. (21.161c) and d are identical with the desired field Eq. (21.159). Before proceeding with the further analysis of these equations, let us note again the close connection of Eq. (21.161d) to the behavior of L under a gauge transformation. If χ is an infinitesimal but arbitrary function on space-time, then by virtue of the current j_μ being conserved L is quasi-invariant under the variation $\delta A_\mu = \partial_\mu \chi$:

$$\delta L = \int d^3x\, j^\mu(\boldsymbol{x},t)\partial_\mu \chi(\boldsymbol{x},t) = -\frac{d}{dt}\int d^3x\, \chi(\boldsymbol{x},t) j_0(\boldsymbol{x},t)\,. \tag{21.162}$$

This fact combines with the complete arbitrariness in the choice of χ and the connection between conservation laws and the action principle (Chapter 4, pages 27 to 30; Chapter 8, pages 86 to 89) to show that what appears on the left-hand side of Eq. (21.161d) must vanish. Thus one of the Lagrangian equations, as well as the absence of an equation for $\ddot{A}_0$, are again related to gauge invariance.

Let us now analyze the Lagrangian equations of motion Eqs. (21.161c) and (21.161d) from the viewpoint of the Lagrangian theory of constraints developed in Chapter 8. (Keep in mind that we have a fourfold infinity of Lagrangian q's here, the four referring to the four components of A_μ and the infinity to all points $\boldsymbol{x}$ in space at a given time t.) The Lagrangian equations have already split into genuine acceleration equations and equations of constraint. The "accelerations" $\ddot{A}_k(\boldsymbol{x},t)$ for each $\boldsymbol{x}$ and for $k=1,2,3$ are specified by Eq. (21.161c) in terms of the "positions" A_μ, the "velocities" $\dot{A}_\mu$ and the sources j_k. On the other hand, Eq. (21.161d) is a constraint of type B because it has an essential dependence on the velocities $\dot{A}_k$. (Actually we have here a continuous family of constraints, one for each point $\boldsymbol{x}$ in space.) According to the general Lagrangian scheme we must take the time derivative of this constraint and ask if on adding it to the already existing acceleration and constraint equations we get more equations for accelerations, or more constraints, or both. In the present case it is clear that no new equations determining accelerations will turn up, because we already have equations for $\ddot{A}_k$, and the time derivative of Eq. (21.161d) will not bring in $\ddot{A}_0$; instead we have (or could have) another constraint. What has to be added to Eqs. (21.161c) and (21.161d) is this:

$$\partial_k \ddot{A}_k - \nabla^2 \dot{A}_0 - \frac{\partial j_0}{\partial t} = 0\,. \tag{21.163}$$

If we eliminate $\ddot{A}_k$ using Eq. (21.161c) we obtain the (possible) constraint:

$$\partial_k(\nabla^2 A_k + \partial_k(\dot{A}_0 - \boldsymbol{\nabla}\cdot\boldsymbol{A}) + j_k) - \nabla^2 \dot{A}_0 - \frac{\partial j_0}{\partial t} = 0\,,$$

that is,

$$\partial^\mu j_\mu = 0\,. \tag{21.164}$$

Therefore the constraint Eq. (21.161d) is automatically preserved in time. The analysis of this Lagrangian system along the lines of Chapter 8 is already complete, and we could state the nature of the time development of the system in this way: choose the variables $A_0(\boldsymbol{x},t)$ for each $\boldsymbol{x}$ as any function of t, in other words choose A_0 as any space-time function; choose initial values $A_k(\boldsymbol{x},0)$ and $\dot{A}_k(\boldsymbol{x},0)$ at time $t=0$ in such a way that the B-type constraint Eq. (21.161d) is obeyed at that time; then the acceleration Eq. (21.161c) will determine $A_k(\boldsymbol{x},t)$ uniquely for all later times, and the B-type constraint Eq. (21.161d) will automatically be satisfied at all times because the external current is conserved.

However, such a description lacks manifest covariance because it singles out the one component A_0 of A_μ, and we wish to use a different approach. Instead of using only those equations that follow from the action principle and permitting A_0 to be an arbitrary space-time function, we introduce a new constraint equation of type B that has the effect essentially of prescribing at each instant of time how A_0 is to be chosen at a succeeding instant of time. Thus we take

Eqs. (21.161c) and (21.161d) that are consequences of the action principle, (call them E, B_1, respectively), add a constraint (B_2) and consider the complete set:

$$\begin{aligned} E: \quad & \ddot{A}_k = \nabla^2 A_k + \partial_k(\dot{A}_0 - \nabla\cdot\mathbf{A}) + j_k\,, \\ B_1: \quad & \partial_k \dot{A}_k - \nabla^2 A_0 - j_0 = 0\,, \\ B_2: \quad & -\dot{A}_0 + \nabla\cdot\mathbf{A} \equiv \partial^\mu A_\mu = 0\,. \end{aligned} \tag{21.165}$$

We must do with B_2 what we already did with B_1, namely add the equation $\dot{B}_2$=0 to the analysis, consider the complete set $E, B_1, B_2, \dot{B}_2$, and ask whether we have ended up with more equations for accelerations and/or more constraints. It is obvious that we do not obtain more constraints because $\dot{B}_2$ involves the acceleration $\ddot{A}_0$; in fact, we now have enough equations to determine all the accelerations. Using the constraints B_1 to simplify the acceleration equations, the complete set takes the form:

$$\begin{aligned} E' \quad & \Box^2 A_\mu \equiv \partial^\nu \partial_\nu A_\mu = -j_\mu\,; \\ B_1 \quad & \partial_k(\partial_0 A_k - \partial_k A_0) - j_0 \equiv -\boldsymbol{\nabla}\cdot\boldsymbol{E} + \rho = 0\,; \\ B_2 \quad & \chi(x) \equiv \partial^\mu A_\mu(x) = 0\,. \end{aligned} \tag{21.166}$$

We can now state the significance of these equations in this way: choose initial values $A_\mu(\boldsymbol{x},0)$ and $\dot{A}_\mu(\boldsymbol{x},0)$ in such a way that B_1 and B_2 are obeyed at time $t=0$. Then the acceleration equations E' permit us to determine $A_\mu(\boldsymbol{x},t)$ for later times, and both constraints B_1 and B_2 are obeyed at all times. In this way manifest covariance has been achieved in the Lagrangian framework.

Though the covariant Eq. (21.166) now has a satisfactory meaning as an initial-value problem, one more step needs to be taken to complete the physical interpretation. It is true that a given set of initial values $A_\mu(\boldsymbol{x},0)$, $\dot{A}_\mu(\boldsymbol{x},0)$ consistent with the constraints B_1 and B_2, will determine a unique solution to the equations E', but two different sets of initial values need not necessarily describe different physical situations. Specifically, we must now add this statement: two solutions $A_\mu(x)$ and $A'_\mu(x)$ of E' whose difference takes the form $A'_\mu(x) - A_\mu(x) = \partial_\mu\Lambda(x)$ with $\Lambda(x)$ obeying $\Box^2\Lambda(x) = 0$ are physically equivalent, because they will lead to the same field strengths $F_{\mu\nu}$. Stated in terms of initial values, two different sets $(A_\mu(\boldsymbol{x},0), \dot{A}_\mu(\boldsymbol{x},0))$ and $(A_k(\boldsymbol{x},0) + \partial_k\Lambda(\boldsymbol{x}), A_0(\boldsymbol{x},0) + \mu(\boldsymbol{x}), \dot{A}_k(\boldsymbol{x},0) + \partial_k\mu(\boldsymbol{x}), \dot{A}_0(\boldsymbol{x},0) + \nabla^2\Lambda(\boldsymbol{x}))$ of initial values are physically equivalent, $\Lambda(\boldsymbol{x})$ and $\mu(\boldsymbol{x})$ being arbitrary.

The development we have just presented is also given in many texts on electromagnetic theory in a much briefer fashion. We have taken more space because we want to make clear the status of each equation, namely whether it is a constraint or not and whether it follows basically from the action principle or not. One can understand the origin of the two constraints B_1, B_2 also from the more usual treatments of this topic. There one says that one-half of the Maxwell equations allow us to represent $F_{\mu\nu}$ in terms of potentials, and the other half then take the form of Eq. (21.159). One then says that, making use of the freedom in the choice of A_μ, we can arrange to have the divergence

$\partial^\mu A_\mu$ always vanishing; this immediately converts Eq. (21.159) into the "wave equation" E' of Eq. (21.166). If this wave equation is used for A_μ, then current conservation tells us that $\partial^\mu A_\mu$ obeys the *homogeneous* wave equation:

$$\Box^2 \partial^\mu A_\mu = 0, \text{ that is, } \Box^2 \chi(x) = 0\,. \tag{21.167}$$

But then, it is enough to ensure that χ and $\partial\chi/\partial t$ vanish at one instant of time, say $t = 0$, and the foregoing equation makes χ vanish for all other times. These two requirements on χ at $t = 0$ are essentially the constraints B_1 and B_2.

We would like to emphasize at this point that this manifestly covariant Lagrangian treatment of the electromagnetic field is in the end not *completely* derivable from an action principle alone. The B-type constraint B_2 had to be introduced "from the outside". This is in contrast to the Coulomb gauge treatment given in the preceding chapter; although manifest covariance was lacking, the complete final set of equations for the physical degrees of freedom is derivable from an action principle using an effective Lagrangian. This is in some ways analogous to the situation encountered in the earlier part of this chapter in the relativistic description of a single particle (see pages 506 to 513). We can use a parameter formalism and derive the equations for a free particle or a charged particle in a given electromagnetic field for instance from a suitable action functional. As long as the parameter θ labeling the trajectory of the particle in space-time is not pinned down, the Lagrangian is necessarily nonstandard, and not all the "accelerations" $(d^2x^\mu(\theta)/d\theta^2)$ can be specified. One possible choice is to specify $x^0(\theta)$ explicitly as a function of θ; for example, θ could be the time variable in some chosen inertial frame; then manifest covariance is lost, but the physical interpretation becomes clear. And at the same time all the basic equations are derivable from an action principle using as Lagrangian (in the free-particle case, for example) the expression in Eq. (20.62). This is somewhat like electrodynamics in the Coulomb gauge. On the other hand, if we want to maintain manifest covariance we can choose θ to be the proper time: then we introduce a new constraint equation (Eq. (21.73)) on $\dot{x}^\mu(\theta)$ and that helps us get covariant equations for all four accelerations. Now, of course, not all the basic equations flow from the action principle. This is somewhat like the treatment of the electromagnetic field given in the preceding few pages. This covariant treatment is called "Lorentz-gauge electrodynamics"; evidently it can be the most convenient form for practical calculations.

The quasi-gauge invariant Lagrangian given in Eq. (21.160) is not a very good one from which to pass to the equivalent Hamiltonian framework, because it would immediately lead to the primary constraint "the momentum canonically conjugate to A_0 vanishes", and that singles out A_0 and again spoils manifest covariance. But we have just seen that the content of Maxwell's equations could be stated in two somewhat different ways: one is to use Eq. (21.159) and whatever consequences follow from it, the other is to use the wave equations E' for all components of A_μ and to supplement this with the two constraints B_1 and B_2, appearing in Eq. (21.166). These two procedures are physically equivalent. (Always the field strengths $F_{\mu\nu}$ are defined to be the curl of A_μ.) Following

Fermi, the equations of the second procedure can also be derived in this way: we choose a new Lagrangian density $\mathcal{L}'(x)$ whose variational equations yield directly the set of wave equations E'; then we add from the outside the constraint B_2 and work out the consequences. This Lagrangian is of standard type and is suitable for passage to the Hamiltonian; after this transcription, the constraint can be expressed in phase-space variables and the consequences worked out along the usual lines. Of course, one obtains equivalence with the Maxwell theory in this approach only at the stage when the constraint is imposed.

We choose, then, the following Lagrangian:

$$\begin{aligned} \mathcal{L}'(x) &= -\frac{1}{2}\partial_\mu A_\nu \partial^\mu A^\nu + A_\mu j^\mu , \\ L' &= \int d^3x\, \mathcal{L}'(x) . \end{aligned} \tag{21.168}$$

This L' is not (quasi) invariant under a gauge transformation. The variational equations we get from it are precisely the equations E':

$$\begin{aligned} \frac{\delta L'}{\delta \dot{A}_\mu} &= \dot{A}^\mu, \ \frac{\delta L'}{\delta A_\mu} = \nabla^2 A^\mu + j^\mu ; \\ \frac{d}{dt}\frac{\delta L'}{\delta \dot{A}_\mu} - \frac{\delta L'}{\delta A_\mu} &= 0 \Rightarrow \Box^2 A_\mu = -j_\mu . \end{aligned} \tag{21.169}$$

These are not yet the Maxwell equations. (For example, they do not require that j_μ be conserved.) We can now pass over to the Hamiltonian. The momentum variables and the fundamental equal time PB's are:

$$\begin{aligned} \pi^\mu(\boldsymbol{x},t) = \text{ conjugate to } A_\mu(\boldsymbol{x},t) &= \frac{\delta L'}{\delta \dot{A}_\mu(\boldsymbol{x},t)} = \dot{A}^\mu(\boldsymbol{x},t) ; \\ \{A_\mu(\boldsymbol{x},t), \pi^\nu(\boldsymbol{y},t)\} &= \delta^\nu_\mu \delta^{(3)}(\boldsymbol{x}-\boldsymbol{y}) , \\ \{A, A\} = \{\pi, \pi\} &= 0 \text{ at equal times} . \end{aligned} \tag{21.170}$$

The Hamiltonian has the form

$$\begin{aligned} H(t) &= \int d^3x\ \mathcal{H}(x) , \\ \mathcal{H}(x) &= \frac{1}{2}(\pi^\mu(x)\pi_\mu(x) + \partial_j A_\mu(x) \partial_j A^\mu(x)) - A^\mu(x) j_\mu(x) . \end{aligned} \tag{21.171}$$

This H may be time dependent because of the presence of j_μ. At this point we get the true Maxwell equations if we impose the constraint $\partial^\mu A_\mu = 0$, which can be stated in phase-space variables as:

$$\chi(x) \equiv \partial_j A_j - \pi_0(x) \approx 0 . \tag{21.172}$$

We have again a continuous family of constraints, one at each point $\boldsymbol{x}$ in space. These are neither primary nor secondary constraints; they are put in from the outside. However, we must follow the general theory and see if they give rise to more constraints. The PB of χ with H (at equal time) is easily worked out, and thereby a new family of constraints is generated:

$$\{\chi(\boldsymbol{x},t), H(t)\} \equiv \chi'(x) = \partial_j \pi_j(x) - \nabla^2 A_0(x) - j_0(x) \approx 0\,. \tag{21.173}$$

Thus the first family of constraints $\chi \approx 0$ is maintained in time if another family of constraints $\chi' \approx 0$ is also imposed. What now of the latter themselves? We have:

$$\begin{aligned} \{\chi'(\boldsymbol{x},t), H(t)\} &= \nabla^2\chi(x) + \partial_j j_j(x)\,, \\ \frac{d\chi'}{dt} \equiv \frac{\partial \chi'}{\partial t} + \{\chi', H\} &\approx 0 \Rightarrow \nabla^2\chi(x) + \partial^\mu j_\mu(x) \approx 0\,. \end{aligned} \tag{21.174}$$

But modulo the constraints $\chi \approx 0$, this condition is met if and only if the current j_μ is conserved. In the Fermi procedure, conservation of the external current is not a necessary condition for the consistency of the Lagrangian equations of motion by themselves, but becomes necessary once we add on the constraints of Eq. (21.172). It is also directly seen from Eq. (21.169) that χ can be zero only if $\partial^\mu j_\mu = 0$ as well. The constraint analysis ends at this stage; the canonical transformation generated by $H(t)$ maps the constrained surface $\chi \approx 0, \chi' \approx 0$ onto itself, and this mapping is the solution of Maxwell's equations.

As long as the source $j_\mu(x)$ is nonzero, we do not have true relativistic invariance. (We have in mind here a *numerically* specified $j_\mu(x)$, not the current of either particles or fields which are themselves dynamical variables.) In the limit $j_\mu = 0$, we deal with the free Maxwell field, and then we can look for the generators of $\mathcal{P}$ characterizing the system. We already have an expression for the Hamiltonian in Eq. (21.171). We get the remaining generators P_j, J_j, and K_j by using the action principle and identifying the constants of motion corresponding to space translations and so on. The free Lagrangian is the expression in Eq. (21.168) with j_μ set equal to zero. (We first obtain all the generators of $\mathcal{P}$ corresponding to this Lagrangian system, and later turn on the constraints). The explicit invariance of the Lagrangian under spatial translations and rotations gives the conservation laws

$$\begin{aligned} \int d^3x \dot{A}^\mu(x)\partial_j A_\mu(x) &= \text{constant}\,, \\ \int d^3x(\dot{A}^\mu(x)(\boldsymbol{x}\times\boldsymbol{\nabla})_j A_\mu(x) + \epsilon_{jkl}\dot{A}_k(x)A_l(x)) &= \text{constant}\,. \end{aligned} \tag{21.175}$$

If we now consider an infinitesimal pure Lorentz transformation and A_μ transforms as a four-vector field under it, we have the variations given in Eq. (20.168):

$$\begin{aligned} \delta A_0 &= \zeta_k A_k + d_\zeta A_0,\ \delta A_k = \zeta_k A_0 + d_\zeta A_k\,, \\ d_\zeta &\equiv t\boldsymbol{\zeta}\cdot\boldsymbol{\nabla} + \boldsymbol{\zeta}\cdot\boldsymbol{x}\frac{\partial}{\partial t}, |\boldsymbol{\zeta}| \ll 1\,. \end{aligned} \tag{21.176}$$

Then the computation of $\delta L'$ is straightforward, and as is expected we find L' to be quasi-invariant:

$$\delta L' = \delta\left(-\frac{1}{2}\int d^3x\, \partial_\mu A_\nu \partial^\mu A^\nu\right) = \frac{d}{dt}\left(-\frac{1}{2}\int d^3x\, \boldsymbol{\zeta}\cdot\boldsymbol{x}\partial_\mu A_\nu \partial^\mu A^\nu\right). \quad (21.177)$$

(Several partial integrations are needed here!) Combining this and Eqs. (21.176) we get the conservation law:

$$\begin{aligned}\int d^3x\, \dot{A}^\mu(x)\delta A_\mu(x) &+ \frac{1}{2}\int d^3x\, \boldsymbol{\zeta}\cdot\boldsymbol{x}\partial_\mu A_\nu(x)\partial^\mu A^\nu(x)\\ =&\zeta_j\int d^3x(\dot{A}_j(x)A_0(x) - \dot{A}_0(x)A_j(x) + t\dot{A}^\mu(x)\partial_j A_\mu(x)\\ &+ \frac{x_j}{2}[\dot{A}^\mu(x)\dot{A}_\mu(x) + \partial_k A^\mu(x)\partial_k A_\mu(x)]) = \text{constant}. \quad (21.178)\end{aligned}$$

We can identify these quantities with the appropriate generators of $\mathcal{P}$. Setting $t = 0$ and replacing $\dot{A}^\mu$ by π^μ, the full set is:

$$\begin{aligned}J_j &= -\int d^3x(\pi^\mu(\boldsymbol{x})(\boldsymbol{x}\times\nabla)_j A_\mu(\boldsymbol{x}) + \epsilon_{jkl}\pi_k(\boldsymbol{x})A_l(\boldsymbol{x})),\\ P_j &= -\int d^3x\, \pi^\mu(\boldsymbol{x})\partial_j A_\mu(\boldsymbol{x}),\\ K_j &= \int d^3x(\pi_j(\boldsymbol{x})A_0(\boldsymbol{x}) - \pi_0(\boldsymbol{x})A_j(\boldsymbol{x}) + \frac{x_j}{2}[\pi^\mu(\boldsymbol{x})\pi_\mu(\boldsymbol{x}) + \partial_k A^\mu(\boldsymbol{x})\partial_k A_\mu(\boldsymbol{x})]),\\ H &= \frac{1}{2}\int d^3x(\pi^\mu(\boldsymbol{x})\pi_\mu(\boldsymbol{x}) + \partial_k A^\mu(\boldsymbol{x})\partial_k A_\mu(\boldsymbol{x})). \quad (21.179)\end{aligned}$$

These are the generators of $\mathcal{P}$ for the system described by the Lagrangian in Eq. (21.168) (with $j_\mu = 0$). To get the realization of $\mathcal{P}$ that describes the free electromagnetic field we must take account of the constraints $\chi(\boldsymbol{x}) \approx 0$. Previously, the analysis of constraints for a Hamiltonian system concerned itself with the question whether a given constraint is preserved in time, and if it is not, what additional constraints must be imposed to ensure this. Now this analysis must obviously be enlarged. Not only must $\chi \approx 0$ be maintained for all time in a given Lorentz frame, it must be maintained under every possible transition to a relativistically equivalent frame. We must ask for the weak vanishing of the PB of $\chi(\boldsymbol{x})$ with every generator of $\mathcal{P}$ appearing in Eq. (21.179), not just with H. One finds again in this analysis that the only new constraints involved are the ones appearing in Eq. (21.173), for with the definitions (at time zero):

$$\chi(\boldsymbol{x}) = \partial_j A_j(\boldsymbol{x}) - \pi_0(\boldsymbol{x}),\ \chi'(\boldsymbol{x}) = \partial_j\pi_j(\boldsymbol{x}) - \nabla^2 A_0(\boldsymbol{x}), \quad (21.180)$$

we find the following set of PB's between χ, χ' and the generators of $\mathcal{P}$:

$$\begin{aligned}\{P_j, \chi(\boldsymbol{x})\} &= \partial_j\chi(\boldsymbol{x}),\ \{J_j, \chi(\boldsymbol{x})\} = (\boldsymbol{x}\times\nabla)_j\chi(\boldsymbol{x});\\ \{P_j, \chi'(\boldsymbol{x})\} &= \partial_j\chi'(\boldsymbol{x}),\ \{J_j, \chi'(\boldsymbol{x})\} = (\boldsymbol{x}\times\nabla)_j\chi'(\boldsymbol{x});\\ \{H, \chi(\boldsymbol{x})\} &= -\chi'(\boldsymbol{x}),\ \{H, \chi'(\boldsymbol{x})\} = -\nabla^2\chi(\boldsymbol{x});\\ \{K_j, \chi(\boldsymbol{x})\} &= -x_j\chi'(\boldsymbol{x}),\ \{K_j, \chi'(\boldsymbol{x})\} = -(\partial_j + x_j\nabla^2)\chi(\boldsymbol{x}). \quad (21.181)\end{aligned}$$

Thus all these brackets vanish modulo the vanishing of χ and χ' themselves; these two families of constraints form a relativistically consistent and invariant set. In this form, the realization of $\mathcal{P}$ that describes the Maxwell field can be described in this way: the set of generators given in Eq. (21.179) gives rise to a canonical realization of $\mathcal{P}$ acting in the ∞^8-dimensional phase space with basic coordinates $\pi^\mu(\boldsymbol{x}), A_\mu(\boldsymbol{x}), \mu = 0, 1, 2, 3$, and $\boldsymbol{x}$ running over all space. This realization of $\mathcal{P}$ leaves invariant the "hypersurface" defined by the ∞^2 constraints $\chi(\boldsymbol{x}) \approx 0, \chi'(\boldsymbol{x}) \approx 0$. Call this hypersurface $\sum$. The restriction of the realization of $\mathcal{P}$ to $\sum$ is not yet the realization of $\mathcal{P}$ corresponding to the Maxwell field because, according to what we said on pages 537 to 538 distinct "points" on $\sum$ may not correspond to distinct physical situations. Thus an alteration of the "coordinates" of a "point" on $\sum$ by the amounts $\triangle A_k(\boldsymbol{x}) = \partial_k \Lambda(\boldsymbol{x})$, $\triangle A_0(\boldsymbol{x}) = \mu(\boldsymbol{x})$, $\triangle \pi_k(\boldsymbol{x}) = \partial_k \mu(\boldsymbol{x})$, $\triangle \pi_0(\boldsymbol{x}) = \nabla^2 \Lambda(\boldsymbol{x})$ leads to a new point on $\sum$ that is physically equivalent to the old one. Here $\Lambda(\boldsymbol{x})$ and $\mu(\boldsymbol{x})$ are arbitrarily chosen numerical functions. Now these changes in π^μ and A_μ are precisely those generated by the constraint functions $\chi(\boldsymbol{x})$ and $\chi'(\boldsymbol{x})$. We easily find that all these constraints are first class:

$$\{\chi(\boldsymbol{x}), \chi(\boldsymbol{y})\} = \{\chi(\boldsymbol{x}), \chi'(\boldsymbol{y})\} = \{\chi'(\boldsymbol{x}), \chi'(\boldsymbol{y})\} = 0\,, \tag{21.182}$$

and if we define G by:

$$G = \int d^3x (\Lambda(\boldsymbol{x})\chi'(\boldsymbol{x}) - \mu(\boldsymbol{x})\chi(\boldsymbol{x}))\,, \tag{21.183}$$

then

$$e^{\tilde{G}}(\pi_k, \pi_0, A_k, A_0) = (\pi_k + \partial_k \mu, \pi_0 + \nabla^2 \Lambda, A_k + \partial_k \Lambda, A_0 + \mu)\,. \tag{21.184}$$

We learn from Eq. (21.182) that the canonical transformations generated by the dynamical variables χ and χ' commute and leave $\sum$ invariant, from Eq. (21.184) that these canonical transformations connect precisely those points on E that are physically equivalent, and from Eq. (21.181) that there is a semi-direct product structure between the generators of $\mathcal{P}$ on the one hand and the dynamical variables $\chi(\boldsymbol{x}), \chi'(\boldsymbol{x})$ on the other. This last statement means that if R is the canonical transformation corresponding to some element of $\mathcal{P}$, then we have relations of the form:

$$RG = G', \; Re^{\tilde{G}}R^{-1} = e^{\tilde{G}'}, \tag{21.185}$$

where G is constructed as in Eq. (21.183) as is G'; the coefficient functions Λ', μ' occurring in G' are determined by Λ, μ. Thus in addition to all the generators of $\mathcal{P}$ being first class,the definition of physical equivalence of points on $\sum$ is relativistically invariant. It is now the restriction of the original realization of $\mathcal{P}$ to $\sum$, plus the further breakup of $\sum$ into subsets consisting of points that are physically equivalent, that describes the free Maxwell field. One might ask: why not introduce suitable new coordinates on $\sum$, deal with just those that distinguish physically different states, and dispense with constraints? That is just what the Coulomb gauge does, and in the process manifest covariance is lost!

The Hamiltonian for the free Maxwell field that appears in Eq. (20.198) is positive definite, whereas that in Eq. (21.179) is not. We have positive contributions for $\mu = 1, 2, 3$ and negative ones for $\mu = 0$. But on the surface $\sum$ there is a cancellation of terms, and H becomes positive definite. We leave it to the reader to check that if $\pi_k(\boldsymbol{x})$ and $A_k(\boldsymbol{x})$ are separated into transverse and longitudinal parts and if the *value* of H is computed using $\chi \approx 0, \chi' \approx 0$, then the positive contributions from the longitudinal parts precisely cancel out the negative ones from the variables π_0, A_0. Effectively, the energy of the field consists of contributions from the transverse fields alone and this is positive.

Finally, it is possible to express the generators of $\mathcal{P}$ in Eq. (21.179) as manifestly covariant integrals of suitable tensorial densities over arbitrary spacelike surfaces. Thus expressions like Eqs. (21.149) and (21.150) can be obtained. The final expressions are:

$$\begin{aligned} P_\mu &= \int d\sigma^\lambda \theta^{(M)}_{\lambda\mu}, \; J_{\mu\nu} = \int d\sigma^\lambda (x_\mu \theta^{(M)}_{\lambda\nu} - x_\nu \theta^{(M)}_{\lambda\mu} - \partial_\lambda A^\alpha (M_{\mu\nu})_{\alpha\beta} A^\beta)\,; \\ \theta^{(M)}_{\lambda\mu} &= \partial_\lambda A^\alpha \partial_\mu A_\alpha + g_{\lambda\mu}\mathcal{L}', \; \mathcal{L}' = -\frac{1}{2}\partial_\mu A_\nu \partial^\mu A^\nu\,; \\ (M_{\mu\nu})_{\alpha\beta} &= g_{\mu\beta} g_{\nu\alpha} - g_{\mu\alpha} g_{\nu\beta}\,. \end{aligned} \tag{21.186}$$

The superscript M on θ refers to the Maxwell field. We leave it to the reader to check these expressions.

Chapter 22

Relativistic Action-at-a-Distance Theories

We have seen that the canonical formalism is incapable of describing a relativistic many-particle system including interactions, if manifest covariance is demanded. (In this chapter the only relativity group we are concerned with is the Poincaré group.) Such dynamical systems must be described by other means. For suitable types of interactions we may still be able to derive the basic equations of the system from an action principle, and in this chapter we study examples of precisely such systems. However, the action functional can no longer be expected to be the time integral of a suitable Lagrangian which is local in time, and hence the basic "equations of motion" also are not local in time. Interactions between particles in the Newtonian nonrelativistic (or better Galilean relativistic) domain are summarized in the notion of a potential that acts nonlocally ("at a distance") in space at a given time. If this is to be generalized to the Poincaré-relativistic domain and the description is manifestly covariant, the nonlocality in space implicit in the notion of a potential automatically implies nonlocality in time. This comes in from the geometry of Lorentz transformations that makes simultaneity frame-dependent; in contrast to a nonrelativistic potential that can act at one instant of time in every frame, a "relativistic potential" acts at unequal times as well.

On the other hand, going back to the no-interaction theorem of the preceding chapter, we must recognize that it applies to those dynamical systems that are composed of *particles alone*. The difficulty is that if the world-line conditions on the particle coordinates are imposed, there is no way in which we could add interaction terms in the Hamiltonian and the generators of pure Lorentz transformations without at the same time upsetting the PB relations reflecting the structure of $\mathcal{P}$. One way of avoiding this problem would be to include new degrees of freedom belonging not to particles but to fields. The fields could carry their own energy momentum, and there could be interaction between particles and fields. Such a combined system could again lead to a realization of the Poincaré group within the canonical framework, with the world-line conditions for the particle positions (and analogous conditions for the fields) being satisfied. What we shall find is that, *under suitable circumstances the key equations describing a relativistic action-at-a-distance theory of particles alone can be reinterpreted as those of a canonical Hamiltonian system of particles and fields with local interactions.*

We would like to make a few remarks on these two forms of describing interaction. From the earliest days of dynamics the question of action at a distance versus action by contact has been discussed. One gets the impression that the question has been decided in favor of action by contact. In pre-Newtonian days all interactions were by contact. To affect an object at a distance, one either pulled it by a string or pushed it by a rod. Newton's theory of gravitation, however, changed this belief, because the sun seemed to affect the motion of the earth from a great distance without any visible string or rod. This gave rise to the possibility of action at a distance. Such a picture continued to suffice for an understanding of electrostatics and magnetostatics. The idea of a field was introduced in gravitation, electrostatics, and magnetostatics to restore action by contact, but these fields, which were "extensions" of the material particles and provided the "invisible strings" between them, were convenient auxiliary quantities that were not essential for the description of static phenomena. With the advent of Maxwell's electrodynamics, however, it appeared as if introduction of the field was essential for the description of observed phenomena, both in virtue of the finiteness of the velocity of propagation of interactions and the existence of radiation. The electromagnetic field had to be viewed as a dynamical system in its own right, with its own degrees of freedom. But trying to introduce interaction between the electromagnetic field on the one hand, and charged particles with degrees of freedom of their own on the other, led to fresh troubles. If elementary classical point charges are to interact with the electromagnetic field in the manner deduced from the Maxwell equations, then there is a serious inconsistency: the point charges must explode. This explosion or divergence is caused by the action of a charged particle on itself; if the field produced by the particle is determined by some solution of Maxwell's equations with the particle as source and if this field produces a force on the particle according to the Lorentz law of force, we end up with an infinite self-force because the field of the particle diverges at the location of that particle. If we wish to remain true to the spirit of this Maxwell-Lorentz theory of fields and particles in mutual interaction, which implies accepting the action of a charged particle on itself, we must make this self-force finite by modifying the theory in the appropriate direction. As we indicate toward the end of this chapter, this can be done by introducing additional fields, the properties of which may be somewhat strange at first sight; these properties are essential if we want a finite and consistent theory.

The following question must also be raised: do the observed phenomena of electromagnetism rule out an action-at-a-distance theory, in which self-effects are explicitly excluded? Could we not reconcile the phenomena by using a suitable physical interpretation of a direct action at a distance among charged particles alone? This question was raised and answered in the affirmative by Wheeler and Feynman. (J.A. Wheeler and R.P. Feynman, *Rev. Mod. Phys.*, **17**, 157 (1945); **21**, 425 (1949)). They showed that, provided certain entities collectively acting as the "absorber" are introduced, all processes involving radiation can be equally well derived from a theory involving direct action at a distance between charged particles. They do avoid the infinities because of the self-forces, but at the price of complicating the effective interaction between the charged particles and a field, if we were to rephrase their theory as an equivalent field theory. In other words, their theory is not equivalent to Maxwell's theory even with the introduction of the absorber. In any case, their work does show that the direct observation of a "propagating influence" such as electromagnetic radiation in no way rules out the possibility of action at a distance.

We consider two examples of interactions, a scalar and a vector one, in the action-at-a-distance framework. In each case we attempt to establish a correspondence with a suitable theory of particles and fields in interaction, a scalar field in one case and a vector field in the other. The latter leads to the Maxwell-Lorentz electrodynamics. Much of this work is formal in nature because we ignore the inconsistencies caused by the appearance of infinite self-action. How to make this finite is indicated later on. This point must be kept in mind during the manipulations needed in comparing a theory of one kind with a theory of the other kind.

The Case of Scalar Interactions

We consider a dynamical system consisting of a number of point particles. We label these particles by the letters $r, s, \ldots$; the quantities pertaining to the rth particle carry r as a subscript. To describe the space-time trajectory of the rth particle we introduce a parameter θ_r that runs continuously and monotonically from $-\infty$ to $+\infty$ as the trajectory is traversed from the infinite past to the infinite future. The space time coordinates of particle r at the point on its trajectory labeled θ_r are written $x_r^\mu(\theta_r)$, and quite often the argument θ_r is omitted. Let $C_1, C_2, \ldots C_r, \ldots$ be a conceivable set of space-time trajectories for the particles $1, 2, \ldots r, \ldots$, and choose a segment on each trajectory; on C_r let the chosen segment correspond to an "initial" value $\theta_r = \sigma_r$ and a "final" value $\theta_r = \rho_r(\rho_r > \sigma_r)$. We now set up an action Φ that is a functional of the trajectories C_r and in addition depends on the set of initial instants σ_r and of final instants ρ_r. If there were no interactions, Φ would simply be a sum of expressions of the form given in Eq. (21.52). We include binary interactions by adding terms depending on pairs of particles; such additions must preserve invariance under changes in choice of the parameters θ_r. With m_r being the rest masses of the particles, we define the action functional

$$\Phi[C_1, C_2, \ldots; \sigma_1, \sigma_2, \ldots; \rho_1, \rho_2, \ldots] = -\sum_r m_r \int_{\sigma_r}^{\rho_r} d\theta_r (-\dot{x}_r^2)^{1/2}$$

$$+ \frac{1}{2} \sum_{r,s} \int_{\sigma_r}^{\rho_r} d\theta_r \int_{\sigma_s}^{\rho_s} d\theta_s (-\dot{x}_r^2)^{1/2} (-\dot{x}_s^2)^{1/2} K_{rs}((x_r - x_s)^2) . \qquad (22.1)$$

The dot on x_r signifies the derivative with respect to θ_r, and $\dot{x}_r^2$ is the Lorentz square of the velocity, $\dot{x}_r^\mu \dot{x}_{r\mu}$. (For each r, θ_r is the argument of $x_{r\mu}$.) The interaction is described by a collection of functions K_{rs} (and obviously $K_{rs} = K_{sr}$) and we take these to depend only on the squared space-time interval between a pair of points, one on each trajectory. Because the dependence of the interaction terms on the velocities $\dot{x}_{r\mu}, \dot{x}_{s\mu}$ is the minimum required to preserve invariance against changes in θ_r, we call this a theory with *scalar interactions*. It is assumed that the functions K_{rs} vanish for spacelike separations of the particles, that is, $K_{rs}(\tau)$=0 for $\tau > 0$. The self-interaction terms in Eq. (22.1), corresponding to $r = s$, are to be interpreted in this sense: we have two parameters θ_r, θ_r' both running independently over the given stretch σ_r to ρ_r of the one world-line C_r, and the corresponding term in Eq. (21.1) is:

$$\frac{1}{2} \int_{\sigma_r}^{\rho_r} d\theta_r \int_{\sigma_r}^{\rho_r} d\theta_r' (-\dot{x}_r'^2)^{1/2} (-\dot{x}_r^2)^{1/2} K_{rr}((x_r - x_r')^2) ,$$

x_r denoting $x_r(\theta_r)$ and x_r' denoting $x_r(\theta_r')$. The derivation of a system of "equations of motion" from a principle of stationary action proceeds now along lines slightly different from the case of Lagrangian mechanics. In the latter case, we could start with the action functional computed for any finite segment of the system trajectory in configuration space, say for the time t in the interval $t_1 \leq t \leq t_2$, and demand the stationarity of this action under variations of just this segment of the trajectory. The resulting equations of motion that are implied for t in the range $t_1 < t < t_2$ are actually independent of the terminal times t_1, t_2; both the Lagrangian and the equations of motion are local in time. It is clear that with the action in Eq. (22.1) we must proceed differently. Let us, to begin with, compute the variation of the "finite" action (that is, the action referring to finite segments of the particle world lines) under a small change in the world lines; the result of this computation proves useful later on. Keeping the terminal instants (σ_r, ρ_r) fixed and changing the dependences of $x_{r\mu}$ on θ_r by infinitesimal functions $\delta x_{r\mu}(\theta_r)$, the change in Φ is seen to be:

$$
\begin{aligned}
\triangle\Phi[C_1,\ldots;\sigma_1,\ldots;\rho_1,\ldots] &= \sum_r \int_{\sigma_r}^{\rho_r} d\theta_r p_{r\mu}\delta\dot{x}_r^\mu \\
&+ \sum_{rs} \int_{\sigma_r}^{\rho_r} d\theta_r \int_{\sigma_s}^{\rho_s} d\theta_s \\
&\times \left(-\frac{p_{r\mu}}{m_r}\delta\dot{x}_r^\mu(-\dot{x}_s^2)^{1/2}K_{rs} + (-\dot{x}_r^2)^{1/2}(-\dot{x}_s^2)^{1/2}\frac{\partial K_{rs}}{\partial x_{r\mu}}\delta x_{r\mu} \right), \\
p_{r\mu} &= \frac{m_r\dot{x}_{r\mu}}{(-\dot{x}_r^2)^{1/2}}\,.
\end{aligned}
\tag{22.2}
$$

(The arguments of the x's and K's are self-evident and need not be indicated explicitly.) As usual, we integrate by parts the terms involving $\delta\dot{x}_{r\mu}$, collect in one place all terms involving $\dot{p}_{r\mu}$, and end up with

$$
\begin{aligned}
\triangle\Phi[C_1,\ldots;\sigma_1\ldots;\rho_1,\ldots] &= \sum_r [p_{r\mu}\delta x_r^\mu]_{\sigma_r}^{\rho_r} \\
&- \sum_{rs} \int_{\sigma_s}^{\rho_s} d\theta_s(-\dot{x}_s^2)^{1/2} \left[\frac{p_{r\mu}}{m_r}\delta x_r^\mu K_{rs}\right]_{\sigma_r}^{\rho_r} \\
&- \sum_r \int_{\sigma_r}^{\rho_r} d\theta_r \left(1 - \frac{1}{m_r}\sum_s \int_{\sigma_s}^{\rho_s} d\theta_s(-\dot{x}_s^2)^{1/2}K_{rs}\right)\dot{p}_{r\mu}\delta x_r^\mu \\
&+ \sum_{rs} \int_{\sigma_r}^{\rho_r} d\theta_r \int_{\sigma_s}^{\rho_s} d\theta_s(-\dot{x}_r^2)^{1/2}(-\dot{x}_s^2)^{1/2}W_{r\mu\nu}\delta x_r^\mu \frac{\partial K_{rs}}{\partial x_{r\nu}}\,, \\
W_{r\mu\nu} &= g_{\mu\nu} - \frac{\dot{x}_{r\mu}\dot{x}_{r\nu}}{\dot{x}_r^2}\,.
\end{aligned}
\tag{22.3}
$$

To impose the principle of stationary action, we proceed as follows. We consider the action corresponding to the entire trajectories of all the particles; that is, we set $\sigma_r = -\infty, \rho_r = +\infty$. We envisage variations $\delta x_{r\mu}(\theta_r)$ that vanish outside some finite region in θ_r, or at any rate vanish as $\theta_r \to \pm\infty$, and then demand that for all such variations $\triangle\Phi$ be zero. This immediately leads to the "equations of motion":

$$
\begin{aligned}
&\left(1-\frac{1}{m_r}\sum_s \int_{-\infty}^{\infty} d\theta_s(-\dot{x}_s^2)^{1/2}K_{rs}\right)\dot{p}_{r\mu} \\
&\quad = (-\dot{x}_r^2)^{1/2}W_{r\mu\nu}\sum_s \int_{-\infty}^{\infty} d\theta_s(-\dot{x}_s^2)^{1/2}\frac{\partial K_{rs}}{\partial x_{r\nu}} \\
&\quad = (-\dot{x}_r^2)^{1/2}W_{r\mu\nu}\frac{\partial}{\partial x_{r\nu}}\sum_s \int_{-\infty}^{\infty} d\theta_s(-\dot{x}_s^2)^{1/2}K_{rs}\,.
\end{aligned}
\tag{22.4}
$$

The argument of K_{rs} in Eq. (22.4) is always $(x_r - x_s)^2$ with $x_{r\mu}$ kept fixed and $x_{s\mu}$ being integrated over. Several comments are now in order:

1. The equations we have obtained are not differential equations of motion but integro-differential equations. We might call them equations for world lines. They are by construction manifestly covariant under the Poincaré group. The analysis of the nature of their solutions is certainly much more complicated than if we were dealing with a system of differential equations, but we do not go into this point any further.

2. It would be wrong to impose the action principle directly on the "finite" action defined in Eq. (22.1) by considering variations $\delta x_{r\mu}$ that vanish outside the finite segments $\sigma_r < \theta_r < \rho_r$, because then the equations for the world lines would, in general, depend on these instants. One must deal with the total action for this purpose. Depending on the functions K_{rs}, it may happen that the total action is not finite; that is, it does not exist. Then it plays only a formal role; all that is used is its form. But as long as we consider variations $\delta x_{r\mu}$ that vanish outside bounded segments of the world lines, the *variation* of the total action presumably has a well-defined meaning.

3. As we have seen in the previous chapter (see Eq. (21.54)), the presence of the elements of the singular matrix $||W_{r\mu\nu}||$ on the righthand side in Eq. (22.4) is quite natural, reflecting the arbitrariness in the choice of the parameters θ_r. A set of covariant equations that specify the four "accelerations" $\dot{p}_{r\mu}$ must be consistent with the fact that these four quantities are not independent but obey $\dot{p}_{r\mu}p_r^{\mu}{=}0$ or $\dot{x}_{r\mu}\dot{p}_r^{\mu}{=}0$, and Eq. (22.4) obeys this requirement. Having obtained the equations of motion, we could set the θ_r equal to the corresponding proper times τ_r; that would remove the factors $(-\dot{x}_r^2)^{1/2}$ everywhere.

4. We assumed that the functions K_{rs} vanished for positive argument. Thus $K_{rs}((x_r - x_s)^2)$ in general does not vanish for positive as well as negative timelike and lightlike separations $(x_r - x_s)^{\mu}$. Going back to Eq. (22.4) we may say that the motion of particle r at a point P_r on its world line is affected by all those points on the world line of particle s that are in the absolute past of P_r as well as those in the absolute future of P_r. These are called retarded and advanced effects, respectively. The interesting point is that when we work from an action principle both kinds of effects are necessarily present, because K_{rs} is a symmetric function of x_r and x_s. A given binary interaction term in the action in Eq. (22.1) contributes to the equation of motion of particle r as well as of particle s. If there is a part of it that describes the retarded effect of s on the motion of r, that same part gives rise to an advanced effect of r on the motion of s.

Let us now turn to the derivation of conservation laws that must follow from the manifest invariance of the action under the transformations of the Poincaré group. As we have seen in the discussion in Chapter 4, we must combine the equations for world lines, the *general* form for the variation $\triangle\Phi$, and the form expected for $\triangle\Phi$ knowing the invariance properties of Φ if the variation $\delta x_{r\mu}$ is that induced by an infinitesimal transformation from $\mathcal{P}$. The slight complication we encounter is caused by the fact that only the variation of the total action leads to the equations for world lines. To save writing, let us at this stage choose each θ_r to be the proper time τ_r of the corresponding particle, and let dots now denote derivatives with respect to proper time. By the very definition of Φ in Eq. (22.1), it is evident that if the variation in the trajectories corresponds to an infinitesimal inhomogeneous Lorentz transformation,

$$\delta x_{r\mu}(\theta_r) = a_\mu + \lambda_{\mu\nu} x_r^\nu(\theta_r), |a| << 1, |\lambda| << 1, \lambda_{\mu\nu} = -\lambda_{\nu\mu}\,, \tag{22.5}$$

and if the σ_r and ρ_r are not altered, then $\triangle\Phi$ given in Eq. (22.3) must vanish:

$$\sum_r \left[\left(1 - \frac{1}{m_r}\sum_s \int_{\sigma_s}^{\rho_s} d\tau_s K_{rs}\right) p_{r\mu}\delta x_r^\mu\right]_{\sigma_r}^{\rho_r}$$
$$-\sum_r \int_{\sigma_r}^{\rho_r} d\tau_r \left(1 - \frac{1}{m_r}\sum_s \int_{\sigma_s}^{\rho_s} d\tau_s K_{rs}\right) \dot{p}_{r\mu}\delta x_r^\mu$$
$$+\sum_{rs} \int_{\sigma_r}^{\rho_r} d\tau_r \int_{\sigma_s}^{\rho_s} d\tau_s W_{r\mu\nu}\delta x_r^\mu \frac{\partial K_{rs}}{\partial x_{r\nu}} = 0 \tag{22.6}$$

To use the equations of motion of Eq. (22.4) to simplify the second term, the former must be written as:

$$\left(1 - \frac{1}{m_r}\sum_s \int_{\sigma_s}^{\rho_s} d\tau_s K_{rs}\right)\dot{p}_{r\mu}$$
$$= \frac{1}{m_r}\sum_s \left(\int_{-\infty}^{\sigma_s} + \int_{\rho_s}^{\infty}\right) d\tau_s K_{rs}\dot{p}_{r\mu} + W_{r\mu\nu}\frac{\partial}{\partial x_{r\nu}}\sum_s \int_{-\infty}^{\infty} d\tau_s K_{rs}\,. \tag{22.7}$$

Then Eq. (22.6) takes the form:

$$\sum_r \left[\left(1 - \frac{1}{m_r}\sum_s \int_{\sigma_s}^{\rho_s} d\tau_s K_{rs}\right) p_{r\mu}\delta x_r^\mu\right]_{\sigma_r}^{\rho_r}$$
$$-\sum_{rs} \frac{1}{m_r}\int_{\sigma_r}^{\rho_r} d\tau_r \left(\int_{-\infty}^{\sigma_s} + \int_{\rho_s}^{\infty}\right) d\tau_s K_{rs}\dot{p}_{r\mu}\delta x_r^\mu$$
$$-\sum_{rs} \int_{\sigma_r}^{\rho_r} d\tau_r \left(\int_{-\infty}^{\sigma_s} + \int_{\rho_s}^{\infty}\right) d\tau_s W_{r\mu\nu}\delta x_r^\mu \frac{\partial K_{rs}}{\partial x_{r\nu}} = 0\,. \tag{22.8}$$

It is now better to integrate the second term here by parts with respect to τ_r so that $\dot{p}_{r\mu}$ does not appear explicitly anywhere. On doing so, the boundary term combines with the first term in Eq. (22.8) to simplify it, whereas the remainder combines with and simplifies the last term in Eq. (22.8):

$$\sum_r \left[\left(1 - \frac{1}{m_r} \sum_s \int_{-\infty}^{\infty} d\tau_s K_{rs} \right) p_{r\mu} \delta x_r^\mu \right]_{\sigma_r}^{\rho_r} - \sum_{rs} \int_{\sigma_r}^{\rho_r} d\tau_r \left(\int_{-\infty}^{\sigma_s} + \int_{\rho_s}^{\infty} \right) d\tau_s \frac{\partial K_{rs}}{\partial x_{r\mu}} \delta x_{r\mu} = 0 \,. \tag{22.9}$$

(Note that $p_{r\mu}\delta\dot{x}_r^\mu = 0$.) The result above is valid for the transformation Eq. (22.5) and for any arbitrarily chosen set of initial instants σ_r and final instants ρ_r. We could interpret it as a conservation law if the left-hand side could be exhibited as some quantity depending on the ρ_r alone minus the same quantity with ρ_r replaced by σ_r for all r. The first term is already in such a form, as is the second term if we suitably split the τ_r integrations:

$$\sum_{rs} \int_{\sigma_r}^{\rho_r} d\tau_r \left(\int_{-\infty}^{\sigma_s} + \int_{\rho_s}^{\infty} \right) d\tau_s \frac{\partial K_{rs}}{\partial x_{r\mu}} \delta x_{r\mu} = \sum_{rs} \left(\int_{\sigma_r}^{\infty} \int_{-\infty}^{\sigma_s} - \int_{\rho_r}^{\infty} \int_{-\infty}^{\sigma_s} + \int_{-\infty}^{\rho_r} \int_{\rho_s}^{\infty} - \int_{-\infty}^{\sigma_r} \int_{\rho_s}^{\infty} \right) d\tau_r d\tau_s \frac{\partial K_{rs}}{\partial x_{r\mu}} \delta x_{r\mu} \,. \tag{22.10}$$

The translation and Lorentz invariances of K_{rs} imply that with δx given by Eq. (22.5),

$$\frac{\partial K_{rs}}{\partial x_{r\mu}} = -\frac{\partial K_{rs}}{\partial x_{s\mu}} \,, \quad \frac{\partial K_{rs}}{\partial x_{r\mu}} \delta x_{r\mu} = -\frac{\partial K_{rs}}{\partial x_{s\mu}} \delta x_{s\mu} \,. \tag{22.11}$$

Because of the summation over r and s, the second and fourth sets of double integrals in Eq. (22.10) cancel one another, and Eq. (22.9) can be stated as:

$$\sum_r \left[(1 - \frac{1}{m_r} \sum_s \int_{-\infty}^{\infty} d\tau_s K_{rs}) p_{r\mu} \delta x_r^\mu \right]_{\tau_r = \rho_r} + \sum_{rs} \int_{\rho_r}^{\infty} d\tau_r \int_{-\infty}^{\rho_s} d\tau_s \frac{\partial K_{rs}}{\partial x_{r\mu}} \delta x_{r\mu} = (\text{same expression evaluated with } \sigma_r \text{ in place of } \rho_r) \,. \tag{22.12}$$

This statement has the form of a conservation law. By substituting for $\delta x_{r\mu}$ we can identify the coefficients of a^μ and $\lambda^{\mu\nu}$ as the total energy-momentum four-vector P_μ and the four-dimensional angular momentum tensor $J_{\mu\nu}$, respectively:

$$
\begin{aligned}
P_\mu(\rho_1 \ldots) = &\sum_r \left[\left(1 - \frac{1}{m_r} \sum_s \int_{-\infty}^{\infty} d\tau_s K_{rs} \right) p_{r\mu} \right]_{\tau_r = \rho_r} \\
&+ \sum_{rs} \int_{\rho_r}^{\infty} d\tau_r \int_{-\infty}^{\rho_s} d\tau_s \frac{\partial K_{rs}}{\partial x_r^\mu}, \\
J_{\mu\nu}(\rho_1 \ldots) = &\sum_r \left[\left(1 - \frac{1}{m_r} \sum_s \int_{-\infty}^{\infty} d\tau_s K_{rs} \right) (x_{r\mu} p_{r\nu} - x_{r\nu} p_{r\mu}) \right]_{\tau_r = \rho_r} \\
&+ \sum_{rs} \int_{\rho_r}^{\infty} d\tau_r \int_{-\infty}^{\rho_s} d\tau_s \left(x_{r\mu} \frac{\partial}{\partial x_r^\nu} - x_{r\nu} \frac{\partial}{\partial x_r^\mu} \right) K_{rs}. \qquad (22.13)
\end{aligned}
$$

The law of conservation of energy-momentum-angular momentum is, then, the statement that for a given state of motion in accord with the equations of motion (Eq. (22.4)), the values of P_μ and $J_{\mu\nu}$ are the same independent of what set of instants on the world lines is chosen to evaluate them. That the quantities P_μ and $J_{\mu\nu}$ are conserved in this sense could also have been shown by direct integration of the equations of motion.

Let us interpret these constants of motion further. It is obvious from the form of P_μ that for a given state of motion the values obtained for P_μ in two different inertial frames are related in the correct way by means of the matrix of the appropriate homogeneous Lorentz transformation. Thus P_μ and $J_{\mu\nu}$ transform correctly. The first term in P_μ includes the individual particle contributions $p_{r\mu}$ to the total energy-momentum of the system, each individual term evaluated at the chosen instant ρ_r on the world line. All other terms in P_μ are caused by the presence of interaction; because of manifest covariance, there is an energy of interaction as well as a momentum (and angular momentum) of interaction. A typical double-integral term in P_μ,

$$
\int_{\rho_r}^{\infty} d\tau_r \int_{-\infty}^{\rho_s} d\tau_s \frac{\partial K_{rs}}{\partial x_r^\mu},
$$

is to be interpreted as "impulse in transit", that is, the energy-momentum in transit between particles r and s. If all the instants ρ_r were chosen so that the corresponding points on the world lines were all relatively spacelike, then only the part of K_{rs} corresponding to *positive* time-difference $(x_r - x_s)^0 > 0$ contributes in the foregoing double integral, and this contribution can be interpreted as the energy-momentum that left particle s *upto* the instant ρ_s and will arrive at r *after* the instant ρ_r. [This can be modified if the events $x_r^\mu(\rho_r)$ are not all relatively spacelike.] There is an extra interaction term in the first term of P_μ itself proportional to $p_{r\mu}$; it is like making the mass of particle r vary during the motion because the "single particle" terms in P_μ can be written as

$$
\left(1 - \frac{1}{m_r} \sum_s \int_{-\infty}^{\infty} d\tau_s K_{rs} \right) p_{r\mu} = \left(m_r - \sum_s \int_{-\infty}^{\infty} d\tau_s K_{rs} \right) \dot{x}_{r\mu}.
$$

We see that in the case of electromagnetic interactions (or vector interactions) such a term is not present.

It should be noted that these expressions for P_μ and $J_{\mu\nu}$ are not unique. It is true that the action functional has led to specific quantities whose values turn out to be independent of the set of instants ρ_r chosen to evaluate them. But we could modify our definition of P_μ, say, by adding to what we already have an extra contribution that transforms as a four-vector, depends on the state of motion, but is explicitly independent of the instants ρ_r. This might be achieved by making it a functional of the *entire* world lines of all the particles. The same comment applies to $J_{\mu\nu}$. In a canonical theory there would be severe restrictions on such additions to P_μ and $J_{\mu\nu}$ because these must obey the PB relations corresponding to the structure of $\mathcal{P}$. In an action-at-a-distance theory, some ambiguity will remain. This point is to be made because when we try to connect the expressions in Eq. (22.13) to those occurring in a canonical theory of particles and fields, extra terms of the type mentioned will appear.

Let us now ask under what circumstances the foregoing theory is "equivalent" to a suitable canonical theory of particles and fields in mutual interaction. "Equivalence" must mean that there is a correspondence between the basic dynamical equations in the two cases, and also between the conserved quantities. This will come about if the functions $K_{rs}((x_r - x_s)^2)$ factorize in their dependence on the particle labels r, s and if the dependence on $(x_r - x_s)^2$ is suitably chosen:

$$K_{rs}((x_r - x_s)^2) = g_r g_s K((x_r - x_s)^2)\,. \tag{22.14}$$

(A sum of such factorized expressions is also allowed and is, in fact, needed for a finite theory, as we indicate later.) The constants g_r are real, and to begin with we choose for K the time-symmetric invariant function $\bar{\Delta}(x;\kappa)$ as follows:

$$\begin{aligned} K((x_r - x_s)^2) =& \bar{\Delta}(x_r - x_s;\kappa),\ (\partial^\mu\partial_\mu - \kappa^2)\bar{\Delta}(x;\kappa) = -\delta^{(4)}(x)\,,\\ \bar{\Delta}(x;\kappa) =& \frac{1}{4\pi}\delta(x^2) - \theta(-x^2)\frac{\kappa}{8\pi\sqrt{-x^2}}J_1(\kappa\sqrt{-x^2})\,. \end{aligned} \tag{22.15}$$

Here J_1 is the Bessel function of order one, θ is the step function that vanishes for negative argument and is unity for positive argument, and $x^2 = x^\mu x_\mu$ as usual. The function $\bar{\Delta}(x;\kappa)$ is actually a function of x^2 as is seen above, and is even under the strong reflection, $x^\mu \to -x^\mu$. It has the property of vanishing whenever x^μ is a spacelike vector. The function $\bar{\Delta}$ is actually one of several important solutions to the inhomogeneous differential equation appearing in Eq. (22.15) and is distinguished by being a real, time-symmetric solution. Other important solutions are the retarded solution $\Delta_{ret}(x;\kappa)$ that vanishes if $x^0 < 0$, and the advanced solution $\Delta_{adv}(x;\kappa)$ that vanishes if $x^0 > 0$. The various solutions of the homogeneous differential equation obtained by omitting the $\delta^{(4)}(x)$ are also important. At the end of this chapter we list these functions and some of their important properties. With the foregoing choice for K_{rs}, let us put the equations of motion Eq. (22.4) into a more suggestive form. For each state of motion of the multiparticle system let us define a quantity $\bar{\phi}(x)$, which is a functional of the set of world lines and a function of x:

$$\bar{\phi}(x) \equiv \sum_s g_s \int_{-\infty}^{\infty} d\tau_s \bar{\Delta}(x - x_s; \kappa) . \qquad (22.16)$$

The functional dependence is left implicit. This "field" $\bar{\phi}(x)$ is defined completely in terms of the particle trajectories and is *not* an independent variable. But it helps us write the equations of motion for the action-at-a-distance theory in the simple form:

$$\left(1 - \frac{g_r}{m_r}\bar{\phi}(x_r)\right) \dot{p}_{r\mu} = g_r(g_{\mu\nu} + \dot{x}_{r\mu}\dot{x}_{r\nu}) \left[\frac{\partial\bar{\phi}(x)}{\partial x_\nu}\right]_{x=x_r} , \qquad (22.17)$$

and it itself obeys:

$$(\partial^\mu\partial_\mu - \kappa^2)\bar{\phi}(x) \equiv j(x) = -\sum_s g_s \int_{-\infty}^{\infty} d\tau_s \delta^{(4)}(x - x_s) . \qquad (22.18)$$

This auxiliary quantity $\bar{\phi}$ depends as a function on x and as a functional on the particle world lines, and only the former dependence is being differentiated in Eqs. (22.17) and (22.18). Equation (22.18) is not really an equation of motion for $\bar{\phi}$, but a fact, and $\bar{\phi}$ is not any solution but precisely the one appearing in Eq. (22.16). However, a system of genuine equations of motion can be obtained in the canonical framework for a real dynamical scalar field ϕ in interaction with a collection of particles, having exactly the form of Eqs. (22.17) and (22.18)!

At this point we must recognize that by choosing $K_{rs} = g_r g_s \bar{\Delta}$, the foregoing equations have become meaningless. This is because $\bar{\Delta}$ is a singular function of x^2 as is evident from Eq. (22.15); going back to the action functional in Eq. (22.1) it follows that the self-interaction terms (terms with $r = s$) diverge, hence there are explicit infinities in the equations of motion. One way out at this stage is to omit all the self-interaction terms in the action functional in Eq. (22.1), but retain the choice made above for the functions K_{rs} for $r \neq s$. This certainly leads to a mathematically well-defined system. (One only has to discount the crossing of the world lines of two distinct particles as being infinitely improbable in four-dimensional space-time.) This is, in fact, the approach taken by many people. But if one tries to express such a theory in terms of some kind of equivalent field theory, one has to do one of two things: corresponding to each particle introduce the field produced by *that* particle and make it act on every *other* particle only, or corresponding to each particle define the field that acts on it as the sum of the fields caused by all the other particles. Thus this field carries the space-time argument x as well as a particle index r, and *there are as many fields as particles.* We wish to deal with a unique field; hence we are forced to include self-interaction terms in the action. To make these terms finite we must choose for K_{rs} a sum of expressions like the one in Eqs. (22.14) and (22.15), arranged so that the singularities near $x^2 = 0$ cancel out. It being *understood that this is to be done*, we now proceed in a formal way to describe the canonical theory of particles and one scalar field in interaction that leads to equations just like Eqs. (22.17) and (22.18).

The new dynamical system is defined by the action functional:

$$\Phi'[\Sigma_1, \Sigma_2] = -\frac{1}{2}\int_{\Sigma_1}^{\Sigma_2} d^4x\{\partial^\mu\phi(x)\partial_\mu\phi(x) + \kappa^2\phi^2(x)\}$$
$$-\sum_r m_r \int_{\Sigma_1}^{\Sigma_2} d\theta_r(-\dot{x}_r^2)^{1/2} + \sum_r g_r \int_{\Sigma_1}^{\Sigma_2} d\theta_r(-\dot{x}_r^2)^{1/2}\phi(x_r)\,. \tag{22.19}$$

The notation is as follows: $\phi(x)$ is a scalar field, the particle variables are as before; $\sum_1$ and $\sum_2$ are any two spacelike surfaces in space-time with $\sum_2$ being later than $\sum_1$ and both extending to infinity in spacelike directions, and $\Phi'[\sum_1, \sum_2]$ is the action contained in the space-time region between $\sum_1$ and $\sum_2$; the integrals over θ_r run over the world line of particle r, the limits on θ_r being the points $\theta_r = \sigma_r$ and $\theta_r = \rho_r$ at which the world line crosses $\sum_1$ and $\sum_2$, respectively. Thus $\Phi'[\sum_1, \sum_2]$ is a functional of the world lines and of the values assigned to $\phi(x)$, and the third term is the interaction term. To derive the equations of motion and the conserved quantities, we compute $\triangle\Phi'$ for a general variation: we keep $\sum_1$ and $\sum_2$ fixed, we make a small variation $\delta\phi(x)$ in $\phi(x)$ at each point x, and a small variation $\delta x_{r\mu}(\theta_r)$ in the world line of particle r. Because $\sum_1$ and $\sum_2$ are fixed, this last variation produces small changes $\delta\sigma_r, \delta\rho_r$ in the values of θ_r at the points of intersection of the world line and the surfaces $\sum_1, \sum_2$. Denoting the contributions to $\triangle\Phi'$ from the three terms in Eq. (22.19) by $\triangle_1\Phi', \triangle_2\Phi'$ and $\triangle_3\Phi'$, we get:

$$\triangle_1\Phi'\,[\Sigma_1, \Sigma_2] = -\int_{\Sigma_1}^{\Sigma_2} d^4x\{\partial_\mu\phi(x)\partial^\mu\delta\phi(x) + \kappa^2\phi(x)\delta\phi(x)\}$$
$$= -\left[\int_{\Sigma} d\sigma^\mu\partial_\mu\phi(x)\delta\phi(x)\right]_{\Sigma_1}^{\Sigma_2}$$
$$+\int_{\Sigma_1}^{\Sigma_2} d^4x\delta\phi(x)(\partial_\mu\partial^\mu - \kappa^2)\phi(x)\,;$$

$$\triangle_2\Phi'\,[\Sigma_1, \Sigma_2] = -\sum_r m_r\left(\int_{\sigma_r+\delta\sigma_r}^{\rho_r+\delta\rho_r} d\theta_r(-\dot{x}_r^2)^{1/2}\left(1 + \frac{\dot{x}_{r\mu}\delta\dot{x}_r^\mu + \ldots}{\dot{x}_r^2}\right)\right.$$
$$\left.-\int_{\sigma_r}^{\rho_r} d\theta_r(-\dot{x}_r^2)^{1/2}\right)$$
$$= -\sum_r m_r\left(\left[(-\dot{x}_r^2)^{1/2}\delta\theta_r\right]_{\Sigma_1}^{\Sigma_2} + \int_{\Sigma_1}^{\Sigma_2} d\theta_r(-\dot{x}_r^2)^{1/2}\frac{\dot{x}_{r\mu}\delta\dot{x}_r^\mu}{\dot{x}_r^2}\right)$$

$$= \left[\sum_r (p_{r\mu} \delta x_r^\mu - m_r(-\dot{x}_r^2)^{1/2} \delta\theta_r) \right]_{\Sigma_1}^{\Sigma_2} - \sum_r \int_{\Sigma_1}^{\Sigma_2} d\theta_r \dot{p}_{r\mu} \delta x_r^\mu ;$$

$$\begin{aligned}
\triangle_3 \Phi' [\Sigma_1, \Sigma_2] &= \sum_r g_r \left(\int_{\sigma_r + \delta\sigma_r}^{\rho_r + \delta\rho_r} d\theta_r (-\dot{x}_r^2)^{1/2} \left(1 + \frac{\dot{x}_{r\mu} \delta \dot{x}_r^\mu}{\dot{x}_r^2} + \ldots \right) \right. \\
&\quad \left. \times (\phi(x_r + \delta x_r) + \delta\phi(x_r)) - \int_{\sigma_r}^{\rho_r} d\theta_r (-\dot{x}_r^2)^{1/2} \phi(x_r) \right) \\
&= \left[\sum_r g_r \phi(x_r) (-\dot{x}_r^2)^{1/2} \delta\theta_r \right]_{\Sigma_1}^{\Sigma_2} \\
&\quad + \sum_r g_r \int_{\Sigma_1}^{\Sigma_2} d\theta_r (-\dot{x}_r^2)^{1/2} \\
&\quad \times \left(\frac{\partial\phi(x_r)}{\partial x_{r\mu}} \delta x_{r\mu} + \delta\phi(x_r) + \frac{\dot{x}_{r\mu} \delta \dot{x}_r^\mu}{\dot{x}_r^2} \phi(x_r) \right) \\
&= \left[\sum_r g_r \phi(x_r) \left((-\dot{x}_r^2)^{1/2} \delta\theta_r - p_{r\mu} \frac{\delta x_r^\mu}{m_r} \right) \right]_{\Sigma_1}^{\Sigma_2} \\
&\quad + \sum_r g_r \int_{\Sigma_1}^{\Sigma_2} d\theta_r \left\{ \phi(x_r) \dot{p}_{r\mu} \frac{\delta x_r^\mu}{m_r} + (-\dot{x}_r^2)^{1/2} \right. \\
&\quad \left. \times \left(\delta\phi(x_r) + W_{r\mu\nu} \delta x_r^\mu \frac{\partial\phi(x_r)}{\partial x_{r\nu}} \right) \right\} .
\end{aligned}$$

We have written $\partial\phi(x_r)/\partial x_{r\nu}$ for what should properly be $[\partial\phi(x)/\partial x_\nu]_{x=x_r}$. Adding these expressions, we can write $\triangle\Phi'$ as a boundary term plus two integrals involving $\delta\phi(x)$ and δx_r^μ, respectively:

$$\begin{aligned}
&\triangle\Phi'[\Sigma_1, \Sigma_2] \\
&= \left[\sum_r \left(1 - \frac{g_r}{m_r} \phi(x_r) \right) (p_{r\mu} \delta x_r^\mu - m_r(-\dot{x}_r^2)^{1/2} \delta\theta_r) - \int_\Sigma d\sigma^\mu \partial_\mu \phi(x) \delta\phi(x) \right]_{\Sigma_1}^{\Sigma_2} \\
&+ \int_{\Sigma_1}^{\Sigma_2} d^4x \delta\phi(x) \left\{ (\partial_\mu \partial^\mu - \kappa^2) \phi(x) + \sum_r g_r \int_{-\infty}^{\infty} d\theta_r (-\dot{x}_r^2)^{1/2} \delta^{(4)}(x - x_r) \right\} \\
&- \sum_r \int_{\Sigma_1}^{\Sigma_2} d\theta_r \delta x_r^\mu \left\{ \left(1 - \frac{g_r}{m_r} \phi(x_r) \right) \dot{p}_{r\mu} - g_r (-\dot{x}_r^2)^{1/2} W_{r\mu\nu} \frac{\partial\phi(x_r)}{\partial x_{r\nu}} \right\} .
\end{aligned} \tag{22.20}$$

We can now impose the action principle and state: for variations about the actual motion $\triangle\Phi'$ consists of boundary terms only. Switching to proper-time variables, we thus get the system of equations of motion:

$$\left(1 - \frac{g_r}{m_r}\phi(x_r)\right)\dot{p}_{r\mu} = g_r(g_{\mu\nu} + \dot{x}_{r\mu}\dot{x}_{r\nu})\frac{\partial\phi(x_r)}{\partial x_{r\nu}}$$
$$(\partial_\mu\partial^\mu - \kappa^2)\phi(x) = -\sum_s g_s \int_{-\infty}^{\infty} d\tau_s \delta^{(4)}(x - x_s) = j(x)\,, \qquad (22.21)$$

to be compared with Eqs. (22.17) and (22.18).

In one sense there is a great difference between this system and the system of particles under direct interaction. In the present case the field carries its own energy and can act as the repository of energy, momentum, and angular momentum. It is subjected to independent variations of its own in obtaining the equations of motion. Despite this difference, which we discuss presently, the similarity between the two systems is more than appears in the equations of motion. The expressions for the conserved quantities can be transformed from the action-by-contact form to the action-at-a-distance form. Let us see how this happens. When the equations of motion Eq. (22.21) are obeyed, we have the formula:

$$\Delta\Phi'[\Sigma_1, \Sigma_2] = \Bigg[\sum_r \left(1 - \frac{g_r}{m_r}\phi(x_r)\right)(p_{r\mu}\delta x_r^\mu - m_r(-\dot{x}_r^2)^{1/2}\delta\theta_r) - \int_\Sigma d\sigma^\mu \partial_\mu\phi(x)\delta\phi(x)\Bigg]_{\Sigma_1}^{\Sigma_2} \qquad (22.22)$$

for a general variation. Now a variation induced by an infinitesimal Poincaré transformation is given by

$$x_{r\mu}(\theta_r) \to x'_{r\mu}(\theta_r) = x_{r\mu}(\theta_r) + a_\mu + \lambda_{\mu\nu}x_r^\nu(\theta_r),\ \delta x_{r\mu}(\theta_r) = a_\mu + \lambda_{\mu\nu}x_r^\nu(\theta_r)\,;$$
$$\phi(x) \to \phi'(x^\mu + a^\mu + \lambda^{\mu\nu}x_\nu) = \phi(x)\,,$$
$$\delta\phi(x) = \phi'(x) - \phi(x) = -(a_\mu + \lambda_{\mu\nu}x^\nu)\partial^\mu\phi(x)\,. \qquad (22.23)$$

(If in the original motion the particle was at $x_{r\mu}$ at the instant θ_r, in the new motion it is at $x_{r\mu} + a_\mu + \lambda_{\mu\nu}x_r^\nu$ at that instant; what value the field possessed at x_μ previously it now possesses at $x_\mu + a_\mu + \lambda_{\mu\nu}x^\nu$). Now from the structure of $\Phi'[\sum_1, \sum_2]$ what do we expect for $\Delta\Phi'[\sum_1, \sum_2]$? Because the first term in Φ' involves the Lagrangian density $\mathcal{L}^{\text{free}}$ for the field ϕ by itself and this density is constructed to be like a scalar field under Poincaré transformations, we expect:

$$\begin{aligned}\Delta_1\Phi' &= \int_{\Sigma_1}^{\Sigma_2} d^4x(\mathcal{L}^{\text{free}}(\phi'(x)) - \mathcal{L}^{\text{free}}(\phi(x)))\\ &= \int_{\Sigma_1}^{\Sigma_2} d^4x(\mathcal{L}^{\text{free}}(\phi(x - a - \lambda x)) - \mathcal{L}^{\text{free}}(\phi(x)))\\ &= \int_{\Sigma_1}^{\Sigma_2} d^4x(-(a_\mu + \lambda_{\mu\nu}x^\nu)\partial^\mu\mathcal{L}^{\text{free}}) = -\Bigg[\int_\Sigma d\sigma^\mu(a_\mu + \lambda_{\mu\nu}x^\nu)\mathcal{L}^{\text{free}}\Bigg]_{\Sigma_1}^{\Sigma_2}.\end{aligned}$$

Similarly, because $\dot{x}_r'^2 = \dot{x}_r^2$ and $\phi'(x_r') = \phi(x_r)$, we have:

$$\begin{aligned}\triangle \int_{\Sigma_1}^{\Sigma_2} d\theta_r (-\dot{x}_r^2)^{1/2} &= \Big[(-\dot{x}_r^2)^{1/2}\delta\theta\Big]_{\Sigma_1}^{\Sigma_2}\,,\\ \triangle \int_{\Sigma_1}^{\Sigma_2} d\theta_r (-\dot{x}_r^2)\phi(x_r) &= \Big[(-\dot{x}_r^2)^{1/2}\phi(x_r)\delta\theta_r\Big]_{\Sigma_1}^{\Sigma_2}\,.\end{aligned}$$

All in all, we expect for $\triangle\Phi'$ the value:

$$\begin{aligned}\triangle\Phi'[\Sigma_1,\Sigma_2] &= \Bigg[\sum_r (g_r\phi(x_r) - m_r)(-\dot{x}_r^2)^{1/2}\delta\theta_r\\ &\quad - \int_\Sigma d\sigma^\mu (a_\mu + \lambda_{\mu\nu}x^\nu)\mathcal{L}^{\text{free}}\Bigg]_{\Sigma_1}^{\Sigma_2}\,. \end{aligned} \tag{22.24}$$

On comparing Eqs. (22.22) and (22.24), the $\delta\theta$ terms cancel and we find that the quantity:

$$\sum_r \left(1 - \frac{g_r}{m_r}\phi(x_r)\right) p_{r\mu}\delta x_r^\mu + \int_\Sigma d\sigma^\lambda (a_\mu + \lambda_{\mu\nu}x^\nu)(g_\lambda^\mu \mathcal{L}^{\text{free}} + \partial_\lambda\phi\partial^\mu\phi)\,,$$

the particle contributions taken at the points at which they cross the surface Σ, is independent of Σ. The coefficient of a^μ gives the conserved energy-momentum P'_μ, that of $\lambda^{\mu\nu}$ gives $J'_{\mu\nu}$:

$$\begin{aligned} P'_\mu &= \sum_r \left(1 - \frac{g_r}{m_r}\phi(x_r)\right) p_{r\mu} + \int_\Sigma d\sigma^\lambda \theta_{\lambda\mu}\,,\\ J'_{\mu\nu} &= \sum_r \left(1 - \frac{g_r}{m_r}\phi(x_r)\right)(x_{r\mu}p_{r\nu} - x_{r\nu}p_{r\mu}) + \int_\Sigma d\sigma^\lambda (x_\mu\theta_{\lambda\nu} - x_\nu\theta_{\lambda\mu})\,,\\ \theta_{\lambda\mu} &= \partial_\lambda\phi\partial_\mu\phi - \frac{1}{2}g_{\lambda\mu}(\partial^\rho\phi\partial_\rho\phi + \kappa^2\phi^2)\,. \end{aligned} \tag{22.25}$$

These characterize the action-by-contact theory. Now let us show in what sense P'_μ is the same as P_μ that appears in Eq. (22.13). For the present theory we have in Eq. (22.21) an equation of motion for the field ϕ as well as for each particle, and in formally solving for ϕ one must specify the boundary conditions properly. Let us choose that solution for ϕ in which ϕ becomes completely fixed as a functional of the particle trajectories by setting:

$$\phi(x) = \sum_r g_r \int_{-\infty}^{\infty} d\tau_r \bar{\triangle}(x - x_r;\kappa)\,. \tag{22.26}$$

Then there is exact agreement between the single-particle terms in P_μ and P'_μ. (Put $K_{rs} = g_r g_s \bar{\Delta}$ in Eq. (22.13)!) Now let us transform the surface term in P'_μ. Because

$$\partial^\lambda \theta_{\lambda\mu} = \partial_\mu \phi(\partial^\lambda \partial_\lambda - \kappa^2)\phi = -\sum_s g_s \int_{-\infty}^{\infty} d\tau_s \delta^{(4)}(x - x_s)\partial_\mu \phi(x)\,, \quad (22.27)$$

use of Gauss' theorem yields

$$\int_{\Sigma} d\sigma^\lambda \theta_{\lambda\mu} = -\int_{-\infty}^{\Sigma} d^4x \sum_s g_s \int_{-\infty}^{\infty} d\tau_s \delta^{(4)}(x - x_s)\partial_\mu \phi(x) + \underset{\Sigma' \to -\infty}{\text{Lt}} \int_{\Sigma'} d\sigma^\lambda \theta_{\lambda\mu}\,. \quad (22.28)$$

With the help of Eq. (22.26), the first term becomes

$$\begin{aligned} -\sum_s g_s \int_{-\infty}^{\Sigma} d\tau_s \frac{\partial \phi(x_s)}{\partial x_s^\mu} &= -\sum_{rs} g_r g_s \int_{-\infty}^{\rho_s} d\tau_s \int_{-\infty}^{\infty} d\tau_r \frac{\partial \bar{\Delta}(x_s - x_r;\kappa)}{\partial x_s^\mu} \\ &= -\sum_{rs} g_r g_s \int_{-\infty}^{\rho_s} d\tau_s \int_{\rho_r}^{\infty} d\tau_r \frac{\partial \bar{\Delta}(x_s - x_r;\kappa)}{\partial x_s^\mu} \\ &= \sum_{rs} g_r g_s \int_{\rho_r}^{\infty} d\tau_r \int_{-\infty}^{\rho_s} d\tau_s \frac{\partial \bar{\Delta}(x_r - x_s;\kappa)}{\partial x_r^\mu}\,. \end{aligned} \quad (22.29)$$

In going from the first line to the second, we made use of the antisymmetry of the integrand in r and s, so that the integration in τ_r over the range $(-\infty, \rho_r)$ dropped out. This result matches exactly with the second term in P_μ in Eq. (22.13); thus with the choice made for $\phi(x)$ in Eq. (22.26), P'_μ and P_μ agree except for the last term in Eq. (22.28). What is this term? To evaluate it we make use of the fact that the invariant function $\bar{\Delta}$ is one-half the sum of the retarded and advanced invariant functions Δ_{ret} and Δ_{adv}, so that Eq. (22.26) can be written as

$$\begin{aligned} \phi(x) &= \frac{1}{2}(\phi_{\text{ret}}(x) + \phi_{\text{adv}}(x))\,, \\ \phi_{\substack{\text{ret}\\ \text{adv}}}(x) &= \sum_r g_r \int_{-\infty}^{\infty} d\tau_r \Delta_{\substack{\text{ret}\\ \text{adv}}}(x - x_r;\kappa)\,. \end{aligned} \quad (22.30)$$

By definition, in the limit $\sum' \to -\infty$, the part ϕ_{ret} in ϕ drops out, and $\theta^{\lambda\mu}$ is formed from ϕ_{adv} alone; hence we have

$$\underset{\Sigma' \to -\infty}{Lt} \int_{\Sigma'} d\sigma^\lambda \theta_{\lambda\mu} = \underset{\Sigma' \to -\infty}{Lt} \frac{1}{4} \int_{\Sigma'} d\sigma^\lambda \cdot \{\partial_\lambda \phi_{\text{adv}} \partial_\mu \phi_{\text{adv}} - \frac{1}{2} g_{\lambda\mu}(\partial^\rho \phi_{\text{adv}} \partial_\rho \phi_{\text{adv}} + \kappa^2 \phi_{\text{adv}}^2)\}\,.$$

By the same token, because $\phi_{\mathrm{adv}} \to 0$ as $\sum' \to +\infty$, we can use Gauss' theorem again to say:

$$\underset{\Sigma' \to -\infty}{Lt} \int_{\Sigma'} d\sigma^{\lambda}\theta_{\lambda\mu} = -\frac{1}{4}\int d^4x\partial^{\lambda}\theta_{\lambda\mu}(\phi_{\mathrm{adv}})$$
$$= -\frac{1}{4}\int d^4x\partial_{\mu}\phi_{\mathrm{adv}}(\partial^{\lambda}\partial_{\lambda} - \kappa^2)\phi_{\mathrm{adv}}$$
$$= \frac{1}{4}\sum_s g_s \int d^4x \int_{-\infty}^{\infty} d\tau_s \delta^{(4)}(x - x_s)\partial_{\mu}\phi_{\mathrm{adv}}(x)$$
$$= \frac{1}{4}\sum_s g_s \int_{-\infty}^{\infty} d\tau_s \frac{\partial\phi_{\mathrm{adv}}(x_s)}{\partial x_s^{\mu}} .$$

We now use Eq. (22.30), together with the relation $\triangle_{\mathrm{adv}}(x;\kappa) = \triangle_{\mathrm{ret}}(-x;\kappa)$ to finally conclude that:

$$\underset{\Sigma' \to -\infty}{Lt} \int_{\Sigma'} d\sigma^{\lambda}\theta_{\lambda\mu} = \frac{1}{4}\sum_{rs} g_r g_s \int_{-\infty}^{\infty} d\tau_s \frac{\partial\triangle_{\mathrm{adv}}(x_s - x_r;\kappa)}{\partial x_s^{\mu}}$$
$$= -\frac{1}{8}\sum_{rs} g_r g_s \int_{-\infty}^{\infty} d\tau_r \int_{-\infty}^{\infty} d\tau_s \frac{\partial\triangle(x_r - x_s;\kappa)}{\partial x_r^{\mu}} ,$$

where $\triangle = \triangle_{\mathrm{ret}} - \triangle_{\mathrm{adv}}$ is a basic solution of the homogeneous differential equation $(\partial^{\lambda}\partial_{\lambda} - \kappa^2)\triangle = 0$; the relation between P_{μ} and P'_{μ} is:

$$P_{\mu} = P'_{\mu} + \frac{1}{8}\sum_{rs} g_r g_s \int_{-\infty}^{\infty} d\tau_r \int_{-\infty}^{\infty} d\tau_s \frac{\partial\triangle(x_r - x_s;\kappa)}{\partial x_r^{\mu}} . \tag{22.31}$$

Keep in mind the fact that this correspondence obtains with the solution of Eq. (22.26) chosen for ϕ in the action-by-contact theory. Now we had already seen that P_{μ} for the action-at-a-distance theory is not completely specified, and the term appearing in Eq. (22.31) is precisely of the form that could not be fixed in defining P_{μ}. It is a vector quantity; it is a functional of the particle world-lines; it is explicitly independent of the instants ρ_r appearing in the definition of P_{μ} in Eq. (22.13), because we integrate over the entire world lines. Therefore, we can maintain that there is an agreement in the two theories as far as their expressions for energy and momentum are concerned, provided always that ϕ is chosen as in Eq. (22.26).

Before asking what happens if Eq. (22.26) is altered, let us develop the canonical aspects of the particle-plus-field theory a little further. For this theory we have the possibility of defining PB relations and constructing a set of generators for $\mathcal{P}$. To do these things, let us go back to the (noncovariant-looking) form in which the time variable is distinguished. (This is not absolutely essential, but it makes the interpretation easier.) Thus instead of several parameters θ_r we have one common time parameter t, and each particle has three coordinates $x_{rj}(t)$; the action $\Phi'[\sum_1, \sum_2]$, with $\sum_1$ and $\sum_2$ taken to be plane surfaces of constant times t_1, t_2, becomes:

$$\Phi'[t_1, t_2] = \int_{t_1}^{t_2} dt L(t),$$

$$L(t) = \int d^3x \frac{1}{2}\{\dot{\phi}(\boldsymbol{x}, t)^2 - \partial_j\phi(\boldsymbol{x}, t)\partial_j\phi(\boldsymbol{x}, t) - \kappa^2\phi(\boldsymbol{x}, t)^2\}$$
$$- \sum_r m_r(1 - \dot{x}_{rj}\dot{x}_{rj})^{1/2} + \sum_r g_r(1 - \dot{x}_{rj}\dot{x}_{rj})^{1/2}\phi(\boldsymbol{x}_r(t), t). \quad (22.32)$$

The basic Lagrangian variables are $x_{rj}(t), j$=1,2,3 and $\phi(\boldsymbol{x}, t)$. Because the symbol $p_{r\mu}$ has already been used to denote the kinetic momentum of particle r, let us write π_{rj} for the momentum canonically conjugate to x_{rj}; the field conjugate to $\phi(\boldsymbol{x}, t)$ is written $\pi(\boldsymbol{x}, t)$. These variables are:

$$\pi_{rj}(t) = \frac{\partial L}{\partial \dot{x}_{rj}(t)} = (m_r - g_r\phi(\boldsymbol{x}_r(t), t))\frac{\dot{x}_{rj}(t)}{(1 - \dot{x}_{rk}\dot{x}_{rk})^{1/2}},$$
$$\pi(\boldsymbol{x}, t) = \frac{\delta L}{\delta\dot{\phi}(\boldsymbol{x}, t)} = \dot{\phi}(\boldsymbol{x}, t). \quad (22.33)$$

We can express the velocity variable $\dot{x}_{rj}$ in terms of π_{rj},

$$\dot{x}_{rj} = \frac{\pi_{rj}}{[\pi_{rk}\pi_{rk} + (m_r - g_r\phi(\boldsymbol{x}_r, t))^2]^{1/2}}. \quad (22.34)$$

We can set up the fundamental equal-time PB's

$$\{x_{rj}(t), \pi_{sk}(t)\} = \delta_{rs}\delta_{jk}, \{\phi(\boldsymbol{x}, t), \pi(\boldsymbol{y}, t)\} = \delta^{(3)}(\boldsymbol{x} - \boldsymbol{y}), \quad (22.35)$$

all other brackets between these variables being defined to vanish. The generators P'_μ and $J'_{\mu\nu}$ that appear in covariant form in Eq. (22.29) can be written in terms of the Hamiltonian variables at time $t = 0$ by choosing the surface $\sum$ to be all space at time zero:

$$J'_j = \sum_r (\boldsymbol{x}_r \times \boldsymbol{\pi}_r)_j - \int d^3x\, \pi(\boldsymbol{x})(\boldsymbol{x} \times \boldsymbol{\nabla})_j\phi(\boldsymbol{x}),$$

$$P'_j = \sum_r \pi_{rj} - \int d^3x\, \pi(\boldsymbol{x})\partial_j\phi(\boldsymbol{x});$$

$$K'_j = \sum_r x_{rj}[\pi_{rk}\pi_{rk} + (m_r - g_r\phi(\boldsymbol{x}_r))^2]^{1/2} + \int d^3x\, x_j\mathcal{H}^{(0)}(\boldsymbol{x}),$$

$$H' = \sum_r [\pi_{rk}\pi_{rk} + (m_r - g_r\phi(\boldsymbol{x}_r))^2]^{1/2} + \int d^3x\, \mathcal{H}^{(0)}(\boldsymbol{x}),$$

$$\mathcal{H}^{(0)}(\boldsymbol{x}) = \frac{1}{2}\{\pi(\boldsymbol{x})^2 + (\boldsymbol{\nabla}\phi(\boldsymbol{x}))^2 + \kappa^2\phi(\boldsymbol{x})^2\}. \quad (22.36)$$

We leave it as an exercise to the reader to check that the Lie bracket relations of $\mathcal{P}$ are obeyed. Let us note that the $E(3)$ generators J'_j and P'_j have taken up kinematic forms, whereas K'_j and H' contain interaction terms. *The world-line conditions of the preceding chapter* (Eq. (21.80)) *are now explicitly satisfied*, guaranteeing that in this canonical theory the particle positions transform in a manifestly covariant way. This happens because for each $r, K'_j - x_{rj}H'$ is independent of π_{rk}. Thus we have avoided the no-interaction theorem of the preceding chapter by allowing for new degrees of freedom that are not particle-like in nature but belong to a field. At the same time, the field ϕ also has a manifestly covariant transformation property because we have no difficulty in verifying the PB relations:

$$\{J'_{\mu\nu}, \phi(x)\} = (x_\mu \partial_\nu - x_\nu \partial_\mu)\phi(x), \ \{P'_\mu, \phi(x)\} = \partial_\mu \phi(x) . \tag{22.37}$$

All this is no surprise because we took a manifestly covariant particle-plus-field theory containing interaction and put it into canonical form.

Let us go back to comparing the action-at-a-distance and action-by-contact theories. Call these systems A and B, respectively. Equation (22.31) shows that for every state of motion of A we can define a state of motion of B such that we have a correspondence between the two sets of conserved quantities. Considering system B by itself, the usual way in which one "solves" the equation of motion for $\phi(x)$ in Eq. (22.21) in such a way as to exhibit the independent degrees of freedom of the field is to use the retarded Green's function $\triangle_{\text{ret}}(x; \kappa)$:

$$\phi(x) = -\int d^4x' \triangle_{\text{ret}}(x - x'; \kappa) j(x') + \phi_{\text{in}}(x) . \tag{22.38}$$

This $\phi_{\text{in}}(x)$ is a solution of the homogeneous field equation and it essentially describes the state of the field in the distant past before interactions took place. The freedom in the choice of $\phi_{\text{in}}(x)$ is simply the freedom we have in choosing the initial configuration of the field, analogous to the freedom we have in setting the configuration of a system of particles at one instant of time in ordinary mechanics. How are we to reconcile the fact that the field contains its own autonomous vibrations independent of particles, with the correspondence between A and B? The resolution of this "dilemma" lies in the observation that as we specifically chose the field to be given by Eq. (22.26), the field had *no* autonomy. It was completely defined in terms of the particle variables. Hence the question of what happens to $\phi(x)$ at infinity, that is the nature of $\phi_{\text{in}}(x)$, cannot be set by us independently of the configuration of particles. In fact, comparing Eq. (22.38) with Eq. (22.26), we have

$$\begin{aligned} \phi_{\text{in}}(x) &= \frac{1}{2}\int d^4x' \triangle(x - x'; \kappa) j(x') = -\frac{1}{2}\sum_s g_s \int_{-\infty}^{\infty} d\tau_s \triangle(x - x_s; \kappa) , \\ \triangle &= \triangle_{\text{ret}} - \triangle_{\text{adv}} . \end{aligned} \tag{22.39}$$

In other words, to obtain a correspondence, we had to arrange that the degrees of freedom of the field did not really show up.

It is only a subset of the possible states of motion of system B that turn up in correspondence with the states of system A. These are those in which the particles of system B neither emit nor absorb energy momentum ("radiation") from the field ϕ but simply redistribute these quantities among themselves by interaction. This is natural because of the way system A is defined; there is nothing to which the particles can give away their energy momentum except other particles. Isolated particles do not, in the very nature of things, lose energy to empty space!

What of the remaining states of motion of system B? These are those in which we can identify a nonvanishing initial and/or final contribution to the total energy momentum of the system because of the field ϕ by itself such that a true exchange of energy momentum has taken place between the particles and the field. Such states of system B can be separated from the previous ones more precisely as follows: analogously to Eq. (22.38), we can define the "outgoing field" ϕ_{out} by

$$\phi(x) = -\int d^4x' \Delta_{\text{adv}}(x - x'; \kappa) j(x') + \phi_{\text{out}}(x) , \tag{22.40}$$

so that in all cases we have the relation

$$\phi_{\text{out}}(x) - \phi_{\text{in}}(x) = -\int d^4x' \Delta(x - x'; \kappa) j(x') . \tag{22.41}$$

If the solution of Eq. (22.26) for $\phi(x)$ is used, it implies

$$\phi_{\text{in}}(x) + \phi_{\text{out}}(x) = 0, \phi_{\text{in}}(x) = \frac{1}{2}\int d^4x' \Delta(x - x'; \kappa) j(x') . \tag{22.42}$$

The more general states of motion of system B are those in which $\phi_{\text{in}}(x) + \phi_{\text{out}}(x)$ is a nonzero solution of the homogeneous field equation, and it is these that at present have no analogues in system A. If we want at all to consider energy leaving a collection of particles in the action-at-a-distance framework, there should be some other particles (to which it has been lost!) that act as absorbers or suppliers of energy. We must divide the particles into two sets, those of interest and the rest that constitute the "absorber." To be able to talk about radiation we must make the following provisions: (1) There should be an arbitrarily large number of absorber particles that are at all times sufficiently far away from the particles of interest. (2) We should compute only the energy, momentum, and angular momentum that are emitted *from* the particles of interest *to* the absorber particles, and the calculation of the radiated quantities of interest takes into account only these. Then the degrees of freedom of the field in system B can be interpreted as being really the degrees of freedom of the myriads of particles constituting the absorber, and the choice of boundary conditions for the field is really a disguised assumption about the configuration of the absorber particles. We may then ask: why bother with the fields? The answer is that the introduction of the field enables us to restore, albeit formally, a local interaction structure; in radiative processes we need not explicitly talk about the absorber but instead replace it by a boundary condition on the field.

Now comes the question of making the theories finite by modifying the previous choice for the interaction kernels K_{rs} of the action-at-a-distance theory.

Maintaining the factorization with respect to r and s as in Eq. (22.14), we now choose for $K(x^2)$ the following linear combination:

$$K((x_r - x_s)^2) = \sum_\alpha \epsilon_\alpha C_\alpha^2 \bar{\Delta}(x_r - x_s; \kappa_\alpha)\,. \tag{22.43}$$

Here ϵ_α are sign factors $(\pm 1), C_\alpha$ are real constants, and κ_α distinct parameters. This is to replace Eq. (22.15), and the ϵ_α and C_α have to be chosen to avoid explicit infinities. The equations of motion of the action-at-a-distance theory, namely Eq. (22.4), now have the form:

$$\left(1 - \frac{g_r}{m_r}\bar{\psi}(x_r)\right) \dot{p}_{r\mu} = g_r(g_{\mu\nu} + \dot{x}_{r\mu}\dot{x}_{r\nu}) \left[\frac{\partial \bar{\psi}(x)}{\partial x_\nu}\right]_{x=x_r}, \tag{22.44}$$

where $\bar{\psi}$ is a linear superposition of auxiliary "fields" $\bar{\phi}_\alpha$ defined in this way:

$$\begin{aligned} \bar{\psi}(x) &= \sum_\alpha \epsilon_\alpha C_\alpha \bar{\phi}_\alpha(x)\,, \\ \bar{\phi}_\alpha(x) &\equiv C_\alpha \sum_s g_s \int_{-\infty}^{\infty} d\tau_s \bar{\Delta}(x - x_s; \kappa_\alpha)\,. \end{aligned} \tag{22.45}$$

As before, $\bar{\psi}$ and $\bar{\phi}_\alpha$ are derived quantities defined completely in terms of the particle world lines. The different $\bar{\phi}_\alpha(x)$ obey:

$$(\partial^\mu \partial_\mu - \kappa_\alpha^2)\bar{\phi}_\alpha(x) = -C_\alpha \sum_s g_s \int_{-\infty}^{\infty} d\tau_s \delta^{(4)}(x - x_s) = C_\alpha j(x)\,. \tag{22.46}$$

We must now choose the ϵ_α and C_α in such a way that both $\bar{\psi}$ and $\partial\bar{\psi}/\partial x_\nu$ remain finite on the world line of each particle, because these appear in Eq. (22.44). The important point to note is that the particle "sees" *only* the linear combination $\sum_\alpha \epsilon_\alpha C_\alpha \bar{\phi}_\alpha(x)$ and *never* the individual terms separately. By examining the form of $\bar{\Delta}(x; \kappa)$ given in Eq. (22.15), these finiteness requirements can be satisfied by imposing two conditions:

$$\sum_\alpha \epsilon_\alpha C_\alpha^2 = 0, \ \sum_\alpha \epsilon_\alpha C_\alpha^2 \kappa_\alpha^2 = 0\,. \tag{22.47}$$

Then both the term $\delta(x^2)$ and the term proportional to $\theta(-x^2)$ [whose first derivative would have produced a $\delta(x^2)$] cancel out in the combination $\sum_\alpha \epsilon_\alpha C_\alpha^2 \bar{\Delta}(x; \kappa_\alpha)$. Taking the "leading" term α=1 to be characterized by $\epsilon_1 = C_1 = 1$, we see that it suffices to include two additional values of α. (At least two are needed.) And of the two sign factors ϵ_2, ϵ_3 at least one must be negative. We can solve Eq. (22.47) to get

$$C_2^2 = \frac{\kappa_3^2 - \kappa_1^2}{\kappa_2^2 - \kappa_3^2}, \ C_3^2 = 1 + C_2^2\,, \tag{22.48}$$

by choosing, for example, $\epsilon_2 = 1, \epsilon_3 = -1$; this is consistent if the κ's are suitably restricted.

An analogous canonical theory of particles and fields in local interaction, if it were to reproduce equations like Eqs. (22.44) and (22.46), must use the action functional

$$\begin{aligned}\Phi'[\Sigma_1, \Sigma_2] = &-\frac{1}{2}\sum_\alpha \epsilon_\alpha \int_{\Sigma_1}^{\Sigma_2} d^4x\{\partial^\mu\phi_\alpha(x)\partial_\mu\phi_\alpha(x) + \kappa_\alpha^2\phi_\alpha^2(x)\} \\ &-\sum_r m_r \int_{\Sigma_1}^{\Sigma_2} d\theta_r(-\dot{x}_r^2)^{1/2} \\ &+\sum_r g_r \int_{\Sigma_1}^{\Sigma_2} d\theta_r(-\dot{x}_r^2)^{1/2}\left(\sum_\alpha \epsilon_\alpha C_\alpha \phi_\alpha(x_r)\right) \end{aligned} \tag{22.49}$$

in place of Eq. (22.19). The point to note is the presence of the sign factors ϵ_α in the free-field terms; those fields for which $\epsilon = -1$ contribute negatively to the energy of the system, in contrast to "normal" fields. But this is the price to pay for a finite theory with local interactions and local self-actions.

Vector-Interactions: Electromagnetism

The treatment of this case is very similar in spirit to what has been done with scalar interactions. The only new point of principle is that in the field formulation the Lorentz condition on the vector potential must be included as a constraint. We simply indicate the principal expressions that replace those of the scalar theory.

The action-at-a-distance theory of particles subject to vector interactions works with the action functional

$$\begin{aligned}\Phi[C_1, C_2, \ldots; \sigma_1, \sigma_2, \ldots; \rho_1, \rho_2, \ldots] = &-\sum_r m_r \int_{\sigma_r}^{\rho_r} d\theta_r(-\dot{x}_r^2)^{1/2} \\ &+\frac{1}{2}\sum_{r,s}\int_{\sigma_r}^{\rho_r} d\theta_r \int_{\sigma_s}^{\rho_s} d\theta_s\, \dot{x}_{r\mu}\dot{x}_s^\mu K_{rs}((x_r - x_s)^2)\,. \end{aligned} \tag{22.50}$$

(Using the same symbols Φ, K_{rs} as before does not lead to any confusion.) The interaction term could be generalised somewhat without essentially complicating the velocity dependence, but we will work with the above simple version. This finite action replaces the one in Eq. (22.1). The invariance under changes in θ_r is maintained, and the structure of the interaction is in some ways simpler than before as far as the velocity dependence goes. The analogue to Eq. (22.3) is now:

$$\triangle\Phi[C_1,\ldots;\sigma_1,\ldots;\rho_1,\ldots] = \sum_r [p_{r\mu}\delta x_r^\mu]_{\sigma_r}^{\rho_r} + \sum_{rs}\int_{\sigma_s}^{\rho_s} d\theta_s \dot{x}_{s\mu}[K_{rs}\delta x_r^\mu]_{\sigma_r}^{\rho_r}$$
$$-\sum_r \int_{\sigma_r}^{\rho_r} d\theta_r \dot{p}_{r\mu}\delta x_r^\mu$$
$$+\sum_{rs}\int_{\sigma_r}^{\rho_r} d\theta_r \int_{\sigma_s}^{\rho_s} d\theta_s \dot{x}_{r\nu}\delta x_{r\mu}\left(\dot{x}_s^\nu \frac{\partial}{\partial x_{r\mu}} - \dot{x}_s^\mu \frac{\partial}{\partial x_{r\nu}}\right) K_{rs}\,. \tag{22.51}$$

This leads to the set of equations for the world lines

$$\dot{p}_{r\mu} = \left[\frac{\partial}{\partial x_r^\mu}\left(\sum_s \int_{-\infty}^{\infty} d\theta_s \dot{x}_{s\nu} K_{rs}\right) - \frac{\partial}{\partial x_r^\nu}\left(\sum_s \int_{-\infty}^{\infty} d\theta_s \dot{x}_{s\mu} K_{rs}\right)\right]\dot{x}_r^\nu\,, \tag{22.52}$$

to be compared with Eq. (22.4). Because the expression in square brackets is anti-symmetric in μ and ν, these equations are compatible with the requirement $\dot{x}_r^\mu p_{r\mu}=0$. The derivation of conservation laws combines the equations of motion with the form invariance of Φ under Poincaré transformations as before, and after some algebra we obtain these expressions for energy, momentum and angular momentum:

$$P_\mu(\rho_1,\ldots) = \sum_r \left[p_{r\mu} + \sum_s \int_{-\infty}^{\infty} d\tau_s \dot{x}_{s\mu} K_{rs}\right]_{\tau_r=\rho_r}$$
$$+\sum_{rs}\int_{\rho_r}^{\infty} d\tau_r \int_{-\infty}^{\rho_s} d\tau_s \dot{x}_{r\nu}\dot{x}_s^\nu \frac{\partial K_{rs}}{\partial x_r^\mu}\,,$$

$$J_{\mu\nu}(\rho_1,\ldots) = \sum_r \left[x_{r\mu}\left(p_{r\nu} + \sum_s \int_{-\infty}^{\infty} d\tau_s \dot{x}_{s\nu} K_{rs}\right)\right.$$
$$\left.-x_{r\nu}\left(p_{r\mu} + \sum_s \int_{-\infty}^{\infty} d\tau_s \dot{x}_{s\mu} K_{rs}\right)\right]_{\tau_r=\rho_r}$$
$$+\sum_{rs}\int_{\rho_r}^{\infty} d\tau_r \int_{-\infty}^{\rho_s} d\tau_s \left[\dot{x}_{r\lambda}\dot{x}_s^\lambda\left(x_{r\mu}\frac{\partial}{\partial x_r^\nu} - x_{r\nu}\frac{\partial}{\partial x_r^\mu}\right)\right.$$
$$\left.+(\dot{x}_{r\mu}\dot{x}_{s\nu} - \dot{x}_{r\nu}\dot{x}_{s\mu})\right] K_{rs}\,. \tag{22.53}$$

Here the interaction contributes a part to the "single-particle" terms as well, but later on we give a form in which these are also explicitly included in the two-particle terms.

To establish contact with a field formalism and in particular with Maxwell-Lorentz electrodynamics, we must make the choice

$$K_{rs}((x_r - x_s)^2) = \frac{e_r e_s}{4\pi}\delta((x_r - x_s)^2)\,. \tag{22.54}$$

$(1/4\pi)\delta(x^2)$ is the value of $\bar{\Delta}(x;\kappa)$ when κ is set equal to zero. This introduces infinities, to be removed in the manner already indicated. Proceeding formally, with this choice for the interaction kernels the equations of motion Eq. (22.52) take the expected form:

$$\dot{p}_{r\mu} = e_r \bar{F}_{\mu\nu}(x_r)\dot{x}_r^\nu \tag{22.55}$$

where the "field-strength" $\bar{F}_{\mu\nu}$ and the related potentials $\bar{A}_\mu$ are all determined by the particle motions as:

$$\begin{aligned} \bar{F}_{\mu\nu}(x) &\equiv \partial_\mu \bar{A}_\nu(x) - \partial_\nu \bar{A}_\mu(x)\,, \\ \bar{A}_\mu(x) &\equiv \frac{1}{4\pi}\sum_s e_s \int_{-\infty}^{\infty} d\tau_s \dot{x}_{s\mu}\delta((x-x_s)^2)\,. \end{aligned} \tag{22.56}$$

By virtue of its definition, $\bar{A}_\mu$ satisfies both the wave equation and the Lorentz gauge condition:

$$\begin{aligned} \partial^\lambda\partial_\lambda \bar{A}_\mu(x) &= -\sum_s e_s \int_{-\infty}^{\infty} d\tau_s \dot{x}_{s\mu}\delta^{(4)}(x-x_s) \equiv -j_\mu(x)\,; \\ \partial^\mu \bar{A}_\mu(x) &= -\frac{1}{4\pi}\sum_s e_s \int_{-\infty}^{\infty} d\tau_s \dot{x}_{s\mu}\frac{\partial}{\partial x_{s\mu}}\delta((x-x_s)^2) \\ &= -\frac{1}{4\pi}\sum_s e_s \int_{-\infty}^{\infty} d\tau_s \frac{d}{d\tau_s}\delta((x-x_s)^2) = 0\,. \end{aligned} \tag{22.57}$$

Thus all the basic equations of the Maxwell-Lorentz theory are present in the action-at-a-distance form. Before proceeding to the corresponding field theory, let us show that with the choice of Eq. (22.54) the interaction energy momentum included with the single-particle contribution to P_μ can be written slightly differently. This is as follows:

$$\begin{aligned} \left[\sum_{rs}\int_{-\infty}^{\infty} d\tau_s \dot{x}_{s\mu}K_{rs}\right]_{\tau_r=\rho_r} &= \frac{1}{4\pi}\sum_{rs} e_r e_s \int_{-\infty}^{\infty} d\tau_s \dot{x}_{s\mu} \\ &\quad \times \int_{-\infty}^{\rho_r} d\tau_r \frac{d}{d\tau_r}\delta((x_r-x_s)^2) \\ &= \frac{1}{2\pi}\sum_{rs} e_r e_s \int_{-\infty}^{\infty} d\tau_s \int_{-\infty}^{\rho_r} d\tau_r \{\dot{x}_{s\mu}\dot{x}_{r\nu}(x_r-x_s)^\nu \\ &\quad + \dot{x}_{r\mu}\dot{x}_{s\nu}(x_r-x_s)^\nu\}\delta'((x_r-x_s)^2)\,. \end{aligned} \tag{22.58}$$

The prime on δ' stands for the derivative of δ with respect to its argument. In arriving at this last form we added on a perfect differential with respect to τ_s which integrates out to zero as the limits on the τ_s integration are $\pm\infty$. Now the purpose of adding on this vanishing term is that we might have an integrand explicitly antisymmetric in r and s. Having achieved this, now the τ_s integration in Eq. (22.58) can be restricted to the range ρ_s to $+\infty$; thus this term can be combined with the last term in P_μ in Eq. (22.53). The single-particle contributions and the interaction terms completely separate to give P_μ the alternative form

$$P_\mu(\rho_1,\ldots) = \sum_r [p_{r\mu}]_{\tau_r=\rho_r} + \frac{1}{2\pi}\sum_{rs} e_r e_s \int_{\rho_r}^{\infty} d\tau_r \int_{-\infty}^{\rho_s} d\tau_s (\dot{x}_{r\nu}\dot{x}_s^\nu (x_r - x_s)_\mu - \dot{x}_{r\mu}\dot{x}_s^\nu (x_r - x_s)_\nu - \dot{x}_{s\mu}\dot{x}_r^\nu (x_r - x_s)_\nu)\delta'((x_r - x_s)^2). \quad (22.59)$$

To get the system of Eqs. (22.55) and (22.57) from a theory of particles and a field, we use the Fermi formulation of electrodynamics in which the Lorentz gauge condition is not obtained from the action principle but is added in separately. In place of the action functional in Eq. (22.19) we have to take

$$\Phi'[\Sigma_1, \Sigma_2] = -\frac{1}{2}\int_{\Sigma_1}^{\Sigma_2} d^4x \partial_\mu A_\nu(x)\partial^\mu A^\nu(x) - \sum_r m_r \int_{\Sigma_1}^{\Sigma_2} d\theta_r (-\dot{x}_r^2)^{1/2} + \sum_r e_r \int_{\Sigma_1}^{\Sigma_2} d\theta_r \dot{x}_{r\mu} A^\mu(x_r). \quad (22.60)$$

The result of a general variation in the particle trajectories and the field A_μ is to produce

$$\begin{aligned}\triangle\Phi'[\Sigma_1, \Sigma_2] = & \Bigg[\sum_r \Big((p_{r\mu} + e_r A_\mu(x_r))\delta x_r^\mu + \big(e_r \dot{x}_{r\mu} A^\mu(x_r) - m_r(-\dot{x}_r^2)^{1/2}\big)\delta\theta_r\Big) - \int_\Sigma d\sigma^\mu \partial_\mu A_\nu(x)\delta A^\nu(x)\Bigg]_{\Sigma_1}^{\Sigma_2} \\ & + \int_{\Sigma_1}^{\Sigma_2} d^4x \delta A^\mu(x)\Big\{\partial^\lambda\partial_\lambda A_\mu(x) + \sum_r e_r \int_{-\infty}^{\infty} d\theta_r \dot{x}_{r\mu}\delta^{(4)}(x - x_r)\Big\} \\ & - \sum_r \int_{\Sigma_1}^{\Sigma_2} d\theta_r \delta x_r^\mu \{\dot{p}_{r\mu} - e_r F_{\mu\nu}(x_r)\dot{x}_r^\nu\}, \\ F_{\mu\nu}(x) \equiv & \partial_\mu A_\nu(x) - \partial_\nu A_\mu(x). \end{aligned} \quad (22.61)$$

This is to be compared with Eq. (22.20). Imposing the action principle, we end up with the equations of motion

$$\begin{aligned}\dot{p}_{r\mu} &= e_r F_{\mu\nu}(x_r)\dot{x}_r^\nu, \\ \partial^\lambda\partial_\lambda A_\mu(x) &= -\sum_r e_r \int_{-\infty}^{\infty} d\theta_r \dot{x}_{r\mu}\delta^{(4)}(x - x_r), \end{aligned} \quad (22.62)$$

which are the desired ones. We must now add the subsidiary condition

$$\chi(x) \equiv \partial^\mu A_\mu(x) = 0; \quad (22.63)$$

and as we have seen in the preceding chapter, because the current of particles, $j_\mu(x)$, is conserved, it suffices for the complete vanishing of $\chi(x)$ at all times that $\chi(\boldsymbol{x}, t)$ and $\dot{\chi}(\boldsymbol{x}, t)$ vanish at one time.

Once the field and particle equations of motion are obeyed, $\triangle\Phi'$ consists of boundary terms alone, and the dynamical quantities P'_μ, $J'_{\mu\nu}$ can be read off by working out the case of the variation produced by an infinitesimal Poincaré transformation. These conserved quantities turn out to have the forms

$$\begin{aligned}
P'_\mu &= \sum_r (p_{r\mu} + e_r A_\mu(x_r)) + \int_\Sigma d\sigma^\lambda \theta_{\lambda\mu} , \\
J'_{\mu\nu} &= \sum_r (x_{r\mu}(p_{r\nu} + e_r A_\nu(x_r)) - x_{r\nu}(p_{r\mu} + e_r A_\mu(x_r))) \\
&\quad + \int_\Sigma d\sigma^\lambda (x_\mu \theta_{\lambda\nu} - x_\nu \theta_{\lambda\mu} + \partial_\lambda A_\mu \cdot A_\nu - \partial_\lambda A_\nu \cdot A_\mu) , \\
\theta_{\lambda\mu} &= \partial_\lambda A_\rho \partial_\mu A^\rho - \frac{1}{2} g_{\lambda\mu} \partial_\sigma A_\rho \partial^\sigma A^\rho .
\end{aligned} \tag{22.64}$$

The particle terms must be evaluated on the same surface on which the integration is done. The characteristic changes as compared to the scalar field are evident (cf. Eq. (22.25)). The preceding form for P'_μ is analogous to the form given in Eq. (22.53) for the action-at-a-distance theory, in that part of the interaction contribution appears added on to the $p_{r\mu}$ in the single-particle term. We can rewrite P'_μ in such a way that formally the single-particle terms consist of $p_{r\mu}$ alone, but we omit the details.

If in the particles-plus-field theory we choose the particular solution

$$A_\mu(x) = \frac{1}{4\pi} \sum_r e_r \int_{-\infty}^{\infty} d\tau_r \dot{x}_{r\mu} \delta((x - x_r)^2) , \tag{22.65}$$

then we find, by suitably splitting up $\delta((x-x')^2)$ into its retarded and advanced parts and using Gauss' theorem, that the two energy-momentum vectors P_μ, P'_μ are related by

$$P_\mu = P'_\mu + \frac{1}{8} \sum_{rs} e_r e_s \int_{-\infty}^{\infty} d\tau_r \int_{-\infty}^{\infty} d\tau_s \dot{x}_{r\nu} \dot{x}_s^\nu \frac{\partial D(x_r - x_s)}{\partial x_r^\mu} , \tag{22.66}$$

$D(x)$ being the function $\triangle(x;\kappa)$ for $\kappa = 0$. This relation between P_μ and P'_μ is the analogue of Eq. (22.31).

To complete the parallel to the scalar case, we must discuss the canonical aspects of the particles-plus-field theory. The action Φ' in Eq. (22.60) has the

Lagrangian

$$
\begin{aligned}
L(t) =& \frac{1}{2}\int d^3x(\dot{A}_\nu(\boldsymbol{x},t)\dot{A}^\nu(\boldsymbol{x},t) - \partial_j A_\nu(\boldsymbol{x},t)\partial_j A^\nu(\boldsymbol{x},t)) \\
& - \sum_r m_r(1-\dot{x}_{rj}\dot{x}_{rj})^{1/2} + \sum_r e_r(A_0(\boldsymbol{x}_r(t),t) + \dot{x}_{rj}A_j(\boldsymbol{x}_r(t),t))\,;
\end{aligned}
\tag{22.67}
$$

thus the canonical conjugates $\pi_{rj}, \pi^\nu(\boldsymbol{x},t)$ to $x_{rj}, A_\nu(\boldsymbol{x},t)$ respectively are

$$
\begin{aligned}
\pi_{rj}(t) &= \frac{m_r\dot{x}_{rj}}{(1-\dot{x}_{rk}\dot{x}_{rk})^{1/2}} + e_r A_j(\boldsymbol{x}_r,t)\,, \\
\pi^\nu(\boldsymbol{x},t) &= \dot{A}^\nu(\boldsymbol{x},t)\,.
\end{aligned}
\tag{22.68}
$$

Solving for $\dot{x}_{rj}$ is easy:

$$
\dot{x}_{rj} = (\pi_{rj} - e_r A_j(\boldsymbol{x}_r,t))/[m_r^2 + (\boldsymbol{\pi}_r - e_r\boldsymbol{A}(\boldsymbol{x}_r,t))^2]^{1/2}\,. \tag{22.69}
$$

Once again the characteristic differences between scalar and vector interactions show up on comparing this with Eq. (22.34). The equal-time PB's that do not vanish are these:

$$
\{x_{rj}(t), \pi_{sk}(t)\} = \delta_{rs}\delta_{jk}\,, \qquad \{A_\nu(\boldsymbol{x},t), \pi^\mu(\boldsymbol{y},t)\} = \delta^\mu_\nu \delta^{(3)}(\boldsymbol{x}-\boldsymbol{y})\,. \tag{22.70}
$$

We can exhibit all the generators of $\mathcal{P}$ for this system in terms of the phase-space variables at time zero; they are directly obtained from Eq. (22.64):

$$
\begin{aligned}
J'_j =& \sum(\boldsymbol{x}_r \times \boldsymbol{\pi}_r)_j - \int d^3x(\pi^\nu(\boldsymbol{x})(\boldsymbol{x}\times\boldsymbol{\nabla})_j A_\nu(\boldsymbol{x}) + \epsilon_{jkl}\pi_k(\boldsymbol{x})A_l(\boldsymbol{x}))\,, \\
P'_j =& \sum_r \pi_{rj} - \int d^3x\, \pi^\nu(\boldsymbol{x})\partial_j A_\nu(\boldsymbol{x})\,; \\
K'_j =& \sum_r x_{rj}([m_r^2 + (\boldsymbol{\pi}_r - e_r\boldsymbol{A}(\boldsymbol{x}_r))^2]^{1/2} - e_r A_0(\boldsymbol{x}_r)) \\
& + \int d^3x(\pi_j(\boldsymbol{x})A_0(\boldsymbol{x}) - \pi_0(\boldsymbol{x})A_j(\boldsymbol{x}) + x_j\mathcal{H}^{(0)}(\boldsymbol{x}))\,, \\
H' =& \sum_r([m_r^2 + (\pi_r - e_r A(\boldsymbol{x}_r))^2]^{1/2} - e_r A_0(\boldsymbol{x}_r)) + \int d^3x\, \mathcal{H}^{(0)}(\boldsymbol{x})\,, \\
\mathcal{H}^{(0)}(\boldsymbol{x}) =& \frac{1}{2}(\pi^\nu(\boldsymbol{x})\pi_\nu(\boldsymbol{x}) + \partial_k A^\nu(\boldsymbol{x})\partial_k A_\nu(\boldsymbol{x}))\,.
\end{aligned}
\tag{22.71}
$$

Again the $E(3)$ generators are kinematic, and K'_j and H' contain interaction. The contributions to the generators from the field by itself agree with what we had obtained in Eq. (21.179). For each r, there is no dependence on π_{rk} in $K'_j - x_{rj}H'$, so the particle world-line conditions are obeyed; similarly for the field variables. It remains to impose the family of constraints (one for each $\boldsymbol{x}$!)

$$
\chi(\boldsymbol{x}) = \partial_j A_j(\boldsymbol{x}) - \pi_0(\boldsymbol{x}) \approx 0\,, \tag{22.72}
$$

and examine the consequences. On taking the PB with H', this $\chi(\boldsymbol{x})$ calls forth a new family of constraints $\chi'(\boldsymbol{x}) \approx 0$ according to

$$\begin{aligned}\{\chi(\boldsymbol{x}), H'\} &= \partial_j \pi_j(\boldsymbol{x}) - \nabla^2 A_0(\boldsymbol{x}) + j^0(\boldsymbol{x}) = \chi'(\boldsymbol{x}) \approx 0\,,\\ j^0(\boldsymbol{x}) &= \sum_r e_r \delta^{(3)}(\boldsymbol{x} - \boldsymbol{x}_r)\,. \end{aligned}\tag{22.73}$$

We leave it to the reader to see that the two sets of constraints $\chi(\boldsymbol{x}) \approx 0, \chi'(\boldsymbol{x}) \approx 0$ are relativistically complete in the sense that the PB's of χ and χ' with all the generators of $\mathcal{P}$ vanish weakly, This generalizes our discussion page 542 of the preceding chapter, especially (21.181), to the interacting case.

It suffices to take a total of three fields $A_{\mu\alpha}(x), \alpha$=1,2,3 to make the theory finite. For α=1, we can assume $\epsilon_1 = C_1 = +1$, and $\kappa_1 = 0$; then the second of Eq. (22.47) forces ϵ_2 and ϵ_3 to be of opposite signs. With $\epsilon_2 = +1$ and $\epsilon_3 = -1$, it is easily checked that Eq. (22.47) can be solved with $\kappa_2^2 > \kappa_3^2$. Then $A_{\mu 1}$ and $A_{\mu 2}$ contribute "normally" to the conserved quantities, $A_{\mu 3}$ "abnormally."

The Invariant Singular Functions

Here let us list the definitions and main relationships among the functions $\bar{\triangle}, \triangle_{\rm ret}, \triangle_{\rm adv}$ and $\triangle$ that we have been using. It is best to start with the function $\triangle$ because all the others can be derived therefrom. For real (positive) $\kappa, \triangle(x;\kappa)$ may be defined by the boundary-value problem

$$(\partial^\mu \partial_\mu - \kappa^2)\triangle(\boldsymbol{x};\kappa) = 0,\ \triangle(\boldsymbol{x},0;\kappa) = 0, \dot{\triangle}(\boldsymbol{x},t;\kappa)\Big|_{t=0} = \delta^{(3)}(\boldsymbol{x})\,. \tag{22.74}$$

We can write the solution in the form

$$\triangle(x;\kappa) = (2\pi)^{-3} \int \frac{d^3k}{\omega} \exp[i\boldsymbol{k}\cdot\boldsymbol{x}] \sin\omega t,\ \omega = +(\boldsymbol{k}^2 + \kappa^2)^{1/2}\,. \tag{22.75}$$

Some relevant relations are

$$\triangle(\boldsymbol{x},t;\kappa) = \triangle(-\boldsymbol{x},t;\kappa) = -\triangle(\boldsymbol{x},-t;\kappa) = \triangle^*(\boldsymbol{x},t;\kappa)\,. \tag{22.76}$$

The fact that $\triangle(\boldsymbol{x};\kappa)$ vanishes if $t = 0$ can be generalized to the relativistically invariant statement that it vanishes if x is spacelike.

The other three functions we are interested in all obey an inhomogeneous equation:

$$(\partial^\mu \partial_\mu - \kappa^2)(\triangle_{\rm ret}(x;\kappa), \triangle_{\rm adv}(x;\kappa), \bar{\triangle}(x;\kappa)) = -\delta^{(4)}(x)\,. \tag{22.77}$$

We can generate these solutions from $\triangle(x;\kappa)$. Let us begin with $\triangle_{\rm ret}$, which is defined as that solution of Eq. (22.77) that vanishes for negative t. Setting

$$\triangle_{\rm ret}(x;\kappa) = \theta(t)\triangle(x;\kappa)\,, \tag{22.78}$$

certainly makes $\triangle_{\text{ret}}$ vanish if $t < 0$. Further,

$$\begin{aligned}\frac{\partial \triangle_{ret}(x;\kappa)}{\partial t} &= \delta(t)\triangle(\boldsymbol{x},0;\kappa)+\theta(t)\dot{\triangle}(\boldsymbol{x},t;\kappa)=\theta(t)\dot{\triangle}(\boldsymbol{x},t;\kappa)\,,\\ \frac{\partial^2 \triangle_{ret}(x;\kappa)}{\partial t^2} &= \delta(t)\dot{\triangle}(\boldsymbol{x},0;\kappa)+\theta(t)\frac{\partial^2\triangle(x;\kappa)}{\partial t^2}\\ &= \delta^{(4)}(x)+\theta(t)\frac{\partial^2\triangle(x:\kappa)}{\partial t^2} \qquad (22.79)\end{aligned}$$

where we used the boundary conditions on $\triangle$, and $\dot{\theta}(t) = \delta(t)$. Therefore,

$$\begin{aligned}(\partial^\mu\partial_\mu-\kappa^2)\triangle_{ret}(x;\kappa) =&\theta(t)(\nabla^2-\kappa^2)\triangle(x;\kappa)-\delta^{(4)}(x)-\theta(t)\frac{\partial^2\triangle(x;\kappa)}{\partial t^2}\\ =&\theta(t)(\partial^\mu\partial_\mu-\kappa^2)\triangle(x;\kappa)-\delta^{(4)}(x)=-\delta^{(4)}(x)\,; \quad (22.80)\end{aligned}$$

thus Eq. (22.78) is justified. In a similar way, $\triangle_{adv}(x:\kappa)$ which is defined to vanish for $t > 0$ is seen to be given by

$$\triangle_{adv}(x;\kappa) = -\theta(-t)\triangle(x;\kappa)\,. \qquad (22.81)$$

Both functions $\triangle_{ret}$ and $\triangle_{adv}$ are real, and between them and $\triangle$ we read off the relations

$$\begin{aligned}\triangle_{ret}(\boldsymbol{x},t;\kappa) &= \triangle_{adv}(\boldsymbol{x},-t;\kappa)\,,\\ \triangle(x;\kappa) &= \triangle_{ret}(x;\kappa)-\triangle_{adv}(x;\kappa)\,. \qquad (22.82)\end{aligned}$$

The functions $\triangle_{ret}$ and $\triangle_{adv}$ share with $\triangle$ the property of vanishing for space-like x, and are likewise even functions of $\boldsymbol{x}$.

The real, time-symmetric solution $\bar{\triangle}(x;\kappa)$ to Eq. (22.77) is defined as just the average of the solutions $\triangle_{ret}$ and $\triangle_{adv}$:

$$\bar{\triangle}(x;\kappa)=\frac{1}{2}(\triangle_{ret}(x;\kappa)+\triangle_{adv}(x;\kappa))=\frac{1}{2}\epsilon(t)\triangle(x;\kappa)\,, \qquad (22.83)$$

where $\epsilon(t)$ is the sign function (±1 for t positive (negative)). Whereas $\triangle, \triangle_{ret}$ and $\triangle_{adv}$ are all functions of x^2 and the sign of $x^0, \bar{\triangle}$ is a function of x^2 alone.

Manifestly covariant integral representations can be given for all these functions, and these integrals can be explicitly evaluated leading to forms such as the one given in Eq. (22.15). All these functions involve Bessel functions. These details may be found in any text on relativistic quantum field theory; hence we do not go into them. There is an exception in case $\kappa = 0$, when very simple expressions exist for these functions, all in terms of the Dirac delta function. Writing $D(x)$, $D_{ret}(x)$, $D_{adv}(x)$ and $\overline{D}(x)$ when $\kappa = 0$, we have:

$$\begin{aligned}D(x) &= \frac{1}{4\pi r}[\delta(r-t)-\delta(r+t)]\,,\\ D_{ret}(x) &= \frac{1}{4\pi r}\delta(r-t),\ D_{adv}(x)=\frac{1}{4\pi r}\delta(r+t)\,,\\ \bar{D}(x) &= \frac{1}{4\pi}\delta(r^2-t^2)=\frac{1}{4\pi}\delta(x^2),\ r=|\boldsymbol{x}|\,. \qquad (22.84)\end{aligned}$$

Thus, in this special case, all these functions vanish away from the light cone.

Chapter 23

Conclusion

In the preceding chapters we have treated various aspects of classical dynamics and have endeavoured to show that the structure of the theory is intimately related to an associative algebra with derivations. Transformations of dynamical systems are intimately associated not only with their time development but also with the "apparent" changes due to a change in the frame of reference. In many ways our approach to classical mechanics in this book has been conditioned by knowledge of developments in quantum mechanics. Without trying to explain in detail the quantum mechanical notions involved, we conclude this book by first discussing some of the analogies as well as the differences between classical mechanics and quantum mechanics, and then we mention the ways in which the latter has "opened our eyes" to aspects of the former.

The basic elements involved in the description of a classical dynamical system are the following: the phase space of the system must be defined, identifying the physical meaning of each independent coordinate of this space; the observables or dynamical variables of the system are all real functions on phase space (or they may be a subset chosen according to some criteria); the set of all possible states is the set of all positive-definite normalized probability distributions (state functions) on phase space. It is presupposed that all these definitions hold relative to some specific observer. The dynamical variables form an associative algebra via pointwise multiplication in phase space; and the PB, of one fixed dynamical variable with all others, acts as a derivation. The PB also converts the set of all dynamical variables into a Lie algebra. The group of transformations that this Lie structure gives rise to, and under which the PB itself is invariant, is the group of (regular) canonical transformations. The importance of these transformations lies in that they represent the changes in the state of the system as perceived by the observer. We are tempted to call a change "real" if it expresses the development of the system by itself in time, the observer remaining unchanged, and "apparent" if a change of observer is involved. Dynamical variables and state functions are connected to results of measurement by the statement that the phase space integral of the product of a dynamical variable and a state function is the expected value of the former in the state corresponding to the latter. The family of all state functions forms a convex set: given a set of state functions, any real linear combination of these with positive coefficients is also a possible state function, provided the sum of the coefficients is unity. The extreme points of this set are, by definition, those state functions that cannot be obtained by such linear combinations from other state functions, and in classical mechanics such extremal or "pure" states correspond to probability distributions in phase space that are concentrated at one point (a delta function distribution). In these pure states, and only in these, every dynamical variable has a precise value; furthermore, the expected value of the product of two variables is the product of the two expected values. This last point must be particularly appreciated for the following reason. There is generally a reluctance to admit that state functions that are not extremals (delta functions) must be taken seriously. One often has the feeling that in dealing with a classical system, even if one has to describe a particular state by means of a "spread-out" probability distribution in phase space, one can suppose that "actually" every dynamical variable has a precise numerical value but one is just not aware of it. This feeling seems reasonable because classical mechanics predicts no limitations on the accuracy with which all the independent variables of a system may be simultaneously measured: thus, one is tempted to say that using a spread-out probability distribution is just the result of a poor measurement. But using such a state function is the only proper way of representing an incomplete measurement, and must be taken as seriously as the representative of any other measurement! If after an incomplete measurement, we were to suppose that "actually" the variable A has the precise value A_0, and B the value B_0, then consistency demands that we also suppose that AB "actually" has the value A_0B_0; but the expected values in general do not behave that way, $\langle AB\rangle \neq \langle A\rangle \langle B\rangle$; thus the suppositions we have made do not correctly represent the knowledge of the system that we have after the incomplete measurement.

The description of a quantum system has some points of similarity with the foregoing. Instead of distributions in the phase space we now deal with the vectors in and density operators on the Hilbert space of the system. (We do not imply a direct correspondence between phase space and Hilbert space!) The observables or dynamical variables are represented by Hermitian operators on the Hilbert space. They form a Lie algebra again, with $-i/\hbar$ times the commutator appearing as the Lie bracket; $\hbar$ is Plancks constant. Though operator multiplication is associative, there is no associative algebra structure for the set of observables by themselves; there is only the Lie structure. This Lie bracket gives rise to the group of unitary transformations on Hilbert space, and it is these transformations that represent changes in the system. States are represented by those Hermitian operators that are positive semidefinite and have trace unity, and the expected value of the observable A in the state ρ is $\langle A\rangle = tr(A\rho)\cdot(tr = \text{trace})$. The representatives ρ of states are called density matrices. The trace operation replaces integration over phase space. Once again the set of all possible density matrices forms a convex set, exactly as in the classical theory. The real differences between classical mechanics and quantum mechanics show up in the nature and properties of the extreme points of this set. Now these points are those operators ρ that are projection operators onto single vectors in the Hilbert space-these are the pure states in quantum mechanics. The specifically quantum mechanical property of these pure states, having no analogue among the pure states of classical mechanics, is the superposition principle, which is the following: given any set of vectors in the Hilbert space, not only does each one determine a pure state via the projection operator onto it, we can also consider any complex linear combination of the given vectors and obtain another pure state by means of the projection onto this resultant vector. This superposition principle operating at the level of vectors in Hilbert space is quite distinct from the process of forming real positive normalized linear combinations of density matrices to get new density matrices; the latter alone has an exact parallel in classical mechanics. The other major difference between classical and quantum mechanics is that now there are no states, not even pure ones, in which every observable can be said to have a definite value. Quantum mechanics predicts definite limitations on the accuracies with which different observables of a system may be simultaneously measured.

Within the framework of quantum mechanics, a clear separation between the dynamical variables viewed as an algebraic system and the states in which expected values are assigned to observables is pretty well forced upon us. We do not think of an observable as always having some value or values with various probabilities. A similar viewpoint can fruitfully be adopted toward classical mechanics. We must pay attention to the *form* of the phase space function that represents an observable, not only to its value in some state. It is these forms that matter in transformation theory.

We have, in this book, presented classical mechanics as a consistent and closed discipline. We have taken some care in demonstrating the intimate relationships between some of the modern developments like the analysis of the Galilei and Poincaré groups and the fundamental concepts of mechanics. These discussions and expositions are within the classical framework. But much of the inspiration for these efforts came from discoveries and developments in quantum theory, even from particle theory. For example, the angular momentum commutation relations in quantum theory had been known for several decades before the corresponding classical Poisson bracket relations were explored. The notion of the Galilei and Poincaré groups as specifying the irreducible systems is quite new to classical mechanics, although they have been known for some time in quantum theory.

Why is this? How does it come about that such basic and natural ideas did not develop in classical mechanics? We offer a hypothesis: Perhaps it is because classical mechanics was visualized in too concrete terms, with much more emphasis on *values* rather than *forms* of things. Familiarity breeds contempt-unless we constantly see more and more in the familiar! In quantum theory we are saved this degradation because things would never quite become familiar! But in both cases we must distinguish the abstract ideal systems from the concrete physical model. We hope that the presentation in this book serves to restore the perspective.

We have deliberately avoided having an extended discussion of the relations between classical mechanics and quantum theory in the previous chapters because all too often a concrete physical system is visualized as a classical mechanical system. In a sense what ought to be discussed is the *interrelationship* of the two theories, rather than consider classical mechanics as a limit of quantum theory or suggest that classical mechanics is not really correct.

It would not be lost on the discerning reader that classical mechanics is a *framework* for describing certain physical systems and their interactions and that it is *not a theory*. This is the same with quantum mechanics. For example, even after some suitable choice of the physical system is made, there is the essential problem of determining the Lagrangian for the system. The choice of the precise form of the Lagrangian is the specification of the interaction. Until this specification is made, we do not have a theory of the interacting system. Classical mechanics provides a framework, a language, for describing this interaction; but the choice of the Lagrangian is as essential a step as the choice of the dynamical system itself

There are, of course, general dynamical principles that may be invoked to limit the possible Lagrangians and, in turn, to arrive at useful conclusions independently of the specific choice of the Lagrangian. We may cite the existence of the constants ("first integrals") of motion of energy, total linear and angular momentum, and the uniform motion of the center of mass (energy) in systems invariant under Galilean (Poincaré) transformations. The most familiar of these is the principle of conservation of energy valid for almost all dynamical systems that we study. In these cases, general principles constrain the possible form of the Lagrangian for an interacting system.

The transformation theory of dynamics does much more than this! We saw, in Chapter 19, how the notion of a "particle" with finite mass and a triplet of canonical variable pairs emerges from an analysis of the Galilean group of transformations and its realizations. But such a particle is the elementary building block for a Newtonian theory. In a sense we may say that *the entities of Newtonian mechanics themselves may be thought of as derived notions starting with abstract notions of transformations.* As we said earlier, this approach to the classical situation is inspired by the way one has come to look upon particles in quantum theory. These relativity transformations are not invariance transformations of a specific dynamical system; rather they carry a state of an abstract system through an entire "orbit" of states. The orbit under the relativity transformations is the direct realization of all the states of the elementary entity: it may be identified with the entity itself. An elementary entity consists of no more than what is revealed by application of all these transformations to it. *The transformation is the system*: the medium is the message.

Our words echo the immortal lines of the sage-poet Ved Vyāas:

vāsanād vāsudevasya vāsitam te jagatrayam
sarvabhūta nivāsosi vāsudeva namostuté.

Freely translated: All the three worlds, past, present and future, are inherent as potentialities in the presiding Intelligence. Worthy of adoration is this abiding entity in all that is static and dynamic.

Looked at from this point of view it is somewhat disappointing that transformation theory does not uniquely determine the interaction structure apart from a finite number of parameters. True, it gives rise to constraints that in turn yield useful theorems, but the interaction structure itself is arbitrary to a large extent. This question was raised almost a century ago by Heinrich Hertz, who pointed out that in mechanics we have to deal with the notions of particles, motion, and forces; he proposed to reduce the third notion to the second so that forces may be derived from motions of particles. Because there were apparently forces where no ordinary intervening particles could be found, it was the proposal of Hertz to introduce "concealed particles" and "concealed motions." Hertz's ideas did not develop to the point where he could formulate a simple satisfactory theory. Apart from his untimely death, from the vantage point of the late twentieth century we can see that the quantum theory was a necessary development before "virtual particles" could be assigned the role of accounting for forces. And what Hertz envisaged found its precise formulation at the hands of Hideki Yukawa.

The reason why quantum theory was necessary for the logical completion of this program can be seen as follows. In classical mechanics there is a fundamental or intrinsic distinction between the two kinds of dynamical entities, particles and fields. They are quite different in character, not only in that a particle has a finite number of degrees of freedom, whereas the field has an infinite number of degrees of freedom, but in a more immediate sense. Particles are localized objects, and even a large collection of them, say, forming a rigid body, can be made to occupy a limited volume of space. Moreover, this condition is a natural state of the system. A field, on the other hand, is extended all over space, and its nature is to spread. In the preceding chapters we found that in studying the dynamics of Galilean and Lorentzian fields there are genuine differences as well as similarities between particles and fields. If we generate a localized excitation in a field, it will automatically spread and propagate. This makes the field a natural device to mediate interactions between particles. A scheme involving motions and only contact interactions between particles themselves is at variance with our experience in the classical domain, but if we include only motions and contact interactions between particles and fields, the scheme is qualitatively satisfactory. We may cite the interaction between charged particles and the electric field, as well as the interaction between particles suspended in a fluid and the turbulent velocity field in the fluid. Hertz did not take this aspect into account. Our study in Chapter 22 exhibits the role of the mediating field in studying interaction at a distance between particles. Fields and particles are distinct types of dynamical systems in classical physics, and the direct interaction between particles does not involve "concealed particles" as Hertz assumed, but rather fields. The situation is quite different in quantum physics: fields have quanta that are particles. Quanta can be exchanged between particles, and Hertz's program can be realized; this was precisely what was achieved by Yukawa in his meson theory of nuclear forces. Thus a search for a principle in which interactions can be understood in terms of motions alone leads towards quantum theory.

This synthesis of the notions of particles and fields has been achieved after some revision of our concept of the nature of particles. Satyendra Nath Bose showed fifty years ago that photons obeyed a particular kind of statistics which enabled us to consider them as levels of excitation of a radiation oscillator. We now extend this to all Bose-Einstein particles, and apply related rules to Fermi-Dirac particles. With this amended status, quantum particles and quantum fields (at least in the limit of no interactions) are two ways of talking about the same system.

There are other chains of ideas that also strongly suggest quantum theory. Dirac has pointed to the principle of limited divisibility, that there is a scale at which division of an object into parts will come to an end. In a sense this may also be seen as the requirement that physical objects possess only finite numbers of degrees of freedom: if they were infinitely divisible, there would be the possibility of infinite numbers of independent component motions. If, in principle, any finite objects must have only a finite number of degrees of freedom, it should not be infinitely divisible.

There is one more context in which classical mechanics and quantum theory share conceptual structure. This is in the problem of measurement where both classical and quantum theories are deficient. The so-called disturbance under measurement and the uncertainty principle in quantum theory notwithstanding, the real problem of measurement is what is called the "reduction of the wave packet" in conventional quantum theory of measurement and its classical counterpart. Where does dynamics end and measurement begin? We find that these are hard and subtle questions that have only been inadequately treated, even more so in classical physics. We believe that they are important and ought to be investigated by physicists.

Classical mechanics is an eternal discipline, where harmony abounds. It is beautiful in the true sense of the term. It is new every time we grow and look at it anew. If we have conveyed a sense of awe and adoration to this eternal beauty, our labor is worthwhile.

Index

The following abbreviations are used whenever convenient:

Dirac bracket - DB Galilei group - G
Generalized Poisson bracket - GPB
Homogeneous Lorentz group - HLG
Lagrange bracket - LB
Poincaré group - P
Poisson bracket - PB
Three-dimensional Euclidean group - $E(3)$
Three-dimensional rotation group - $R(3)$
Three-dimensional translation group - T_3

Printed in the United States
By Bookmasters

Printed in the United States
By Bookmasters